S0-ASM-390

*Multivariable Calculus
with Linear Algebra
and Series*

JAMES
GERBER

Multivariable Calculus with Linear Algebra and Series

—————————WILLIAM F. TRENCH AND BERNARD KOLMAN

Drexel University

BETHEL
COLLEGE

ACADEMIC PRESS
New York and London

Copyright © 1972, by Academic Press, Inc.

All rights reserved.
No part of this book may be reproduced in any form,
by photostat, microfilm, retrieval system, or any
other means, without written permission from
the publishers.

ACADEMIC PRESS, INC.
111 Fifth Avenue, New York, New York 10003

United Kingdom Edition published by
ACADEMIC PRESS, INC. (LONDON) LTD.
24/28 Oval Road, London NW1 7DD

Library of Congress Catalog Card Number: 70-182649
AMS (MOS) 1970 Subject Classification: 26A60

Printed in the United States of America

To Lucille and Lillie

26 X/6

Contents

vii

Preface

This book is an expansion of the authors' *Elementary Multivariable Calculus*. It presents a modern, but not extreme, treatment of linear algebra, the calculus of several variables, and series at a level appropriate for the sophomore science, engineering, or mathematics major who has completed a standard first-year calculus course. Our choice of topics has been influenced, but not dictated by, the recommendations of the Committee on Undergraduate Preparation in Mathematics. The main emphasis throughout the book is on maintaining a sophomore level of understanding. Thus, we have not hesitated to omit proofs of difficult theorems, preferring instead to illustrate their meanings with numerous examples. In particular, there are few "epsilon–delta" arguments in this book. Most theorems and definitions are followed by worked-out illustrative examples.

By way of introduction to the various concepts of the multivariable calculus, we usually recall the corresponding notion from the one-dimensional calculus. Simple notions are presented before their generalizations (for example, real-valued functions are introduced before vector-valued functions, rather than as special cases of the latter) and examples illustrate each theorem and definition. Each section contains, in addition to an ample number of routine exercises, a set of theoretical exercises designed to fill the gaps in proofs and extend results obtained in the text. The latter can also be used to raise the level of the book. Answers to selected exercises are given in the back of the book.

Chapter 1 discusses linear equations and matrices, including determinants. Chapter 2, on vector spaces and linear transformations, includes eigenvalues and eigenvectors. Chapter 3 discusses vector analysis and analytic geometry in R^3. It also includes sections on curves and surfaces. Chapter 4 covers the differential calculus of real-valued functions of n variables from a modern point of view. The differential is treated as a linear transformation, and concepts from linear algebra are used as an integral part of the discussion. Chapter 5 treats vector-valued functions as ordered m-tuples of real-valued functions, so that the results of Chapter 2 can be directly applied. Chapter 6 deals with integration (line, surface, and multiple integrals) and includes a discussion of Green's and Stokes's

Theorems and the Divergence Theorem. Chapter 7 deals with infinite sequences, infinite series, and power series in one variable.

Chapters 3–6 can be used for a one-semester course for students who have already taken linear algebra as well as first-year calculus. The entire book can be used as the text for a combined, full-year course in linear algebra and sophomore calculus, including a brief treatment of sequences and series.

Acknowledgments

We wish to express our thanks to Miss Susan R. Gershuni, who typed most of the manuscript, and to Mrs. Marjorie M. Bawduniak, who also helped with the typing; to Albert J. Herr and Robert L. Higgins, who provided solutions and corrections to some of the problems; to Mrs. Anna Hernandez for the art work; and to the staff of Academic Press, for their interest, encouragement, and cooperation.

We also wish to thank three unknown reviewers for their helpful suggestions for improving the manuscript.

Chapter 1

Linear Equations and Matrices

1.1 Linear Systems and Matrices

Many problems in mathematics and the sciences involve *linear* relationships between two sets of variables $y_1, \ldots, y_n$ and $x_1, \ldots, x_n$; that is, relationships of the form

$$
\begin{aligned}
y_1 &= a_{11}x_1 + a_{12}x_2 + \cdots + a_{1n}x_n, \\
y_2 &= a_{21}x_1 + a_{22}x_2 + \cdots + a_{2n}x_n, \\
&\ \vdots \\
y_m &= a_{m1}x_1 + a_{m2}x_2 + \cdots + a_{mn}x_n,
\end{aligned}
$$

(1)

where the a_{ij} $(1 \leq i \leq m, 1 \leq j \leq n)$ are constants. In a typical situation $a_{11}, \ldots, a_{mn}$ and $y_1, \ldots, y_m$ are known, and it is required to find $x_1, \ldots, x_n$ such that all of the equations in (1) are satisfied. Then (1) is called *a system of m linear equations in n unknowns*, or simply a *linear* system. An ordered n-tuple $(x_1, \ldots, x_n)$ which satisfies (1) is *a solution of* (1). When $n = 2, 3$, or 4 we shall often write the unknowns as x, y, z, and w, to avoid unnecessary subscripts.

The student has no doubt encountered linear systems before, and can solve them by the *method of elimination*, if m and n are small.

Example 1.1 Consider the system of two equations in two unknowns x and y:

$$
\begin{aligned}
x - 2y &= 3, \\
2x + y &= 1.
\end{aligned}
$$

(2)

Subtracting twice the first equation from the second yields

$$5y = -5,$$

which does not involve x; thus x has been *eliminated*. Clearly $y = -1$, and the first equation of (2) yields

$$x = 3 + 2y = 3 + 2(-1) = 1.$$

Thus $(x, y) = (1, -1)$ is a solution of (2). In this example, there is only one solution, and we say that the solution is *unique*.

Example 1.2 Consider the system of three equations in three unknowns x, y, and z:

$$\begin{aligned} x + y - z &= 4, \\ 3x - 2y + 2z &= -3, \\ 4x + 2y - 3z &= 11. \end{aligned}$$ (3)

To eliminate x we subtract three times the first equation from the second and four times the first from the third. This yields two equations in y and z:

$$\begin{aligned} -5y + 5z &= -15, \\ -2y + z &= -5, \end{aligned}$$

which can be further simplified by multiplying the first equation by $-\frac{1}{5}$:

$$\begin{aligned} y - z &= 3, \\ -2y + z &= -5. \end{aligned}$$ (4)

Adding twice the first equation to the second eliminates y:

$$-z = 1.$$

Combining this with the first equations of (3) and (4), we obtain a new system

$$\begin{aligned} x + y - z &= 4, \\ y - z &= 3, \\ z &= -1. \end{aligned}$$ (5)

Because of the way we obtained (5) from (3), the two systems have the same solutions. The solution of (5) is unique and is obtained by solving successively for z, y, and x:

$$z = -1,$$

$$y = 3 + z = 3 - 1 = 2,$$

$$x = 4 - y + z = 4 - 2 - 1 = 1.$$

Thus, (3) has the unique solution $(x, y, z) = (1, 2, -1)$.

Example 1.3 Consider the system

(6)
$$x + y = 2,$$

$$2x + 2y = 5.$$

Multiplying the first equation by 2 yields

$$2x + 2y = 4,$$

which contradicts the second. Therefore (6) has no solution.

Example 1.4 Consider the system

(7)
$$x + y - z = 1,$$

$$x - y + z = 3.$$

Adding the two equations yields $2x = 4$, or $x = 2$. However, y and z cannot be obtained separately from (7), since both equations are satisfied if $x = 2$ and $y - z = -1$. Thus $(2, z - 1, z)$ is a solution for any z, and we see that a linear system may have more than one solution.

Matrices

The method of elimination is adequate for the solution of linear systems when m and n are small. However, more sophisticated methods are required if m and n are large, and for theoretical investigations. For this reason, we turn to the theory of matrices.

We shall assume that all quantities appearing below are real numbers (which we also call *scalars*), although much that we say also holds for complex numbers.

Definition 1.1 An $m \times n$ (read "m by n") matrix $\mathbf{A}$ is a rectangular array of mn real numbers arranged in m rows and n columns:

$$(8) \qquad \mathbf{A} = \begin{bmatrix} a_{11} & a_{12} & \cdots & a_{1n} \\ a_{21} & a_{22} & \cdots & a_{2n} \\ \vdots & \vdots & & \vdots \\ a_{m1} & a_{m2} & \cdots & a_{mn} \end{bmatrix}.$$

We call $(a_{i1}, a_{i2}, \ldots, a_{in})$ the ith row of $\mathbf{A}$ $(1 \leq i \leq m)$ and

$$\begin{bmatrix} a_{1j} \\ a_{2j} \\ \vdots \\ a_{mj} \end{bmatrix}$$

the jth column $(1 \leq j \leq n)$. The number a_{ij} which is in the ith row and jth column of $\mathbf{A}$ is called *the i, jth element of $\mathbf{A}$*, and we shall often abbreviate (8) as

$$\mathbf{A} = [a_{ij}].$$

If $m = n$, so that $\mathbf{A}$ is square, we shall say that $\mathbf{A}$ is a *square matrix of order n*. In this case, the elements $a_{11}, a_{22}, \ldots, a_{nn}$ form the *main diagonal* of $\mathbf{A}$. If $\mathbf{A}$ is square and $a_{ij} = 0$ for $i \neq j$, then $\mathbf{A}$ is said to be a *diagonal* matrix.

We emphasize that a matrix is an ordered array; that is,

$$\begin{bmatrix} 1 & 2 \\ 3 & 4 \end{bmatrix} \neq \begin{bmatrix} 1 & 3 \\ 2 & 4 \end{bmatrix}$$

even though the elements of the two matrices form the same set of numbers. Two $m \times n$ matrices $\mathbf{A} = [a_{ij}]$ and $\mathbf{B} = [b_{ij}]$ are equal if and only if $a_{ij} = b_{ij}$ $(1 \leq i \leq m, 1 \leq j \leq n)$.

Note that the definition of equality of matrices $\mathbf{A}$ and $\mathbf{B}$ requires that $\mathbf{A}$ and $\mathbf{B}$ have the same number of rows and the same number of columns; thus, if $\mathbf{A}$ is $m \times n$ and $\mathbf{B}$ is $m_1 \times n_1$, with either $m_1 \neq m$ or $n_1 \neq n$, then $\mathbf{A}$ and $\mathbf{B}$ cannot be equal.

Example 1.5 Let

$$\mathbf{A} = \begin{bmatrix} 1 & 2 & -1 \\ 0 & 2 & 3 \end{bmatrix}, \qquad \mathbf{B} = \begin{bmatrix} 1 & 2 & 3 \\ -1 & -2 & 0 \\ 3 & 2 & 4 \end{bmatrix}.$$

$$C = [1 \quad 2 \quad 3], \qquad D = \begin{bmatrix} 1 \\ -1 \\ 2 \end{bmatrix},$$

and

$$E = [2].$$

Then A is a 2×3 matrix with

$$a_{11} = 1, \qquad a_{12} = 2, \qquad a_{13} = -1,$$

$$a_{21} = 0, \qquad a_{22} = 2, \qquad a_{23} = 3.$$

B is a square matrix of order 3, with diagonal elements $b_{11} = 1$, $b_{22} = -2$, and $b_{33} = 4$. C is a 1×3 matrix and D is a 3×1 matrix; E is a 1×1 matrix. Also,

$$F = \begin{bmatrix} -1 & 0 & 0 \\ 0 & 2 & 0 \\ 0 & 0 & 3 \end{bmatrix} \quad \text{and} \quad G = \begin{bmatrix} -1 & 0 \\ 0 & 0 \end{bmatrix}$$

are diagonal matrices.

Definition 1.2 (Matrix Addition)   The *sum* of two $m \times n$ matrices $A = [a_{ij}]$ and $B = [b_{ij}]$ is defined by

$$A + B = C = [c_{ij}],$$

where

$$c_{ij} = a_{ij} + b_{ij} \qquad (1 \le i \le m, \quad 1 \le j \le n).$$

Example 1.6   If

$$A = \begin{bmatrix} 1 & 2 & -3 \\ 3 & -2 & 0 \end{bmatrix} \quad \text{and} \quad B = \begin{bmatrix} -1 & 2 & 3 \\ 2 & 1 & 3 \end{bmatrix},$$

then

$$A + B = \begin{bmatrix} 1-1 & 2+2 & -3+3 \\ 3+2 & -2+1 & 0+3 \end{bmatrix} = \begin{bmatrix} 0 & 4 & 0 \\ 5 & -1 & 3 \end{bmatrix}.$$

Note that the sum of two matrices is defined if and only if they have the same number of rows and the same number of columns. Therefore, whenever we write $\mathbf{A} + \mathbf{B}$, it is to be understood that $\mathbf{A}$ and $\mathbf{B}$ are both $m \times n$ matrices.

Theorem 1.1   (a) Matrix addition is *commutative*; that is

$$\mathbf{A} + \mathbf{B} = \mathbf{B} + \mathbf{A}.$$

(b) Matrix addition is *associative*; that is

$$\mathbf{A} + (\mathbf{B} + \mathbf{C}) = (\mathbf{A} + \mathbf{B}) + \mathbf{C}.$$

(c) There exists a unique $m \times n$ matrix $\mathbf{0}$, called the $m \times n$ *zero matrix*, such that

(9) $$\mathbf{A} + \mathbf{0} = \mathbf{A}$$

for all $m \times n$ matrices $\mathbf{A}$.

(d) For each matrix $\mathbf{A}$ there exists a unique matrix $-\mathbf{A}$ such that

$$\mathbf{A} + (-\mathbf{A}) = \mathbf{0}.$$

Proof   (a) The i, jth elements of $\mathbf{A} + \mathbf{B}$ and $\mathbf{B} + \mathbf{A}$ are $a_{ij} + b_{ij}$ and $b_{ij} + a_{ij}$, respectively. Since

$$a_{ij} + b_{ij} = b_{ij} + a_{ij} \qquad (1 \leq i \leq m, \quad 1 \leq j \leq n),$$

it follows that $\mathbf{A} + \mathbf{B} = \mathbf{B} + \mathbf{A}$.

(b) Exercise T-1.

(c) The matrix $\mathbf{0}$ whose elements are all zero satisfies (9). Conversely, if

$$\mathbf{A} + \mathbf{B} = \mathbf{A},$$

then

$$a_{ij} + b_{ij} = a_{ij} \qquad (1 \leq i \leq m, \quad 1 \leq j \leq n),$$

which implies that

$$b_{ij} = 0 \qquad (1 \leq i \leq m, \quad 1 \leq j \leq n);$$

hence $\mathbf{B} = \mathbf{0}$.

(d) We leave it to the student (Exercise T-2) to show that

$$-\mathbf{A} = [-a_{ij}].$$

Example 1.7 If

$$A = \begin{bmatrix} 1 & 2 & -1 \\ 3 & 2 & 0 \end{bmatrix} \quad \text{and} \quad B = \begin{bmatrix} 2 & -1 & 3 \\ -2 & 1 & 4 \end{bmatrix},$$

then

$$A + B = \begin{bmatrix} 1+2 & 2-1 & -1+3 \\ 3-2 & 2+1 & 0+4 \end{bmatrix} = \begin{bmatrix} 3 & 1 & 2 \\ 1 & 3 & 4 \end{bmatrix}$$

and

$$B + A = \begin{bmatrix} 2+1 & -1+2 & 3-1 \\ -2+3 & 1+2 & 4+0 \end{bmatrix} = \begin{bmatrix} 3 & 1 & 2 \\ 1 & 3 & 4 \end{bmatrix};$$

hence $A + B = B + A$, in agreement with (a).

Example 1.8 Let

$$A = \begin{bmatrix} 1 & -2 & 3 \\ -1 & 2 & 4 \\ 3 & 2 & -1 \end{bmatrix}, \quad B = \begin{bmatrix} 0 & 2 & 1 \\ 2 & 3 & 0 \\ 0 & 1 & 2 \end{bmatrix}, \quad \text{and} \quad C = \begin{bmatrix} 1 & -2 & -1 \\ 0 & 2 & 1 \\ 3 & 2 & 3 \end{bmatrix}.$$

Then

$$A + B = \begin{bmatrix} 1 & 0 & 4 \\ 1 & 5 & 4 \\ 3 & 3 & 1 \end{bmatrix},$$

and

$$(10) \quad (A + B) + C = \begin{bmatrix} 1+1 & 0+(-2) & 4+(-1) \\ 1+0 & 5+2 & 4+1 \\ 3+3 & 3+2 & 1+3 \end{bmatrix} = \begin{bmatrix} 2 & -2 & 3 \\ 1 & 7 & 5 \\ 6 & 5 & 4 \end{bmatrix}.$$

On the other hand,

$$B + C = \begin{bmatrix} 1 & 0 & 0 \\ 2 & 5 & 1 \\ 3 & 3 & 5 \end{bmatrix}$$

and

$$(11) \quad \mathbf{A} + (\mathbf{B} + \mathbf{C}) = \begin{bmatrix} 1+1 & (-2)+0 & 3+0 \\ (-1)+2 & 2+5 & 4+1 \\ 3+3 & 2+3 & (-1)+5 \end{bmatrix}$$

$$= \begin{bmatrix} 2 & -2 & 3 \\ 1 & 7 & 5 \\ 6 & 5 & 4 \end{bmatrix}.$$

Comparing (10) and (11) confirms (b) for this special case.

For every pair of positive integers m and n, there is an $m \times n$ zero matrix. One can usually decide from the context which zero matrix is under discussion at any given time. When this is not the case, we shall denote the $m \times n$ zero matrix by $_m\mathbf{0}_n$.

Example 1.9 The 2×3 zero matrix is

$$_2\mathbf{0}_3 = \begin{bmatrix} 0 & 0 & 0 \\ 0 & 0 & 0 \end{bmatrix}.$$

If

$$\mathbf{A} = \begin{bmatrix} 1 & -1 & 2 \\ 3 & 2 & -1 \end{bmatrix},$$

then

$$\mathbf{A} + {}_2\mathbf{0}_3 = \begin{bmatrix} 1+0 & -1+0 & 2+0 \\ 3+0 & 2+0 & -1+0 \end{bmatrix} = \begin{bmatrix} 1 & -1 & 2 \\ 3 & 2 & -1 \end{bmatrix} = \mathbf{A}.$$

Moreover (from Exercise T-2),

$$-\mathbf{A} = \begin{bmatrix} -1 & 1 & -2 \\ -3 & -2 & 1 \end{bmatrix}.$$

Following the usual practice for ordinary addition, we write

$$\mathbf{A} + (-\mathbf{B}) = \mathbf{A} - \mathbf{B}.$$

Example 1.10 Let

$$\mathbf{A} = \begin{bmatrix} 1 & 0 \\ 2 & 1 \end{bmatrix} \quad \text{and} \quad \mathbf{B} = \begin{bmatrix} 3 & 0 \\ 4 & -5 \end{bmatrix};$$

then

$$A - B = \begin{bmatrix} 1 - 3 & 0 - 0 \\ 2 - 4 & 1 + 5 \end{bmatrix} = \begin{bmatrix} -2 & 0 \\ -2 & 6 \end{bmatrix}.$$

Definition 1.3 (Scalar Multiplication) If $A = [a_{ij}]$ is an $m \times n$ matrix and c is a scalar, then we define the $m \times n$ matrix $B = cA$ by $B = [b_{ij}]$, where

$$b_{ij} = ca_{ij} \qquad (1 \leq i \leq m, \quad 1 \leq j \leq n).$$

We say that B is a *scalar multiple* of A.

Example 1.11 Let $c = -2$ and

$$A = \begin{bmatrix} 1 & -3 \\ 2 & 4 \\ -1 & 2 \end{bmatrix};$$

then

$$cA = \begin{bmatrix} -2 & 6 \\ -4 & -8 \\ 2 & -4 \end{bmatrix}.$$

It can be shown (Exercise T-2) that

$$-A = (-1)A.$$

Theorem 1.2 If A and B are $m \times n$ matrices and r and s are scalars, then

(a) $r(sA) = (rs)A$;
(b) $(r + s)A = rA + sA$;
(c) $r(A + B) = rA + rB$.

We leave the proof to the student (Exercise T-3).

Example 1.12 Let $r = 4$, $s = -2$,

$$A = \begin{bmatrix} 1 & 2 \\ -2 & 3 \end{bmatrix} \qquad \text{and} \qquad B = \begin{bmatrix} 3 & -1 \\ 2 & 0 \end{bmatrix}.$$

Then

$$r(s\mathbf{A}) = 4((-2)\mathbf{A}) = 4\begin{bmatrix} (-2)1 & (-2)2 \\ (-2)(-2) & (-2)3 \end{bmatrix}$$

$$= 4\begin{bmatrix} -2 & -4 \\ 4 & -6 \end{bmatrix} = \begin{bmatrix} 4(-2) & 4(-4) \\ 4(4) & 4(-6) \end{bmatrix}$$

$$= \begin{bmatrix} -8 & -16 \\ 16 & -24 \end{bmatrix} = -8\begin{bmatrix} 1 & 2 \\ -2 & 3 \end{bmatrix}$$

$$= (4(-2))\begin{bmatrix} 1 & 2 \\ -2 & 3 \end{bmatrix} = (rs)\mathbf{A},$$

$$r\mathbf{A} + s\mathbf{A} = \begin{bmatrix} 4 & 8 \\ -8 & 12 \end{bmatrix} + \begin{bmatrix} -2 & -4 \\ 4 & -6 \end{bmatrix}$$

$$= \begin{bmatrix} 2 & 4 \\ -4 & 6 \end{bmatrix} = 2\begin{bmatrix} 1 & 2 \\ -2 & 3 \end{bmatrix} = (r+s)\mathbf{A},$$

and

$$r\mathbf{A} + r\mathbf{B} = \begin{bmatrix} 4 & 8 \\ -8 & 12 \end{bmatrix} + \begin{bmatrix} 12 & -4 \\ 8 & 0 \end{bmatrix}$$

$$= \begin{bmatrix} 16 & 4 \\ 0 & 12 \end{bmatrix} = 4\begin{bmatrix} 4 & 1 \\ 0 & 3 \end{bmatrix} = 4\begin{bmatrix} 1+3 & 2+(-1) \\ (-2)+2 & 3+0 \end{bmatrix}$$

$$= 4\left(\begin{bmatrix} 1 & 2 \\ -2 & 3 \end{bmatrix} + \begin{bmatrix} 3 & -1 \\ 2 & 0 \end{bmatrix}\right) = r(\mathbf{A} + \mathbf{B}).$$

Summation Notation

Before defining the product of two matrices, it is convenient to introduce a compact notation for sums.

Definition 1.4 If $m < n$ and $\alpha_m, \alpha_{m+1}, \ldots, \alpha_n$ are scalars, the symbol $\sum\limits_{i=m}^{n} \alpha_i$ will be interpreted as

$$\sum_{i=m}^{n} \alpha_i = \alpha_m + \alpha_{m+1} + \cdots + \alpha_n.$$

Here $\sum$ is called the *summation symbol*, i is the *index of summation*, and m and n are the *limits of summation*.

There is an obvious analogy between this terminology and that of the integral calculus, in which we write

$$\int_{x=a}^{b} f(x)\ dx,$$

where $\int$ is the integral (summation) sign, x is the variable of integration, and a and b are the limits of integration.

Example 1.13 Let $\alpha_3 = 1$, $\alpha_4 = 4$, $\alpha_5 = 3$, and $\alpha_6 = -1$. Then

$$\sum_{i=3}^{3} \alpha_i = \alpha_3 = 1,$$

$$\sum_{i=3}^{4} \alpha_i = \alpha_3 + \alpha_4 = 5,$$

$$\sum_{i=3}^{5} \alpha_i = \alpha_3 + \alpha_4 + \alpha_5 = 8,$$

and

$$\sum_{i=3}^{6} \alpha_i = \alpha_3 + \alpha_4 + \alpha_5 + \alpha_6 = 7.$$

Example 1.14 If $\alpha_1 = \alpha_2 = \cdots = \alpha_n = 1$, then

$$\sum_{i=1}^{n} \alpha_i = \sum_{i=1}^{n} 1 = \underbrace{1 + 1 + \cdots + 1}_{n \text{ times}} = n.$$

If $\alpha_i = i$, then

$$\sum_{i=1}^{n} \alpha_i = \sum_{i=1}^{n} i = 1 + 2 + \cdots + n = \frac{n(n+1)}{2}$$

(Exercise T-4).

Example 1.15 If $a_1, \ldots, a_n$ and $b_1, \ldots, b_n$ are real numbers, then

$$\sum_{i=1}^{n} a_i b_i = a_1 b_1 + a_2 b_2 + \cdots + a_n b_n.$$

We leave the verification of the following properties of the summation symbol to the student (Exercise T-5):

$$\sum_{i=1}^{n} (a_i + b_i)c_i = \sum_{i=1}^{n} a_i c_i + \sum_{i=1}^{n} b_i c_i,$$

$$\sum_{i=1}^{n} (ca_i) = c \sum_{i=1}^{n} a_i.$$

If more than one index is present, we must be careful to note which is the summation index. If

$$\mathbf{A} = \begin{bmatrix} a_{11} & a_{12} & \cdots & a_{1n} \\ a_{21} & a_{22} & \cdots & a_{2n} \\ \vdots & \vdots & & \vdots \\ a_{m1} & a_{m2} & \cdots & a_{mn} \end{bmatrix},$$

then

(12) $$r_i = \sum_{j=1}^{n} a_{ij}$$

is the sum of the elements in the ith row, while

(13) $$c_j = \sum_{i=1}^{m} a_{ij}$$

is the sum of the elements in the jth column.

Given two indices, one can sum first with respect to one and then with respect to the other to obtain a *double* sum; thus

(14) $$\sum_{i=1}^{m} (\sum_{j=1}^{n} a_{ij}) = S$$

is the sum of all the elements of $\mathbf{A}$, calculated by first adding the elements in each row to obtain the subtotals $r_1, \ldots, r_m$ in (12), and then adding them together. The same result can be accomplished by adding the elements in each column to obtain $c_1, \ldots, c_n$ as in (13), and then adding them together; thus

(15) $$\sum_{j=1}^{n} (\sum_{i=1}^{m} a_{ij}) = S.$$

Comparing (14) and (15) and dropping the parentheses yields

$$\sum_{i=1}^{m} \sum_{j=1}^{n} a_{ij} = \sum_{j=1}^{n} \sum_{i=1}^{m} a_{ij},$$

which says that a double sum may be evaluated by summing first on either index.

Example 1.16 Let

$$\mathbf{A} = \begin{bmatrix} 1 & 2 & 3 & 4 \\ 1 & 0 & 3 & 1 \end{bmatrix};$$

then

$$r_1 = \sum_{j=1}^{4} a_{1j} = 1 + 2 + 3 + 4 = 10,$$

$$r_2 = \sum_{j=1}^{4} a_{2j} = 1 + 0 + 3 + 1 = 5,$$

and

(16) $$S = \sum_{i=1}^{2} \sum_{j=1}^{4} a_{ij} = \sum_{i=1}^{2} r_i = 10 + 5 = 15.$$

Reversing the order of summation, we obtain

$$c_1 = \sum_{i=1}^{2} a_{i1} = 1 + 1 = 2,$$

$$c_2 = \sum_{i=1}^{2} a_{i2} = 2 + 0 = 2,$$

$$c_3 = \sum_{i=1}^{2} a_{i3} = 3 + 3 = 6,$$

$$c_4 = \sum_{i=1}^{2} a_{i4} = 4 + 1 = 5,$$

and

$$S = \sum_{j=1}^{4} \sum_{i=1}^{2} a_{ij} = \sum_{j=1}^{4} c_j = 2 + 2 + 6 + 5 = 15,$$

which agrees with (16).

Matrix Multiplication

Definition 1.5 (Matrix Multiplication)   If $A = [a_{ij}]$ is an $m \times p$ matrix and $B = [b_{ij}]$ is a $p \times n$ matrix, then $C = AB$ is the $m \times n$ matrix with i, jth element

$$c_{ij} = a_{i1}b_{1j} + a_{i2}b_{2j} + \cdots + a_{ip}b_{pj}$$

$$= \sum_{k=1}^{p} a_{ik}b_{kj} \qquad (1 \leq i \leq m, \quad 1 \leq j \leq n).$$

According to this definition the i, jth element of the product AB is obtained by multiplying each element in the ith row of A by the corresponding element in the jth column of B and adding the resulting products, as depicted in Figure 1.1. Thus AB is defined if and only if A has as many columns as B has rows. Consequently, the product AB of an $m \times p$ matrix A and a $q \times n$ matrix B is defined if and only if $p = q$ (in which case AB is $m \times n$), while BA is defined if and only if $n = m$ (in which case BA is $q \times p$).

Example 1.17   Let

$$A = \begin{bmatrix} 4 & -1 & 2 \\ 3 & 2 & -4 \end{bmatrix} \quad \text{and} \quad B = \begin{bmatrix} 2 & 2 \\ 3 & 0 \\ -1 & 3 \end{bmatrix};$$

then

$$c_{11} = a_{11}b_{11} + a_{12}b_{21} + a_{13}b_{31}$$
$$= (4)(2) + (-1)(3) + 2(-1) = 3,$$
$$c_{12} = a_{11}b_{12} + a_{12}b_{22} + a_{13}b_{32}$$
$$= 4(2) + (-1)(0) + (2)(3) = 14,$$
$$c_{21} = a_{21}b_{11} + a_{22}b_{21} + a_{23}b_{31}$$
$$= (3)(2) + (2)(3) + (-4)(-1) = 16,$$

and

$$c_{22} = a_{21}b_{12} + a_{22}b_{22} + a_{23}b_{32}$$
$$= (3)(2) + (2)(0) + (-4)(3) = -6.$$

Thus

$$AB = C = \begin{bmatrix} 3 & 14 \\ 16 & -6 \end{bmatrix}.$$

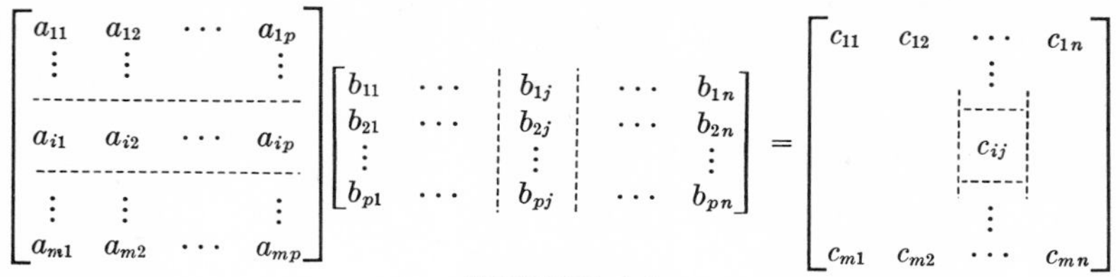

$$\begin{bmatrix} a_{11} & a_{12} & \cdots & a_{1p} \\ \vdots & \vdots & & \vdots \\ \hline a_{i1} & a_{i2} & \cdots & a_{ip} \\ \hline \vdots & \vdots & & \vdots \\ a_{m1} & a_{m2} & \cdots & a_{mp} \end{bmatrix} \begin{bmatrix} b_{11} & \cdots & b_{1j} & \cdots & b_{1n} \\ b_{21} & \cdots & b_{2j} & \cdots & b_{2n} \\ \vdots & & \vdots & & \vdots \\ b_{p1} & \cdots & b_{pj} & \cdots & b_{pn} \end{bmatrix} = \begin{bmatrix} c_{11} & c_{12} & \cdots & c_{1n} \\ & & \vdots & \\ & & c_{ij} & \\ & & \vdots & \\ c_{m1} & c_{m2} & \cdots & c_{mn} \end{bmatrix}$$

FIGURE 1.1

For this example, **BA** is also defined:

$$\mathbf{BA} = \begin{bmatrix} 14 & 2 & -4 \\ 12 & -3 & 6 \\ 5 & 7 & -14 \end{bmatrix}.$$

Example 1.18 Let

$$\mathbf{A} = \begin{bmatrix} 2 & 1 & 2 \\ 0 & 1 & 2 \end{bmatrix} \quad \text{and} \quad \mathbf{B} = \begin{bmatrix} 1 & -1 & 0 \\ 1 & 0 & 1 \\ 2 & 2 & 1 \end{bmatrix};$$

then

$$\mathbf{AB} = \begin{bmatrix} 7 & 2 & 3 \\ 5 & 4 & 3 \end{bmatrix},$$

while **BA** is not defined.

If a and b are real numbers, then $ab = ba$; that is, ordinary multiplication is *commutative*. The last two examples show that this is not the case for matrix multiplication. In Example 1.17, both **AB** and **BA** are defined, but they are not equal, while in Example 1.18, the former is defined, but the latter is not. However, matrix multiplication exhibits other familiar properties of multiplication, as the next two theorems show.

Theorem 1.3 If **A**, **B**, and **C** are $m \times p$, $p \times q$, and $q \times n$ matrices, respectively, then

(17) $$(\mathbf{AB})\mathbf{C} = \mathbf{A}(\mathbf{BC});$$

that is, matrix multiplication is *associative*.

Proof. First let us be sure that the meaning of (17) is clear. The product on the left is calculated by first forming the $m \times q$ matrix $\mathbf{D} = \mathbf{AB}$, and then multiplying $\mathbf{D}$ on the right by $\mathbf{C}$. The resulting product is an $m \times n$ matrix. On the right-hand side of (17), we obtain first the $p \times n$ matrix $\mathbf{F} = \mathbf{BC}$, and then multiply it on the left by $\mathbf{A}$. The result is again an $m \times n$ matrix. We must show that the two $m \times n$ matrices in (17) are equal.

Let $\mathbf{D} = [d_{ij}]$ and $\mathbf{F} = [f_{ij}]$. By definition,

$$d_{ik} = \sum_{r=1}^{p} a_{ir}b_{rk} \qquad (1 \le i \le m, \quad 1 \le k \le q),$$

and the i, jth element of $(\mathbf{AB})\mathbf{C} = \mathbf{DC}$ is

$$\sum_{k=1}^{q} d_{ik}c_{kj} = \sum_{k=1}^{q} \left(\sum_{r=1}^{p} a_{ir}b_{rk} \right) c_{kj} \qquad (1 \le i \le m, \quad 1 \le j \le n).$$

Changing the order of summation, it can be seen that this is equal to

$$\sum_{r=1}^{p} a_{ir} \left(\sum_{k=1}^{q} b_{rk}c_{kj} \right) = \sum_{r=1}^{p} a_{ir} f_{rj} \qquad (1 \le i \le m, \quad 1 \le j \le n),$$

which is the i, jth element of $\mathbf{AF} = \mathbf{A}(\mathbf{BC})$. This completes the proof.

Example 1.19 Let

$$\mathbf{A} = \begin{bmatrix} 1 & 2 & -1 & 3 \\ 2 & 0 & 0 & 2 \end{bmatrix}, \quad \mathbf{B} = \begin{bmatrix} 1 & 0 & 2 \\ -1 & 1 & 3 \\ 2 & 2 & -1 \\ 3 & 0 & 2 \end{bmatrix}, \quad \text{and} \quad \mathbf{C} = \begin{bmatrix} 1 & -1 \\ 2 & 2 \\ 3 & 0 \end{bmatrix}.$$

Then

$$\mathbf{AB} = \mathbf{D} = \begin{bmatrix} 6 & 0 & 15 \\ 8 & 0 & 8 \end{bmatrix},$$

$$\mathbf{BC} = \mathbf{F} = \begin{bmatrix} 7 & -1 \\ 10 & 3 \\ 3 & 2 \\ 9 & -3 \end{bmatrix},$$

and

$$(\mathbf{AB})\mathbf{C} = \mathbf{A}(\mathbf{BC}) = \begin{bmatrix} 51 & -6 \\ 32 & -8 \end{bmatrix}.$$

Theorem 1.4 Suppose $\mathbf{A}$ is an $m \times p$ matrix, $\mathbf{B}$ and $\mathbf{C}$ are $p \times q$ matrices, $\mathbf{D}$ and $\mathbf{E}$ are $v \times p$ matrices, and t is a scalar. Then

(a) $\mathbf{A}(\mathbf{B} + \mathbf{C}) = \mathbf{AB} + \mathbf{AC}$;
(b) $(\mathbf{D} + \mathbf{E})\mathbf{C} = \mathbf{DC} + \mathbf{EC}$;
(c) $\mathbf{A}(t\mathbf{B}) = t(\mathbf{AB}) = (t\mathbf{A})\mathbf{B}$.

We leave the proof to the student (Exercise T-6).

Symmetric Matrices

Definition 1.6 The *transpose* of an $m \times n$ matrix $\mathbf{A} = [a_{ij}]$ is the $n \times m$ matrix $\mathbf{A}^{\mathrm{T}} = [a_{ij}^{\mathrm{T}}]$, whose i,jth element is given by

$$a_{ij}^{\mathrm{T}} = a_{ji} \qquad (1 \le i \le n, \quad 1 \le j \le m).$$

Thus, the ith row of $\mathbf{A}$ and the ith column of $\mathbf{A}^{\mathrm{T}}$ have the same elements. The same is true of the jth column of $\mathbf{A}$ and jth row of $\mathbf{A}^{\mathrm{T}}$.

Example 1.20 Let

$$\mathbf{A} = \begin{bmatrix} 1 & 3 \\ 2 & 2 \\ -1 & 4 \end{bmatrix}, \quad \mathbf{B} = \begin{bmatrix} 1 & 2 & 3 \\ 2 & -3 & 4 \\ 3 & 4 & 5 \end{bmatrix}, \quad \text{and} \quad \mathbf{C} = \begin{bmatrix} 0 & 1 & 2 \\ -1 & 0 & 3 \\ -2 & -3 & 0 \end{bmatrix};$$

then

$$\mathbf{A}^{\mathrm{T}} = \begin{bmatrix} 1 & 2 & -1 \\ 3 & 2 & 4 \end{bmatrix}, \quad \mathbf{B}^{\mathrm{T}} = \begin{bmatrix} 1 & 2 & 3 \\ 2 & -3 & 4 \\ 3 & 4 & 5 \end{bmatrix}, \quad \text{and} \quad \mathbf{C}^{\mathrm{T}} = \begin{bmatrix} 0 & -1 & -2 \\ 1 & 0 & -3 \\ 2 & 3 & 0 \end{bmatrix}.$$

Theorem 1.5 If $\mathbf{A}$ and $\mathbf{B}$ are matrices and r is a scalar, then

(a) $(\mathbf{A}^{\mathrm{T}})^{\mathrm{T}} = \mathbf{A}$;
(b) $(\mathbf{A} + \mathbf{B})^{\mathrm{T}} = \mathbf{A}^{\mathrm{T}} + \mathbf{B}^{\mathrm{T}}$ if $\mathbf{A}$ and $\mathbf{B}$ are both $m \times n$;
(c) $(r\mathbf{A})^{\mathrm{T}} = r\mathbf{A}^{\mathrm{T}}$;
(d) $(\mathbf{AB})^{\mathrm{T}} = \mathbf{B}^{\mathrm{T}}\mathbf{A}^{\mathrm{T}}$ if $\mathbf{A}$ is $m \times p$ and $\mathbf{B}$ is $p \times n$.

Proof. We prove only (d), leaving the rest of the proof for the student (Exercise T-7). The i, jth element of **AB** is

$$c_{ij} = \sum_{k=1}^{p} a_{ik}b_{kj};$$

hence the i, jth element of $(\mathbf{AB})^{\mathrm{T}}$ is

$$c_{ij}^{\mathrm{T}} = c_{ji} = \sum_{k=1}^{p} a_{jk}b_{ki} = \sum_{k=1}^{p} b_{ki}a_{jk} = \sum_{k=1}^{p} b_{ik}^{\mathrm{T}}a_{kj}^{\mathrm{T}},$$

which is the i, jth element of $\mathbf{B}^{\mathrm{T}}\mathbf{A}^{\mathrm{T}}$. This proves (d).

Example 1.21 Let

$$\mathbf{A} = \begin{bmatrix} 1 & 2 \\ -2 & 3 \\ 0 & 1 \end{bmatrix} \quad \text{and} \quad \mathbf{B} = \begin{bmatrix} 2 & 1 & -1 & 1 \\ 3 & 2 & 1 & 0 \end{bmatrix};$$

then

$$\mathbf{A}^{\mathrm{T}} = \begin{bmatrix} 1 & -2 & 0 \\ 2 & 3 & 1 \end{bmatrix}, \quad \mathbf{B}^{\mathrm{T}} = \begin{bmatrix} 2 & 3 \\ 1 & 2 \\ -1 & 1 \\ 1 & 0 \end{bmatrix},$$

$$\mathbf{AB} = \begin{bmatrix} 8 & 5 & 1 & 1 \\ 5 & 4 & 5 & -2 \\ 3 & 2 & 1 & 0 \end{bmatrix}, \quad \text{and} \quad \mathbf{B}^{\mathrm{T}}\mathbf{A}^{\mathrm{T}} = \begin{bmatrix} 8 & 5 & 3 \\ 5 & 4 & 2 \\ 1 & 5 & 1 \\ 1 & -2 & 0 \end{bmatrix},$$

so that $(\mathbf{AB})^{\mathrm{T}} = \mathbf{B}^{\mathrm{T}}\mathbf{A}^{\mathrm{T}}$.

Definition 1.7 A matrix **A** such that

$$\mathbf{A}^{\mathrm{T}} = \mathbf{A}$$

is said to be *symmetric*. That is, a symmetric matrix **A** is a square matrix, say of order n, such that

$$a_{ij} = a_{ji} \qquad (1 \leq i, \ j \leq n).$$

The matrix **B** of Example 1.20 is symmetric. A *diagonal* matrix such as

$$\begin{bmatrix} 4 & 0 & 0 \\ 0 & -1 & 0 \\ 0 & 0 & 2 \end{bmatrix}$$

is clearly symmetric. The $n \times n$ diagonal matrix

$$\mathbf{I}_n = \begin{bmatrix} 1 & 0 & \cdots & 0 \\ 0 & 1 & \cdots & 0 \\ \vdots & \vdots & & \vdots \\ 0 & 0 & & 1 \end{bmatrix}$$

all of whose diagonal elements equal 1, is called the *identity matrix* of order n; thus

$$\mathbf{I}_2 = \begin{bmatrix} 1 & 0 \\ 0 & 1 \end{bmatrix}, \ \mathbf{I}_3 = \begin{bmatrix} 1 & 0 & 0 \\ 0 & 1 & 0 \\ 0 & 0 & 1 \end{bmatrix}, \ldots.$$

If **A** is an $m \times n$ matrix, then

$$\mathbf{I}_m \mathbf{A} = \mathbf{A} \mathbf{I}_n = \mathbf{A}$$

(Exercise T-11).

When it is unnecessary to specify the order of the identity matrix, or when it is clear from the context, we shall write simply **I** rather than $\mathbf{I}_n$.

Linear Systems in Matrix Form

Using the definition of matrix multiplication, we can rewrite the linear system (1) as

$$\begin{bmatrix} y_1 \\ y_2 \\ \vdots \\ y_m \end{bmatrix} = \begin{bmatrix} a_{11} & a_{12} & \cdots & a_{1n} \\ a_{21} & a_{22} & \cdots & a_{2n} \\ \vdots & \vdots & & \vdots \\ a_{m1} & a_{m2} & \cdots & a_{mn} \end{bmatrix} \begin{bmatrix} x_1 \\ x_2 \\ \vdots \\ x_n \end{bmatrix},$$

where we regard

$$\mathbf{X} = \begin{bmatrix} x_1 \\ x_2 \\ \vdots \\ x_n \end{bmatrix}$$

as an $n \times 1$ matrix,

$$
Y = \begin{bmatrix} y_1 \\ y_2 \\ \vdots \\ y_m \end{bmatrix}
$$

as an $m \times 1$ matrix, and

$$
A = \begin{bmatrix} a_{11} & a_{12} & \cdots & a_{1n} \\ a_{21} & a_{22} & \cdots & a_{2n} \\ \vdots & \vdots & & \vdots \\ a_{m1} & a_{m2} & \cdots & a_{mn} \end{bmatrix}
$$

as an $m \times n$ matrix. We call A the *coefficient matrix* of (1). In terms of these matrices, we can write (1) more briefly as

$$
Y = AX.
$$

For numerical work, it is also convenient to introduce the *augmented matrix* of (1), which is obtained by adjoining Y to A as an additional column. We denote the augmented matrix by $[A \vdots Y]$; thus

$$
[A \vdots Y] = \begin{bmatrix} a_{11} & a_{12} & \cdots & a_{1n} & y_1 \\ a_{21} & a_{22} & \cdots & a_{2n} & y_2 \\ \vdots & \vdots & & \vdots & \vdots \\ a_{m1} & a_{m2} & \cdots & a_{mn} & y_m \end{bmatrix}.
$$

Here, the dotted vertical line has no significance other than to indicate that the last column of $[A \vdots Y]$ is not part of the coefficient matrix.

Example 1.22 The system

$$
x + y - z = 2,
$$

$$
3x + 2z = -1,
$$

$$
4x + 2y - 3z = 0,
$$

can be written as

$$
AX = Y,
$$

with

$$
A = \begin{bmatrix} 1 & 1 & -1 \\ 3 & 0 & 2 \\ 4 & 2 & -3 \end{bmatrix}, \qquad Y = \begin{bmatrix} 2 \\ -1 \\ 0 \end{bmatrix}, \qquad \text{and} \qquad X = \begin{bmatrix} x \\ y \\ z \end{bmatrix}.
$$

Its augmented matrix is

$$[\mathbf{A} \vdots \mathbf{Y}] = \begin{bmatrix} 1 & 1 & -1 & 2 \\ 3 & 0 & 2 & -1 \\ 4 & 2 & -3 & 0 \end{bmatrix}.$$

EXERCISES 1.1

1. Solve by elimination:

(a) $x + y = -1$, (b) $x + y + 2z = -3$,

$2x - y = 7$; $2x - y + z = -3$,

$x - 2y - 3z = 6$.

2. Solve by elimination:

(a) $2x + y = 2$, (b) $x + y - z = -3$,

$3x - 2y = -11$; $2x - 2y + 3z = 1$,

$2x + y + 4z = 10$.

3. Solve for x and y:

(a) $\begin{bmatrix} 2 & -3 \\ 2x & 3y \end{bmatrix} = \begin{bmatrix} 2 & -3 \\ 4 & -6 \end{bmatrix}$;

(b) $\begin{bmatrix} 3 & 4 \\ 2x + y & 2y - x \end{bmatrix} = \begin{bmatrix} 3 & 4 \\ 5 & 5 \end{bmatrix}$.

4. Solve for x and y:

(a) $\begin{bmatrix} 2x & 3y \\ -2 & 1 \end{bmatrix} = \begin{bmatrix} 4x & 9 \\ -2 & 1 \end{bmatrix}$;

(b) $\begin{bmatrix} x + 3 & 4 \\ -2 & y - 2 \end{bmatrix} = \begin{bmatrix} y + 5 & 4 \\ -2 & -x + 2 \end{bmatrix}$.

In Exercises 5 and 6 let

$$A = \begin{bmatrix} 1 & 2 & 3 \\ -1 & 2 & 1 \\ 3 & 2 & 1 \end{bmatrix}, \quad B = \begin{bmatrix} 1 & 2 \\ 0 & 1 \end{bmatrix}, \quad C = \begin{bmatrix} 3 & 2 \\ 2 & -1 \end{bmatrix},$$

$$D = \begin{bmatrix} 1 & 3 & 2 \\ -1 & 2 & 2 \end{bmatrix}, \quad E = \begin{bmatrix} 0 & 2 & 1 \\ -1 & 2 & 1 \\ 0 & 1 & 1 \end{bmatrix},$$

$$F = \begin{bmatrix} 0 & 1 & 2 \\ 1 & -2 & 0 \end{bmatrix}, \quad G = \begin{bmatrix} 0 & 2 \\ 1 & 3 \end{bmatrix}.$$

5. Compute, if defined:

   (a) 2A;   (b) A + B;   (c) A − E;   (d) 2B + 3C;

   (e) 2D − E.

6. Compute, if defined:

   (a) B − C;   (b) 2D − 3F;   (c) B + (C + G);

   (d) A − D;   (e) B − F + C.

7. Verify Theorem 1.1 for

$$A = \begin{bmatrix} 1 & 2 & 3 \\ -2 & 1 & 3 \end{bmatrix}, \quad B = \begin{bmatrix} -1 & 4 & 2 \\ 3 & 1 & 2 \end{bmatrix}, \quad C = \begin{bmatrix} 0 & 1 & 1 \\ 0 & 1 & 2 \end{bmatrix}.$$

8. Verify Theorem 1.2 for

$$A = \begin{bmatrix} 1 & 0 & 1 \\ 2 & 2 & -3 \end{bmatrix}, \quad B = \begin{bmatrix} -3 & 2 & 5 \\ 4 & 4 & -2 \end{bmatrix}, \quad r = 2 \text{ and } s = -3.$$

9. (a) Compute $\sum_{i=1}^{5} \alpha_i$, where $\alpha_i = 2i - 1$.

   (b) Compute $\sum_{i=1}^{7} \alpha_i$, where $\alpha_i = 2i$.

10. Let $\alpha_i = i + 1$ and $\beta_i = 2i - 1$. Compute

    (a) $\sum\limits_{i=1}^{5} (2\alpha_i + \beta_i)$;   (b) $\sum\limits_{i=1}^{5} (\alpha_i - 3\beta_i)$.

11. Let

$$\mathbf{A} = [a_{ij}] = \begin{bmatrix} 1 & 2 & 3 & 4 \\ 4 & 2 & -1 & 3 \\ 2 & 4 & 5 & 1 \\ 4 & 3 & -2 & 1 \end{bmatrix}.$$

    Compute

    (a) $c_j = \sum\limits_{i=1}^{4} a_{ij} \quad (1 \le j \le 4)$;   (b) $r_i = \sum\limits_{j=1}^{4} a_{ij}$;

    (c) $\sum\limits_{j=1}^{4} c_j$;   (d) $\sum\limits_{i=1}^{4} r_i$.

12. Repeat Exercise 11 for

$$\mathbf{A} = \begin{bmatrix} 1 & 1 & -1 & 2 \\ -2 & 1 & 1 & 3 \\ 4 & -1 & 5 & 4 \\ 2 & 0 & 0 & 1 \end{bmatrix}.$$

In Exercises 13, 14, and 15 let

$$\mathbf{A} = \begin{bmatrix} 1 & 2 & 2 \\ -1 & 3 & 1 \end{bmatrix}, \quad \mathbf{B} = \begin{bmatrix} 2 & 2 \\ 3 & -1 \end{bmatrix}, \quad \mathbf{C} = \begin{bmatrix} 1 & 2 \\ 2 & 3 \\ -1 & 1 \end{bmatrix},$$

$$\mathbf{D} = \begin{bmatrix} 1 & 0 \\ 2 & -1 \end{bmatrix}, \quad \mathbf{E} = \begin{bmatrix} 1 & 1 \\ 2 & 1 \\ 3 & -1 \end{bmatrix}, \quad \mathbf{F} = \begin{bmatrix} 1 & 2 & -1 \\ 0 & 1 & 1 \end{bmatrix}.$$

13. Compute, if defined:

    (a) **AB**;   (b) **BA**;   (c) **AC** − 2**B**;

    (d) **AE** + 4**F**;   (e) **BD** + **AE**;   (f) **ACD**.

14. Compute, if defined:

    (a) (**A** + **F**)**E**;   (b) **B**(**C** + **D**);   (c) **A**(2**C**) − 5**C**(2**D**);

    (d) (**B** + **D**)**AE**;   (e) 2**DF** + 4**CE**.

15. Compute, if defined:

    (a) (**A** + **F**)$^\mathrm{T}$, **A**$^\mathrm{T}$ + **F**$^\mathrm{T}$;   (b) (3**A**)$^\mathrm{T}$, 3**A**$^\mathrm{T}$;   (c) (**AC**)$^\mathrm{T}$, **C**$^\mathrm{T}$**A**$^\mathrm{T}$;

    (d) (2**A** + 3**B**)$^\mathrm{T}$;   (e) (**A** + **F**)$^\mathrm{T}$**B**.

16. If

$$\mathbf{A} = \begin{bmatrix} 0 & 1 \\ 0 & 0 \end{bmatrix} \quad \text{and} \quad \mathbf{B} = \begin{bmatrix} 1 & 0 \\ 0 & 0 \end{bmatrix},$$

    show that **AB** = **0**.

17. Let

$$\mathbf{A} = \begin{bmatrix} 0 & 1 \\ 0 & 0 \end{bmatrix}, \quad \mathbf{B} = \begin{bmatrix} 1 & 0 \\ 0 & 0 \end{bmatrix}, \quad \text{and} \quad \mathbf{C} = \begin{bmatrix} 1 & 1 \\ 1 & 1 \end{bmatrix}.$$

    Show that **AC** = **BC**.

18. If

$$\mathbf{A} = \begin{bmatrix} 1 & 2 \\ 3 & 1 \end{bmatrix} \quad \text{and} \quad \mathbf{B} = \begin{bmatrix} -1 & 2 \\ 3 & 1 \end{bmatrix},$$

    show that **AB** ≠ **BA**.

In Exercises 19 and 20, write each system of equations in matrix form and find its coefficient matrix and augmented matrix.

19. (a) $x + y = 10$,   (b) $x - y + 2z = 5$,

        $2x - 3y = 5$; $2x + 3y - z = -3$,

                        $4x + 2y + 3z = 8$;

    (c) $x - y + 2z - w = 4$,

        $2x + 2y + 3z = 8$.

20.   (a) $3x - y = 5,$ (b) $x + y - 2z = 4,$
          $3x + 3y = 8;$ $x - y = 2;$

     (c) $x + 2y + w = 4,$
          $2x - y + z = -2,$
          $y - 2z + 2w = 6.$

THEORETICAL EXERCISES

T-1.   Show that $(\mathbf{A} + \mathbf{B}) + \mathbf{C} = \mathbf{A} + (\mathbf{B} + \mathbf{C})$ whenever $\mathbf{A}$, $\mathbf{B}$, and $\mathbf{C}$ are $m \times n$ matrices.

T-2.   If $\mathbf{A} = [a_{ij}]$ show that $-\mathbf{A} = [-a_{ij}] = (-1)\mathbf{A}$.

T-3.   Prove Theorem 1.2.

T-4.   Show that

$$\sum_{i=1}^{n} i = \frac{n(n + 1)}{2} \,.$$

T-5.   Show that

$$\sum_{i=1}^{n} (a_i + b_i) c_i = \sum_{i=1}^{n} a_i c_i + \sum_{i=1}^{n} b_i c_i$$

and

$$\sum_{i=1}^{n} (ca_i) = c \sum_{i=1}^{n} a_i.$$

T-6.   Prove Theorem 1.4.

T-7.   Prove (a), (b), and (c) of Theorem 1.5.

T-8.   A *skew symmetric* matrix is a square matrix $\mathbf{A} = [a_{ij}]$ such that $\mathbf{A}^{\mathrm{T}} = -\mathbf{A}$. Show that $a_{ij} = -a_{ji}$ for a skew symmetric matrix, and that $a_{ii} = 0$.

T-9.   *Prove:* If $\mathbf{A}$ is both symmetric and skew symmetric, then $\mathbf{A} = \mathbf{0}$.

T-10. *Prove:* An arbitrary $n \times n$ matrix can be written uniquely in the form $\mathbf{A} = \mathbf{B} + \mathbf{C}$, where $\mathbf{B}$ is symmetric and $\mathbf{C}$ is skew symmetric.

T-11. *Prove:* If **A** is an $m \times n$ matrix, then $\mathbf{I}_m\mathbf{A} = \mathbf{AI}_n = \mathbf{A}$.

T-12. (a) If **A** has a row of zeros, show that **AB** has a row of zeros.
      (b) If **B** has a column of zeros, show that **AB** has a column of zeros.

T-13. *Prove:* If **A** is symmetric, then $\mathbf{A}^\mathsf{T}$ is symmetric.

T-14. *Prove:* The sum and product of diagonal matrices are diagonal.

T-15. Suppose **A** and **B** are symmetric.
      (a) Show that $\mathbf{A} + \mathbf{B}$ is symmetric.
      (b) Show that **AB** is symmetric if and only if $\mathbf{AB} = \mathbf{BA}$.

T-16. A matrix $\mathbf{A} = [a_{ij}]$ is *upper triangular* if $a_{ij} = 0$ for $i > j$. Show that the sum and product of upper triangular matrices are upper triangular.

1.2 Solution of Equations

In Section 1.1, we associated the augmented matrix

$$[\mathbf{A} \mid \mathbf{Y}] = \begin{bmatrix} a_{11} & a_{12} & \cdots & a_{1n} & y_1 \\ a_{21} & a_{22} & \cdots & a_{2n} & y_2 \\ \vdots & \vdots & & \vdots & \vdots \\ a_{m1} & a_{m2} & \cdots & a_{mn} & y_m \end{bmatrix}$$

with the linear system

(1)
$$\begin{aligned} a_{11}x_1 + a_{12}x_2 + \cdots + a_{1n}x_n &= y_1 \\ a_{21}x_1 + a_{22}x_2 + \cdots + a_{2n}x_n &= y_2 \\ &\vdots \\ a_{m1}x_1 + a_{m2}x_2 + \cdots + a_{mn}x_n &= y_n. \end{aligned}$$

Thus, every linear system has an augmented matrix and, conversely, any matrix with more than one column is the augmented matrix of a linear system. The advantage of the augmented matrix is its brevity. Properly interpreted, it has the same meaning as system (1), but requires less writing. In this section, we consider matrix operations that provide efficient methods for the solution of linear systems, and are also of interest in other applications.

Elementary Row Operations

The method of elimination, as described in Section 1.1, is essentially a method by which a linear system such as

$$x + y - z = 4,$$
$$(2) \qquad 3x - 2y + 2z = -3,$$
$$4x + 2y - 3z = 11,$$

is transformed into a simple system such as

$$x + y - z = 4,$$
$$(3) \qquad y - z = 3,$$
$$z = -1,$$

by means of operations which guarantee that (2) and (3) have the same solutions. The operations by which this was accomplished in Example 1.2 were of two kinds: multiplication of an equation by a nonzero constant, and addition of a multiple of one equation to another. In some problems, it is convenient to employ a third operation which does not change the solutions of a linear system; namely, the interchange of two equations.

The effects of these operations on the corresponding augmented matrices can be described in terms of the matrix operations defined next. However, these operations are applicable to arbitrary matrices which are not necessarily being considered as augmented matrices of linear systems.

Definition 2.1 The following operations on the rows of an $m \times n$ matrix $\mathbf{A} = [a_{ij}]$ are called *elementary row operations:*

(a) Multiplication of the ith row by a nonzero constant c; that is, replacing $a_{i1}, \ldots, a_{in}$ by $ca_{i1}, \ldots, ca_{in}$.

(b) Adding d times the rth row to the sth row, where $r \neq s$; that is, replacing $a_{s1}, \ldots, a_{sn}$ by $a_{s1} + da_{r1}, \ldots, a_{sn} + da_{rn}$.

(c) Interchanging the rth row and the sth row; that is, replacing $a_{r1}, \ldots, a_{rn}$ by $a_{s1}, \ldots, a_{sn}$, and replacing $a_{s1}, \ldots, a_{sn}$ by $a_{r1}, \ldots, a_{rn}$.

Example 2.1 Consider the matrix

$$\mathbf{A} = \begin{bmatrix} 2 & 2 & 4 & 6 \\ 0 & 0 & 1 & 2 \\ 3 & 4 & 7 & 10 \end{bmatrix}.$$

Multiplying the first row of **A** by $\frac{1}{2}$ produces

$$\mathbf{A}_1 = \begin{bmatrix} 1 & 1 & 2 & 3 \\ 0 & 0 & 1 & 2 \\ 3 & 4 & 7 & 10 \end{bmatrix}.$$

Adding -3 times the first row of $\mathbf{A}_1$ to its third row produces

$$\mathbf{A}_2 = \begin{bmatrix} 1 & 1 & 2 & 3 \\ 0 & 0 & 1 & 2 \\ 0 & 1 & 1 & 1 \end{bmatrix}.$$

Interchanging the second and third rows of $\mathbf{A}_2$ produces

$$\mathbf{A}_3 = \begin{bmatrix} 1 & 1 & 2 & 3 \\ 0 & 1 & 1 & 1 \\ 0 & 0 & 1 & 2 \end{bmatrix}.$$

Definition 2.2 An $m \times n$ matrix **B** is *row equivalent* to an $m \times n$ matrix **A** if there is a sequence of matrices $\mathbf{A}_0, \mathbf{A}_1, \ldots, \mathbf{A}_N$ such that $\mathbf{A} = \mathbf{A}_0, \mathbf{B} = \mathbf{A}_N$, and, for each $i = 1, \ldots, N - 1$, $\mathbf{A}_i$ is obtained by applying an elementary row operation to $\mathbf{A}_{i-1}$.

Example 2.2 The matrix

$$\mathbf{B} = \begin{bmatrix} 1 & 1 & 2 & 3 \\ 0 & 1 & 1 & 1 \\ 0 & 0 & 1 & 2 \end{bmatrix}$$

is row equivalent to

$$\mathbf{A} = \begin{bmatrix} 2 & 2 & 4 & 6 \\ 0 & 0 & 1 & 2 \\ 3 & 4 & 7 & 10 \end{bmatrix},$$

since we exhibited a sequence $\mathbf{A}, \mathbf{A}_1, \mathbf{A}_2, \mathbf{B}$ in Example 2.1 which satisfies the requirements of Definition 2.2.

Theorem 2.1

(a) A matrix $\mathbf{A}$ is row equivalent to itself.
(b) If $\mathbf{B}$ is row equivalent to $\mathbf{A}$, then $\mathbf{A}$ is row equivalent to $\mathbf{B}$.
(c) If $\mathbf{A}$ is row equivalent to $\mathbf{B}$ and $\mathbf{B}$ is row equivalent to $\mathbf{C}$, then $\mathbf{A}$ is row equivalent to $\mathbf{C}$.

We leave the proof of this theorem to the student (Exercise T-1, Section 1.3).

Because of part (b) of Theorem 2.1, we shall often replace the statement "$\mathbf{B}$ is row equivalent to $\mathbf{A}$," by "$\mathbf{A}$ and $\mathbf{B}$ are row equivalent."

Theorem 2.2   Let the augmented matrices of the two linear systems

$$
\begin{aligned}
a_{11}x_1 + a_{12}x_2 + \cdots + a_{1n}x_n &= y_1, \\
a_{21}x_1 + a_{22}x_2 + \cdots + a_{2n}x_n &= y_2, \\
&\;\;\vdots \\
a_{m1}x_1 + a_{m2}x_2 + \cdots + a_{mn}x_n &= y_m,
\end{aligned}
$$

(4)

and

$$
\begin{aligned}
b_{11}x_1 + b_{12}x_2 + \cdots + b_{1n}x_n &= z_1, \\
b_{21}x_2 + b_{22}x_2 + \cdots + b_{2n}x_n &= z_2, \\
&\;\;\vdots \\
b_{m1}x_1 + b_{m2}x_2 + \cdots + b_{mn}x_n &= z_m,
\end{aligned}
$$

(5)

be row equivalent. Then every solution of (4) is a solution of (5), and every solution of (5) is a solution of (4).

Proof.   Denote the augmented matrices of (4) and (5) by $[\mathbf{A} \mid \mathbf{Y}]$ and $[\mathbf{B} \mid \mathbf{Z}]$, respectively. By Definition 2.2, there is a sequence of matrices $[\mathbf{A} \mid \mathbf{Y}] = [\mathbf{A}_0 \mid \mathbf{Y}_0], [\mathbf{A}_1 \mid \mathbf{Y}_1], \ldots, [\mathbf{A}_N \mid \mathbf{Y}_N] = [\mathbf{B} \mid \mathbf{Z}]$ such that, for $i = 1, \ldots, N$, $[\mathbf{A}_i \mid \mathbf{Y}_i]$ is obtained by applying an elementary row operation to $[\mathbf{A}_{i-1} \mid \mathbf{Y}_{i-1}]$. Since the elementary row operations were defined precisely so that the corresponding linear systems

$$\mathbf{A}_{i-1}\mathbf{X} = \mathbf{Y}_{i-1} \quad \text{and} \quad \mathbf{A}_i\mathbf{X} = \mathbf{Y}_i \quad (i = 1, 2, \ldots, N)$$

would have the same solutions (Exercise T-1), the theorem follows.

Example 2.3   We saw in Example 2.2 that the augmented matrices of the systems

$$
\begin{aligned}
2x + 2y + 4z &= 6, \\
z &= 2, \\
3x + 4y + 7z &= 10,
\end{aligned}
$$

(6)

and

$$
\begin{aligned}
x + y + 2z &= 3, \\
y + z &= 1, \\
z &= 2,
\end{aligned}
$$
(7)

are row equivalent. Theorem 2.2 implies that every solution of (6) is a solution of (7), and conversely. In this case there is only one solution, $(x, y, z) = (0, -1, 2)$, as can be seen by solving (7) for z, y, and x, in that order.

It should be clear to the student that the transformation of the augmented matrix of (6) into the augmented matrix of (7) (see Example 2.1) is equivalent to solving (6) by the method of elimination.

Matrices in Row Echelon and Reduced Row Echelon Form

Definition 2.3   An $m \times n$ matrix **A** is said to be *in row echelon form* if

(a) each of the first k rows $(1 \leq k \leq m)$ has at least one nonzero element and rows $k + 1$ through m (if $k < m$) contain only zero elements;

(b) counting from left to right, the first nonzero element in each of the first k rows is a 1; and

(c) if $k \geq 2$ and, for $1 \leq i \leq k$, the first 1 in the ith row appears in the j_ith column, then $j_1 < j_2 < \cdots < j_k$.

If, in addition to (a), (b), and (c), the j_ith column $(i = 1, \ldots, k)$ contains only one nonzero element (the 1 required by (c)), then **A** is said to be *in reduced row echelon form*.

Example 2.4   The matrices

$$
\mathbf{A} = \begin{bmatrix} 1 & 1 & 3 & 4 \\ 0 & 1 & 2 & 1 \\ 0 & 0 & 1 & 1 \end{bmatrix},
$$

and

$$
\mathbf{B} = \begin{bmatrix} 1 & 0 & 2 & 0 & 4 & 4 \\ 0 & 0 & 1 & 0 & 1 & 1 \\ 0 & 0 & 0 & 0 & 0 & 1 \\ 0 & 0 & 0 & 0 & 0 & 0 \end{bmatrix}
$$

are in row echelon form. In $\mathbf{A}$, $k = 3$, $j_1 = 1$, $j_2 = 2$, and $j_3 = 3$; in $\mathbf{B}$, $k = 3$, $j_1 = 1$, $j_2 = 3$, and $j_3 = 6$.

The matrices

$$\mathbf{C} = \begin{bmatrix} 1 & 1 & 0 & 0 \\ 0 & 0 & 0 & 0 \\ 0 & 1 & 0 & 2 \end{bmatrix},$$

$$\mathbf{D} = \begin{bmatrix} 2 & 2 & 0 & 0 \\ 0 & 1 & 0 & 2 \\ 0 & 0 & 0 & 0 \end{bmatrix},$$

and

$$\mathbf{E} = \begin{bmatrix} 0 & 1 & 0 & 2 \\ 1 & 1 & 0 & 0 \\ 0 & 0 & 0 & 0 \end{bmatrix}$$

are not in row echelon form, since they violate (a), (b), and (c), respectively. However, $\mathbf{C}$, $\mathbf{D}$, and $\mathbf{E}$ are all row equivalent to

$$\mathbf{F} = \begin{bmatrix} 1 & 1 & 0 & 0 \\ 0 & 1 & 0 & 2 \\ 0 & 0 & 0 & 0 \end{bmatrix}$$

(verify!), which is in row echelon form.

Example 2.5 The matrices

$$\mathbf{A}_1 = \begin{bmatrix} 1 & 0 & 0 & 2 \\ 0 & 1 & 0 & -1 \\ 0 & 0 & 1 & 1 \end{bmatrix},$$

$$\mathbf{B}_1 = \begin{bmatrix} 1 & 0 & 0 & 0 & 2 & 0 \\ 0 & 0 & 1 & 0 & 1 & 0 \\ 0 & 0 & 0 & 0 & 0 & 1 \end{bmatrix},$$

and

$$\mathbf{F}_1 = \begin{bmatrix} 1 & 0 & 0 & -2 \\ 0 & 1 & 0 & 2 \\ 0 & 0 & 0 & 0 \end{bmatrix}$$

are in reduced row echelon form.

From Examples 1.2 and 2.3, it can be seen that solving a linear system by the method of elimination is equivalent to finding a matrix in row echelon form which is row equivalent to the augmented matrix of the system. The next theorem guarantees that this can always be done.

Theorem 2.3 Any nonzero $m \times n$ matrix $\mathbf{A} = [a_{ij}]$ is row equivalent to a matrix $\mathbf{B}$ in row echelon form.

Proof. Suppose the leftmost nonzero element of $\mathbf{A}$ appears in the j_1th column. If $a_{1j_1} = 0$, then let $\mathbf{B} = [b_{ij}]$ be obtained from interchanging the first and i_1th rows of $\mathbf{A}$, where i_1 is chosen so that $a_{i_1 j_1} \neq 0$. If $a_{1j_1} \neq 0$, let $\mathbf{B} = \mathbf{A}$. In either case, $b_{1j_1} \neq 0$. Now divide the first row of $\mathbf{B}$ by b_{1j_1} and then subtract b_{rj_1} times the resulting first row from the rth row ($r = 2, \ldots, m$). This produces a matrix $\mathbf{A}_1$ whose first j_1 columns contain only one nonzero element, a 1 in the $(1, j_1)$-position. If $m \geq 2$ let $\mathbf{C}$ be the $(m-1) \times n$ matrix obtained by deleting the first row from $\mathbf{A}_1$. If $\mathbf{C} = \mathbf{0}$, we are finished; if $\mathbf{C} \neq \mathbf{0}$, then treat $\mathbf{C}$ as we have just treated $\mathbf{A}$, while carrying along the first row of $\mathbf{A}_1$ unaffected (see Examples 2.6 and 2.7). Continuing in this way, we eventually arrive at a matrix in row echelon form which is row equivalent to $\mathbf{A}$.

Example 2.6 To obtain a matrix in row echelon form which is row equivalent to

$$\mathbf{A} = \begin{bmatrix} 0 & 0 & 2 & -1 \\ 0 & 2 & -1 & 1 \\ 0 & 6 & 1 & 1 \\ 0 & 0 & 1 & 3 \end{bmatrix},$$

we interchange the first and second rows:

$$\begin{bmatrix} 0 & 2 & -1 & 1 \\ 0 & 0 & 2 & -1 \\ 0 & 6 & 1 & 1 \\ 0 & 0 & 1 & 3 \end{bmatrix}.$$

Then multiply the first row by $\frac{1}{2}$:

$$\begin{bmatrix} 0 & 1 & -\frac{1}{2} & \frac{1}{2} \\ 0 & 0 & 2 & -1 \\ 0 & 6 & 1 & 1 \\ 0 & 0 & 1 & 3 \end{bmatrix}.$$

Next subtract six times the first row from the third row:

$$(8) \qquad \mathbf{A_1} = \begin{bmatrix} 0 & 1 & -\frac{1}{2} & \frac{1}{2} \\ 0 & 0 & 2 & -1 \\ 0 & 0 & 4 & -2 \\ 0 & 0 & 1 & 3 \end{bmatrix}.$$

The matrix $\mathbf{C}$ mentioned in the proof of Theorem 2.3 is obtained by deleting the first row from $\mathbf{A_1}$:

$$\mathbf{C} = \begin{bmatrix} 0 & 0 & 2 & -1 \\ 0 & 0 & 4 & -2 \\ 0 & 0 & 0 & 3 \end{bmatrix}.$$

We now continue to apply row operations to (8) which do not, however, involve the first row. (Thus, we are actually operating on the rows of $\mathbf{C}$.)

Multiplying the second row of (8) by $\frac{1}{2}$ yields

$$\begin{bmatrix} 0 & 1 & -\frac{1}{2} & \frac{1}{2} \\ 0 & 0 & 1 & -\frac{1}{2} \\ 0 & 0 & 4 & -2 \\ 0 & 0 & 1 & 3 \end{bmatrix}.$$

Subtracting four times the second row from the third row and one times the second row from the fourth row yields

$$\begin{bmatrix} 0 & 1 & -\frac{1}{2} & \frac{1}{2} \\ 0 & 0 & 1 & -\frac{1}{2} \\ 0 & 0 & 0 & 0 \\ 0 & 0 & 0 & \frac{7}{2} \end{bmatrix}.$$

Interchanging the third and fourth rows yields

$$\begin{bmatrix} 0 & 1 & -\frac{1}{2} & \frac{1}{2} \\ 0 & 0 & 1 & -\frac{1}{2} \\ 0 & 0 & 0 & \frac{7}{2} \\ 0 & 0 & 0 & 0 \end{bmatrix},$$

and multiplying the third row by $\frac{2}{7}$ yields

$$\begin{bmatrix} 0 & 1 & -\frac{1}{2} & \frac{1}{2} \\ 0 & 0 & 1 & -\frac{1}{2} \\ 0 & 0 & 0 & 1 \\ 0 & 0 & 0 & 0 \end{bmatrix},$$

which is in row echelon form, and is row equivalent to **A**.

It is not necessary to rewrite the entire matrix after each elementary row operation; in fact, the work should be arranged so that successive matrices which are actually written out differ by more than one elementary row operation, except possibly near the end of the calculation. We illustrate this in the next example.

Example 2.7   Let

$$\mathbf{A} = \begin{bmatrix} 1 & 1 & 2 & 1 \\ 2 & 3 & 5 & 3 \\ 2 & 2 & 4 & 2 \\ 3 & 4 & 7 & 4 \\ 4 & 5 & 10 & 7 \end{bmatrix}.$$

By subtracting suitable multiples of the first row from the other rows, we find that **A** is row equivalent to

$$\begin{bmatrix} 1 & 1 & 2 & 1 \\ 0 & 1 & 1 & 1 \\ 0 & 0 & 0 & 0 \\ 0 & 1 & 1 & 1 \\ 0 & 1 & 2 & 3 \end{bmatrix}.$$

By subtracting the second row of this matrix from its last two rows and rearranging the rows of the result, we find that **A** is row equivalent to

$$\begin{bmatrix} 1 & 1 & 2 & 1 \\ 0 & 1 & 1 & 1 \\ 0 & 0 & 1 & 2 \\ 0 & 0 & 0 & 0 \\ 0 & 0 & 0 & 0 \end{bmatrix}.$$

Theorem 2.4   Every nonzero matrix **A** is row equivalent to a unique matrix **C** in reduced row echelon form.

The proof of this theorem is quite technical, and we omit it. The method for finding **C** is straightforward: First find a matrix **B** in row echelon form which is row equivalent to **A**. Then apply elementary row operations to **B** to produce zeros above the leading 1 in each row of **B**.

The next example illustrates this procedure.

Example 2.8 Let

$$
\mathbf{A} = \begin{bmatrix} 1 & 1 & 2 & 1 & 3 & 1 \\ 2 & 2 & 5 & 4 & 7 & 0 \\ 1 & 1 & 2 & 2 & 5 & 1 \\ 1 & 1 & 3 & 3 & 5 & 0 \end{bmatrix};
$$

then **A** is row equivalent to

$$
\mathbf{B}_1 = \begin{bmatrix} 1 & 1 & 2 & 1 & 3 & 1 \\ 0 & 0 & 1 & 2 & 1 & -2 \\ 0 & 0 & 0 & 1 & 2 & 0 \\ 0 & 0 & 0 & 0 & 1 & 1 \end{bmatrix}
$$

(verify!), which is in row echelon form. Subtracting twice the second row of $\mathbf{B}_1$ from the first row yields

$$
\mathbf{B}_2 = \begin{bmatrix} 1 & 1 & 0 & -3 & 1 & 5 \\ 0 & 0 & 1 & 2 & 1 & -2 \\ 0 & 0 & 0 & 1 & 2 & 0 \\ 0 & 0 & 0 & 0 & 1 & 1 \end{bmatrix}.
$$

Adding three times the third row of $\mathbf{B}_2$ to its first row and subtracting twice the third row from the second yields

$$
\mathbf{B}_3 = \begin{bmatrix} 1 & 1 & 0 & 0 & 7 & 5 \\ 0 & 0 & 1 & 0 & -3 & -2 \\ 0 & 0 & 0 & 1 & 2 & 0 \\ 0 & 0 & 0 & 0 & 1 & 1 \end{bmatrix}.
$$

Subtracting suitable multiples of the last row from the others yields

$$\mathbf{B}_4 = \begin{bmatrix} 1 & 1 & 0 & 0 & 0 & -2 \\ 0 & 0 & 1 & 0 & 0 & 1 \\ 0 & 0 & 0 & 1 & 0 & -2 \\ 0 & 0 & 0 & 0 & 1 & 1 \end{bmatrix},$$

which is in reduced row echelon form, and is row equivalent to $\mathbf{A}$.

The next three examples illustrate the use of row echelon and reduced row echelon matrices in the solution of linear systems. The student should verify the row equivalence of the matrices in each example.

Example 2.9 The augmented matrix of the system

(9)
$$\begin{aligned} x + y - z &= 4, \\ 3x - 2y + 2z &= -3, \\ 4x + 2y - 3z &= 11, \end{aligned}$$

is

$$\begin{bmatrix} 1 & 1 & -1 & 4 \\ 3 & -2 & 2 & -3 \\ 4 & 2 & -3 & 11 \end{bmatrix},$$

which is row equivalent to

(10)
$$\begin{bmatrix} 1 & 1 & -1 & 4 \\ 0 & 1 & -1 & 3 \\ 0 & 0 & 1 & -1 \end{bmatrix}.$$

The system with this augmented matrix,

$$\begin{aligned} x + y - z &= 4, \\ y - z &= 3, \\ z &= -1, \end{aligned}$$

has the same solutions as (9). These three equations can be solved successively, as in Example 1.2, to obtain $z = -1$, $y = 2$, $x = 1$. The solution can also be obtained by showing that (10) is row equivalent to

$$\begin{bmatrix} 1 & 0 & 0 & 1 \\ 0 & 1 & 0 & 2 \\ 0 & 0 & 1 & -1 \end{bmatrix},$$

which is in reduced row echelon form, and represents the very simple system

$$x = 1,$$
$$y = 2,$$
$$z = -1.$$

Example 2.10 We saw in Example 2.8, that the augmented matrix of the system

$$x_1 + x_2 + 2x_3 + x_4 + 3x_5 = 1,$$
$$2x_1 + 2x_2 + 5x_3 + 4x_4 + 7x_5 = 0,$$
(11)
$$x_1 + x_2 + 2x_3 + 2x_4 + 5x_5 = 1,$$
$$x_1 + x_2 + 3x_3 + 3x_4 + 5x_5 = 0,$$

is row equivalent to

$$\begin{bmatrix} 1 & 1 & 0 & 0 & 0 & -2 \\ 0 & 0 & 1 & 0 & 0 & 1 \\ 0 & 0 & 0 & 1 & 0 & -2 \\ 0 & 0 & 0 & 0 & 1 & 1 \end{bmatrix},$$

which is the augmented matrix of the system

$$x_1 + x_2 = -2,$$
$$x_3 = 1,$$
$$x_4 = -2,$$
$$x_5 = 1.$$

According to Theorem 2.2, this system has the same solutions as (11).

We conclude that (11) has more than one solution; in fact $(x_1, x_2, x_3, x_4, x_5)$ is a solution of (11) if and only if

$$x_1 = r,$$

$$x_2 = -2 - r,$$

$$x_3 = 1,$$

$$x_4 = -2,$$

$$x_5 = 1,$$

where r is an arbitrary real number.

Example 2.11 The augmented matrix of the system

(12)
$$x + y - z + w = 4,$$
$$2x - y + z - w = -1,$$
$$3x + y + 2z + 4w = 0,$$

is row equivalent to

$$\begin{bmatrix} 1 & 0 & 0 & 0 & 1 \\ 0 & 1 & 0 & 2 & 1 \\ 0 & 0 & 1 & 1 & -2 \end{bmatrix},$$

which is the augmented matrix of the system

$$x = 1,$$

$$y + 2w = 1,$$

$$z + w = -2;$$

therefore (x, y, z, w) is a solution of (12) if and only if

$$x = 1,$$

$$y = 1 - 2r,$$

$$z = -2 - r,$$

$$w = r,$$

where r is an arbitrary real number.

Example 2.12   The augmented matrix of the system

$$x + y - z = 4,$$

(13)
$$x + y + 2z = 2,$$

$$2x + 2y + z = 5,$$

is

$$\begin{bmatrix} 1 & 1 & -1 & 4 \\ 1 & 1 & 2 & 2 \\ 2 & 2 & 1 & 5 \end{bmatrix},$$

which is row equivalent to

$$\begin{bmatrix} 1 & 1 & -1 & 4 \\ 0 & 0 & 1 & -\frac{2}{3} \\ 0 & 0 & 0 & 1 \end{bmatrix}.$$

The last row of this matrix stands for the equation

$$0 \cdot x + 0 \cdot y + 0 \cdot z = 1,$$

which is not satisfied for any (x, y, z). Therefore, (13) has no solutions.

This example suggests a necessary and sufficient condition for a linear system to have no solution.

Theorem 2.5   The system $\mathbf{AX} = \mathbf{Y}$ of m equations in n unknowns has no solution if and only if its augmented matrix is row equivalent to a matrix containing a row whose first n elements are zero, and whose $(n + 1)$th element is nonzero.

We leave the proof of this theorem to the student (Exercise T-2).

Homogeneous Systems

A *homogeneous* linear system is a linear system of the form

$$a_{11}x_1 + a_{12}x_2 + \cdots + a_{1n}x_n = 0,$$

(14)
$$a_{21}x_1 + a_{22}x_2 + \cdots + a_{2n}x_n = 0,$$

$$\vdots$$

$$a_{m1}x_1 + a_{m2}x_2 + \cdots + a_{mn}x_n = 0,$$

or, more briefly,

$$\mathbf{AX} = \mathbf{0}.$$

Clearly, (14) has the solution

$$x_1 = x_2 = \cdots = x_n = 0,$$

which is called the *trivial* solution. A solution of (14) for which at least one x_i is nonzero is said to be *nontrivial*. An interesting question is: Under what conditions does (14) have nontrivial solutions? The next theorem gives a partial answer to this question.

Theorem 2.6 A homogeneous system of m equations in n unknowns always has a nontrivial solution if the number of unknowns exceeds the number of equations; that is, if $n > m$.

Proof. Suppose $[\mathbf{B} \mathbin{\vdots} \mathbf{0}]$ is in reduced row echelon form, and row equivalent to $[\mathbf{A} \mathbin{\vdots} \mathbf{0}]$. In the notation of Definition 2.3, let the first 1's in each of the k nonzero rows of $[\mathbf{B} \mathbin{\vdots} \mathbf{0}]$ appear in columns $j_1, \ldots, j_k$. Since $k \leq m < n$, the unknowns $x_{j_1}, \ldots, x_{j_k}$ are determined by the remaining unknowns, which can be assigned arbitrary values.

Example 2.13 The augmented matrix of the homogeneous system

$$x + 2y - z + w = 0,$$

(15) $$x + y + w = 0,$$

$$x + 4y - 3z + w = 0,$$

is row equivalent to

$$\begin{bmatrix} 1 & 0 & 1 & 1 & 0 \\ 0 & 1 & -1 & 0 & 0 \\ 0 & 0 & 0 & 0 & 0 \end{bmatrix},$$

which is associated with the system

$$x + z + w = 0,$$

$$y - z = 0.$$

This system can be solved for x and y in terms of z and w:

$$x = -z - w,$$

(16)

$$y = z.$$

By taking $z = 1$ and $w = 0$ (for example), we find from (16) that $(x, y, z, w) = (-1, 1, 1, 0)$ is a solution of (15). By taking $z = 2$ and $w = 1$, we obtain $(x, y, z, w) = (-3, 2, 2, 1)$, another solution.

EXERCISES 1.2

1. If

$$\mathbf{A} = \begin{bmatrix} 1 & 0 & 1 & 2 \\ 2 & -1 & 3 & 1 \\ 4 & 1 & 2 & -3 \end{bmatrix},$$

find the matrices obtained by performing the following elementary row operations on $\mathbf{A}$:

(a) Interchange the first and third rows.
(b) Add -2 times the second row to the third row.
(c) Multiply the first row by 3.

2. Let

$$\mathbf{A} = \begin{bmatrix} 1 & 2 & 1 \\ 0 & 1 & 2 \\ 1 & 1 & 2 \\ 2 & -1 & 3 \end{bmatrix}.$$

(a) Find a matrix $\mathbf{B}$ in row echelon form which is row equivalent to $\mathbf{A}$.
(b) Find a matrix $\mathbf{C}$ in reduced row echelon form which is row equivalent to $\mathbf{A}$.

3. Repeat Exercise 2 for

$$A = \begin{bmatrix} 1 & -1 & 2 & 0 \\ 0 & 2 & -1 & 3 \\ -1 & 2 & 4 & 3 \end{bmatrix}.$$

4. Which of the following matrices are in row echelon form? Which is in reduced row echelon form?

$$A = \begin{bmatrix} 1 & 0 & 1 & 2 & 1 \\ 0 & 0 & 1 & 0 & 0 \\ 0 & 0 & 0 & 0 & 0 \end{bmatrix},$$

$$B = \begin{bmatrix} 1 & 0 & 1 & 2 & 1 \\ 0 & 0 & 2 & 0 & 1 \\ 0 & 0 & 0 & 0 & 0 \end{bmatrix},$$

$$C = \begin{bmatrix} 0 & 1 & 2 & 0 & 1 & 2 \\ 0 & 0 & 1 & 2 & 3 & 1 \\ 0 & 0 & 0 & 1 & 2 & 1 \\ 0 & 0 & 0 & 0 & 0 & 1 \end{bmatrix},$$

$$D = \begin{bmatrix} 1 & 0 & 0 \\ 0 & 1 & 0 \\ 0 & 0 & 1 \end{bmatrix},$$

$$E = \begin{bmatrix} 0 & 0 & 1 & 2 & 1 & 0 \\ 0 & 0 & 1 & 2 & 3 & 0 \\ 0 & 1 & 2 & 1 & 2 & 1 \end{bmatrix},$$

$$\mathbf{F} = \begin{bmatrix} 0 & 0 & 1 & 0 & 0 & 2 \\ 0 & 0 & 0 & 1 & 0 & 1 \\ 0 & 0 & 0 & 0 & 1 & 3 \end{bmatrix}.$$

In Exercises 5 through 7 find all solutions of the given systems of equations.

5. (a)
$$x + y - z = -2,$$
$$x - 2y + 2z = 7,$$
$$2x - y - z = 1;$$

(b)
$$x + y + z = 1,$$
$$2x - 3y + 2z = 2,$$
$$3x - y - z = 3,$$
$$4x - y + 4z = 4,$$
$$x + 2y - 3z = 2;$$

(c)
$$x - y + z + w = 2,$$
$$x + y - z - w = 4,$$
$$3x + y - z - w = 10.$$

6. (a)
$$x + y - z = 4,$$
$$x - 2y + 2z = 7,$$
$$2x - y + z = 3;$$

(b)
$$x - y + z + w = 2,$$
$$x + y - 2z - w = 1,$$
$$2x - 3y + z - w = 3,$$
$$2x - 4y + 5z + 4w = 5,$$
$$4x - y - 3z - 3w = 5;$$

(c)
$$x - y + 2z = 1,$$
$$2x + y - 3z = -5,$$
$$x - y - z = -2.$$

7. (a)
$$x + y - z = 4,$$
$$x - 2y + 2z = 7,$$
$$x + 4y - 4z = 1;$$

(b)
$$x - y + z + w = 2,$$
$$x + 2y - 3z - w = 3,$$
$$3x - z - w = 4;$$

(c)
$$x + 2y - z = -4,$$
$$2x + 2y + 3z = 9,$$
$$x + y - z = -3.$$

In Exercises 8 through 10, find all values of the parameter a such that the given system of equations has (i) a unique solution; (ii) no solution; (iii) infinitely many solutions.

8.
$$x + y - z = 3,$$
$$x - 2y + 3z = 4,$$
$$x + y + (a^2 - 10)z = a.$$

9.
$$x - y = a,$$
$$x + (a^2 - 5)y = 2.$$

10.
$$x + y + z = 1,$$
$$2x - 3y + 4z = 4,$$
$$2x + 2y + (a^2 + 1)z = a + 1.$$

In Exercises 11 through 14, find all solutions of the system of equations with the given augmented matrices.

11. (a) $\begin{bmatrix} 1 & 0 & 1 & 2 & 3 \\ 2 & 1 & 2 & -1 & 4 \\ 0 & -1 & 0 & 6 & 5 \end{bmatrix}$ (b) $\begin{bmatrix} 1 & 2 & 3 & 1 \\ -2 & 2 & 3 & 4 \\ 1 & 8 & 12 & 5 \end{bmatrix}$

12. (a) $\begin{bmatrix} 1 & 2 & 1 & 2 & 2 \\ 1 & 1 & 2 & 0 & -2 \\ 2 & 1 & 3 & -1 & -4 \\ 3 & 1 & 2 & 1 & 2 \\ 5 & 7 & 8 & 4 & -2 \end{bmatrix}$ (b) $\begin{bmatrix} 1 & 2 & 1 & 3 & 4 \\ 4 & 2 & -3 & 1 & 5 \\ -1 & 4 & 4 & 8 & 7 \end{bmatrix}$

13. (a) $\begin{bmatrix} 1 & 2 & -1 & 0 \\ -2 & 1 & 2 & 0 \\ -3 & -1 & 3 & 0 \\ -5 & 2 & 4 & 0 \end{bmatrix}$ (b) $\begin{bmatrix} 1 & 3 & 1 & 2 & 1 & 0 \\ 2 & 3 & -1 & 2 & 2 & 0 \\ 3 & 2 & 0 & 1 & 3 & 0 \end{bmatrix}$

14. (a) $\begin{bmatrix} 1 & 2 & 1 & 0 \\ 1 & 1 & 2 & 0 \\ 1 & 2 & 3 & 0 \end{bmatrix}$ (b) $\begin{bmatrix} 1 & 2 & 1 & 3 & 0 \\ 2 & 1 & -1 & 2 & 0 \\ 3 & -1 & 2 & -2 & 0 \end{bmatrix}$

THEORETICAL EXERCISES

T-1. *Prove:* If the augmented matrix $[A_1 \mid Y_1]$ is obtained by applying an elementary row operation to $[A \mid Y]$, then $AX = Y$ and $A_1X_1 = Y_1$ have the same solutions.

T-2. Prove Theorem 2.5.

T-3. *Prove:* If A and B are row equivalent, then the systems of equations $AX = 0$ and $BX = 0$ have the same solutions.

T-4. Let

$$A = \begin{bmatrix} a & b \\ c & d \end{bmatrix}.$$

Show that A is row equivalent to I_2 if and only if $ad - bc \neq 0$.

T-5. Let

$$A = \begin{bmatrix} a & b \\ c & d \end{bmatrix}.$$

Show that the system $AX = 0$ has only the trivial solution if and only if $ad - bc \neq 0$.

T-6. *Prove:* If a matrix A is row equivalent to 0, then $A = 0$.

1.3 The Inverse of a Matrix

Corresponding to every nonzero real number a, there is a real number a^{-1} such that $a^{-1}a = aa^{-1} = 1$. To find the solution of

(1) $$ax = y,$$

where y is given, we multiply both sides of (1) by a^{-1}:

(2) $$a^{-1}(ax) = a^{-1}y.$$

Since

$$a^{-1}(ax) = (a^{-1}a)x = 1 \cdot x = x,$$

we conclude from (2) that the solution of (1) is $x = a^{-1}y$.

In this section, we consider the analogous problem for square matrices.

Definition 3.1 Let $\mathbf{A}$ be an $n \times n$ matrix. If there is an $n \times n$ matrix $\mathbf{B}$ such that

$$(3) \qquad\qquad \mathbf{AB} = \mathbf{BA} = \mathbf{I},$$

we say that $\mathbf{A}$ is *nonsingular*, and that $\mathbf{B}$ is an *inverse* of $\mathbf{A}$. If no such matrix $\mathbf{B}$ exists, we say that $\mathbf{A}$ is *singular*.

Notice that this definition requires that $\mathbf{A}$ be square. We state this property as part of the definition for emphasis only. In fact, it is implied by (3), since if $\mathbf{AB}$ and $\mathbf{BA}$ are both defined and equal, then $\mathbf{A}$ and $\mathbf{B}$ must be square and of the same order (Exercise T-1).

Example 3.1 Let

$$\mathbf{A} = \begin{bmatrix} 3 & 1 \\ 1 & 2 \end{bmatrix};$$

Since

$$\begin{bmatrix} \frac{2}{5} & -\frac{1}{5} \\ -\frac{1}{5} & \frac{3}{5} \end{bmatrix}\begin{bmatrix} 3 & 1 \\ 1 & 2 \end{bmatrix} = \begin{bmatrix} 3 & 1 \\ 1 & 2 \end{bmatrix}\begin{bmatrix} \frac{2}{5} & -\frac{1}{5} \\ -\frac{1}{5} & \frac{3}{5} \end{bmatrix} = \begin{bmatrix} 1 & 0 \\ 0 & 1 \end{bmatrix},$$

it follows that

$$\mathbf{B} = \begin{bmatrix} \frac{2}{5} & -\frac{1}{5} \\ -\frac{1}{5} & \frac{3}{5} \end{bmatrix}$$

is an inverse of $\mathbf{A}$. Hence $\mathbf{A}$ is nonsingular.

Example 3.2 Let

$$\mathbf{A} = \begin{bmatrix} 1 & 2 \\ 2 & 4 \end{bmatrix};$$

if $\mathbf{A}$ has an inverse

$$\mathbf{B} = \begin{bmatrix} b_{11} & b_{12} \\ b_{21} & b_{22} \end{bmatrix}$$

then

$$(4) \qquad \begin{bmatrix} 1 & 2 \\ 2 & 4 \end{bmatrix}\begin{bmatrix} b_{11} & b_{12} \\ b_{21} & b_{22} \end{bmatrix} = \begin{bmatrix} 1 & 0 \\ 0 & 1 \end{bmatrix}.$$

Equating the elements of the first column of the product on the left to the corresponding elements on the right yields

$$b_{11} + 2b_{21} = 1,$$
$$2b_{11} + 4b_{21} = 0.$$

This system has no solution; therefore we conclude that no 2×2 matrix **B** can satisfy (4), and that **A** is singular.

Theorem 3.1   If a matrix has an inverse, then the inverse is unique.

   Proof.   Suppose **B** and **C** are inverses of **A**. Then

$$\mathbf{BA} = \mathbf{AC} = \mathbf{I},$$

and

$$\mathbf{B} = \mathbf{BI} = \mathbf{B}(\mathbf{AC}) = (\mathbf{BA})\mathbf{C} = \mathbf{IC} = \mathbf{C},$$

which completes the proof.

   Henceforth, we shall denote the inverse of **A** by $\mathbf{A}^{-1}$; thus

(5)
$$\mathbf{AA}^{-1} = \mathbf{A}^{-1}\mathbf{A} = \mathbf{I}$$

if $\mathbf{A}^{-1}$ exists.

Theorem 3.2

   (a)   If **A** is nonsingular, then $\mathbf{A}^{-1}$ is nonsingular and

(6)
$$(\mathbf{A}^{-1})^{-1} = \mathbf{A}.$$

   (b)   If **A** and **B** are nonsingular, then **AB** is nonsingular, and

(7)
$$(\mathbf{AB})^{-1} = \mathbf{B}^{-1}\mathbf{A}^{-1}.$$

   (c)   If **A** is nonsingular, then $\mathbf{A}^\mathrm{T}$ is nonsingular and

$$(\mathbf{A}^\mathrm{T})^{-1} = (\mathbf{A}^{-1})^\mathrm{T}.$$

   Proof.   (a)   By Definition 3.1 (with **A** replaced by $\mathbf{A}^{-1}$), it follows that $\mathbf{A}^{-1}$ is nonsingular if and only if there is a matrix **B** such that

(8)
$$\mathbf{A}^{-1}\mathbf{B} = \mathbf{BA}^{-1} = \mathbf{I}.$$

From (5), it follows that (8) is satisfied with $\mathbf{B} = \mathbf{A}$; thus **A** is an inverse of $\mathbf{A}^{-1}$. The uniqueness of the inverse implies (6).

   (b)   Since

$$(\mathbf{AB})(\mathbf{B}^{-1}\mathbf{A}^{-1}) = \mathbf{A}(\mathbf{BB}^{-1})\mathbf{A}^{-1} = \mathbf{AIA}^{-1} = \mathbf{AA}^{-1} = \mathbf{I}$$

and

$$(\mathbf{B}^{-1}\mathbf{A}^{-1})(\mathbf{AB}) = \mathbf{B}^{-1}(\mathbf{A}^{-1}\mathbf{A})\mathbf{B} = (\mathbf{B}^{-1}\mathbf{I})\mathbf{B} = \mathbf{B}^{-1}\mathbf{B} = \mathbf{I},$$

AB is nonsingular. The uniqueness of the inverse implies (7).

   We leave the proof of (c) to the student (Exercise T-2).

   By repeated application of (b), we obtain the following corollary.

Corollary 3.1 If $\mathbf{A}_1, \mathbf{A}_2, \ldots, \mathbf{A}_k$ are nonsingular $n \times n$ matrices, then the product $\mathbf{A}_1\mathbf{A}_2, \ldots, \mathbf{A}_k$ is also nonsingular, and

$$(\mathbf{A}_1\mathbf{A}_2 \cdots \mathbf{A}_k)^{-1} = \mathbf{A}_k^{-1}\mathbf{A}_{k-1}^{-1} \cdots \mathbf{A}_1^{-1}.$$

Elementary Matrices

We shall now discuss elementary matrices, which can be used to find the inverse of an arbitrary nonsingular matrix.

Definition 3.2 An $n \times n$ *elementary matrix* is a matrix obtained by performing an elementary row operation on the $n \times n$ identity matrix.

Example 3.3 The matrices

$$\mathbf{E}_1 = \begin{bmatrix} 0 & 0 & 1 \\ 0 & 1 & 0 \\ 1 & 0 & 0 \end{bmatrix}, \quad \mathbf{E}_2 = \begin{bmatrix} 1 & 0 & 0 \\ 0 & -3 & 0 \\ 0 & 0 & 1 \end{bmatrix}, \quad \text{and} \quad \mathbf{E}_3 = \begin{bmatrix} 1 & 0 & 0 \\ 0 & 1 & 0 \\ -4 & 0 & 1 \end{bmatrix}$$

are elementary matrices: $\mathbf{E}_1$ is obtained from $\mathbf{I}_3$ by interchanging the latter's first and third rows, $\mathbf{E}_2$ by multiplying the second row of $\mathbf{I}_3$ by -3, and $\mathbf{E}_3$ by adding -4 times the first row of $\mathbf{I}_3$ to its third row.

Theorem 3.3 Suppose $\mathbf{A}$ is an $m \times n$ matrix, and let $\mathbf{B}$ be obtained by applying an elementary row operation to $\mathbf{A}$. Let $\mathbf{E}$ be the elementary matrix obtained by applying the same operation to $\mathbf{I}_m$. Then $\mathbf{B} = \mathbf{E}\mathbf{A}$.

We leave the proof of this theorem to the student (Exercise T-3).

Example 3.4 Let

$$\mathbf{A} = \begin{bmatrix} 1 & 2 & -1 & 3 \\ 4 & 2 & -3 & 1 \\ 2 & 5 & 4 & 2 \end{bmatrix}, \qquad \mathbf{B} = \begin{bmatrix} 2 & 5 & 4 & 2 \\ 4 & 2 & -3 & 1 \\ 1 & 2 & -1 & 3 \end{bmatrix},$$

$$\mathbf{C} = \begin{bmatrix} 1 & 2 & -1 & 3 \\ -8 & -4 & 6 & -2 \\ 2 & 5 & 4 & 2 \end{bmatrix}, \quad \text{and} \quad \mathbf{D} = \begin{bmatrix} -11 & -4 & 8 & 0 \\ 4 & 2 & -3 & 1 \\ 2 & 5 & 4 & 2 \end{bmatrix};$$

thus **B**, **C**, and **D** are each obtained by applying an elementary row operation to **A**. The same row operations applied to $\mathbf{I}_3$ yield

$$\mathbf{E}_b = \begin{bmatrix} 0 & 0 & 1 \\ 0 & 1 & 0 \\ 1 & 0 & 0 \end{bmatrix}, \quad \mathbf{E}_c = \begin{bmatrix} 1 & 0 & 0 \\ 0 & -2 & 0 \\ 0 & 0 & 1 \end{bmatrix}, \quad \text{and} \quad \mathbf{E}_d = \begin{bmatrix} 1 & -3 & 0 \\ 0 & 1 & 0 \\ 0 & 0 & 1 \end{bmatrix},$$

respectively. It is easily verified that $\mathbf{E}_b\mathbf{A} = \mathbf{B}$, $\mathbf{E}_c\mathbf{A} = \mathbf{C}$, and $\mathbf{E}_d\mathbf{A} = \mathbf{D}$, as implied by Theorem 3.3.

Theorem 3.4 Two $m \times n$ matrices **A** and **B** are row equivalent if and only if

$$\text{(9)} \qquad \mathbf{B} = \mathbf{E}_N \cdots \mathbf{E}_1\mathbf{A},$$

where $\mathbf{E}_1, \ldots, \mathbf{E}_N$ are $m \times m$ elementary matrices.

Proof. By Definition 2.2 and Theorem 3.3, **A** and **B** are row equivalent if and only if there is a sequence of matrices $\mathbf{A}_0, \ldots, \mathbf{A}_N$ such that

$$\text{(10)} \qquad \mathbf{A} = \mathbf{A}_0, \qquad \mathbf{B} = \mathbf{A}_N, \qquad \mathbf{A}_i = \mathbf{E}_i\mathbf{A}_{i-1} \qquad (1 \le i \le N),$$

where $\mathbf{E}_i$ is a suitable elementary matrix. However, (10) is just another way of writing (9). Hence, the proof is complete.

Example 3.5 Let

$$\mathbf{A} = \begin{bmatrix} 1 & 2 & -1 & 0 \\ -2 & 3 & 1 & -3 \\ 4 & 1 & 2 & -1 \end{bmatrix}$$

and

$$\mathbf{B} = \begin{bmatrix} 4 & 1 & 2 & -1 \\ -2 & 3 & 1 & -3 \\ 9 & 4 & 3 & -2 \end{bmatrix};$$

that is, **B** is obtained by interchanging the first and third rows of **A** and then adding twice the first row of the resulting matrix to its third row. The first operation corresponds to the elementary matrix

$$\mathbf{E}_1 = \begin{bmatrix} 0 & 0 & 1 \\ 0 & 1 & 0 \\ 1 & 0 & 0 \end{bmatrix},$$

and the second to

$$\mathbf{E}_2 = \begin{bmatrix} 1 & 0 & 0 \\ 0 & 1 & 0 \\ 2 & 0 & 1 \end{bmatrix};$$

therefore, Theorem 3.4 implies that

$$\mathbf{B} = \mathbf{E}_2\mathbf{E}_1\mathbf{A},$$

which the student should verify.

Theorem 3.5 An elementary matrix $\mathbf{E}$ has an inverse $\mathbf{E}^{-1}$ which is also an elementary matrix.

Proof. Suppose $\mathbf{E}$ is obtained by interchanging the rth and sth row of $\mathbf{I}$. From Theorem 3.3, multiplying $\mathbf{E}$ on the left by itself interchanges the rth and sth rows of $\mathbf{E}$, which simply restores them to their original positions in $\mathbf{I}$. Thus $\mathbf{EE} = \mathbf{I}$ and it follows that $\mathbf{E} = \mathbf{E}^{-1}$.

Next, suppose $\mathbf{E}$ and $\mathbf{F}$ are obtained by multiplying the rth row of $\mathbf{I}$ by c and $1/c$, respectively, where $c \neq 0$. It follows from Theorem 3.3 that $\mathbf{EF} = \mathbf{FE} = \mathbf{I}$, so that $\mathbf{E}^{-1} = \mathbf{F}$.

We leave it to the student (Exercise T-4) to verify that if E results from adding c times the ith row of $\mathbf{I}$ to the sth row $(r \neq s)$, then $\mathbf{E}^{-1}$ is the elementary matrix obtained by subtracting c times the rth row from the sth row.

Example 3.6 The inverses of the elementary matrices of Example 3.4 are

$$\mathbf{E}_b^{-1} = \begin{bmatrix} 0 & 0 & 1 \\ 0 & 1 & 0 \\ 1 & 0 & 0 \end{bmatrix}, \quad \mathbf{E}_c^{-1} = \begin{bmatrix} 1 & 0 & 0 \\ 0 & -\frac{1}{2} & 0 \\ 0 & 0 & 1 \end{bmatrix}, \quad \text{and} \quad \mathbf{E}_d^{-1} = \begin{bmatrix} 1 & 3 & 0 \\ 0 & 1 & 0 \\ 0 & 0 & 1 \end{bmatrix}.$$

(Verify.)

Lemma 3.1   Except for I_n, every $n \times n$ matrix in reduced row echelon form has a row of zeros, and is consequently singular.

We leave the proof of this theorem to the student (Exercises T-5 and T-6).

Theorem 3.6   An $n \times n$ matrix **A** is nonsingular if and only if it can be written as a product of elementary matrices.

Proof.   Let **C** be the unique $n \times n$ matrix in reduced row echelon form which is row equivalent to **A** (Theorem 2.4). By Theorem 3.4,

$$(11) \qquad\qquad C = E_N E_{N-1} \cdots E_1 A$$

for suitable elementary matrices $E_1, \ldots, E_N$. If **A** is nonsingular, then so is **C** (Corollary 3.1), and Lemma 3.1 implies that $C = I$. Thus, (11) becomes

$$I = E_N E_{N-1} \cdots E_1 A$$

or

$$A = E_1^{-1} E_2^{-1} \cdots E_N^{-1}.$$

The matrices on the right-hand side are elementary matrices (Theorem 3.5), so that this completes the proof in one direction. For the converse, we have only to observe that since elementary matrices are nonsingular, any product of elementary matrices is also nonsingular.

Corollary 3.2   An $n \times n$ matrix is singular if and only if it is row equivalent to a matrix, in reduced row echelon form, which has a row of zeros. Consequently, an $n \times n$ matrix is nonsingular if and only if it is row equivalent to I_n.

We leave the proof of this corollary to the student (Exercise T-7).

According to Definition 3.1, **B** is the inverse of **A** if and only if both $AB = I$ and $BA = I$. The next theorem shows that it is not necessary to assume that both of these equations hold, since each implies the other.

Theorem 3.7   Suppose **A** and **B** are both $n \times n$ matrices and either

$$(12) \qquad\qquad AB = I$$

or

$$(13) \qquad\qquad BA = I.$$

Then **A** and **B** are nonsingular and $B = A^{-1}$.

Proof. Suppose (12) holds and let **C** be as defined by (11); that is, **C** is in reduced row echelon form and row equivalent to **A**. If **A** is singular, then **C** has a row of zeros (Corollary 3.2). From the definition of matrix multiplication, **CB** also has a row of zeros, and is consequently singular (Corollary 3.2). However, from (11) and (12),

$$\mathbf{CB} = (\mathbf{E}_N\mathbf{E}_{N-1}\cdots\mathbf{E}_1\mathbf{A})\mathbf{B}$$

$$= (\mathbf{E}_N\mathbf{E}_{N-1}\cdots\mathbf{E}_1)\mathbf{AB}$$

$$= (\mathbf{E}_N\mathbf{E}_{N-1}\cdots\mathbf{E}_1)\mathbf{I}$$

$$= \mathbf{E}_N\mathbf{E}_{N-1}\cdots\mathbf{E}_1,$$

which implies that **CB** is nonsingular (Theorem 3.6), a contradiction. Therefore, we conclude that **A** is nonsingular. Now it follows from (12) that

$$\mathbf{B} = \mathbf{IB} = (\mathbf{A}^{-1}\mathbf{A})\mathbf{B} = \mathbf{A}^{-1}(\mathbf{AB}) = \mathbf{A}^{-1}\mathbf{I} = \mathbf{A}^{-1},$$

which completes the proof under assumption (12). A similar proof with **A** and **B** interchanged establishes the result under assumption (13).

Finding the Inverse of a Matrix

We now give a method for determining whether an $n \times n$ matrix **A** is nonsingular and, if so, for finding $\mathbf{A}^{-1}$.

Theorem 3.8 Let **A** be an $n \times n$ matrix and form the $n \times 2n$ auxiliary matrix

$$\mathbf{A}^* = [\mathbf{A} \mid \mathbf{I}_n]$$

whose first n columns are those of **A** and whose last n columns are those of $\mathbf{I}_n$. Let $\mathbf{C}^*$ be the matrix in reduced row echelon form that is row equivalent to $\mathbf{A}^*$. Then **A** is nonsingular if and only if $\mathbf{C}^*$ is of the form

(14) $$\mathbf{C}^* = [\mathbf{I}_n \mid \mathbf{B}],$$

in which case $\mathbf{B} = \mathbf{A}^{-1}$.

Proof. Let $\mathbf{E}_1, \ldots, \mathbf{E}_N$ be elementary $n \times n$ matrices such that

$$\mathbf{C}^* = \mathbf{E}_N\mathbf{E}_{N-1}\cdots\mathbf{E}_1\mathbf{A}^*.$$

Then

(15) $$\mathbf{C}^* = [\mathbf{E}_N\mathbf{E}_{N-1}\cdots\mathbf{E}_1\mathbf{A} \mid \mathbf{E}_N\mathbf{E}_{N-1}\cdots\mathbf{E}_1].$$

Since $\mathbf{C}^*$ is in reduced row echelon form, so is the matrix formed from its first n columns; since this matrix is row equivalent to $\mathbf{A}$, it follows from Corollary 3.2 that $\mathbf{A}$ is nonsingular if and only if

$$(16) \qquad \mathbf{E}_N\mathbf{E}_{N-1}\cdots\mathbf{E}_1\mathbf{A} = \mathbf{I}_n.$$

If (16) is satisfied, then Theorem 3.7 implies that

$$\mathbf{A}^{-1} = \mathbf{E}_N\mathbf{E}_{N-1}\cdots\mathbf{E}_1;$$

therefore (15) can be rewritten as

$$\mathbf{C}^* = [\mathbf{I}_n \mid \mathbf{A}^{-1}],$$

which completes the proof.

Example 3.7 Let

$$\mathbf{A} = \begin{bmatrix} 1 & 0 & 1 \\ 1 & 1 & 0 \\ 0 & 1 & 1 \end{bmatrix};$$

then

$$\mathbf{A}^* = \begin{bmatrix} 1 & 0 & 1 & \vdots & 1 & 0 & 0 \\ 1 & 1 & 0 & \vdots & 0 & 1 & 0 \\ 0 & 1 & 1 & \vdots & 0 & 0 & 1 \end{bmatrix}.$$

We find $\mathbf{C}^*$ by the following steps:

(a) Subtract the first row from the second:

$$\begin{bmatrix} 1 & 0 & 1 & \vdots & 1 & 0 & 0 \\ 0 & 1 & -1 & \vdots & -1 & 1 & 0 \\ 0 & 1 & 1 & \vdots & 0 & 0 & 1 \end{bmatrix}.$$

(b) Subtract the second row from the third:

$$\begin{bmatrix} 1 & 0 & 1 & \vdots & 1 & 0 & 0 \\ 0 & 1 & -1 & \vdots & -1 & 1 & 0 \\ 0 & 0 & 2 & \vdots & 1 & -1 & 1 \end{bmatrix}.$$

(c)   Divide the third row by 2:

$$\left[\begin{array}{ccc:ccc} 1 & 0 & 1 & 1 & 0 & 0 \\ 0 & 1 & -1 & -1 & 1 & 0 \\ 0 & 0 & 1 & \frac{1}{2} & -\frac{1}{2} & \frac{1}{2} \end{array}\right].$$

(d)   Subtract the third row from the first:

$$\left[\begin{array}{ccc:ccc} 1 & 0 & 0 & \frac{1}{2} & \frac{1}{2} & -\frac{1}{2} \\ 0 & 1 & -1 & -1 & 1 & 0 \\ 0 & 0 & 1 & \frac{1}{2} & -\frac{1}{2} & \frac{1}{2} \end{array}\right].$$

(e)   Finally, add the third row to the second to obtain

$$\mathbf{C}^* = \left[\begin{array}{ccc:ccc} 1 & 0 & 0 & \frac{1}{2} & \frac{1}{2} & -\frac{1}{2} \\ 0 & 1 & 0 & -\frac{1}{2} & \frac{1}{2} & \frac{1}{2} \\ 0 & 0 & 1 & \frac{1}{2} & -\frac{1}{2} & \frac{1}{2} \end{array}\right],$$

which is of the form (14). Hence

(17)
$$\mathbf{A}^{-1} = \left[\begin{array}{ccc} \frac{1}{2} & \frac{1}{2} & -\frac{1}{2} \\ -\frac{1}{2} & \frac{1}{2} & \frac{1}{2} \\ \frac{1}{2} & -\frac{1}{2} & \frac{1}{2} \end{array}\right].$$

Example 3.8   Let

$$\mathbf{A} = \left[\begin{array}{ccc} 1 & 0 & 1 \\ 1 & 1 & 0 \\ 0 & 1 & -1 \end{array}\right];$$

then

$$\mathbf{A}^* = \left[\begin{array}{ccc:ccc} 1 & 0 & 1 & 1 & 0 & 0 \\ 1 & 1 & 0 & 0 & 1 & 0 \\ 0 & 1 & -1 & 0 & 0 & 1 \end{array}\right].$$

To find **C*** we proceed as follows:

(a)   Subtract the first row from the second:

$$\begin{bmatrix} 1 & 0 & 1 & \vdots & 1 & 0 & 0 \\ 0 & 1 & -1 & \vdots & -1 & 1 & 0 \\ 0 & 1 & -1 & \vdots & 0 & 0 & 1 \end{bmatrix}.$$

(b)   Subtract the second row from the third:

(18)
$$\begin{bmatrix} 1 & 0 & 1 & \vdots & 1 & 0 & 0 \\ 0 & 1 & -1 & \vdots & -1 & 1 & 0 \\ 0 & 0 & 0 & \vdots & 1 & -1 & 1 \end{bmatrix}.$$

This is not **C***, but there is no point in proceeding further, since it is clear from (18) that the last row of **C*** will have three leading zeros. Consequently, **C*** cannot be of the form (14), and therefore **A** is singular.

Using the Inverse to Solve Linear Systems

A system of n linear equations in n unknowns,

$$a_{11}x_1 + a_{12}x_2 + \cdots + a_{1n}x_n = y_1,$$
$$a_{21}x_1 + a_{22}x_2 + \cdots + a_{2n}x_n = y_2,$$
$$\vdots$$
$$a_{n1}x_1 + a_{n2}x_2 + \cdots + a_{nn}x_n = y_n,$$

or, more briefly,

(19) $$\mathbf{AX} = \mathbf{Y},$$

may fail to have a solution for a given **Y**, or have more than one solution. However, neither of these situations can arise if $\mathbf{A}^{-1}$ exists.

Theorem 3.9   The system (19) of n equations in n unknowns has a solution for every given $n \times 1$ matrix **Y** if and only if **A** is nonsingular. Moreover, if **A** is nonsingular, then

(20) $$\mathbf{X} = \mathbf{A}^{-1}\mathbf{Y}$$

is the unique solution of (19).

Proof. If $\mathbf{A}$ is nonsingular, then (20) satisfies (19), as can be verified by direct substitution of (20) into (19). On the other hand, if $\mathbf{X}$ is a solution of (19) and $\mathbf{A}$ is nonsingular, then

$$\mathbf{A}^{-1}\mathbf{Y} = \mathbf{A}^{-1}(\mathbf{AX}) = (\mathbf{A}^{-1}\mathbf{A})\mathbf{X} = \mathbf{IX} = \mathbf{X};$$

thus (20) is the only solution of (19).

Now suppose (19) has a solution for every $\mathbf{Y}$. We want to show that $\mathbf{A}$ is nonsingular. Let $\mathbf{Y}_1, \mathbf{Y}_2, \ldots, \mathbf{Y}_n$ be defined by

$$\mathbf{Y}_1 = \begin{bmatrix} 1 \\ 0 \\ \vdots \\ 0 \end{bmatrix}, \ \mathbf{Y}_2 = \begin{bmatrix} 0 \\ 1 \\ 0 \\ \vdots \\ 0 \end{bmatrix}, \ldots, \mathbf{Y}_n = \begin{bmatrix} 0 \\ \vdots \\ 0 \\ 1 \end{bmatrix};$$

that is, $\mathbf{Y}_n$ is the nth column of $\mathbf{I}_n$. Let

$$\mathbf{X}_1 = \begin{bmatrix} b_{11} \\ b_{21} \\ \vdots \\ b_{n1} \end{bmatrix}, \ \mathbf{X}_2 = \begin{bmatrix} b_{12} \\ b_{22} \\ \vdots \\ b_{n2} \end{bmatrix}, \ldots, \mathbf{X}_n = \begin{bmatrix} b_{1n} \\ b_{2n} \\ \vdots \\ b_{nn} \end{bmatrix}$$

be respective solutions of

$$\mathbf{AX}_j = \mathbf{Y}_j \qquad (1 \leq j \leq n),$$

which exist by hypothesis. Finally, let $\mathbf{B}$ be the matrix that has $\mathbf{X}_1, \ldots, \mathbf{X}_n$ as columns:

$$\mathbf{B} = \begin{bmatrix} b_{11} & b_{12} & \cdots & b_{1n} \\ b_{21} & b_{22} & \cdots & b_{2n} \\ \vdots & \vdots & & \vdots \\ b_{n1} & b_{n2} & \cdots & b_{nn} \end{bmatrix}.$$

Then

$$\mathbf{AB} = \mathbf{I}$$

(verify!), and Theorem 3.7 implies that $\mathbf{B} = \mathbf{A}^{-1}$. This completes the proof of the theorem.

Corollary 3.3 If $\mathbf{A}$ is a nonsingular $n \times n$ matrix, then the homogeneous system

$$\mathbf{AX} = \mathbf{0}$$

of n equations in n unknowns has only the trivial solution.

Example 3.9 The matrix

$$\mathbf{A} = \begin{bmatrix} 1 & 0 & 1 \\ 1 & 1 & 0 \\ 0 & 1 & 1 \end{bmatrix}$$

was shown in Example 3.7 to have inverse (17). Therefore, the solution of

$$x_1 + x_3 = y_1,$$
$$x_1 + x_2 = y_2,$$
$$x_2 + x_3 = y_3$$

is

$$\begin{bmatrix} x_1 \\ x_2 \\ x_3 \end{bmatrix} = A^{-1} \begin{bmatrix} y_1 \\ y_2 \\ y_3 \end{bmatrix},$$

or

$$x_1 = \tfrac{1}{2}(y_1 + y_2 - y_3),$$
$$x_2 = \tfrac{1}{2}(-y_1 + y_2 + y_3),$$
$$x_3 = \tfrac{1}{2}(y_1 - y_2 + y_3).$$

Example 3.10 The matrix

$$\mathbf{A} = \begin{bmatrix} 1 & 0 & 1 \\ 1 & 1 & 0 \\ 0 & 1 & -1 \end{bmatrix}$$

was shown to be singular in Example 3.8; therefore Theorem 3.9 implies that

$$x_1 + x_3 = y_1,$$

(21)
$$x_1 + x_2 = y_2,$$

$$x_2 - x_3 = y_3$$

fails to have solutions for some triples (y_1, y_2, y_3). We can obtain more specific information by considering the augmented matrix of (21),

$$\begin{bmatrix} 1 & 0 & 1 & y_1 \\ 1 & 1 & 0 & y_2 \\ 0 & 1 & -1 & y_3 \end{bmatrix},$$

which is row equivalent to

$$\begin{bmatrix} 1 & 0 & 1 & y_1 \\ 0 & 1 & -1 & y_2 - y_1 \\ 0 & 0 & 0 & y_3 - y_2 + y_1 \end{bmatrix}.$$

Since the last row of this matrix represents the equation

$$0 \cdot x_1 + 0 \cdot x_2 + 0 \cdot x_3 = y_3 - y_2 + y_1,$$

it follows that (21) has no solution unless

$$y_3 - y_2 + y_1 = 0,$$

in which case

$$x_1 = y_1 - r,$$
$$x_2 = y_2 - y_1 + r,$$
$$x_3 = r$$

is a solution, for any real number r.

EXERCISES 1.3

In Exercises 1 through 4 work directly from Definition 3.1.

1. Show that

$$\mathbf{A} = \begin{bmatrix} 1 & 2 \\ 2 & 4 \end{bmatrix}$$

is singular.

2. Show that

$$A = \begin{bmatrix} 1 & 2 & 3 \\ -1 & 1 & 2 \\ 0 & 3 & 5 \end{bmatrix}$$

is singular.

3. Find the inverse of

$$A = \begin{bmatrix} 1 & 1 \\ 3 & 2 \end{bmatrix}.$$

4. Find the inverse of

$$A = \begin{bmatrix} 1 & 0 & 1 \\ 1 & 1 & 0 \\ 0 & 1 & 1 \end{bmatrix}.$$

5. Write the elementary matrices corresponding to the following elementary row operations on a $4 \times n$ matrix.

(a) Adding -3 times the fourth row to the third row.
(b) Multiplying the second row by 4.
(c) Interchanging the first and third rows.

6. Let

$$A = \begin{bmatrix} 1 & 2 & -1 & 0 & -1 \\ 3 & 4 & 2 & 1 & 0 \\ -2 & 3 & 5 & -2 & 1 \end{bmatrix}.$$

Perform the following elementary row operations on A to obtain B, C, and D. Then verify that $B = E_b A$, $C = E_c A$, and $D = E_d A$, where E_b, E_c, and E_d are the elementary matrices corresponding to given row operations.

(a) B is obtained by multiplying the third row of A by -3.
(b) C is obtained by interchanging the second and third rows of A.
(c) D is obtained by adding 2 times the first row of A to its third row.

7. Find the inverses of the following elementary matrices.

(a) $\mathbf{E}_1 = \begin{bmatrix} 1 & 0 & 0 \\ 0 & 0 & 1 \\ 0 & 1 & 0 \end{bmatrix}$; (b) $\mathbf{E}_2 = \begin{bmatrix} 1 & 0 & -4 \\ 0 & 1 & 0 \\ 0 & 0 & 1 \end{bmatrix}$;

(c) $\mathbf{E}_3 = \begin{bmatrix} 1 & 0 & 0 \\ 0 & 5 & 0 \\ 0 & 0 & 1 \end{bmatrix}$.

8. Show that

$$\mathbf{A} = \begin{bmatrix} 0 & 5 & 0 \\ 1 & 0 & -4 \\ 0 & 0 & 1 \end{bmatrix}$$

is nonsingular and write it as a product of elementary matrices.

9. Show that

$$\mathbf{A} = \begin{bmatrix} 2 & 3 & -2 \\ 0 & -1 & 5 \\ 0 & -2 & 4 \end{bmatrix}$$

is nonsingular and write it as a product of elementary matrices.

In Exercises 10 through 15, find matrices in reduced row echelon form which are row equivalent to the given matrices, and find the inverses of those which are nonsingular.

10. (a) $\begin{bmatrix} 2 & 2 & 3 \\ 3 & 0 & 2 \\ 1 & 0 & 1 \end{bmatrix}$; (b) $\begin{bmatrix} 2 & 3 & -2 \\ 6 & 0 & -3 \\ 2 & -6 & -1 \end{bmatrix}$; (c) $\begin{bmatrix} 4 & 2 \\ -2 & -1 \end{bmatrix}$.

11.　(a)
$$\begin{bmatrix} 4 & 2 & 3 & 3 \\ 2 & 1 & 2 & 1 \\ 3 & 2 & -1 & 6 \\ -1 & -2 & 3 & -6 \end{bmatrix};$$
　(b)
$$\begin{bmatrix} 4 & 3 \\ 3 & 4 \end{bmatrix};$$

　(c)
$$\begin{bmatrix} 3 & 2 & 0 & 1 \\ 1 & 0 & 0 & 1 \\ 0 & 2 & 3 & 2 \\ 1 & 3 & -2 & 1 \end{bmatrix}.$$

12.　(a)
$$\begin{bmatrix} 1 & 2 & 3 \\ 2 & 1 & -5 \\ 4 & 5 & 1 \end{bmatrix};$$
　(b)
$$\begin{bmatrix} 1 & 3 \\ 2 & 5 \end{bmatrix};$$
　(c)
$$\begin{bmatrix} 1 & 1 & 2 \\ 2 & 3 & 4 \\ 3 & 2 & 1 \end{bmatrix}.$$

13.　(a)
$$\begin{bmatrix} 1 & 1 & 2 & -1 \\ 2 & 2 & -1 & 3 \\ -1 & 4 & 1 & 2 \\ 3 & 1 & 0 & 1 \end{bmatrix};$$
　(b)
$$\begin{bmatrix} 2 & 3 & 0 \\ 2 & 2 & 2 \\ 1 & -1 & 5 \end{bmatrix}.$$

　(c)
$$\begin{bmatrix} 2 & 2 & 1 \\ 1 & -1 & 2 \\ 3 & 2 & -1 \end{bmatrix}.$$

14.　(a)
$$\begin{bmatrix} 1 & 2 & 3 & 1 \\ -1 & 2 & 1 & 3 \\ 2 & 4 & 1 & -3 \\ 5 & 4 & 7 & -3 \end{bmatrix};$$
　(b)
$$\begin{bmatrix} 4 & 0 & 1 \\ 2 & 2 & -1 \\ 1 & 2 & 1 \end{bmatrix};$$

(c) $\begin{bmatrix} 2 & 3 & 1 \\ 3 & 1 & 1 \\ -2 & 2 & -1 \end{bmatrix}$.

15. (a) $\begin{bmatrix} -1 & 1 \\ 2 & 3 \end{bmatrix}$; (b) $\begin{bmatrix} 1 & 2 & 1 \\ 3 & -1 & 2 \\ 2 & 1 & 3 \end{bmatrix}$; (c) $\begin{bmatrix} 3 & 1 & 5 \\ -1 & 2 & -4 \\ 2 & 3 & 3 \end{bmatrix}$.

16. Solve each of the following systems by means of Theorem 3.9.

(a) $x + z = 1,$ (b) $2x + 3y = 0,$

$x + y = 2,$ $2x + y = 2.$

$y + z = -1;$

17. Solve each of the following systems by means of Theorem 3.9.

(a) $x + z = 2,$ (b) $x - y = 1,$

$x + 2y + z = -1,$ $x + y = 0.$

$y + 3z = 1;$

18. Each of the following is the coefficient matrix of a homogeneous system $\mathbf{AX} = \mathbf{0}$. Which systems have only the trivial solution?

(a) $\begin{bmatrix} 0 & 2 & 1 & 3 \\ 2 & -1 & 3 & 4 \\ -2 & 1 & 5 & 2 \\ 0 & 1 & 0 & 2 \end{bmatrix}$; (b) $\begin{bmatrix} 3 & -2 & 4 \\ -2 & 3 & -1 \\ 2 & 4 & 8 \end{bmatrix}$;

(c) $\begin{bmatrix} 4 & 2 & -1 \\ 2 & 3 & 0 \\ 1 & 2 & -1 \end{bmatrix}$.

19. Repeat Exercise 18 for

(a) $\begin{bmatrix} 1 & 2 \\ -3 & 4 \end{bmatrix}$; (b) $\begin{bmatrix} 3 & 2 & 5 \\ -2 & 3 & -12 \\ 1 & 4 & -1 \end{bmatrix}$;

(c) $\begin{bmatrix} 3 & 2 & 1 \\ 2 & 1 & 1 \\ -1 & -2 & 2 \end{bmatrix}$.

THEORETICAL EXERCISES

T-1. *Prove:* If $\mathbf{AB} = \mathbf{BA}$ then $\mathbf{A}$ and $\mathbf{B}$ are both square, and of the same order.

T-2. Show that if $\mathbf{A}$ is nonsingular then so is $\mathbf{A^T}$, and $(\mathbf{A^T})^{-1} = (\mathbf{A^{-1}})^{\mathbf{T}}$. Then show that the inverse of a nonsingular symmetric matrix is symmetric.

T-3. Prove Theorem 3.3.

T-4. *Prove:* If $\mathbf{E}$ is the elementary matrix obtained by adding c times the ith row of $\mathbf{I}_n$ to the jth row $(i \neq j)$, and $\mathbf{F}$ is the elementary matrix obtained by subtracting c times the ith row from the jth row of $\mathbf{I}_n$, then $\mathbf{F} = \mathbf{E}^{-1}$.

T-5. *Prove:* An $n \times n$ matrix in reduced row echelon form either has a row of zeros or is equal to $\mathbf{I}_n$.

T-6. *Prove:* A square matrix $\mathbf{A}$ which has a row of zeros is singular. (*Hint:* Show that $\mathbf{AB}$ has a row of zeros, for any $\mathbf{B}$.)

T-7. Prove Corollary 3.2.

T-8. Let

$$\mathbf{A} = \begin{bmatrix} a & b \\ c & d \end{bmatrix}.$$

Show that **A** is nonsingular if and only if $ad - bc \neq 0$. If this condition is satisfied, find $\mathbf{A}^{-1}$.

T-9. *Prove:* Two $n \times n$ matrices **A** and **B** are row equivalent if and only if there is a nonsingular matrix **P** such that $\mathbf{B} = \mathbf{PA}$.

T-10. Prove Theorem 2.1, Section 1.2.

1.4 Determinants

The second-order determinant $D = D(a, b, c, d)$ is written as

$$D = D(a, b, c, d) = \begin{vmatrix} a & b \\ c & d \end{vmatrix}$$

and defined by

$$D(a, b, c, d) = ad - bc.$$

It is important to notice that the value of D depends upon the order in which a, b, c, and d appear. For example,

$$D(c, d, a, b) = \begin{vmatrix} c & d \\ a & b \end{vmatrix} = cb - da = -D(a, b, c, d).$$

Second-order determinants occur in the solution of pairs of linear equations in two unknowns. Consider the system

(1)
$$a_{11}x_1 + a_{12}x_2 = y_1,$$
$$a_{21}x_1 + a_{22}x_2 = y_2.$$

Subtracting a_{12} times the second equation from a_{22} times the first yields

(2)
$$(a_{22}a_{11} - a_{12}a_{21})x_1 = a_{22}y_1 - a_{12}y_2,$$

and subtracting a_{21} times the first from a_{11} times the second yields

(3)
$$(a_{11}a_{22} - a_{21}a_{12})x_2 = a_{11}y_2 - a_{21}y_1.$$

From (2) and (3), it can be seen that (1) has a solution for every given y_1 and y_2 if and only if

(4)
$$\begin{vmatrix} a_{11} & a_{12} \\ a_{21} & a_{22} \end{vmatrix} = a_{11}a_{22} - a_{12}a_{21} \neq 0,$$

in which case the solution is

$$(5) \qquad x_1 = \frac{\begin{vmatrix} y_1 & a_{12} \\ y_2 & a_{22} \end{vmatrix}}{\begin{vmatrix} a_{11} & a_{12} \\ a_{21} & a_{22} \end{vmatrix}} , \qquad x_2 = \frac{\begin{vmatrix} a_{11} & y_1 \\ a_{21} & y_2 \end{vmatrix}}{\begin{vmatrix} a_{11} & a_{12} \\ a_{21} & a_{22} \end{vmatrix}} .$$

The determinant (4) is called the *determinant* of the matrix

$$\mathbf{A} = \begin{bmatrix} a_{11} & a_{12} \\ a_{21} & a_{22} \end{bmatrix}$$

and is denoted by det $\mathbf{A}$.

In this section, we shall define and develop the properties of determinants of arbitrary order. To do this properly, we must briefly discuss permutations.

Permutations

Definition 4.1 For each $n \geq 1$, let S_n be the set of integers $1, 2, \ldots, n$. Any arrangement $(j_1, j_2, \ldots, j_n)$ of the elements of S_n is called a *permutation* of S_n. Thus, $(j_1, j_2, \ldots, j_n)$ is a permutation of S_n if and only if each j_r is an integer between 1 and n, and $j_r \neq j_s$ if $r \neq s$.

To count the distinct permutations of S_n we observe that there are n ways to choose j_1, then $n - 1$ ways to choose j_2, then $n - 2$ ways to choose j_3, and so on. Since distinct combinations of choices result in distinct permutations, there are

$$n \cdot (n - 1) \cdots 2 \cdot 1 = n!$$

distinct permutations of S_n.

Example 4.1 There is only $1! = 1$ permutation of S_1; namely, (1). There are $2! = 2$ permutations of S_2: $(1, 2)$ and $(2, 1)$. There are $3! = 6$ permutations of S_3: $(1, 2, 3)$, $(2, 3, 1)$, $(3, 1, 2)$, $(1, 3, 2)$, $(3, 2, 1)$, and $(2, 1, 3)$.

Definition 4.2 Let $(j_1, j_2, \ldots, j_n)$ be a permutation of S_n, and suppose $r < s$. If $j_r > j_s$, we say that the ordered pair (j_r, j_s) is an *inversion*. The permutation $(j_1, j_2, \ldots, j_n)$ is *even* or *odd* according to whether its total number of inversions is even or odd.

Example 4.2 The permutation $(1, 2)$ of S_2 is even, since it has no inversions; $(2, 1)$ is odd, since it has one inversion.

Example 4.3 Consider the six permutations of S_3:

$(1, 2, 3)$—no inversions—even;

$(2, 3, 1)$—inversions $(2, 1)$, $(3, 1)$—even;

$(3, 1, 2)$—inversions $(3, 1)$, $(3, 2)$—even;

$(1, 3, 2)$—inversion $(3, 2)$—odd;

$(3, 2, 1)$—inversions $(3, 2)$, $(3, 1)$, $(2, 1)$—odd;

$(2, 1, 3)$—inversion $(2, 1)$—odd.

Lemma 4.1 Suppose a permutation P has m inversions, and let a new permutation P_1 be obtained by interchanging two integers in P. Then the number of inversions in P_1 differs from m by an odd integer.

We leave the proof of this lemma to the student (Exercise T-1).

Example 4.4 The permutation $(1, 3, 2, 5, 4)$ has two inversions. Interchanging the second and third entries yields $(1, 2, 3, 5, 4)$, which has one inversion. Interchanging the first and last entry yields $(4, 3, 2, 5, 1)$, which has seven inversions.

Determinants of nth order

Definition 4.3 The *nth order determinant*

$$D = \begin{vmatrix} a_{11} & a_{12} & \cdots & a_{1n} \\ a_{21} & a_{22} & \cdots & a_{2n} \\ \vdots & \vdots & & \vdots \\ a_{n1} & a_{n2} & \cdots & a_{nn} \end{vmatrix},$$

where the a_{ij} are real numbers, is defined by

$$D = \sum e(j_1, j_2, \ldots, j_n) a_{1j_1} a_{2j_2} \cdots a_{nj_n},$$

where the sum is taken over all permutations of S_n (that is, it includes exactly one term for each permutation), and

$$e(j_1, j_2, \ldots, j_n) = \begin{cases} +1 & \text{if} \quad (j_1, j_2, \ldots, j_n) \text{ is even,} \\ -1 & \text{if} \quad (j_1, j_2, \ldots, j_n) \text{ is odd.} \end{cases}$$

We shall often associate D with the matrix

$$\mathbf{A} = \begin{bmatrix} a_{11} & a_{12} & \cdots & a_{1n} \\ a_{21} & a_{22} & \cdots & a_{2n} \\ \vdots & \vdots & & \vdots \\ a_{n1} & a_{n2} & \cdots & a_{nn} \end{bmatrix},$$

in which case we write

$$D = \det \mathbf{A}.$$

We shall refer to the rows and columns of $\mathbf{A}$ as the *rows and columns of D*.

Example 4.5   Let

$$D = \begin{vmatrix} a_{11} & a_{12} \\ a_{21} & a_{22} \end{vmatrix} ;$$

since $(1, 2)$ and $(2, 1)$ are, respectively, even and odd permutations of S_2, Definition 4.2 yields

$$D = a_{11}a_{22} - a_{12}a_{21},$$

which agrees with (4). A more specific example is

$$\begin{vmatrix} 1 & 2 \\ 3 & 4 \end{vmatrix} = (1)(4) - (2)(3) = -2.$$

Example 4.6   Definition 4.3 and the results of Example 4.3 yield, for $n = 3$,

$$(6) \qquad D = \begin{vmatrix} a_{11} & a_{12} & a_{13} \\ a_{21} & a_{22} & a_{23} \\ a_{31} & a_{32} & a_{33} \end{vmatrix}$$

$$= a_{11}a_{22}a_{33} + a_{12}a_{23}a_{31} + a_{13}a_{21}a_{32}$$

$$- a_{11}a_{23}a_{32} - a_{13}a_{22}a_{31} - a_{12}a_{21}a_{33}.$$

A more specific example is

$$\begin{vmatrix} 1 & 2 & 3 \\ 1 & 0 & -1 \\ 1 & 1 & 2 \end{vmatrix} = (1)(0)(2) + (2)(-1)(1) + (3)(1)(1)$$
$$- (1)(-1)(1) - (3)(0)(1) - (2)(1)(2)$$
$$= -2.$$

Evaluating determinants directly from Definition 4.3 is tedious even for small n. Theorems 4.1 through 4.7 lead to practical methods for finding the value of a determinant.

Theorem 4.1   The value of a determinant is not changed if its respective rows and columns are interchanged; that is,

$$(7) \qquad \begin{vmatrix} a_{11} & a_{21} & \cdots & a_{n1} \\ a_{12} & a_{22} & \cdots & a_{n2} \\ \vdots & \vdots & & \vdots \\ a_{1n} & a_{2n} & \cdots & a_{nn} \end{vmatrix} = \begin{vmatrix} a_{11} & a_{12} & \cdots & a_{1n} \\ a_{21} & a_{22} & \cdots & a_{2n} \\ \vdots & \vdots & & \vdots \\ a_{n1} & a_{n2} & \cdots & a_{nn} \end{vmatrix} ,$$

or, in terms of the matrices associated with the two sides of (7),

$$\det \mathbf{A}^{\mathsf{T}} = \det \mathbf{A}.$$

Proof.   A complete proof of this theorem requires more of a digression into the theory of permutations than we wish to make. Therefore, we give only enough of the proof to make the result plausible.

We start by rewriting the left side of (7) as

$$D_1 = \begin{vmatrix} b_{11} & b_{12} & \cdots & b_{1n} \\ b_{21} & b_{22} & \cdots & b_{2n} \\ \vdots & \vdots & & \vdots \\ b_{n1} & b_{n2} & \cdots & b_{nn} \end{vmatrix} ,$$

where

$$(8) \qquad b_{ij} = a_{ji} \qquad (1 \le i, \ j \le n).$$

From Definition 4.3,

$$D_1 = \sum e(k_1, k_2, \ldots, k_n) b_{1k_1} b_{2k_2} \cdots b_{nk_n},$$

which, from (8), can be rewritten as

$$(9) \qquad D_1 = \sum e(k_1, k_2, \ldots, k_n) a_{k_1 1} a_{k_2 2} \cdots a_{k_n n}.$$

The factors in a typical term of (9) can be rearranged so that their row indices are in order reading from left to right; that is,

$$(10) \qquad a_{k_1 1} a_{k_2 2} \cdots a_{k_n n} = a_{1 j_1} a_{2 j_2} \cdots a_{n j_n},$$

where the permutation $(j_1, j_2, \ldots, j_n)$ is the *inverse* of $(k_1, k_2, \ldots, k_n)$. We shall not discuss the inverse of a permutation, but let it suffice to say that a permutation and its inverse are both even or both odd; consequently,

$$e(k_1, k_2, \ldots, k_n) = e(j_1, j_2, \ldots, j_n).$$

Substituting this and (10) into (9) yields

$$D_1 = \sum e(j_1, j_2, \ldots, j_n) a_{1j_1} a_{2j_2} \cdots a_{nj_n} = D,$$

which completes the proof.

Since $D_1 = D$, (9) tells us that we could just as well have stated the definition of an nth order determinant with the column indices, rather than the row indices, in natural order.

Example 4.7   The determinant

$$D_1 = \begin{vmatrix} a_{11} & a_{21} & a_{31} \\ a_{12} & a_{22} & a_{32} \\ a_{13} & a_{23} & a_{33} \end{vmatrix}$$

is obtained by interchanging the rows and columns of (6) in Example 4.6. From (9),

$$D_1 = a_{11}a_{22}a_{33} + a_{21}a_{32}a_{13} + a_{31}a_{12}a_{23}$$

$$- a_{11}a_{32}a_{23} - a_{31}a_{22}a_{13} - a_{21}a_{12}a_{33}.$$

Rearranging the factors in each term so that the row indices are in natural order yields

$$D_1 = a_{11}a_{22}a_{33} + a_{13}a_{21}a_{32} + a_{12}a_{23}a_{31}$$

$$- a_{11}a_{23}a_{32} - a_{13}a_{22}a_{31} - a_{12}a_{21}a_{33},$$

which is the same as the right side of (6).

Theorem 4.2   If two rows (columns) of a determinant are interchanged, then the determinant changes sign.

Proof.   Let

$$(11) \qquad D = \begin{vmatrix} a_{11} & a_{12} & \cdots & a_{1n} \\ a_{21} & a_{22} & \cdots & a_{2n} \\ \vdots & \vdots & & \vdots \\ a_{n1} & a_{n2} & \cdots & a_{nn} \end{vmatrix}$$

and

$$D_1 = \begin{vmatrix} b_{11} & b_{12} & \cdots & b_{1n} \\ b_{21} & b_{22} & \cdots & b_{2n} \\ \vdots & \vdots & & \vdots \\ b_{n1} & b_{n2} & \cdots & b_{nn} \end{vmatrix},$$

where

$$b_{ij} = a_{ij} \qquad (i \neq r, \quad i \neq s),$$

$$b_{rj} = a_{sj},$$

$$b_{sj} = a_{rj};$$

thus, D_1 is obtained by interchanging the rth and sth rows of D. Assume that $1 \leq r < s \leq n$. Then

$$(12) \quad D_1 = \sum e(k_1, \ldots, k_r, \ldots, k_s, \ldots, k_n) b_{1k_1} \cdots b_{rk_r} \cdots b_{sk_s} \cdots b_{nk_n}$$

$$= \sum e(k_1, \ldots, k_r, \ldots, k_s, \ldots, k_n) a_{1k_1} \cdots a_{sk_r} \cdots a_{rk_s} \cdots a_{nk_n}.$$

The row indices in the product

$$a_{1k_1} \cdots a_{sk_r} \cdots a_{rk_s} \cdots a_{nk_n}$$

are not in natural order, but can be made so by interchanging a_{sk_r} and a_{rk_s}. This changes neither the value of the product nor the permutation with which it is associated in (12); hence (12) can be rewritten as

$$D_1 = \sum e(k_1, \ldots, k_r, \ldots, k_s, \ldots, k_n) a_{1k_1} \cdots a_{rk_s} \cdots a_{sk_r} \cdots a_{nk_n}.$$

From Definition 4.3,

$$D = \sum e(k_1, \ldots, k_s, \ldots, k_r, \ldots, k_n) a_{1k_1} \cdots a_{rk_s} \cdots a_{sk_r} \cdots a_{nk_n}.$$

According to Lemma 4.1,

$$e(k_1, \ldots, k_s, \ldots, k_r, \ldots, k_n) = -e(k_1, \ldots, k_r, \ldots, k_s, \ldots, k_n),$$

since the two permutations in this equation differ only by the interchange of k_r and k_s. Hence $D_1 = -D$. This completes the proof of the theorem for interchange of rows; we leave the proof for the interchange of columns to the student (Exercise T-2).

Example 4.8 Let $n = 3$ and suppose D_1 results from interchanging the first and third rows of (6); thus

$$D_1 = \begin{vmatrix} b_{11} & b_{12} & b_{13} \\ b_{21} & b_{22} & b_{23} \\ b_{31} & b_{32} & b_{33} \end{vmatrix},$$

where

(13) $\qquad b_{1j} = a_{3j}, \qquad b_{2j} = a_{2j}, \qquad b_{3j} = a_{1j} \qquad (j = 1, 2, 3).$

Then

$$D_1 = b_{11}b_{22}b_{33} + b_{12}b_{23}b_{31} + b_{13}b_{21}b_{32}$$
$$- b_{11}b_{23}b_{32} - b_{13}b_{22}b_{31} - b_{12}b_{21}b_{33},$$

which, from (13), can be rewritten as

$$D_1 = a_{13}a_{22}a_{31} + a_{11}a_{23}a_{32} + a_{12}a_{21}a_{33}$$
$$- a_{12}a_{23}a_{31} - a_{11}a_{22}a_{33} - a_{13}a_{21}a_{32}.$$

Comparing each term on the right with the corresponding term in (6) yields $D_1 = -D$.

Theorem 4.3 If two rows (columns) of a determinant D are equal, then $D = 0$.

Proof. Interchanging the identical rows (columns) produces D itself. However, Theorem 4.2 implies that this changes the sign of D. Hence $D = -D$, which implies that $D = 0$.

Theorem 4.4 If each element of a row (column) of a determinant is multiplied by a constant c, then the value of the determinant is also multiplied by c.

Proof. Let D be defined by (11) and

$$D_1 = \begin{vmatrix} a_{11} & a_{12} & \cdots & a_{1n} \\ \vdots & \vdots & & \vdots \\ ca_{r1} & ca_{r2} & \cdots & ca_{rn} \\ \vdots & & & \\ a_{r1} & a_{n2} & \cdots & a_{nn} \end{vmatrix} ;$$

then

$$D_1 = \sum e(j_1, j_2, \ldots, j_n)a_{1j_1}\cdots(ca_{rj_r})\cdots a_{nj_n}$$
$$= c \sum e(j_1, j_2, \ldots, j_n)a_{1j_1}\cdots a_{rj_r}\cdots a_{nj_n}$$
$$= cD.$$

Corollary 4.1 If a determinant D has a row (column) of zeros, then $D = 0$.

Theorem 4.5 The value of a determinant is unchanged if a multiple of one of its rows (columns) is added to a different row (column).

Proof. Let D be defined by (11) and

$$
D_1 = \begin{vmatrix} a_{11} & a_{12} & \cdots & a_{1n} \\ \vdots & \vdots & & \vdots \\ a_{r1} + ca_{s1} & a_{r2} + ca_{s2} & \cdots & a_{rn} + ca_{sn} \\ \vdots & \vdots & & \vdots \\ a_{n1} & a_{n2} & \cdots & a_{nn} \end{vmatrix} ;
$$

that is, the elements of D_1 in all rows except the rth are the same as the corresponding elements of D, while those in the rth row are as shown, with $s \neq r$ and c a constant. By Definition 4.3,

$$
D_1 = \sum e(j_1, j_2, \ldots, j_n) a_{1j_1} \cdots (a_{rj_r} + ca_{sj_r}) \cdots a_{nj_n}
$$

and therefore

$$
(14) \qquad D_1 = \sum e(j_1, j_2, \ldots, j_n) a_{1j_1} \cdots a_{rj_r} \cdots a_{nj_n}
$$
$$
+ c \sum e(j_1, j_2, \ldots, j_n) a_{1j_1} \cdots a_{sj_r} \cdots a_{nj_n}.
$$

The first sum on the right side of (14) is D, while the second is the value of a determinant whose rth and sth rows are identical. The second sum vanishes, by Theorem 4.3. Consequently, $D_1 = D$, which completes the proof.

Example 4.9 Consider

$$
D_1 = \begin{vmatrix} a_{11} & a_{12} & a_{13} \\ a_{21} + ca_{31} & a_{22} + ca_{32} & a_{22} + ca_{33} \\ a_{31} & a_{32} & a_{33} \end{vmatrix}
$$
$$
= a_{11}(a_{22} + ca_{32})a_{33} + a_{12}(a_{23} + ca_{33})a_{31}
$$
$$
+ a_{13}(a_{21} + ca_{31})a_{32} - a_{11}(a_{23} + ca_{33})a_{32}
$$
$$
- a_{13}(a_{22} + ca_{32})a_{31} - a_{12}(a_{21} + ca_{31})a_{33}
$$
$$
= [a_{11}a_{22}a_{33} + a_{12}a_{23}a_{31} + a_{13}a_{21}a_{32}
$$
$$
- a_{11}a_{23}a_{32} - a_{13}a_{22}a_{31} - a_{12}a_{21}a_{33}]
$$
$$
+ c[a_{11}a_{32}a_{33} + a_{12}a_{33}a_{31} + a_{13}a_{31}a_{32}
$$
$$
- a_{11}a_{33}a_{32} - a_{13}a_{32}a_{31} - a_{12}a_{31}a_{33}].
$$

The first bracketed expression is the value of D as given in (6), and the second is zero.

A matrix $\mathbf{A} = [a_{ij}]$ is in *lower triangular form* if $a_{ij} = 0$ for $1 \leq i < j \leq n$, or in *upper triangular form* if $a_{ij} = 0$ for $1 \leq j < i \leq n$ (see Exercise T-16 of Section 1.1). A matrix in either of these forms is said to be *triangular*.

The next theorem follows directly from Definition 4.3 (Exercise T-21).

Theorem 4.6 The determinant of a triangular matrix $\mathbf{A} = [a_{ij}]$ is equal to the product of the diagonal elements of A; thus,

$$\det \mathbf{A} = a_{11}a_{22}\cdots a_{nn}.$$

Theorems 4.2 through 4.6 are useful for evaluating determinants, as we shall illustrate in the next three examples.

Example 4.10 Let

$$D = \begin{vmatrix} 1 & 2 & 3 \\ 1 & 0 & -1 \\ 1 & 1 & 2 \end{vmatrix}.$$

Subtracting the first row from the second and third does not change the value of D (Theorem 4.5); hence

$$D = \begin{vmatrix} 1 & 2 & 3 \\ 0 & -2 & -4 \\ 0 & -1 & -1 \end{vmatrix}.$$

From Theorem 4.4, this can be rewritten as

$$D = (-2) \begin{vmatrix} 1 & 2 & 3 \\ 0 & 1 & 2 \\ 0 & -1 & -1 \end{vmatrix}.$$

Adding the second row to the third yields

$$D = (-2) \begin{vmatrix} 1 & 2 & 3 \\ 0 & 1 & 2 \\ 0 & 0 & 1 \end{vmatrix},$$

which is in upper triangular form. Now Theorem 4.6 implies that

$$D = (-2)(1)(1)(1) = -2,$$

which agrees with the result obtained in Example 4.6.

Example 4.11 Consider

$$D = \begin{vmatrix} 6 & 12 & 18 & 12 \\ 1 & 2 & 3 & 4 \\ 2 & 4 & 9 & 3 \\ 1 & 4 & 12 & 4 \end{vmatrix}.$$

Factoring a 6 out of the first row yields

$$D = 6 \begin{vmatrix} 1 & 2 & 3 & 2 \\ 1 & 2 & 3 & 4 \\ 2 & 4 & 9 & 3 \\ 1 & 4 & 12 & 4 \end{vmatrix}.$$

Taking common factors out of the second and third columns yields

$$D = (6)(2)(3) \begin{vmatrix} 1 & 1 & 1 & 2 \\ 1 & 1 & 1 & 4 \\ 2 & 2 & 3 & 3 \\ 1 & 2 & 4 & 4 \end{vmatrix}.$$

Subtracting the first column from the second and third columns, and twice the first column from the fourth column, yields

$$D = 36 \begin{vmatrix} 1 & 0 & 0 & 0 \\ 1 & 0 & 0 & 2 \\ 2 & 0 & 1 & -1 \\ 1 & 1 & 3 & 2 \end{vmatrix}.$$

Interchanging the second and fourth columns yields

$$D = -36 \begin{vmatrix} 1 & 0 & 0 & 0 \\ 1 & 2 & 0 & 0 \\ 2 & -1 & 1 & 0 \\ 1 & 2 & 3 & 1 \end{vmatrix}$$

(Theorem 4.2), which is in lower triangular form; hence

$$D = -36\,(1)\,(2)\,(1)\,(1) = -72.$$

Example 4.12 Let

$$D = \begin{vmatrix} 1 & x & x^2 \\ 1 & y & y^2 \\ 1 & z & z^2 \end{vmatrix}.$$

From Theorem 4.3, $D = 0$ if $x = y$, $x = z$, or $y = z$. Using the other theorems given earlier, we find that

$$D = \begin{vmatrix} 1 & x & x^2 \\ 0 & y - x & y^2 - x^2 \\ 0 & z - x & z^2 - x^2 \end{vmatrix}$$

$$= (y - x)\,(z - x) \begin{vmatrix} 1 & x & x^2 \\ 0 & 1 & y + x \\ 0 & 1 & z + x \end{vmatrix}$$

$$= (y - x)\,(z - x) \begin{vmatrix} 1 & x & x^2 \\ 0 & 1 & y + x \\ 0 & 0 & z - y \end{vmatrix}$$

$$= (y - x)\,(z - x)\,(z - y) \begin{vmatrix} 1 & x & x^2 \\ 0 & 1 & y + x \\ 0 & 0 & 1 \end{vmatrix}$$

$$= (y - x)\,(z - x)\,(z - y).$$

Expansion by Cofactors

Definition 4.4 Let

$$(15) \qquad D = \begin{vmatrix} a_{11} & a_{12} & \cdots & a_{1n} \\ a_{21} & a_{22} & \cdots & a_{2n} \\ \vdots & \vdots & & \vdots \\ a_{n1} & a_{n2} & \cdots & a_{nn} \end{vmatrix} .$$

For each i, j $(1 \leq i, j \leq n)$, the sum of all terms in the expansion

$$(16) \qquad D = \sum e(j_1, j_2, \ldots, j_n) a_{1j_1} a_{2j_2} \cdots a_{nj_n}$$

which multiply a_{ij} is called the *cofactor* of a_{ij}, and is denoted by A_{ij}.

Example 4.13 Let $n = 3$. By inspection of (6) in Example 4.6, we find that

$$A_{11} = a_{22}a_{33} - a_{23}a_{32} = \begin{vmatrix} a_{22} & a_{23} \\ a_{32} & a_{33} \end{vmatrix},$$

$$A_{12} = a_{23}a_{31} - a_{21}a_{33} = - \begin{vmatrix} a_{21} & a_{23} \\ a_{31} & a_{33} \end{vmatrix},$$

and

$$A_{13} = a_{21}a_{32} - a_{22}a_{31} = \begin{vmatrix} a_{21} & a_{22} \\ a_{31} & a_{32} \end{vmatrix}.$$

Theorem 4.7 Let D be defined by (15). Then:

(a) $\quad D = a_{i1}A_{i1} + a_{i2}A_{i2} + \cdots + a_{in}A_{in} \qquad (1 \leq i \leq n);$

(b) $\quad D = a_{1j}A_{1j} + a_{2j}A_{2j} + \cdots + a_{nj}A_{nj} \qquad (1 \leq j \leq n);$

(c) $\quad 0 = a_{k1}A_{i1} + a_{k2}A_{i2} + \cdots + a_{kn}A_{in} \qquad (k \neq i);$

(d) $\quad 0 = a_{1k}A_{1j} + a_{2k}A_{2j} + \cdots + a_{nk}A_{nj} \qquad (k \neq j).$

Proof. By definition, $a_{i1}A_{i1}$ is the sum of all those terms in (16) for which $j_i = 1$, $a_{i2}A_{i2}$ is the sum of all terms for which $j_i = 2$, and so forth. Since every term in (16) contains exactly one factor from the ith row, it follows that every term is contained in exactly one of the sums $a_{i1}A_{i1}, \ldots,$ $a_{in}A_{in}$. This proves (a). To prove (c), define D_1 to be the determinant obtained by replacing the ith row of D by its kth row $(k \neq i)$. The cofactors of the elements in the ith row of D_1 are the same as those of D; hence, applying (a) to D_1 yields

$$D_1 = a_{k1}A_{i1} + a_{k2}A_{i2} + \cdots + a_{kn}A_{in}.$$

However, D_1 has two identical rows (the ith and kth); hence $D_1 = 0$, which proves (c).

We leave the proof of (b) and (d) to the student (Exercise T-4).

Evaluating D by means of (a) is called *expansion in terms of the ith row;* using (b) is called *expansion in terms of the jth column.*

The following theorem, which we state without proof, provides a convenient way to compute the cofactors of a determinant.

Theorem 4.8 The cofactor of a_{ij} in (15) is given by

(17) $$A_{ij} = (-1)^{i+j}M_{ij},$$

where M_{ij}, the *minor* of a_{ij}, is the determinant obtained by deleting the ith row and jth column of D.

The relationship (17) between minors and cofactors can be easily remembered by visualizing an $n \times n$ "checkerboard" with alternating plus and minus signs, starting with a plus in the upper left hand corner. Figure 4.1 depicts the situation for $n = 3$ and $n = 4$.

FIGURE 4.1

Example 4.14 The cofactors of

(18) $$D = \begin{vmatrix} 1 & 2 & 3 \\ 1 & 0 & -1 \\ 1 & 1 & 2 \end{vmatrix}$$

are

$$A_{11} = \begin{vmatrix} 0 & -1 \\ 1 & 2 \end{vmatrix} = 1, \qquad A_{12} = -\begin{vmatrix} 1 & -1 \\ 1 & 2 \end{vmatrix} = -3,$$

$$A_{13} = \begin{vmatrix} 1 & 0 \\ 1 & 1 \end{vmatrix} = 1, \qquad A_{21} = -\begin{vmatrix} 2 & 3 \\ 1 & 2 \end{vmatrix} = -1,$$

$$A_{22} = \begin{vmatrix} 1 & 3 \\ 1 & 2 \end{vmatrix} = -1, \qquad A_{23} = -\begin{vmatrix} 1 & 2 \\ 1 & 1 \end{vmatrix} = 1,$$

$$A_{31} = \begin{vmatrix} 2 & 3 \\ 0 & -1 \end{vmatrix} = -2, \qquad A_{32} = -\begin{vmatrix} 1 & 3 \\ 1 & -1 \end{vmatrix} = 4,$$

$$A_{33} = \begin{vmatrix} 1 & 2 \\ 1 & 0 \end{vmatrix} = -2.$$

Expanding D according to the elements of its first row yields

$$D = (1)A_{11} + (2)A_{12} + (3)A_{13}$$
$$= (1)(1) + 2(-3) + (3)(1) = -2,$$

which agrees with the result of Example 4.6. Expanding D by the elements of its second column yields

$$D = (2)A_{12} + (0)A_{22} + (1)A_{32}$$
$$= (2)(-3) + (0)(-1) + (1)(4) = -2.$$

Finally, multiplying the elements of the third column by the cofactors of the first column and adding yields

$$3A_{11} + (-1)A_{21} + 2A_{31} = (3)(1) + (-1)(-1) + (2)(-2) = 0,$$

as implied by (d) of Theorem 4.7.

Theorem 4.7 can be used in conjunction with Theorems 4.2 through 4.6 to reduce the order of the determinant to be evaluated. This is accomplished by manipulating the rows and columns of the given determinant in such a way as to produce a new determinant with at most one nonzero element in some row or column. The next two examples illustrate the procedure.

Example 4.15 Subtracting the first row of (18) from the other two yields

$$D = \begin{vmatrix} 1 & 2 & 3 \\ 0 & -2 & -4 \\ 0 & -1 & -1 \end{vmatrix},$$

which can be expanded on the elements of the first row to yield

$$D = (1) \begin{vmatrix} -2 & -4 \\ -1 & -1 \end{vmatrix} = (1)(2-4) = -2.$$

Example 4.16 Let

$$D = \begin{vmatrix} 1 & 1 & 2 & 2 \\ 1 & 4 & 4 & 4 \\ 1 & 1 & 2 & 4 \\ 2 & 3 & 4 & 3 \end{vmatrix};$$

subtracting twice the first column from the third yields

$$D = \begin{vmatrix} 1 & 1 & 0 & 2 \\ 1 & 4 & 2 & 4 \\ 1 & 1 & 0 & 4 \\ 2 & 3 & 0 & 3 \end{vmatrix},$$

which can be expanded on the elements of the third column to yield

$$D = (-2) \begin{vmatrix} 1 & 1 & 2 \\ 1 & 1 & 4 \\ 2 & 3 & 3 \end{vmatrix}.$$

Now subtract the first column from the second:

$$D = (-2) \begin{vmatrix} 1 & 0 & 2 \\ 1 & 0 & 4 \\ 2 & 1 & 3 \end{vmatrix},$$

and expand on the elements of the second column to obtain

$$D = (-2)(-1) \begin{vmatrix} 1 & 2 \\ 1 & 4 \end{vmatrix}$$

$$= (-2)(-1)(4 - 2) = 4.$$

Cramer's Rule and Matrix Inversion

At the beginning of this section, we saw how second order determinants occur in solving a pair of equations in two unknowns. We now generalize this idea to a system of n equations in n unknowns.

Consider the system

$$(19) \qquad \begin{aligned} a_{11}x_1 + a_{12}x_2 + \cdots + a_{1n}x_n &= y_1, \\ a_{21}x_1 + a_{22}x_2 + \cdots + a_{2n}x_n &= y_2, \\ &\vdots \\ a_{n1}x_1 + a_{n2}x_2 + \cdots + a_{nn}x_n &= y_n, \end{aligned}$$

or, more briefly,

$$(20) \qquad \sum_{j=1}^{n} a_{ij}x_j = y_i \qquad (1 \leq i \leq n).$$

Let r be a fixed integer $(1 \leq r \leq n)$ and multiply both sides of (20) by A_{ir}, the cofactor of a_{ir} in the determinant $D = \det[a_{ij}]$. Adding the resulting equations yields

$$\sum_{i=1}^{n} A_{ir} \left(\sum_{j=1}^{n} a_{ij}x_j \right) = \sum_{i=1}^{n} y_i A_{ir}.$$

Changing the order of summation on the left yields

$$(21) \qquad \sum_{j=1}^{n} \left(\sum_{i=1}^{n} a_{ij}A_{ir} \right) x_j = \sum_{i=1}^{n} y_i A_{ir} \qquad (1 \leq r \leq n).$$

From Theorem 4.7, (b) and (d),

$$\sum_{i=1}^{n} a_{ij}A_{ir} = \begin{cases} D & \text{if } j = r, \\ 0 & \text{if } j \neq r, \end{cases}$$

and (21) reduces to

$$Dx_r = \sum_{i=1}^{n} y_i A_{ir} \qquad (1 \leq r \leq n).$$

The right side of this equation can be represented as a certain determinant D_r, as defined in the next theorem. We leave the remaining details of the proof of this theorem to the student (Exercise T-5).

Theorem 4.9 (Cramer's Rule) The system (19) of n equations in n unknowns has a solution for every given

$$\mathbf{Y} = \begin{bmatrix} y_1 \\ y_2 \\ \vdots \\ y_n \end{bmatrix}$$

if and only if $D = \det [a_{ij}] \neq 0$, in which case the solution is unique and is given by

$$x_r = \frac{D_r}{D} \qquad (1 \leq r \leq n),$$

where D_r is the determinant obtained by replacing the rth column of D by $\mathbf{Y}$.

We derived this result for $n = 2$ at the beginning of this section, where we found the solution of (1) to be given by (5), provided the determinant (4) is nonzero.

Example 4.17 Consider the system

$$x + 2y + 3z = 1,$$
(22) $$x - z = -1,$$
$$x + y + 2z = 0,$$

which has the coefficient matrix

$$\mathbf{A} = \begin{bmatrix} 1 & 2 & 3 \\ 1 & 0 & -1 \\ 1 & 1 & 2 \end{bmatrix}$$

with $D = \det \mathbf{A} = -2$ (Example 4.6). According to Cramer's rule the unique solution of (22) is given by

$$x = -\frac{1}{2}\begin{vmatrix} 1 & 2 & 3 \\ -1 & 0 & -1 \\ 0 & 1 & 2 \end{vmatrix} = -1,$$

$$y = -\frac{1}{2}\begin{vmatrix} 1 & 1 & 3 \\ 1 & -1 & -1 \\ 1 & 0 & 2 \end{vmatrix} = 1,$$

$$z = -\frac{1}{2}\begin{vmatrix} 1 & 2 & 1 \\ 1 & 0 & -1 \\ 1 & 1 & 0 \end{vmatrix} = 0.$$

The practical value of Cramer's rule as a method for solving linear systems is limited, since it requires the computation of $n + 1$ determinants of order n. (The methods of Section 1.2 are more efficient for numerical work.) However, Cramer's rule has interesting theoretical consequences. The following theorem is an example.

Theorem 4.10 An $n \times n$ matrix $\mathbf{A}$ is nonsingular if and only if $D = \det \mathbf{A} \neq 0$, in which case

(23)
$$\mathbf{A}^{-1} = \frac{1}{D} \operatorname{adj} \mathbf{A},$$

where $\operatorname{adj} \mathbf{A}$ (the *adjoint* of $\mathbf{A}$) is the $n \times n$ matrix defined by

$$\operatorname{adj} \mathbf{A} = \begin{bmatrix} A_{11} & A_{21} & \cdots & A_{n1} \\ A_{12} & A_{22} & \cdots & A_{n2} \\ \vdots & \vdots & & \vdots \\ A_{1n} & A_{2n} & \cdots & A_{nn} \end{bmatrix};$$

that is, $\operatorname{adj} \mathbf{A}$ is obtained by replacing each element of $\mathbf{A}$ by its cofactor, and then taking the transpose.

Proof. That $D \neq 0$ is a necessary and sufficient condition for $\mathbf{A}$ to be nonsingular follows from Theorems 3.9 and 4.9. We leave it to the student to verify (23) by showing that

$$(24) \qquad \left(\frac{1}{D} \operatorname{adj} \mathbf{A}\right) \mathbf{A} = \mathbf{A}\left(\frac{1}{D} \operatorname{adj} \mathbf{A}\right) = \mathbf{I}.$$

(Exercise T-6).

Example 4.18 Let

$$\mathbf{A} = \begin{bmatrix} 1 & 2 & 3 \\ 1 & 0 & -1 \\ 1 & 1 & 2 \end{bmatrix};$$

then from (23),

$$(25) \qquad \mathbf{A}^{-1} = \frac{1}{D}\begin{bmatrix} A_{11} & A_{21} & A_{31} \\ A_{12} & A_{22} & A_{32} \\ A_{13} & A_{23} & A_{33} \end{bmatrix}.$$

Substituting the results of Example 4.14 into (25) yields

$$\mathbf{A}^{-1} = -\frac{1}{2}\begin{bmatrix} 1 & -1 & -2 \\ -3 & -1 & 4 \\ 1 & 1 & -2 \end{bmatrix}$$

$$= \begin{bmatrix} -\frac{1}{2} & \frac{1}{2} & 1 \\ \frac{3}{2} & \frac{1}{2} & -2 \\ -\frac{1}{2} & -\frac{1}{2} & 1 \end{bmatrix}.$$

The Determinant of a Product of Two Matrices

We conclude this section by showing that the determinant of the product of two $n \times n$ matrices equals the product of the determinants.

Lemma 4.2 If **E** is an elementary $n \times n$ matrix and **B** is an arbitrary $n \times n$ matrix, then

$$\det \mathbf{EB} = (\det \mathbf{E})(\det \mathbf{B}).$$

We leave the proof of this lemma to the student (Exercise T-7). By repeated application of this result, we obtain the following lemma (Exercise T-8).

Lemma 4.3 If $\mathbf{E}_1, \ldots, \mathbf{E}_k$ are elementary $n \times n$ matrices and B is an arbitrary $n \times n$ matrix, then

$$\det \mathbf{E}_1 \mathbf{E}_2 \cdots \mathbf{E}_k \mathbf{B} = (\det \mathbf{E}_1) \cdots (\det \mathbf{E}_k) \det \mathbf{B}$$

$$= \det(\mathbf{E}_1 \cdots \mathbf{E}_k) \det \mathbf{B}.$$

Theorem 4.11 If **A** and **B** are $n \times n$ matrices, then

(26) $$\det \mathbf{AB} = (\det \mathbf{A})(\det \mathbf{B}).$$

Proof. Let **A** be row equivalent to the matrix **C** in reduced row-echelon form; by Theorem 3.4, we can write

$$\mathbf{A} = \mathbf{E}_1 \mathbf{E}_2 \cdots \mathbf{E}_k \mathbf{C},$$

where $\mathbf{E}_1, \ldots, \mathbf{E}_k$ are elementary matrices. Now

$$\mathbf{AB} = (\mathbf{E}_1 \cdots \mathbf{E}_k)\mathbf{CB},$$

and, by Lemma 4.3,

(27) $$\det \mathbf{AB} = \det(\mathbf{E}_1 \cdots \mathbf{E}_k) \det(\mathbf{CB}).$$

If **A** is nonsingular, then **C** = **I** (Corollary 3.2), and (27) reduces to (26), since $\mathbf{A} = \mathbf{E}_1 \cdots \mathbf{E}_k$. If **A** is singular, then **C** has a row of zeros (Corollary 3.2), and so does **CB** (Exercise T-12 of Section 1.2). Hence, det **CB** = 0, and (27) implies that det **AB** = 0. Since det **A** = 0 (Theorem 4.10), we conclude that (26) also holds in this case. This completes the proof.

Corollary 4.2 If **A** is a nonsingular matrix, then

$$\det \mathbf{A}^{-1} = \frac{1}{\det \mathbf{A}}.$$

Proof. Since $\mathbf{AA}^{-1} = \mathbf{I}$, $(\det \mathbf{A})(\det \mathbf{A}^{-1}) = \det \mathbf{I} = 1$.

EXERCISES 1.4

1. List the inversions in the following permutations of S_5:

   (a) $(4, 2, 1, 3, 5)$;   (b) $(5, 1, 3, 2, 4)$;

   (c) $(3, 2, 1, 5, 4)$;   (d) $(1, 4, 2, 3, 5)$.

2. List the inversions in the following permutations of S_5:

   (a) $(2, 1, 3, 4, 5)$;   (b) $(3, 5, 2, 1, 4)$;

   (c) $(4, 3, 2, 1, 5)$;   (d) $(5, 2, 3, 1, 4)$.

3. Which of the following permutations of S_5 are even?

   (a) $(3, 1, 2, 4, 5)$;   (b) $(1, 5, 4, 3, 2)$;

   (c) $(2, 3, 5, 4, 1)$;   (d) $(4, 5, 2, 1, 3)$.

4. Which of the following permutations of S_5 are odd?

   (a) $(4, 5, 3, 1, 2)$;   (b) $(2, 3, 1, 5, 4)$;

   (c) $(4, 5, 3, 2, 1)$;   (d) $(3, 2, 4, 1, 5)$.

5. Verify that the number of inversions in the following pairs of permutations of S_6 differ by an odd integer.

   (a) $(1, 2, 3, 4, 5, 6)$ and $(4, 2, 3, 1, 5, 6)$;

   (b) $(3, 4, 5, 1, 2, 6)$ and $(3, 5, 4, 1, 2, 6)$;

   (c) $(6, 5, 4, 3, 1, 2)$ and $(2, 5, 4, 3, 1, 6)$.

6. Evaluate the following determinants directly from Definition 4.3.

   (a) $\begin{vmatrix} -3 & 2 & 3 \\ 0 & 4 & -5 \\ 0 & 0 & 2 \end{vmatrix}$;   (b) $\begin{vmatrix} 4 & 0 & 0 \\ 5 & -2 & 0 \\ 2 & 3 & -3 \end{vmatrix}$;

   (c) $\begin{vmatrix} 4 & 0 & 0 \\ 0 & 5 & 0 \\ 0 & 0 & -3 \end{vmatrix}$.

7. Evaluate the following determinants directly from Definition 4.3.

(a) $\begin{vmatrix} 4 & 2 \\ 3 & -1 \end{vmatrix}$; (b) $\begin{vmatrix} -4 & -2 & 3 \\ 0 & 1 & 2 \\ 3 & -4 & 2 \end{vmatrix}$;

(c) $\begin{vmatrix} 2 & 1 & 0 & 3 \\ 1 & 0 & 1 & 3 \\ 3 & 0 & -1 & 4 \\ 5 & -2 & 0 & 1 \end{vmatrix}$.

8. If

$$\mathbf{A} = \begin{bmatrix} 4 & 3 & 2 \\ 1 & 2 & 3 \\ -1 & 0 & 1 \end{bmatrix},$$

verify that $\det \mathbf{A} = \det \mathbf{A}^{\mathrm{T}}$.

9. If

$$D = \begin{vmatrix} a_1 & a_2 & a_3 \\ b_1 & b_2 & b_3 \\ c_1 & c_2 & c_3 \end{vmatrix} = 5,$$

determine the value of

$$D_1 = \begin{vmatrix} a_1 & a_2 & a_3 \\ b_1 + 2c_1 & b_2 + 2c_2 & b_3 + 2c_3 \\ c_1 & c_2 & c_3 \end{vmatrix},$$

$$D_2 = \begin{vmatrix} a_1 & 2a_2 & a_3 \\ b_1 & 2b_2 & b_3 \\ c_1 & 2c_2 & c_3 \end{vmatrix},$$

and

$$D_3 = \begin{vmatrix} a_3 & a_2 & a_1 \\ b_3 & b_2 & b_1 \\ c_3 & c_2 & c_1 \end{vmatrix}.$$

10. If

$$D = \begin{vmatrix} a_1 & a_2 & a_3 \\ b_1 & b_2 & b_3 \\ c_1 & c_2 & c_3 \end{vmatrix} = -1$$

determine the value of

$$D_1 = \begin{vmatrix} a_1 + 2b_1 & b_1 - c_1 & 4c_1 \\ a_2 + 2b_2 & b_2 - c_2 & 4c_2 \\ a_3 + 2b_3 & b_3 - c_3 & 4c_3 \end{vmatrix}$$

and

$$D_2 = \begin{vmatrix} a_1 & a_2 & a_3 \\ b_1 + 2a_1 & b_2 + 2a_1 & b_3 + 2a_3 \\ 2a_1 & 2a_2 & 2a_3 \end{vmatrix}.$$

Evaluate the determinants in Exercises 11 through 13 by means of Theorems 4.1 through 4.6.

11. (a) $\begin{vmatrix} -3 & 2 \\ 5 & 1 \end{vmatrix}$; (b) $\begin{vmatrix} 0 & 2 & 3 \\ 0 & 1 & 5 \\ 2 & 1 & 2 \end{vmatrix}$; (c) $\begin{vmatrix} -1 & 2 & 3 & 0 \\ -1 & 2 & 1 & 0 \\ 4 & 5 & 6 & 2 \\ 1 & 0 & 0 & 2 \end{vmatrix}.$

12. (a) $\begin{vmatrix} 0 & 1 \\ 1 & 2 \end{vmatrix}$; (b) $\begin{vmatrix} -1 & 0 & -1 \\ 2 & 0 & 3 \\ -3 & 2 & -3 \end{vmatrix}$;

(c) $\begin{vmatrix} 4 & 1 & 2 & 5 \\ 1 & 0 & 0 & 2 \\ 3 & 0 & 1 & 2 \\ 0 & 2 & -3 & 4 \end{vmatrix}$.

13. (a) $\begin{vmatrix} 1 & 2 & 1 & 0 \\ 0 & 1 & -3 & 2 \\ -1 & -1 & 2 & 1 \\ 2 & 5 & 5 & 7 \end{vmatrix}$; (b) $\begin{vmatrix} 1 & 2 & 0 & -1 \\ 0 & 1 & 0 & 2 \\ 2 & 5 & 0 & 1 \\ 1 & 3 & 0 & 3 \end{vmatrix}$;

(c) $\begin{vmatrix} 1 & 2 & 3 & 4 \\ 1 & 3 & 4 & 5 \\ 2 & 5 & 7 & 10 \\ 1 & 3 & 4 & 7 \end{vmatrix}$.

14. Compute all cofactors of

$$D = \begin{vmatrix} 1 & 0 & 1 \\ 1 & 1 & 0 \\ 0 & 1 & 1 \end{vmatrix} ,$$

and verify (a) and (c) of Theorem 4.7.

15. Compute all cofactors of

$$D = \begin{vmatrix} 1 & 0 & 1 \\ 2 & 1 & 1 \\ 0 & 1 & 0 \end{vmatrix} ,$$

and verify (b) and (d) of Theorem 4.7.

16.   Using Theorem 4.7, evaluate the following determinants.

(a) $\begin{vmatrix} 1 & 2 & 0 \\ -3 & 4 & 1 \\ 2 & 1 & 0 \end{vmatrix}$;   (b) $\begin{vmatrix} 5 & 1 & 2 \\ -1 & 2 & 3 \\ 0 & 1 & 2 \end{vmatrix}$;   (c) $\begin{vmatrix} 0 & 0 & 1 \\ -1 & 2 & 3 \\ 1 & 1 & 2 \end{vmatrix}$.

17.   Using Theorem 4.7, evaluate the following determinants.

(a) $\begin{vmatrix} 3 & 2 & 1 \\ 0 & 2 & 1 \\ 1 & 1 & 0 \end{vmatrix}$;   (b) $\begin{vmatrix} 1 & 1 & 0 \\ 2 & 3 & 4 \\ 5 & 6 & 4 \end{vmatrix}$;   (c) $\begin{vmatrix} -1 & 2 & 3 \\ 2 & 1 & 5 \\ 2 & 3 & 0 \end{vmatrix}$.

18.   Evaluate the following determinants by any convenient method.

(a) $\begin{vmatrix} 1 & 2 & 0 & 0 \\ -2 & 3 & 0 & 0 \\ 0 & 0 & 2 & 3 \\ 0 & 0 & -1 & 2 \end{vmatrix}$;   (b) $\begin{vmatrix} 1 & 2 & 2 & 3 \\ -2 & 3 & 1 & 2 \\ 0 & 0 & -2 & 3 \\ 0 & 0 & 5 & 1 \end{vmatrix}$.

19.   Evaluate the following determinants by any convenient method.

(a) $\begin{vmatrix} 1 & 2 & 3 & 4 & 1 \\ 0 & 0 & 1 & 0 & 2 \\ 0 & 0 & 1 & 1 & -1 \\ 2 & 0 & 0 & 1 & 5 \\ 1 & 0 & 2 & 1 & 0 \end{vmatrix}$;   (b) $\begin{vmatrix} 1 & 2 & 1 & 0 & 0 \\ 0 & 1 & -3 & 2 & 0 \\ -1 & -1 & 2 & 1 & 0 \\ 2 & 5 & 5 & 7 & 0 \\ 0 & 0 & 0 & 0 & 1 \end{vmatrix}$.

20.   Solve, by Cramer's rule:

$$x + y - z = 5,$$
$$2x + y - 3z = 10,$$
$$3x + 4y + 5z = 1.$$

21.   Solve, by Cramer's rule:

$$x - y + z - 2w = 2,$$
$$2x + y - 3z + 3w = 1,$$
$$3x + 2y + w = 7,$$
$$2x + y - z = 2.$$

22.   Solve, by Cramer's rule:

$$2x - y + z = 7,$$
$$3x + 2y + z = 3,$$
$$2x - y + 3z = 9.$$

23.   Find the inverse of

$$\mathbf{A} = \begin{bmatrix} 1 & 0 & 1 \\ 1 & 1 & 0 \\ 0 & 1 & 1 \end{bmatrix}$$

by the method of Theorem 4.10. (*Hint:* Use the results of Exercise 14.)

24.   Find the inverse of

$$\mathbf{A} = \begin{bmatrix} 1 & 0 & 1 \\ 2 & 1 & 1 \\ 0 & 1 & 0 \end{bmatrix}$$

by the method of Theorem 4.10. (*Hint:* Use the results of Exercise 15.)

In Exercises 25 through 28, invert the given matrices using the method of Theorem 4.10.

25.   (a) $\begin{bmatrix} 1 & -2 \\ 3 & 4 \end{bmatrix}$;   (b) $\begin{bmatrix} 4 & 2 & 1 \\ 3 & -1 & 2 \\ 0 & 1 & 2 \end{bmatrix}$.

26.   (a) $\begin{bmatrix} 0 & 1 \\ 2 & 3 \end{bmatrix}$;   (b) $\begin{bmatrix} 0 & 1 & 2 \\ 0 & -1 & 2 \\ 2 & 3 & 1 \end{bmatrix}$.

27.   (a) $\begin{bmatrix} 2 & 2 \\ 0 & 3 \end{bmatrix}$;   (b) $\begin{bmatrix} 1 & 0 & 3 \\ -1 & 1 & 4 \\ 2 & 2 & 1 \end{bmatrix}$.

28.   (a) $\begin{bmatrix} 3 & 2 \\ 2 & 4 \end{bmatrix}$;   (b) $\begin{bmatrix} 1 & 0 & 2 \\ -1 & 1 & 2 \\ 0 & 2 & 2 \end{bmatrix}$.

29.   Verify that $\det(\mathbf{AB}) = (\det \mathbf{A})(\det \mathbf{B})$:

(a) $\mathbf{A} = \begin{bmatrix} 1 & 2 \\ 3 & -4 \end{bmatrix}$, $\mathbf{B} = \begin{bmatrix} -2 & 3 \\ 5 & 1 \end{bmatrix}$;

(b) $\mathbf{A} = \begin{bmatrix} 1 & -2 & 3 \\ 0 & 1 & -3 \\ 2 & 3 & 2 \end{bmatrix}$, $\mathbf{B} = \begin{bmatrix} 4 & -1 & 2 \\ 5 & 0 & 1 \\ 0 & 0 & -3 \end{bmatrix}$.

30.   If $\det \mathbf{A} = 3$, find $\det \mathbf{A}^3$.

31.   If $\det \mathbf{A} = -3$, find $\det \mathbf{A}^{-1}$.

THEORETICAL EXERCISES

T-1.   Prove Lemma 1. (*Hint:* First show that the interchange of adjacent integers in a permutation changes the number of inversions by ± 1. Then show that the interchange of any two integers can be accomplished by an odd number of interchanges of adjacent integers.)

T-2.   *Prove:* If two columns of a determinant are interchanged, the determinant changes sign. (*Hint:* Use Theorem 4.1 and the proof already given for the corresponding statement about rows in Theorem 4.2.)

T-3.   *Prove:* If $\mathbf{A}$ is an $n \times n$ matrix and c is a real number, then $\det(c\mathbf{A}) = c^n \det \mathbf{A}$.

T-4. Prove parts (b) and (d) of Theorem 4.7.

T-5. Complete the proof of Theorem 4.9.

T-6. Using Theorem 4.7, verify Eq. (24).

T-7. Prove Lemma 4.2. (*Hint:* Consider each of the three types of elementary matrices separately, and use Theorems 3.3, 4.2, 4.4, and 4.5.)

T-8. Prove Lemma 4.3 by repeated application of Lemma 4.2.

T-9. *Prove:* If $\mathbf{A}$ is a nonsingular $n \times n$ matrix then

$$\det(\text{adj } \mathbf{A}) = (\det \mathbf{A})^{n-1}.$$

T-10. If $\mathbf{A}$ is an $n \times n$ matrix, show that the homogeneous system $\mathbf{AX} = \mathbf{0}$ has a nontrivial solution if and only if $\det \mathbf{A} = 0$.

T-11. If $\mathbf{AB} = \mathbf{I}$, show that $\det \mathbf{A} \neq 0$ and $\det \mathbf{B} \neq 0$.

T-12. If $\det(\mathbf{AB}) = 0$, show that either $\det \mathbf{A} = 0$ or $\det \mathbf{B} = 0$.

T-13. Show that $\det(\mathbf{AB}) = \det(\mathbf{BA})$.

T-14. Use Theorem 4.10 to show that $\begin{bmatrix} a & b \\ c & d \end{bmatrix}$ is nonsingular if and only if $ad - bc \neq 0$.

T-15. Show that a triangular matrix is nonsingular if and only if every entry on its main diagonal is nonzero.

T-16. If $\mathbf{A}^2 = \mathbf{A}$, show that either $\mathbf{A}$ is singular or $\det \mathbf{A} = 1$.

T-17. Show that $\det(\mathbf{A}^\mathsf{T}\mathbf{B}^\mathsf{T}) = (\det \mathbf{A})(\det \mathbf{B})$.

T-18. If $\mathbf{A} = \mathbf{A}^{-1}$, show that $\det \mathbf{A} = \pm 1$.

T-19. Compute the inverse of the matrix in Example 4.12.

T-20. If $\mathbf{A}$ is an $n \times n$ matrix and t is a real number, show that $\det(\mathbf{A} - t\mathbf{I})$ is a polynomial in t of degree n.

T-21. Prove Theorem 4.6 directly from Definition 4.3.

Chapter 2

Vector Spaces and Linear Transformations

2.1 Vector Spaces

The student has probably dealt with vectors in his previous calculus and physics courses. In this section, we introduce the idea of a vector space, develop some of the consequences of the definition, and consider several examples.

Definition 1.1   A *real vector space*, or *vector space over the real numbers*, is a set $S = \{\mathbf{U}, \mathbf{V}, \mathbf{W}, \ldots\}$ of elements called *vectors* on which two operations, *vector addition* and *scalar multiplication*, are defined. That is, to each pair of elements $\mathbf{U}$ and $\mathbf{V}$ in S, there corresponds a unique *vector sum* $\mathbf{U} + \mathbf{V}$, and to each $\mathbf{U}$ in S and real number a, there corresponds a vector $a\mathbf{U}$, called a *scalar multiple of* $\mathbf{U}$. Moreover, if $\mathbf{U}$, $\mathbf{V}$, and $\mathbf{W}$ are vectors and a and b are real numbers, then:

(a) $\mathbf{U} + \mathbf{V} = \mathbf{V} + \mathbf{U}$. (Vector addition is commutative.)
(b) $\mathbf{U} + (\mathbf{V} + \mathbf{W}) = (\mathbf{U} + \mathbf{V}) + \mathbf{W}$. (Vector addition is associative.)
(c)   There is a *zero vector* $\mathbf{0}$ in S, such that

$$\mathbf{U} + \mathbf{0} = \mathbf{U}$$

for all $\mathbf{U}$ in S.
(d)   Each $\mathbf{U}$ in S has an *additive inverse*, denoted by $-\mathbf{U}$, such that

$$\mathbf{U} + (-\mathbf{U}) = \mathbf{0}.$$

(e) $a(b\mathbf{U}) = (ab)\mathbf{U}.$
(f) $(a + b)\mathbf{U} = a\mathbf{U} + b\mathbf{U}.$
(g) $a(\mathbf{U} + \mathbf{V}) = a\mathbf{U} + a\mathbf{V}.$
(h) $1\mathbf{U} = \mathbf{U}.$

We shall usually refer to a real vector space simply as a *vector space*. Following the usual practice for the real numbers, we shall write $\mathbf{U} + (-\mathbf{V})$ more simply as $\mathbf{U} - \mathbf{V}$. Finally, we shall often refer to real numbers as *scalars*.

There are two harmless inconsistencies in the way that we indicate vector addition and scalar multiplication. The juxtaposition of b and $\mathbf{U}$ on the left side of (e) indicates scalar multiplication, while the juxtaposition of a and b on the right indicates ordinary multiplication of two real numbers. Also, the plus sign on the left side of (f) indicates ordinary addition of two real numbers, while the one on the right indicates vector addition. We could avoid these inconsistencies by introducing new symbols, such as $\oplus$ and $\odot$, to indicate vector addition and scalar multiplication. Then (f) would be written as

$$(a + b) \odot \mathbf{U} = a \odot \mathbf{U} \oplus b \odot \mathbf{U}.$$

However, this notation is cumbersome and unnecessary, since it is always possible to distinguish between the two kinds of addition and multiplication. Clearly, if a and b are real numbers and $\mathbf{U}$ and $\mathbf{V}$ are vectors, then $a + b$ indicates ordinary addition, while $a\mathbf{U} + b\mathbf{V}$ indicates vector addition. Also, $(ab)\mathbf{V}$ indicates ordinary multiplication of a by b, and scalar multiplication of $\mathbf{V}$ by ab.

Example 1.1   The real numbers, with vector addition and scalar multiplication defined to coincide with ordinary addition and multiplication, respectively, form a real vector space (Exercise 1). We denote this vector space by R.

Example 1.2   The set of $m \times n$ matrices, with the operations of matrix addition (Definition 1.2, Chapter 1) and scalar multiplication (Definition 1.3, Chapter 1) is a vector space over the reals, as we verified in Theorems 1.1 and 1.2 of Chapter 1.

Example 1.3   The set of $n \times 1$ matrices

(1)
$$\mathbf{X} = \begin{bmatrix} x_1 \\ x_2 \\ \vdots \\ x_n \end{bmatrix},$$

with vector addition defined by

$$\begin{bmatrix} x_1 \\ x_2 \\ \vdots \\ x_n \end{bmatrix} + \begin{bmatrix} y_1 \\ y_2 \\ \vdots \\ y_n \end{bmatrix} = \begin{bmatrix} x_1 + y_1 \\ x_2 + y_2 \\ \vdots \\ x_n + y_n \end{bmatrix},$$

and scalar multiplication defined by

$$a \begin{bmatrix} x_1 \\ x_2 \\ \vdots \\ x_n \end{bmatrix} = \begin{bmatrix} ax_1 \\ ax_2 \\ \vdots \\ ax_n \end{bmatrix}$$

is a real vector space, which we denote by R^n. An important special case of the spaces considered in Example 1.2, it is the vector space that occurs most frequently throughout this book. Elements of R^n are called *n-vectors;* in fact, throughout much of this book (from Chapter 3 onward), we shall call them simply *vectors.* The scalars $x_1, x_2, \ldots, x_n$ are called the *first, second, $\ldots$, n*th components of **X**, respectively.

The zero vector in R^n is the *n*-vector

$$\mathbf{0} = \begin{bmatrix} 0 \\ 0 \\ \vdots \\ 0 \end{bmatrix},$$

and, if **X** is defined by (1), then

$$-\mathbf{X} = \begin{bmatrix} -x_1 \\ -x_2 \\ \vdots \\ -x_n \end{bmatrix} = (-1) \begin{bmatrix} x_1 \\ x_2 \\ \vdots \\ x_n \end{bmatrix} = (-1)\mathbf{X}.$$

Example 1.4 In R^3, let

$$\mathbf{X} = \begin{bmatrix} 1 \\ -2 \\ 3 \end{bmatrix} \quad \text{and} \quad \mathbf{Y} = \begin{bmatrix} 0 \\ 2 \\ -4 \end{bmatrix};$$

then

$$\mathbf{X} + \mathbf{Y} = \begin{bmatrix} 1 + 0 \\ -2 + 2 \\ 3 - 4 \end{bmatrix} = \begin{bmatrix} 1 \\ 0 \\ -1 \end{bmatrix},$$

$$-\mathbf{Y} = \begin{bmatrix} 0 \\ -2 \\ 4 \end{bmatrix},$$

$$\mathbf{X} - \mathbf{Y} = \begin{bmatrix} 1 \\ -4 \\ 7 \end{bmatrix},$$

and

$$2\mathbf{X} = \begin{bmatrix} 2 \\ -4 \\ 6 \end{bmatrix}.$$

In Section 2.5, we shall consider R^n in detail.

Example 1.5 Let S be the set of all real-valued functions defined on $(-\infty, \infty)$. If f and g are in S, and a is a real number, we define $f + g$ and cf by

(2) $$(f + g)(t) = f(t) + g(t)$$

and

(3) $$(cf)(t) = cf(t)$$

for $-\infty < t < \infty$. With these definitions, S is a vector space (Exercise 3).

Example 1.6 Let P be the set of polynomials of the form

(4) $$p(t) = a_0 + a_1t + \cdots + a_nt^n,$$

where

$$a_0, a_1, \ldots, a_n$$

are arbitrary real numbers. If c is a real number, then

$$(cp)(t) = (ca_0) + (ca_1)t + \cdots + (ca_n)t^n,$$

and if

$$q(t) = b_0 + b_1 t + \cdots + b_m t^m$$

with $m \geq n$, then

$$(p + q)(t) = (a_0 + b_0) + (a_1 + b_1)t + \cdots + (a_m + b_m)t^n,$$

where, if $m > n$, we define $a_r = 0$ for $n + 1 \leq r \leq m$. If $n > m$ we define

$$(p + q)(t) = (a_0 + b_0) + (a_1 + b_1)t + \cdots + (a_n + b_n)t^n,$$

with $b_r = 0$ for $m + 1 \leq r \leq n$. Thus, if

(5) $$p(t) = 1 + 2t + 4t^3,$$

and

(6) $$q(t) = 2 + 5t^2 + t^3 + t^4,$$

then

$$\begin{aligned}(p + q)(t) &= (1 + 2) + (2 + 0)t + (0 + 5)t^2 \\ &\quad + (4 + 1)t^3 + (0 + 1)t^4 \\ &= 3 + 2t + 5t^2 + 5t^3 + t^4,\end{aligned}$$

and

$$(4p)(t) = 4 + 8t + 16t^3.$$

The *degree* of the polynomial (4) is the highest power of t that appears on the right side of (4) with a nonzero coefficient. Thus, (5) and (6) are of degree three and four, respectively, while

$$r(t) = 2$$

is of degree zero.

If p and q have degree n or less, then $p + q$ cannot be of degree greater than n. However, we cannot conclude from this that the degree of $p + q$ is n or less, because if $p = -q$, then $p + q$ is the zero polynomial, which has no nonzero coefficients and, therefore, no degree. We remove this inconvenience by arbitrarily assigning the degree $-\infty$ to the zero polynomial (-1 would do as well), and agreeing that $-\infty < n$ for every $n \geq 0$. With this agreement, we can say that if p and q are of degree n or less, then so is $p + q$.

Example 1.7   The set P_n of polynomials of degree n or less, with vector addition and scalar multiplication a defined in Example 1.6, is a vector space, given our agreement that the zero polynomial is in P_n for all $n \geq 0$ (Exercise 4).

Properties of Vector Spaces

We have seen several examples of mathematical structures which have the properties required in Definition 1.1. Although each of these structures is in some sense different from the others, they must all share any property which follows directly from Definition 1.1. The next theorem gives some examples of such properties.

Theorem 1.1   Any vector space S has the following properties.

(i)   The zero vector **0** is unique.

(ii)   For each **U** in S, $-$**U** is unique.

(iii) 0**U** = **0** for every **U** in S.

(iv) a**0** = **0** for every scalar a.

(v) (-1)**U** = $-$**U**.

(vi)   If c**U** = **0** then $c = 0$ or **U** = **0**.

Proof of (i).   Suppose $\mathbf{0}_1$ and $\mathbf{0}_2$ are both zero vectors; then $\mathbf{U} + \mathbf{0}_1 = \mathbf{U}$ and $\mathbf{V} + \mathbf{0}_2 = \mathbf{V}$ for every **U** and **V** in S. Taking $\mathbf{U} = \mathbf{0}_2$ and $\mathbf{V} = \mathbf{0}_1$ yields the two equations $\mathbf{0}_2 + \mathbf{0}_1 = \mathbf{0}_2$ and $\mathbf{0}_1 + \mathbf{0}_2 = \mathbf{0}_1$. From (a), the left sides of these equations are equal, and it follows that $\mathbf{0}_1 = \mathbf{0}_2$.

Proof of (ii).   Suppose $\mathbf{V}_1$ and $\mathbf{V}_2$ are both additive inverses of **U**; thus $\mathbf{U} + \mathbf{V}_1 = \mathbf{U} + \mathbf{V}_2 = \mathbf{0}$. Then

$$(7) \qquad \mathbf{V}_1 = \mathbf{V}_1 + \mathbf{0} = \mathbf{V}_1 + (\mathbf{U} + \mathbf{V}_2) = (\mathbf{V}_1 + \mathbf{U}) + \mathbf{V}_2$$

$$= (\mathbf{U} + \mathbf{V}_1) + \mathbf{V}_2 = \mathbf{0} + \mathbf{V}_2 = \mathbf{V}_2 + \mathbf{0} = \mathbf{V}_2;$$

that is, $\mathbf{V}_1 = \mathbf{V}_2$, which completes the proof.

Each of the equalities in (7) depends on the definitions of $\mathbf{V}_1$ and $\mathbf{V}_2$, or on one of the properties of Definition 1.1.

Proof of (iii).   Let $\mathbf{V} = -(0\mathbf{U})$; then

$$\mathbf{0} = 0\mathbf{U} + \mathbf{V} = (0 + 0)\mathbf{U} + \mathbf{V}$$

$$= (0\mathbf{U} + 0\mathbf{U}) + \mathbf{V} = 0\mathbf{U} + (0\mathbf{U} + \mathbf{V})$$

$$= 0\mathbf{U} + \mathbf{0} = 0\mathbf{U},$$

which proves (iii).

1. Additive Inverse
2. Property f. + Associative
3. Prop. of Add opposite
4. Prop. of Zero

Proof of (iv).   Exercise T-1.

Proof of (v).

$$(-1)\mathbf{U} + \mathbf{U} = (-1)\mathbf{U} + 1\mathbf{U} = (-1 + 1)\mathbf{U} = 0\mathbf{U};$$

hence, from (iii),

$$(-1)\mathbf{U} + \mathbf{U} = \mathbf{0},$$

and (ii) implies (iv).

Proof of (vi).   Suppose $c\mathbf{U} = \mathbf{0}$ and $c \neq 0$. Then

$$\mathbf{U} = 1\mathbf{U} = \left(\frac{c}{c}\right)\mathbf{U} = \frac{1}{c}\,(c\mathbf{U}) = \frac{1}{c}\,\mathbf{0} = \mathbf{0},$$

where the last equality follows from (iv).

Corollary 1.1   If $\mathbf{U}$, $\mathbf{V}$, and $\mathbf{W}$ are vectors and $\mathbf{U} + \mathbf{V} = \mathbf{U} + \mathbf{W}$, then $\mathbf{V} = \mathbf{W}$.

Proof.   Exercise T-3.

Corollary 1.2   If $\mathbf{U} \neq \mathbf{0}$ and $a\mathbf{U} = b\mathbf{U}$, then $a = b$.

Proof.   Exercise T-4.

So far, vector addition is defined only for pairs of vectors. Consequently, an expression such as $\mathbf{U} + \mathbf{V} + \mathbf{W}$ is as yet undefined. We remedy this by defining

(8) $$\mathbf{U} + \mathbf{V} + \mathbf{W} = (\mathbf{U} + \mathbf{V}) + \mathbf{W}$$

or

(9) $$\mathbf{U} + \mathbf{V} + \mathbf{W} = \mathbf{U} + (\mathbf{V} + \mathbf{W}).$$

These are equivalent definitions, since (b) of Definition 1.1 guarantees that the right sides of (8) and (9) are equal.

Having defined the sum of any three vectors, we can define the sum of any four by

(10) $$\mathbf{U}_1 + \mathbf{U}_2 + \mathbf{U}_3 + \mathbf{U}_4 = (\mathbf{U}_1 + \mathbf{U}_2 + \mathbf{U}_3) + \mathbf{U}_4$$

or

(11) $$\mathbf{U}_1 + \mathbf{U}_2 + \mathbf{U}_3 + \mathbf{U}_4 = \mathbf{U}_1 + (\mathbf{U}_2 + \mathbf{U}_3 + \mathbf{U}_4),$$

since the right-hand sides are equal (Exercise T-5). Continuing in this way, suppose we have defined the sum of $N - 1$ vectors ($N \geq 3$). Then we can define the sum of N vectors as

$$\mathbf{U}_1 + \mathbf{U}_2 + \cdots + \mathbf{U}_N = \mathbf{U}_1 + (\mathbf{U}_2 + \cdots + \mathbf{U}_N)$$

or

$$\mathbf{U}_1 + \mathbf{U}_2 + \cdots + \mathbf{U}_N = (\mathbf{U}_1 + \cdots + \mathbf{U}_{N-1}) + \mathbf{U}_N,$$

since the two definitions are equivalent.

Subspaces of a Vector Space

Definition 1.2 A *subspace* of a vector space S is a collection T of vectors in S such that:

(a) If $\mathbf{U}$ and $\mathbf{V}$ are in T, then $\mathbf{U} + \mathbf{V}$ is in T.
(b) If $\mathbf{U}$ is in T and a is a scalar, then $a\mathbf{U}$ is in T.

Example 1.8 Every vector space S has at least two subspaces: S itself and the subspace consisting of only the zero vector (Exercise 6). We denote the latter by $\{\mathbf{0}\}$. A subspace different from $\{\mathbf{0}\}$ will be called a *nonzero* subspace.

Example 1.9 The set of matrices of the form

$$\begin{bmatrix} 0 & a_{12} & a_{13} \\ 0 & a_{22} & a_{23} \\ 0 & a_{32} & a_{33} \end{bmatrix}$$

is a subspace of the vector space of 3×3 matrices. (Verify.)

Example 1.10 In R^3, the set of all vectors of the form

$$X = \begin{bmatrix} x_1 \\ c \\ x_3 \end{bmatrix},$$

where c is a fixed real number, is a subspace if and only if $c = 0$. (Verify.)

Example 1.11 The vector space P of polynomials (Example 1.6) is a subspace of the vector space S of real valued functions defined on $(-\infty, \infty)$

(Example 1.5). Examples of other subspaces of S are the collection of functions in S which vanish at a fixed $t = t_0$, and the collection of continuous functions in S (Exercise T-7).

Example 1.12 For each $n \geq 0$ the space P_n of polynomials of degree less than or equal to n (Example 1.7) is a subspace of P. Moreover, if $m \leq n$, then P_m is a subspace of P_n. The set of polynomials of exact degree n is not a subspace of P. (Why not?)

Example 1.13 Let T be the set of solutions of the homogeneous system $\mathbf{AX} = \mathbf{0}$ of m equations in n unknowns. Then T is a subspace of R^n, since $\mathbf{AX_1} = \mathbf{0}$ and $\mathbf{AX_2} = \mathbf{0}$ imply that

$$\mathbf{A}(\mathbf{X_1} + \mathbf{X_2}) = \mathbf{AX_1} + \mathbf{AX_2} = \mathbf{0} + \mathbf{0} = \mathbf{0}$$

and

$$\mathbf{A}(c\mathbf{X_1}) = c(\mathbf{AX_1}) = c\mathbf{0} = \mathbf{0},$$

for any scalar c. However, the solutions of $\mathbf{AX} = \mathbf{Y}$ do not form a subspace of R^n if $\mathbf{Y} \neq \mathbf{0}$. (Why not?)

Example 1.14 Let $\mathbf{U}$ and $\mathbf{V}$ be fixed elements of a vector space S and define T to be the collection of vectors in S which can be written in the form

$$\mathbf{W} = a\mathbf{U} + b\mathbf{V},$$

where a and b are arbitrary real numbers. If c is a scalar and $\mathbf{W}$ is in T, then

$$c\mathbf{W} = c(a\mathbf{U} + b\mathbf{V})$$
$$= c(a\mathbf{U}) + c(b\mathbf{V})$$
$$= (ca)\mathbf{U} + (cb)\mathbf{V};$$

hence, $c\mathbf{W}$ is in S. If $\mathbf{W_1} = a_1\mathbf{U} + b_1\mathbf{V}$ and $\mathbf{W_2} = a_2\mathbf{U} + b_2\mathbf{V}$ are two vectors in T, then

$$\text{(12)} \qquad \mathbf{W_1} + \mathbf{W_2} = (a_1\mathbf{U} + b_1\mathbf{V}) + (a_2\mathbf{U} + b_2\mathbf{V})$$
$$= (a_1 + a_2)\mathbf{U} + (b_1 + b_2)\mathbf{V};$$

hence $\mathbf{W_1} + \mathbf{W_2}$ is also in T. We conclude that T is a subspace of S.

The next theorem follows directly from Definitions 1.1 and 1.2 (Exercise T-9).

Theorem 1.2 A subset T of a real vector space S is a subspace of S if and only if T is itself a vector space with respect to the vector addition and scalar multiplication defined in S.

EXERCISES 2.1

1. Verify that the real numbers, with the usual addition and multiplication operations, form a real vector space.

2. Verify in detail that R^n (Example 1.3) is a real vector space.

3. Let S be the set of all real valued functions defined on an interval I (which may be infinite), with vector addition defined by

$$(f + g)(t) = f(t) + g(t)$$

and scalar multiplication defined by

$$(cf)(t) = cf(t)$$

for all t in I. Show that S is a vector space.

4. Verify that P_n (Example 1.7) is a vector space.

5. Let S be the set of all ordered pairs of real numbers, with addition defined by $(a, b) + (c, d) = (a + c, b + d)$, and scalar multiplication by $\alpha(a, b) = (a, \alpha b)$. Show that S is not a vector space.

6. Show that the set $\{\mathbf{0}\}$, which contains only the zero vector of a vector space S, is a subspace of S.

7. State which of the following are subspaces of R^3: the set of vectors of the form

$$(a) \begin{bmatrix} x \\ y \\ 1 \end{bmatrix}; \quad (b) \begin{bmatrix} x \\ 0 \\ z \end{bmatrix}; \quad (c) \begin{bmatrix} x \\ y \\ x + y \end{bmatrix}; \quad (d) \begin{bmatrix} x \\ y \\ x + 1 \end{bmatrix}.$$

8. State which of the following are subspaces of R^3: the set of vectors of the form

$$(a) \begin{bmatrix} x \\ x \\ 1 \end{bmatrix}; \quad (b) \begin{bmatrix} x \\ x \\ 0 \end{bmatrix}; \quad (c) \begin{bmatrix} x \\ y \\ |z| \end{bmatrix}; \quad (d) \begin{bmatrix} x^2 \\ y \\ z \end{bmatrix}.$$

9. State which of the following are subspaces of the vector space of $n \times n$ matrices: the set of (a) symmetric matrices; (b) singular matrices; (c) nonsingular matrices; (d) diagonal matrices.

10. State which of the following are subspaces of the vector space of $n \times n$ matrices: the set of (a) matrices whose diagonal elements sum to zero; (b) triangular matrices; (c) upper triangular matrices; (d) matrices in row-echelon form.

11. State which of the following are subspaces of the vector space P (Example 1.6): the set of (a) even polynomials; (b) polynomials of degree 2; (c) polynomials with no real roots; (d) polynomials which vanish on a certain set of points $\{t_1, t_2, \ldots, t_n\}$.

12. State which of the following are subspaces of the set S defined in Exercise 3, with $I = (-\infty, \infty)$: the set of (a) even functions; (b) odd functions; (c) nonnegative functions; (d) functions such that $f(0) + f(1) = f(2)$; (e) functions such that $f(0) = 0$; (f) functions such that $f(0) + f(1) = f(2) + 1$.

13. Verify that the set of solutions of

$$2x - 3y + 4z + w = 0,$$
$$3x - 4y + 2z - w = 0,$$

is a subspace of R^4.

14. Verify that the set of vectors of the form

$$a \begin{bmatrix} 1 \\ 2 \\ -1 \\ 2 \end{bmatrix} + b \begin{bmatrix} 0 \\ 1 \\ 2 \\ -3 \end{bmatrix},$$

where a and b are arbitrary scalars, is a subspace of R^4.

THEORETICAL EXERCISES

T-1. Prove (iv) of Theorem 1.1.

T-2. *Prove:* $-(-\mathbf{U}) = \mathbf{U}$.

T-3. Prove Corollary 1.1.

T-4. Prove Corollary 1.2.

T-5. Verify that Eqs. (10) and (11) are equivalent definitions of the sum of four vectors.

T-6. Verify Eq. (12) in detail, stating at each step which of the properties in Definition 1.1 is being invoked.

T-7. Let $C[a, b]$ denote the set of functions which are continuous on the interval $I = [a, b]$, and define vector addition and multiplication as in Exercise 3. Show that $C[a, b]$ is a vector space.

T-8. For each $n \geq 1$, let $C^n[a, b]$ be the set of functions which are n times differentiable on $[a, b]$. Show that $C^n[a, b]$ is a subspace of $C[a, b]$ (Exercise T-8).

T-9. Prove Theorem 1.2.

T-10. Let $p_0, p_1, \ldots, p_n$ be functions defined on $[a, b]$. Show that the set of functions in $C^n[a, b]$ which satisfy

$$p_0 \frac{d^n f}{dx^n} + p_1 \frac{d^{n-1} f}{dx^{n-1}} + \cdots + p_n f = 0$$

is a subspace of $C^n[a, b]$.

2.2 Linear Independence and Bases

In this section, we consider several ideas that are important for the study of vector spaces.

Definition 2.1 A vector **U** is said to be a *linear combination* of vectors $\mathbf{U}_1, \ldots, \mathbf{U}_m$ if it can be written in the form

$$\mathbf{U} = c_1 \mathbf{U}_1 + \cdots + c_m \mathbf{U}_m,$$

where $c_1, \ldots, c_n$ are scalars.

Example 2.1 In R^4 let

$$\mathbf{X}_1 = \begin{bmatrix} 1 \\ 0 \\ 2 \\ -1 \end{bmatrix}, \quad \mathbf{X}_2 = \begin{bmatrix} 2 \\ 1 \\ -3 \\ 1 \end{bmatrix}, \quad \mathbf{X}_3 = \begin{bmatrix} 0 \\ 1 \\ 2 \\ -1 \end{bmatrix}, \quad \text{and} \quad \mathbf{X} = \begin{bmatrix} 5 \\ 1 \\ -6 \\ 2 \end{bmatrix};$$

then $\mathbf{X}$ is a linear combination of $\mathbf{X}_1$, $\mathbf{X}_2$, and $\mathbf{X}_3$, since

$$\mathbf{X} = \mathbf{X}_1 + 2\mathbf{X}_2 - \mathbf{X}_3.$$

Definition 2.2   Let $\mathbf{U}_1, \ldots, \mathbf{U}_m$ be fixed vectors in a vector space S. The *subspace of S spanned* by $\mathbf{U}_1, \ldots, \mathbf{U}_m$ is the set of all linear combinations of $\mathbf{U}_1, \ldots, \mathbf{U}_m$. It is denoted by $S[\mathbf{U}_1, \ldots, \mathbf{U}_m]$. Thus $\mathbf{U}$ is in $S[\mathbf{U}_1, \ldots, \mathbf{U}_m]$ if and only if it can be written as

$$\mathbf{U} = c_1\mathbf{U}_1 + \cdots + c_m\mathbf{U}_m$$

for some choice of scalars $c_1, \ldots, c_m$.

The student should verify that $S[\mathbf{U}_1, \ldots, \mathbf{U}_m]$ is a subspace of S (Exercise T-1).

Example 2.2   The subspace of R^3 spanned by

$$\mathbf{X}_1 = \begin{bmatrix} 1 \\ 0 \\ 0 \end{bmatrix}, \qquad \mathbf{X}_2 = \begin{bmatrix} 0 \\ 0 \\ 1 \end{bmatrix}, \qquad \text{and} \qquad \mathbf{X}_3 = \begin{bmatrix} 1 \\ 0 \\ 1 \end{bmatrix}$$

is the set of vectors of the form

$$\mathbf{X} = \begin{bmatrix} x_1 \\ 0 \\ x_3 \end{bmatrix}.$$

Each such $\mathbf{X}$ can be written in infinitely many ways as a linear combination of $\mathbf{X}_1$, $\mathbf{X}_2$, and $\mathbf{X}_3$; thus

$$\begin{bmatrix} x_1 \\ 0 \\ x_3 \end{bmatrix} = c_1\mathbf{X}_1 + c_2\mathbf{X}_2 + c_3\mathbf{X}_3,$$

provided

$$c_1 + c_3 = x_1, \qquad c_2 + c_3 = x_3.$$

Here, any one of c_1, c_2, or c_3 can be chosen arbitrarily, and then the other two are determined uniquely. Note that since each vector

$$\mathbf{X} = \begin{bmatrix} x_1 \\ 0 \\ x_3 \end{bmatrix}$$

is of the form $\mathbf{X} = x_1\mathbf{X}_1 + x_3\mathbf{X}_2$, the pair $\{\mathbf{X}_1, \mathbf{X}_2\}$ already spans the subspace. In fact, any pair chosen from $\mathbf{X}_1$, $\mathbf{X}_2$, and $\mathbf{X}_3$ already spans $S[\mathbf{X}_1, \mathbf{X}_2, \mathbf{X}_3]$ (Exercise T-2).

Example 2.3 The vectors

$$\mathbf{E}_1 = \begin{bmatrix} 1 \\ 0 \\ 0 \end{bmatrix}, \quad \mathbf{E}_2 = \begin{bmatrix} 0 \\ 1 \\ 0 \end{bmatrix}, \quad \mathbf{E}_3 = \begin{bmatrix} 0 \\ 0 \\ 1 \end{bmatrix}$$

span R^3, since

$$\begin{bmatrix} x_1 \\ x_2 \\ x_3 \end{bmatrix} = x_1\mathbf{E}_1 + x_2\mathbf{E}_2 + x_3\mathbf{E}_3;$$

thus $S[\mathbf{E}_1, \mathbf{E}_2, \mathbf{E}_3] = R^3$. Similarly,

$$(1) \qquad \mathbf{E}_1 = \begin{bmatrix} 1 \\ 0 \\ \vdots \\ 0 \end{bmatrix}, \quad \mathbf{E}_2 = \begin{bmatrix} 0 \\ 1 \\ 0 \\ \vdots \\ 0 \end{bmatrix}, \ldots, \quad \mathbf{E}_n = \begin{bmatrix} 0 \\ \vdots \\ 0 \\ 1 \end{bmatrix}$$

span R^n, since

$$\begin{bmatrix} x_1 \\ \vdots \\ x_n \end{bmatrix} = x_1\mathbf{E}_1 + \cdots + x_n\mathbf{E}_n.$$

Example 2.4 The polynomials $p_1(t) = t^2 + t + 1$, $p_2(t) = t^2 - 1$, and $p_3(t) = 1$ span P_2, the vector space of polynomials of degree less than or equal to 2. To verify this, we must show that any polynomial

(2) $$p(t) = a + bt + ct^2$$

in P_2 can be written as

(3)
$$p(t) = \alpha p_1(t) + \beta p_2(t) + \gamma p_3(t)$$
$$= \alpha(t^2 + t + 1) + \beta(t^2 - 1) + \gamma$$

for suitable constants α, β, and γ. Equating the coefficients of like powers of t in (2) and (3) yields

$$\alpha + \beta = c,$$
$$\alpha = b,$$
$$\alpha - \beta + \gamma = a,$$

which has the unique solution

$$\alpha = b,$$
$$\beta = c - b,$$
$$\gamma = a - 2b + c;$$

thus, if $p(t)$ is defined by (2), then

$$p(t) = bp_1(t) + (c - b)p_2(t) + (a - 2b + c)p_3(t).$$

Example 2.5 The polynomials $p_1(t) = t^2 + 1$ and $p_2(t) = t - 1$ do not span P_2, since any linear combination of p_1 and p_2 is of the form

(4)
$$p(t) = \alpha(t^2 + 1) + \beta(t - 1)$$
$$= \alpha t^2 + \beta t + \alpha - \beta.$$

Equating coefficients of powers of t in (2) and (4) yields

$$\alpha = c,$$
$$\beta = b,$$
$$\alpha - \beta = a;$$

therefore, p is not in $S[p_1, p_2]$ unless $a = c - b$.

Linear Dependence

Definition 2.3 The vectors $\mathbf{U}_1, \ldots, \mathbf{U}_m$ are said to be *linearly dependent* if there are scalars $c_1, \ldots, c_m$, not all zero, such that

$$(5) \qquad c_1\mathbf{U}_1 + \cdots + c_m\mathbf{U}_m = \mathbf{0}.$$

Otherwise, $\mathbf{U}_1, \ldots, \mathbf{U}_m$ are said to be *linearly independent*; in this case, (5) implies that $c_1 = \cdots = c_m = 0$. We shall also say that the set $\{\mathbf{U}_1, \ldots, \mathbf{U}_m\}$ is linearly dependent or independent, as the case may be.

Example 2.6 The vectors $\mathbf{X}_1$, $\mathbf{X}_2$, and $\mathbf{X}_3$ defined in Example 2.2 are linearly dependent, since

$$\mathbf{X}_1 + \mathbf{X}_2 - \mathbf{X}_3 = \mathbf{0}.$$

However, any pair chosen from these three vectors is linearly independent. (Exercise T-2.)

Example 2.7 The vectors $\mathbf{E}_1$, $\mathbf{E}_2$, and $\mathbf{E}_3$ of Example 2.3 are linearly independent, since

$$c_1\begin{bmatrix} 1 \\ 0 \\ 0 \end{bmatrix} + c_2\begin{bmatrix} 0 \\ 1 \\ 0 \end{bmatrix} + c_3\begin{bmatrix} 0 \\ 0 \\ 1 \end{bmatrix} = \begin{bmatrix} c_1 \\ c_2 \\ c_3 \end{bmatrix} = \begin{bmatrix} 0 \\ 0 \\ 0 \end{bmatrix}$$

if and only if $c_1 = c_2 = c_3 = 0$. Similarly, $\mathbf{E}_1, \ldots, \mathbf{E}_n$, as defined by (1), are linearly independent in R^n.

Example 2.8 The polynomials p_1, p_2, and p_3 of Example 2.4 are linearly independent, for if

$$\alpha(t^2 + t + 1) + \beta(t^2 - 1) + \gamma = 0$$

for all t, then the coefficients of 1, t, and t^2 on the left must vanish. Hence

$$\alpha + \beta = 0,$$

$$\alpha = 0,$$

$$\alpha - \beta + \gamma = 0,$$

which has the unique solution $\alpha = \beta = \gamma = 0$.

The last two examples illustrate the basic method for proving that a set of vectors $\mathbf{U}_1, \ldots, \mathbf{U}_m$ is linearly independent: assume that (5) holds for some choice of $c_1, \ldots, c_m$, and then deduce that $c_1 = \cdots = c_m = 0$.

Example 2.9 To test the polynomials $p_1(t) = t^2 + t + 1$, $p_2(t) = t - 1$, $p_3(t) = t^2 + 1$, and $p_4(t) = 3t - 2$ for linear dependence, we observe that

$$\alpha(t^2 + t + 1) + \beta(t - 1) + \gamma(t^2 + 1) + \delta(3t - 2) = 0$$

for all t if and only if

$$\alpha + \gamma = 0,$$

$$\alpha + \beta + 3\delta = 0,$$

$$\alpha - \beta + \gamma - 2\delta = 0.$$

This system has the solution $(\alpha, \beta, \gamma, \delta) = (1, 2, -1, -1)$; hence

$$p_1(t) + 2p_2(t) - p_3(t) - p_4(t) = 0$$

for all t, and p_1, p_2, p_3, and p_4 are linearly dependent.

Example 2.10 To test the vectors

$$\mathbf{X}_1 = \begin{bmatrix} 1 \\ 0 \\ 1 \end{bmatrix}, \qquad \mathbf{X}_2 = \begin{bmatrix} 1 \\ 0 \\ -1 \end{bmatrix}, \qquad \text{and} \qquad \mathbf{X}_3 = \begin{bmatrix} 3 \\ 0 \\ 1 \end{bmatrix},$$

for linear dependence, we observe that

$$a \begin{bmatrix} 1 \\ 0 \\ 1 \end{bmatrix} + b \begin{bmatrix} 1 \\ 0 \\ -1 \end{bmatrix} + c \begin{bmatrix} 3 \\ 0 \\ 1 \end{bmatrix} = \begin{bmatrix} 0 \\ 0 \\ 0 \end{bmatrix}$$

if and only if

$$a + b + 3c = 0,$$

$$a - b + c = 0.$$

This system has the solution $a = -2$, $b = -1$, $c = 1$. Hence, $\mathbf{X}_1$, $\mathbf{X}_2$, and $\mathbf{X}_3$ are linearly dependent, since

$$-2\mathbf{X}_1 - \mathbf{X}_2 + \mathbf{X}_3 = \mathbf{0}.$$

Example 2.11 Any set of vectors that contains $\mathbf{0}$ must be linearly dependent; if $\mathbf{U}_i = \mathbf{0}$ in Definition 2.3, then (5) is satisfied with $c_i = 1$ and $c_j = 0$ for $j \neq i$.

Theorem 2.1 The vectors $\mathbf{U}_1, \ldots, \mathbf{U}_m$ are linearly dependent if and only if one of them is a linear combination of the others.

Proof. If $V = \{\mathbf{U}_1, \ldots, \mathbf{U}_m\}$ is a linearly dependent set, then (5) holds with some $c_i \neq 0$; hence we can divide by c_i and solve for $\mathbf{U}_i$ as a linear combination of the remaining vectors in V. Conversely, if $\mathbf{U}_i$ is a linear combination of the other members of V, then (5) holds for some $c_1, \ldots, c_m$ with $c_i = -1$, and V is linearly dependent.

Theorem 2.2 If $V = \{\mathbf{U}_1, \ldots, \mathbf{U}_m\}$ is a set of vectors, not all zero, then it is possible to choose linearly independent vectors from V which span the same subspace as V.

Proof. If V is linearly independent, there is nothing to prove. Now suppose V is linearly dependent. By Theorem 2.2, one of the $\mathbf{U}_i$ is a linear combination of the others. We can assume that $\mathbf{U}_m$ is a linear combination of $\mathbf{U}_1, \ldots, \mathbf{U}_{m-1}$. (If this is not the case, we simply renumber the elements of V to make it so.) Then

$$\mathbf{U}_m = a_1\mathbf{U}_1 + \cdots + a_{m-1}\mathbf{U}_{m-1},$$

and if

$$\mathbf{U} = c_1\mathbf{U}_1 + \cdots + c_m\mathbf{U}_m,$$

then $\mathbf{U}$ can be written as a linear combination of $\mathbf{U}_1, \ldots, \mathbf{U}_{m-1}$:

$$\mathbf{U} = (c_1 + a_1 c_m)\mathbf{U}_1 + \cdots + (c_{m-1} + a_{m-1}c_m)\mathbf{U}_{m-1}.$$

Therefore,

$$S[\mathbf{U}_1, \ldots, \mathbf{U}_{m-1}] = S[\mathbf{U}_1, \ldots, \mathbf{U}_m].$$

If $\mathbf{U}_1, \ldots, \mathbf{U}_{m-1}$ are linearly independent, we are finished. If they are not, we repeat the argument, eliminating one $\mathbf{U}_i$ each time, until those remaining are linearly independent. This must occur after fewer than m steps, since some $\mathbf{U}_i \neq 0$.

Example 2.12 In Example 2.2, $V = \{\mathbf{X}_1, \mathbf{X}_2, \mathbf{X}_3\}$ is a linearly dependent set. Three linearly independent subsets of V span $S[\mathbf{X}_1, \mathbf{X}_2, \mathbf{X}_3]$; namely, $\{\mathbf{X}_1, \mathbf{X}_2\}$, $\{\mathbf{X}_1, \mathbf{X}_3\}$, and $\{\mathbf{X}_2, \mathbf{X}_3\}$ (Exercise T-2).

Example 2.13 The vectors

$$\mathbf{X}_1 = \begin{bmatrix} 2 \\ 0 \\ -3 \end{bmatrix}, \quad \mathbf{X}_2 = \begin{bmatrix} 0 \\ 0 \\ 1 \end{bmatrix}, \quad \mathbf{X}_3 = \begin{bmatrix} 1 \\ 0 \\ 0 \end{bmatrix}, \quad \text{and} \quad \mathbf{X}_4 = \begin{bmatrix} 1 \\ 0 \\ 1 \end{bmatrix}$$

are linearly dependent in R^3, since

$$\mathbf{X}_1 + 2\mathbf{X}_2 - 3\mathbf{X}_3 + \mathbf{X}_4 = 0.$$

Solving this for $\mathbf{X}_4$ yields

(6) $$\mathbf{X}_4 = -\mathbf{X}_1 - 2\mathbf{X}_2 + 3\mathbf{X}_3.$$

Now suppose

(7) $$\mathbf{X} = c_1\mathbf{X}_1 + c_2\mathbf{X}_2 + c_3\mathbf{X}_3 + c_4\mathbf{X}_4$$

is an arbitrary vector in $H = S[\mathbf{X}_1, \mathbf{X}_2, \mathbf{X}_3, \mathbf{X}_4]$. Substituting (6) into (7) yields

(8) $$\mathbf{X} = c_1\mathbf{X}_1 + c_2\mathbf{X}_2 + c_3\mathbf{X}_3 + c_4(-\mathbf{X}_1 - 2\mathbf{X}_2 + 3\mathbf{X}_3)$$
$$= (c_1 - c_4)\mathbf{X}_1 + (c_2 - 2c_4)\mathbf{X}_2 + (c_3 + 3c_4)\mathbf{X}_3,$$

which implies that $H = S[\mathbf{X}_1, \mathbf{X}_2, \mathbf{X}_3]$. However, $\mathbf{X}_1$, $\mathbf{X}_2$, and $\mathbf{X}_3$ are also linearly dependent, since

$$\mathbf{X}_3 = \tfrac{1}{2}\mathbf{X}_1 + \tfrac{3}{2}\mathbf{X}_2.$$

Substituting this into (8) yields

$$\mathbf{X} = (c_1 - c_4)\mathbf{X}_1 + (c_2 - 2c_4)\mathbf{X}_2 + (c_3 + 3c_4)\left(\frac{1}{2}\mathbf{X}_1 + \frac{3}{2}\mathbf{X}_2\right)$$

$$= \left(c_1 + \frac{c_3}{2} + \frac{c_4}{2}\right)\mathbf{X}_1 + \left(c_2 + \frac{3}{2}c_3 + \frac{5}{2}c_4\right)\mathbf{X}_2.$$

Therefore $H = S[\mathbf{X}_1, \mathbf{X}_2]$; moreover $\mathbf{X}_1$ and $\mathbf{X}_2$ are linearly independent, since

$$a_1\mathbf{X}_1 + a_2\mathbf{X}_2 = 0$$

implies that

$$2a_1 = 0,$$

$$-3a_1 + a_2 = 0,$$

which has the unique solution $a_1 = a_2 = 0$. Thus, we have found two linearly independent vectors in $V = \{\mathbf{X}_1, \mathbf{X}_2, \mathbf{X}_3, \mathbf{X}_4\}$ that span the same subspace as V. This can be accomplished in other ways; we leave it to the student to verify that any pair of vectors from V is linearly independent and spans H.

Definition 2.4 A set $B = \{\mathbf{U}_1, \ldots, \mathbf{U}_m\}$ of elements in a vector space S is a *basis* for S if B is linearly independent and spans S.

Example 2.14 We saw in Example 2.7 that $\{\mathbf{E}_1, \ldots, \mathbf{E}_n\}$, as defined by (1), is a basis for R^n.

In Example 2.13, we showed that $\mathbf{X}_1$ and $\mathbf{X}_2$ form a basis for H.

The polynomials $p_0(t) = 1$, $p_1(t) = t$, and $p_2(t) = t^2$ form a basis for P_2. (Verify.)

Theorem 2.3 If $B = \{\mathbf{U}_1, \ldots, \mathbf{U}_m\}$ is a basis for a vector space S, then every vector in S can be written uniquely as a linear combination of $\mathbf{U}_1, \ldots, \mathbf{U}_m$.

Proof. Since B is a basis for S, any vector $\mathbf{U}$ in S can be written as

$$(9) \qquad \mathbf{U} = c_1 \mathbf{U}_1 + \cdots + c_m \mathbf{U}_m$$

for some choice of the scalars $c_1, \ldots, c_m$. Suppose also that

$$(10) \qquad \mathbf{U} = d_1 \mathbf{U}_1 + \cdots + d_m \mathbf{U}_m;$$

subtracting (9) from (10) yields

$$\mathbf{0} = (d_1 - c_1)\mathbf{U}_1 + \cdots + (d_m - c_m)\mathbf{U}_m,$$

and the linear independence of B implies that

$$d_1 - c_1 = \cdots = d_m - c_m = 0,$$

which completes the proof.

Theorem 2.4 If $B = \{\mathbf{U}_1, \ldots, \mathbf{U}_m\}$ is a basis for a vector space S, then no linearly independent set of vectors in S contains more than m vectors, and every basis contains exactly m vectors.

Proof. Let $B_1 = \{\mathbf{V}_1, \ldots, \mathbf{V}_k\}$ be a linearly independent set of vectors in S and suppose $k > m$. Since B is a basis for S, we can write

$$\mathbf{V}_1 = a_1 \mathbf{U}_1 + \cdots + a_m \mathbf{U}_m,$$

where $a_i \neq 0$ for some i. Suppose $a_m \neq 0$ (if this is not so, simply renumber the elements of B to make it so); then

$$\mathbf{U}_m = \frac{\mathbf{V}_1}{a_m} - \frac{1}{a_m}(a_1 \mathbf{U}_1 + \cdots + a_{m-1}\mathbf{U}_{m-1})$$

and $[\mathbf{U}_1, \ldots, \mathbf{U}_{m-1}, \mathbf{V}_1]$ spans S. Therefore, we can write

$$\mathbf{V}_2 = b_1 \mathbf{U}_1 + \cdots + b_{m-1}\mathbf{U}_{m-1} + c\mathbf{V}_1$$

where $b_i \neq 0$ for at least one i, $1 \leq i \leq m - 1$, since $b_1 = \cdots = b_{m-1} = 0$

would imply that $\mathbf{V}_2 = c\mathbf{V}_1$, a contradiction. Again we assume (possibly after renumbering $\mathbf{U}_1, \ldots, \mathbf{U}_{m-1}$) that $b_{m-1} \neq 0$; then

$$\mathbf{U}_{m-1} = \frac{\mathbf{V}_2}{b_{m-1}} - \frac{1}{b_{m-1}}\,(b_1\mathbf{U}_1 + \cdots + b_{m-2}\mathbf{U}_{m-2} + c\mathbf{V}_1).$$

This implies that $\{\mathbf{U}_1, \ldots, \mathbf{U}_{m-2}, \mathbf{V}_1, \mathbf{V}_2\}$ spans S. Carrying out this argument m times, we conclude that $\{\mathbf{V}_1, \ldots, \mathbf{V}_m\}$ spans S. Since $k > m$, we can write

$$\mathbf{V}_{m+1} = d_1\mathbf{V}_1 + \cdots + d_m\mathbf{V}_m,$$

which contradicts the assumption that B_1 is a linearly independent set. Hence $k \leq m$. If B_1 is also a basis for S, the same argument with B and B_1 interchanged implies that $k \geq m$. Hence two bases have the same number of elements, and the proof is complete.

Definition 2.5   A vector space S which has a basis consisting of a finite number of vectors is said to be *finite-dimensional*. In particular, if a basis contains m elements, then S is *m-dimensional*, and we write dim $S = m$. A vector space consisting of a zero vector alone is *zero-dimensional*, and a nonzero vector space that does not have a finite basis is *infinite-dimensional*.

Example 2.15   The vector space R^n is n-dimensional, with basis $\mathbf{E}_1, \ldots,$ $\mathbf{E}_n$ as defined by (1). This basis is called the *natural basis* for R^n.

The vector space P_m of polynomials of degree less than or equal to m has dimension $m + 1$, since the polynomials $p_0(t) = 1$, $p_1(t) = 1, \ldots, p_m(t) = t^m$ form a basis for P_m.

The vector space P of all polynomials is infinite-dimensional, because if $q_0, \ldots, q_k$ is a finite set of polynomials and d is the largest of their degrees, then no polynomial of degree greater than d can be written as a linear combination of $q_0, \ldots, q_k$. Hence, no finite set of polynomials can be a basis for P.

Theorem 2.5   Every nonzero subspace T of a finite-dimensional vector space S has a basis, and dim $T \leq$ dim S.

Proof.   Let dim $S = n$. A linearly independent set of vectors in T is also linearly independent in S. Therefore, Theorem 2.4 implies that no linearly independent set in T contains more than n elements. If m is the maximum number of vectors in any linearly independent set in T, then any linearly independent set in T with m elements is a basis for T. We leave the proof of this statement to the student (Exercise T-3).

Example 2.16 The set T of vectors of the form

$$\begin{bmatrix} a \\ b \\ a + b \\ a - b \end{bmatrix}$$

is a two-dimensional subspace of R^4, since

$$\begin{bmatrix} a \\ b \\ a + b \\ a - b \end{bmatrix} = a \begin{bmatrix} 1 \\ 0 \\ 1 \\ 1 \end{bmatrix} + b \begin{bmatrix} 0 \\ 1 \\ 1 \\ -1 \end{bmatrix},$$

and the vectors on the right are linearly independent.

The next theorem states that any collection of linearly independent vectors in a finite-dimensional space can be included in a basis for the space.

Theorem 2.6 If $m < n$ and $\mathbf{U}_1, \ldots, \mathbf{U}_m$ are linearly independent vectors in an n-dimensional vector space S, then there exist vectors $\mathbf{V}_1, \ldots, \mathbf{V}_{n-m}$ such that $\{\mathbf{U}_1, \ldots, \mathbf{U}_m, \mathbf{V}_1, \ldots, \mathbf{V}_{n-m}\}$ is a basis for S.

Proof. Since $\mathbf{U}_1, \ldots, \mathbf{U}_m$ is not a basis for S, there exists a vector $\mathbf{V}_1$ which is not a linear combination of $\mathbf{U}_1, \ldots, \mathbf{U}_m$. Hence $\mathbf{U}_1, \ldots, \mathbf{U}_m, \mathbf{V}_1$ are linearly independent. If $n = m + 1$, we are finished, by Theorem 2.4. If $n > m + 1$ there is a vector $\mathbf{V}_2$ which is not a linear combination of $\mathbf{U}_1, \ldots, \mathbf{U}_m, \mathbf{V}_1$. Hence $\mathbf{U}_1, \ldots, \mathbf{U}_m, \mathbf{V}_1, \mathbf{V}_2$ are linearly independent. By carrying out this argument $n - m$ times, we arrive at a set $\{\mathbf{U}_1, \ldots, \mathbf{U}_m, \mathbf{V}_1, \ldots, \mathbf{V}_{n-m}\}$ of n linearly independent vectors, which must be a basis for S (Exercise T-6).

Bases in R^n

We now give a useful necessary and sufficient condition for a set of vectors to be a basis for R^n.

Theorem 2.7 The n-vectors

$$(11) \qquad \mathbf{X}_1 = \begin{bmatrix} a_{11} \\ a_{21} \\ \vdots \\ a_{n1} \end{bmatrix}, \ \mathbf{X}_2 = \begin{bmatrix} a_{12} \\ a_{22} \\ \vdots \\ a_{n2} \end{bmatrix}, \dots, \ \mathbf{X}_n = \begin{bmatrix} a_{1n} \\ a_{2n} \\ \vdots \\ a_{nn} \end{bmatrix}$$

form a basis for R^n if and only if the matrix

$$(12) \qquad \mathbf{A} = \begin{bmatrix} a_{11} & a_{12} & \cdots & a_{1n} \\ a_{21} & a_{22} & \cdots & a_{2n} \\ \vdots & \vdots & & \vdots \\ a_{n1} & a_{n2} & \cdots & a_{nn} \end{bmatrix},$$

whose jth column is the vector $\mathbf{X}_j$, is nonsingular.

Proof. The vectors $\mathbf{X}_1, \dots, \mathbf{X}_n$ form a basis for R^n if and only if every

$$\mathbf{X} = \begin{bmatrix} x_1 \\ \vdots \\ x_n \end{bmatrix}$$

can be expressed in the form

$$(13) \qquad \mathbf{X} = c_1\mathbf{X}_1 + c_2\mathbf{X}_2 + \cdots + c_n\mathbf{X}_n$$

(Exercise T-5). By equating components on both sides of (13), we find that it is equivalent to

$$(14) \qquad \begin{aligned} x_1 &= a_{11}c_1 + a_{12}c_2 + \cdots + a_{1n}c_n, \\ x_2 &= a_{21}c_1 + a_{22}c_2 + \cdots + a_{2n}c_n, \\ &\vdots \\ x_n &= a_{n1}c_1 + a_{n2}c_2 + \cdots + a_{nn}c_n. \end{aligned}$$

Thus, $\{\mathbf{X}_1, \dots, \mathbf{X}_n\}$ is a basis for R^n if and only if the linear system (14) has a solution $(c_1, \dots, c_n)$ for every given $\mathbf{X}$, which is equivalent to the nonsingularity of A (Theorem 3.9, Chapter 1). This completes the proof.

From this theorem and Theorem 4.10, Chapter 1, we obtain the following corollary.

Corollary 2.1 The vectors $\mathbf{X}_1, \dots, \mathbf{X}_n$ defined by (11) are linearly independent, and therefore form a basis for R^n if and only if the determinant

$$\begin{vmatrix} a_{11} & a_{12} & \cdots & a_{1n} \\ a_{21} & a_{22} & \cdots & a_{2n} \\ \vdots & \vdots & & \vdots \\ a_{n1} & a_{n2} & \cdots & a_{nn} \end{vmatrix}$$

is nonzero.

Theorem 2.7 and Corollary 2.1 are useful because they allow us to use the computational techniques of Chapter 1 to determine first whether $\{X_1, \ldots, X_n\}$ is a basis for R^n, and if it is, to find $c_1, \ldots, c_n$ in (13).

Example 2.17 The vectors

$$X_1 = \begin{bmatrix} 1 \\ 1 \\ 2 \end{bmatrix}, \qquad X_2 = \begin{bmatrix} 0 \\ 1 \\ 1 \end{bmatrix}, \qquad \text{and} \qquad X_3 = \begin{bmatrix} 1 \\ 2 \\ 3 \end{bmatrix}$$

do not form a basis for R^3, since

$$\begin{vmatrix} 1 & 0 & 1 \\ 1 & 1 & 2 \\ 2 & 1 & 3 \end{vmatrix} = 0.$$

(Verify.)

Example 2.18 The vectors

$$X_1 = \begin{bmatrix} 1 \\ 1 \\ 0 \end{bmatrix}, \qquad X_2 = \begin{bmatrix} 0 \\ 1 \\ 1 \end{bmatrix}, \qquad \text{and} \qquad X_3 = \begin{bmatrix} 1 \\ 0 \\ 1 \end{bmatrix}$$

form a basis for R^3, since we have previously shown (Example 3.7, Chapter 1) that

$$A = \begin{bmatrix} 1 & 0 & 1 \\ 1 & 1 & 0 \\ 0 & 1 & 1 \end{bmatrix}$$

is nonsingular, with inverse

$$
\mathbf{A}^{-1} = \begin{bmatrix} \frac{1}{2} & \frac{1}{2} & -\frac{1}{2} \\ -\frac{1}{2} & \frac{1}{2} & \frac{1}{2} \\ \frac{1}{2} & -\frac{1}{2} & \frac{1}{2} \end{bmatrix}.
$$

Let

$$
\mathbf{X} = \begin{bmatrix} 2 \\ 1 \\ 3 \end{bmatrix};
$$

to find c_1, c_2, and c_3 such that

$$
\mathbf{X} = c_1\mathbf{X}_1 + c_2\mathbf{X}_2 + c_3\mathbf{X}_3,
$$

we must solve

$$
\begin{bmatrix} 1 & 0 & 1 \\ 1 & 1 & 0 \\ 0 & 1 & 1 \end{bmatrix} \begin{bmatrix} c_1 \\ c_2 \\ c_3 \end{bmatrix} = \begin{bmatrix} 2 \\ 1 \\ 3 \end{bmatrix}.
$$

Multiplying both sides by $\mathbf{A}^{-1}$ yields

$$
\begin{bmatrix} c_1 \\ c_2 \\ c_3 \end{bmatrix} = \begin{bmatrix} 0 \\ 1 \\ 2 \end{bmatrix};
$$

hence

$$
\mathbf{X} = \mathbf{X}_2 + 2\mathbf{X}_3.
$$

Corollary 2.2   The vectors $\mathbf{X}_1,\ldots,\mathbf{X}_n$ defined by (11) form a basis for R^n if and only if the same is true of the vectors

$$
\mathbf{Y}_1 = \begin{bmatrix} a_{11} \\ a_{12} \\ \vdots \\ a_{1n} \end{bmatrix}, \ \mathbf{Y}_2 = \begin{bmatrix} a_{21} \\ a_{22} \\ \vdots \\ a_{2n} \end{bmatrix}, \ldots, \ \mathbf{Y}_n = \begin{bmatrix} a_{n1} \\ a_{n2} \\ \vdots \\ a_{nn} \end{bmatrix},
$$

which are simply the columns of $\mathbf{A}^{\mathrm{T}}$, where $\mathbf{A}$ is defined by (12).

We leave the proof of this corollary to the student (Exercise T-8).

Example 2.19 From Example 2.17, the vectors

$$\begin{bmatrix} 1 \\ 1 \\ 2 \end{bmatrix}, \quad \begin{bmatrix} 0 \\ 1 \\ 1 \end{bmatrix}, \quad \text{and} \quad \begin{bmatrix} 1 \\ 2 \\ 3 \end{bmatrix}$$

must be linearly dependent. (Verify.) From Example 2.18, the vectors

$$\begin{bmatrix} 1 \\ 1 \\ 0 \end{bmatrix}, \quad \begin{bmatrix} 0 \\ 1 \\ 1 \end{bmatrix}, \quad \text{and} \quad \begin{bmatrix} 1 \\ 0 \\ 1 \end{bmatrix}$$

form a basis for R^3.

EXERCISES 2.2

1. Find which of the following vectors are linear combinations of

$$\mathbf{X}_1 = \begin{bmatrix} 1 \\ -1 \\ 3 \end{bmatrix}, \quad \mathbf{X}_2 = \begin{bmatrix} -1 \\ 0 \\ 2 \end{bmatrix}, \quad \text{and} \quad \mathbf{X}_3 = \begin{bmatrix} 0 \\ -1 \\ 5 \end{bmatrix}.$$

(a) $\begin{bmatrix} 9 \\ 2 \\ 14 \end{bmatrix}$; (b) $\begin{bmatrix} 1 \\ -3 \\ 13 \end{bmatrix}$; (c) $\begin{bmatrix} 0 \\ 1 \\ 2 \end{bmatrix}$; (d) $\begin{bmatrix} -1 \\ -2 \\ 12 \end{bmatrix}$.

2. Write the following vectors as linear combinations of

$$\mathbf{X}_1 = \begin{bmatrix} 1 \\ 0 \\ 3 \end{bmatrix}, \quad \mathbf{X}_2 = \begin{bmatrix} -1 \\ 2 \\ -3 \end{bmatrix}, \quad \text{and} \quad \mathbf{X}_3 = \begin{bmatrix} 0 \\ 1 \\ 2 \end{bmatrix}.$$

(a) $\begin{bmatrix} 1 \\ 3 \\ 5 \end{bmatrix}$;   (b) $\begin{bmatrix} -2 \\ 6 \\ 1 \end{bmatrix}$;   (c) $\begin{bmatrix} -1 \\ 4 \\ 5 \end{bmatrix}$;   (d) $\begin{bmatrix} 0 \\ 3 \\ 2 \end{bmatrix}$.

3. Find which of the following polynomials are linear combinations of

$p_1(t) = 3t^2 + 2t - 1$, $p_2(t) = t^2 - 1$,   and $p_3(t) = t + 1$.

(a) $t^2 + 3t + 2$;   (b) $t^2 + 2t$;   (c) t;   (d) $t^2 - t + 2$.

4. If

$$\mathbf{X_1} = \begin{bmatrix} 1 \\ -2 \\ 3 \end{bmatrix}, \qquad \mathbf{X_2} = \begin{bmatrix} -2 \\ 3 \\ -4 \end{bmatrix}, \qquad \text{and} \qquad \mathbf{X_3} = \begin{bmatrix} 4 \\ -5 \\ 6 \end{bmatrix},$$

show that $S[\mathbf{X_1}, \mathbf{X_2}, \mathbf{X_3}] = S[\mathbf{X_1}, \mathbf{X_2}]$. (It is sufficient to show that $\mathbf{X_3}$ is a linear combination of $\mathbf{X_1}$ and $\mathbf{X_2}$. Why?)

5. Let $\mathbf{X_1}$, $\mathbf{X_2}$, and $\mathbf{X_3}$ be as defined in Exercise 4. If

$$\mathbf{Y_1} = \begin{bmatrix} -1 \\ 1 \\ -1 \end{bmatrix} \qquad \text{and} \qquad \mathbf{Y_2} = \begin{bmatrix} -3 \\ 5 \\ -7 \end{bmatrix},$$

show that $S[\mathbf{X_1}, \mathbf{X_2}, \mathbf{X_3}] = S[\mathbf{Y_1}, \mathbf{Y_2}]$.

6. State which sets span R^2:

(a) $\begin{bmatrix} 1 \\ 1 \end{bmatrix}$, $\begin{bmatrix} 0 \\ 1 \end{bmatrix}$;   (b) $\begin{bmatrix} 1 \\ -1 \end{bmatrix}$, $\begin{bmatrix} 2 \\ -2 \end{bmatrix}$, $\begin{bmatrix} 0 \\ 0 \end{bmatrix}$;

(c) $\begin{bmatrix} 0 \\ 1 \end{bmatrix}$, $\begin{bmatrix} 1 \\ 0 \end{bmatrix}$, $\begin{bmatrix} 1 \\ 1 \end{bmatrix}$.

7. State which sets span R^3:

(a) $\begin{bmatrix} 1 \\ 0 \\ 1 \end{bmatrix}$, $\begin{bmatrix} 1 \\ 1 \\ 1 \end{bmatrix}$;   (b) $\begin{bmatrix} 0 \\ 0 \\ 0 \end{bmatrix}$, $\begin{bmatrix} 1 \\ 2 \\ 1 \end{bmatrix}$, $\begin{bmatrix} 1 \\ -1 \\ 2 \end{bmatrix}$, $\begin{bmatrix} 0 \\ 0 \\ 1 \end{bmatrix}$;

(c) $\begin{bmatrix} 1 \\ -1 \\ 2 \end{bmatrix}, \begin{bmatrix} 1 \\ 2 \\ 3 \end{bmatrix}, \begin{bmatrix} 1 \\ 4 \\ -2 \end{bmatrix}$; (d) $\begin{bmatrix} 1 \\ 0 \\ 0 \end{bmatrix}, \begin{bmatrix} 1 \\ -1 \\ 2 \end{bmatrix}, \begin{bmatrix} 0 \\ 1 \\ 0 \end{bmatrix}, \begin{bmatrix} 1 \\ 1 \\ 1 \end{bmatrix}$.

8. State which sets span P_1:

    (a) $t - 1, t + 1, 1$;   (b) $0, t - 1$;   (c) $t - 2, t$;
    (d) $t - 3, 2t - 6$.

9. State which sets span P_2:

    (a) $t - 1, t^2 - 1, t$;   (b) $t^2 - 1, t$;   (c) $2t^2, t + 1, t - 1$;
    (d) $t - 1, t + 1, t^2$.

In Exercises 10 through 13, determine which of the given sets of vectors are linearly dependent. For those that are, express one vector in the set as a linear combination of the others.

10. (a) $\begin{bmatrix} 1 \\ 2 \\ 1 \end{bmatrix}, \begin{bmatrix} 0 \\ 1 \\ 1 \end{bmatrix}, \begin{bmatrix} 1 \\ 0 \\ 1 \end{bmatrix}$;   (b) $\begin{bmatrix} 3 \\ 1 \\ 1 \end{bmatrix}, \begin{bmatrix} 2 \\ 2 \\ -3 \end{bmatrix}, \begin{bmatrix} -4 \\ 0 \\ -5 \end{bmatrix}$;

    (c) $\begin{bmatrix} 1 \\ 2 \\ -1 \end{bmatrix}, \begin{bmatrix} 3 \\ 2 \\ 4 \end{bmatrix}$;   (d) $\begin{bmatrix} 2 \\ -1 \\ 3 \end{bmatrix}, \begin{bmatrix} 8 \\ 4 \\ 4 \end{bmatrix}, \begin{bmatrix} 4 \\ 6 \\ -2 \end{bmatrix}$.

11. (a) $\begin{bmatrix} 0 \\ 0 \\ 0 \end{bmatrix}, \begin{bmatrix} 1 \\ 1 \\ 2 \end{bmatrix}, \begin{bmatrix} -2 \\ 3 \\ 4 \end{bmatrix}$;   (b) $\begin{bmatrix} -1 \\ 2 \\ -2 \end{bmatrix}, \begin{bmatrix} 3 \\ 4 \\ 2 \end{bmatrix}, \begin{bmatrix} 0 \\ 0 \\ 1 \end{bmatrix}$;

    (c) $\begin{bmatrix} 1 \\ 2 \\ -3 \end{bmatrix}, \begin{bmatrix} -1 \\ -3 \\ 1 \end{bmatrix}, \begin{bmatrix} -1 \\ 1 \\ 9 \end{bmatrix}$;   (d) $\begin{bmatrix} 1 \\ 0 \\ 2 \end{bmatrix}, \begin{bmatrix} 0 \\ 1 \\ 2 \end{bmatrix}, \begin{bmatrix} -1 \\ 2 \\ 0 \end{bmatrix}$.

12. (a) $t^2 + t, t - 1, 2t + 2$;   (b) $t, t - 1, t^2$;
    (c) $t^2 - 1, t + 1, t^2 - t, t^2 + t$;
    (d) $t^2 + t + 1, -2t^2 - t, t^2 + 2t + 3$.

13. (a) $\cos^2 t,\ \sin^2 t,\ \cos 2t$;
    (b) $t,\ t^2,\ e^t$;
    (c) $\sin t,\ e^t,\ \cos t$;
    (d) $\tan^2 t,\ \sec^2 t,\ 1$.

14. Which sets in Exercise 6 are bases for R^2?

15. Which sets in Exercise 7 are bases for R^3?

16. Which sets in Exercise 8 are bases for P_1?

17. Which sets in Exercise 9 are bases for P_2?

18. Express $\mathbf{Y} = \begin{bmatrix} 2 \\ 3 \end{bmatrix}$ in terms of the following bases for R^2:

    (a) $\mathbf{X}_1 = \begin{bmatrix} 1 \\ 2 \end{bmatrix}$, $\mathbf{X}_2 = \begin{bmatrix} 1 \\ -3 \end{bmatrix}$;   (b) $\mathbf{X}_1 = \begin{bmatrix} 0 \\ 1 \end{bmatrix}$, $\mathbf{X}_2 = \begin{bmatrix} 1 \\ 1 \end{bmatrix}$;

    (c) $\mathbf{X}_1 = \begin{bmatrix} -1 \\ 2 \end{bmatrix}$, $\mathbf{X}_2 = \begin{bmatrix} -1 \\ 1 \end{bmatrix}$;   (d) $\mathbf{X}_1 = \begin{bmatrix} 0 \\ 1 \end{bmatrix}$, $\mathbf{X}_2 = \begin{bmatrix} 1 \\ 0 \end{bmatrix}$.

19. Express $\mathbf{Y} = \begin{bmatrix} 2 \\ 1 \\ 3 \end{bmatrix}$ in terms of the following bases for R^3:

    (a) $\mathbf{X}_1 = \begin{bmatrix} 1 \\ 1 \\ 1 \end{bmatrix}$, $\mathbf{X}_2 = \begin{bmatrix} 1 \\ 2 \\ 3 \end{bmatrix}$, $\mathbf{X}_3 = \begin{bmatrix} 0 \\ 1 \\ 0 \end{bmatrix}$;

    (b) $\mathbf{X}_1 = \begin{bmatrix} 1 \\ 2 \\ 1 \end{bmatrix}$, $\mathbf{X}_2 = \begin{bmatrix} 2 \\ 1 \\ 3 \end{bmatrix}$, $\mathbf{X}_3 = \begin{bmatrix} 1 \\ 0 \\ 1 \end{bmatrix}$;

    (c) $\mathbf{X}_1 = \begin{bmatrix} 1 \\ 2 \\ 1 \end{bmatrix}$, $\mathbf{X}_2 = \begin{bmatrix} 1 \\ 1 \\ 4 \end{bmatrix}$, $\mathbf{X}_3 = \begin{bmatrix} 1 \\ 5 \\ 1 \end{bmatrix}$.

20. Find a basis for the subspace of R^3 spanned by

(a) $\mathbf{X}_1 = \begin{bmatrix} 1 \\ 2 \\ 3 \end{bmatrix}$, $\mathbf{X}_2 = \begin{bmatrix} 2 \\ 0 \\ 1 \end{bmatrix}$, $\mathbf{X}_3 = \begin{bmatrix} 3 \\ 2 \\ 4 \end{bmatrix}$;

(b) $\mathbf{X}_1 = \begin{bmatrix} 1 \\ -1 \\ 2 \end{bmatrix}$, $\mathbf{X}_2 = \begin{bmatrix} 4 \\ 4 \\ 7 \end{bmatrix}$, $\mathbf{X}_3 = \begin{bmatrix} 1 \\ 2 \\ 1 \end{bmatrix}$, $\mathbf{X}_4 = \begin{bmatrix} 2 \\ 3 \\ 4 \end{bmatrix}$.

21. Find a basis for the subspace of P_2 spanned by

(a) $p_1(t) = t^2 + 1, p_2(t) = t^2 - 1, p_3(t) = t^2$;
(b) $p_1(t) = 2t^2 + t - 1, p_2(t) = 2t + 2, p_3(t) = t^2 - 1,$
$p_4(t) = 3t + 3.$

22. Use Corollary 2.1 to verify that

$$\begin{bmatrix} 1 \\ 1 \\ 2 \end{bmatrix}, \quad \begin{bmatrix} 1 \\ 2 \\ 3 \end{bmatrix}, \quad \text{and} \quad \begin{bmatrix} -1 \\ 2 \\ 0 \end{bmatrix}$$

form a basis for R^3.

23. Find a basis for R^3 that contains $\begin{bmatrix} 1 \\ -1 \\ 2 \end{bmatrix}$.

24. In the space of real-valued functions defined for all x, find a basis for $S[f_1, f_2, f_3]$, where

$$f_1(t) = \cos^2 t, \qquad f_2(t) = \sin^2 t, \qquad \text{and} \qquad f_3(t) = \cos 2t.$$

25. Find a basis for R^3 which contains

$$\begin{bmatrix} 1 \\ 0 \\ 1 \end{bmatrix} \qquad \text{and} \qquad \begin{bmatrix} 1 \\ 2 \\ -3 \end{bmatrix}.$$

THEORETICAL EXERCISES

T-1. Verify that $S[\mathbf{U}_1, \ldots, \mathbf{U}_m]$ (Definition 2.2) is a subspace of S.

T-2. Show that any pair of the vectors

$$\mathbf{X}_1 = \begin{bmatrix} 1 \\ 0 \\ 0 \end{bmatrix}, \qquad \mathbf{X}_2 = \begin{bmatrix} 0 \\ 0 \\ 1 \end{bmatrix}, \qquad \mathbf{X}_3 = \begin{bmatrix} 1 \\ 0 \\ 1 \end{bmatrix}$$

is linearly independent and spans $S[\mathbf{X}_1, \mathbf{X}_2, \mathbf{X}_3]$.

T-3. Suppose no linearly independent set of vectors in a nonzero vector space S has more than m elements. Show that any set of m linearly independent vectors is a basis for S.

T-4. *Prove:* If T is a subspace of a finite-dimensional vector space S and dim T = dim S, then $T = S$.

T-5. *Prove:* If vectors $\mathbf{U}_1, \ldots, \mathbf{U}_n$ span an n-dimensional vector space S, then they must be linearly independent and, consequently, form a basis for S.

T-6. *Prove:* If $\mathbf{U}_1, \ldots, \mathbf{U}_n$ are n linearly independent vectors in an n-dimensional vector space S, then they must span S and, consequently, form a basis for S.

T-7. Let $\mathbf{U}_1, \mathbf{U}_2, \ldots, \mathbf{U}_n$ be elements of a vector space S. Show that the dimension of $S[\mathbf{U}_1, \ldots, \mathbf{U}_n]$ equals the number of linearly independent vectors among $\mathbf{U}_1, \ldots, \mathbf{U}_n$.

T-8. Prove Corollary 2.2. (*Hint:* Recall that det $\mathbf{A}^\mathrm{T}$ = det $\mathbf{A}$.)

T-9. Let S_1 and S_2 be finite sets of vectors in a vector space and let S_1 be a subset of S_2. *Prove:*

(a) If S_1 is linearly dependent, so is S_2.
(b) If S_2 is linearly independent, so is S_1.

2.3 Linear Transformations

A *linear transformation* from a vector space S to a vector space T is a rule that assigns elements of S to elements of T in a way that preserves vector addition and scalar multiplication. We make this precise in the following definition.

Definition 3.1 Let S and T be real vector spaces. A *linear transformation* **L** *from S into T* is a rule by which every vector **U** in S is associated with a unique *image vector* **L(U)** in T, in such a way that:

(a) **L(U + V) = L(U) + L(V)** for every **U** and **V** in S.
(b) **L**$(c$**U**$) = c$**L(U)** for every **U** in S and scalar c.

We denote this by

$$\mathbf{L}\colon S \to T.$$

Example 3.1 Let **L**: $R^3 \to R^2$ be defined by

$$\mathbf{L}\left(\begin{bmatrix} x \\ y \\ z \end{bmatrix}\right) = \begin{bmatrix} x \\ y \end{bmatrix}.$$

This is a linear transformation, since

$$\mathbf{L}\left(\begin{bmatrix} x_1 \\ y_1 \\ z_1 \end{bmatrix} + \begin{bmatrix} x_2 \\ y_2 \\ z_2 \end{bmatrix}\right) = \mathbf{L}\left(\begin{bmatrix} x_1 + x_2 \\ y_1 + y_2 \\ z_1 + z_2 \end{bmatrix}\right) = \begin{bmatrix} x_1 + x_2 \\ y_1 + y_2 \end{bmatrix}$$

$$= \begin{bmatrix} x_1 \\ y_1 \end{bmatrix} + \begin{bmatrix} x_2 \\ y_2 \end{bmatrix} = \mathbf{L}\left(\begin{bmatrix} x_1 \\ y_1 \\ z_1 \end{bmatrix}\right) + \mathbf{L}\left(\begin{bmatrix} x_2 \\ y_2 \\ z_2 \end{bmatrix}\right),$$

and

$$\mathbf{L}\left(c\begin{bmatrix} x \\ y \\ z \end{bmatrix}\right) = \mathbf{L}\left(\begin{bmatrix} cx \\ cy \\ cz \end{bmatrix}\right) = \begin{bmatrix} cx \\ cy \end{bmatrix} = c\begin{bmatrix} x \\ y \end{bmatrix} = c\mathbf{L}\left(\begin{bmatrix} x \\ y \\ z \end{bmatrix}\right).$$

Example 3.2 Let **L**: $R^n \to R^m$ be defined by **L(X) = AX**, where **A** is an $m \times n$ matrix. If **X** and **Y** are in R^n, then

$$\mathbf{L(X + Y) = A(X + Y) = AX + AY = L(X) + L(Y)}$$

and

$$\mathbf{L}(c\mathbf{X}) = \mathbf{A}(c\mathbf{X}) = c(\mathbf{AX}) = c\mathbf{L}(\mathbf{X})$$

for every scalar c. Hence, $\mathbf{L}$ is a linear transformation.

The transformation defined in Example 3.1 can be written in this form, with $m = 2$, $n = 3$, and

$$\mathbf{A} = \begin{bmatrix} 1 & 0 & 0 \\ 0 & 1 & 0 \end{bmatrix},$$

since

$$\mathbf{A} \begin{bmatrix} x \\ y \\ z \end{bmatrix} = \begin{bmatrix} 1 & 0 & 0 \\ 0 & 1 & 0 \end{bmatrix} \begin{bmatrix} x \\ y \\ z \end{bmatrix} = \begin{bmatrix} x \\ y \end{bmatrix}.$$

Example 3.3   Let $\mathbf{L}: P \to R$ be defined by

$$\mathbf{L}(p) = \int_0^1 p(t) \, dt.$$

From known properties of integration,

$$\mathbf{L}(p + q) = \int_0^1 [p(t) + q(t)] \, dt = \int_0^1 p(t) \, dt + \int_0^1 q(t) \, dt$$

$$= \mathbf{L}(p) + \mathbf{L}(q)$$

if p and q are polynomials, and

$$\mathbf{L}(cp) = \int_0^1 cp(t) \, dt = c \int_0^1 p(t) \, dt = c\mathbf{L}(p)$$

if p is a polynomial and c is a scalar. Hence, $\mathbf{L}$ is a linear transformation.

Example 3.4   Let $\mathbf{L}: P_n \to P_{n-1}$ be defined by

$$\mathbf{L}(p) = \frac{dp}{dt}.$$

Then,

$$\mathbf{L}(p + q) = \frac{d}{dt}(p + q) = \frac{dp}{dt} + \frac{dq}{dt} = \mathbf{L}(p) + \mathbf{L}(q)$$

if p and q are in P_n, and

$$\mathbf{L}(cp) = c\frac{dp}{dt} = c\mathbf{L}(p)$$

if p is in P_n and c is a scalar. Hence, $\mathbf{L}$ is a linear transformation.

Example 3.5 Let

$$\mathbf{L}\left(\begin{bmatrix} x \\ y \end{bmatrix}\right) = \begin{bmatrix} x+1 \\ y \end{bmatrix}.$$

Then

$$\mathbf{L}\left(\begin{bmatrix} x_1 \\ y_1 \end{bmatrix} + \begin{bmatrix} x_2 \\ y_2 \end{bmatrix}\right) = \mathbf{L}\left(\begin{bmatrix} x_1 + x_2 \\ y_1 + y_2 \end{bmatrix}\right) = \begin{bmatrix} x_1 + x_2 + 1 \\ y_1 + y_2 \end{bmatrix},$$

while

$$\mathbf{L}\left(\begin{bmatrix} x_1 \\ y_1 \end{bmatrix}\right) + \mathbf{L}\left(\begin{bmatrix} x_2 \\ y_2 \end{bmatrix}\right) = \begin{bmatrix} x_1 + 1 \\ y_1 \end{bmatrix} + \begin{bmatrix} x_2 + 1 \\ y_2 \end{bmatrix} = \begin{bmatrix} x_1 + x_2 + 2 \\ y_1 + y_2 \end{bmatrix};$$

hence

$$\mathbf{L}\left(\begin{bmatrix} x_1 \\ y_1 \end{bmatrix} + \begin{bmatrix} x_2 \\ y_2 \end{bmatrix}\right) \neq \mathbf{L}\left(\begin{bmatrix} x_1 \\ y_1 \end{bmatrix}\right) + \mathbf{L}\left(\begin{bmatrix} x_2 \\ y_2 \end{bmatrix}\right),$$

and therefore $\mathbf{L}$ is not a linear transformation.

Theorem 3.1 If $\mathbf{L}\colon S \to T$ is a linear transformation, then

$$\mathbf{L}(a_1\mathbf{U}_1 + \cdots + a_n\mathbf{U}_n) = a_1\mathbf{L}(\mathbf{U}_1) + \cdots + a_n\mathbf{L}(\mathbf{U}_n)$$

for any vectors $\mathbf{U}_1, \ldots, \mathbf{U}_n$ in S and scalars $c_1, \ldots, c_n$.

We leave the proof of this theorem to the student (Exercise T-1).

Definition 3.2 A linear transformation $\mathbf{L}\colon S \to T$ is *one-to-one* if $\mathbf{L}(\mathbf{U})$ and $\mathbf{L}(\mathbf{V})$ are distinct vectors in T whenever $\mathbf{U}$ and $\mathbf{V}$ are distinct vectors in S. Equivalently, $\mathbf{L}$ is one-to-one if $\mathbf{L}(\mathbf{U}) = \mathbf{L}(\mathbf{V})$ implies that $\mathbf{U} = \mathbf{V}$.

Example 3.6 The linear transformation $\mathbf{L}\colon R^2 \to R^2$ defined by

$$\mathbf{L}\left(\begin{bmatrix} x \\ y \end{bmatrix}\right) = \begin{bmatrix} x+y \\ x \end{bmatrix}$$

is one-to-one, because if

$$L\left(\begin{bmatrix} x_1 \\ y_1 \end{bmatrix}\right) = L\left(\begin{bmatrix} x_2 \\ y_2 \end{bmatrix}\right)$$

then,

$$x_1 + y_1 = x_2 + y_2,$$

$$x_1 = x_2,$$

which implies that

$$\begin{bmatrix} x_1 \\ y_1 \end{bmatrix} = \begin{bmatrix} x_2 \\ y_2 \end{bmatrix}$$

Example 3.7   The linear transformation $L\colon R^3 \to R^2$ defined by

$$L\left(\begin{bmatrix} x \\ y \\ z \end{bmatrix}\right) = \begin{bmatrix} x \\ y \end{bmatrix}$$

is clearly not one-to-one, since, for example,

$$L\left(\begin{bmatrix} x \\ y \\ 1 \end{bmatrix}\right) = L\left(\begin{bmatrix} x \\ y \\ 2 \end{bmatrix}\right).$$

When dealing with two vector spaces S and T, it is necessary to recognize that the zero vectors of the two spaces are distinct. We could emphasize this by denoting them by $\mathbf{0}_S$ and $\mathbf{0}_T$, respectively. However, the subscripts are cumbersome and can be omitted without loss of clarity since it is always possible to distinguish between the two zero vectors from the context. For example, if $L\colon S \to T$, then each of the equations

(1) $L(U) = 0,$

(2) $L(0) = V,$

and

(3) $L(0) = 0$

has only one possible interpretation: in (1), $\mathbf{0} = \mathbf{0}_T$; in (2), $\mathbf{0} = \mathbf{0}_S$; on the left side of (3), $\mathbf{0} = \mathbf{0}_S$, and on the right, $\mathbf{0} = \mathbf{0}_T$.

Definition 3.3 If $\mathbf{L}: S \to T$ is a linear transformation, then the *kernel* of $\mathbf{L}$, denoted by ker $\mathbf{L}$, is the set of all vectors $\mathbf{U}$ in S such that $\mathbf{L(U)} = \mathbf{0}$.

The kernel of a linear transformation $\mathbf{L}$ is also called the *null space* of $\mathbf{L}$.

Example 3.8 For $\mathbf{L}: R^3 \to R^2$ defined by

$$\mathbf{L}\left(\begin{bmatrix} x \\ y \\ z \end{bmatrix}\right) = \begin{bmatrix} x \\ y \end{bmatrix},$$

ker $\mathbf{L}$ consists of all vectors in R^3 of the form

$$\begin{bmatrix} 0 \\ 0 \\ z \end{bmatrix}.$$

For $\mathbf{L}: P_n \to P_{n-1}$ defined by $\mathbf{L}(p) = dp/dt$, ker $\mathbf{L}$ consists of the constant polynomials.

For $\mathbf{L}: R^2 \to R^2$ defined by

$$\mathbf{L}\left(\begin{bmatrix} x \\ y \end{bmatrix}\right) = \begin{bmatrix} x + y \\ x \end{bmatrix},$$

ker $\mathbf{L}$ contains only the zero vector $\begin{bmatrix} 0 \\ 0 \end{bmatrix}$. (Verify.)

Theorem 3.2 The kernel of a linear transformation $\mathbf{L}: S \to T$ is a subspace of S.

Proof. First, we observe that $\mathbf{L(0)} = \mathbf{0}$ (Exercise T-2), so that ker $\mathbf{L}$ has at least one element. If $\mathbf{U}$ and $\mathbf{V}$ are in ker $\mathbf{L}$, then

$$\mathbf{L(U + V)} = \mathbf{L(U)} + \mathbf{L(V)} = \mathbf{0} + \mathbf{0} = \mathbf{0}$$

and

$$\mathbf{L}(c\mathbf{U}) = c\mathbf{L(U)} = c\mathbf{0} = \mathbf{0};$$

thus $\mathbf{U} + \mathbf{V}$ and $c\mathbf{U}$ are in ker $\mathbf{L}$. It follows that ker $\mathbf{L}$ is a subspace of $\mathbf{L}$.

Example 3.9   The kernels of the first two linear transformations discussed in Example 3.8 are one dimensional subspaces of R^3 and P_n, respectively, while the kernel of the third is the zero subspace of R^2. The kernel of $\mathbf{L}\colon R^3 \to R^3$, defined by

$$\mathbf{L}\left(\begin{bmatrix} x \\ y \\ z \end{bmatrix}\right) = \begin{bmatrix} z \\ z \\ z \end{bmatrix},$$

is the two dimensional subspace of R^3 spanned by

$$\begin{bmatrix} 1 \\ 0 \\ 0 \end{bmatrix} \quad \text{and} \quad \begin{bmatrix} 0 \\ 1 \\ 0 \end{bmatrix}.$$

(Verify.)

Theorem 3.3   A linear transformation $\mathbf{L}\colon S \to T$ is one-to-one if and only if ker $\mathbf{L} = \{\mathbf{0}\}$.

Proof.   Suppose ker $\mathbf{L} = \{\mathbf{0}\}$ and $\mathbf{L}(\mathbf{U}) = \mathbf{L}(\mathbf{V})$; then

$$\mathbf{L}(\mathbf{U} - \mathbf{V}) = \mathbf{L}(\mathbf{U}) - \mathbf{L}(\mathbf{V}) = \mathbf{0}.$$

Thus, $\mathbf{U} - \mathbf{V}$ is in ker $\mathbf{L}$, which implies that $\mathbf{U} - \mathbf{V} = \mathbf{0}$; that is, $\mathbf{U} = \mathbf{V}$. For the converse, suppose $\mathbf{L}$ is one-to-one and $\mathbf{L}(\mathbf{U}) = \mathbf{0}$. Since $\mathbf{L}(\mathbf{0}) = \mathbf{0}$ (Exercise T-2), $\mathbf{L}(\mathbf{U}) = \mathbf{L}(\mathbf{0})$, and it follows that $\mathbf{U} = \mathbf{0}$. Thus, ker $\mathbf{L} = \{\mathbf{0}\}$, which completes the proof.

Definition 3.4   Let $\mathbf{L}\colon S \to T$ be a linear transformation. The set of vectors in T which are images of vectors in S is called the *range* of $\mathbf{L}$, and is denoted by range $\mathbf{L}$. Thus, a vector $\mathbf{V}$ in T is in range $\mathbf{L}$ if and only if there is a vector $\mathbf{U}$ in S such that $\mathbf{V} = \mathbf{L}(\mathbf{U})$. If range $\mathbf{L} = T$ then $\mathbf{L}$ is said to be *onto* T.

Theorem 3.4   The range of a linear transformation $\mathbf{L}\colon S \to T$ is a subspace of T.

We leave the proof of this theorem to the student (Exercise T-3).

If $\mathbf{L}: R^n \to R^m$ is defined by

(4)
$$\mathbf{L(X)} = \mathbf{AX},$$

where $\mathbf{A}$ is an $m \times n$ matrix, then range $\mathbf{L}$ is the set of vectors $\mathbf{Y}$ in R^n for which the linear system

$$\mathbf{AX} = \mathbf{Y}$$

has a solution, and ker $\mathbf{L}$ is the space of solutions of the homogeneous system

$$\mathbf{AX} = \mathbf{0}.$$

Example 3.10 Let $\mathbf{L}: R^3 \to R^2$ be defined by

$$\mathbf{L}\left(\begin{bmatrix} x \\ y \\ z \end{bmatrix}\right) = \begin{bmatrix} x + y \\ x - y \end{bmatrix},$$

which can be written in the form (4) as

$$\mathbf{L}\left(\begin{bmatrix} x \\ y \\ z \end{bmatrix}\right) = \begin{bmatrix} 1 & 1 & 0 \\ 1 & -1 & 0 \end{bmatrix}\begin{bmatrix} x \\ y \\ z \end{bmatrix}.$$

If $\begin{bmatrix} u \\ v \end{bmatrix}$ is an arbitrary vector in R^3 then

$$\mathbf{L}\left(\begin{bmatrix} \dfrac{u + v}{2} \\ \dfrac{u - v}{2} \\ z \end{bmatrix}\right) = \begin{bmatrix} u \\ v \end{bmatrix}$$

for any z; hence range $\mathbf{L} = R^2$, and $\mathbf{L}$ is onto. Since

$$x + y = 0,$$

$$x - y = 0,$$

only if $x = y = 0$, ker $\mathbf{L}$ is the one-dimensional subspace of R^3 consisting of vectors of the form

$$\begin{bmatrix} 0 \\ 0 \\ z \end{bmatrix}.$$

Example 3.11   Let $\mathbf{L}: R^3 \to R^2$ be defined by

$$\mathbf{L}\left(\begin{bmatrix} x \\ y \\ z \end{bmatrix}\right) = \begin{bmatrix} x \\ -x \end{bmatrix}.$$

The range of $\mathbf{L}$ is the one-dimensional subspace of R^2 spanned by $\begin{bmatrix} 1 \\ -1 \end{bmatrix}$.

The kernel of $\mathbf{L}$ is the two-dimensional subspace of R^3 spanned by

$$\begin{bmatrix} 0 \\ 1 \\ 0 \end{bmatrix} \quad \text{and} \quad \begin{bmatrix} 0 \\ 0 \\ 1 \end{bmatrix}.$$

In each of the last two examples, the dimensions of ker $\mathbf{L}$ and range $\mathbf{L}$ add up to the dimension of the space on which $\mathbf{L}$ is defined. This is true in general, as shown by the next theorem.

Theorem 3.5   Let $\mathbf{L}: S \to T$ be a linear transformation and suppose dim $S = n$. Then

$$(5) \qquad\qquad \dim(\ker \mathbf{L}) + \dim(\text{range } \mathbf{L}) = n.$$

Proof.   Let $m = \dim(\ker \mathbf{L})$. If $m = n$, then ker $\mathbf{L} = S$ (Exercise T-4, Section 2.2), which means that $\mathbf{L}(\mathbf{U}) = \mathbf{0}$ for every $\mathbf{U}$ in S; thus range $\mathbf{L} = \{\mathbf{0}\}$, $\dim(\text{range } \mathbf{L}) = 0$, and (5) holds.

Now suppose $1 \le m < n$ and let $\{\mathbf{U}_1, \ldots, \mathbf{U}_m\}$ be a basis for ker $\mathbf{L}$. By Theorem 2.6 (Section 2.2), there are vectors $\mathbf{V}_1, \ldots, \mathbf{V}_{n-m}$ such that

$$(6) \qquad\qquad \{\mathbf{U}_1, \ldots, \mathbf{U}_m, \mathbf{V}_1, \ldots, \mathbf{V}_{n-m}\}$$

is a basis for S. Any vector $\mathbf{W}$ in range $\mathbf{L}$ can be written as $\mathbf{W} = \mathbf{L}(\mathbf{U})$ for some $\mathbf{U}$ in S, or, since (6) is a basis for S, in the form

$$(7) \quad \mathbf{W} = \mathbf{L}(c_1\mathbf{U}_1 + \cdots + c_m\mathbf{U}_m + c_{m+1}\mathbf{V}_1 + \cdots + c_n\mathbf{V}_{n-m})$$

$$= \mathbf{L}(c_1\mathbf{U}_1 + \cdots + c_m\mathbf{U}_m) + c_{m+1}\mathbf{L}(\mathbf{V}_1) + \cdots + c_n\mathbf{L}(\mathbf{V}_{n-m}).$$

Since $c_1\mathbf{U}_1 + \cdots + c_m\mathbf{U}_m$ is in ker $\mathbf{L}$,

$$\mathbf{L}(c_1\mathbf{U}_1 + \cdots + c_m\mathbf{U}_m) = \mathbf{0};$$

hence, from (7),

$$\mathbf{W} = c_{m+1}\mathbf{L}(\mathbf{V}_1) + \cdots + c_n\mathbf{L}(\mathbf{V}_{n-m}),$$

which means that $\{\mathbf{L}(\mathbf{V}_1), \ldots, \mathbf{L}(\mathbf{V}_{n-m})\}$ spans range $\mathbf{L}$. Moreover, these vectors are linearly independent, since if

$$(8) \quad a_1\mathbf{L}(\mathbf{V}_1) + \cdots + a_{n-m}\mathbf{L}(\mathbf{V}_{n-m}) = \mathbf{0},$$

then

$$\mathbf{L}(a_1\mathbf{V}_1 + \cdots + a_{n-m}\mathbf{V}_{n-m}) = \mathbf{0},$$

which implies that $a_1\mathbf{V}_1 + \cdots + a_n\mathbf{V}_{n-m}$ is in ker $\mathbf{L}$; consequently it can be written as a linear combination of $\mathbf{U}_1, \ldots, \mathbf{U}_m$, which means that

$$(9) \quad a_1\mathbf{V}_1 + \cdots + a_{n-m}\mathbf{V}_{n-m} + b_1\mathbf{U}_1 + \cdots + b_m\mathbf{U}_m = \mathbf{0}$$

for some choice of scalars $b_1, \ldots, b_m$. However, since (6) is a basis for S, the vectors on the left side of (9) are linearly independent; hence the scalars in (9) vanish. Thus, we have deduced from (8) that $a_1 = \cdots = a_{n-m} = 0$, which means that $\mathbf{L}(\mathbf{V}_1), \ldots, \mathbf{L}(\mathbf{V}_{n-m})$ are linearly independent vectors in T. Since they span range $\mathbf{L}$, dim range $\mathbf{L} = n - m$ and (5) is verified.

In the remaining case ker $\mathbf{L} = \{\mathbf{0}\}$. We leave it to the student to show that, in this case, dim range $\mathbf{L} = n$ (Exercise T-8), which satisfies (5).

Isomorphisms

Definition 3.5 A linear transformation which is one-to-one and onto is called an *isomorphism*. If $\mathbf{L}: S \to T$ is an isomorphism, then S is said to be *isomorphic* to T.

Example 3.12   The linear transformation $\mathbf{L} \colon R_n \to R^n$ defined by

$$\mathbf{L} \colon ([x_1, x_2, \ldots, x_n]) = \begin{bmatrix} x_1 \\ x_2 \\ \vdots \\ x_n \end{bmatrix}$$

is one-to-one and onto; hence R_n is isomorphic to R^n.

Example 3.13   Let $p_1(t) = 1$, $p_2(t) = t, \ldots, p_n(t) = t^{n-1}$. The linear transformation $\mathbf{L} \colon P_{n-1} \to R^n$ defined by

$$\mathbf{L}(a_1 p_1 + a_2 p_2 + \cdots + a_n p_n) = \begin{bmatrix} a_1 \\ a_2 \\ \vdots \\ a_n \end{bmatrix}$$

is an isomorphism. Since this follows from the next theorem, we shall not carry out the proof for this special case.

Theorem 3.6   Any n-dimensional vector space S is isomorphic to R^n.

Proof.   Let $\mathbf{U}_1, \ldots, \mathbf{U}_n$ be a basis for S, and define $\mathbf{L} \colon S \to R^n$ by

$$(10) \qquad \mathbf{L}(\mathbf{U}) = \begin{bmatrix} \beta_1 \\ \beta_2 \\ \vdots \\ \beta_n \end{bmatrix},$$

where $\beta_1, \beta_2, \ldots, \beta_n$ are the unique scalars such that

$$\mathbf{U} = \beta_1 \mathbf{U}_1 + \beta \mathbf{U}_2 + \cdots + \beta_n \mathbf{U}_n$$

(Theorem 2.3, Section 2.2). If c is a scalar, then

$$c\mathbf{U} = (c\beta_1)\mathbf{U}_1 + (c\beta_2)\mathbf{U}_2 + \cdots + (c\beta_n)\mathbf{U}_n$$

and the definition (10) yields

$$\mathbf{L}(c\mathbf{U}) = \begin{bmatrix} c\beta_1 \\ c\beta_2 \\ \vdots \\ c\beta_n \end{bmatrix} = c \begin{bmatrix} \beta_1 \\ \beta_2 \\ \vdots \\ \beta_n \end{bmatrix} = c\mathbf{L}(\mathbf{U}).$$

If

$$\mathbf{V} = \gamma_1 \mathbf{U}_1 + \gamma_2 \mathbf{U}_2 + \cdots + \gamma_n \mathbf{U}_n,$$

then

$$\mathbf{U} + \mathbf{V} = (\beta_1 + \gamma_1)\mathbf{U}_1 + (\beta_2 + \gamma_2)\mathbf{U}_2 + \cdots + (\beta_n + \gamma_n)\mathbf{U}_n,$$

and (10) yields

$$\mathbf{L}(\mathbf{U} + \mathbf{V}) = \begin{bmatrix} \beta_1 + \gamma_1 \\ \beta_2 + \gamma_2 \\ \vdots \\ \beta_n + \gamma_n \end{bmatrix} = \begin{bmatrix} \beta_1 \\ \beta_2 \\ \vdots \\ \beta_n \end{bmatrix} + \begin{bmatrix} \gamma_1 \\ \gamma_2 \\ \vdots \\ \gamma_n \end{bmatrix}$$

$$= \mathbf{L}(\mathbf{U}) + \mathbf{L}(\mathbf{V}).$$

Therefore, $\mathbf{L}$ is a linear transformation. We leave it to the student to verify that $\mathbf{L}$ is one-to-one and onto, and therefore an isomorphism (Exercise T-5).

Coordinates with Respect to a Basis

A natural by product of the proof of Theorem 3.6 is the notion of a coordinate vector, defined next.

Definition 3.6   Let $B = \{\mathbf{U}_1, \mathbf{U}_2, \ldots, \mathbf{U}_n\}$ be a basis for an n-dimensional vector space S, and let

$$\mathbf{U} = \beta_1\mathbf{U}_1 + \beta_2\mathbf{U}_2 + \cdots + \beta_n\mathbf{U}_n$$

be an arbitrary vector in S. Then the *coordinate vector* $(\mathbf{U})_B$ *of the vector* $\mathbf{U}$ *with respect to the basis B* is the n-vector

$$(\mathbf{U})_B = \begin{bmatrix} \beta_1 \\ \beta_2 \\ \vdots \\ \beta_n \end{bmatrix}.$$

The components of $(\mathbf{U})_B$ are called the *coordinates* of $\mathbf{U}$ with respect to B.

Example 3.14   If $B = \{\mathbf{U}_1, \mathbf{U}_2, \ldots, \mathbf{U}_n\}$ is a basis for S, then

$$(\mathbf{U}_1)_B = \mathbf{E}_1, \ (\mathbf{U}_2)_B = \mathbf{E}_2, \ldots, \ (\mathbf{U}_n)_B = \mathbf{E}_n,$$

where $\{\mathbf{E}_1, \ldots, \mathbf{E}_n\}$ is the natural basis for R^n.

Example 3.15   Let B_0 be the natural basis for R^n. Then $(\mathbf{X})_{B_0} = \mathbf{X}$ for every $\mathbf{X}$ in R^n; moreover, B_0 is the only basis for R^n for which this is true (Exercise T-6).

Example 3.16 We have seen that if

$$\mathbf{X}_1 = \begin{bmatrix} 1 \\ 1 \\ 0 \end{bmatrix}, \qquad \mathbf{X}_2 = \begin{bmatrix} 0 \\ 1 \\ 1 \end{bmatrix}, \qquad \text{and} \qquad \mathbf{X}_3 = \begin{bmatrix} 1 \\ 0 \\ 1 \end{bmatrix},$$

then,

$$B = \{\mathbf{X}_1, \mathbf{X}_2, \mathbf{X}_3\}$$

is a basis for R^3 (Example 2.17, Section 2.2). To find the coordinate vector of

$$\mathbf{X} = \begin{bmatrix} x \\ y \\ z \end{bmatrix}$$

with respect to this basis, we must find β_1, β_2, and β_3 such that

$$\begin{bmatrix} x \\ y \\ z \end{bmatrix} = \beta_1 \begin{bmatrix} 1 \\ 1 \\ 0 \end{bmatrix} + \beta_2 \begin{bmatrix} 0 \\ 1 \\ 1 \end{bmatrix} + \beta_3 \begin{bmatrix} 1 \\ 0 \\ 1 \end{bmatrix}.$$

In terms of components, this is equivalent to the linear system

$$(11) \qquad \begin{bmatrix} 1 & 0 & 1 \\ 1 & 1 & 0 \\ 0 & 1 & 1 \end{bmatrix} \begin{bmatrix} \beta_1 \\ \beta_2 \\ \beta_3 \end{bmatrix} = \begin{bmatrix} x \\ y \\ z \end{bmatrix}.$$

The unique solution of (11) is given by

$$\begin{bmatrix} \beta_1 \\ \beta_2 \\ \beta_3 \end{bmatrix} = \begin{bmatrix} \frac{1}{2} & \frac{1}{2} & -\frac{1}{2} \\ -\frac{1}{2} & \frac{1}{2} & \frac{1}{2} \\ \frac{1}{2} & -\frac{1}{2} & \frac{1}{2} \end{bmatrix} \begin{bmatrix} x \\ y \\ z \end{bmatrix}$$

(see Examples 3.7 and 3.9, Section 1.3); therefore,

$$(\mathbf{X})_B = \begin{bmatrix} \beta_1 \\ \beta_2 \\ \beta_3 \end{bmatrix} = \frac{1}{2} \begin{bmatrix} x + y - z \\ -x + y + z \\ x - y + z \end{bmatrix}.$$

Example 3.17 Consider the basis $B = \{p_1, p_2, p_3\}$ for P_2, where $p_1(t) = 1$, $p_2(t) = t$, and $p_3(t) = t^2$. If

(12) $$p(t) = 1 + 2t + 3t^2,$$

then

$$(p)_B = \begin{bmatrix} 1 \\ 2 \\ 3 \end{bmatrix}.$$

Another basis for P_2 is $C = \{q_1, q_2, q_3\}$, where $q_1(t) = t^2 + t + 1$, $q_2(t) = t^2 - 1$, and $q_3(t) = 1$ (Examples 2.4 and 2.8, Section 2.2). In terms of this basis, the polynomial (12) can be written as

$$p = 2q_1 + q_2;$$

hence,

$$(p)_C = \begin{bmatrix} 2 \\ 1 \\ 0 \end{bmatrix}.$$

Theorem 3.7 Let B be a basis for a vector space S. If $\mathbf{U}$ and $\mathbf{V}$ are vectors in S, then

$$(\mathbf{U} + \mathbf{V})_B = (\mathbf{U})_B + (\mathbf{V})_B,$$

and if c is a scalar, then

$$(c\mathbf{U})_B = c(\mathbf{U})_B.$$

More generally, if $\mathbf{U}_1, \ldots, \mathbf{U}_m$ are vectors in S and $c_1, \ldots, c_n$ are scalars, then

$$(c_1\mathbf{U}_1 + \cdots + c_m\mathbf{U}_m)_B = c_1(\mathbf{U}_1)_B + \cdots + c_m(\mathbf{U}_m)_B.$$

Proof.   This theorem follows from the proof of Theorem 3.6, where it was shown that taking coordinates with respect to a basis B defines an isomorphism from S onto R^n, for some n. We leave the details of the proof to the student (Exercise T-7).

Matrix Representation of a Linear Transformation

All linear transformations from one finite-dimensional vector space to another can be viewed as being essentially from R^n to R^m for some n and m, as we shall now see.

Theorem 3.8   Let S and T be finite-dimensional vector spaces with bases $B = \{\mathbf{U}_1, \ldots, \mathbf{U}_n\}$ and $C = \{\mathbf{V}_1, \ldots, \mathbf{V}_m\}$, respectively. Suppose

$$(13) \qquad \mathbf{U} = \beta_1 \mathbf{U}_1 + \beta_2 \mathbf{U}_2 + \cdots + \beta_n \mathbf{U}_n$$

is an arbitrary vector in S and

$$\mathbf{L}(\mathbf{U}) = \gamma_1 \mathbf{V}_1 + \gamma_2 \mathbf{V}_2 + \cdots + \gamma_m \mathbf{V}_m$$

is its image in T; thus

$$(14) \qquad (\mathbf{U})_B = \begin{bmatrix} \beta_1 \\ \beta_2 \\ \vdots \\ \beta_n \end{bmatrix} \quad \text{and} \quad (\mathbf{L}(\mathbf{U}))_C = \begin{bmatrix} \gamma_1 \\ \gamma_2 \\ \vdots \\ \gamma_m \end{bmatrix}.$$

Then

$$(15) \qquad (\mathbf{L}(\mathbf{U}))_C = \mathbf{A}(\mathbf{U})_B,$$

where $\mathbf{A}$ is the $m \times n$ matrix whose jth column is $(\mathbf{L}(\mathbf{U}_j))_C$, the coordinate vector of $\mathbf{L}(\mathbf{U}_j)$ with respect to C.

Proof.   Let

$$\mathbf{L}(\mathbf{U}_j) = a_{1j} \mathbf{V}_1 + a_{2j} \mathbf{V}_2 + \cdots + a_{mj} \mathbf{V}_m$$

thus

$$(16) \qquad (\mathbf{L}(\mathbf{U}_j))_C = \begin{bmatrix} a_{1j} \\ a_{2j} \\ \vdots \\ a_{mj} \end{bmatrix}$$

If $\mathbf{U}$ is given by (13), then

$$\mathbf{L}(\mathbf{U}) = \beta_1 \mathbf{L}(\mathbf{U}_1) + \beta_2 \mathbf{L}(\mathbf{U}_2) + \cdots + \beta_n \mathbf{L}(\mathbf{U}_n),$$

and, from Theorem 3.7,

$$(\mathbf{L}(\mathbf{U}))_C = \beta_1(\mathbf{L}(\mathbf{U}_1))_C + \beta_2(\mathbf{L}(\mathbf{U}_2))_C + \cdots + \beta_n(\mathbf{L}(\mathbf{U}_n))_C.$$

From (14) and (16), this can be rewritten as

$$
\begin{bmatrix} \gamma_1 \\ \gamma_2 \\ \vdots \\ \gamma_m \end{bmatrix} = \beta_1 \begin{bmatrix} a_{11} \\ a_{21} \\ \vdots \\ a_{m1} \end{bmatrix} + \beta_2 \begin{bmatrix} a_{12} \\ a_{22} \\ \vdots \\ a_{m2} \end{bmatrix} + \cdots + \beta_n \begin{bmatrix} a_{1n} \\ a_{2n} \\ \vdots \\ a_{mn} \end{bmatrix},
$$

$$
= \begin{bmatrix} \beta_1 a_{11} + \beta_2 a_{12} + \cdots + \beta_n a_{1n} \\ \beta_1 a_{21} + \beta_2 a_{22} + \cdots + \beta_n a_{2n} \\ \vdots \\ \beta_1 a_{m1} + \beta_2 a_{m2} + \cdots + \beta_n a_{mn} \end{bmatrix},
$$

or in matrix form as

$$
\begin{bmatrix} \gamma_1 \\ \gamma_2 \\ \vdots \\ \gamma_m \end{bmatrix} = \begin{bmatrix} a_{11} & a_{12} & \cdots & a_{1n} \\ a_{21} & a_{22} & \cdots & a_{2n} \\ \vdots & \vdots & & \vdots \\ a_{m1} & a_{m2} & \cdots & a_{mn} \end{bmatrix} \begin{bmatrix} \beta_1 \\ \beta_2 \\ \vdots \\ \beta_n \end{bmatrix},
$$

which completes the proof.

Definition 3.7 The matrix $\mathbf{A}$ of Theorem 3.8 is said to be *the matrix of* $\mathbf{L}$ *with respect to the bases B and C*, and (15) is said to *represent* $\mathbf{L}$ *with respect to B and C*.

Example 3.18 Let $\mathbf{L}: R^2 \to R^3$ be defined by

$$
(17) \qquad \mathbf{L}\left(\begin{bmatrix} x_1 \\ x_2 \end{bmatrix}\right) = \begin{bmatrix} x_1 + x_2 \\ x_1 - x_2 \\ x_2 \end{bmatrix}
$$

$$
= \begin{bmatrix} y_1 \\ y_2 \\ y_3 \end{bmatrix}.
$$

The representation of $\mathbf{L}$ with respect to the natural bases in R^2 and R^3 is simply the matrix equation that expresses

$$\begin{bmatrix} y_1 \\ y_2 \\ y_3 \end{bmatrix}$$

in terms of $\begin{bmatrix} x_1 \\ x_2 \end{bmatrix}$; by inspection of (17), this is

(18)
$$\begin{bmatrix} y_1 \\ y_2 \\ y_3 \end{bmatrix} = \begin{bmatrix} 1 & 1 \\ 1 & -1 \\ 0 & 1 \end{bmatrix} \begin{bmatrix} x_1 \\ x_2 \end{bmatrix}.$$

This result is consistent with Theorem 3.8, since the first column of the 2×3 matrix in (18) is $\mathbf{L}\left(\begin{bmatrix} 1 \\ 0 \end{bmatrix}\right)$ and the second is $\mathbf{L}\left(\begin{bmatrix} 0 \\ 1 \end{bmatrix}\right)$.

Example 3.19 Let us find the representation of the linear transformation (17) in terms of the bases

$$B = \left\{ \begin{bmatrix} 1 \\ 1 \end{bmatrix}, \begin{bmatrix} 0 \\ 1 \end{bmatrix} \right\} \quad \text{and} \quad C = \left\{ \begin{bmatrix} 1 \\ 0 \\ 1 \end{bmatrix}, \begin{bmatrix} 0 \\ 1 \\ 1 \end{bmatrix}, \begin{bmatrix} 0 \\ 0 \\ 1 \end{bmatrix} \right\}.$$

Since

$$\mathbf{L}\left(\begin{bmatrix} 1 \\ 1 \end{bmatrix}\right) = \begin{bmatrix} 2 \\ 0 \\ 1 \end{bmatrix} = 2\begin{bmatrix} 1 \\ 0 \\ 1 \end{bmatrix} + 0\begin{bmatrix} 0 \\ 1 \\ 1 \end{bmatrix} - 1\begin{bmatrix} 0 \\ 0 \\ 1 \end{bmatrix}$$

and

$$\mathbf{L}\left(\begin{bmatrix} 0 \\ 1 \end{bmatrix}\right) = \begin{bmatrix} 1 \\ -1 \\ 1 \end{bmatrix} = 1\begin{bmatrix} 1 \\ 0 \\ 1 \end{bmatrix} - 1\begin{bmatrix} 0 \\ 1 \\ 1 \end{bmatrix} + 1\begin{bmatrix} 0 \\ 0 \\ 1 \end{bmatrix},$$

the required representation is

$$
(19) \qquad
\begin{bmatrix} \gamma_1 \\ \gamma_2 \\ \gamma_3 \end{bmatrix}
=
\begin{bmatrix} 2 & 1 \\ 0 & -1 \\ -1 & 1 \end{bmatrix}
\begin{bmatrix} \beta_1 \\ \beta_2 \end{bmatrix},
$$

by which we mean that (19) associates the vector

$$
(20) \qquad \mathbf{U} = \beta_1 \begin{bmatrix} 1 \\ 1 \end{bmatrix} + \beta_2 \begin{bmatrix} 0 \\ 1 \end{bmatrix}
$$

in R^2 with

$$
(21) \qquad \mathbf{V} = \mathbf{L}(\mathbf{U}) = \gamma_1 \begin{bmatrix} 1 \\ 0 \\ 1 \end{bmatrix} + \gamma_2 \begin{bmatrix} 0 \\ 1 \\ 1 \end{bmatrix} + \gamma_3 \begin{bmatrix} 0 \\ 0 \\ 1 \end{bmatrix}
$$

in R^3.

We emphasize that (18) and (19) are different ways of representing the *same* linear transformation. To check this, let us compute $\mathbf{L}\left(\begin{bmatrix} 2 \\ 1 \end{bmatrix} \right)$ in two ways. From (18),

$$
(22) \qquad \mathbf{L}\left(\begin{bmatrix} 2 \\ 1 \end{bmatrix} \right) = \begin{bmatrix} 3 \\ 1 \\ 1 \end{bmatrix}.
$$

To obtain $\mathbf{L}\left(\begin{bmatrix} 2 \\ 1 \end{bmatrix} \right)$ from (19), we must first express $\begin{bmatrix} 2 \\ 1 \end{bmatrix}$ in the form (20). The result is

$$
\begin{bmatrix} 2 \\ 1 \end{bmatrix} = 2 \begin{bmatrix} 1 \\ 1 \end{bmatrix} - 1 \begin{bmatrix} 0 \\ 1 \end{bmatrix};
$$

hence, from (19),

$$
\begin{bmatrix} \gamma_1 \\ \gamma_2 \\ \gamma_3 \end{bmatrix}
=
\begin{bmatrix} 2 & 1 \\ 0 & -1 \\ -1 & 1 \end{bmatrix}
\begin{bmatrix} 2 \\ -1 \end{bmatrix}
=
\begin{bmatrix} 3 \\ 1 \\ -3 \end{bmatrix},
$$

and from (21)

$$\mathbf{L(U)} = 3\begin{bmatrix} 1 \\ 0 \\ 1 \end{bmatrix} + \begin{bmatrix} 0 \\ 1 \\ 1 \end{bmatrix} - 3\begin{bmatrix} 0 \\ 0 \\ 1 \end{bmatrix} = \begin{bmatrix} 3 \\ 1 \\ 1 \end{bmatrix},$$

which agrees with (22).

Example 3.20   Let $\mathbf{L}: R^2 \to R^3$ be defined in terms of the bases

$$B = \left\{ \begin{bmatrix} 1 \\ 1 \end{bmatrix}, \begin{bmatrix} 1 \\ -1 \end{bmatrix} \right\} \qquad \text{and} \qquad C = \left\{ \begin{bmatrix} 1 \\ 1 \\ 0 \end{bmatrix}, \begin{bmatrix} 0 \\ 1 \\ 1 \end{bmatrix}, \begin{bmatrix} 1 \\ 0 \\ 1 \end{bmatrix} \right\}$$

by

$$(23) \qquad \begin{bmatrix} \gamma_1 \\ \gamma_2 \\ \gamma_3 \end{bmatrix} = \begin{bmatrix} 1 & -1 \\ 0 & 1 \\ 2 & 0 \end{bmatrix} \begin{bmatrix} \beta_1 \\ \beta_2 \end{bmatrix};$$

that is,

$$(24) \qquad \mathbf{L}\left(\beta_1 \begin{bmatrix} 1 \\ 1 \end{bmatrix} + \beta_2 \begin{bmatrix} 1 \\ -1 \end{bmatrix} \right) = \gamma_1 \begin{bmatrix} 1 \\ 1 \\ 0 \end{bmatrix} + \gamma_2 \begin{bmatrix} 0 \\ 1 \\ 1 \end{bmatrix} + \gamma_3 \begin{bmatrix} 1 \\ 0 \\ 1 \end{bmatrix}.$$

To represent $\mathbf{L}$ with respect to the natural bases in R^2 and R^3, we must find the coordinates of

$$\mathbf{L}\left(\begin{bmatrix} 1 \\ 0 \end{bmatrix} \right) \qquad \text{and} \qquad \mathbf{L}\left(\begin{bmatrix} 0 \\ 1 \end{bmatrix} \right)$$

with respect to the natural bases for R^3. Since

$$\begin{bmatrix} 1 \\ 0 \end{bmatrix} = \frac{1}{2}\begin{bmatrix} 1 \\ 1 \end{bmatrix} + \frac{1}{2}\begin{bmatrix} 1 \\ -1 \end{bmatrix},$$

we substitute $\beta_1 = \beta_2 = \frac{1}{2}$ in (23) and find that $\gamma_1 = 0$, $\gamma_2 = \frac{1}{2}$, and $\gamma_3 = 1$; hence, from (24)

$$
(25) \qquad \mathbf{L}\left(\begin{bmatrix} 1 \\ 0 \end{bmatrix} \right) = 0 \begin{bmatrix} 1 \\ 1 \\ 0 \end{bmatrix} + \frac{1}{2} \begin{bmatrix} 0 \\ 1 \\ 1 \end{bmatrix} + 1 \begin{bmatrix} 1 \\ 0 \\ 1 \end{bmatrix} = \begin{bmatrix} 1 \\ \frac{1}{2} \\ \frac{3}{2} \end{bmatrix}.
$$

Similarly,

$$
\begin{bmatrix} 0 \\ 1 \end{bmatrix} = \frac{1}{2} \begin{bmatrix} 1 \\ 1 \end{bmatrix} - \frac{1}{2} \begin{bmatrix} 1 \\ -1 \end{bmatrix}
$$

so that $\beta_1 = \frac{1}{2}$ and $\beta_2 = -\frac{1}{2}$; from (23) and (24),

$$
(26) \qquad \mathbf{L}\left(\begin{bmatrix} 0 \\ 1 \end{bmatrix} \right) = 1 \begin{bmatrix} 1 \\ 1 \\ 0 \end{bmatrix} - \frac{1}{2} \begin{bmatrix} 0 \\ 1 \\ 1 \end{bmatrix} + 1 \begin{bmatrix} 1 \\ 0 \\ 1 \end{bmatrix} = \begin{bmatrix} 2 \\ \frac{1}{2} \\ \frac{1}{2} \end{bmatrix}.
$$

From (25) and (26), the required representation is

$$
\begin{bmatrix} y_1 \\ y_2 \\ y_3 \end{bmatrix} = \begin{bmatrix} 1 & 2 \\ \frac{1}{2} & \frac{1}{2} \\ \frac{3}{2} & \frac{1}{2} \end{bmatrix} \begin{bmatrix} x_1 \\ x_2 \end{bmatrix}.
$$

If $\mathbf{L}$ is a linear transformation of an n-dimensional vector space S into itself $(S = T)$, then we can let $B = C$ in Definition 3.7. In this case, we speak of the *matrix of* $\mathbf{L}$, and the *representation of* $\mathbf{L}$, with respect to B. If $B = \{\mathbf{U}_1, \ldots, \mathbf{U}_n\}$ then we can write

$$
\mathbf{U} = \beta_1\mathbf{U}_1 + \beta_2\mathbf{U}_2 + \cdots + \beta_n\mathbf{U}_n,
$$

and

$$
\mathbf{L}(\mathbf{U}) = \gamma_1\mathbf{U}_1 + \gamma_2\mathbf{U}_2 + \cdots + \gamma_n\mathbf{U}_n,
$$

and the representation of $\mathbf{L}$ with respect to B is

$$
\begin{bmatrix} \gamma_1 \\ \gamma_2 \\ \vdots \\ \gamma_n \end{bmatrix} = \mathbf{A} \begin{bmatrix} \beta_1 \\ \beta_2 \\ \vdots \\ \beta_n \end{bmatrix}
$$

where **A** is the $n \times n$ matrix whose jth column is the coordinate vector of $\mathbf{L}(\mathbf{U}_j)$ with respect to B. It is not necessary to prove this, since it is a special case of Theorem 3.8.

Example 3.21 Let $\mathbf{L} : P_2 \to P_2$ be defined by $\mathbf{L}(p) = dp/dt$, and take $B = \{p_1, p_2, p_3\}$, where $p_1(t) = 1$, $p_2(t) = t$, and $p_3(t) = t^2$. Then,

$$\mathbf{L}(p_1) = 0 = 0 \cdot p_1 + 0 \cdot p_2 + 0 \cdot p_3,$$

$$\mathbf{L}(p_2) = 1 = 1 \cdot p_1 + 0 \cdot p_2 + 0 \cdot p_3,$$

$$\mathbf{L}(p_3) = 2t = 0 \cdot p_1 + 2 \cdot p_2 + 0 \cdot p_3.$$

Therefore,

$$\mathbf{L}(\beta_1 p_1 + \beta_2 p_2 + \beta_3 p_3) = \gamma_1 p_1 + \gamma_2 p_2 + \gamma_3 p_3,$$

where

$$\begin{bmatrix} \gamma_1 \\ \gamma_2 \\ \gamma_3 \end{bmatrix} = \begin{bmatrix} 0 & 1 & 0 \\ 0 & 0 & 2 \\ 0 & 0 & 0 \end{bmatrix} \begin{bmatrix} \beta_1 \\ \beta_2 \\ \beta_3 \end{bmatrix}.$$

(Verify this directly.)

Example 3.22 Consider the transformation $\mathbf{L} : R^2 \to R^2$ defined by

$$(27) \qquad \begin{bmatrix} y_1 \\ y_2 \end{bmatrix} = \mathbf{L}\left(\begin{bmatrix} x_1 \\ x_2 \end{bmatrix} \right) = \begin{bmatrix} x_1 + 3x_2 \\ -x_1 + x_2 \end{bmatrix}.$$

Since

$$\mathbf{L}\left(\begin{bmatrix} 1 \\ 0 \end{bmatrix} \right) = \begin{bmatrix} 1 \\ -1 \end{bmatrix} \quad \text{and} \quad \mathbf{L}\left(\begin{bmatrix} 0 \\ 1 \end{bmatrix} \right) = \begin{bmatrix} 3 \\ 1 \end{bmatrix},$$

the representation of $\mathbf{L}$ with respect to the natural basis

$$\left\{ \begin{bmatrix} 1 \\ 0 \end{bmatrix}, \begin{bmatrix} 0 \\ 1 \end{bmatrix} \right\}$$

is

$$\begin{bmatrix} y_1 \\ y_2 \end{bmatrix} = \begin{bmatrix} 1 & 3 \\ -1 & 1 \end{bmatrix} \begin{bmatrix} y_1 \\ y_2 \end{bmatrix},$$

which is easily verified by inspection of (27).

Example 3.23 To represent (27) with respect to the basis

$$B = \left\{ \begin{bmatrix} 1 \\ 1 \end{bmatrix}, \begin{bmatrix} 1 \\ -1 \end{bmatrix} \right\},$$

we express

$$\mathbf{L}\left(\begin{bmatrix} 1 \\ 1 \end{bmatrix} \right) \quad \text{and} \quad \mathbf{L}\left(\begin{bmatrix} 1 \\ -1 \end{bmatrix} \right)$$

in terms of B:

$$\mathbf{L}\left(\begin{bmatrix} 1 \\ 1 \end{bmatrix} \right) = \begin{bmatrix} 4 \\ 0 \end{bmatrix} = 2 \begin{bmatrix} 1 \\ 1 \end{bmatrix} + 2 \begin{bmatrix} 1 \\ -1 \end{bmatrix},$$

$$\mathbf{L}\left(\begin{bmatrix} 1 \\ -1 \end{bmatrix} \right) = \begin{bmatrix} -2 \\ -2 \end{bmatrix} = -2 \begin{bmatrix} 1 \\ 1 \end{bmatrix} + 0 \begin{bmatrix} 1 \\ -1 \end{bmatrix}.$$

Therefore,

$$\mathbf{L}\left(\beta_1 \begin{bmatrix} 1 \\ 1 \end{bmatrix} + \beta_2 \begin{bmatrix} 1 \\ -1 \end{bmatrix} \right) = \gamma_1 \begin{bmatrix} 1 \\ 1 \end{bmatrix} + \gamma_2 \begin{bmatrix} 1 \\ -1 \end{bmatrix},$$

where

$$\begin{bmatrix} \gamma_1 \\ \gamma_2 \end{bmatrix} = \begin{bmatrix} 2 & -2 \\ 2 & 0 \end{bmatrix} \begin{bmatrix} \beta_1 \\ \beta_2 \end{bmatrix}.$$

EXERCISES 2.3

1. State whether the following define linear transformations.

(a) $\quad \mathbf{L}\left(\begin{bmatrix} x \\ y \\ z \end{bmatrix} \right) = \begin{bmatrix} x + y \\ x - y \\ z + x \end{bmatrix} \qquad (\mathbf{L}: R^3 \rightarrow R^3).$

(b) $\quad \mathbf{L}\left(\begin{bmatrix} x \\ y \end{bmatrix} \right) = \begin{bmatrix} x + y \\ y - x \end{bmatrix} \qquad (\mathbf{L}: R^2 \rightarrow R^2).$

(c) $L(at^2 + bt + c) = 2at + b$ $(L: P_2 \rightarrow P_1)$.

(d) $L\left(\begin{bmatrix} x \\ y \end{bmatrix}\right) = \begin{bmatrix} 2 & -1 \\ 3 & 4 \end{bmatrix}\begin{bmatrix} x \\ y \end{bmatrix}$ $(L: R^2 \rightarrow R^2)$.

(e) $L(p) = tp(t) + 1$ $(L: P_1 \rightarrow P_2)$.

2.   State whether the following define linear transformations.

(a) $L\left(\begin{bmatrix} x \\ y \end{bmatrix}\right) = \begin{bmatrix} y \\ x \end{bmatrix}$ $(L: R^2 \rightarrow R^2)$.

(b) $L\left(\begin{bmatrix} x \\ y \\ z \end{bmatrix}\right) = \begin{bmatrix} x^2 - y^2 \\ x^2 + y^2 \\ x - z \end{bmatrix}$ $(L: R^3 \rightarrow R^3)$.

(c) $L(p) = p(2)$ $(L: P^2 \rightarrow R)$.

(d) $L\left(\begin{bmatrix} x \\ y \end{bmatrix}\right) = \begin{bmatrix} 1 & -3 \\ 2 & 1 \end{bmatrix}\begin{bmatrix} x \\ y \end{bmatrix} + \begin{bmatrix} 3 \\ 4 \end{bmatrix}$ $(L: R^2 \rightarrow R^2)$.

(e) $L(p) = tp(t) + p(0)$ $(L: P_1 \rightarrow P_2)$.

In Exercises 3 through 9, (a) find a basis for ker L; (b) find a basis for range L (see Exercise T-8); (c) state whether L is one-to-one; (d) state whether L is onto; (e) verify Theorem 3.5.

3. $L\left(\begin{bmatrix} x \\ y \\ z \end{bmatrix}\right) = \begin{bmatrix} x + y \\ x - z \\ y + z \end{bmatrix}$ $(L: R^3 \rightarrow R^3)$.

4. $L(at^2 + bt + c) = 2at + b$ $(L: P_2 \rightarrow P_1)$.

5. $L(p) = tp'(t) + p(0)$ $(L: P_2 \rightarrow P_2)$.

6. $L(p) = \int_0^1 p(t)\, dt$ $(L: P_2 \rightarrow R)$.

7. $L\left(\begin{bmatrix} x \\ y \\ z \end{bmatrix}\right) = \begin{bmatrix} 1 & 2 & -1 \\ -1 & 2 & 0 \end{bmatrix} \begin{bmatrix} x \\ y \\ z \end{bmatrix}$ $(L: R^3 \to R^3)$.

8. $L(p) = tp(t)$ $(L: P_2 \to P_3)$.

9. $L\left(\begin{bmatrix} x \\ y \\ z \end{bmatrix}\right) = \begin{bmatrix} x + y \\ x - y \\ z \end{bmatrix}$ $(L: R^3 \to R^3)$.

10. Find the coordinate vectors of the following vectors with respect to the basis

$$B = \left\{ \begin{bmatrix} 1 \\ 0 \\ 1 \end{bmatrix}, \begin{bmatrix} 1 \\ 1 \\ 1 \end{bmatrix}, \begin{bmatrix} 1 \\ 2 \\ 0 \end{bmatrix} \right\}$$

for R^3.

(a) $\begin{bmatrix} 1 \\ 0 \\ 1 \end{bmatrix}$; (b) $\begin{bmatrix} 1 \\ 0 \\ 0 \end{bmatrix}$; (c) $\begin{bmatrix} 0 \\ 0 \\ 0 \end{bmatrix}$; (d) $\begin{bmatrix} -2 \\ 1 \\ 0 \end{bmatrix}$.

11. Find the coordinate vectors of the following polynomials with respect to the basis

$$B = \{t^2 + 1, t - 1, t + 1\} \quad \text{for} \quad P_2.$$

(a) $t^2 + 1$; (b) $t^2 - 2t + 1$; (c) $2t + 1$;
(d) $3t^2 - 2t - 1$.

12. Find the matrix of $L: R^2 \to R^3$, defined by

$$L\left(\begin{bmatrix} x \\ y \end{bmatrix}\right) = \begin{bmatrix} x + y \\ x - y \\ 2x + 3y \end{bmatrix},$$

with respect to the natural bases for R^2 and R^3, and also with respect to the bases

$$B = \left\{ \begin{bmatrix} 1 \\ 1 \end{bmatrix}, \begin{bmatrix} 1 \\ -1 \end{bmatrix} \right\} \quad \text{and} \quad C = \left\{ \begin{bmatrix} 1 \\ 0 \\ 1 \end{bmatrix}, \begin{bmatrix} 1 \\ 1 \\ 1 \end{bmatrix}, \begin{bmatrix} 1 \\ 2 \\ 3 \end{bmatrix} \right\} .$$

Compute $\mathbf{L}\left(\begin{bmatrix} 2 \\ 3 \end{bmatrix} \right)$, using both representations.

13.   Find the matrix of $\mathbf{L}: R^3 \to R^2$, defined by

$$\mathbf{L}\left(\begin{bmatrix} x \\ y \\ z \end{bmatrix} \right) = \begin{bmatrix} x - y + z \\ y + z \end{bmatrix} ,$$

with respect to the natural bases for R^3 and R^2, and also with respect to the bases

$$B = \left\{ \begin{bmatrix} 1 \\ 0 \\ 1 \end{bmatrix}, \begin{bmatrix} 1 \\ 1 \\ 1 \end{bmatrix}, \begin{bmatrix} 1 \\ 2 \\ 3 \end{bmatrix} \right\} \quad \text{and} \quad C = \left\{ \begin{bmatrix} -1 \\ 1 \end{bmatrix}, \begin{bmatrix} 2 \\ 1 \end{bmatrix} \right\} .$$

Compute $\mathbf{L}\left(\begin{bmatrix} 1 \\ -1 \\ 2 \end{bmatrix} \right)$, using both representations.

14.   Suppose the matrix of $\mathbf{L}: R^2 \to R^3$ with respect to the bases

$$B = \left\{ \begin{bmatrix} 1 \\ 1 \end{bmatrix}, \begin{bmatrix} 0 \\ 1 \end{bmatrix} \right\} \quad \text{and} \quad C = \left\{ \begin{bmatrix} 1 \\ 1 \\ 1 \end{bmatrix}, \begin{bmatrix} 1 \\ 0 \\ 2 \end{bmatrix}, \begin{bmatrix} 1 \\ 0 \\ 1 \end{bmatrix} \right\}$$

is

$$A = \begin{bmatrix} 1 & 1 \\ -2 & 1 \\ 3 & 0 \end{bmatrix}.$$

What is the matrix of **L** with respect to the natural bases for R^2 and R^3?

15. Find the matrix of $\mathbf{L}: P_1 \to P_2$, defined by $\mathbf{L}(p) = tp(t)$, with respect to the bases $B = \{t, 1\}$ and $C = \{t^2, t, 1\}$.

16. Find the matrix of $\mathbf{L}: P_2 \to P_1$, defined by $\mathbf{L}(p) = p'(t) + tp''(t)$, with respect to the bases $B = \{t^2 + 1, t - 1, t + 2\}$ and $C = \{t - 1, t + 1\}$.

17. Represent $\mathbf{L}: R^3 \to R^3$, defined by

$$\mathbf{L}\left(\begin{bmatrix} x \\ y \\ z \end{bmatrix}\right) = \begin{bmatrix} 2x - y \\ x + 2y \\ x + y + z \end{bmatrix},$$

with respect to the natural basis for R^3, and with respect to the basis

$$B = \left\{ \begin{bmatrix} 1 \\ 0 \\ 1 \end{bmatrix}, \begin{bmatrix} 0 \\ 1 \\ 1 \end{bmatrix}, \begin{bmatrix} 1 \\ 1 \\ 0 \end{bmatrix} \right\}.$$

Compute $\mathbf{L}\left(\begin{bmatrix} 2 \\ -1 \\ 3 \end{bmatrix}\right)$, using both representations.

18. If the matrix of $\mathbf{L}: R^3 \to R^3$ with respect to the basis

$$B = \left\{ \begin{bmatrix} 1 \\ 0 \\ 1 \end{bmatrix}, \begin{bmatrix} 0 \\ 1 \\ 1 \end{bmatrix}, \begin{bmatrix} 1 \\ 1 \\ 0 \end{bmatrix} \right\}$$

is

$$A = \begin{bmatrix} 2 & 0 & 1 \\ -1 & 0 & 1 \\ 1 & 0 & 0 \end{bmatrix},$$

then what is the matrix of L with respect to the natural basis for R^3?

THEORETICAL EXERCISES

T-1. Prove Theorem 3.1.

T-2. *Prove:* If L is a linear transformation, then $L(0) = 0$.

T-3. Prove Theorem 3.4.

T-4. Let L be a linear transformation from S to T, and suppose $\{U_1, U_2, \ldots, U_n\}$ is a basis for S. Show that $\{L(U_1), L(U_2), \ldots, L(U_n)\}$ spans range L.

T-5. Prove that the linear transformation defined in Theorem 3.6 is an isomorphism.

T-6. Let $B = \{X_1, X_2, \ldots, X_n\}$ be a basis for R^n and

$$X = \begin{bmatrix} x_1 \\ x_2 \\ \vdots \\ x_n \end{bmatrix} = \beta_1 X_1 + \beta_2 X_2 + \cdots + \beta_n X_n;$$

thus,

$$(X)_B = \begin{bmatrix} \beta_1 \\ \beta_2 \\ \vdots \\ \beta_n \end{bmatrix}.$$

Prove: $(X)_B = X$ for every X in R^n if and only if $X_1 = E_1$, $X_2 = E_2, \ldots, X_n = E_n$.

T-7. Prove Theorem 3.7.

T-8. Complete the proof of Theorem 3.5 for the case where ker $L = 0$.

T-9. Exhibit an isomorphism from the vector space of $m \times n$ matrices onto R^{mn}.

T-10. Let $\mathbf{L}: S \to T$, where dim $S =$ dim $T = n$. Show that $\mathbf{L}$ is one-to-one if and only if $\mathbf{L}$ is onto.

T-11. Show that $\mathbf{L}: S \to T$ is one-to-one if and only if the image of any linearly independent set in S is a linearly independent set in T.

T-12. *Prove:* Vectors $\mathbf{U}_1, \ldots, \mathbf{U}_k$ in an n-dimensional space S are linearly independent if and only if their coordinate vectors with respect to any basis for S are linearly independent in R^n.

T-13. Let $B = \{\mathbf{U}_1, \ldots, \mathbf{U}_n\}$ and $C = \{\mathbf{V}_1, \ldots, \mathbf{V}_n\}$ be two bases for a vector space S, and let $\mathbf{W}$ be an arbitrary vector in S. Show that

$$(\mathbf{W})_B = \mathbf{A}(\mathbf{W})_C,$$

where the jth column of $\mathbf{A}$ is $(\mathbf{V}_j)_C$.

2.4 Rank of a Matrix

We saw in Section 2.3 that every linear transformation $\mathbf{L}: R^n \to R^m$ can be represented as

(1) $$\mathbf{L}(\mathbf{X}) = \mathbf{A}\mathbf{X},$$

where $\mathbf{A}$ is an $m \times n$ matrix. In Chapter 1 we developed methods for solving a linear system

(2) $$\mathbf{A}\mathbf{X} = \mathbf{Y}.$$

These problems are clearly related, since (2) has a solution if and only if $\mathbf{Y}$ is in range $\mathbf{L}$. We shall now consider this relationship.

Definition 4.1 The *null space* of an $m \times n$ matrix $\mathbf{A}$ is the collection of n-vectors $\mathbf{X}$ that satisfy

$$\mathbf{A}\mathbf{X} = \mathbf{0};$$

thus it is the kernel of the linear transformation (1). The dimension of the null space of $\mathbf{A}$ is the *nullity* of $\mathbf{A}$. It is denoted by $N(\mathbf{A})$.

The range of the linear transformation (1) is the subspace of R^m spanned by the columns of $\mathbf{A}$, since

$$\mathbf{L(X)} = \begin{bmatrix} a_{11} & a_{12} & \cdots & a_{1n} \\ a_{21} & a_{22} & \cdots & a_{2n} \\ \vdots & \vdots & & \vdots \\ a_{m1} & a_{m2} & \cdots & a_{mn} \end{bmatrix} \begin{bmatrix} x_1 \\ x_2 \\ \vdots \\ x_n \end{bmatrix}$$

$$= x_1 \begin{bmatrix} a_{11} \\ a_{21} \\ \vdots \\ a_{m1} \end{bmatrix} + x_2 \begin{bmatrix} a_{12} \\ a_{22} \\ \vdots \\ a_{m2} \end{bmatrix} + \cdots + x_n \begin{bmatrix} a_{1n} \\ a_{2n} \\ \vdots \\ a_{mn} \end{bmatrix}.$$

It can be seen from the last member of this equation that an m-vector $\mathbf{Y}$ is in range $\mathbf{L}$ if and only if $\mathbf{Y}$ is a linear combination of the columns of $\mathbf{A}$. This motivates the following definition.

Definition 4.2 The subspace of R^m spanned by the columns of an $m \times n$ matrix $\mathbf{A}$ is called the *column space of* $\mathbf{A}$. The dimension of the column space of $\mathbf{A}$ is the *column rank of* $\mathbf{A}$.

It is also useful to make corresponding definitions for the rows of $\mathbf{A}$.

Definition 4.3 The subspace of R_n spanned by the rows of an $m \times n$ matrix $\mathbf{A}$ is called the *row space of* $\mathbf{A}$. The dimension of the row space is the *row rank of* $\mathbf{A}$.

Example 4.1 Consider the linear transformation $\mathbf{L}: R^6 \to R^4$ defined by

$$\mathbf{L(X)} = \mathbf{AX} = \begin{bmatrix} 1 & 1 & 2 & 1 & 3 & 1 \\ 2 & 2 & 5 & 4 & 7 & 0 \\ 1 & 1 & 2 & 2 & 5 & 1 \\ 1 & 1 & 3 & 3 & 5 & 0 \end{bmatrix} \begin{bmatrix} x_1 \\ x_2 \\ x_3 \\ x_4 \\ x_5 \\ x_6 \end{bmatrix}.$$

From Example 2.8, Section 1.2, **A** is row equivalent to

$$\mathbf{B} = \begin{bmatrix} 1 & 1 & 0 & 0 & 0 & -2 \\ 0 & 0 & 1 & 0 & 0 & 1 \\ 0 & 0 & 0 & 1 & 0 & -2 \\ 0 & 0 & 0 & 0 & 1 & 1 \end{bmatrix}.$$

Since $\mathbf{AX} = \mathbf{0}$ and $\mathbf{BX} = \mathbf{0}$ have the same solutions (Theorem 2.2, Section 1.2), and the solutions of the latter are the 6-vectors such that

$$x_1 = -x_2 + 2x_6,$$

$$x_3 = -x_6,$$

$$x_4 = 2x_6,$$

$$x_5 = -x_6,$$

it follows that the null space of **A** consists of vectors of the form

$$\mathbf{X} = x_2 \begin{bmatrix} -1 \\ 1 \\ 0 \\ 0 \\ 0 \\ 0 \end{bmatrix} + x_6 \begin{bmatrix} 2 \\ 0 \\ -1 \\ 2 \\ -1 \\ 1 \end{bmatrix},$$

where x_2 and x_6 are arbitrary. Thus $N(\mathbf{A}) = 2$ and, from Theorem 3.5, Section 2.3, the column rank of **A** (which is the dimension of range **A**) is 4. Therefore, the column space of **A** is R^4.

The row and column ranks of a matrix are equal, as we shall now show.

Lemma 4.1   If an $m \times n$ matrix **A** has column rank n, then its row rank is also n.

Proof.   Let

$$\mathbf{A} = \begin{bmatrix} a_{11} & a_{12} & \cdots & a_{1n} \\ a_{21} & a_{22} & \cdots & a_{2n} \\ \vdots & & & \\ a_{m1} & a_{m2} & \cdots & a_{mn} \end{bmatrix}$$

have n linearly independent columns, and denote its row rank by r. Then

(3) $$r \le n \le m,$$

since the number of vectors in a linearly independent set cannot exceed the dimension of the space. Interchanging the rows of a matrix does not change its row rank; hence we may assume that the first r rows of $\mathbf{A}$ form a basis for the row space. If $r < n$, then the columns of

$$\mathbf{A}_1 = \begin{bmatrix} a_{11} & a_{12} & \cdots & a_{1n} \\ a_{21} & a_{22} & \cdots & a_{2n} \\ \vdots & \vdots & & \vdots \\ a_{r1} & a_{r2} & \cdots & a_{rn} \end{bmatrix}$$

are linearly dependent, since no subset of R^r containing more than r elements can be linearly independent. Therefore, there are constants $c_1, c_2, \ldots, c_n$, not all zero, such that

(4) $$c_1 a_{i1} + c_2 a_{i2} + \cdots + c_n a_{in} = 0 \qquad (1 \le i \le r).$$

Rows $r + 1, \ldots, m$ of $\mathbf{A}$ can be written as linear combinations of rows 1 through r, since the latter form a basis for the row space of $\mathbf{A}$. Consequently,

$$[a_{i1}, a_{i2}, \ldots, a_{in}] = d_{i1}[a_{11}, a_{12}, \ldots, a_{1n}] + d_{i2}[a_{21}, a_{22}, \ldots, a_{2n}]$$
$$+ \cdots + d_{ir}[a_{r1}, a_{r2}, \ldots, a_{rn}] \qquad (r + 1 \le i \le n)$$

for suitable constants $\{d_{ij}\}$. By writing this equation in terms of components and using (4), we can infer that

(5) $$c_1 a_{i1} + c_2 a_{i2} + \cdots + c_n a_{in} = 0 \qquad (r + 1 \le i \le n).$$

(Verify.) Now (4) and (5) imply that

$$c_1 \begin{bmatrix} a_{11} \\ a_{21} \\ \vdots \\ a_{n1} \end{bmatrix} + c_2 \begin{bmatrix} a_{12} \\ a_{22} \\ \vdots \\ a_{n2} \end{bmatrix} + \cdots + c_n \begin{bmatrix} a_{1n} \\ a_{2n} \\ \vdots \\ a_{nn} \end{bmatrix} = \begin{bmatrix} 0 \\ 0 \\ \vdots \\ 0 \end{bmatrix},$$

which is false, since $c_1, \ldots, c_n$ are not all zero and the vectors on the left (the columns of $\mathbf{A}$) are linearly independent. This contradiction stems from the assumption that $r < n$. Hence, $r \geq n$. This together with (3) yields $r = n$, which completes the proof of the lemma.

Theorem 4.1 The row and column ranks of a matrix are equal.

Proof. Let the row and column ranks of $\mathbf{A}$ be r and s, respectively. If $r = 0$ or $s = 0$, then $\mathbf{A} = \mathbf{0}$ and the conclusion holds; hence, we may assume that r and s are both positive. For convenience, assume that the first s columns of $\mathbf{A}$ are linearly independent. Then the row rank of

$$\mathbf{A}_1 = \begin{bmatrix} a_{11} & a_{12} & \cdots & a_{1s} \\ a_{21} & a_{22} & \cdots & a_{2s} \\ \vdots & \vdots & & \vdots \\ a_{m1} & a_{m2} & \cdots & a_{ms} \end{bmatrix}$$

is also s, from Lemma 4.1. Again for convenience, assume that the first s rows of $\mathbf{A}_1$ are linearly independent in R_s. Then, surely the first s rows of $\mathbf{A}$ are linearly independent in R_n (Exercise T-1). Therefore, $r \geq s$. We have now shown that

(6)   row rank of $\mathbf{A} \geq$ column rank of $\mathbf{A}$

for any matrix $\mathbf{A}$. On replacing $\mathbf{A}$ by $\mathbf{A}^\mathrm{T}$ in (6), we find that

(7)   row rank of $\mathbf{A}^\mathrm{T} \geq$ column rank of $\mathbf{A}^\mathrm{T}$.

The row and column ranks of $\mathbf{A}$ equal, respectively, the column and row ranks of $\mathbf{A}^\mathrm{T}$ (Exercise T-2); hence, (7) yields

(8)   column rank of $\mathbf{A} \geq$ row rank of $\mathbf{A}$.

Now, (6) and (8) imply that the row and column ranks of $\mathbf{A}$ are equal, which completes the proof.

Definition 4.4 The *rank* of a matrix $\mathbf{A}$, denoted by $R(\mathbf{A})$, is the common value of its row and column ranks.

The next theorem can be obtained by applying Theorem 3.5 of Section 2.3 to the linear transformation defined by $\mathbf{L}(\mathbf{X}) = \mathbf{AX}$ (Exercise T-3).

Theorem 4.2 If $\mathbf{A}$ is an $m \times n$ matrix, then

$$R(\mathbf{A}) + N(\mathbf{A}) = n.$$

Definition 4.5 A matrix obtained by crossing out selected rows, selected columns, or selected rows and columns of a matrix **A** is called a *submatrix* *of* **A**. The determinant of a square submatrix of **A** is called a *subdeterminant* *of* **A**.

Example 4.2 Consider the matrix **A** of Example 4.1. Crossing out the second and sixth columns of **A** yields the submatrix

$$
\begin{bmatrix}
1 & 2 & 1 & 3 \\
2 & 5 & 4 & 7 \\
1 & 2 & 2 & 5 \\
1 & 3 & 3 & 5
\end{bmatrix}.
$$

Crossing out the last row of **A** yields the submatrix

$$
\begin{bmatrix}
1 & 1 & 2 & 1 & 3 & 1 \\
2 & 2 & 5 & 4 & 7 & 0 \\
1 & 1 & 2 & 2 & 5 & 1
\end{bmatrix}.
$$

Crossing out the second and sixth columns and the last row of **A** yields the submatrix

$$
\begin{bmatrix}
1 & 2 & 1 & 3 \\
2 & 5 & 4 & 7 \\
1 & 2 & 2 & 5
\end{bmatrix}.
$$

The submatrices

$$
\begin{bmatrix}
1 & 1 \\
2 & 2
\end{bmatrix}
\quad \text{and} \quad
\begin{bmatrix}
2 & 1 \\
5 & 4
\end{bmatrix}
$$

have determinants 0 and 3, respectively.

Theorem 4.3 The rank of a nonzero matrix **A** equals the largest integer s such that **A** has a nonzero subdeterminant of order s.

Proof. Let **A** be an $m \times n$ matrix with rank r, and suppose **B** is an $r \times n$ submatrix of **A** whose r rows are linearly independent; then **B** has rank r. By Theorem 4.1, $\mathbf{B}$ has r linearly independent columns. Let $\mathbf{C}$ be the $r \times r$ submatrix consisting of r linearly independent columns of $\mathbf{B}$. Then $\mathbf{C}$ is a submatrix of $\mathbf{A}$, and det $\mathbf{C} \neq 0$ (Corollary 2.1, Section 2.2). Therefore,

$$(9) \qquad\qquad r \leq s.$$

On the other hand, if $\mathbf{D}$ is an $s \times s$ submatrix of $\mathbf{A}$ with nonzero determinant, then the columns of $\mathbf{D}$ are linearly independent in R^s (Corollary 2.1, Section 2.2). The columns in $\mathbf{A}$ from which the columns of $\mathbf{D}$ were obtained must also be linearly independent in R^m (Exercise T-1), and therefore $s \leq r$. This and (9) imply that $r = s$, which completes the proof.

Theorem 4.4   Row equivalent matrices have identical row spaces.

Proof.   If $\mathbf{A}$ and $\mathbf{B}$ are row equivalent, then each row of $\mathbf{B}$ is a linear combination of the rows of $\mathbf{A}$. Hence, any linear combination of the rows of $\mathbf{B}$ can be written as a linear combination of the rows of $\mathbf{A}$, which means that the row space of $\mathbf{A}$ contains the row space of $\mathbf{B}$. A similar argument with $\mathbf{A}$ and $\mathbf{B}$ interchanged implies that the row space of $\mathbf{B}$ contains that of $\mathbf{A}$. Hence, the two row spaces are identical.

Lemma 4.2   The rank of a matrix $\mathbf{B}$ in row echelon form equals the number of nonzero rows in $\mathbf{B}$.

Proof.   Suppose $\mathbf{B}$ has exactly r nonzero rows and the leading ones in rows $1, 2, \ldots, r$ occur in columns $j_1 < j_2 < \cdots < j_r$. Let $\mathbf{B}_1$ be the submatrix consisting of these columns, and let $\mathbf{B}_2$ be the submatrix consisting of the first r rows of $\mathbf{B}_1$. Then $\mathbf{B}_2$ is an $r \times r$ submatrix of $\mathbf{B}$ in upper triangular form, with ones on its diagonal. Therefore, det $\mathbf{B}_2 \neq 0$, and $R(\mathbf{B}) \geq r$. Since any submatrix of $\mathbf{B}$ with more than r rows must contain a row of zeros, no $s \times s$ subdeterminant of $\mathbf{B}$ can be nonzero if $s > r$. Consequently, $R(\mathbf{B}) = r$, which completes the proof.

Example 4.3   To find the row rank of

$$\mathbf{A} = \begin{bmatrix} 1 & 1 & 0 & -1 \\ 0 & 2 & 1 & 3 \\ 1 & 1 & 0 & -1 \\ 1 & 1 & 0 & -1 \end{bmatrix},$$

we note that $\mathbf{A}$ is row equivalent to

$$\begin{bmatrix} 1 & 1 & 0 & -1 \\ 0 & 1 & \frac{1}{2} & \frac{3}{2} \\ 0 & 0 & 0 & 0 \\ 0 & 0 & 0 & 0 \end{bmatrix}.$$

Therefore, $R(\mathbf{A}) = 2$. This can also be obtained by computing the column rank of $\mathbf{A}$, which is the same as the row rank of

$$\mathbf{A}^{\mathrm{T}} = \begin{bmatrix} 1 & 0 & 1 & 1 \\ 1 & 2 & 1 & 1 \\ 0 & 1 & 0 & 0 \\ -1 & 3 & -1 & -1 \end{bmatrix}$$

(Exercise T-2). Since $\mathbf{A}^{\mathrm{T}}$ is row equivalent to

$$\begin{bmatrix} 1 & 0 & 1 & 1 \\ 0 & 1 & 0 & 0 \\ 0 & 0 & 0 & 0 \\ 0 & 0 & 0 & 0 \end{bmatrix},$$

the column rank of $\mathbf{A}$ is also two, as expected. We can also verify Theorem 4.3 by observing that

$$\begin{vmatrix} 1 & 1 \\ 0 & 2 \end{vmatrix} = 2 \neq 0,$$

while all subdeterminants of $\mathbf{A}$ of order greater than 2 have value zero.

Theorem 4.5   Let $\mathbf{A}$ be row equivalent to a matrix $\mathbf{B}$ in row echelon form, with exactly r nonzero rows. Then $R(\mathbf{A}) = r$, and if the leading ones in the nonzero rows of **B** occur in columns $j_1 < j_2 < \cdots < j_r$, then columns $j_1, j_2, \ldots, j_r$ of **A** form a basis for the column space of **A**.

Proof. From Theorem 4.4, $R(\mathbf{A}) = R(\mathbf{B})$, and from Lemma 4.2, $R(\mathbf{B}) = r$. Therefore, $R(\mathbf{A}) = r$. Let $\mathbf{A}_1$ and $\mathbf{B}_1$ be the matrices consisting of columns $j_1, j_2, \ldots, j_r$ of **A** and **B**, respectively. Then $\mathbf{A}_1$ is row equivalent to $\mathbf{B}_1$ (Exercise T-4) and therefore $R(\mathbf{A}_1) = R(\mathbf{B}_1) = r$. Consequently, the columns of $\mathbf{A}_1$ are linearly independent, and therefore form a basis for the column space of **A**.

Example 4.4 Suppose we wish to find bases for the kernel and range of the linear transformation $\mathbf{L}: R^4 \to R^3$ defined by

$$\mathbf{L}(\mathbf{X}) = \mathbf{AX} = \begin{bmatrix} 1 & 2 & 0 & -1 \\ 2 & 3 & -1 & 2 \\ -1 & -3 & -1 & 5 \end{bmatrix} \begin{bmatrix} x_1 \\ x_2 \\ x_3 \\ x_4 \end{bmatrix}.$$

The matrix

$$\mathbf{B} = \begin{bmatrix} 1 & 0 & -2 & 7 \\ 0 & 1 & 1 & -4 \\ 0 & 0 & 0 & 0 \end{bmatrix},$$

which is in row echelon form, is row equivalent to **A**. (Verify.) Hence, $\mathbf{AX} = \mathbf{0}$ if and only if $\mathbf{BX} = \mathbf{0}$, which is true if and only if

$$x_1 = 2x_3 - 7x_4,$$

$$x_2 = -x_3 + 4x_4.$$

Thus, **X** is in ker **L** if and only if it is of the form

$$\mathbf{X} = \begin{bmatrix} 2x_3 - 7x_4 \\ -x_3 + 4x_4 \\ x_3 \\ x_4 \end{bmatrix} = x_3 \begin{bmatrix} 2 \\ -1 \\ 1 \\ 0 \end{bmatrix} + x_4 \begin{bmatrix} -7 \\ 4 \\ 0 \\ 1 \end{bmatrix},$$

where x_3 and x_4 are arbitrary. Therefore,

$$\begin{bmatrix} 2 \\ -1 \\ 1 \\ 0 \end{bmatrix} \quad \text{and} \quad \begin{bmatrix} -7 \\ 4 \\ 0 \\ 1 \end{bmatrix}$$

form a basis for ker **L**. Since the leading ones in **B** occur in its first two columns, the first two columns of **A** form a basis for range **L**. Of course, as the next theorem shows, these are not the only columns of **A** with this property.

Theorem 4.6 Let $R(\mathbf{A}) = r$, let **B** be row equivalent to **A** and in reduced row echelon form, and let $\mathbf{B_1}$ be the matrix obtained by deleting the zero rows of **B**. Then columns $p_1 < p_2 < \cdots < p_r$ of **A** form a basis for the column space of **A** if and only if the $r \times r$ subdeterminant formed from columns $p_1, \ldots, p_r$ of $\mathbf{B_1}$ is nonzero.

We leave the proof of this theorem to the student (Exercise T-5).

Theorem 4.6 provides a method for finding bases for the subspace spanned by any finite set of vectors in R^n.

Example 4.5 To find a basis for $S[\mathbf{X_1}, \mathbf{X_2}, \mathbf{X_3}, \mathbf{X_4}]$, where

$$\mathbf{X_1} = \begin{bmatrix} 1 \\ 2 \\ -2 \\ 0 \end{bmatrix}, \quad \mathbf{X_2} = \begin{bmatrix} 0 \\ -1 \\ 2 \\ 2 \end{bmatrix}, \quad \mathbf{X_3} = \begin{bmatrix} 1 \\ 3 \\ -4 \\ -2 \end{bmatrix}, \quad \text{and} \quad \mathbf{X_4} = \begin{bmatrix} 0 \\ 1 \\ -2 \\ 2 \end{bmatrix},$$

we consider the matrix

$$\mathbf{A} = \begin{bmatrix} 1 & 0 & 1 & 0 \\ 2 & -1 & 3 & 1 \\ -2 & 2 & -4 & -2 \\ 0 & 2 & -2 & 2 \end{bmatrix},$$

which is row equivalent to

$$\mathbf{B} = \begin{bmatrix} 1 & 0 & 1 & 0 \\ 0 & 1 & -1 & -1 \\ 0 & 0 & 0 & 1 \\ 0 & 0 & 0 & 0 \end{bmatrix},$$

a matrix in row echelon form. Thus, $S[\mathbf{X}_1, \mathbf{X}_2, \mathbf{X}_3, \mathbf{X}_4]$ is a three-dimensional subspace of R^4; to choose bases for it from among $\mathbf{X}_1$, $\mathbf{X}_2$, $\mathbf{X}_3$, and $\mathbf{X}_4$, we examine the 3×3 subdeterminants of

$$\mathbf{B}_1 = \begin{bmatrix} 1 & 0 & 1 & 0 \\ 0 & 1 & -1 & -1 \\ 0 & 0 & 0 & 1 \end{bmatrix}.$$

Since

$$\begin{vmatrix} 0 & 1 & 0 \\ 1 & -1 & -1 \\ 0 & 0 & 1 \end{vmatrix}, \quad \begin{vmatrix} 1 & 1 & 0 \\ 0 & -1 & -1 \\ 0 & 0 & 1 \end{vmatrix}, \quad \text{and} \quad \begin{vmatrix} 1 & 0 & 0 \\ 0 & 1 & -1 \\ 0 & 0 & 1 \end{vmatrix}$$

are nonzero, while

$$\begin{vmatrix} 1 & 0 & 1 \\ 0 & 1 & -1 \\ 0 & 0 & 0 \end{vmatrix} = 0,$$

we conclude that $\{\mathbf{X}_2, \mathbf{X}_3, \mathbf{X}_4\}$, $\{\mathbf{X}_1, \mathbf{X}_3, \mathbf{X}_4\}$, and $\{\mathbf{X}_1, \mathbf{X}_2, \mathbf{X}_4\}$ are bases for $S[\mathbf{X}_1, \mathbf{X}_2, \mathbf{X}_3, \mathbf{X}_4]$, but $\{\mathbf{X}_1, \mathbf{X}_2, \mathbf{X}_3\}$ is not.

Example 4.6 Suppose we wish to find bases for the subspace of R^3 spanned by $\mathbf{X}_1$, $\mathbf{X}_2$, $\mathbf{X}_3$, $\mathbf{X}_4$, the columns of

$$\mathbf{A} = \begin{bmatrix} 1 & 2 & -1 & 2 \\ 0 & 0 & 1 & -1 \\ 1 & 2 & 1 & 0 \end{bmatrix}.$$

Since **A** is row equivalent to

$$\mathbf{B} = \begin{bmatrix} 1 & 2 & 0 & 1 \\ 0 & 0 & 1 & -1 \\ 0 & 0 & 0 & 0 \end{bmatrix},$$

we examine the 2×2 subdeterminants of

$$B_1 = \begin{bmatrix} 1 & 2 & 0 & 1 \\ 0 & 0 & 1 & -1 \end{bmatrix}.$$

All 2×2 subdeterminants of B_1 are nonzero except for $\begin{vmatrix} 1 & 2 \\ 0 & 0 \end{vmatrix}$; therefore,

any pair $\{\mathbf{X}_i, \mathbf{X}_j\}$ $(i < j)$, other than $\{\mathbf{X}_1, \mathbf{X}_2\}$, form a basis for $S[\mathbf{X}_1, \mathbf{X}_2, \mathbf{X}_3, \mathbf{X}_4]$.

Theorem 4.7   Let $k \leq n$ and suppose $\mathbf{X}_1, \mathbf{X}_2, \ldots, \mathbf{X}_k$ are n-vectors. Define **A** to be the $n \times k$ matrix which has $\mathbf{X}_1, \mathbf{X}_2, \ldots, \mathbf{X}_k$ as columns. Let $\mathbf{B} = [b_{ij}]$ be any matrix in row echelon form which is row equivalent to **A**. Then $\mathbf{X}_1, \mathbf{X}_2, \ldots, \mathbf{X}_k$ are linearly independent if and only if $b_{11} = b_{22} = \cdots = b_{kk} = 1$.

We leave the proof of this theorem to the student (Exercise T-6).

Example 4.7   To test the vectors

$$\mathbf{X}_1 = \begin{bmatrix} 1 \\ 2 \\ -1 \\ 3 \end{bmatrix}, \quad \mathbf{X}_2 = \begin{bmatrix} 1 \\ 3 \\ -4 \\ 0 \end{bmatrix}, \quad \text{and} \quad \mathbf{X}_3 = \begin{bmatrix} 0 \\ 2 \\ 1 \\ 0 \end{bmatrix}$$

for linear independence, we form the matrix

$$\mathbf{A} = \begin{bmatrix} 1 & 1 & 0 \\ 2 & 3 & 2 \\ -1 & -4 & 1 \\ 3 & 0 & 0 \end{bmatrix},$$

which is row equivalent to

$$B = \begin{bmatrix} 1 & 1 & 0 \\ 0 & 1 & 2 \\ 0 & 0 & 1 \\ 0 & 0 & 0 \end{bmatrix}$$

in row echelon form. Since $b_{11} = b_{22} = b_{33} = 1$, $\mathbf{X}_1$, $\mathbf{X}_2$, and $\mathbf{X}_3$ are linearly independent.

EXERCISES 2.4

In Exercises 1 through 3 find a basis for the null space of the given matrix.

1. $\begin{bmatrix} 1 & 2 & 3 & 4 \\ 1 & 9 & 10 & 10 \\ -2 & 3 & 1 & -2 \\ -3 & 8 & 5 & 0 \end{bmatrix}$.

2. $\begin{bmatrix} 1 & -1 & 2 & 3 & 5 \\ 0 & 1 & 0 & 0 & 1 \\ 2 & 1 & 4 & -1 & 2 \end{bmatrix}$.

3. $\begin{bmatrix} 1 & 2 & -1 & 0 & 1 \\ 0 & 4 & -2 & -2 & 3 \\ 3 & 2 & -1 & 2 & 0 \end{bmatrix}$.

4. Find the dimension of the null space of

$$A = \begin{bmatrix} -1 & 2 & 3 & -1 \\ 6 & -3 & -4 & 1 \\ 4 & 1 & 2 & -1 \end{bmatrix}.$$

5. Find the dimension of the null space of

$$A = \begin{bmatrix} 1 & 2 & -1 & 3 & -2 \\ 0 & 1 & 0 & 0 & 1 \\ 2 & 1 & -1 & 3 & 4 \\ 0 & 2 & 3 & 4 & 1 \end{bmatrix}.$$

In Exercises 6 through 9, find bases for ker **L** and range **L**, where $\mathbf{L}(\mathbf{X}) = \mathbf{AX}$.

6. $A = \begin{bmatrix} 1 & 1 & 2 & 2 & 3 \\ 2 & 1 & 1 & 0 & 1 \\ 3 & -1 & 0 & -4 & -4 \end{bmatrix}.$

7. $A = \begin{bmatrix} 1 & 1 & -1 & 3 & 5 \\ -1 & 2 & 4 & 3 & 1 \\ 0 & 1 & 1 & 2 & 2 \\ 2 & -3 & -7 & -4 & 0 \end{bmatrix}.$

8. $A = \begin{bmatrix} 1 & 2 & -3 \\ 1 & 1 & 1 \\ -1 & 2 & 2 \end{bmatrix}.$

9. $A = \begin{bmatrix} 1 & 0 & 2 & 1 \\ 2 & 1 & 3 & 0 \\ -1 & 0 & -2 & -1 \\ 2 & 3 & 1 & -4 \end{bmatrix}.$

In Exercises 10 through 13, find the rank of **A** by computing the dimensions of its column and row spaces, and by finding a nonzero determinant of largest order.

10. $\mathbf{A} = \begin{bmatrix} 1 & 2 & 1 & 2 \\ 2 & 0 & 1 & 1 \\ -1 & 1 & 2 & -2 \end{bmatrix}$.

11. $\mathbf{A} = \begin{bmatrix} -2 & -1 & -5 & 0 & 2 \\ 3 & 2 & 8 & -1 & -2 \\ -1 & 1 & -1 & -2 & 4 \\ 2 & 4 & 8 & -6 & 4 \end{bmatrix}$.

12. $\mathbf{A} = \begin{bmatrix} 2 & 0 & 3 \\ 0 & 1 & 1 \\ -1 & 2 & -1 \end{bmatrix}$.

13. $\mathbf{A} = \begin{bmatrix} 2 & 0 & 4 & 1 & 3 \\ -1 & 1 & 10 & 5 & 4 \\ 2 & 2 & 4 & 0 & 2 \\ -4 & -1 & 1 & 3 & -1 \\ 3 & 2 & -7 & -1 & 2 \end{bmatrix}$.

14. Find a basis for $S[\mathbf{X}_1, \mathbf{X}_2, \mathbf{X}_3, \mathbf{X}_4]$, where

$$\mathbf{X}_1 = \begin{bmatrix} 2 \\ -1 \\ 3 \\ 1 \end{bmatrix}, \quad \mathbf{X}_2 = \begin{bmatrix} 4 \\ -5 \\ -5 \\ -5 \end{bmatrix}, \quad \mathbf{X}_3 = \begin{bmatrix} -1 \\ 2 \\ 4 \\ 3 \end{bmatrix}, \quad \text{and} \quad \mathbf{X}_4 = \begin{bmatrix} 5 \\ -4 \\ 2 \\ -1 \end{bmatrix}.$$

15. Find a basis for $S[\mathbf{X}_1, \mathbf{X}_2, \mathbf{X}_3, \mathbf{X}_4, \mathbf{X}_5]$ where

$$\mathbf{X}_1 = \begin{bmatrix} 4 \\ -1 \\ 2 \end{bmatrix}, \quad \mathbf{X}_2 = \begin{bmatrix} 0 \\ 5 \\ 0 \end{bmatrix}, \quad \mathbf{X}_3 = \begin{bmatrix} -2 \\ 3 \\ -1 \end{bmatrix},$$

$$\mathbf{X}_4 = \begin{bmatrix} 1 \\ 3 \\ -1 \end{bmatrix}, \quad \text{and} \quad \mathbf{X}_5 = \begin{bmatrix} 1 \\ 1 \\ 2 \end{bmatrix}.$$

16. Are the following vectors linearly independent in R^3?

$$\mathbf{X}_1 = \begin{bmatrix} 1 \\ -1 \\ 2 \end{bmatrix}, \quad \mathbf{X}_2 = \begin{bmatrix} 2 \\ -1 \\ 3 \end{bmatrix}.$$

17. Are the following vectors linearly independent in R^5?

$$\mathbf{X}_1 = \begin{bmatrix} 1 \\ 2 \\ -1 \\ 3 \\ 4 \end{bmatrix}, \quad \mathbf{X}_2 = \begin{bmatrix} -2 \\ 4 \\ -3 \\ -1 \\ 5 \end{bmatrix}, \quad \text{and} \quad \mathbf{X}_3 = \begin{bmatrix} -3 \\ 2 \\ -2 \\ -4 \\ 1 \end{bmatrix}.$$

THEORETICAL EXERCISES

T-1. Suppose a matrix $\mathbf{A}_1$ is obtained by deleting selected columns of a matrix $\mathbf{A}$. Let the rows of $\mathbf{A}_1$ be linearly independent. Show that the rows of $\mathbf{A}$ are also linearly independent. Also, prove the corresponding statement with "rows" and "columns" interchanged.

T-2. *Prove:* The row and column ranks of $\mathbf{A}$ equal the column and row ranks, respectively, of $\mathbf{A}^{\mathrm{T}}$.

T-3. Prove Theorem 4.2.

T-4. Let $\mathbf{A}_1$ and $\mathbf{B}_1$ consist of columns $j_1, j_2, \ldots, j_r$ of $\mathbf{A}$ and $\mathbf{B}$, respectively. *Prove:* If $\mathbf{A}$ and $\mathbf{B}$ are row equivalent, so are $\mathbf{A}_1$ and $\mathbf{B}_1$.

T-5. Prove Theorem 4.6.

T-6. Prove Theorem 4.7.

T-7. Prove directly from the definition of linear independence: If **A** is in row echelon form, then the nonzero rows of **A** form a basis for its row space.

T-8. *Prove:* The system $\mathbf{AX} = \mathbf{Y}$ has a solution if and only if **A** and $[\mathbf{A} \mid \mathbf{Y}]$ have the same rank. (*Hint:* See the remark preceding Definition 4.2.)

T-9. *Prove:* The determinant of an $n \times n$ matrix **A** is nonzero if and only if $R(\mathbf{A}) = n$.

T-10. An $n \times n$ matrix **A** is singular if and only if $R(\mathbf{A}) < n$.

START

2.5 More about R^n

When dealing with vectors in R^2 and R^3, we shall usually write

$$
\mathbf{X} = \begin{bmatrix} x \\ y \end{bmatrix} \quad \text{and} \quad \mathbf{X} = \begin{bmatrix} x \\ y \\ z \end{bmatrix}
$$

to avoid unnecessary subscripts. This notation also suggests the familiar and natural correspondence between vectors in R^2 and R^3 and points in the plane and space, respectively. For example, let us recall the situation in the plane. We choose an arbitrary point O, the *origin*, and draw through it two perpendicular lines (Fig. 5.1), the *x-axis* and the *y-axis*, assigning a positive direction to each. Through an arbitrary point P in the plane we draw lines perpendicular to the x- and y-axes and denote the points where

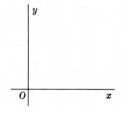

FIGURE 5.1

these lines intersect the axes by P' and P'', respectively (Fig. 5.2). Then the directed distance from P' to O (positive in Fig. 5.2a, negative in Fig. 5.2b) is the x-coordinate of P, and the directed distance from P'' to O (positive in Figs. 5.2a and 5.2b) is the y-coordinate of P. We say that P *has coordinates* (x, y) and write $P = (x, y)$.

Thus there is a one-to-one correspondence between vectors in R^2 and points in the plane:

$$\begin{bmatrix} x \\ y \end{bmatrix} \leftrightarrow (x, y).$$

In fact, we shall identify vectors and points and henceforth view R^2 both as a vector space and as the collection of points in the plane. More generally

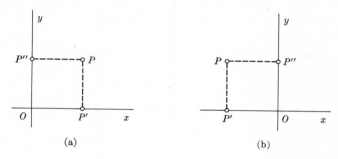

(a)   (b)

FIGURE 5.2

we shall think of R^n as a vector space on the one hand, and as a collection of points on the other. We shall not always distinguish between

$$\mathbf{X} = \begin{bmatrix} x_1 \\ \vdots \\ x_n \end{bmatrix},$$

and

$$\mathbf{X} = (x_1, \ldots, x_n);$$

both will stand for vectors and points in R^n. However, we shall usually employ the vertical display when we wish to emphasize that $\mathbf{X}$ is a vector and the horizontal when this is not necessary.

When it is necessary to distinguish between the vector $\mathbf{X}$ and the point $\mathbf{X}$ we shall refer to the latter as the point P with coordinates $(x_1, \ldots, x_n)$.

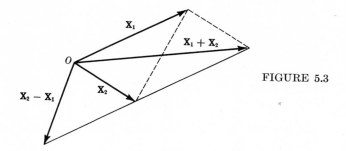

FIGURE 5.3

If **X** and **Y** are points in R^n then the vector $\mathbf{X} - \mathbf{Y}$ is called the *position vector of* **X** *with respect to* **Y**; in the special case $\mathbf{Y} = \mathbf{0}$ we say that **X** is the position vector of the point $(x_1, \ldots, x_n)$ (with respect to the origin).

We associate with each vector in R^2 the directed line segment from the origin to the point **X**. We assume that the student is familiar with the parallelogram law for vector addition in R^2, which says that if $\mathbf{X}_1$ and $\mathbf{X}_2$ are two vectors such that their associated line segments are nonparallel (thus $\mathbf{X}_1$ and $\mathbf{X}_2$ are linearly independent), then $\mathbf{X}_2 + \mathbf{X}_1$ is associated with one diagonal of the parallelogram in Fig. 5.3, and $\mathbf{X}_2 - \mathbf{X}_1$ is associated with a line segment through the origin and parallel to the other diagonal.

We cannot visualize R^n if $n > 3$. Nevertheless, important geometrical ideas such as distance, direction, line, and plane can be generalized to R^n.

In R^2 the distance from the origin to a point P with coordinates (x, y), is, according to the Pythagorean theorem,

$$|\overrightarrow{OP}| = \sqrt{x^2 + y^2}$$

(Fig. 5.4). This is also the *length of the vector* $\begin{bmatrix} x \\ y \end{bmatrix}$. We generalize this to define distance and length in R^n.

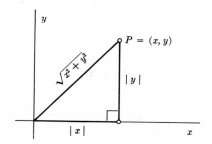

FIGURE 5.4

Inner Product

Definition 5.1 The *length*, or *magnitude*, of the vector

$$X = \begin{bmatrix} x_1 \\ \vdots \\ x_n \end{bmatrix}$$

and the *distance from the origin to the point* $X = (x_1, \ldots, x_n)$ are both defined by

$$|X| = \sqrt{x_1^2 + \cdots + x_n^2}.$$

If Y is a second point in R^n then the *distance between the points* X *and* Y is defined to be the length of the vector $X - Y$:

$$|X - Y| = \sqrt{(x_1 - y_1)^2 + \cdots + (x_n - y_n)^2}.$$

Example 5.1 If

$$X = \begin{bmatrix} 1 \\ 2 \\ -1 \\ 3 \end{bmatrix} \quad \text{and} \quad Y = \begin{bmatrix} 3 \\ 0 \\ 2 \\ -2 \end{bmatrix},$$

then

$$|X| = \sqrt{1^2 + 2^2 + (-1)^2 + 3^2} = \sqrt{15},$$

$$|Y| = \sqrt{3^2 + 0^2 + 2^2 + (-2)^2} = \sqrt{17},$$

and

$$|X - Y| = \sqrt{(1 - 3)^2 + (2 - 0)^2 + (-1 - 2)^2 + (3 + 2)^2} = \sqrt{42}.$$

The angle between two nonzero vectors

$$X_1 = \begin{bmatrix} x_1 \\ y_1 \end{bmatrix} \quad \text{and} \quad X_2 = \begin{bmatrix} x_2 \\ y_2 \end{bmatrix}$$

in R^2 is defined to be the angle between the directed line segments from the origin to the points X_1 and X_2 (Fig. 5.5). Applying the law of cosines to the

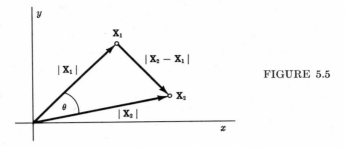

FIGURE 5.5

triangle in Fig. 5.5 yields

(1) $\qquad |\mathbf{X}_2 - \mathbf{X}_1|^2 = |\mathbf{X}_1|^2 + |\mathbf{X}_2|^2 - 2|\mathbf{X}_1||\mathbf{X}_2|\cos\theta.$

But

$$|\mathbf{X}_2 - \mathbf{X}_1|^2 = (x_2 - x_1)^2 + (y_2 - y_1)^2$$
$$= x_2^2 + y_2^2 + x_1^2 + y_1^2 - 2(x_1x_2 + y_1y_2)$$
$$= |\mathbf{X}_2|^2 + |\mathbf{X}_1|^2 - 2(x_1x_2 + y_1y_2).$$

Substituting this in (1) yields

(2) $$\cos\theta = \frac{x_1x_2 + y_1y_2}{|\mathbf{X}_1||\mathbf{X}_2|}.$$

To define the angle between two vectors in R^n, we need the notion of the inner product of two n-vectors.

Definition 5.2 The *inner product* of two n-vectors

$$\mathbf{X} = \begin{bmatrix} x_1 \\ \vdots \\ x_n \end{bmatrix} \quad \text{and} \quad \mathbf{Y} = \begin{bmatrix} y_1 \\ \vdots \\ y_n \end{bmatrix}$$

is defined by

$$\mathbf{X}\cdot\mathbf{Y} = x_1y_1 + \cdots + x_ny_n.$$

In R^2 and R^3 the inner product is also called the *scalar product* or *dot product*.

Example 5.2 For $\mathbf{X}$ and $\mathbf{Y}$ as defined in Example 5.1,

$$\mathbf{X}\cdot\mathbf{Y} = (1)(3) + (2)(0) + (-1)(2) + (3)(-2)$$
$$= -5.$$

The next theorem follows directly from the definition of inner product.

Theorem 5.1   If **X**, **Y**, and **Z** are n-vectors and c is a scalar, then

(a) $\quad\quad\quad \mathbf{X} \cdot \mathbf{Y} = \mathbf{Y} \cdot \mathbf{X};$

(b) $\quad \mathbf{X} \cdot (\mathbf{Y} + \mathbf{Z}) = \mathbf{X} \cdot \mathbf{Y} + \mathbf{X} \cdot \mathbf{Z};$

(c) $\quad\quad (c\mathbf{X}) \cdot \mathbf{Y} = \mathbf{X} \cdot (c\mathbf{Y}) = c(\mathbf{X} \cdot \mathbf{Y});$

(d) $\quad\quad\quad \mathbf{X} \cdot \mathbf{X} = |\,\mathbf{X}\,|^2 \geq 0$, with equality if and only if $\mathbf{X} = \mathbf{0}$.

We leave the proof to the student (Exercise T-1).

Schwarz's inequality, which we prove next, occurs in many branches of mathematics. It is also known as Cauchy's inequality, or as the Cauchy–Schwarz inequality.

Theorem 5.2 (Schwarz's Inequality)   If **X** and **Y** are n-vectors, then

(3) $$|\,\mathbf{X} \cdot \mathbf{Y}\,| \leq |\,\mathbf{X}\,|\,|\,\mathbf{Y}\,|,$$

with equality if and only if one of the vectors is a scalar multiple of the other.

(Note that on the left $|\,|$ stands for the absolute value of a scalar, while on the right it stands for the length of a vector. We shall use the same symbol for these two quantities because it is convenient to do so, and it is always possible from the context to decide which meaning is intended.)

Proof.   If $\mathbf{X} = \mathbf{0}$, then $\mathbf{X} \cdot \mathbf{Y} = |\,\mathbf{X}\,|\,|\,\mathbf{Y}\,| = 0$, and $\mathbf{X} = 0\mathbf{Y}$ and we are finished. Now assume **X** and **Y** are nonzero and let a be a real parameter. From the properties listed in Theorem 5.1,

(4) $$0 \leq |\,\mathbf{X} - a\mathbf{Y}\,|^2 = (\mathbf{X} - a\mathbf{Y}) \cdot (\mathbf{X} - a\mathbf{Y})$$

$$= |\,\mathbf{X}\,|^2 - 2a(\mathbf{X} \cdot \mathbf{Y}) + |\,\mathbf{Y}\,|^2 a^2.$$

For fixed **X** and **Y** the last expression in (4) is a quadratic polynomial $p(a)$ which is nonnegative for all real values of a, and therefore cannot have distinct real roots. (If it had distinct roots, it would have to assume negative values between them, as shown in Fig. 5.6.) Since the roots of $p(a)$

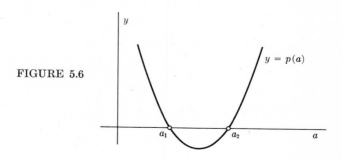

FIGURE 5.6

are, by the quadratic formula,

$$a = (\mathbf{X} \cdot \mathbf{Y}) \pm \sqrt{(\mathbf{X} \cdot \mathbf{Y})^2 - |\mathbf{X}|^2 |\mathbf{Y}|^2},$$

this means that

$$(\mathbf{X} \cdot \mathbf{Y})^2 - |\mathbf{X}|^2 |\mathbf{Y}|^2 \leq 0,$$

which implies (3). If equality occurs in (3) then $p(a)$ has the repeated real root $a_0 = (\mathbf{X} \cdot \mathbf{Y})$ and, from (4), $|\mathbf{X} - a_0 \mathbf{Y}| = 0$ and $\mathbf{X} = a_0 \mathbf{Y}$. We leave it to the student to show that if $\mathbf{X} = a\mathbf{Y}$ for some real a, then the equality occurs in (3).

Example 5.3   If

$$\mathbf{X} = \begin{bmatrix} 2 \\ -1 \\ 3 \\ 4 \end{bmatrix}, \qquad \mathbf{Y} = \begin{bmatrix} 1 \\ 3 \\ -1 \\ 0 \end{bmatrix}, \qquad \text{and} \qquad \mathbf{Z} = \begin{bmatrix} 4 \\ -2 \\ 6 \\ 8 \end{bmatrix},$$

then

$$|\mathbf{X}| = \sqrt{30}, \qquad |\mathbf{Y}| = \sqrt{11}, \qquad |\mathbf{Z}| = \sqrt{120},$$

and

$$\mathbf{X} \cdot \mathbf{Y} = -4, \qquad \mathbf{X} \cdot \mathbf{Z} = 60.$$

Thus

$$|\mathbf{X} \cdot \mathbf{Y}| = 4 < \sqrt{330} = |\mathbf{X}| |\mathbf{Y}|$$

and

$$|\mathbf{X} \cdot \mathbf{Z}| = |60| = \sqrt{3600} = |\mathbf{X}| |\mathbf{Z}|.$$

From the last equality we know that $\mathbf{Z}$ is a scalar multiple of $\mathbf{X}$; in fact $\mathbf{Z} = 2\mathbf{X}$.

In R^2 the points $\mathbf{X}$, $\mathbf{Y}$ and $\mathbf{X} + \mathbf{Y}$ are vertices of a parallelogram with sides $O\mathbf{X}$ and $O\mathbf{Y}$ as shown in Fig. 5.7; thus there is a triangle with sides of

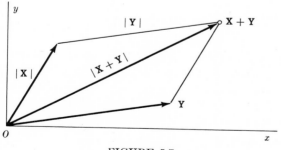

FIGURE 5.7

length $|\mathbf{X}|$, $|\mathbf{Y}|$, and $|\mathbf{X} + \mathbf{Y}|$. Since the length of a side of a triangle cannot exceed the sum of lengths of the other two, it follows that

$$|\mathbf{X} + \mathbf{Y}| \leq |\mathbf{X}| + |\mathbf{Y}|.$$

That this inequality holds in R^n for any n is shown in the next theorem.

Theorem 5.3 (Triangle Inequality.) If $\mathbf{X}$ and $\mathbf{Y}$ are n-vectors then

(5) $$|\mathbf{X} + \mathbf{Y}| \leq |\mathbf{X}| + |\mathbf{Y}|,$$

and equality holds if and only if $\mathbf{X} = \mathbf{0}$ or $\mathbf{Y} = a\mathbf{X}$ with $a \geq 0$.

Proof. From (4) with $a = -1$,

$$|\mathbf{X} + \mathbf{Y}|^2 = |\mathbf{X}|^2 + 2(\mathbf{X} \cdot \mathbf{Y}) + |\mathbf{Y}|^2$$

and Schwarz's inequality implies that

$$|\mathbf{X} + \mathbf{Y}|^2 \leq |\mathbf{X}|^2 + 2|\mathbf{X}||\mathbf{Y}| + |\mathbf{Y}|^2$$
$$= (|\mathbf{X}| + |\mathbf{Y}|)^2,$$

from which (5) can be obtained by taking square roots. We leave the proof of the assertion about equality to the student.

Example 5.4 Let $\mathbf{X}$, $\mathbf{Y}$, and $\mathbf{Z}$ be as defined in Example 5.3. Then

$$|\mathbf{X} + \mathbf{Y}| = \sqrt{33} < \sqrt{30} + \sqrt{11} = |\mathbf{X}| + |\mathbf{Y}|$$

and

$$|\mathbf{X} + \mathbf{Z}| = \sqrt{270} = \sqrt{30} + \sqrt{120} = |\mathbf{X}| + |\mathbf{Z}|.$$

(Note that $\mathbf{Z} = 2\mathbf{X}$.)

Definition 5.3 The *cosine of the angle* between two nonzero n-vectors $\mathbf{X}$ and $\mathbf{Y}$ is defined by

(6) $$\cos \theta = \frac{\mathbf{X} \cdot \mathbf{Y}}{|\mathbf{X}||\mathbf{Y}|}.$$

This definition generalizes (2), to which it reduces when $n = 2$; it is also equivalent to the usual geometric definition for $n = 3$. If we were not sure that the magnitude of the right side of (6) is bounded by unity for any n-vectors $\mathbf{X}$ and $\mathbf{Y}$, then we would not have the right to write it as $\cos \theta$; however, Schwarz's inequality resolves any doubts on that score.

In R^2 two vectors $\mathbf{X}$ and $\mathbf{Y}$ are parallel if and only if $|\cos \theta| = 1$; they are perpendicular, or orthogonal, if and only if $\cos \theta = 0$. This geometric language is useful in R^n.

Definition 5.4 The n-vectors **X** and **Y** said to be *parallel* if $|\mathbf{X} \cdot \mathbf{Y}| = |\mathbf{X}||\mathbf{Y}|$, and *in the same direction* if $\mathbf{X} \cdot \mathbf{Y} = |\mathbf{X}||\mathbf{Y}|$. They are said to be *orthogonal* if $\mathbf{X} \cdot \mathbf{Y} = 0$.

From Theorem 5.2, two n-vectors are parallel if and only if one is a scalar multiple of the other, and in the same direction if and only if the scalar is nonnegative. The zero vector is both parallel and perpendicular to every vector.

Example 5.5 If

$$
\mathbf{X} = \begin{bmatrix} 1 \\ 0 \\ 1 \\ 0 \end{bmatrix} \quad \text{and} \quad \mathbf{Y} = \begin{bmatrix} 1 \\ 0 \\ 0 \\ 1 \end{bmatrix},
$$

then

$$
|\mathbf{X}| = \sqrt{2}, \qquad |\mathbf{Y}| = \sqrt{2}, \qquad \mathbf{X} \cdot \mathbf{Y} = 1,
$$

and

$$
\cos \theta = \tfrac{1}{2}.
$$

The vector

$$
\mathbf{Z} = \begin{bmatrix} 0 \\ 1 \\ 0 \\ 0 \end{bmatrix}
$$

is orthogonal to **X** and **Y**, since

$$
\mathbf{X} \cdot \mathbf{Z} = \mathbf{Y} \cdot \mathbf{Z} = 0,
$$

and

$$
\mathbf{W} = \begin{bmatrix} 2 \\ 0 \\ 2 \\ 0 \end{bmatrix}
$$

is in the same direction as **X**, since

$$\mathbf{W} \cdot \mathbf{X} = 4 = \sqrt{2} \sqrt{8} = |\mathbf{W}| \, |\mathbf{X}|;$$

of course, $\mathbf{W} = 2\mathbf{X}$.

Example 5.6 A *unit vector* **U** is a vector of unit length:

$$|\mathbf{U}| = 1.$$

If **X** is an arbitrary nonzero vector then

$$\mathbf{U} = \left[\frac{1}{|\mathbf{X}|}\right]\mathbf{X} = \frac{\mathbf{X}}{|\mathbf{X}|}$$

is the unique unit vector in the direction of **X**. For example, if

$$\mathbf{X} = \begin{bmatrix} 1 \\ -1 \\ 2 \\ 3 \end{bmatrix}, \quad \text{then} \quad \mathbf{U} = \frac{1}{\sqrt{15}}\begin{bmatrix} 1 \\ -1 \\ 2 \\ 3 \end{bmatrix}$$

is the unit vector in the direction of **X**. We say that **U** is obtained by *normalizing* **X**.

Orthonormal Bases in R^n

Definition 5.5 A set $V = \{\mathbf{X}_1, \ldots, \mathbf{X}_k\}$ of n-vectors in R^n is said to be *orthogonal* if any two distinct vectors in V are orthogonal; that is, if $\mathbf{X}_i \cdot \mathbf{X}_j = 0$ for $i \neq j$. An orthogonal set of unit vectors is said to be *orthonormal*.

Example 5.7 The set

$$V = \left\{ \begin{bmatrix} 1 \\ 0 \\ 1 \end{bmatrix}, \begin{bmatrix} 0 \\ 1 \\ 0 \end{bmatrix}, \begin{bmatrix} -1 \\ 0 \\ 1 \end{bmatrix} \right\}$$

is orthogonal in R^3. Normalizing the vectors in V yields the orthonormal set

$$U = \left\{ \begin{bmatrix} \dfrac{1}{\sqrt{2}} \\ 0 \\ \dfrac{1}{\sqrt{2}} \end{bmatrix}, \begin{bmatrix} 0 \\ 1 \\ 0 \end{bmatrix}, \begin{bmatrix} -\dfrac{1}{\sqrt{2}} \\ 0 \\ \dfrac{1}{\sqrt{2}} \end{bmatrix} \right\}.$$

Example 5.8 In R^3, the natural basis

$$B = \left\{ \begin{bmatrix} 1 \\ 0 \\ 0 \end{bmatrix}, \begin{bmatrix} 0 \\ 1 \\ 0 \end{bmatrix}, \begin{bmatrix} 0 \\ 0 \\ 1 \end{bmatrix} \right\}$$

is an orthonormal set. More generally, $\mathbf{E}_1, \ldots, \mathbf{E}_n$ is an orthonormal set in R^n.

Theorem 5.4 An orthogonal set $V = \{\mathbf{X}_1, \ldots, \mathbf{X}_k\}$ of nonzero vectors in R^n is linearly independent.

Proof. Suppose

(7) $$a_1\mathbf{X}_1 + a_2\mathbf{X}_2 + \cdots + a_k\mathbf{X}_k = \mathbf{0}.$$

Taking inner product on both sides with $\mathbf{X}_i$ yields

$$(a_1\mathbf{X}_1 + a_2\mathbf{X}_2 + \cdots + a_n\mathbf{X}_n) \cdot \mathbf{X}_i = \mathbf{0} \cdot \mathbf{X}_i = 0,$$

which, from (b) and (c) of Theorem 5.1, can be rewritten as

(8) $$a_1(\mathbf{X}_1 \cdot \mathbf{X}_i) + \cdots + a_i(\mathbf{X}_i \cdot \mathbf{X}_i) + \cdots + a_n(\mathbf{X}_n \cdot \mathbf{X}_i) = 0.$$

Since $\mathbf{X}_i \cdot \mathbf{X}_j = 0$ if $i \neq j$, (8) reduces to

$$0 = a_i(\mathbf{X}_i \cdot \mathbf{X}_i) = a_i \, |\, \mathbf{X}_i\, |^2,$$

which, from (d) of Theorem 5.1, implies that $a_i = 0$, because $|\, \mathbf{X}_i\, | \neq 0$. Since this argument holds for $i = 1, \ldots, n$, we have inferred from (7) that $a_1 = a_2 = \cdots = a_n = 0$. Consequently, V is a linearly independent set.

Corollary 5.1 An orthonormal set of vectors in R^n is linearly independent. Consequently, if H is an m-dimensional subspace of R^n, then no orthonormal set of vectors in H can contain more than m elements, and any orthonormal set in H which contains m elements is a basis for H.

We leave the proof of this corollary to the student (Exercise T-2).

Theorem 5.5 Let $\{\mathbf{X}_1, \ldots, \mathbf{X}_k\}$ be a linearly independent set of vectors in R^n, and let $H = S[\mathbf{X}_1, \ldots, \mathbf{X}_k]$. Then H has an orthonormal basis $B = \{\mathbf{Y}_1, \ldots, \mathbf{Y}_k\}$.

Proof. We first normalize $\mathbf{X}_1$, which is nonzero, by hypothesis; thus

$$\mathbf{Y}_1 = \frac{\mathbf{X}_1}{|\mathbf{X}_1|}$$

is a unit vector which spans $S[\mathbf{X}_1]$. If $k = 1$, we are finished; if $k > 1$, define

$$(9) \qquad\qquad \mathbf{Z}_2 = \mathbf{X}_2 - (\mathbf{X}_2 \cdot \mathbf{Y}_1)\mathbf{Y}_1;$$

then

$$\mathbf{Z}_2 \cdot \mathbf{Y}_1 = \mathbf{X}_2 \cdot \mathbf{Y}_1 - (\mathbf{X}_2 \cdot \mathbf{Y}_1)(\mathbf{Y}_1 \cdot \mathbf{Y}_1)$$

$$= \mathbf{X}_2 \cdot \mathbf{Y}_1 - \mathbf{X}_2 \cdot \mathbf{Y}_1 = 0,$$

which means that $\mathbf{Z}_2$ is orthogonal to $\mathbf{Y}_1$. Moreover, $\mathbf{Z}_2$ is a nontrivial linear combination of $\mathbf{X}_1$ and $\mathbf{X}_2$, and consequently nonzero. Therefore, $\mathbf{Y}_1$ and

$$\mathbf{Y}_2 = \frac{\mathbf{Z}_2}{|\mathbf{Z}_2|}$$

form an orthonormal set in $S[\mathbf{X}_1, \mathbf{X}_2]$.

If $k = 2$, we are finished. If $k > 2$, suppose $2 \leq m \leq k - 1$ and that we have found orthonormal vectors $\mathbf{Y}_1, \ldots, \mathbf{Y}_m$ in $S[\mathbf{X}_1, \ldots, \mathbf{X}_m]$. Define

$$(10) \qquad \mathbf{Z}_{m+1} = \mathbf{X}_{m+1} - (\mathbf{X}_{m+1} \cdot \mathbf{Y}_1)\mathbf{Y}_1 - \cdots - (\mathbf{X}_{m+1} \cdot \mathbf{Y}_m)\mathbf{Y}_m.$$

Then

$$\mathbf{Z}_{m+1} \cdot \mathbf{Y}_i = (\mathbf{X}_{m+1} \cdot \mathbf{Y}_i) - (\mathbf{X}_{m+1} \cdot \mathbf{Y}_i)(\mathbf{Y}_i \cdot \mathbf{Y}_i) = 0,$$

since $\mathbf{Y}_i \cdot \mathbf{Y}_j = 0$ if $i \neq j$ and $\mathbf{Y}_i \cdot \mathbf{Y}_j = 1$. Consequently, $\mathbf{Z}_{m+1}$ is orthogonal to $\mathbf{Y}_1, \ldots, \mathbf{Y}_m$. Being a nontrivial linear combination of $\mathbf{X}_1, \ldots, \mathbf{X}_{m+1}$ (remember that $\mathbf{Y}_1, \ldots, \mathbf{Y}_m$ are in $S[\mathbf{X}_1, \ldots, \mathbf{X}_m]$), it is also nonzero; hence we can define the unit vector

$$\mathbf{Y}_{m+1} = \frac{\mathbf{Z}_{m+1}}{|\mathbf{Z}_{m+1}|}.$$

Now $\{\mathbf{Y}_1, \ldots, \mathbf{Y}_{m+1}\}$ is an orthonormal set in $S[\mathbf{X}_1, \ldots, \mathbf{X}_{m+1}]$. Continuing in this way leads to an orthonormal set $\mathbf{Y}_1, \ldots, \mathbf{Y}_k$ in $S[\mathbf{X}_1, \ldots, \mathbf{X}_k]$. From Corollary 5.1, the vectors $\mathbf{Y}_1, \ldots, \mathbf{Y}_k$ form a basis for $S[\mathbf{X}_1, \ldots, \mathbf{X}_k]$. This completes the proof of the theorem.

The method used in this proof to obtain $\{\mathbf{Y}_1, \ldots, \mathbf{Y}_k\}$ from $\{\mathbf{X}_1, \ldots, \mathbf{X}_k\}$ is known as the *Gram–Schmidt process*.

Corollary 5.2 Every subspace of R^n has an orthonormal basis.

Example 5.9 Consider the subspace of R^4 spanned by

$$\mathbf{X}_1 = \begin{bmatrix} 1 \\ 0 \\ 1 \\ 0 \end{bmatrix}, \qquad \mathbf{X}_2 = \begin{bmatrix} 0 \\ 1 \\ 1 \\ 0 \end{bmatrix}, \qquad \text{and} \qquad \mathbf{X}_3 = \begin{bmatrix} 0 \\ 0 \\ -1 \\ 0 \end{bmatrix}.$$

Normalizing $\mathbf{X}_1$ yields

$$\mathbf{Y}_1 = \frac{1}{\sqrt{2}} \begin{bmatrix} 1 \\ 0 \\ 1 \\ 0 \end{bmatrix},$$

a unit vector that spans $S[\mathbf{X}_1]$. Since $\mathbf{X}_2 \cdot \mathbf{Y}_1 = \dfrac{1}{\sqrt{2}}$, it follows from (9) that

$$\mathbf{Z}_2 = \mathbf{X}_2 - \frac{1}{\sqrt{2}} \mathbf{Y}_1 = \begin{bmatrix} -\frac{1}{2} \\ 1 \\ \frac{1}{2} \\ 0 \end{bmatrix}$$

is orthogonal to $\mathbf{Y}_1$. (Verify.) Normalizing $\mathbf{Z}_2$ yields

$$\mathbf{Y}_2 = \frac{\mathbf{Z}_2}{|\mathbf{Z}_2|} = \frac{1}{\sqrt{6}} \begin{bmatrix} -1 \\ 2 \\ 1 \\ 0 \end{bmatrix}.$$

Thus, $\mathbf{Y}_1$ and $\mathbf{Y}_2$ are orthonormal vectors in $S[\mathbf{X}_1, \mathbf{X}_2]$. Since $\mathbf{X}_3 \cdot \mathbf{Y}_1 = -\dfrac{1}{\sqrt{2}}$

and $\mathbf{X}_3 \cdot \mathbf{Y}_2 = -\dfrac{1}{\sqrt{6}}$, it follows from (10), with $m = 2$, that

$$\mathbf{Z}_3 = \mathbf{X}_3 + \frac{1}{\sqrt{2}}\,\mathbf{Y}_1 + \frac{1}{\sqrt{6}}\,\mathbf{Y}_2 = \begin{bmatrix} \frac{1}{3} \\ \frac{1}{3} \\ -\frac{1}{3} \\ 0 \end{bmatrix}$$

is orthogonal to $\mathbf{Y}_1$ and $\mathbf{Y}_2$ (verify), and normalizing $\mathbf{Z}_3$ yields

$$\mathbf{Y}_3 = \frac{1}{\sqrt{3}} \begin{bmatrix} 1 \\ 1 \\ -1 \\ 0 \end{bmatrix}.$$

Thus, $\{\mathbf{Y}_1, \mathbf{Y}_2, \mathbf{Y}_3\}$ is an orthonormal basis for $S[\mathbf{X}_1, \mathbf{X}_2, \mathbf{X}_3]$. The student should verify this by expressing $\mathbf{X}_1, \mathbf{X}_2, \mathbf{X}_3$ in terms of $\{\mathbf{Y}_1, \mathbf{Y}_2, \mathbf{Y}_3\}$.

If $B = \{\mathbf{X}_1, \ldots, \mathbf{X}_n\}$ is a basis for R^n and $\mathbf{X}$ is an arbitrary vector, then the coefficients $a_1, \ldots, a_n$ in

$$\mathbf{X} = a_1\mathbf{X}_1 + a_2\mathbf{X}_2 + \cdots + a_n\mathbf{X}_n$$

are obtained by solving a system of n equations in n unknowns. (See the proof of Theorem 2.7.) However, this computation is considerably simplified if B is an orthonormal basis.

Theorem 5.6 Let $B = \{\mathbf{Y}_1, \ldots, \mathbf{Y}_n\}$ be an orthonormal basis for R^n, and let $\mathbf{X}$ be an arbitrary n-vector. Then

$$\mathbf{X} = a_1\mathbf{Y}_1 + a_2\mathbf{Y}_2 + \cdots + a_n\mathbf{Y}_n,$$

where

$$a_i = (\mathbf{X} \cdot \mathbf{Y}_i) \qquad (1 \le i \le n).$$

We leave the proof to the student (Exercise T-3).

EXERCISES 2.5

1. Find the lengths of the following vectors.

(a) $\begin{bmatrix} 1 \\ 2 \\ 3 \\ 4 \end{bmatrix}$;
(b) $\begin{bmatrix} 0 \\ 0 \\ 0 \end{bmatrix}$;
(c) $\begin{bmatrix} 1 \\ 0 \\ 1 \end{bmatrix}$;
(d) $\begin{bmatrix} 1 \\ -2 \\ -3 \\ -3 \end{bmatrix}$.

2. Find the lengths of the following vectors.

(a) $\begin{bmatrix} 1 \\ -1 \\ 2 \\ -3 \end{bmatrix}$;
(b) $\begin{bmatrix} 0 \\ 1 \\ -2 \end{bmatrix}$;
(c) $\begin{bmatrix} 1 \\ 1 \\ 3 \\ -2 \\ 4 \end{bmatrix}$;
(d) $\begin{bmatrix} 0 \\ 0 \\ 2 \\ -1 \\ -2 \end{bmatrix}$.

3. Find the distance between the following pairs of points.
    (a) $\mathbf{X} = (1, 2, 3, -1)$, $\mathbf{Y} = (-2, -3, 4, 2)$;
    (b) $\mathbf{X} = (1, 2, 3)$, $\mathbf{Y} = (3, 2, -5)$;
    (c) $\mathbf{X} = (-1, -2, -3, 4)$, $\mathbf{Y} = (4, 5, 6, 3)$.

4. Find the distance between the following pairs of points.
    (a) $\mathbf{X} = (1, 2, -1, 2)$, $\mathbf{Y} = (2, -1, 4, 3)$;
    (b) $\mathbf{X} = (-1, 4, 5, 0)$, $\mathbf{Y} = (0, 2, 3, 4)$;
    (c) $\mathbf{X} = (1, 3, 2)$, $\mathbf{Y} = (4, -1, 3)$.

5. Find $\mathbf{X} \cdot \mathbf{Y}$:

(a) $\mathbf{X} = \begin{bmatrix} 1 \\ 2 \\ -3 \\ -2 \end{bmatrix}$, $\mathbf{Y} = \begin{bmatrix} -2 \\ -3 \\ 4 \\ -2 \end{bmatrix}$;
(b) $\mathbf{X} = \begin{bmatrix} 0 \\ 1 \\ 0 \end{bmatrix}$, $\mathbf{Y} = \begin{bmatrix} 1 \\ 0 \\ 0 \end{bmatrix}$;

(c) $\mathbf{X} = \begin{bmatrix} 2 \\ 2 \\ -3 \\ 1 \end{bmatrix}$, $\mathbf{Y} = \begin{bmatrix} 3 \\ -4 \\ 2 \\ -2 \end{bmatrix}$; (d) $\mathbf{X} = \begin{bmatrix} 1 \\ 1 \\ 1 \\ 3 \end{bmatrix}$, $\mathbf{Y} = \begin{bmatrix} 2 \\ -1 \\ 3 \\ 4 \end{bmatrix}$.

6. Find $\mathbf{X} \cdot \mathbf{Y}$:

(a) $\mathbf{X} = \begin{bmatrix} -1 \\ 2 \\ 3 \\ -2 \end{bmatrix}$, $\mathbf{Y} = \begin{bmatrix} 1 \\ 1 \\ 0 \\ 2 \end{bmatrix}$; (b) $\mathbf{X} = \begin{bmatrix} 1 \\ 0 \\ 2 \end{bmatrix}$, $\mathbf{Y} = \begin{bmatrix} 2 \\ -1 \\ 2 \end{bmatrix}$;

(c) $\mathbf{X} = \begin{bmatrix} 2 \\ 3 \\ 4 \\ -1 \end{bmatrix}$, $\mathbf{Y} = \begin{bmatrix} -3 \\ -2 \\ 1 \\ 0 \end{bmatrix}$; (d) $\mathbf{X} = \begin{bmatrix} 0 \\ 0 \\ 1 \\ 2 \end{bmatrix}$, $\mathbf{Y} = \begin{bmatrix} 1 \\ 2 \\ -1 \\ 3 \end{bmatrix}$.

7. Find $|\mathbf{X} - \mathbf{Y}|$ for each pair of vectors in Exercise 5.

8. Find $|\mathbf{X} - \mathbf{Y}|$ for each pair of vectors in Exercise 6.

9. Find the cosine of the angle between $\mathbf{X}$ and $\mathbf{Y}$ for each pair of vectors in Exercise 5.

10. Find the cosine of the angle between $\mathbf{X}$ and $\mathbf{Y}$ for each pair of vectors in Exercise 6.

11. Verify Schwarz's inequality for

(a) $\mathbf{X} = \begin{bmatrix} 1 \\ 2 \\ 3 \end{bmatrix}$, $\mathbf{Y} = \begin{bmatrix} -1 \\ 2 \\ -1 \end{bmatrix}$; (b) $\mathbf{X} = \begin{bmatrix} 1 \\ -1 \\ 2 \end{bmatrix}$, $\mathbf{Y} = \begin{bmatrix} 3 \\ -3 \\ 6 \end{bmatrix}$.

12.   Verify the triangle inequality for

(a) $\mathbf{X} = \begin{bmatrix} 1 \\ 2 \\ 3 \end{bmatrix}$, $\mathbf{Y} = \begin{bmatrix} -1 \\ 2 \\ -1 \end{bmatrix}$;   (b) $\mathbf{X} = \begin{bmatrix} 1 \\ 3 \\ 1 \end{bmatrix}$, $\mathbf{Y} = \begin{bmatrix} -1 \\ 0 \\ 2 \end{bmatrix}$.

13.   Find a unit vector in the direction of $\mathbf{X}$:

(a) $\mathbf{X} = \begin{bmatrix} 1 \\ 2 \end{bmatrix}$;   (b) $\mathbf{X} = \begin{bmatrix} 2 \\ -3 \\ 4 \end{bmatrix}$;

(c) $\mathbf{X} = \begin{bmatrix} 1 \\ 2 \\ 1 \\ -1 \end{bmatrix}$;   (d) $\mathbf{X} = \begin{bmatrix} 0 \\ 0 \\ 1 \\ -5 \end{bmatrix}$.

14.   Find a unit vector in the direction of $\mathbf{X}$:

(a) $\mathbf{X} = \begin{bmatrix} -2 \\ 3 \end{bmatrix}$;   (b) $\mathbf{X} = \begin{bmatrix} 1 \\ -2 \\ 3 \\ 4 \end{bmatrix}$;

(c) $\mathbf{X} = \begin{bmatrix} 2 \\ -3 \\ 4 \\ 1 \end{bmatrix}$;   (d) $\mathbf{X} = \begin{bmatrix} 1 \\ 2 \\ -1 \end{bmatrix}$.

15. State which of the following are orthogonal sets.

(a) $\begin{bmatrix} 1 \\ 1 \\ 2 \\ 0 \end{bmatrix}$, $\begin{bmatrix} -1 \\ -1 \\ -1 \\ 1 \end{bmatrix}$, $\begin{bmatrix} -1 \\ 1 \\ 0 \\ 0 \end{bmatrix}$;   (b) $\begin{bmatrix} 1 \\ 1 \\ 0 \\ 0 \end{bmatrix}$, $\begin{bmatrix} 0 \\ 0 \\ 1 \\ 1 \end{bmatrix}$, $\begin{bmatrix} 0 \\ 1 \\ 0 \\ 0 \end{bmatrix}$;

(c) $\begin{bmatrix} 1 \\ 0 \\ 0 \\ 0 \end{bmatrix}$, $\begin{bmatrix} 0 \\ 1 \\ -1 \\ 1 \end{bmatrix}$, $\begin{bmatrix} 0 \\ -1 \\ -1 \\ 0 \end{bmatrix}$.

16. State which of the following are orthonormal sets.

(a) $\begin{bmatrix} \dfrac{1}{\sqrt{3}} \\[6pt] \dfrac{1}{\sqrt{3}} \\[6pt] \dfrac{1}{\sqrt{3}} \end{bmatrix}$, $\begin{bmatrix} 0 \\[6pt] \dfrac{1}{\sqrt{2}} \\[6pt] -\dfrac{1}{\sqrt{2}} \end{bmatrix}$, $\begin{bmatrix} 0 \\ 1 \\ 0 \end{bmatrix}$;   (b) $\begin{bmatrix} \dfrac{1}{\sqrt{6}} \\[6pt] \dfrac{1}{\sqrt{6}} \\[6pt] \dfrac{2}{\sqrt{6}} \end{bmatrix}$, $\begin{bmatrix} -\dfrac{1}{\sqrt{3}} \\[6pt] -\dfrac{1}{\sqrt{3}} \\[6pt] \dfrac{1}{\sqrt{2}} \end{bmatrix}$, $\begin{bmatrix} -\dfrac{1}{\sqrt{2}} \\[6pt] \dfrac{1}{\sqrt{2}} \\[6pt] 0 \end{bmatrix}$;

(c) $\begin{bmatrix} 2 \\ -2 \\ 1 \\ 1 \end{bmatrix}$, $\begin{bmatrix} 2 \\ 1 \\ -2 \\ 0 \end{bmatrix}$, $\begin{bmatrix} 1 \\ 2 \\ 2 \\ 0 \end{bmatrix}$.

17. Find an orthonormal basis for the subspace of R^3 spanned by

$\begin{bmatrix} 1 \\ 0 \\ 1 \end{bmatrix}$ and $\begin{bmatrix} 0 \\ 1 \\ 1 \end{bmatrix}$.

18. Find an orthonormal basis for the subspace of R^4 spanned by

$$\begin{bmatrix} 0 \\ 0 \\ 1 \\ 0 \end{bmatrix}, \quad \begin{bmatrix} 1 \\ 1 \\ 1 \\ 0 \end{bmatrix}, \quad \text{and} \quad \begin{bmatrix} 1 \\ 0 \\ 1 \\ 1 \end{bmatrix}.$$

19. Find an orthonormal basis for the subspace of R^4 spanned by

$$\begin{bmatrix} 1 \\ 1 \\ 0 \\ 1 \end{bmatrix}, \quad \begin{bmatrix} 2 \\ 0 \\ 1 \\ 2 \end{bmatrix}, \quad \text{and} \quad \begin{bmatrix} 1 \\ -1 \\ 1 \\ -1 \end{bmatrix}.$$

20. Find an orthonormal basis for the subspace of R^4 spanned by

$$\begin{bmatrix} 1 \\ 1 \\ -1 \\ 0 \end{bmatrix}, \quad \begin{bmatrix} 2 \\ 0 \\ 1 \\ 0 \end{bmatrix}, \quad \text{and} \quad \begin{bmatrix} 1 \\ 0 \\ 0 \\ 0 \end{bmatrix}.$$

21. Express the following vectors as linear combinations of the orthonormal set

$$\mathbf{X}_1 = \begin{bmatrix} \dfrac{1}{\sqrt{6}} \\[2mm] \dfrac{1}{\sqrt{6}} \\[2mm] \dfrac{2}{\sqrt{6}} \end{bmatrix}, \quad \mathbf{X}_2 = \begin{bmatrix} \dfrac{1}{\sqrt{2}} \\[2mm] -\dfrac{1}{\sqrt{2}} \\[2mm] 0 \end{bmatrix}, \quad \mathbf{X}_3 = \begin{bmatrix} \dfrac{1}{\sqrt{3}} \\[2mm] \dfrac{1}{\sqrt{3}} \\[2mm] -\dfrac{1}{\sqrt{3}} \end{bmatrix}.$$

(a) $\begin{bmatrix} 1 \\ 2 \\ 3 \end{bmatrix}$;   (b) $\begin{bmatrix} 2 \\ -1 \\ 2 \end{bmatrix}$;   (c) $\begin{bmatrix} 1 \\ -1 \\ 2 \end{bmatrix}$;   (d) $\begin{bmatrix} 1 \\ 1 \\ 2 \end{bmatrix}$.

22. Express the following vectors as linear combinations of the orthonormal set

$$\mathbf{X}_1 = \begin{bmatrix} \dfrac{1}{\sqrt{3}} \\[6pt] \dfrac{1}{\sqrt{2}} \\[6pt] 0 \\[6pt] \dfrac{1}{\sqrt{3}} \end{bmatrix}, \quad \mathbf{X}_2 = \begin{bmatrix} \dfrac{1}{\sqrt{2}} \\[6pt] -\dfrac{1}{\sqrt{2}} \\[6pt] 0 \\[6pt] 0 \end{bmatrix}, \quad \mathbf{X}_3 = \begin{bmatrix} \dfrac{1}{\sqrt{6}} \\[6pt] \dfrac{1}{\sqrt{6}} \\[6pt] 0 \\[6pt] -\dfrac{2}{\sqrt{6}} \end{bmatrix}, \quad \mathbf{X}_4 = \begin{bmatrix} 0 \\[6pt] 0 \\[6pt] 1 \\[6pt] 0 \end{bmatrix}.$$

(a) $\begin{bmatrix} 1 \\ -1 \\ 0 \\ 1 \end{bmatrix}$;   (b) $\begin{bmatrix} 3 \\ 1 \\ 0 \\ -1 \end{bmatrix}$;   (c) $\begin{bmatrix} 0 \\ 0 \\ 0 \\ 1 \end{bmatrix}$;   (d) $\begin{bmatrix} 3 \\ 2 \\ -1 \\ 1 \end{bmatrix}$.

THEORETICAL EXERCISES

T-1.   Prove Theorem 5.1.

T-2.   Prove Corollary 5.1.

T-3.   Prove Theorem 5.6.

T-4.   *Prove:* If $\mathbf{Y}$ is orthogonal to $\mathbf{X}_1, \ldots, \mathbf{X}_m$, then $\mathbf{Y}$ is orthogonal to $S[\mathbf{X}_1, \ldots, \mathbf{X}_m]$.

T-5.   Let $\mathbf{X}_0$ be a fixed vector in R^n. Prove that the set of vectors orthogonal to $\mathbf{X}_0$ is a subspace of R^n.

T-6.   (a)   *Prove:* If $\mathbf{X} \cdot \mathbf{Y} = 0$ for all $\mathbf{X}$ in R^n, then $\mathbf{Y} = \mathbf{0}$.
(b)   *Prove:* If $\mathbf{X} \cdot \mathbf{Y} = \mathbf{X} \cdot \mathbf{Z}$ for all $\mathbf{X}$ in R^n, then $\mathbf{Y} = \mathbf{Z}$.

T-7. Show that

$$|\mathbf{X}_1 + \cdots + \mathbf{X}_n| \leq |\mathbf{X}_1| + \cdots + |\mathbf{X}_n|$$

and give conditions for equality.

T-8. Show that $|\mathbf{X} + \mathbf{Y}|^2 = |\mathbf{X}|^2 + |\mathbf{Y}|^2$ if and only if $\mathbf{X} \cdot \mathbf{Y} = 0$.

T-9. Show that

(a) $(\mathbf{X} + \alpha\mathbf{Y}) \cdot \mathbf{Z} = \mathbf{X} \cdot \mathbf{Z} + \alpha\mathbf{Y} \cdot \mathbf{Z}$.
(b) $\mathbf{X} \cdot (\alpha\mathbf{Y}) = \alpha\mathbf{X} \cdot \mathbf{Y}$.
(c) $(\mathbf{X} + \mathbf{Y}) \cdot \alpha\mathbf{Z} = \alpha\mathbf{X} \cdot \mathbf{Z} + \alpha\mathbf{Y} \cdot \mathbf{Z}$.

T-10. Show that $\mathbf{X} \cdot \mathbf{Y} = \mathbf{X}^\mathsf{T}\mathbf{Y}$.

T-11. Define the distance $d(\mathbf{U}, \mathbf{V})$ between n-vectors $\mathbf{U}$ and $\mathbf{V}$ as $d(\mathbf{U}, \mathbf{V}) = |\mathbf{U} - \mathbf{V}|$. Show that

(a) $d(\mathbf{U}, \mathbf{V}) \geq 0$.
(b) $d(\mathbf{U}, \mathbf{V}) = 0$ if and only if $\mathbf{U} = \mathbf{V}$.
(c) $d(\mathbf{U}, \mathbf{V}) = d(\mathbf{V}, \mathbf{U})$.
(d) $d(\mathbf{U}, \mathbf{V}) \leq d(\mathbf{U}, \mathbf{W}) + d(\mathbf{W}, \mathbf{V})$, for any $\mathbf{W}$.

2.6 Eigenvalues and Eigenvectors

In this section, all matrices (other than n-vectors) are square. If $\mathbf{A}$ is an $n \times n$ matrix, then

$$\mathbf{Y} = \mathbf{AX}$$

defines a linear transformation from R^n into itself. In many applications, it is useful to know which n-vectors $\mathbf{X}$, if any, are such that $\mathbf{X}$ and $\mathbf{AX}$ are parallel.

Definition 6.1 A real number λ is called an *eigenvalue* of the $n \times n$ matrix $\mathbf{A}$ if there is a nonzero n-vector $\mathbf{X}$ such that

$$(1) \qquad\qquad \mathbf{AX} = \lambda\mathbf{X}.$$

An n-vector with this property is said to be an *eigenvector of $\mathbf{A}$ associated with λ.*

Eigenvalues are also called *proper values*, or *characteristic values* of $\mathbf{A}$. Correspondingly, eigenvectors are also called *proper vectors*, or *characteristic vectors*.

Example 6.1 The identity matrix has exactly one eigenvalue, $\lambda = 1$, and every nonzero n-vector $\mathbf{X}$ is an associated eigenvector, since

(2) $$\mathbf{IX} = (1)\mathbf{X}.$$

Notice that $\mathbf{X} = \mathbf{0}$ is *not* an eigenvector of $\mathbf{I}$, even though it satisfies (2), since eigenvectors must be nonzero, by definition.

Example 6.2 If

$$\mathbf{A} = \begin{bmatrix} 0 & 1 \\ 1 & 0 \end{bmatrix},$$

then

$$\mathbf{A}\begin{bmatrix} 1 \\ 1 \end{bmatrix} = \begin{bmatrix} 0 & 1 \\ 1 & 0 \end{bmatrix}\begin{bmatrix} 1 \\ 1 \end{bmatrix} = \begin{bmatrix} 1 \\ 1 \end{bmatrix} = (1)\begin{bmatrix} 1 \\ 1 \end{bmatrix}$$

and

$$\mathbf{A}\begin{bmatrix} 1 \\ -1 \end{bmatrix} = \begin{bmatrix} 0 & 1 \\ 1 & 0 \end{bmatrix}\begin{bmatrix} 1 \\ -1 \end{bmatrix} = \begin{bmatrix} -1 \\ 1 \end{bmatrix} = (-1)\begin{bmatrix} 1 \\ -1 \end{bmatrix};$$

hence $\lambda_1 = 1$ and $\lambda_2 = -1$ are eigenvalues of $\mathbf{A}$ with corresponding eigenvectors

$$\mathbf{X}_1 = \begin{bmatrix} 1 \\ 1 \end{bmatrix} \quad \text{and} \quad \mathbf{X}_2 = \begin{bmatrix} 1 \\ -1 \end{bmatrix}.$$

Since (1) can be rewritten as

(3) $$(\mathbf{A} - \lambda\mathbf{I})\mathbf{X} = \mathbf{0},$$

it follows that the eigenvectors of $\mathbf{A}$ corresponding to an eigenvalue λ, along with the zero vector, form a subspace of R^n (Exercise T-1). We call this subspace the *eigenspace of $\mathbf{A}$ corresponding to* λ.

Definition 6.2 The *characteristic polynomial* of an $n \times n$ matrix $\mathbf{A} = [a_{ij}]$ is the nth degree polynomial

(4) $$p(\lambda) = \det(\mathbf{A} - \lambda\mathbf{I})$$

$$= \begin{vmatrix} a_{11} - \lambda & a_{12} & \cdots & a_{1n} \\ a_{21} & a_{22} - \lambda & \cdots & a_{2n} \\ \vdots & \vdots & & \\ a_{1n} & a_{2n} & \cdots & a_{nn} - \lambda \end{vmatrix}.$$

The *characteristic equation* of **A** is

(5) $$p(\lambda) = 0.$$

It can be verified directly from Definition 4.3, Section 1.4, that p as defined by (4) is in fact a polynomial of degree n (Exercise T-2).

Theorem 6.1 A real number λ is an eigenvalue of **A** if and only if it satisfies the characteristic equation (5).

Proof. For a given λ, (3) is simply a linear system of n equations in n unknowns, with determinant $p(\lambda)$. Such a system has a nontrivial solution ($\mathbf{X} \neq \mathbf{0}$) if and only if its determinant vanishes (Exercise T-10, Section 1.4), which yields the conclusion of the theorem.

Example 6.3 Let

$$\mathbf{A} = \begin{bmatrix} 0 & 1 \\ 1 & 0 \end{bmatrix};$$

then

$$\mathbf{A} - \lambda\mathbf{I} = \begin{bmatrix} -\lambda & 1 \\ 1 & -\lambda \end{bmatrix}$$

and

$$p(\lambda) = \begin{vmatrix} -\lambda & 1 \\ 1 & -\lambda \end{vmatrix} = \lambda^2 - 1.$$

Thus the eigenvalues of **A** are $\lambda_1 = 1$ and $\lambda_2 = -1$, as we saw by inspection in Example 6.2. To find the eigenvectors associated with λ_1, we consider the system

$$(\mathbf{A} - \lambda\mathbf{I})\mathbf{X} = \mathbf{0}$$

with, specifically, $\lambda = \lambda_1 = 1$:

$$(\mathbf{A} - \mathbf{I})\mathbf{X} = \mathbf{0},$$

or

$$\begin{bmatrix} -1 & 1 \\ 1 & -1 \end{bmatrix} \begin{bmatrix} x \\ y \end{bmatrix} = \begin{bmatrix} 0 \\ 0 \end{bmatrix}.$$

The solutions of this system are the vectors $\begin{bmatrix} x \\ y \end{bmatrix}$ such that $x = y$; thus the eigenvectors of **A** corresponding to $\lambda = 1$ are the nonzero multiples of

$$\mathbf{X}_1 = \begin{bmatrix} 1 \\ 1 \end{bmatrix}.$$

Similarly, to find the eigenvectors associated with $\lambda_2 = -1$, we consider the system

$$(\mathbf{A} + \mathbf{I})\mathbf{X} = 0,$$

or

$$\begin{bmatrix} 1 & 1 \\ 1 & 1 \end{bmatrix}\begin{bmatrix} x \\ y \end{bmatrix} = \begin{bmatrix} 0 \\ 0 \end{bmatrix},$$

whose solution space is spanned by $\begin{bmatrix} 1 \\ -1 \end{bmatrix}$; hence any nonzero multiple of

$\begin{bmatrix} 1 \\ -1 \end{bmatrix}$ is an eigenvector of **A** corresponding to $\lambda_2 = -1$.

Example 6.4   The characteristic polynomial of

$$\mathbf{A} = \begin{bmatrix} 0 & -1 \\ 1 & 0 \end{bmatrix}$$

is

$$p(\lambda) = \begin{vmatrix} -\lambda & -1 \\ 1 & -\lambda \end{vmatrix} = \lambda^2 + 1,$$

which has no real roots; hence **A** has no eigenvalues.

Since

$$\begin{vmatrix} r & s \\ r & -s \end{vmatrix} = -2rs,$$

eigenvectors corresponding to the distinct eigenvalues of **A** in Example 6.3 are linearly independent (Corollary 2.1, Section 2.2). This is always true, as we shall now prove.

Theorem 6.2   Let $\lambda_1, \ldots, \lambda_k$ be distinct eigenvalues of a matrix **A**, with associated eigenvectors $\mathbf{X}_1, \ldots, \mathbf{X}_k$. Then $\mathbf{X}_1, \ldots, \mathbf{X}_k$ are linearly independent.

Proof. Suppose $\mathbf{X}_1, \ldots, \mathbf{X}_k$ are linearly dependent. Then there is a smallest integer $j \leq k$ such that $\mathbf{X}_1, \ldots, \mathbf{X}_j$ are linearly dependent. Moreover, $j \geq 2$ (why?) and $\mathbf{X}_j$ must be a linear combination of $\mathbf{X}_1, \ldots, \mathbf{X}_{j-1}$, which are linearly independent (Exercise T-3); thus

$$(6) \qquad \mathbf{X}_j = c_1 \mathbf{X}_1 + \cdots + c_{j-1} \mathbf{X}_{j-1}.$$

Now

$$\mathbf{A}\mathbf{X}_j = c_1 \mathbf{A}\mathbf{X}_1 + \cdots + c_{j-1} \mathbf{A}\mathbf{X}_{j-1},$$

which can be rewritten as

$$(7) \qquad \lambda_j \mathbf{X}_j = c_1 \lambda_1 \mathbf{X}_1 + \cdots + c_{-1} \lambda_{j-1} \mathbf{X}_{j-1},$$

by definition of $\mathbf{X}_1, \ldots, \mathbf{X}_j$. Multiplying (6) by λ_j and subtracting the result from (7) yields

$$\mathbf{0} = c_1 (\lambda_1 - \lambda_j) \mathbf{X}_1 + \cdots + c_{j-1} (\lambda_{j-1} - \lambda_j) \mathbf{X}_{j-1},$$

which implies that

$$(8) \qquad c_1 (\lambda_1 - \lambda_j) = \cdots = c_{j-1} (\lambda_{j-1} - \lambda_j) = 0,$$

because of the linear independence of $\mathbf{X}_1, \ldots, \mathbf{X}_{j-1}$. Since $\lambda_i \neq \lambda_j$ if $1 \leq i \leq j - 1$, (8) yields

$$c_1 = \cdots = c_{j-1} = 0,$$

so that, from (6), $\mathbf{X}_j = \mathbf{0}$. However, this is a contradiction, since $\mathbf{X}_j$ is an eigenvector of $\mathbf{A}$, and consequently nonzero. Therefore, $\mathbf{X}_1, \ldots, \mathbf{X}_k$ are linearly independent, which completes the proof.

The next theorem follows immediately from Theorem 6.2, since any set of n linearly independent n-vectors is a basis for R^n.

Theorem 6.3 If an $n \times n$ matrix $\mathbf{A}$ has n distinct eigenvalues, then there is a basis for R^n consisting of eigenvectors of $\mathbf{A}$.

Example 6.5 Let

$$\mathbf{A} = \begin{bmatrix} 1 & 0 & 0 \\ 0 & 1 & 1 \\ 0 & 1 & 1 \end{bmatrix}.$$

The characteristic polynomial of **A** is

$$p(\lambda) = \begin{vmatrix} 1 - \lambda & 0 & 0 \\ 0 & 1 - \lambda & 1 \\ 0 & 1 & 1 - \lambda \end{vmatrix}$$

$$= -\lambda(\lambda - 1)(\lambda - 2);$$

hence, the eigenvalues of **A** are $\lambda_1 = 0$, $\lambda_2 = 1$, and $\lambda_3 = 2$. The associated eigenvectors satisfy the systems

$$\mathbf{AX} = \begin{bmatrix} 1 & 0 & 0 \\ 0 & 1 & 1 \\ 0 & 1 & 1 \end{bmatrix} \begin{bmatrix} x \\ y \\ z \end{bmatrix} = \begin{bmatrix} 0 \\ 0 \\ 0 \end{bmatrix},$$

$$(\mathbf{A} - \mathbf{I})\mathbf{X} = \begin{bmatrix} 0 & 0 & 0 \\ 0 & 0 & 1 \\ 0 & 1 & 0 \end{bmatrix} \begin{bmatrix} x \\ y \\ z \end{bmatrix} = \begin{bmatrix} 0 \\ 0 \\ 0 \end{bmatrix},$$

and

$$(\mathbf{A} - 2\mathbf{I})\mathbf{X} = \begin{bmatrix} -1 & 0 & 0 \\ 0 & -1 & 1 \\ 0 & 1 & -1 \end{bmatrix} \begin{bmatrix} x \\ y \\ z \end{bmatrix} = \begin{bmatrix} 0 \\ 0 \\ 0 \end{bmatrix},$$

respectively. By solving these systems, we find eigenvectors

$$\mathbf{X}_1 = \begin{bmatrix} 0 \\ 1 \\ -1 \end{bmatrix}, \qquad \mathbf{X}_2 = \begin{bmatrix} 1 \\ 0 \\ 0 \end{bmatrix}, \qquad \text{and} \qquad \mathbf{X}_3 = \begin{bmatrix} 0 \\ 1 \\ 1 \end{bmatrix}.$$

Corollary 2.1 implies that these vectors form a basis for R^3, since

$$\begin{vmatrix} 0 & 1 & 0 \\ 1 & 0 & 1 \\ -1 & 0 & 1 \end{vmatrix} = -2.$$

If the characteristic polynomial of an $n \times n$ matrix **A** has nonreal roots, then no basis for R^n consists entirely of eigenvectors of **A**. We omit the proof of this statement, which is beyond the scope of this book. (See Example 6.4.)

An eigenvalue $\lambda = \lambda_1$ is said to be of *multiplicity* $k \geq 1$ if $(\lambda - \lambda_1)^k$ divides $p(\lambda)$ but, $(\lambda - \lambda_1)^{k+1}$ does not. If the characteristic polynomial of an $n \times n$ matrix **A** has only real roots, but some are of multiplicity greater than one, then it may or may not be possible to construct a basis for R^n consisting of eigenvectors of **A**, as we shall see in the next two examples.

Example 6.6   Consider

$$\mathbf{A} = \begin{bmatrix} 0 & 0 & 1 \\ 0 & 2 & 0 \\ 0 & 0 & 2 \end{bmatrix},$$

which has the characteristic polynomial

$$\begin{vmatrix} -\lambda & 0 & 1 \\ 0 & 2-\lambda & 0 \\ 0 & 0 & 2-\lambda \end{vmatrix} = -\lambda(\lambda - 2)^2;$$

hence, its eigenvalues are $\lambda_1 = 0$, of multiplicity 1, and $\lambda_2 = 2$, of multiplicity 2. The eigenspace corresponding to $\lambda_1 = 0$ consists of the solutions of

$$\mathbf{AX} = \begin{bmatrix} 0 & 0 & 1 \\ 0 & 2 & 0 \\ 0 & 0 & 2 \end{bmatrix} \begin{bmatrix} x \\ y \\ z \end{bmatrix} = \begin{bmatrix} 0 \\ 0 \\ 0 \end{bmatrix},$$

which are of the form $\begin{bmatrix} r \\ 0 \\ 0 \end{bmatrix}$, with r arbitrary; hence the eigenspace is spanned by

$$\mathbf{X_1} = \begin{bmatrix} 1 \\ 0 \\ 0 \end{bmatrix}.$$

The eigenspace corresponding to $\lambda_2 = 2$ consists of the solutions of

$$(\mathbf{A} - 2\mathbf{I})\mathbf{X} = \begin{bmatrix} -2 & 0 & 1 \\ 0 & 0 & 0 \\ 0 & 0 & 0 \end{bmatrix} \begin{bmatrix} x \\ y \\ z \end{bmatrix} = \begin{bmatrix} 0 \\ 0 \\ 0 \end{bmatrix},$$

which satisfy

$$-2x + z = 0,$$

so that $x = r$ and $y = s$ can be chosen arbitrarily, and then $z = 2r$ is determined. Thus, the solutions are of the form

$$\mathbf{X} = \begin{bmatrix} r \\ s \\ 2r \end{bmatrix} = r \begin{bmatrix} 1 \\ 0 \\ 2 \end{bmatrix} + s \begin{bmatrix} 0 \\ 1 \\ 0 \end{bmatrix};$$

hence

$$\mathbf{X}_2 = \begin{bmatrix} 0 \\ 1 \\ 0 \end{bmatrix} \quad \text{and} \quad \mathbf{X}_3 = \begin{bmatrix} 1 \\ 0 \\ 2 \end{bmatrix}$$

span the eigenspace corresponding to $\lambda_2 = 2$. Thus $\{\mathbf{X}_1, \mathbf{X}_2, \mathbf{X}_3\}$ is a basis for R^3. (Verify.)

Example 6.7 The matrix

$$\mathbf{A} = \begin{bmatrix} 0 & 0 & 1 \\ 0 & 1 & 1 \\ 0 & 1 & 1 \end{bmatrix}$$

has characteristic polynomial

$$\begin{vmatrix} -\lambda & 0 & 1 \\ 0 & 1-\lambda & 1 \\ 0 & 1 & 1-\lambda \end{vmatrix} = -\lambda^2(\lambda - 2).$$

Its eigenvalues are $\lambda_1 = 2$, of multiplicity 1, and $\lambda_2 = \lambda_3 = 0$, of multiplicity 2. The eigenspace corresponding to $\lambda_1 = 2$ consists of solutions of

$$(\mathbf{A} - 2\mathbf{I})\mathbf{X} = \begin{bmatrix} -2 & 0 & 1 \\ 0 & -1 & 1 \\ 0 & 1 & -1 \end{bmatrix} \begin{bmatrix} x \\ y \\ z \end{bmatrix} = \begin{bmatrix} 0 \\ 0 \\ 0 \end{bmatrix},$$

and is spanned by

$$\mathbf{X}_1 = \begin{bmatrix} 1 \\ 2 \\ 2 \end{bmatrix}.$$

The eigenspace corresponding to $\lambda_2 = 0$ consists of solutions of

$$\mathbf{A}\mathbf{X} = \begin{bmatrix} 0 & 0 & 1 \\ 0 & 1 & 1 \\ 0 & 1 & 1 \end{bmatrix} \begin{bmatrix} x \\ y \\ z \end{bmatrix},$$

and is spanned by

$$\mathbf{X}_2 = \begin{bmatrix} 1 \\ 0 \\ 0 \end{bmatrix}.$$

Consequently, $\mathbf{A}$ has only two linearly independent eigenvectors, and no basis for R^3 can be constructed from eigenvectors of $\mathbf{A}$.

Eigenvectors of Symmetric Matrices

The eigenvectors of a symmetric $n \times n$ matrix always span R^n, as we shall now see. We state the next theorem without proof.

Theorem 6.4   The characteristic polynomial of an $n \times n$ symmetric matrix $\mathbf{A}$ is of the form

$$p(\lambda) = (-1)^n (\lambda - \lambda_1)^{n_1} (\lambda - \lambda_2)^{n_2} \cdots (\lambda - \lambda_k)^{n_k},$$

where $\lambda_1, \lambda_2, \ldots, \lambda_k$ are real numbers and

$$(9) \qquad n_1 + n_2 + \cdots + n_k = n.$$

Moreover, the eigenspaces corresponding to $\lambda_1, \lambda_2, \ldots, \lambda_k$ are of dimensions $n_1, n_2, \ldots, n_k$.

Lemma 6.1   Eigenvectors associated with distinct eigenvalues of a symmetric matrix **A** are orthogonal.

Proof.   First, we observe that if **X** and **Y** are n-vectors, then $\mathbf{X} \cdot \mathbf{Y} = \mathbf{X}^T \mathbf{Y}$. (Verify.) Now suppose **A** is symmetric and $\mathbf{A}\mathbf{X}_i = \lambda_i \mathbf{X}_i$ for $i = 1, 2$. Then

$$(10) \qquad \lambda_1 (\mathbf{X}_1 \cdot \mathbf{X}_2) = (\lambda_1 \mathbf{X}_1) \cdot \mathbf{X}_2 = (\mathbf{A}\mathbf{X}_1) \cdot \mathbf{X}_2$$
$$= (\mathbf{A}\mathbf{X}_1)^T \mathbf{X}_2 = \mathbf{X}_1^T \mathbf{A}^T \mathbf{X}_2.$$

Since **A** is symmetric, $\mathbf{A}^T = \mathbf{A}$, and we may continue the string of equalities (10) with

$$(11) \qquad \mathbf{X}_1^T \mathbf{A}^T \mathbf{X}_2 = \mathbf{X}_1^T \mathbf{A}\mathbf{X}_2 = (\mathbf{X}_1) \cdot (\mathbf{A}\mathbf{X}_2)$$
$$= \mathbf{X}_1 \cdot (\lambda_2 \mathbf{X}_2) = \lambda_2 (\mathbf{X}_1 \cdot \mathbf{X}_2).$$

From (10) and (11),

$$(12) \qquad \lambda_1 (\mathbf{X}_1 \cdot \mathbf{X}_2) = \lambda_2 (\mathbf{X}_1 \cdot \mathbf{X}_2).$$

If $\lambda_1 \neq \lambda_2$, then (12) implies that $\mathbf{X}_1 \cdot \mathbf{X}_2 = 0$, which completes the proof.

Example 6.8   Consider the symmetric matrix

$$\mathbf{A} = \begin{bmatrix} 1 & 0 & 0 \\ 0 & 1 & -1 \\ 0 & -1 & 1 \end{bmatrix},$$

with characteristic polynomial

$$p(\lambda) = \begin{bmatrix} 1 - \lambda & 0 & 0 \\ 0 & 1 - \lambda & -1 \\ 0 & -1 & 1 - \lambda \end{bmatrix}$$
$$= -\lambda(\lambda - 1)(\lambda - 2).$$

The eigenvalues of $\mathbf{A}$ are $\lambda_1 = 0$, $\lambda_2 = 1$, and $\lambda_3 = 2$, with corresponding eigenvectors

$$\mathbf{X}_1 = \begin{bmatrix} 0 \\ 1 \\ 1 \end{bmatrix}, \quad \mathbf{X}_2 = \begin{bmatrix} 1 \\ 0 \\ 0 \end{bmatrix}, \quad \text{and} \quad \mathbf{X}_3 = \begin{bmatrix} 0 \\ 1 \\ -1 \end{bmatrix},$$

which are orthogonal, as implied by Lemma 6.1.

Theorem 6.5 If $\mathbf{A}$ is an $n \times n$ symmetric matrix, then there is an orthonormal basis for R^n consisting of eigenvectors of $\mathbf{A}$.

Proof. Let $\lambda_1, \ldots, \lambda_k$ be the distinct eigenvalues of $\mathbf{A}$, of multiplicities $n_1, n_2, \ldots, n_k$, respectively. By Theorem 6.4 and Corollary 5.2, Section 2.5, the associated eigenspaces have orthonormal bases $B_1, \ldots, B_k$, consisting of $n_1, \ldots, n_k$ vectors, respectively. Since vectors within a given B_i are orthogonal by definition, and vectors in B_i and B_j $(i \neq j)$ are orthogonal by Lemma 6.1, it follows from (9) that $B_1, \ldots, B_k$ contain exactly n orthonormal vectors, all eigenvectors of $\mathbf{A}$, which form a basis for R^n.

Example 6.9 The characteristic polynomial of

$$\mathbf{A} = \begin{bmatrix} 0 & 1 & 1 \\ 1 & 0 & 1 \\ 1 & 1 & 0 \end{bmatrix}$$

is

$$p(\lambda) = \begin{vmatrix} -\lambda & 1 & 1 \\ 1 & -\lambda & 1 \\ 1 & 1 & -\lambda \end{vmatrix} = -(\lambda - 2)(\lambda + 1)^2;$$

hence the eigenvalues of $\mathbf{A}$ are $\lambda_1 = 2$, of multiplicity 1, and $\lambda_2 = -1$, of multiplicity 2. The eigenspace associated with $\lambda_1 = 2$ is spanned by

$$\mathbf{X}_1 = \begin{bmatrix} 1 \\ 1 \\ 1 \end{bmatrix}.$$

The eigenvectors associated with $\lambda_2 = -1$ are the nontrivial solutions of

$$\begin{bmatrix} 1 & 1 & 1 \\ 1 & 1 & 1 \\ 1 & 1 & 1 \end{bmatrix} \begin{bmatrix} x \\ y \\ z \end{bmatrix} = \begin{bmatrix} 0 \\ 0 \\ 0 \end{bmatrix}.$$

This system has two linearly independent solutions

$$\mathbf{X}_2 = \begin{bmatrix} 1 \\ -1 \\ 0 \end{bmatrix} \quad \text{and} \quad \mathbf{X}_3 = \begin{bmatrix} 1 \\ 0 \\ -1 \end{bmatrix},$$

which are both orthogonal to $\mathbf{X}_1$. Since $\mathbf{X}_2 \cdot \mathbf{X}_3 \neq 0$, we use the Gram–Schmidt process to find an orthonormal basis for $S[\mathbf{X}_2, \mathbf{X}_3]$:

$$\mathbf{Y}_2 = \frac{\mathbf{X}_2}{|\mathbf{X}_2|} = \frac{1}{\sqrt{2}} \begin{bmatrix} 1 \\ -1 \\ 0 \end{bmatrix};$$

$$\mathbf{X}_3 \cdot \mathbf{Y}_2 = \frac{1}{\sqrt{2}};$$

$$\mathbf{Z}_3 = \mathbf{X}_3 - (\mathbf{X}_3 \cdot \mathbf{Y}_2)\mathbf{Y}_2 = \begin{bmatrix} \frac{1}{2} \\ \frac{1}{2} \\ -1 \end{bmatrix};$$

$$\mathbf{Y}_3 = \frac{\mathbf{Z}_3}{|\mathbf{Z}_3|} = \frac{1}{\sqrt{6}} \begin{bmatrix} 1 \\ 1 \\ -2 \end{bmatrix}.$$

Now $\mathbf{Y}_2$ and $\mathbf{Y}_3$ form an orthonormal basis for the eigenspace associated with $\lambda_2 = -1$. Normalizing $\mathbf{X}_1$ produces

$$\mathbf{Y}_1 = \frac{1}{\sqrt{3}} \begin{bmatrix} 1 \\ 1 \\ 1 \end{bmatrix},$$

and $\{Y_1, Y_2, Y_3\}$ is an orthonormal basis for R^3, comprised of eigenvectors of $\mathbf{A}$.

Example 6.10 The characteristic polynomial of

$$\mathbf{A} = \begin{bmatrix} 0 & 0 & 0 \\ 0 & 1 & 1 \\ 0 & 1 & 1 \end{bmatrix}$$

is

$$P(\lambda) = \begin{vmatrix} -\lambda & 0 & 0 \\ 0 & 1-\lambda & 1 \\ 0 & 1 & 1-\lambda \end{vmatrix} = -\lambda^2(\lambda - 2);$$

hence the eigenvalues of $\mathbf{A}$ are $\lambda_1 = 0$, of multiplicity 2, and $\lambda_2 = 2$, of multiplicity 1. The eigenvectors associated with $\lambda_1 = 0$ are the nontrivial solutions of

$$\begin{bmatrix} 0 & 0 & 0 \\ 0 & 1 & 1 \\ 0 & 1 & 1 \end{bmatrix} \begin{bmatrix} x \\ y \\ z \end{bmatrix} = \begin{bmatrix} 0 \\ 0 \\ 0 \end{bmatrix}.$$

This system has two linearly independent solutions,

$$\mathbf{X}_1 = \begin{bmatrix} 1 \\ 0 \\ 0 \end{bmatrix}, \qquad \mathbf{X}_2 = \begin{bmatrix} 0 \\ 1 \\ -1 \end{bmatrix}.$$

It happens that these vectors are already orthogonal.

The eigenspace associated with $\lambda_2 = 2$ is spanned by

$$\mathbf{X}_3 = \begin{bmatrix} 0 \\ 1 \\ 1 \end{bmatrix},$$

which is orthogonal to $\mathbf{X}_1$ and $\mathbf{X}_2$, as implied by Lemma 6.1. Normalizing

$\mathbf{X}_1$, $\mathbf{X}_2$, $\mathbf{X}_3$ yields the orthonormal basis

$$\mathbf{Y}_1 = \begin{bmatrix} 1 \\ 0 \\ 0 \end{bmatrix}, \qquad \mathbf{Y}_2 = \frac{1}{\sqrt{2}} \begin{bmatrix} 0 \\ 1 \\ -1 \end{bmatrix}, \qquad \mathbf{Y}_3 = \frac{1}{\sqrt{2}} \begin{bmatrix} 0 \\ 1 \\ 1 \end{bmatrix}$$

for R^3.

The next theorem presents a useful consequence of Theorem 6.5.

Theorem 6.6 Let $\mathbf{A}$ be an $n \times n$ symmetric matrix and suppose $B = \{\mathbf{Y}_1, \ldots, \mathbf{Y}_n\}$ is an orthonormal basis for R^n consisting of eigenvectors of $\mathbf{A}$; that is,

$$(13) \qquad \mathbf{AY}_i = \lambda_i \mathbf{Y}_i \qquad (1 \le i \le n).$$

If $\lambda_i \neq 0$ for $i = 1, \ldots, n$, then the system

$$(14) \qquad \mathbf{AX} = \mathbf{Y}$$

has the unique solution

$$(15) \qquad \mathbf{X} = \frac{(\mathbf{Y} \cdot \mathbf{Y}_1)}{\lambda_1} \mathbf{Y}_1 + \frac{(\mathbf{Y} \cdot \mathbf{Y}_2)}{\lambda_2} \mathbf{Y}_2 + \cdots + \frac{(\mathbf{Y} \cdot \mathbf{Y}_n)}{\lambda_n} \mathbf{Y}_n$$

for any given $\mathbf{Y}$. If $\lambda_1 = \cdots = \lambda_k = 0$ and $\lambda_{k+1}, \ldots, \lambda_n$ are all nonzero, then (14) has a solution if and only if

$$\mathbf{Y} \cdot \mathbf{Y}_i = 0 \qquad (1 \le i \le k),$$

in which case $\mathbf{X}$ is a solution if and only if

$$\mathbf{X} = a_1 \mathbf{Y}_1 + \cdots + a_k \mathbf{Y}_k + \frac{\mathbf{Y} \cdot \mathbf{Y}_{k+1}}{\lambda_{k+1}} \mathbf{Y}_{k+1} + \cdots + \frac{\mathbf{Y} \cdot \mathbf{Y}_n}{\lambda_n} \mathbf{Y}_n,$$

where $a_1, \ldots, a_k$ are arbitrary.

Proof. Let

$$(16) \qquad \mathbf{X} = a_1 \mathbf{Y}_1 + a_2 \mathbf{Y}_2 + \cdots + a_n \mathbf{Y}_n,$$

where $a_1, a_2, \ldots, a_n$ are to be determined. From Theorem 5.6 of Section 2.5,

$$(17) \qquad \mathbf{Y} = (\mathbf{Y} \cdot \mathbf{Y}_1) \mathbf{Y}_1 + (\mathbf{Y} \cdot \mathbf{Y}_2) \mathbf{Y}_2 + \cdots + (\mathbf{Y} \cdot \mathbf{Y}_n) \mathbf{Y}_n.$$

Substituting (16) and (17) into (14) yields

$$a_1 \mathbf{AY}_1 + a_2 \mathbf{AY}_2 + \cdots + a_n \mathbf{AY}_n = (\mathbf{Y} \cdot \mathbf{Y}_1) \mathbf{Y}_1 + (\mathbf{Y} \cdot \mathbf{Y}_2) \mathbf{Y}_2 + \cdots + (\mathbf{Y} \cdot \mathbf{Y}_n) \mathbf{Y}_n,$$

which, from (13), can be rewritten as

$$a_1\lambda_1\mathbf{Y}_1 + a_2\lambda_2\mathbf{Y}_2 + \cdots + a_n\lambda_n\mathbf{Y}_n = (\mathbf{Y}\cdot\mathbf{Y}_1)\mathbf{Y}_1 + (\mathbf{Y}\cdot\mathbf{Y}_2)\mathbf{Y}_2 + \cdots + (\mathbf{Y}\cdot\mathbf{Y}_n)\mathbf{Y}_n.$$

Equating coefficients of $\mathbf{Y}_1, \ldots, \mathbf{Y}_n$ yields

$$(18) \qquad a_i\lambda_i = (\mathbf{Y}\cdot\mathbf{Y}_i) \qquad (1 \leq i \leq n).$$

If $\lambda_i \neq 0$, $i = 1, \ldots, n$, then (15) follows immediately. We leave the rest of the proof to the student (Exercise T-4).

Example 6.11 In Example 6.9 we found the orthonormal basis

$$\mathbf{Y}_1 = \frac{1}{\sqrt{3}}\begin{bmatrix} 1 \\ 1 \\ 1 \end{bmatrix}, \qquad \mathbf{Y}_2 = \frac{1}{\sqrt{2}}\begin{bmatrix} 1 \\ -1 \\ 0 \end{bmatrix}, \qquad \mathbf{Y}_3 = \frac{1}{\sqrt{6}}\begin{bmatrix} 1 \\ 1 \\ -2 \end{bmatrix}$$

for R^3, consisting of eigenvectors of

$$\mathbf{A} = \begin{bmatrix} 0 & 1 & 1 \\ 1 & 0 & 1 \\ 1 & 1 & 0 \end{bmatrix},$$

associated, respectively, with the eigenvalues, $\lambda_1 = 2$, $\lambda_2 = \lambda_3 = -1$; thus

$$(19) \qquad \mathbf{A}\mathbf{Y}_1 = 2\mathbf{Y}_1, \qquad \mathbf{A}\mathbf{Y}_2 = -\mathbf{Y}_2, \qquad \mathbf{A}\mathbf{Y}_3 = -\mathbf{Y}_3.$$

Suppose we wish to solve

$$\mathbf{A}\mathbf{X} = \mathbf{Y}$$

with

$$\mathbf{Y} = \begin{bmatrix} 2 \\ -1 \\ 3 \end{bmatrix}.$$

Since the eigenvalues of $\mathbf{A}$ are all nonzero, Theorem 6.6 yields the solution

$$\mathbf{X} = a_1\mathbf{Y}_1 + a_2\mathbf{Y}_2 + a_3\mathbf{Y}_3,$$

where, from (18) and (19),

$$a_1 = \frac{\mathbf{Y}\cdot\mathbf{Y}_1}{2} = \frac{2}{\sqrt{3}}, \qquad a_2 = \frac{\mathbf{Y}\cdot\mathbf{Y}_2}{(-1)} = -\frac{3}{\sqrt{2}}, \qquad a_3 = \frac{\mathbf{Y}\cdot\mathbf{Y}_3}{(-1)} = \frac{5}{\sqrt{6}};$$

thus

$$\mathbf{X} = \frac{2}{\sqrt{3}}\,\mathbf{Y}_1 - \frac{3}{\sqrt{2}}\,\mathbf{Y}_2 + \frac{5}{\sqrt{6}}\,\mathbf{Y}_3$$

$$= \frac{2}{3}\begin{bmatrix} 1 \\ 1 \\ 1 \end{bmatrix} - \frac{3}{2}\begin{bmatrix} 1 \\ -1 \\ 0 \end{bmatrix} + \frac{5}{6}\begin{bmatrix} 1 \\ 1 \\ -2 \end{bmatrix} = \begin{bmatrix} 0 \\ 3 \\ -1 \end{bmatrix}.$$

The case where 0 is an eigenvalue of **A** is considered in Exercise 22.

Diagonalizable Matrices

Definition 6.3 A matrix **B** is said to be *similar* to **A** if there is a nonsingular matrix **P** such that

$$\mathbf{B} = \mathbf{P}^{-1}\mathbf{A}\mathbf{P}.$$

Example 6.12 Let

$$\mathbf{A} = \begin{bmatrix} 2 & 1 \\ 3 & -1 \end{bmatrix} \quad \text{and} \quad \mathbf{P} = \begin{bmatrix} -1 & 1 \\ 1 & 0 \end{bmatrix}.$$

Then

$$\mathbf{P}^{-1} = \begin{bmatrix} 0 & 1 \\ 1 & 1 \end{bmatrix}$$

and

$$\mathbf{B} = \mathbf{P}^{-1}\mathbf{A}\mathbf{P} = \begin{bmatrix} 0 & 1 \\ 1 & 1 \end{bmatrix}\begin{bmatrix} 2 & 1 \\ 3 & -1 \end{bmatrix}\begin{bmatrix} -1 & 1 \\ 1 & 0 \end{bmatrix}$$

$$= \begin{bmatrix} -4 & 3 \\ -5 & 5 \end{bmatrix}$$

is similar to **A**.

Theorem 6.7

(a) **A** is similar to **A**.
(b) If **B** is similar to **A**, then **A** is similar to **B**.
(c) If **B** is similar to **A** and **C** is similar to **B**, then **C** is similar to **A**.

We leave the proof to the student (Exercise T-5).

Part (b) allows us to say simply "**A** and **B** are similar," rather than "**A** is similar to **B**" or "**B** is similar to **A**."

The next theorem, which we state without proof, interprets similarity of matrices in terms of transformations from R^n to R^n.

Theorem 6.8 Suppose **A** is the matrix of **L**: $R^n \to R^n$ with respect to the natural basis for R^n; thus

$$\mathbf{L}(\mathbf{X}) = \mathbf{AX}.$$

Then

$$\mathbf{B} = \mathbf{P}^{-1}\mathbf{AP}$$

is the matrix of **L** with respect to the basis $\{\mathbf{X}_1, \dots, \mathbf{X}_n\}$, where $\mathbf{X}_j$ is the jth column of **P**.

Definition 6.4 A matrix **A** is said to be *diagonalizable* if it is similar to a diagonal matrix.

Theorem 6.9 A matrix **A** is diagonalizable if and only if it has n linearly independent eigenvectors, in which case **A** is similar to a diagonal matrix **D** whose diagonal elements are the eigenvalues of **A**.

Proof. First we observe that

(20) $$\mathbf{AP} = \mathbf{PD},$$

where **D** is the diagonal matrix

$$\mathbf{D} = \begin{bmatrix} \lambda_1 & 0 & \cdots & 0 \\ 0 & \lambda_2 & \cdots & 0 \\ \vdots & \vdots & & \\ 0 & 0 & \cdots & \lambda_n \end{bmatrix}$$

and the columns of **P** are nonzero n-vectors, if and only if the jth column of **P** is an eigenvector of **A** corresponding to λ_j, for $j = 1, 2, \dots, n$ (Exercise T-6). Moreover, **P** is nonsingular if and only if its columns form a basis for R^n (Corollary 2.1, Section 2.2). Since **A** and **D** are similar if and only if there is a nonsingular matrix **P** such that (20) holds, the proof is complete.

Example 6.13 We saw in Example 6.2 that

$$\mathbf{A} = \begin{bmatrix} 0 & 1 \\ 1 & 0 \end{bmatrix}$$

has eigenvectors

$$\mathbf{X_1} = \begin{bmatrix} 1 \\ 1 \end{bmatrix} \quad \text{and} \quad \mathbf{X_2} = \begin{bmatrix} 1 \\ -1 \end{bmatrix},$$

corresponding to $\lambda_1 = 1$ and $\lambda_2 = -1$, respectively. Let $\mathbf{P}$ be the matrix whose columns are $\mathbf{X_1}$ and $\mathbf{X_2}$:

$$\mathbf{P} = \begin{bmatrix} 1 & 1 \\ 1 & -1 \end{bmatrix}.$$

Then

$$\mathbf{P^{-1}} = \begin{bmatrix} \frac{1}{2} & \frac{1}{2} \\ \frac{1}{2} & -\frac{1}{2} \end{bmatrix}$$

and

$$\mathbf{P^{-1}AP} = \begin{bmatrix} \frac{1}{2} & \frac{1}{2} \\ \frac{1}{2} & -\frac{1}{2} \end{bmatrix} \begin{bmatrix} 0 & 1 \\ 1 & 0 \end{bmatrix} \begin{bmatrix} 1 & 1 \\ 1 & -1 \end{bmatrix} = \begin{bmatrix} 1 & 0 \\ 0 & -1 \end{bmatrix},$$

as is implied by Theorem 6.9.

Example 6.14   From Example 6.5, if

$$\mathbf{A} = \begin{bmatrix} 1 & 0 & 0 \\ 0 & 1 & 1 \\ 0 & 1 & 1 \end{bmatrix} \quad \text{and} \quad \mathbf{P} = \begin{bmatrix} 0 & 1 & 0 \\ 1 & 0 & 0 \\ -1 & 0 & 1 \end{bmatrix},$$

then

$$\mathbf{P^{-1}AP} = \begin{bmatrix} 0 & 0 & 0 \\ 0 & 1 & 0 \\ 0 & 0 & 2 \end{bmatrix}$$

(Exercise 23). From Example 6.6, if

$$\mathbf{A} = \begin{bmatrix} 0 & 0 & 1 \\ 0 & 2 & 0 \\ 0 & 0 & 2 \end{bmatrix} \quad \text{and} \quad \mathbf{P} = \begin{bmatrix} 1 & 0 & 1 \\ 0 & 1 & 0 \\ 0 & 0 & 2 \end{bmatrix},$$

then

$$\mathbf{P^{-1}AP} = \begin{bmatrix} 0 & 0 & 0 \\ 0 & 2 & 0 \\ 0 & 0 & 2 \end{bmatrix}.$$

(Exercise 24). From Example 6.7,

$$\mathbf{A} = \begin{bmatrix} 0 & 0 & 1 \\ 0 & 1 & 1 \\ 0 & 1 & 1 \end{bmatrix}$$

is not diagonalizable.

Orthogonal Matrices

It follows from Theorems 6.5 and 6.9 that a symmetric matrix is always diagonalizable; moreover, if **A** is symmetric, we can write

$$\mathbf{D} = \mathbf{P^{-1}AP},$$

where **D** is diagonal and the columns of **P** form an orthonormal basis for R^n.

Definition 6.5 An $n \times n$ matrix **P** is said to be an *orthogonal matrix* if its columns form an orthonormal basis for R^n.

Example 6.15 Examples of orthogonal matrices are

$$\mathbf{P_1} = \begin{bmatrix} \dfrac{1}{\sqrt{3}} & \dfrac{1}{\sqrt{2}} & \dfrac{1}{\sqrt{6}} \\[3ex] \dfrac{1}{\sqrt{3}} & -\dfrac{1}{\sqrt{2}} & \dfrac{1}{\sqrt{6}} \\[3ex] \dfrac{1}{\sqrt{3}} & 0 & -\dfrac{2}{\sqrt{6}} \end{bmatrix}$$

and

$$\mathbf{P}_2 = \begin{bmatrix} 1 & 0 & 0 \\ 0 & \dfrac{1}{\sqrt{2}} & \dfrac{1}{\sqrt{2}} \\ 0 & -\dfrac{1}{\sqrt{2}} & \dfrac{1}{\sqrt{2}} \end{bmatrix},$$

whose columns are the eigenvectors found in Examples 6.9 and 6.10, respectively.

Theorem 6.10 An orthogonal matrix $\mathbf{P}$ is nonsingular, and its inverse equals its transpose; thus

$$(21) \qquad\qquad \mathbf{P}^{-1} = \mathbf{P}^{\mathrm{T}}.$$

Conversely, if $\mathbf{P}$ satisfies (21), then $\mathbf{P}$ is an orthogonal matrix.

Proof. Let $\mathbf{P} = [p_{ij}]$, where

$$\mathbf{X}_1 = \begin{bmatrix} p_{11} \\ p_{21} \\ \vdots \\ p_{n1} \end{bmatrix}, \ \mathbf{X}_2 = \begin{bmatrix} p_{12} \\ p_{22} \\ \vdots \\ p_{n2} \end{bmatrix}, \ldots, \ \mathbf{X}_n = \begin{bmatrix} p_{1n} \\ p_{2n} \\ \vdots \\ p_{nn} \end{bmatrix}$$

are orthonormal vectors. Suppose

$$\mathbf{P}^{\mathrm{T}}\mathbf{P} = [c_{ij}];$$

then

$$c_{ij} = \sum_{k=1}^{n} p_{ik}^{\mathrm{T}} p_{kj}$$

$$= \sum_{k=1}^{n} p_{ki} p_{kj}$$

$$= \mathbf{X}_i \cdot \mathbf{X}_j.$$

Since $\mathbf{X}_1, \ldots, \mathbf{X}_n$ are orthonormal, it follows that

$$c_{ij} = \begin{cases} 0 & \text{if} \quad i \neq j, \\ 1 & \text{if} \quad i = j. \end{cases}$$

Therefore, $\mathbf{P^TP} = \mathbf{I}$, which proves (21). We leave the proof of the converse to the student (Exercise T-7).

The student should verify Theorem 6.10 for $\mathbf{P}_1$ and $\mathbf{P}_2$ as defined in Example 6.15.

The next theorem now follows from Theorems 6.5, 6.9, and 6.10.

Theorem 6.11 If $\mathbf{A}$ is a symmetric matrix, then there is an orthogonal matrix $\mathbf{P}$ such that

$$\mathbf{P^TAP} = \mathbf{D},$$

where $\mathbf{D}$ is diagonal and the elements on the diagonal of $\mathbf{D}$ are the eigenvalues of $\mathbf{A}$.

Orthogonal matrices are associated with rotations of coordinate systems. We shall discuss this in Section 3.5. For now we apply them to the study of quadratic forms.

Quadratic Forms

Definition 6.6 The double sum

(22)
$$Q(\mathbf{X}) = \sum_{i=1}^{n} \sum_{j=1}^{n} a_{ij}x_ix_j,$$

where $\mathbf{A} = [a_{ij}]$ is symmetric, is called a *quadratic form in n variables*, and $\mathbf{A}$ is said to be the *matrix of Q*.

We observe that (22) can also be written as

(23)
$$Q(\mathbf{X}) = \mathbf{X^TAX},$$

where

$$\mathbf{X} = \begin{bmatrix} x_1 \\ \vdots \\ x_n \end{bmatrix}.$$

Example 6.16 The matrix of the quadratic form in two variables,

(24)
$$Q(\mathbf{X}) = 2x_1^2 - 4x_1x_2 + x_2^2$$

$$= 2x_1x_1 - 2x_1x_2 - 2x_2x_1 + x_2x_2$$

is

$$\mathbf{A} = \begin{bmatrix} 2 & -2 \\ -2 & 1 \end{bmatrix},$$

and (24) can also be written as

$$Q(\mathbf{X}) = \begin{bmatrix} x_1 & x_2 \end{bmatrix} \begin{bmatrix} 2 & -2 \\ -2 & 1 \end{bmatrix} \begin{bmatrix} x_1 \\ x_2 \end{bmatrix}.$$

Example 6.17 The matrix of the quadratic form in three variables,

$$(25) \qquad Q(\mathbf{X}) = 2x_1x_2 + 2x_2x_3 + 2x_1x_3$$

is

$$\mathbf{A} = \begin{bmatrix} 0 & 1 & 1 \\ 1 & 0 & 1 \\ 1 & 1 & 0 \end{bmatrix},$$

and (25) can also be written as

$$Q(\mathbf{X}) = \begin{bmatrix} x_1 & x_2 & x_3 \end{bmatrix} \begin{bmatrix} 0 & 1 & 1 \\ 1 & 0 & 1 \\ 1 & 1 & 0 \end{bmatrix} \begin{bmatrix} x_1 \\ x_2 \\ x_3 \end{bmatrix}.$$

If we introduce new variables $y_1, \ldots, y_n$ into (22) by means of the substitution

$$\begin{bmatrix} x_1 \\ x_2 \\ \vdots \\ x_n \end{bmatrix} = \mathbf{P} \begin{bmatrix} y_1 \\ y_2 \\ \vdots \\ y_n \end{bmatrix},$$

then (22) becomes

$$Q(\mathbf{X}) = Q(\mathbf{PY}) = (\mathbf{PY})^{\mathrm{T}} \mathbf{A}(\mathbf{PY})$$

$$= \mathbf{Y}^{\mathrm{T}}(\mathbf{P}^{\mathrm{T}}\mathbf{AP})\mathbf{Y},$$

which is a quadratic form in $y_1, \ldots, y_n$, with associated matrix $\mathbf{P}^{\mathrm{T}}\mathbf{AP}$. This leads to the following theorem.

Theorem 6.12 Let $\mathbf{P}$ be an orthogonal matrix that diagonalizes the symmetric matrix $\mathbf{A}$ in (23). Then

$$(26) \qquad Q(\mathbf{PY}) = \lambda_1 y_1^2 + \lambda_2 y_2^2 + \cdots + \lambda_n y_n^2.$$

We leave the proof to the student (Exercise T-8).

Example 6.18 The matrix of

$$Q(\mathbf{X}) = 3x_1^2 + 2x_2^2 + 3x_3^2 + 2x_1x_3$$

is

$$\mathbf{A} = \begin{bmatrix} 3 & 0 & 1 \\ 0 & 2 & 0 \\ 1 & 0 & 3 \end{bmatrix},$$

which has eigenvalues $\lambda_1 = \lambda_2 = 2$ and $\lambda_3 = 4$, with associated orthonormal eigenvectors

$$\mathbf{X}_1 = \begin{bmatrix} 0 \\ 1 \\ 0 \end{bmatrix}, \qquad \mathbf{X}_2 = \begin{bmatrix} \frac{1}{\sqrt{2}} \\ 0 \\ -\frac{1}{\sqrt{2}} \end{bmatrix}, \qquad \mathbf{X}_3 = \begin{bmatrix} \frac{1}{\sqrt{2}} \\ 0 \\ \frac{1}{\sqrt{2}} \end{bmatrix}.$$

(Verify.) Now define the orthogonal matrix

$$\mathbf{P} = \begin{bmatrix} 0 & \frac{1}{\sqrt{2}} & \frac{1}{\sqrt{2}} \\ 1 & 0 & 0 \\ 0 & -\frac{1}{\sqrt{2}} & \frac{1}{\sqrt{2}} \end{bmatrix},$$

and introduce new variables y_1, y_2, y_3 by

$$\mathbf{X} = \mathbf{PY},$$

or

$$x_1 = \frac{y_2 + y_3}{\sqrt{2}}, \qquad x_2 = y_1, \qquad x_3 = -\frac{1}{\sqrt{2}}(y_2 - y_3).$$

According to (26),

$$Q(\mathbf{PY}) = 2y_1^2 + 2y_2^2 + 4y_3^2.$$

The student should verify this directly.

EXERCISES 2.6

Find the characteristic polynomials of the matrices in Exercises 1 through 8. Also, find all eigenvalues and give bases for the corresponding eigenspaces.

1. $\begin{bmatrix} 2 & 4 \\ 4 & 2 \end{bmatrix}$. 2. $\begin{bmatrix} 1 & -3 \\ 1 & 1 \end{bmatrix}$. 3. $\begin{bmatrix} -1 & 2 & 3 \\ 0 & 1 & 6 \\ 0 & 0 & -2 \end{bmatrix}$.

4. $\begin{bmatrix} -1 & 0 & 0 \\ 3 & 2 & 0 \\ 4 & -1 & 2 \end{bmatrix}$. 5. $\begin{bmatrix} -4 & -2 & 1 \\ -2 & -1 & 2 \\ -3 & -2 & 0 \end{bmatrix}$.

6. $\begin{bmatrix} 2 & 2 & 0 \\ 5 & -1 & 3 \\ 0 & 0 & 0 \end{bmatrix}$. 7. $\begin{bmatrix} 2 & 3 & 3 & 4 \\ -1 & 0 & -3 & 2 \\ -1 & -3 & 0 & -2 \\ 2 & 0 & 0 & 1 \end{bmatrix}$.

8. $\begin{bmatrix} 1 & 0 & 1 \\ 0 & 1 & 0 \\ -1 & 0 & 1 \end{bmatrix}$.

In Exercises 9 through 13 find, if possible, a basis for R^3 consisting of eigenvectors of the given matrices.

9. $\begin{bmatrix} 0 & 0 & 1 \\ 0 & 1 & 0 \\ 0 & 0 & 1 \end{bmatrix}$. 10. $\begin{bmatrix} 1 & 0 & 0 \\ 2 & -1 & -6 \\ 2 & 0 & -2 \end{bmatrix}$. 11. $\begin{bmatrix} -3 & 5 & -1 \\ 0 & 2 & 4 \\ 0 & 4 & 2 \end{bmatrix}$.

12. $\begin{bmatrix} 0 & 0 & 0 \\ 0 & 1 & 0 \\ 2 & 2 & 1 \end{bmatrix}$. 13. $\begin{bmatrix} -2 & 0 & 0 \\ 3 & 1 & 0 \\ 4 & 2 & 5 \end{bmatrix}$.

In Exercises 14 through 21 find orthonormal bases for R^3 (R^4 in Exercise 20) consisting of eigenvectors of the given matrices.

14. $\begin{bmatrix} 8 & -2 & 0 \\ -2 & 9 & -2 \\ 0 & -2 & 10 \end{bmatrix}$.

15. $\begin{bmatrix} -1 & -4 & -8 \\ -4 & -7 & 4 \\ -8 & 4 & -1 \end{bmatrix}$.

16. $\begin{bmatrix} 0 & 2 & 2 \\ 2 & 0 & 2 \\ 2 & 2 & 0 \end{bmatrix}$.

17. $\begin{bmatrix} 0 & 3 & 0 \\ 3 & 0 & 4 \\ 0 & 4 & 0 \end{bmatrix}$.

18. $\begin{bmatrix} 1 & 1 & 0 \\ 1 & 1 & 0 \\ 0 & 0 & 1 \end{bmatrix}$.

19. $\begin{bmatrix} -1 & 2 & 2 \\ 2 & 1 & 2 \\ 2 & 2 & -1 \end{bmatrix}$.

20. $\begin{bmatrix} 2 & -1 & 0 & 0 \\ -1 & 2 & 0 & 0 \\ 0 & 0 & -1 & 0 \\ 0 & 0 & 0 & 2 \end{bmatrix}$.

21. Find an orthonormal basis for R^3 consisting of eigenvectors of

$$A = \begin{bmatrix} 1 & -1 & 2 \\ -1 & 1 & 2 \\ 2 & 2 & 2 \end{bmatrix}.$$

Then solve the system $AX = Y$ by the method of Example 6.11, where

(a) $Y = \begin{bmatrix} 2 \\ 1 \\ -1 \end{bmatrix}$;   (b) $Y = \begin{bmatrix} 3 \\ 2 \\ -1 \end{bmatrix}$;   (c) $Y = \begin{bmatrix} 1 \\ -1 \\ 0 \end{bmatrix}$.

22. Find an orthonormal basis for R^3 consisting of eigenvectors of

$$A = \begin{bmatrix} 1 & -5 & 2 \\ -5 & 1 & 2 \\ 2 & 2 & -2 \end{bmatrix}.$$

Show that $AX = Y$ has a solution if and only if Y is orthogonal to

$$X_0 = \begin{bmatrix} 1 \\ 1 \\ 2 \end{bmatrix}.$$

Then solve $AX = Y$ by the method Example 6.11, where

(a) $Y = \begin{bmatrix} 1 \\ -1 \\ 0 \end{bmatrix}$;   (b) $Y = \begin{bmatrix} -2 \\ 0 \\ 1 \end{bmatrix}$;   (c) $Y = \begin{bmatrix} 1 \\ 1 \\ -1 \end{bmatrix}.$

23. *Verify:* If

$$A = \begin{bmatrix} 1 & 0 & 0 \\ 0 & 1 & 1 \\ 0 & 1 & 1 \end{bmatrix} \quad \text{and} \quad P = \begin{bmatrix} 0 & 1 & 0 \\ 1 & 0 & 0 \\ -1 & 0 & 1 \end{bmatrix},$$

then

$$P^{-1}AP = \begin{bmatrix} 0 & 0 & 0 \\ 0 & 1 & 0 \\ 0 & 0 & 2 \end{bmatrix}.$$

24. *Verify:* If

$$A = \begin{bmatrix} 0 & 0 & 1 \\ 0 & 2 & 0 \\ 0 & 0 & 2 \end{bmatrix} \quad \text{and} \quad P = \begin{bmatrix} 1 & 0 & 1 \\ 0 & 1 & 0 \\ 0 & 0 & 2 \end{bmatrix},$$

then

$$\mathbf{P^{-1}AP} = \begin{bmatrix} 0 & 0 & 0 \\ 0 & 2 & 0 \\ 0 & 0 & 2 \end{bmatrix}.$$

For each matrix $\mathbf{A}$ in Exercises 25 through 29, find an orthogonal matrix $\mathbf{P}$ such that $\mathbf{P^TAP}$ is diagonal.

25. $\begin{bmatrix} 0 & 1 \\ 1 & 0 \end{bmatrix}.$

26. $\begin{bmatrix} -3 & 0 & -4 \\ 0 & 5 & 0 \\ -4 & 0 & 3 \end{bmatrix}.$

27. $\begin{bmatrix} 1 & 0 & 0 \\ 0 & 2 & 3 \\ 0 & 3 & 2 \end{bmatrix}.$

28. $\begin{bmatrix} 3 & 0 & 0 \\ 0 & -4 & 3 \\ 0 & 3 & 4 \end{bmatrix}.$

29. $\begin{bmatrix} -1 & 1 & -2 \\ 1 & -1 & -2 \\ -2 & -2 & -2 \end{bmatrix}.$

In Exercises 30 through 34, find a matrix $\mathbf{P}$ such that $Q(\mathbf{PY})$ can be expressed in the form

$$Q(\mathbf{PY}) = \lambda_1 y_1^2 + \lambda_2 y_2^2 + \lambda_3 y_3^2.$$

30. $Q(\mathbf{X}) = x_1^2 + 7x_2^2 + x_3^2 + 8x_1x_2 + 16x_1x_3 - 8x_2x_3.$

31. $Q(\mathbf{X}) = 6x_1x_2 + 8x_2x_3.$

32. $Q(\mathbf{X}) = x_1^2 + x_2^2 + x_3^2 - 4x_1x_2 - 4x_1x_3 - 4x_2x_3.$

33. $Q(\mathbf{X}) = x_1^2 + x_2^2 + 2x_3^2 - 2x_1x_2 + 4x_1x_3 + 4x_2x_3.$

34. $Q(\mathbf{X}) = -3x_1^2 + 5x_2^2 + 3x_3^2 - 8x_1x_3.$

THEORETICAL EXERCISES

T-1. *Prove:* If λ is an eigenvalue of the $n \times n$ matrix $\mathbf{A}$, then the solutions of $\mathbf{AX} = \lambda\mathbf{X}$ form a subspace of R^n.

T-2. Using the definition of a determinant (Definition 4.3, Section 1.4), show that $\det(\mathbf{A} - \lambda\mathbf{I})$ is a polynomial of degree n if $\mathbf{A}$ is an $n \times n$ matrix.

T-3.   *Prove:* If $\mathbf{X}_1, \ldots, \mathbf{X}_{j-1}$ are linearly independent and $\mathbf{X}_1, \ldots, \mathbf{X}_{j-1}, \mathbf{X}_j$ are linearly dependent, then $\mathbf{X}_j$ is a linear combination of $\mathbf{X}_1, \ldots, \mathbf{X}_{j-1}$.

T-4.   Complete the proof of Theorem 6.6.

T-5.   Prove Theorem 6.7.

T-6.   Let $\mathbf{A}$ and $\mathbf{P}$ be $n \times n$ matrices, and suppose the columns of $\mathbf{P}$ are nonzero n-vectors. Show that $\mathbf{AP} = \mathbf{PD}$, where $\mathbf{D} = [d_{ij}]$ is a diagonal matrix, if and only if $d_{11}, \ldots, d_{nn}$ are eigenvalues of $\mathbf{A}$ and, for $j = 1, \ldots, n$, the jth column of $\mathbf{P}$ is an eigenvector of $\mathbf{A}$ corresponding to the eigenvalue d_{jj}.

T-7.   *Prove:* If $\mathbf{P}^{-1} = \mathbf{P}^{\mathrm{T}}$, then $\mathbf{P}$ is an orthogonal matrix.

T-8.   Prove Theorem 6.12.

T-9.   Show that similar matrices have the same characteristic polynomial. (What then can be said about the eigenvalues of similar matrices?)

T-10. Show that the eigenvalues of a triangular matrix are its diagonal elements.

T-11. *Prove:* If λ is an eigenvalue of $\mathbf{A}$, then λ^r is an eigenvalue of $\mathbf{A}^r$.

T-12. Show that $\det \mathbf{A}$ is the product of the eigenvalues of $\mathbf{A}$.

T-13. *Prove:* If $\mathbf{A}$ is an orthogonal matrix, then $\det \mathbf{A} = \pm 1$.

T-14. *Prove:* The product of orthogonal matrices is orthogonal.

T-15. Let $\mathbf{A}$ be symmetric and $\mathbf{P}$ orthogonal. Show that $\mathbf{P}^{-1}\mathbf{AP}$ is symmetric.

T-16. *Prove:* If $\mathbf{A}$ is an $n \times n$ matrix with n mutually orthogonal eigenvectors, then $\mathbf{A}$ is symmetric.

3

Vectors and Analytic Geometry

3.1 Lines and Planes

The only subspaces of R^3, other than the zero subspace and R^3 itself, are those of dimensions one and two. By definition, a one-dimensional subspace consists of all multiples of a fixed nonzero vector, and a two-dimensional subspace consists of all linear combinations of two linearly independent vectors. These subspaces have familiar geometric interpretations.

Lines

Suppose H is a one-dimensional subspace of R^3. Then any nonzero vector

$$
\mathbf{U} = \begin{bmatrix} u \\ v \\ w \end{bmatrix}
$$

in H is a basis for H; thus $\mathbf{X}$ is in H if and only if

(1) $$\mathbf{X} = t\mathbf{U} \qquad (-\infty < t < \infty).$$

The points whose position vectors with respect to the origin are in H lie on a *line L through the origin in the direction of* $\mathbf{U}$; (1) is a parametric equation of this line. In general the *line L through* $\mathbf{X}_0 = (x_0, y_0, z_0)$ *parallel to* $\mathbf{U}$ consists of the points whose position vectors with respect to $\mathbf{X}_0$ are in H (Fig. 1.1); it is given parametrically by

$$\mathbf{X} = \mathbf{X}_0 + t\mathbf{U}$$

or in terms of components by

$$
\begin{aligned}
x &= x_0 + tu, \\
y &= y_0 + tv, \\
z &= z_0 + tw.
\end{aligned}
$$

(2)

If u, v, and w are all nonzero, then we can solve each equation in (2) for t and equate the results to obtain the *symmetric* form of the equation of L:

(3)
$$\frac{x - x_0}{u} = \frac{y - y_0}{v} = \frac{z - z_0}{w}.$$

If one or two of the components of $\mathbf{U}$ vanish, then (3) indicates division by zero, which is never legitimate. Nevertheless it is convenient to adopt

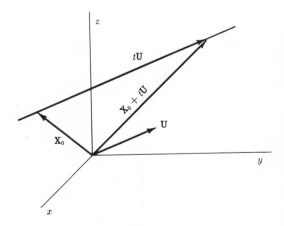

FIGURE 1.1

the convention that (3) stands for the line through $\mathbf{X}_0$ in the direction of $\mathbf{U}$ if at least one of the denominators is nonzero.

Thus

$$\frac{x - x_0}{0} = \frac{y - y_0}{v} = \frac{z - z_0}{w}$$

is the equation of a line along which x is constant ($x = x_0$) and, if u and v are both nonzero,

$$\frac{y - y_0}{v} = \frac{z - z_0}{w}.$$

Example 1.1 The line L through the origin in the direction of

$$\mathbf{U} = \begin{bmatrix} 1 \\ 2 \\ -1 \end{bmatrix}$$

is given parametrically by

$$\begin{bmatrix} x \\ y \\ z \end{bmatrix} = t \begin{bmatrix} 1 \\ 2 \\ -1 \end{bmatrix} \qquad (-\infty < t < \infty);$$

its symmetric equation is

$$x = \tfrac{1}{2}y = -z.$$

Example 1.2 The set of points

$$\mathbf{X} = \begin{bmatrix} 3 \\ 1 \\ -2 \end{bmatrix} + t \begin{bmatrix} 0 \\ -1 \\ 2 \end{bmatrix} \qquad (-\infty < t < \infty)$$

is the line L through $\mathbf{X}_0 = (3, 1, -2)$ in the direction of

$$\mathbf{U} = \begin{bmatrix} 0 \\ -1 \\ 2 \end{bmatrix};$$

its symmetric equation is

$$\frac{x-3}{0} = \frac{(y+1)}{-1} = \frac{z+2}{2}.$$

The vectors from $\mathbf{X}_0$ to points on L are the elements of the subspace spanned by $\mathbf{U}$; that is, a point $\mathbf{X}$ is on L if and only if the vector $\mathbf{X} - \mathbf{X}_0$ is parallel to $\mathbf{U}$.

Definition 1.1 Let $\mathbf{X}_0 = (\bar{x}_1, \ldots, \bar{x}_n)$ and

$$\mathbf{U} = \begin{bmatrix} u_1 \\ \vdots \\ u_n \end{bmatrix} \neq 0.$$

The line L through $\mathbf{X}_0$ in the direction of $\mathbf{U}$ is the set of points given parametrically by

$$(4) \qquad \mathbf{X} = \mathbf{X}_0 + t\mathbf{U} \qquad (-\infty < t < \infty),$$

or in symmetric form by

$$(5) \qquad \frac{x_1 - \bar{x}_1}{u_1} = \frac{x_2 - \bar{x}_2}{u_2} = \cdots = \frac{x_n - \bar{x}_n}{u_n}.$$

If the components of $\mathbf{U}$ are all nonzero then (5) can be obtained by rewriting (4) as

$$x_i = \bar{x}_i + tu_i \qquad (1 \le i \le n),$$

solving each equation for t, and equating the results. If one or more of the components of $\mathbf{U}$ vanishes then (5) must be interpreted as symbolically representing the line defined by (4).

Example 1.3 In R^5 the line through $\mathbf{X}_0 = (2, 3, -1, 4, 5)$ in the direction of

$$\mathbf{U} = \begin{bmatrix} 1 \\ 2 \\ -1 \\ 3 \\ 4 \end{bmatrix} \qquad \text{is given parametrically by} \qquad \mathbf{X} = \begin{bmatrix} 2 \\ 3 \\ -1 \\ 4 \\ 5 \end{bmatrix} + t \begin{bmatrix} 1 \\ 2 \\ -1 \\ 3 \\ 4 \end{bmatrix},$$

and in symmetric form by

$$\frac{x_1 - 2}{1} = \frac{x_2 - 3}{2} = \frac{(x_3 + 1)}{-1} = \frac{x_4 - 4}{3} = \frac{x_5 - 5}{4} .$$

Example 1.4 The symmetric equation

$$\frac{x_1 - 1}{0} = \frac{x_2 - 3}{4} = \frac{x_3 + 5}{0} = \frac{x_4 - 5}{2} = \frac{x_5}{1}$$

represents the line in R^5 through $\mathbf{X_0} = (1, 3, -5, 5, 0)$ in the direction of

$$\mathbf{U} = \begin{bmatrix} 0 \\ 4 \\ 0 \\ 2 \\ 1 \end{bmatrix}.$$

A line is uniquely determined by requiring that it pass through two points. Thus, if $\mathbf{X_0}$ and $\mathbf{X_1}$ are distinct points on L, then L is the line through $\mathbf{X_0}$ in the direction of $\mathbf{X_1} - \mathbf{X_0}$. It is given parametrically by

(6) $$\mathbf{X} = \mathbf{X_0} + t(\mathbf{X_1} - \mathbf{X_0}) ;$$

in Fig. 1.2 this line is depicted in R^3.

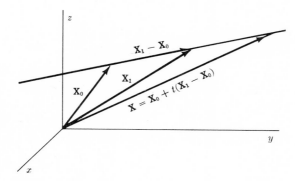

FIGURE 1.2

Example 1.5 In R^4 let $\mathbf{X}_0 = (1, 2, -1, 3)$, and $\mathbf{X}_1 = (4, 5, -2, 1)$. Then

$$\mathbf{X}_1 - \mathbf{X}_0 = \begin{bmatrix} 3 \\ 3 \\ -1 \\ -2 \end{bmatrix};$$

hence the line through $\mathbf{X}_0$ and $\mathbf{X}_1$ is given by

$$\mathbf{X} = \begin{bmatrix} 1 \\ 2 \\ -1 \\ 3 \end{bmatrix} + t \begin{bmatrix} 3 \\ 3 \\ -1 \\ -2 \end{bmatrix}.$$

Clearly any line L can be represented in infinitely many ways in parametric (or symmetric) form: its equation can be written in the form (4), where $\mathbf{X}_0$ in any point on L and $\mathbf{U}_0$ is any vector in the direction of L, or in the form (6), where $\mathbf{X}_1$ and $\mathbf{X}_0$ are any two distinct points on L. We shall give criteria for deciding when lines represented by distinct equations are identical. However, before doing this it is convenient to consider other relationships between lines.

Definition 1.2 If $\mathbf{X}_0$ is common to lines L_1 and L_2, then the lines are said to *intersect* at $\mathbf{X}_0$, which is called their *point of intersection*.

Definition 1.3 If nonzero vectors in the directions of lines L_1 and L_2 are parallel, then L_1 and L_2 are said to be *parallel*.

Definition 1.4 If nonzero vectors in the directions of lines L_1 and L_2 are orthogonal, then L_1 and L_2 are said to be *perpendicular*, or *orthogonal*.

Definition 1.5 Lines L_1 and L_2 are said to be *skew* if they are not parallel and do not intersect.

Theorem 1.1 Let L_1 and L_2 be given parametrically by

(7) $$\mathbf{X} = \mathbf{X}_0 + s\mathbf{U}$$

and

(8)
$$\mathbf{X} = \mathbf{X}_1 + t\mathbf{V}.$$

Then:

(a) L_1 and L_2 are parallel if and only if $\mathbf{U} = a\mathbf{V}$ for some scalar a.
(b) L_1 and L_2 are perpendicular if and only if $\mathbf{U} \cdot \mathbf{V} = 0$.
(c) L_1 and L_2 intersect if and only if $\mathbf{X}_1 - \mathbf{X}_0$ is a linear combination of $\mathbf{U}$ and $\mathbf{V}$.

Proof. Assertions (a) and (b) follow from Definitions 1.3 and 1.4. For (c), suppose L_1 and L_2 intersect at $\bar{\mathbf{X}}$; then there are scalars s_0 and t_0 such that

(9)
$$\bar{\mathbf{X}} = \mathbf{X}_0 + s_0\mathbf{U} = \mathbf{X}_1 + t_0\mathbf{V}.$$

Thus

(10)
$$\mathbf{X}_1 - \mathbf{X}_0 = s_0\mathbf{U} - t_0\mathbf{V}$$

is a linear combination of $\mathbf{U}$ and $\mathbf{V}$. Conversely, if $\mathbf{X}_1 - \mathbf{X}_0$ is of the form (10), then the last two members of (9) are equal, and the common point that they represent is on both lines. This completes the proof of Theorem 1.1.

Example 1.6 The equations

$$\mathbf{X} = \mathbf{X}_0 + s\mathbf{U} \qquad\qquad \mathbf{X} = \mathbf{X}_1 + t\mathbf{V}$$

$$= \begin{bmatrix} -1 \\ 0 \\ 2 \end{bmatrix} + s\begin{bmatrix} 1 \\ 2 \\ 0 \end{bmatrix} \quad \text{and} \quad = \begin{bmatrix} 3 \\ 1 \\ 2 \end{bmatrix} + t\begin{bmatrix} 2 \\ 4 \\ 0 \end{bmatrix}$$

define parallel lines, since $\mathbf{V} = 2\mathbf{U}$.

Example 1.7 The equations

$$\mathbf{X} = \mathbf{X}_0 + s\mathbf{U} \qquad\qquad \mathbf{X} = \mathbf{X}_1 + t\mathbf{V}$$

$$= \begin{bmatrix} -1 \\ 0 \\ 0 \end{bmatrix} + s\begin{bmatrix} 1 \\ 2 \\ 0 \end{bmatrix} \quad \text{and} \quad = \begin{bmatrix} 0 \\ 0 \\ z_0 \end{bmatrix} + t\begin{bmatrix} 1 \\ 0 \\ 2 \end{bmatrix}$$

are not parallel, since $\mathbf{U}$ and $\mathbf{V}$ are linearly independent. They intersect if and only if

$$\begin{bmatrix} -1 \\ 0 \\ 0 \end{bmatrix} + s \begin{bmatrix} 1 \\ 2 \\ 0 \end{bmatrix} = \begin{bmatrix} 0 \\ 0 \\ z_0 \end{bmatrix} + t \begin{bmatrix} 1 \\ 0 \\ 2 \end{bmatrix}$$

has a solution $s = s_0$ and $t = t_0$. The conditions on (s_0, t_0) are

$$s_0 - t_0 = 1,$$

$$2s_0 = 0,$$

$$2t_0 + z_0 = 0.$$

Since the first two equations imply that $s_0 = 0$ and $t_0 = -1$, this system has a solution if and only if $z_0 = 2$; thus the lines intersect if and only if $z_0 = 2$, in which case the point of intersection is $\mathbf{X_0} = (-1, 0, 0)$.

Theorem 1.2 The lines L_1 and L_2 defined by (7) and (8) are identical if and only if $\mathbf{X_1} - \mathbf{X_0}$ and $\mathbf{U}$ are both parallel to $\mathbf{V}$.

We leave the proof of this theorem to the student (Exercise T-1).

Planes

Now suppose H is the two-dimensional subspace of R^3 consisting of vectors of the form

(11) $$\mathbf{X} = s\mathbf{U} + t\mathbf{V} \qquad (-\infty < s, t < \infty),$$

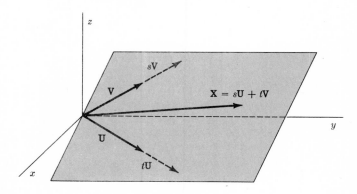

FIGURE 1.3

where **U** and **V** are linearly independent. The points whose position vectors with respect to the origin are in H form a plane through the origin (Fig. 1.3); (11) is a parametric equation of this plane, which we call the *plane of* **U** *and* **V**.

In this text we shall not use parametric equations for planes. It is intuitively evident that a point in R^3 is in the plane shown in Fig. 1.2 if and only if its position vector with respect to the origin is perpendicular to **N**. We shall use this fact to derive a nonparametric equation for the plane defined by (11).

The vector

$$\mathbf{N} = \begin{bmatrix} u_2 v_3 - u_3 v_2 \\ u_3 v_1 - u_1 v_3 \\ u_1 v_2 - u_2 v_1 \end{bmatrix}$$

is perpendicular to both **U** and **V**. This is easily checked by noting that if

$$\mathbf{W} = \begin{bmatrix} w_1 \\ w_2 \\ w_3 \end{bmatrix}$$

is an arbitrary vector then

$$\mathbf{W} \cdot \mathbf{N} = \begin{vmatrix} w_1 & w_2 & w_3 \\ u_1 & u_2 & u_3 \\ v_1 & v_2 & v_3 \end{vmatrix},$$

as can be seen by expanding the determinant with respect to its first row. Hence

$$\mathbf{U} \cdot \mathbf{N} = \mathbf{V} \cdot \mathbf{N} = 0,$$

because a determinant with two identical rows vanishes. Furthermore, **N** = **0** would imply that **U** and **V** are parallel (Exercise T-2), contrary to assumption; hence **N** $\neq$ **0**.

The vector **N** is called the *cross product* of **U** and **V** and is usually denoted by

$$\mathbf{N} = \mathbf{U} \times \mathbf{V}.$$

We shall discuss it further in Section 3.2, but for now we regard it merely as a particular vector perpendicular to **U** and **V**. Since it is also perpendicular to any linear combination of **U** and **V** (Exercise T-5), we say it is *perpendicular to the plane of* **U** *and* **V**.

Example 1.8 Let

$$\mathbf{U} = \begin{bmatrix} 2 \\ 3 \\ 1 \end{bmatrix} \quad \text{and} \quad \mathbf{V} = \begin{bmatrix} 3 \\ 1 \\ 0 \end{bmatrix}.$$

Then

$$\mathbf{N} = \mathbf{U} \times \mathbf{V}$$

$$= \begin{bmatrix} (3)(0) - (1)(1) \\ (1)(3) - (2)(0) \\ (2)(1) - (3)(3) \end{bmatrix} = \begin{bmatrix} -1 \\ 3 \\ -7 \end{bmatrix},$$

and by direct computation $\mathbf{N} \cdot \mathbf{U} = \mathbf{N} \cdot \mathbf{V} = 0$.

Theorem 1.3 A vector **X** is in the subspace of R^3 spanned by two linearly independent vectors **U** and **V** if and only if $\mathbf{X} \cdot (\mathbf{U} \times \mathbf{V}) = 0$.

Proof. If **U** and **V** are linearly independent then **U**, **V**, and $\mathbf{N} = \mathbf{U} \times \mathbf{V}$ are also linearly independent (Exercise T-3); therefore they form a basis for R^3. An arbitrary **X** in R^3 can be written in the form

$$\mathbf{X} = s\mathbf{U} + t\mathbf{V} + a\mathbf{N},$$

where s, t, and a are scalars. Thus

$$\mathbf{X} \cdot \mathbf{N} = s\mathbf{U} \cdot \mathbf{N} + t\mathbf{V} \cdot \mathbf{N} + a\mathbf{N} \cdot \mathbf{N}$$

$$= a \mid \mathbf{N} \mid^2.$$

Thus $\mathbf{X} \cdot \mathbf{N} = 0$ if and only if $a = 0$; that is, if and only if **X** is in the subspace spanned by **U** and **V**.

We now define planes in R^n.

Definition 1.6 If **N** is a nonzero n-vector, then the set of points in $R^n (n \geq 3)$ which satisfy

$$\mathbf{X} \cdot \mathbf{N} = 0$$

is called *the plane through the origin normal to* **N**. More generally, the set of points $\mathbf{X}$ which satisfy

(12)
$$(\mathbf{X} - \mathbf{X}_0) \cdot \mathbf{N} = 0$$

for a fixed $\mathbf{X}_0$ is called the *plane through* $\mathbf{X}_0$ *normal to* $\mathbf{N}$.

Example 1.9 In R^4 the set of points whose coordinates satisfy

$$2(x_1 - 1) + (x_2 - 3) - 3x_3 + 3(x_4 + 2) = 0$$

is the plane through $\mathbf{X}_0 = (1, 3, 0, -2)$ normal to

$$\mathbf{N} = \begin{bmatrix} 2 \\ 1 \\ -3 \\ 3 \end{bmatrix}.$$

The equation of a plane can also be written as

$$\mathbf{X} \cdot \mathbf{N} = d,$$

where d is a constant. Comparing this equation with (12) shows that $d = \mathbf{X}_0 \cdot \mathbf{N}$, where $\mathbf{X}_0$ is any point on the plane.

Example 1.10 The equation of the plane in Example 1.9 can be written as

$$2x_1 + x_2 - 3x_3 + 3x_4 = -1.$$

In Section 3.2 we shall consider planes in R^3 in more detail. We conclude our general discussion with the following theorem:

Theorem 1.4 A plane H through the origin of R^n is an $(n - 1)$-dimensional subspace of R^n.

Proof. Let $\mathbf{N} \neq \mathbf{0}$ be normal to H and suppose $V = \{\mathbf{X}_1, \ldots, \mathbf{X}_n\}$ is a basis for R^n. Then $\mathbf{N}$ is a linear combination of $\mathbf{X}_1, \ldots, \mathbf{X}_n$:

$$\mathbf{N} = a_1 \mathbf{X}_1 + \cdots + a_n \mathbf{X}_n.$$

We can assume, possibly at the expense of renumbering the basis vectors, that $a_n \neq 0$. Then

$$\mathbf{X}_n = \frac{1}{a_n} (\mathbf{N} - a_1 \mathbf{X}_1 - \cdots - a_{n-1} \mathbf{X}_{n-1}),$$

and $V' = \{\mathbf{X}_1, \ldots, \mathbf{X}_{n-1}, \mathbf{N}\}$ spans R^n; hence V' is linearly independent and therefore a basis for R^n. Now consider the vectors

(13) $$\mathbf{Y}_i = \mathbf{X}_i - \frac{(\mathbf{X}_i \cdot \mathbf{N})}{|\mathbf{N}|^2}\,\mathbf{N} \qquad (1 \le i \le n-1);$$

direct calculation shows that

(14) $$\mathbf{Y}_i \cdot \mathbf{N} = 0,$$

and $V'' = \{\mathbf{Y}_1, \ldots, \mathbf{Y}_{n-1}, \mathbf{N}\}$ is linearly independent, for if

$$b_1\mathbf{Y}_1 + \cdots + b_{n-1}\mathbf{Y}_{n-1} + b_n\mathbf{N} = \mathbf{0},$$

then from (13),

$$b_1\mathbf{X}_1 + \cdots + b_{n-1}\mathbf{X}_{n-1} + \left(b_n - b_1\frac{(\mathbf{X}_1 \cdot \mathbf{N})}{|\mathbf{N}|^2} - \cdots - b_{n-1}\frac{(\mathbf{X}_{n-1} \cdot \mathbf{N})}{|\mathbf{N}|^2}\right)\mathbf{N} = \mathbf{0}$$

and the linear independence of V' implies that $b_1 = \cdots = b_n = 0$.

Now, $\mathbf{X}$ is in H if and only if

$$\mathbf{X} \cdot \mathbf{N} = 0;$$

on the other hand, an arbitrary $\mathbf{X}$ in R^n can be written as

(15) $$\mathbf{X} = a_1\mathbf{Y}_1 + \cdots + a_{n-1}\mathbf{Y}_{n-1} + d\mathbf{N},$$

and therefore, from (14),

$$\mathbf{X} \cdot \mathbf{N} = d\,|\mathbf{N}|^2.$$

Thus $\mathbf{X}$ is in H if and only if $d = 0$ in (15). This implies that H is an $(n-1)$-dimensional subspace of R^n.

EXERCISES 3.1

1. State which of the following points are on the line

$$\mathbf{X} = \begin{bmatrix} 2 \\ -1 \\ -2 \\ 3 \end{bmatrix} + t\begin{bmatrix} 0 \\ -1 \\ 4 \\ -3 \end{bmatrix}.$$

(a) $\begin{bmatrix} 2 \\ -3 \\ 6 \\ -3 \end{bmatrix}$;   (b) $\begin{bmatrix} 2 \\ 4 \\ 4 \\ 2 \end{bmatrix}$;   (c) $\begin{bmatrix} 2 \\ 1 \\ -10 \\ 9 \end{bmatrix}$;   (d) $\begin{bmatrix} 2 \\ -2 \\ 2 \\ 0 \end{bmatrix}$.

2.   State which of the following points are on the line

$$\frac{x-2}{3} = \frac{y-4}{-2} = \frac{z+5}{4}.$$

(a) $\begin{bmatrix} -1 \\ 6 \\ -9 \end{bmatrix}$;   (b) $\begin{bmatrix} 5 \\ 2 \\ -1 \end{bmatrix}$;   (c) $\begin{bmatrix} 1 \\ 2 \\ -1 \end{bmatrix}$;   (d) $\begin{bmatrix} 5 \\ 2 \\ 0 \end{bmatrix}$.

3.   Write the equations of the following lines in symmetric form.

(a) $\mathbf{X} = \begin{bmatrix} 2 \\ -1 \\ 3 \\ 3 \end{bmatrix} + t \begin{bmatrix} 0 \\ -1 \\ 4 \\ -3 \end{bmatrix}$;   (b) $\mathbf{X} = \begin{bmatrix} 2 \\ 1 \\ 0 \end{bmatrix} + t \begin{bmatrix} 1 \\ -1 \\ 2 \end{bmatrix}$.

4.   Write parametric equations for the following lines.

(a) $\dfrac{x-2}{4} = \dfrac{y+2}{0} = \dfrac{z-3}{-2}$;

(b) $\dfrac{x_1-4}{0} = \dfrac{x_2+3}{3} = \dfrac{x_3-2}{2} = \dfrac{x_4-3}{-3}$.

5.   Find the equation of the line through $\mathbf{X}_0 = (1, 2, -1)$ in the direction of

$$\mathbf{U} = \begin{bmatrix} 4 \\ -2 \\ 5 \end{bmatrix}.$$

6.   Repeat Exercise 5 for $\mathbf{X}_0 = (3, 2, 1, 0)$ and

$$\mathbf{U} = \begin{bmatrix} 0 \\ 4 \\ 0 \\ -3 \end{bmatrix}.$$

7. Find the equation of the line through $\mathbf{X}_1 = (2, 1, 0, 3)$, and $\mathbf{X}_2 = (-1, -1, 2, 0)$.

8. Repeat Exercise 7 for $\mathbf{X}_1 = (-3, 1, 0)$ and $\mathbf{X}_2 = (3, 4, -1)$.

9. Find the equation of the line through $(4, -2, 5, 2)$ parallel to the line

$$\mathbf{X} = \begin{bmatrix} 1 \\ -2 \\ 3 \\ 2 \end{bmatrix} + t \begin{bmatrix} 0 \\ 1 \\ -2 \\ 4 \end{bmatrix}.$$

10. Find the equation of the line through $(3, -1, 2)$ and parallel to the line

$$\frac{x-2}{-2} = \frac{y+3}{1} = \frac{z-3}{0}.$$

11. Are the points $(1, -2, 3)$, $(2, -3, 5)$, and $(0, -1, 1)$ on the same line?

12. Are the points $(0, 1, 1, -2)$, $(3, 2, -3, 1)$, and $(6, 3, -8, 4)$ on the same line?

13. Find the point of intersection of the lines

$$\mathbf{X} = \begin{bmatrix} 2 \\ 2 \\ 2 \end{bmatrix} + s \begin{bmatrix} 1 \\ -1 \\ 2 \end{bmatrix} \quad \text{and} \quad \mathbf{X} = \begin{bmatrix} 3 \\ 11 \\ -1 \end{bmatrix} + t \begin{bmatrix} 3 \\ -1 \\ 5 \end{bmatrix}.$$

14. Find the point of intersection of the lines

$$\mathbf{X} = \begin{bmatrix} 1 \\ 3 \\ -1 \\ 2 \end{bmatrix} + s \begin{bmatrix} 0 \\ 1 \\ 2 \\ -3 \end{bmatrix} \quad \text{and} \quad \mathbf{X} = \begin{bmatrix} 2 \\ 1 \\ 4 \\ 5 \end{bmatrix} + t \begin{bmatrix} 1 \\ -2 \\ 5 \\ 3 \end{bmatrix}.$$

15. Show that the lines

$$\mathbf{X} = \begin{bmatrix} 2 \\ 3 \\ 1 \\ 4 \end{bmatrix} + s \begin{bmatrix} -2 \\ 1 \\ 0 \\ 1 \end{bmatrix} \quad \text{and} \quad \mathbf{X} = \begin{bmatrix} 0 \\ 1 \\ 2 \\ 3 \end{bmatrix} + t \begin{bmatrix} -1 \\ 3 \\ 2 \\ 1 \end{bmatrix}$$

are skew.

16. Are the lines

$$\mathbf{X} = \begin{bmatrix} 1 \\ 2 \\ -1 \end{bmatrix} + s \begin{bmatrix} 1 \\ -2 \\ 3 \end{bmatrix} \quad \text{and} \quad \mathbf{X} = \begin{bmatrix} 0 \\ 2 \\ 3 \end{bmatrix} + t \begin{bmatrix} 0 \\ 3 \\ 2 \end{bmatrix}$$

perpendicular?

17. Are the lines

$$\mathbf{X} = \begin{bmatrix} 0 \\ 2 \\ 1 \\ 4 \end{bmatrix} + s \begin{bmatrix} 1 \\ -1 \\ -2 \\ 2 \end{bmatrix} \quad \text{and} \quad \mathbf{X} = \begin{bmatrix} 3 \\ 1 \\ -2 \\ -4 \end{bmatrix} + t \begin{bmatrix} 4 \\ 2 \\ 2 \\ -1 \end{bmatrix}$$

perpendicular?

18. Show that

$$\mathbf{X} = \begin{bmatrix} 0 \\ 1 \\ -2 \end{bmatrix} + s \begin{bmatrix} 1 \\ -3 \\ 4 \end{bmatrix} \quad \text{and} \quad \mathbf{X} = \begin{bmatrix} 3 \\ -8 \\ 10 \end{bmatrix} + t \begin{bmatrix} 3 \\ -9 \\ 12 \end{bmatrix}$$

define the same line.

19. Show that

$$\mathbf{X} = \begin{bmatrix} 1 \\ -2 \\ 3 \\ -1 \end{bmatrix} + s \begin{bmatrix} 0 \\ 1 \\ 2 \\ -3 \end{bmatrix} \quad \text{and} \quad \mathbf{X} = \begin{bmatrix} 1 \\ 0 \\ 7 \\ -7 \end{bmatrix} + t \begin{bmatrix} 0 \\ 2 \\ 4 \\ -6 \end{bmatrix}$$

define the same line.

20. Are the lines

$$\mathbf{X} = \begin{bmatrix} 2 \\ -1 \\ 3 \end{bmatrix} + s \begin{bmatrix} 5 \\ 3 \\ -2 \end{bmatrix} \quad \text{and} \quad \mathbf{X} = \begin{bmatrix} 4 \\ -2 \\ 6 \end{bmatrix} + t \begin{bmatrix} -10 \\ -6 \\ 4 \end{bmatrix}$$

parallel?

21. Are the lines

$$\mathbf{X} = \begin{bmatrix} 1 \\ -2 \\ 3 \\ 4 \end{bmatrix} + s \begin{bmatrix} 1 \\ -2 \\ -3 \\ 0 \end{bmatrix} \quad \text{and} \quad \mathbf{X} = \begin{bmatrix} 3 \\ 2 \\ -5 \\ 1 \end{bmatrix} + t \begin{bmatrix} -2 \\ 4 \\ 6 \\ 0 \end{bmatrix}$$

parallel?

22. Find the equation of the line through $(1, 0, 3)$, parallel to the line determined by $(3, 1, -2)$ and $(2, 1, -3)$.

23. Let L be defined by

$$\mathbf{X} = \begin{bmatrix} 1 \\ 0 \\ 2 \end{bmatrix} + t \begin{bmatrix} 3 \\ -1 \\ 2 \end{bmatrix}.$$

Find the equation of the line through $(3, 5, 3)$ that intersects L at right angles.

24. Find the cosine of the angle of intersection between the lines defined in Exercise 13.

25. Find the cosine of the angle of intersection between the lines defined in Exercise 14.

26. If

$$\mathbf{U} = \begin{bmatrix} 1 \\ 3 \\ -4 \end{bmatrix} \quad \text{and} \quad \mathbf{V} = \begin{bmatrix} -2 \\ 0 \\ 3 \end{bmatrix},$$

find $\mathbf{N} = \mathbf{U} \times \mathbf{V}$ and verify that $\mathbf{N}$ is perpendicular to $\mathbf{U}$ and $\mathbf{V}$.

27. Repeat Exercise 26 for

$$\mathbf{U} = \begin{bmatrix} -1 \\ 0 \\ 2 \end{bmatrix} \quad \text{and} \quad \mathbf{V} = \begin{bmatrix} 0 \\ -3 \\ -4 \end{bmatrix}.$$

28. Find the equation of the plane through $(4, -2, 1)$, perpendicular to

$$\mathbf{N} = \begin{bmatrix} 3 \\ -2 \\ 4 \end{bmatrix}.$$

29. Find the equation of the plane through $(1, -2, 3, 0)$, orthogonal to

$$\mathbf{N} = \begin{bmatrix} 2 \\ -1 \\ 4 \\ -5 \end{bmatrix}.$$

30. Find the equation of the plane that contains the points $(1, 3, 2)$, $(-1, 2, 3)$, and $(0, 2, 1)$.

31. Find the equation of the plane that contains $(1, -1, 2)$ and the line

$$\mathbf{X} = \begin{bmatrix} 4 \\ -1 \\ 2 \end{bmatrix} + t \begin{bmatrix} 0 \\ 2 \\ 1 \end{bmatrix}.$$

32. Find the equation of the plane that contains the lines defined in Exercise 13.

33. Find the line through $(1, 3, -2, 4)$, perpendicular to the plane defined by
$$2x_1 + 3x_2 + 3x_3 - 2x_4 = 5.$$

34. Find the equation of the plane through $(3, -2, 4, 2)$ and perpendicular to the line determined by $(1, 0, 2, -3)$ and $(3, -2, 4, 2)$.

THEORETICAL EXERCISES

T-1. Prove Theorem 1.2.

T-2. Prove: If $\mathbf{U} \times \mathbf{V} = \mathbf{0}$ then $\mathbf{U}$ and $\mathbf{V}$ are parallel.

T-3. Prove: If $\mathbf{U}$ and $\mathbf{V}$ are linearly independent, then $\mathbf{U}$, $\mathbf{V}$ and $\mathbf{U} \times \mathbf{V}$ form a basis for R^3.

T-4. (Cross product in R^n). Let $n \geq 2$ and

$$\mathbf{U}_1 = \begin{bmatrix} u_{11} \\ u_{12} \\ \vdots \\ u_{1n} \end{bmatrix}, \qquad \mathbf{U}_2 = \begin{bmatrix} u_{21} \\ u_{22} \\ \vdots \\ u_{2n} \end{bmatrix}, \ldots, \qquad \mathbf{U}_{n-1} = \begin{bmatrix} u_{n-1,1} \\ u_{n-1,2} \\ \vdots \\ u_{n-1,n} \end{bmatrix},$$

and define

$$\mathbf{N} = \begin{bmatrix} v_1 \\ v_2 \\ \vdots \\ v_n \end{bmatrix},$$

where v_i is the cofactor of x_i in the determinant

$$D = \begin{bmatrix} x_1 & x_2 \cdots x_n \\ u_{11} & u_{12} \cdots u_{1n} \\ \vdots & \vdots \\ u_{n-1,1} & u_{n-2,2} \cdots u_{n-1,n} \end{bmatrix}.$$

(a) Show that $\mathbf{N} \cdot \mathbf{U}_i = 0$ $(1 \leq i \leq n-1)$.

(b) Show that $\mathbf{N} = \mathbf{0}$ if and only if $\mathbf{U}_1, \ldots, \mathbf{U}_{n-1}$ are linearly dependent.

T-5.   Show that if **X** is perpendicular to each of $\mathbf{X}_1$, $\mathbf{X}_2$, ..., $\mathbf{X}_k$, then **X** is perpendicular to every vector in $S[\mathbf{X}_1, \mathbf{X}_2, \ldots, \mathbf{X}_k]$.

3.2   Vectors in R^3

Because of its special importance as a mathematical model of the three-dimensional world in which we live, we pay particular attention to R^3. Throughout this section we shall assume that all vectors, lines, and planes are in R^3. It is important to remember this, because some of our statements will be incorrect or meaningless in R^n for $n \neq 3$.

Coplanar and Collinear Vectors

Vectors **U**, **V**, and **W** are linearly dependent if there are constants a, b, and c, not all zero, such that

$$a\mathbf{U} + b\mathbf{V} + c\mathbf{W} = 0.$$

If $c \neq 0$, this equation can be rewritten as

$$\mathbf{W} = s\mathbf{U} + t\mathbf{V},$$

from which it follows that directed line segments drawn from the origin to represent **U**, **V**, and **W** lie in a plane (Fig. 2.1). Therefore we say that three vectors are *coplanar* if they are linearly dependent, and *noncoplanar* if they are not.

Two vectors **U** and **V** are linearly dependent if and only if line segments drawn from the origin in the directions of **U** and **V** are on the same line; hence two vectors are said to be *collinear* if they are parallel, and *noncollinear* if they are not.

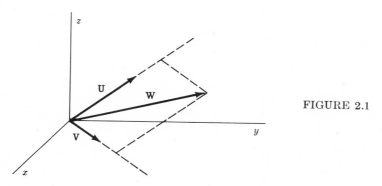

FIGURE 2.1

The Natural Basis for R^3

The natural basis for R^3 is

$$\mathbf{E_1} = \begin{bmatrix} 1 \\ 0 \\ 0 \end{bmatrix}, \qquad \mathbf{E_2} = \begin{bmatrix} 0 \\ 1 \\ 0 \end{bmatrix}, \qquad \text{and} \qquad \mathbf{E_3} = \begin{bmatrix} 0 \\ 0 \\ 1 \end{bmatrix}.$$

It is traditional to denote these vectors as $\mathbf{E_1} = \mathbf{i}$, $\mathbf{E_2} = \mathbf{j}$, and $\mathbf{E_3} = \mathbf{k}$; then an arbitrary vector

$$(1) \qquad\qquad \mathbf{X} = \begin{bmatrix} x \\ y \\ z \end{bmatrix}$$

can be written as

$$\mathbf{X} = x\mathbf{i} + y\mathbf{j} + z\mathbf{k}.$$

We shall use this representation for vectors in R^3 when it is convenient; however, we shall also continue to use (1).

Example 2.1

$$\begin{bmatrix} 1 \\ 2 \\ 3 \end{bmatrix} = \mathbf{i} + 2\mathbf{j} + 3\mathbf{k}, \qquad\qquad \begin{bmatrix} -3 \\ 0 \\ 1 \end{bmatrix} = -3\mathbf{i} + \mathbf{k},$$

and

$$\begin{bmatrix} \frac{2}{3} \\ 5 \\ -1 \end{bmatrix} = \tfrac{2}{3}\mathbf{i} + 5\mathbf{j} - \mathbf{k}.$$

The vectors $\mathbf{i}$, $\mathbf{j}$, and $\mathbf{k}$ can be represented as directed line segments of unit length along the positive directions of the x-, y-, and z-axes, respectively (Fig. 2.2). They form an orthonormal basis for R^3; that is,

$$\mathbf{i} \cdot \mathbf{i} = \mathbf{j} \cdot \mathbf{j} = \mathbf{k} \cdot \mathbf{k} = 1$$

and

$$\mathbf{i} \cdot \mathbf{j} = \mathbf{j} \cdot \mathbf{k} = \mathbf{k} \cdot \mathbf{i} = 0.$$

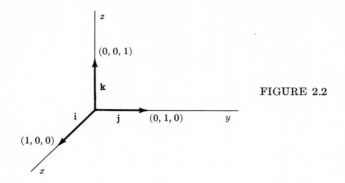

FIGURE 2.2

Cross Product of Two Vectors

Definition 2.1   The cross product $\mathbf{U} \times \mathbf{V}$ of two vectors $\mathbf{U} = u_1\mathbf{i} + u_2\mathbf{j} + u_3\mathbf{k}$ and $\mathbf{V} = v_1\mathbf{i} + v_2\mathbf{j} + v_3\mathbf{k}$ is defined by

$$(2) \qquad \mathbf{U} \times \mathbf{V} = (u_2v_3 - u_3v_2)\mathbf{i} + (u_3v_1 - u_1v_3)\mathbf{j} + (u_1v_2 - u_2v_1)\mathbf{k}.$$

It can be represented symbolically as a determinant,

$$(3) \quad \mathbf{U} \times \mathbf{V} = \begin{vmatrix} \mathbf{i} & \mathbf{j} & \mathbf{k} \\ u_1 & u_2 & u_3 \\ v_1 & v_2 & v_3 \end{vmatrix} = \begin{vmatrix} u_2 & u_3 \\ v_2 & v_3 \end{vmatrix}\mathbf{i} - \begin{vmatrix} u_1 & u_3 \\ v_1 & v_3 \end{vmatrix}\mathbf{j} + \begin{vmatrix} u_1 & u_2 \\ v_1 & v_2 \end{vmatrix}\mathbf{k};$$

thus (2) is the formal expansion of (3) according to the elements of its first row.

Example 2.2   Let $\mathbf{U} = \mathbf{i} + \mathbf{j} - 2\mathbf{k}$, $\mathbf{V} = \mathbf{i} + 2\mathbf{j} + 3\mathbf{k}$, and $\mathbf{W} = 2\mathbf{i} + 4\mathbf{j} + 6\mathbf{k}$. Then

$$\mathbf{U} \times \mathbf{V} = \begin{vmatrix} \mathbf{i} & \mathbf{j} & \mathbf{k} \\ 1 & 1 & -2 \\ 1 & 2 & 3 \end{vmatrix} = 7\mathbf{i} - 5\mathbf{j} + \mathbf{k}$$

and

$$\mathbf{V} \times \mathbf{W} = \begin{vmatrix} \mathbf{i} & \mathbf{j} & \mathbf{k} \\ 1 & 2 & 3 \\ 2 & 4 & 6 \end{vmatrix} = \mathbf{0}.$$

Example 2.3   It is easily verified that

$$\mathbf{i} \times \mathbf{i} = \mathbf{j} \times \mathbf{j} = \mathbf{k} \times \mathbf{k} = 0;$$

(4) $$\mathbf{i} \times \mathbf{j} = \mathbf{k}, \quad \mathbf{j} \times \mathbf{k} = \mathbf{i}, \quad \mathbf{k} \times \mathbf{i} = \mathbf{j}$$

and

(5) $$\mathbf{j} \times \mathbf{i} = -\mathbf{k}, \quad \mathbf{k} \times \mathbf{j} = -\mathbf{i}, \quad \mathbf{i} \times \mathbf{k} = -\mathbf{j}.$$

To remember (4) and (5), it is helpful to refer to Fig. 2.3: if, when reading the two basis vectors in the order in which they appear in the cross product, one reads clockwise in Fig. 2.3, then the cross product is the third basis vector; if one reads counterclockwise, then the cross product is the negative of the third vector.

FIGURE 2.3

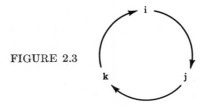

Parts (a) and (b) of the next theorem were established in Section 3.1. The remaining parts follow from the definition of $\mathbf{U} \times \mathbf{V}$; we leave it to the student to verify them.

Theorem 2.1   If $\mathbf{U}$, $\mathbf{V}$, and $\mathbf{W}$ are vectors then:

(a) $\mathbf{U} \times \mathbf{V} = 0$ if and only if $\mathbf{U}$ and $\mathbf{V}$ are parallel.

(b) If $\mathbf{U}$ and $\mathbf{V}$ are linearly independent then $(\mathbf{U} \times \mathbf{V}) \cdot \mathbf{W} = 0$ if and only if $\mathbf{W}$ is in the plane of $\mathbf{U}$ and $\mathbf{V}$; in particular $\mathbf{U} \times \mathbf{V}$ is perpendicular to $\mathbf{U}$ and $\mathbf{V}$.

(c) $\mathbf{V} \times \mathbf{U} = -\mathbf{U} \times \mathbf{V}.$

(d) $\mathbf{U} \times (\mathbf{V} + \mathbf{W}) = \mathbf{U} \times \mathbf{V} + \mathbf{U} \times \mathbf{W}.$

(e) $c(\mathbf{U} \times \mathbf{V}) = (c\mathbf{U}) \times \mathbf{V} = \mathbf{U} \times (c\mathbf{V}).$

The Right-Hand Rule

From (3) and the identity

$$(u_2 v_3 - u_3 v_2)^2 + (u_3 v_1 - u_1 v_3)^2 + (u_1 v_2 - u_2 v_1)^2$$
$$= (u_1{}^2 + u_2{}^2 + u_3{}^2)(v_1{}^2 + v_2{}^2 + v_3{}^2) - (u_1 v_1 + u_2 v_2 + u_3 v_3)^2$$

SKIP 258

it follows that

$$|\mathbf{U} \times \mathbf{V}|^2 = |\mathbf{U}|^2|\mathbf{V}|^2 - (\mathbf{U} \cdot \mathbf{V})^2$$

(6)
$$= |\mathbf{U}|^2|\mathbf{V}|^2\left(1 - \frac{(\mathbf{U} \cdot \mathbf{V})^2}{|\mathbf{U}|^2|\mathbf{V}|^2}\right)$$

$$= |\mathbf{U}|^2|\mathbf{V}|^2(1 - \cos^2\theta),$$

where θ is the angle between $\mathbf{U}$ and $\mathbf{V}$. Of course there are two such angles, as shown in Fig. 2.4. (We ignore the infinitely many additional ones that could be obtained by adding and subtracting multiples of 2π to those shown in Fig. 2.4). We remove this ambiguity by stipulating that θ is always chosen to be the smaller angle between $\mathbf{U}$ and $\mathbf{V}$; then $0 \le \theta \le \pi$, $\sin\theta \ge 0$ and (6) implies that

$$|\mathbf{U} \times \mathbf{V}| = |\mathbf{U}||\mathbf{V}|\sin\theta.$$

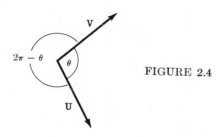

FIGURE 2.4

From Theorem 2.1, if $\mathbf{U} \times \mathbf{V} \ne \mathbf{0}$ then $\mathbf{U} \times \mathbf{V}$ and $-\mathbf{U} \times \mathbf{V}$ are normal to the plane of $\mathbf{U}$ and $\mathbf{V}$ and in opposite directions. There is a convenient geometric rule for distinguishing between them. In Fig. 2.1 we have labeled the coordinate axes so that if a right-hand screw is turned in the direction required to rotate the x-axis through a right angle into coincidence with the y-axis, then the screw will move in the positive z-direction (Fig. 2.5). A coordinate system with this property is said to be *right handed*; by contrast, the system shown in Fig. 2.6 is *left handed*.

Henceforth we shall require that our rectangular coordinate system in R^3 be right handed. With this convention the direction of $\mathbf{U} \times \mathbf{V}$ is that in which a right hand screw perpendicular to the plane of $\mathbf{U}$ and $\mathbf{V}$ would move if it were rotated through an acute angle from $\mathbf{U}$ to $\mathbf{V}$ (Fig. 2.7). From Example 2.3 it is easy to verify this statement if $\mathbf{U}$ and $\mathbf{V}$ are distinct elements of the natural basis for R^3, and a proof is indicated in Exercise T-3

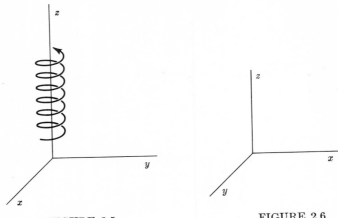

FIGURE 2.5   FIGURE 2.6

for the case in which $\mathbf{U}$ and $\mathbf{V}$ lie in one of the coordinate planes. A general proof, which we only sketch here, can be obtained as follows:

(a) If $\mathbf{U}$ and $\mathbf{V}$ are linearly independent vectors and θ is the acute angle between them, define $\mathbf{N}(\mathbf{U}, \mathbf{V})$ to be the vector of length $|\mathbf{U}| \, |\mathbf{V}| \sin \theta$, perpendicular to the plane of $\mathbf{U}$ and $\mathbf{V}$ and in the direction toward which a right-hand screw would move if rotated through the angle θ from $\mathbf{U}$ to $\mathbf{V}$. (Fig. 2.7). If $\mathbf{U}$ and $\mathbf{V}$ are parallel, define $\mathbf{N}(\mathbf{U}, \mathbf{V}) = 0$. From this definition and Example 2.3 it can be shown that if $\mathbf{U}$ and $\mathbf{V}$ are vectors in the natural basis $\{\mathbf{i}, \mathbf{j}, \mathbf{k}\}$, then

$$\mathbf{N}(\mathbf{U}, \mathbf{V}) = \mathbf{U} \times \mathbf{V}.$$

(b) Show geometrically that

$$\mathbf{N}(\mathbf{U}, \mathbf{V} + \mathbf{W}) = \mathbf{N}(\mathbf{U}, \mathbf{V}) + \mathbf{N}(\mathbf{U}, \mathbf{W})$$

and

$$\mathbf{N}(\mathbf{U}, \mathbf{V}) = -N(\mathbf{V}, \mathbf{U}).$$

(c) Use the results of (a) and (b) to show that

$$\mathbf{N}(\mathbf{U}, \mathbf{V}) = \mathbf{N}(u_1\mathbf{i} + u_2\mathbf{j} + u_3\mathbf{k}, v_1\mathbf{i} + v_2\mathbf{j} + v_3\mathbf{k})$$

$$= (u_2v_3 - u_3v_2)\mathbf{i} + (u_3v_1 - u_1v_3)\mathbf{j} + (u_1v_2 - u_2v_1)\mathbf{k}$$

$$= \mathbf{U} \times \mathbf{V}$$

for all $\mathbf{U}$ and $\mathbf{V}$.

The cross product and scalar product have applications in analytic geometry. We shall present a few of these.

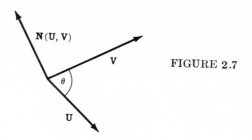

FIGURE 2.7

Area of a Triangle and a Parallelogram

The area A_T of a triangle with base b and height h is

$$A_T = \tfrac{1}{2}bh.$$

Consider the triangle with vertices $\mathbf{X}_1$, $\mathbf{X}_2$, and $\mathbf{X}_3$ (Fig. 2.8); if we take the line segment connecting $\mathbf{X}_1$ and $\mathbf{X}_2$ to be its base then

$$b = |\,\mathbf{X}_2 - \mathbf{X}_1\,|$$

and

$$h = |\,\mathbf{X}_3 - \mathbf{X}_1\,|\sin\theta,$$

where θ is the angle between $\mathbf{X}_1 - \mathbf{X}_2$ and $\mathbf{X}_1 - \mathbf{X}_3$; thus

$$A_T = \tfrac{1}{2}\,|\,\mathbf{X}_2 - \mathbf{X}_1\,|\,|\,\mathbf{X}_3 - \mathbf{X}_1\,|\sin\theta$$

$$= \tfrac{1}{2}\,|\,(\mathbf{X}_2 - \mathbf{X}_1)\,\times\,(\mathbf{X}_3 - \mathbf{X}_1)\,|.$$

The area A_P of the parallelogram with adjacent sides $\mathbf{X}_2 - \mathbf{X}_1$ and $\mathbf{X}_3 - \mathbf{X}_1$ is $2A_T$; hence

$$A_P = |\,(\mathbf{X}_2 - \mathbf{X}_1)\,\times\,(\mathbf{X}_3 - \mathbf{X}_1)\,|.$$

Example 2.4 If $\mathbf{X}_1 = (1, 2, 3)$, $\mathbf{X}_2 = (2, -1, -3)$, and $\mathbf{X}_3 = (0, 1, 2)$,

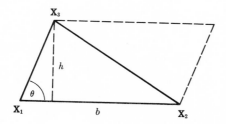

FIGURE 2.8

then

$$X_2 - X_1 = i - 3j - 6k,$$

$$X_3 - X_1 = -(i + j + k),$$

and

$$(X_2 - X_1) \times (X_3 - X_1) = \begin{vmatrix} i & j & k \\ 1 & -3 & -6 \\ -1 & -1 & -1 \end{vmatrix}$$

$$= -3i + 7j - 4k;$$

thus the area of the triangle with vertices X_1, X_2, and X_3 is

$$A_T = \tfrac{1}{2}\sqrt{(-3)^2 + 7^2 + (-4)^2} = \tfrac{1}{2}\sqrt{74}.$$

The Equation of a Plane through Three Points

The plane in R^3 which contains three noncollinear points X_1, X_2, and X_3 (Fig. 2.9) has the equation

(7) $$N \cdot (X - X_1) = 0,$$

where N is perpendicular to $X_2 - X_1$ and $X_3 - X_1$ (since X_2 and X_3 satisfy (7)). Therefore we can take

$$N = (X_2 - X_1) \times (X_3 - X_1).$$

Example 2.5   Let $X_1 = (1, 2, 1)$, $X_2 = (2, -1, 3)$, and $X_3 = (0, 4, 3)$. Then

$$X_2 - X_1 = i - 3j + 2k,$$

$$X_3 - X_1 = -i + 2j + 2k,$$

and

$$(X_2 - X_1) \times (X_3 - X_1) = -10i - 4j - k;$$

FIGURE 2.9

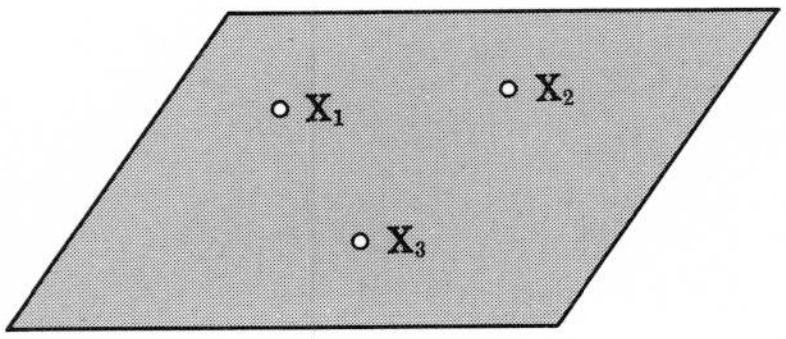

thus the equation of the plane is

$$10(x - 1) + 4(y - 2) + (z - 1) = 0$$

or

$$10x + 4y + z = 19.$$

The Equation of a Line through a Given Point and Intersecting a Given Line at Right Angles

Let a line L_0 be given by

$$\mathbf{X} = \mathbf{X}_0 + t\mathbf{U}_0$$

and suppose we wish to find the equation of the line L_1 that passes through a point $\mathbf{X}_1$ and intersects L_0 at right angles (Fig. 2.10). Let $\mathbf{X}_2$ be the unknown point of intersection of L_1 and L_0 and define $\mathbf{U}_1 = \mathbf{X}_2 - \mathbf{X}_1$. From Fig. 2.10,

(8) $$\mathbf{U}_1 = \mathbf{X}_0 - \mathbf{X}_1 + t_0\mathbf{U}_0$$

and therefore

(9) $$\mathbf{U}_1 \cdot \mathbf{U}_0 = (\mathbf{X}_0 - \mathbf{X}_1) \cdot \mathbf{U}_0 + t_0 \mid \mathbf{U}_0 \mid^2.$$

Since $\mathbf{U}_1$ and $\mathbf{U}_0$ are perpendicular, $\mathbf{U}_1 \cdot \mathbf{U}_0 = 0$, and (9) yields

$$t_0 = \frac{(\mathbf{X}_1 - \mathbf{X}_0) \cdot \mathbf{U}_0}{\mid \mathbf{U}_0 \mid^2}.$$

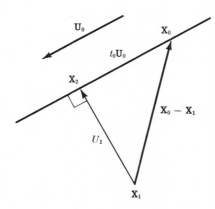

FIGURE 2.10

To obtain $\mathbf{U}_1$, substitute this in (8); then the equation of L_1 is

$$\mathbf{X} = \mathbf{X}_0 + s(a\mathbf{U}_1)$$

where a is any convenient nonzero scalar.

Example 2.6   Let it be required to find the equation of the line L_1 that passes through $\mathbf{X}_1 = (3, 5, 3)$ and intersects the line given by

$$\mathbf{X} = \begin{bmatrix} 1 \\ 0 \\ 2 \end{bmatrix} + t \begin{bmatrix} 3 \\ -1 \\ 2 \end{bmatrix}$$

at right angles. Thus $\mathbf{X}_0 = (1, 0, 2)$,

$$\mathbf{U}_0 = 3\mathbf{i} - \mathbf{j} + 2\mathbf{k},$$

$$\mathbf{X}_1 - \mathbf{X}_0 = 2\mathbf{i} + 5\mathbf{j} + \mathbf{k},$$

$$t_0 = \frac{(2\mathbf{i} + 5\mathbf{j} + \mathbf{k}) \cdot (3\mathbf{i} - \mathbf{j} + 2\mathbf{k})}{|\, 3\mathbf{i} - \mathbf{j} + 2\mathbf{k}\,|} = \frac{3}{14},$$

$$\mathbf{U}_1 = -(2\mathbf{i} + 5\mathbf{j} + \mathbf{k}) + \tfrac{3}{14}(3\mathbf{i} - \mathbf{j} + 2\mathbf{k})$$

$$= -\frac{(19\mathbf{i} + 73\mathbf{j} + 8\mathbf{k})}{14},$$

and L_1 is given by

$$\mathbf{X} = \begin{bmatrix} 1 \\ 0 \\ 2 \end{bmatrix} + s \begin{bmatrix} 19 \\ 73 \\ 8 \end{bmatrix}.$$

Line of Intersection of Two Planes

Let Γ_1 and Γ_2 be planes defined by

(10)
$$\mathbf{X} \cdot \mathbf{N}_1 = d_1$$

and

(11)
$$\mathbf{X} \cdot \mathbf{N}_2 = d_2.$$

If $\mathbf{N}_1$ and $\mathbf{N}_2$ are parallel, we say that Γ_1 *and* Γ_2 *are parallel* (Fig. 2.11). Parallel planes do not intersect unless $d_2\mathbf{N}_1 = d_1\mathbf{N}_2$, in which case they are identical (Exercise T-4).

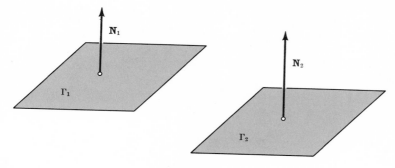

FIGURE 2.11

If Γ_1 and Γ_2 are not parallel, they intersect in a line. This is intuitively evident from Fig. 2.12. To prove it we observe that if $\mathbf{N}_1$ and $\mathbf{N}_2$ are not parallel then $\mathbf{N}_1$, $\mathbf{N}_2$ and $\mathbf{N}_1 \times \mathbf{N}_2$ form a basis for R^3, and any $\mathbf{X}$ in R^3 can be written as

$$\mathbf{X} = a\mathbf{N}_1 + b\mathbf{N}_2 + t(\mathbf{N}_1 \times \mathbf{N}_2)$$

for suitable scalars a, b, and t. Now $\mathbf{X}$ satisfies (10) and (11) if and only if

(12)
$$a \mid \mathbf{N}_1 \mid^2 + b(\mathbf{N}_1 \cdot \mathbf{N}_2) = d_1,$$
$$a(\mathbf{N}_1 \cdot \mathbf{N}_2) + b \mid \mathbf{N}_2 \mid^2 = d_2.$$

The determinant of this pair of equations in a and b is

$$\mid \mathbf{N}_1 \mid^2 \mid \mathbf{N}_2 \mid^2 - (\mathbf{N}_1 \cdot \mathbf{N}_2)^2 = \mid \mathbf{N}_1 \times \mathbf{N}_2 \mid^2 \neq 0 \qquad \text{(see (6))},$$

and therefore the system (12) has a unique solution. If

$$\bar{\mathbf{X}} = a\mathbf{N}_1 + b\mathbf{N}_2,$$

where a and b satisfy (12), then

$$\mathbf{X} = \bar{\mathbf{X}} + t(\mathbf{N}_1 \times \mathbf{N}_2)$$

is the equation of the required line of intersection. Of course, the equation of this line could just as well be written as

$$\mathbf{X} = \mathbf{X}_0 + at(\mathbf{N}_1 \times \mathbf{N}_2)$$

where $\mathbf{X}_0$ is any point common to Γ_1 and Γ_2, and a is any nonzero scalar. (Why?)

Example 2.7 The planes given by

(13)
$$\mathbf{N}_1 \cdot \mathbf{X} = 2x - 3y + 3z = 15$$

and

(14)
$$\mathbf{N}_2 \cdot \mathbf{X} = x + 2y - z = 4$$

are not parallel; hence they intersect in a line L which is parallel to

$$\mathbf{N_1} \times \mathbf{N_2} = -3\mathbf{i} + 5\mathbf{j} + 7\mathbf{k}.$$

To find the equation of the line we need only find a point common to the two planes, which can be done in this case by setting $z = 0$ in (13) and (14) and solving the resulting equations for x and y; this yields the common point $\mathbf{X_0} = (6, -1, 0)$. Thus the line of intersection is given by

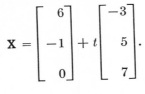

$$\mathbf{X} = \begin{bmatrix} 6 \\ -1 \\ 0 \end{bmatrix} + t \begin{bmatrix} -3 \\ 5 \\ 7 \end{bmatrix}.$$

FIGURE 2.12

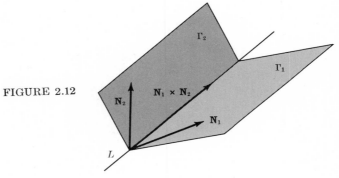

Distance from a Point to a Line

Suppose we wish to find the distance d from the point $\mathbf{X_1}$ to the line L given by

$$\mathbf{X} = \mathbf{X_0} + t\mathbf{U}.$$

Let θ be the angle between $\mathbf{U}$ and $\mathbf{X_1} - \mathbf{X_0}$ (Fig. 2.13); then

$$d = |\mathbf{X_1} - \mathbf{X_0}| \sin \theta$$
$$= \frac{|\mathbf{U} \times (\mathbf{X_1} - \mathbf{X_0})|}{|\mathbf{U}|}.$$

If L passes through the origin we may take $\mathbf{X_0} = \mathbf{0}$ to obtain

$$d = \frac{|\mathbf{U} \times \mathbf{X_1}|}{|\mathbf{U}|};$$

we shall use this result in Section 6.4.

Example 2.8   Let $\mathbf{X}_1 = (-2, 1, 0)$ and let L be given by

$$\mathbf{X} = \mathbf{X}_0 + t\mathbf{U} = \begin{bmatrix} -1 \\ 0 \\ 2 \end{bmatrix} + t \begin{bmatrix} 3 \\ 4 \\ 0 \end{bmatrix};$$

then

$$\mathbf{U} \times (\mathbf{X}_1 - \mathbf{X}_0) = -8\mathbf{i} + 6\mathbf{j} + 7\mathbf{k},$$
$$| \, \mathbf{U} \times (\mathbf{X}_1 - \mathbf{X}_0) \, | = \sqrt{149},$$
$$| \, \mathbf{U} \, | = 5,$$

and the distance from $\mathbf{X}_1$ to L is

$$d = \frac{\sqrt{149}}{5}.$$

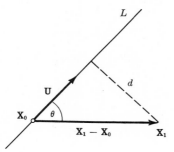

FIGURE 2.13

Distance between Two Lines

The distance d between two lines L_1 and L_2 given by

(15)
$$\mathbf{X} = \mathbf{X}_1 + t\mathbf{U}_1$$

and

(16)
$$\mathbf{X} = \mathbf{X}_2 + s\mathbf{U}_2$$

is defined to be the minimum value attained by $| \, \mathbf{Y}_1 - \mathbf{Y}_2 \, |$ as $\mathbf{Y}_1$ and $\mathbf{Y}_2$ are allowed to vary over L_1 and L_2. If this minimum is attained with $\mathbf{Y}_1 = \bar{\mathbf{Y}}_1$ and $\mathbf{Y}_2 = \bar{\mathbf{Y}}_2$, then $\bar{\mathbf{Y}}_1 - \bar{\mathbf{Y}}_2$ is perpendicular to L_1 and L_2. (See Fig. 2.14 and Exercise T-5). Thus, if L_1 and L_2 are not parallel,

(17)
$$\bar{\mathbf{Y}}_2 - \bar{\mathbf{Y}}_1 = c(\mathbf{U}_1 \times \mathbf{U}_2),$$

since any vector perpendicular to L_1 and L_2 is parallel to $\mathbf{U}_1 \times \mathbf{U}_2$. By taking inner products of both sides of (17) with $\mathbf{U}_1 \times \mathbf{U}_2$, we find that

$$(18) \qquad c = \frac{(\bar{\mathbf{Y}}_2 - \bar{\mathbf{Y}}_1) \cdot (\mathbf{U}_1 \times \mathbf{U}_2)}{|\mathbf{U}_1 \times \mathbf{U}_2|^2} .$$

Since $\bar{\mathbf{Y}}_1$ and $\bar{\mathbf{Y}}_2$ are on L_1 and L_2, (15) and (16) imply that $\bar{\mathbf{Y}}_2 - \bar{\mathbf{Y}}_1$ is of the form

$$\bar{\mathbf{Y}}_2 - \bar{\mathbf{Y}}_1 = \mathbf{X}_2 - \mathbf{X}_1 + \bar{s}\mathbf{U}_2 - \bar{t}\mathbf{U}_1;$$

hence

$$(\bar{\mathbf{Y}}_2 - \bar{\mathbf{Y}}_1) \cdot (\mathbf{U}_1 \times \mathbf{U}_2) = (\mathbf{X}_2 - \mathbf{X}_1) \cdot (\mathbf{U}_1 \times \mathbf{U}_2)$$

and from (18),

$$c = \frac{(\mathbf{X}_2 - \mathbf{X}_1) \cdot (\mathbf{U}_1 \times \mathbf{U}_2)}{|\mathbf{U}_1 \times \mathbf{U}_2|^2} .$$

It now follows from (17) that

$$(19) \qquad d = |c| \, |\mathbf{U}_1 \times \mathbf{U}_2| = \frac{|(\mathbf{X}_2 - \mathbf{X}_1) \cdot (\mathbf{U}_1 \times \mathbf{U}_2)|}{|\mathbf{U}_1 \times \mathbf{U}_2|} .$$

If L_1 and L_2 are parallel, then the distance between them is given by

$$d = \frac{|\mathbf{U}_1 \times (\mathbf{X}_2 - \mathbf{X}_1)|}{|\mathbf{U}_1|}$$

(Exercise T-7).

Example 2.9   Let L_1 and L_2 be given by

$$\mathbf{X} = \mathbf{X}_1 + t\mathbf{U}_1 = \begin{bmatrix} -1 \\ 0 \\ 2 \end{bmatrix} + t \begin{bmatrix} 1 \\ -1 \\ 0 \end{bmatrix}$$

and

$$\mathbf{X} = \mathbf{X}_2 + s\mathbf{U}_2 = \begin{bmatrix} 0 \\ -1 \\ 0 \end{bmatrix} + s \begin{bmatrix} 1 \\ 0 \\ -1 \end{bmatrix};$$

then

$$\mathbf{X}_2 - \mathbf{X}_1 = \mathbf{i} - \mathbf{j} - 2\mathbf{k},$$

$$\mathbf{U}_1 \times \mathbf{U}_2 = \mathbf{i} + \mathbf{j} + \mathbf{k},$$

and

$$d = \frac{2}{\sqrt{3}}.$$

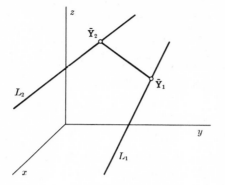

FIGURE 2.14

Distance from a Point to a Plane

Suppose we wish to find the distance d from a point $\mathbf{X}_1$ to a plane Γ defined by

$$\mathbf{N} \cdot (\mathbf{X} - \mathbf{X}_0) = 0.$$

Then

$$d = |\mathbf{X}_1 - \mathbf{X}_0| \, |\cos \theta|$$

where θ is the angle between $\mathbf{X}_1 - \mathbf{X}_0$ and $\mathbf{N}$ (Fig. 2.15); thus

$$\cos \theta = \frac{(\mathbf{X}_1 - \mathbf{X}_0) \cdot \mathbf{N}}{|\mathbf{X}_1 - \mathbf{X}_0| \, |\mathbf{N}|}$$

and

$$d = \frac{|(\mathbf{X}_1 - \mathbf{X}_0) \cdot \mathbf{N}|}{|\mathbf{N}|}.$$

Example 2.10 Let Γ be given by $2x - 3y + 4z = 6$, and $X_1 = (1, -2, 3)$. Then

$$\mathbf{N} = 2\mathbf{i} - 3\mathbf{j} + 4\mathbf{k};$$

therefore $\mathbf{X}_1 \cdot \mathbf{N} = 20$ and $\mathbf{X}_0 \cdot \mathbf{N} = 6$ for any $\mathbf{X}_0$ on Γ. Thus

$$d = \frac{20 - 6}{\sqrt{29}} = \frac{14}{\sqrt{29}}.$$

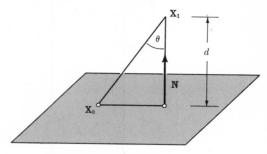

FIGURE 2.15

Scalar Triple Product

If **U**, **V**, and **W** are vectors in R^3, then $\mathbf{U} \cdot (\mathbf{V} \times \mathbf{W})$ is called a *scalar triple product*. Let $\mathbf{U} = u_1\mathbf{i} + u_2\mathbf{j} + u_3\mathbf{k}$, $\mathbf{V} = v_1\mathbf{i} + v_2\mathbf{j} + v_3\mathbf{k}$, and $\mathbf{W} = w_1\mathbf{i} + w_2\mathbf{j} + w_3\mathbf{k}$; since

$$(20) \qquad \mathbf{V} \times \mathbf{W} = (v_2w_3 - v_3w_2)\mathbf{i} + (v_3w_1 - v_1w_3)\mathbf{j} + (v_1w_2 - v_2w_1)\mathbf{k},$$

it follows that

$$\mathbf{U} \cdot (\mathbf{V} \times \mathbf{W}) = u_1(v_2w_3 - v_3w_2) + u_2(v_3w_1 - v_1w_3) + u_3(v_1w_2 - v_2w_1),$$

which can also be written as

$$(21) \qquad \mathbf{U} \cdot (\mathbf{V} \times \mathbf{W}) = \begin{vmatrix} u_1 & u_2 & u_3 \\ v_1 & v_2 & v_3 \\ w_1 & w_2 & w_3 \end{vmatrix}.$$

Transposition (interchange) of two rows of a determinant changes its sign. Since such a transposition is equivalent to interchanging the corresponding vectors on the left side of (21), it follows that

$$\mathbf{U} \cdot (\mathbf{V} \times \mathbf{W}) = \mathbf{V} \cdot (\mathbf{W} \times \mathbf{U}) = \mathbf{W} \cdot (\mathbf{U} \times \mathbf{V})$$
$$= -\mathbf{V} \cdot (\mathbf{U} \times \mathbf{W}) = -\mathbf{W} \cdot (\mathbf{V} \times \mathbf{U}) = -\mathbf{U} \cdot (\mathbf{W} \times \mathbf{V})$$

(Exercise T-10).

Example 2.11 Let

$$\mathbf{U} = \mathbf{i} + \mathbf{j} - \mathbf{k},$$
$$\mathbf{V} = 2\mathbf{i} + \mathbf{j} - 2\mathbf{k},$$
$$\mathbf{W} = \mathbf{i} - \mathbf{j};$$

then

(22) $\mathbf{V} \times \mathbf{W} = -2\mathbf{i} - 2\mathbf{j} - 3\mathbf{k}$

and

$$\mathbf{U} \cdot (\mathbf{V} \times \mathbf{W}) = (\mathbf{i} + \mathbf{j} + \mathbf{k}) \cdot (-2\mathbf{i} - 2\mathbf{j} - 3\mathbf{k}) = -1.$$

Also,

$$\mathbf{W} \times \mathbf{U} = \mathbf{i} + \mathbf{j} + 2\mathbf{k}$$

and

$$\mathbf{V} \cdot (\mathbf{W} \times \mathbf{U}) = (2\mathbf{i} + \mathbf{j} - 2\mathbf{k}) \cdot (\mathbf{i} + \mathbf{j} + 2\mathbf{k}) = -1$$

$$= \mathbf{U} \cdot (\mathbf{V} \times \mathbf{W}).$$

However,

$$\mathbf{V} \cdot (\mathbf{U} \times \mathbf{W}) = (2\mathbf{i} + \mathbf{j} - 2\mathbf{k}) \cdot (-\mathbf{i} - \mathbf{j} - 2\mathbf{k}) = 1$$

$$= -\mathbf{U} \cdot (\mathbf{V} \times \mathbf{W}).$$

Example 2.12 Consider a parallelepiped P with a vertex at the origin and let the edges of P which meet at the origin be associated with vectors **U**, **V**, and **W** as shown in Fig. 2.16. The volume v of P can be obtained by multiplying the area of the face containing **V** and **W**, which we have shown to be $|\mathbf{V} \times \mathbf{W}|$, by the distance d from this face to the face parallel to it. From Fig. 2.16,

$$d = |\mathbf{U}| |\cos \theta|,$$

where θ is the angle between **U** and $\mathbf{V} \times \mathbf{W}$; hence

$$v = |\mathbf{V} \times \mathbf{W}| |\mathbf{U}| |\cos \theta| = |\mathbf{U} \cdot (\mathbf{V} \times \mathbf{W})|.$$

For example, to find the volume of the parallelepiped P with vertices at $(0, 0, 0)$, $(1, 2, 1)$, $(2, -1, 1)$, and $(1, -2, 1)$, we take $\mathbf{U} = \mathbf{i} + 2\mathbf{j} + \mathbf{k}$,

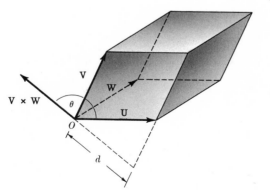

FIGURE 2.16

$V = 2\mathbf{i} - \mathbf{j} + \mathbf{k}$, and $W = \mathbf{i} - 2\mathbf{j} + \mathbf{k}$. Then

$$\mathbf{V} \times \mathbf{W} = \mathbf{i} - \mathbf{j} - 3\mathbf{k}$$

and

$$\mathbf{U} \cdot (\mathbf{V} \times \mathbf{W}) = (\mathbf{i} + 2\mathbf{j} + \mathbf{k}) \cdot (\mathbf{i} - \mathbf{j} - 3\mathbf{k}) = -4;$$

hence the volume of P is

$$v = |\mathbf{U} \cdot (\mathbf{V} \times \mathbf{W})| = 4.$$

Vector Triple Product

Let $\mathbf{U}, \mathbf{V},$ and $\mathbf{W}$ be as defined in our discussion of the scalar triple product. Then $\mathbf{U} \times (\mathbf{V} \times \mathbf{W})$ is called a *vector triple product*. It can be expanded as follows:

$$(23) \quad \mathbf{U} \times (\mathbf{V} \times \mathbf{W}) = u_1(\mathbf{i} \times (\mathbf{V} \times \mathbf{W})) + u_2(\mathbf{j} \times (\mathbf{V} \times \mathbf{W}))$$
$$+ u_3(\mathbf{k} \times (\mathbf{V} \times \mathbf{W}));$$

from (20),

$$\mathbf{i} \times (\mathbf{V} \times \mathbf{W}) = (v_2w_1 - v_1w_2)\mathbf{j} + (v_3w_1 - v_1w_3)\mathbf{k}$$
$$= w_1(v_1\mathbf{i} + v_2\mathbf{j} + v_3\mathbf{k}) - v_1(w_1\mathbf{i} + w_2\mathbf{j} + w_3\mathbf{k})$$
$$= w_1\mathbf{V} - v_1\mathbf{W}.$$

Similarly,

$$\mathbf{j} \times (\mathbf{V} \times \mathbf{W}) = w_2\mathbf{V} - v_2\mathbf{W}$$

and

$$\mathbf{k} \times (\mathbf{V} \times \mathbf{W}) = w_3\mathbf{V} - v_3\mathbf{W}.$$

Substituting these three results into (23) yields

$$\mathbf{U} \times (\mathbf{V} \times \mathbf{W}) = (u_1w_1 + u_2w_2 + u_3w_3)\mathbf{V} - (u_1v_1 + u_2v_2 + u_3v_3)\mathbf{W};$$

thus

$$(24) \qquad \mathbf{U} \times (\mathbf{V} \times \mathbf{W}) = (\mathbf{U} \cdot \mathbf{W})\mathbf{V} - (\mathbf{U} \cdot \mathbf{V})\mathbf{W}.$$

Remembering this equation is made easier by observing that $\mathbf{U} \times (\mathbf{V} \times \mathbf{W})$ must be a linear combination of $\mathbf{V}$ and $\mathbf{W}$ because it is orthogonal to $\mathbf{V} \times \mathbf{W}$.

The triple vector product $(\mathbf{U} \times \mathbf{V}) \times \mathbf{W}$ is in general different from (24); in fact,

$$(\mathbf{U} \times \mathbf{V}) \times \mathbf{W} = -\mathbf{W} \times (\mathbf{U} \times \mathbf{V}) = \mathbf{W} \times (\mathbf{V} \times \mathbf{U}),$$

which can be expanded by interchanging $\mathbf{U}$ and $\mathbf{W}$ in (24). The result is

(25) $$(\mathbf{U} \times \mathbf{V}) \times \mathbf{W} = (\mathbf{W} \cdot \mathbf{U})\mathbf{V} - (\mathbf{W} \cdot \mathbf{V})\mathbf{U}$$

Example 2.13 Let $\mathbf{U}$, $\mathbf{V}$, and $\mathbf{W}$ be as defined in Example 2.10. From (22),

$$\mathbf{U} \times (\mathbf{V} \times \mathbf{W}) = \begin{vmatrix} \mathbf{i} & \mathbf{j} & \mathbf{k} \\ 1 & 1 & -1 \\ -2 & -2 & -3 \end{vmatrix} = -5\mathbf{i} + 5\mathbf{j}.$$

We can obtain the same result from (24); since $\mathbf{U} \cdot \mathbf{W} = 0$, and $\mathbf{U} \cdot \mathbf{V} = 5$,

$$\mathbf{U} \times (\mathbf{V} \times \mathbf{W}) = 0\mathbf{V} - 5\mathbf{W} = -5\mathbf{i} + 5\mathbf{j}.$$

From (25) we find that

$$(\mathbf{U} \times \mathbf{V}) \times \mathbf{W} = 0\mathbf{V} - 1\mathbf{U} = -\mathbf{i} - \mathbf{j} + \mathbf{k}.$$

Sept. 29

EXERCISES 3.2 # 4, 8, 15, 18

1. State which of the following sets of vectors are coplanar.

(a) $\begin{bmatrix} 0 \\ 7 \\ 5 \end{bmatrix}$, $\begin{bmatrix} -3 \\ 5 \\ -2 \end{bmatrix}$, $\begin{bmatrix} 2 \\ -1 \\ 3 \end{bmatrix}$; (b) $\begin{bmatrix} 1 \\ 2 \\ 3 \end{bmatrix}$, $\begin{bmatrix} 0 \\ 1 \\ 2 \end{bmatrix}$, $\begin{bmatrix} 1 \\ 0 \\ 2 \end{bmatrix}$;

(c) $\begin{bmatrix} -3 \\ -2 \\ 1 \end{bmatrix}$, $\begin{bmatrix} 1 \\ 0 \\ -1 \end{bmatrix}$, $\begin{bmatrix} 5 \\ 3 \\ -2 \end{bmatrix}$.

2. Write in terms of $\mathbf{i}$, $\mathbf{j}$, and $\mathbf{k}$:

(a) $\begin{bmatrix} 0 \\ -2 \\ \frac{2}{5} \end{bmatrix}$; (b) $\begin{bmatrix} \frac{4}{5} \\ -7 \\ 5 \end{bmatrix}$; (c) $\begin{bmatrix} 2 \\ 0 \\ -4 \end{bmatrix}$; (d) $\begin{bmatrix} 0 \\ 0 \\ -5 \end{bmatrix}$.

3. Write in "column" form:
   (a) $2\mathbf{i} - 3\mathbf{j}$;   (b) $\mathbf{i} - \frac{2}{3}\mathbf{j} + \frac{4}{7}\mathbf{k}$;
   (c) $2\mathbf{j}$;   (d) $3\mathbf{i} - \mathbf{j} + \mathbf{k}$.

4. Compute $\mathbf{W} = \mathbf{U} \times \mathbf{V}$ and verify that $\mathbf{W}$ is perpendicular to $\mathbf{U}$ and $\mathbf{V}$:

   (a) $\mathbf{U} = \begin{bmatrix} 1 \\ 0 \\ 3 \end{bmatrix}$, $\mathbf{V} = \begin{bmatrix} 1 \\ 2 \\ 3 \end{bmatrix}$;   (b) $\mathbf{U} = \mathbf{V} = \begin{bmatrix} 1 \\ 2 \\ 3 \end{bmatrix}$;

   (c) $\mathbf{U} = \begin{bmatrix} 1 \\ 2 \\ 3 \end{bmatrix}$, $\mathbf{V} = -\mathbf{U}$;   (d) $\mathbf{U} = \begin{bmatrix} -1 \\ 5 \\ 2 \end{bmatrix}$, $\mathbf{V} = \begin{bmatrix} 2 \\ -1 \\ 3 \end{bmatrix}$.

5. Repeat Exercise 4 for the following pairs of vectors.

   (a) $\mathbf{U} = 2\mathbf{i} - 3\mathbf{j} + 4\mathbf{k}$,   (b) $\mathbf{U} = 4\mathbf{i}$,
        $\mathbf{V} = 2\mathbf{i} - \mathbf{j}$; $\mathbf{V} = 3\mathbf{i} + 2\mathbf{j} - 5\mathbf{k}$;
   (c) $\mathbf{U} = 4\mathbf{i} + 3\mathbf{j} - \mathbf{k}$,   (d) $\mathbf{U} = 2\mathbf{i} - \mathbf{j}$,
        $\mathbf{V} = \mathbf{i} + \mathbf{j} - 3\mathbf{k}$; $\mathbf{V} = -3\mathbf{j} + 5\mathbf{k}$.

6. Verify the cross products in Example 3.3.

7. Verify (c), (d), and (e) of Theorem 3.1 for

   $$\mathbf{U} = \begin{bmatrix} 1 \\ 2 \\ 3 \end{bmatrix}, \quad \mathbf{V} = \begin{bmatrix} -1 \\ 3 \\ 4 \end{bmatrix}, \quad \mathbf{W} = \begin{bmatrix} 0 \\ 2 \\ -1 \end{bmatrix}, \quad \text{and} \quad c = -3.$$

8. Verify that $|\mathbf{U} \times \mathbf{V}| = |\mathbf{U}||\mathbf{V}| \sin \theta$ for the pairs of vectors in Exercise 4.

9. Find the area of the triangle with vertices $\mathbf{X}_1 = (1, 0, 2)$, $\mathbf{X}_2 = (2, -1, 3)$, and $\mathbf{X}_3 = (4, -3, 2)$.

10. Find the area of the triangle with vertices $\mathbf{X}_1$, $\mathbf{X}_2$, and $\mathbf{X}_3$ such that $\mathbf{X}_2 - \mathbf{X}_1 = 3\mathbf{i} + \mathbf{j} - 2\mathbf{k}$, and $\mathbf{X}_3 - \mathbf{X}_1 = -\mathbf{i} + 3\mathbf{k}$.

11. Find the area of the triangle with vertices $\mathbf{X}_1$, $\mathbf{X}_2$, and $\mathbf{X}_3$, where $\mathbf{X}_1 = (-1, 0, 2)$, $\mathbf{X}_2 = (4, 1, 3)$, and $\mathbf{X}_3 - \mathbf{X}_1 = \mathbf{i} + 2\mathbf{j} - 3\mathbf{k}$.

12. Find the area of the parallelogram with vertices at $\mathbf{X}_1$, $\mathbf{X}_2$, $\mathbf{X}_3$, and $\mathbf{X}_4$, where $\mathbf{X}_2 - \mathbf{X}_1 = 3\mathbf{i} - 2\mathbf{j} + \mathbf{k}$, and $\mathbf{X}_3 - \mathbf{X}_1 = \mathbf{i} + \mathbf{j} - 3\mathbf{k}$.

13. Find the plane determined by the points $(1, 2, 3)$, $(-2, 1, 4)$, and $(0, 1, 5)$.

14. Find the plane determined by the points $(1, -1, 2)$, $(3, -2, 5)$, and $(0, 0, 2)$.

15. Find the equation of the line through $(1, -1, 3)$ which intersects the line

$$\mathbf{X} = \begin{bmatrix} 1 \\ 3 \\ 5 \end{bmatrix} + t \begin{bmatrix} 1 \\ -1 \\ 3 \end{bmatrix}$$

at right angles.

16. Find the equation of the line through $(-2, 3, 4)$ which intersects the line

$$\frac{x - 3}{0} = \frac{y - 5}{2} = \frac{z - 4}{-3}$$

at right angles.

17. Verify that the planes $x + 2y + 3z = -5$ and $2x + 4y + 6z = 8$ are parallel.

18. Find the equation of the line of intersection of the planes $2x + 3y - z = 8$ and $-x + y + 2z = 6$.

19. Repeat Exercise 18 for the planes $2x - y + 3z = 15$ and $x + y - z = 5$.

20. Find the distance d from $\mathbf{X}_1 = (2, 3, -1)$ to the line L defined by

$$\mathbf{X} = \begin{bmatrix} 0 \\ 1 \\ 2 \end{bmatrix} + t \begin{bmatrix} 1 \\ 2 \\ -3 \end{bmatrix}$$

and find $\bar{\mathbf{X}}$ on L such that $|\bar{\mathbf{X}} - \mathbf{X}_1| = d$.

21. Find the distance d from the point $\mathbf{X}_1 = (1, 1, 3)$ to the line L defined by

$$\frac{x - 3}{2} = \frac{y + 5}{0} = \frac{z - 2}{-3},$$

and find $\bar{\mathbf{X}}$ on L such that $|\bar{\mathbf{X}} - \mathbf{X}_1| = d$.

22. Find the distance between the lines

$$\mathbf{X} = \begin{bmatrix} 1 \\ -2 \\ 3 \end{bmatrix} + s \begin{bmatrix} 2 \\ 2 \\ -1 \end{bmatrix} \quad \text{and} \quad \mathbf{X} = \begin{bmatrix} -2 \\ 3 \\ 4 \end{bmatrix} + t \begin{bmatrix} 2 \\ 3 \\ -2 \end{bmatrix}.$$

23. Find the distance between the lines

$$\mathbf{X} = \begin{bmatrix} 1 \\ 2 \\ -1 \end{bmatrix} + s \begin{bmatrix} 2 \\ -1 \\ 1 \end{bmatrix} \quad \text{and} \quad \mathbf{X} = \begin{bmatrix} 2 \\ 3 \\ 4 \end{bmatrix} + t \begin{bmatrix} -4 \\ 2 \\ -6 \end{bmatrix}.$$

24. Find the distance between the lines

$$\frac{x-3}{2} = \frac{y-4}{0} = \frac{z+5}{-2}$$

and

$$\frac{x+1}{3} = \frac{y-2}{-3} = \frac{z+1}{2}.$$

25. Find the distance from $(3, -1, 2)$ to the plane $2x + 3y - z = 5$.

26. Find the distance from $(2, 1, -3)$ to the plane $x - 2y + z = 2$.

27. Find the distance between the parallel planes $2x + y - z = 8$ and $2x + y - z = 12$.

28. Repeat Exercise 27 for the planes $x + 3y - 5z = 14$ and $2x + 6y - 10z = 12$.

29. Find the plane containing the line

$$\mathbf{X} = \begin{bmatrix} 4 \\ -2 \\ 5 \end{bmatrix} + t \begin{bmatrix} 1 \\ 3 \\ 5 \end{bmatrix}$$

and parallel to the line

$$\mathbf{X} = \begin{bmatrix} 2 \\ 3 \\ 4 \end{bmatrix} + s \begin{bmatrix} 2 \\ 4 \\ -3 \end{bmatrix}.$$

30. Find the plane containing the line

$$X = \begin{bmatrix} 1 \\ 2 \\ -3 \end{bmatrix} + t \begin{bmatrix} 0 \\ 2 \\ -3 \end{bmatrix}$$

and parallel to the line

$$X = \begin{bmatrix} 0 \\ 2 \\ 1 \end{bmatrix} + s \begin{bmatrix} -1 \\ 3 \\ 2 \end{bmatrix}.$$

31. Find the plane determined by the lines

$$X = \begin{bmatrix} 1 \\ 3 \\ -1 \end{bmatrix} + t \begin{bmatrix} 0 \\ 1 \\ 2 \end{bmatrix} \quad \text{and} \quad X = \begin{bmatrix} 2 \\ 1 \\ -5 \end{bmatrix} + s \begin{bmatrix} 1 \\ 2 \\ 4 \end{bmatrix}.$$

32. Find the plane passing through $(1, 2, -1)$ and $(2, 3, 4)$ and parallel to the line

$$X = \begin{bmatrix} 1 \\ 3 \\ 2 \end{bmatrix} + t \begin{bmatrix} 5 \\ 2 \\ -1 \end{bmatrix}.$$

33. Find the plane through $(1, 3, 4)$ and $(-5, 2, 1)$ and parallel to the line

$$\frac{x-2}{3} = \frac{y-5}{0} = \frac{z+5}{-2}.$$

34. Find the plane through the line

$$X = \begin{bmatrix} 2 \\ 3 \\ -1 \end{bmatrix} + t \begin{bmatrix} 1 \\ 2 \\ -1 \end{bmatrix}$$

and perpendicular to the plane $3x + 2y - 5z = 8$.

35. Find all planes through the line

$$\frac{x-2}{3} = \frac{y+2}{-2} = \frac{z-5}{0}.$$

36. Find the plane through the line

$$\mathbf{X} = \begin{bmatrix} 1 \\ 2 \\ 3 \end{bmatrix} + t \begin{bmatrix} 1 \\ 2 \\ -1 \end{bmatrix}$$

and through the point $(2, 3, -1)$.

37. Find the equation of the plane through $(2, 3, -4)$ perpendicular to the planes $x + 2y - 3z = 5$, and $x - y + z = 7$.

38. Verify (21) for

$$\mathbf{U} = \begin{bmatrix} 1 \\ 2 \\ 3 \end{bmatrix}, \qquad \mathbf{V} = \begin{bmatrix} -1 \\ 2 \\ 5 \end{bmatrix}, \qquad \text{and} \qquad \mathbf{W} = \begin{bmatrix} 2 \\ 3 \\ 5 \end{bmatrix}.$$

39. Find the volume of the parallelepiped with edges

$$\mathbf{U} = \begin{bmatrix} 1 \\ 2 \\ 3 \end{bmatrix}, \qquad \mathbf{V} = \begin{bmatrix} 3 \\ -2 \\ 4 \end{bmatrix}, \qquad \text{and} \qquad \mathbf{W} = \begin{bmatrix} 4 \\ 2 \\ 5 \end{bmatrix}.$$

40. Compute $\mathbf{U} \times (\mathbf{V} \times \mathbf{W})$:

(a) $\mathbf{U} = \begin{bmatrix} 1 \\ 2 \\ -1 \end{bmatrix}, \qquad \mathbf{V} = \begin{bmatrix} 2 \\ 3 \\ 4 \end{bmatrix}, \qquad \mathbf{W} = \begin{bmatrix} 1 \\ 2 \\ 3 \end{bmatrix};$

(b) $\mathbf{U} = \begin{bmatrix} 1 \\ 3 \\ 2 \end{bmatrix}, \qquad \mathbf{V} = \begin{bmatrix} 0 \\ 1 \\ 0 \end{bmatrix}, \qquad \mathbf{W} = \begin{bmatrix} 0 \\ 1 \\ 1 \end{bmatrix}.$

41. Compute $\mathbf{U} \times (\mathbf{V} \times \mathbf{W})$:

(a) $\mathbf{U} = \begin{bmatrix} 1 \\ -3 \\ -2 \end{bmatrix}$, $\mathbf{V} = \begin{bmatrix} 1 \\ 0 \\ 1 \end{bmatrix}$, $\mathbf{W} = \begin{bmatrix} 2 \\ 5 \\ 7 \end{bmatrix}$;

(b) $\mathbf{U} = \begin{bmatrix} 0 \\ -1 \\ 2 \end{bmatrix}$, $\mathbf{V} = \begin{bmatrix} 3 \\ 2 \\ 0 \end{bmatrix}$, $\mathbf{W} = \begin{bmatrix} 5 \\ 2 \\ -1 \end{bmatrix}$.

THEORETICAL EXERCISES

T-1. Prove Theorem 2.1.

T-2. Verify the "right-hand rule" for the direction of $\mathbf{U} \times \mathbf{V}$ if $\mathbf{U}$ and $\mathbf{V}$ are distinct elements of the natural basis for R^3.

T-3. Verify the "right-hand rule" for the direction of $\mathbf{U} \times \mathbf{V}$ if $\mathbf{U}$ and $\mathbf{V}$ lie in one of the coordinate planes. Use the results of Exercise T-2.

T-4. Prove : if the planes defined by $\mathbf{X} \cdot \mathbf{N}_1 = d_1$ and $\mathbf{X} \cdot \mathbf{N}_2 = d_2$ are identical, then $d_1 \mathbf{N}_2 = d_2 \mathbf{N}_1$.

T-5. Let L be a line, let $\mathbf{X}_1$ be a point not on L, and let $\mathbf{X}_0$ be the point on L closest to $\mathbf{X}_1$. Show that the line connecting $\mathbf{X}_0$ and $\mathbf{X}_1$ is perpendicular to L.

T-6. Let L_1 and L_2 be nonintersecting parallel lines, and let the distance between them be $d = | \bar{\mathbf{Y}}_1 - \bar{\mathbf{Y}}_2 |$, where $\bar{\mathbf{Y}}_1$ is on L_1 and $\bar{\mathbf{Y}}_2$ is on L_2. Show that $\bar{\mathbf{Y}}_1 - \bar{\mathbf{Y}}_2$ is perpendicular to both L_1 and L_2.

T-7. Show that the distance between two parallel lines L_1 and L_2 is given by

$$d = \frac{| \mathbf{U} \times (\mathbf{X}_2 - \mathbf{X}_1) |}{| \mathbf{U} |},$$

where $\mathbf{U}$ is any nonzero vector parallel to L_1 and L_2, $\mathbf{X}_1$ is on L_1, and $\mathbf{X}_2$ is on L_2.

T-8. Show that (19) yields $d = 0$ if L_1 and L_2 intersect.

T-9. Suppose $\mathbf{U} \neq \mathbf{0}$ and $\mathbf{X}_0$ is a point in R^3. What is the set of points $\mathbf{X}$ which satisfy $\mathbf{U} \times (\mathbf{X} - \mathbf{X}_0) = \mathbf{0}$?

T-10. Verify that

$$\mathbf{U} \cdot (\mathbf{V} \times \mathbf{W}) = \mathbf{V} \cdot (\mathbf{W} \times \mathbf{U}) = \mathbf{W} \cdot (\mathbf{U} \times \mathbf{V})$$
$$= -\mathbf{V} \cdot (\mathbf{U} \times \mathbf{W}) = -\mathbf{W} \cdot (\mathbf{V} \times \mathbf{U}) = -\mathbf{U} \cdot (\mathbf{W} \times \mathbf{V}).$$

T-11. Prove that

$$|\mathbf{U} \times \mathbf{V}|^2 + (\mathbf{U} \cdot \mathbf{V})^2 = |\mathbf{U}|^2 |\mathbf{V}|^2.$$

T-12. Prove that

(a) $\mathbf{U} \times (\mathbf{V} + \mathbf{W}) = \mathbf{U} \times \mathbf{V} + \mathbf{U} \times \mathbf{W}$;
(b) $(\mathbf{U} + \mathbf{V}) \times \mathbf{W} = \mathbf{U} \times \mathbf{W} + \mathbf{V} \times \mathbf{W}$;
(c) $\mathbf{U} \times \mathbf{V} = -(\mathbf{V} \times \mathbf{U})$.

T-13. If $\mathbf{U} \cdot \mathbf{V} = 0$ and $\mathbf{U} \times \mathbf{V} = \mathbf{0}$, show that $\mathbf{U} = \mathbf{0}$ or $\mathbf{V} = \mathbf{0}$.

T-14. Prove the *Jacobi identity*:

$$(\mathbf{U} \times \mathbf{V}) \times \mathbf{W} + (\mathbf{V} \times \mathbf{W}) \times \mathbf{U} + (\mathbf{W} \times \mathbf{U}) \times \mathbf{V} = \mathbf{0}.$$

3.3 Motion in R^3

In this section we discuss the use of vectors to study the motion of a particle in R^3. Since the emphasis is on applications rather than mathematical rigor, we shall be informal here; for example, we use terms like "curve" and "function" even though they are not defined until later, relying on the student's understanding of them on the basis of his first calculus course.

Position, Velocity, and Acceleration Vectors

The position of a particle moving in R^3 can be represented as a vector-valued function of time t:

$$\mathbf{X}(t) = \begin{bmatrix} x(t) \\ y(t) \\ z(t) \end{bmatrix} = x(t)\mathbf{i} + y(t)\mathbf{j} + z(t)\mathbf{k}.$$

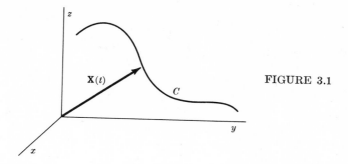

FIGURE 3.1

As t varies the particle moves along a curve C in R^3 (Fig. 3.1); we say that C is defined parametrically by $\mathbf{X} = \mathbf{X}(t)$.

If x, y, and z are continuous functions of t, we say that C is a *continuous curve*; physically this means that the particle does not undergo any sudden jumps.

The *velocity* of the particle is defined to be the vector valued function

$$\mathbf{V}(t) = \mathbf{X}'(t) = x'(t)\mathbf{i} + y'(t)\mathbf{j} + z'(t)\mathbf{k},$$

provided $x'(t)$, $y'(t)$, and $z'(t)$ exist. It has an important geometric interpretation. Suppose $h \neq 0$ and consider the vector

$$\frac{\mathbf{X}(t+h) - \mathbf{X}(t)}{h} = \frac{x(t+h) - x(t)}{h}\mathbf{i} + \frac{y(t+h) - y(t)}{h}\mathbf{j}$$

$$+ \frac{z(t+h) - z(t)}{h}\mathbf{k},$$

which is parallel to the secant vector $\mathbf{X}(t+h) - \mathbf{X}(t)$ (Fig. 3.2). Its components approach the components of $\mathbf{V}(t)$ as h approaches zero; therefore, if $\mathbf{V}(t) \neq 0$, its direction is that approached by the secant vector as h approaches zero; that is, $\mathbf{V}(t)$ is tangent to C at $\mathbf{X} = \mathbf{X}(t)$. The direction of $\mathbf{V}(t)$ is the *instantaneous direction of motion* and $|\mathbf{V}(t)|$ is the *instantaneous speed*.

The *acceleration vector* is defined by

$$\mathbf{A}(t) = \mathbf{V}'(t) = \mathbf{X}''(t) = x''(t)\mathbf{i} + y''(t)\mathbf{j} + z''(t)\mathbf{k},$$

provided $x''(t)$, $y''(t)$, and $z''(t)$ exist.

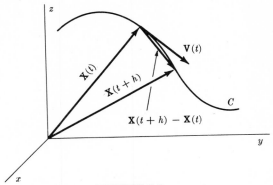

FIGURE 3.2

Example 3.1 Let the motion of a particle be described by

$$\mathbf{X}(t) = a \cos t \, \mathbf{i} + a \sin t \, \mathbf{j} + bt\mathbf{k} \qquad (-\infty < t < \infty);$$

then the curve traversed by the particle is a helix (Fig. 3.3). The velocity vector is

$$\mathbf{V}(t) = -a \sin t \, \mathbf{i} + a \cos t \, \mathbf{j} + b\mathbf{k}$$

and the speed is constant, since

$$| \mathbf{V}(t) | = \sqrt{(-a \sin t)^2 + (a \cos t)^2 + b^2} = \sqrt{a^2 + b^2}.$$

The acceleration vector is

$$\mathbf{A}(t) = -a \cos t \, \mathbf{i} - a \sin t \, \mathbf{j}.$$

The position, velocity and acceleration vectors are examples of *vector functions* of t. We can also define a general vector function $\mathbf{F}$ by

$$\mathbf{F}(t) = f(t)\mathbf{i} + g(t)\mathbf{j} + h(t)\mathbf{k},$$

where f, g, and h are scalar functions of t, and its derivatives by

$$\mathbf{F}'(t) = f'(t)\mathbf{i} + g'(t)\mathbf{j} + h'(t)\mathbf{k},$$

$$\mathbf{F}''(t) = f''(t)\mathbf{i} + g''(t)\mathbf{j} + h''(t)\mathbf{k},$$

etc., so long as f, g, and h possess the derivatives shown on the right.

If ρ is a scalar function of time then the vector function $\rho\mathbf{F}$ is defined by

(1) $$(\rho\mathbf{F})(t) = \rho(t)\mathbf{F}(t),$$

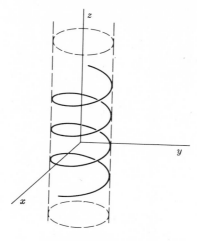

FIGURE 3.3

and if

$$\mathbf{F}_1(t) = f_1(t)\mathbf{i} + g_1(t)\mathbf{j} + h_1(t)\mathbf{k},$$
$$\mathbf{F}_2(t) = f_2(t)\mathbf{i} + g_2(t)\mathbf{j} + h_2(t)\mathbf{k},$$

then we define the sum $\mathbf{F}_1 + \mathbf{F}_2$ by

$$(\mathbf{F}_1 + \mathbf{F}_2)(t) = \mathbf{F}_1(t) + \mathbf{F}_2(t).$$

Similarly, we define $\mathbf{F}_1 \cdot \mathbf{F}_2$ and $\mathbf{F}_1 \times \mathbf{F}_2$ by

(2) $$(\mathbf{F}_1 \cdot \mathbf{F}_2)(t) = \mathbf{F}_1(t) \cdot \mathbf{F}_2(t)$$

and

$$(\mathbf{F}_1 \times \mathbf{F}_2)(t) = \mathbf{F}_1(t) \times \mathbf{F}_2(t).$$

Theorem 3.1 Let $\mathbf{F}_1$ and $\mathbf{F}_2$ be vector functions and ρ be a scalar function of t. Then

(a) $(\mathbf{F}_1 + \mathbf{F}_2)'(t) = \mathbf{F}_1'(t) + \mathbf{F}_2'(t)$;
(b) $(\rho\mathbf{F}_1)'(t) = (\rho'\mathbf{F}_1 + \rho\mathbf{F}_1')(t)$;
(c) $(\mathbf{F}_1 \cdot \mathbf{F}_2)'(t) = (\mathbf{F}_1' \cdot \mathbf{F}_2 + \mathbf{F}_1 \cdot \mathbf{F}_2')(t)$,

and

(d) $(\mathbf{F}_1 \times \mathbf{F}_2)'(t) = (\mathbf{F}_1' \times \mathbf{F}_2 + \mathbf{F}_1 \times \mathbf{F}_2')(t)$,

provided the derivatives on the right hand sides exist.

Proof.   We prove (b) and (c), and leave the rest for the student. From (1),

$$(\rho \mathbf{F})'(t) = \frac{d}{dt} \left[\rho(t) \mathbf{F}(t) \right]$$

$$= \left[\frac{d}{dt} \rho(t) f(t) \right] \mathbf{i} + \left[\frac{d}{dt} \rho(t) g(t) \right] \mathbf{j} + \left[\frac{d}{dt} \rho(t) h(t) \right] \mathbf{k}$$

$$= \left[\rho'(t) f(t) + \rho(t) f'(t) \right] \mathbf{i} + \left[\rho'(t) g(t) + \rho(t) g'(t) \right] \mathbf{j}$$

$$+ \left[\rho'(t) h(t) + \rho(t) h'(t) \right] \mathbf{k}$$

$$= \rho'(t) \left[f(t) \mathbf{i} + g(t) \mathbf{j} + h(t) \mathbf{k} \right]$$

$$+ \rho(t) \left[f'(t) \mathbf{i} + g'(t) \mathbf{j} + h'(t) \mathbf{k} \right]$$

$$= \rho'(t) \mathbf{F}(t) + \rho(t) \mathbf{F}'(t)$$

$$= (\rho' \mathbf{F} + \rho \mathbf{F}')(t),$$

which proves (b). From (2)

$$(\mathbf{F}_1 \cdot \mathbf{F}_2)'(t) = \frac{d}{dt} \left(f_1(t) f_2(t) + g_1(t) g_2(t) + h_1(t) h_2(t) \right)$$

$$= f_1'(t) f_2(t) + g_1'(t) g_2(t) + h_1'(t) h_2(t)$$

$$+ f_1(t) f_2'(t) + g_1(t) g_2'(t) + h_1(t) h_2'(t)$$

$$= \mathbf{F}_1'(t) \cdot \mathbf{F}_2(t) + \mathbf{F}_1(t) \cdot \mathbf{F}_2'(t)$$

$$= (\mathbf{F}_1' \cdot \mathbf{F}_2 + \mathbf{F}_1 \cdot \mathbf{F}_2')(t),$$

which proves (c).

Example 3.2   Let

$$\mathbf{F}_1(t) = \cos t \, \mathbf{i} + \sin t \, \mathbf{j} + t \mathbf{k},$$

$$\mathbf{F}_2(t) = t \mathbf{i} + t^2 \mathbf{j} + t^3 \mathbf{k},$$

and

$$\rho(t) = e^{t^2};$$

then

$$\mathbf{F}_1'(t) = -\sin t \, \mathbf{i} + \cos t \, \mathbf{j} + \mathbf{k},$$

$$\mathbf{F}_2'(t) = \mathbf{i} + 2t \mathbf{j} + 3t^2 \mathbf{k},$$

and

$$\rho'(t) = 2t e^{t^2}.$$

We illustrate Theorem 3.1 by verifying (a) through (d):

(a) $(\mathbf{F_1} + \mathbf{F_2})(t) = (\cos t + t)\mathbf{i} + (\sin t + t^2)\mathbf{j} + (t + t^3)\mathbf{k}$

and

$$(\mathbf{F_1} + \mathbf{F_2'})(t) = (-\sin t + 1)\mathbf{i} + (\cos t + 2t)\mathbf{j} + (1 + 3t^2)\mathbf{k}$$
$$= (-\sin t\,\mathbf{i} + \cos t\,\mathbf{j} + \mathbf{k}) + (\mathbf{i} + 2t\mathbf{j} + 3t^2\mathbf{k})$$
$$= \mathbf{F_1'}(t) + \mathbf{F_2'}(t).$$

(b) $(\rho\mathbf{F_1})(t) = (e^{t^2}\cos t)\mathbf{i} + (e^{t^2}\sin t)\mathbf{j} + (e^{t^2}t)\mathbf{k}$

and

$$(\rho\mathbf{F_1})'(t) = (2te^{t^2}\cos t - e^{t^2}\sin t)\mathbf{i} + (2te^{t^2}\sin t + e^{t^2}\cos t)\mathbf{j}$$
$$+ (2t^2e^{t^2} + e^{t^2})\mathbf{k}$$
$$= 2te^{t^2}(\cos t\,\mathbf{i} + \sin t\,\mathbf{j} + t\mathbf{k})$$
$$+ e^{t^2}(-\sin t\,\mathbf{i} + \cos t\,\mathbf{j} + \mathbf{k})$$
$$= \rho'(t)\mathbf{F_1}(t) + \rho(t)\mathbf{F_1'}(t)$$
$$= (\rho'\mathbf{F_1} + \rho\mathbf{F_1'})(t).$$

(c) $(\mathbf{F_1}\cdot\mathbf{F_2})(t) = t\cos t + t^2\sin t + t^4$

and

(3) $(\mathbf{F_1}\cdot\mathbf{F_2})'(t) = (\cos t - t\sin t) + (2t\sin t + t^2\cos t) + 4t^3$
$$= \cos t + t\sin t + t^2\cos t + 4t^3.$$

On the other hand,

$$(\mathbf{F_1'}\cdot\mathbf{F_2} + \mathbf{F_1}\cdot\mathbf{F_2'})(t) = \mathbf{F_1'}(t)\cdot\mathbf{F_2}(t) + \mathbf{F_1}(t)\cdot\mathbf{F_2'}(t)$$
$$= (-t\sin t + t^2\cos t + t^3) + (\cos t + 2t\sin t + 3t^3)$$
$$= \cos t + t\sin t + t^2\cos t + 4t^3,$$

which agrees with (3).

(d) $(\mathbf{F_1} \times \mathbf{F_2})(t) = \begin{vmatrix} \mathbf{i} & \mathbf{j} & \mathbf{k} \\ \cos t & \sin t & t \\ t & t^2 & t^3 \end{vmatrix}$

$$= (t^3\sin t - t^3)\mathbf{i} + (t^2 - t^3\cos t)\mathbf{j}$$
$$+ (t^2\cos t - t\sin t)\mathbf{k}$$

and

(4)　　$(F_1 \times F_2)'(t) = (3t^2 \sin t + t^3 \cos t - 3t^2)\mathbf{i}$

$+ (2t - 3t^2 \cos t + t^3 \sin t)\mathbf{j}$

$+ (2t \cos t - t^2 \sin t - \sin t - t \cos t)\mathbf{k}.$

On the other hand,

(5)　　$(F_1' \times F_2)(t) = \begin{vmatrix} \mathbf{i} & \mathbf{j} & \mathbf{k} \\ -\sin t & \cos t & 1 \\ t & t^2 & t^3 \end{vmatrix}$

$= (t^3 \cos t - t^2)\mathbf{i} + (t + t^3 \sin t)\mathbf{j}$

$+ (-t^2 \sin t - t \cos t)\mathbf{k}$

and

(6)　　$(F_1 \times F_2')(t) = \begin{vmatrix} \mathbf{i} & \mathbf{j} & \mathbf{k} \\ \cos t & \sin t & t \\ 1 & 2t & 3t^2 \end{vmatrix}$

$= (3t^2 \sin t - 2t^2)\mathbf{i} + (t - 3t^2 \cos t)\mathbf{j}$

$+ (2t \cos t - \sin t)\mathbf{k}.$

Adding the right hand sides of (4) and (5) yields the right hand side of (6).

Example 3.3　　To calculate the derivative of $|F(t)|$ we write

$$|F(t)|^2 = F(t) \cdot F(t)$$

and differentiate both sides:

$$2|F(t)|\frac{d}{dt}|F(t)| = F'(t) \cdot F(t) + F(t) \cdot F'(t)$$

$$= 2F(t) \cdot F'(t).$$

Thus, if $F(t) \neq 0$,

(7)　　$$\frac{d}{dt}|F(t)| = \frac{F'(t) \cdot F(t)}{|F(t)|}.$$

If $|F(t)|$ is constant then $F(t)$ is perpendicular to its derivative, since (7) becomes

$$0 = F'(t) \cdot F(t).$$

For example, if

$$\mathbf{F}(t) = \cos t\,\mathbf{i} + \sin t\,\mathbf{j} + \mathbf{k},$$

then

$$\mathbf{F}'(t) = -\sin t\,\mathbf{i} + \cos t\,\mathbf{j}$$

and

$$\mathbf{F}(t)\cdot\mathbf{F}'(t) = -\cos t \sin t + \sin t \cos t = 0,$$

which is to be expected, since $|\mathbf{F}(t)| \equiv \sqrt{2}$.

Example 3.4   If $\mathbf{F}(t) \neq \mathbf{0}$, define the unit vector

$$\mathbf{U}(t) = \frac{\mathbf{F}(t)}{|\mathbf{F}(t)|};$$

then, from (b) of Theorem 3.1,

$$\mathbf{U}'(t) = \frac{\mathbf{F}'(t)}{|\mathbf{F}(t)|} + \left[\frac{d}{dt}\left(\frac{1}{|\mathbf{F}(t)|}\right)\right]\mathbf{F}(t)$$

$$= \frac{\mathbf{F}'(t)}{|\mathbf{F}(t)|} - \left[\frac{d}{dt}|\mathbf{F}(t)|\right]\frac{\mathbf{F}(t)}{|\mathbf{F}(t)|^2}.$$

From (7) this can be rewritten as

$$\mathbf{U}'(t) = \frac{\mathbf{F}'(t)}{|\mathbf{F}(t)|} - \frac{(\mathbf{F}(t)\cdot\mathbf{F}'(t))}{|\mathbf{F}(t)|^2}\mathbf{F}(t)$$

$$= \frac{|\mathbf{F}(t)|^2\,\mathbf{F}'(t) - (\mathbf{F}(t)\cdot\mathbf{F}'(t))\mathbf{F}(t)}{|\mathbf{F}(t)|^3},$$

which can be expressed as a triple product:

(8)
$$\frac{d}{dt}\frac{\mathbf{F}(t)}{|\mathbf{F}(t)|} = \frac{\mathbf{F}(t) \times (\mathbf{F}'(t) \times \mathbf{F}(t))}{|\mathbf{F}(t)|^3}.$$

Unit Tangent, Normal, and Binormal

Let us again consider a particle moving along a curve C. We have seen that its velocity vector $\mathbf{V}(t)$ is tangent to C at $\mathbf{X} = \mathbf{X}(t)$, provided $\mathbf{V}(t) \neq \mathbf{0}$ (which we assume throughout this section). The *unit tangent vector to C at* $\mathbf{X} = \mathbf{X}(t)$ is defined by

(9)
$$\mathbf{T}(t) = \frac{\mathbf{V}(t)}{|\mathbf{V}(t)|}.$$

From (8)

(10)
$$\mathbf{T}'(t) = \frac{\mathbf{V}(t) \times (\mathbf{A}(t) \times \mathbf{V}(t)}{|\mathbf{V}(t)|^3} .$$

If $\mathbf{A}(t)$ and $\mathbf{V}(t)$ are noncollinear then we can define $\mathbf{N}(t)$, the *unit principal normal vector* $\mathbf{N}$, to be the unit vector in the direction of $\mathbf{T}'(t)$; thus

(11)
$$\mathbf{N}(t) = \frac{\mathbf{T}'(t)}{|\mathbf{T}'(t)|}$$
$$= \frac{\mathbf{V}(t) \times (\mathbf{A}(t) \times \mathbf{V}(t))}{|\mathbf{V}(t)| |\mathbf{A}(t) \times \mathbf{V}(t)|} .$$

The unit binormal vector to C at $\mathbf{X} = \mathbf{X}(t)$ is defined by

$$\mathbf{B}(t) = \mathbf{T}(t) \times \mathbf{N}(t);$$

from (9) and (11), it is given explicitly by

(12)
$$\mathbf{B}(t) = \frac{\mathbf{V}(t) \times \mathbf{A}(t)}{|\mathbf{V}(t) \times \mathbf{A}(t)|}$$

(Exercise T-2).

The plane Γ_t through $\mathbf{X} = \mathbf{X}(t)$ that contains $\mathbf{A}(t)$ and $\mathbf{V}(t)$ is the *osculating plane at* $\mathbf{X} = \mathbf{X}(t)$. From (9) and (11), Γ_t contains $\mathbf{T}(t)$ and $\mathbf{N}(t)$. The situation is shown in Fig. 3.4, where the solid curve is C and the dotted curve is the projection of C on Γ_t. If we consider Γ_t to be the plane of the page, then $\mathbf{B}(t)$ is perpendicular to the page and pointed toward the reader.

The equation of Γ_t is, according to Eq. (12), Section 3.1, and Eq. (12) above,

$$(\mathbf{X} - \mathbf{X}(t)) \cdot \mathbf{B}(t) = 0.$$

Example 3.5   Let us calculate $\mathbf{T}$, $\mathbf{N}$, and $\mathbf{B}$ and find the equation of the osculating plane to the helix defined by

$$\mathbf{X}(t) = \cos t\, \mathbf{i} + \sin t\, \mathbf{j} + e^t \mathbf{k}$$

at $\mathbf{X}(0) = (1, 0, 1)$. The velocity and acceleration vectors are

$$\mathbf{V}(t) = -\sin t\, \mathbf{i} + \cos t\, \mathbf{j} + e^t \mathbf{k}$$

and

$$\mathbf{A}(t) = -\cos t\, \mathbf{i} - \sin t\, \mathbf{j} + e^t \mathbf{k};$$

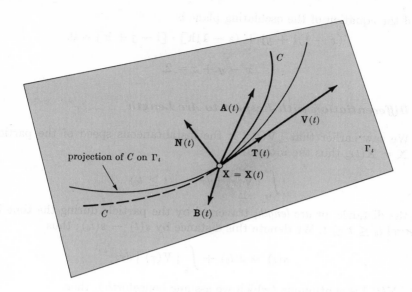

FIGURE 3.4

thus $\mathbf{V}(0) = \mathbf{j} + \mathbf{k}$ and $\mathbf{A}(0) = -\mathbf{i} + \mathbf{k}$. From (9),

$$\mathbf{T}(0) = \frac{\mathbf{V}(0)}{|\mathbf{V}(0)|} = \frac{1}{\sqrt{2}}\,(\mathbf{j} + \mathbf{k}).$$

Since

$$\mathbf{V}(0) \times (\mathbf{A}(0) \times \mathbf{V}(0)) = (\mathbf{j} + \mathbf{k}) \times [(-\mathbf{i} + \mathbf{k}) \times (\mathbf{j} + \mathbf{k})]$$

$$= (\mathbf{j} + \mathbf{k}) \times (-\mathbf{i} + \mathbf{j} - \mathbf{k})$$

$$= -2\mathbf{i} - \mathbf{j} + \mathbf{k},$$

it follows from (11) that

$$\mathbf{N}(0) = \frac{1}{\sqrt{6}}\,(-2\mathbf{i} - \mathbf{j} + \mathbf{k}).$$

Now, from (12),

$$\mathbf{B}(0) = \frac{1}{\sqrt{2}}\,(\mathbf{j} + \mathbf{k}) \times \frac{1}{\sqrt{6}}\,(-2\mathbf{i} - \mathbf{j} + \mathbf{k})$$

$$= \frac{1}{\sqrt{3}}\,(\mathbf{i} - \mathbf{j} + \mathbf{k}),$$

and the equation of the osculating plane is

$$[(x-1)\mathbf{i} + y\mathbf{j} + (z-1)\mathbf{k}] \cdot [\mathbf{i} - \mathbf{j} + \mathbf{k}] = 0,$$

or

$$x - y + z = 2.$$

Differentiation with Respect to Arc Length

We saw earlier that $|\mathbf{V}(t)|$ is the instantaneous speed of the particle at $\mathbf{X} = \mathbf{X}(t)$; thus the integral

$$\int_{t_0}^{t} |\mathbf{V}(\tau)| \, d\tau \qquad (t \geq t_0)$$

is the distance, or *arc length*, traversed by the particle during the time interval $t_0 \leq t \leq t$. We denote this distance by $s(t) - s(t_0)$; thus

$$s(t) = s(t_0) + \int_{t_0}^{t} |\mathbf{V}(\tau)| \, d\tau.$$

If $|\mathbf{V}(t)|$ is continuous (which we assume henceforth), then

(13) $$s'(t) = |\mathbf{V}(t)|.$$

The derivative of T with respect to arc length is defined by

$$\frac{d\mathbf{T}}{ds}(t) = \lim_{h \to 0} \frac{\mathbf{T}(t+h) - \mathbf{T}(t)}{s(t+h) - s(t)},$$

which can be rewritten as

$$\frac{d\mathbf{T}}{ds}(t) = \lim_{h \to 0} \frac{\dfrac{\mathbf{T}(t+h) - \mathbf{T}(t)}{h}}{\dfrac{s(t+h) - s(t)}{h}}$$

$$= \frac{\mathbf{T}'(t)}{|\mathbf{V}(t)|};$$

thus, from (10) and (11),

$$\frac{d\mathbf{T}}{ds}(t) = \frac{\mathbf{V}(t) \times (\mathbf{A}(t) \times \mathbf{V}(t))}{|\mathbf{V}(t)|^4} = \frac{|\mathbf{V}(t) \times \mathbf{A}(t)|}{|\mathbf{V}(t)|^3} \mathbf{N}(t).$$

Hence $(d\mathbf{T}/ds)(t)$ is a vector in the direction of $\mathbf{N}(t)$. We call its magnitude the *curvature* of C at $\mathbf{X} = \mathbf{X}(t)$ and denote it by $\kappa(t)$:

$$\kappa(t) = \frac{|\mathbf{V}(t) \times \mathbf{A}(t)|}{|\mathbf{V}(t)|^3}.$$

The *radius of curvature* of C at $\mathbf{X} = \mathbf{X}(t)$ is defined by

$$\rho(t) = \frac{1}{\kappa(t)}.$$

Example 3.6 For the helix considered in Example 3.1 the curvature is constant, since

(14)
$$\mathbf{V}(t) \times \mathbf{A}(t) = \begin{vmatrix} \mathbf{i} & \mathbf{j} & \mathbf{k} \\ -a\sin t & a\cos t & b \\ -a\cos t & -a\sin t & 0 \end{vmatrix}$$

$$= ab\sin t\, \mathbf{i} - ab\cos t\, \mathbf{j} + a^2\mathbf{k}$$

and

$$\kappa(t) = \frac{a\sqrt{a^2 + b^2}}{(\sqrt{a^2 + b^2})^3}.$$

$$= \frac{a}{a^2 + b^2}.$$

Thus the radius of curvatures is also constant:

(15)
$$\rho(t) = \frac{a^2 + b^2}{a}.$$

If $b = 0$, the helix reduces to a circle of radius a in the xy-plane; thus it follows from (15) that the radius of curvature of a circle is equal to its radius.

Example 3.7 From (9) and (13),

$$\mathbf{V}(t) = s'(t)\mathbf{T}(t);$$

hence, utilizing (11),

$$\mathbf{A}(t) = s''(t)\mathbf{T}(t) + s'(t)\mathbf{T}'(t)$$

$$= s''(t)\mathbf{T}(t) + s'(t)\,|\,\mathbf{T}'(t)\,|\,\mathbf{N}(t).$$

According to (10), and because $\mathbf{V}(t)$ and $\mathbf{A}(t) \times \mathbf{V}(t)$ are perpendicular,

$$|\,\mathbf{T}'(t)\,| = \frac{|\,\mathbf{A}(t) \times \mathbf{V}(t)\,|}{|\,\mathbf{V}(t)\,|^2} = \frac{|\,\mathbf{A}(t) \times \mathbf{V}(t)\,|}{|\,\mathbf{V}(t)\,|^3}\,|\,\mathbf{V}(t)\,|$$

$$= \kappa(t)\,|\,\mathbf{V}(t)\,|$$

$$= \frac{s'(t)}{\rho(t)};$$

thus $\mathbf{A}(t)$ can be written as

$$\mathbf{A}(t) = \mathbf{A}_N(t) + \mathbf{A}_T(t),$$

where

$$\mathbf{A}_T(t) = s''(t)\mathbf{T}(t)$$

and

$$\mathbf{A}_N(t) = \frac{(s'(t))^2}{\rho(t)} \mathbf{N}(t)$$

are the *tangential* and *normal* components of acceleration, respectively.

The derivative of the unit binormal with respect to arc length is defined by

$$(16) \qquad \frac{d\mathbf{B}}{ds}(t) = \lim_{h \to 0} \frac{\mathbf{B}(t+h) - \mathbf{B}(t)}{s(t+h) - s(t)}$$

$$= \lim_{h \to 0} \frac{\dfrac{\mathbf{B}(t+h) - \mathbf{B}(t)}{h}}{\dfrac{s(t+h) - s(t)}{h}}$$

$$= \frac{\mathbf{B}'(t)}{|\mathbf{V}(t)|}.$$

Differentiating (12) with respect to t and using (8) with $\mathbf{F} = \mathbf{V} \times \mathbf{A}$, noting that

$$(\mathbf{V} \times \mathbf{A})' = (\mathbf{V} \times \mathbf{A}') + (\mathbf{V}' \times \mathbf{A})$$

$$= (\mathbf{V} \times \mathbf{A}') + (\mathbf{A} \times \mathbf{A}) = \mathbf{V} \times \mathbf{A}',$$

we find that

$$\mathbf{B}'(t) = \frac{(\mathbf{V}(t) \times \mathbf{A}(t)) \times [(\mathbf{V}(t) \times \mathbf{A}'(t)) \times (\mathbf{V}(t) \times \mathbf{A}(t))]}{|\mathbf{V}(t) \times \mathbf{A}(t)|^3}$$

$$= \frac{(\mathbf{V}(t) \times \mathbf{A}(t)) \times [(\mathbf{A}(t) \times \mathbf{V}(t)) \cdot \mathbf{A}'(t)]\mathbf{V}(t)}{|\mathbf{V}(t) \times \mathbf{A}(t)|^3}$$

$$= \frac{(\mathbf{A}(t) \times \mathbf{V}(t)) \cdot \mathbf{A}'(t)}{|\mathbf{V}(t) \times \mathbf{A}(t)|^3} [\mathbf{V}(t) \times (\mathbf{A}(t) \times \mathbf{V}(t))].$$

From this and (11) and (16),

$$\frac{d\mathbf{B}}{ds}(t) = -\frac{(\mathbf{V}(t) \times \mathbf{A}(t)) \cdot \mathbf{A}'(t)}{|\mathbf{V}(t) \times \mathbf{A}(t)|^2} \mathbf{N}(t).$$

Thus $(d\mathbf{B}/ds)(t)$ is parallel to $\mathbf{N}(t)$. The quantity

$$\tau(t) \;=\; -\frac{|\,\mathbf{V}(t)\,|\,(\mathbf{V}(t) \times \mathbf{A}(t)) \cdot \mathbf{A}'(t)}{|\,\mathbf{V}(t) \times \mathbf{A}(t)\,|^2}$$

is called the *torsion of* C at $\mathbf{X} = \mathbf{X}(t)$; it is identically zero if and only if C lies in a plane.

Example 4.8   For the helix of Example 4.1,

and
$$\mathbf{A}'(t) = a \sin t\,\mathbf{i} - a \cos t\,\mathbf{j}$$

$$\mathbf{V}(t) \times \mathbf{A}(t) = ab \sin t\,\mathbf{i} - ab \cos t\,\mathbf{j} + a^2\mathbf{k}$$

(see (14)). Therefore

$$\tau(t) \;=\; \frac{-a^2b}{a^4 + a^2b^2}$$

$$=\; \frac{-b}{a^2 + b^2}.$$

If $b = 0$, then C lies in the xy-plane and $\tau(t) \equiv 0$.

Oct. 6
#2, 4, 8, 12, 14, 16, 18, 20, 23

EXERCISES 3.3

In Exercises 1 through 4 find the velocity and acceleration vectors and the cosine of the angle between them. Also, find the instantaneous speed and direction of motion.

1. $\mathbf{X}(t) = t^3\mathbf{i} + t^2\mathbf{j} + \mathbf{k}$.

2. $\mathbf{X}(t) = \cos t\,\mathbf{i} + \sin t\,\mathbf{j} + \mathbf{k}$.

3. $\mathbf{X}(t) = e^t\mathbf{i} + \cos t\,\mathbf{j} + t^3\mathbf{k}$.

4. $\mathbf{X}(t) = (t^2 - t)\mathbf{i} + (t^3 + t^2)\mathbf{j}$.

5.   Verify Theorem 3.1 for $\mathbf{F}_1(t) = e^t\mathbf{i} + e^{-t}\mathbf{j}$, $\mathbf{F}_2(t) = \cos t\,\mathbf{j} + \sin t\,\mathbf{k}$, and $\rho(t) = e^t$.

6.   Verify Theorem 3.1 for $\mathbf{F}_1(t) = t\mathbf{i} + t^2\mathbf{j} + t^3\mathbf{k}$, $\mathbf{F}_2(t) = t^3\mathbf{i} + t^2\mathbf{j} + t\mathbf{k}$, and $\rho(t) = t$.

In Exercises 7 through 11 compute

$$\frac{d}{dt} \mid \mathbf{F}(t) \mid \qquad \text{and} \qquad \frac{d}{dt}\left(\frac{\mathbf{F}(t)}{\mid \mathbf{F}(t) \mid}\right)$$

directly and also by means of Eqs. (7) and (8), respectively.

7. $\mathbf{F}(t) = (t + 1)\mathbf{i} + (t - 1)\mathbf{j} + t^2\mathbf{k}$.

8. $\mathbf{F}(t) = e^t\mathbf{i} + e^{-t}\mathbf{j}$.

9. $\mathbf{F}(t) = e^t \sin t \, \mathbf{i} + e^t \cos t \, \mathbf{k}$.

10. $\mathbf{F}(t) = \mathbf{i} - \mathbf{j} + t\mathbf{k}$.

11. $\mathbf{F}(t) = \sqrt{2}t\mathbf{i} - t\mathbf{j} + \mathbf{k}$.

In Exercises 12 through 15 compute $\mathbf{T}(t_0)$, $\mathbf{N}(t_0)$, and $\mathbf{B}(t_0)$ and find the equation of the osculating plane to the curve at $\mathbf{X}(t_0)$.

12. $\mathbf{X}(t) = 2t\mathbf{i} + t^2\mathbf{j} + t^3\mathbf{k}; \qquad t_0 = 1$.

13. $\mathbf{X}(t) = \sin t \, \mathbf{i} + \cos t \, \mathbf{j} + 2\mathbf{k}; \qquad t_0 = \dfrac{\pi}{3}$.

14. $\mathbf{X}(t) = e^t \sin t \, \mathbf{i} + e^t \cos t \, \mathbf{j} + e^t\mathbf{k}; \qquad t_0 = 0$.

15. $\mathbf{X}(t) = t^2\mathbf{i} + (t + 1)\mathbf{j} + t\mathbf{k}; \qquad t_0 = 0$.

In Exercises 16 through 18 find the arc length of the curve defined by $\mathbf{X} = \mathbf{X}(t)$ $(t_0 \le t \le t_1)$.

16. $\mathbf{X}(t) = \cos 3t \, \mathbf{i} + \sin 3t \, \mathbf{j} + 4t\mathbf{k}; \qquad 0 \le t \le 2\pi$.

17. $\mathbf{X}(t) = t\mathbf{i} + \dfrac{t^2}{2}\mathbf{j} + \dfrac{2\sqrt{2}}{3}t^{3/2}\mathbf{k}; \qquad 1 \le t \le 2$.

18. $\mathbf{X}(t) = e^t\mathbf{i} + \sqrt{2}t\mathbf{j} - e^{-t}\mathbf{k}; \qquad 0 \le t \le 1$.

In Exercises 19 through 22 compute the curvature, radius of curvature, and torsion of the curve defined by $\mathbf{X} = \mathbf{X}(t)$ at the point $\mathbf{X}(t_0)$. Also find the tangential and normal components of acceleration at $t = t_0$.

19. $\mathbf{X}(t) = \cos t \, \mathbf{i} + \sin t \, \mathbf{j} + t\mathbf{k}; \qquad t_0 = \dfrac{\pi}{3}$.

20. $\mathbf{X}(t) = e^t \sin t \, \mathbf{i} + e^{-t} \cos t \, \mathbf{j} + e^{-t}\mathbf{k}; \qquad t_0 = 0$.

21. $\mathbf{X}(t) = t\mathbf{i} + t^2\mathbf{j} + (t^2 + 1)\mathbf{k}; \qquad t_0 = 0$.

22. $\mathbf{X}(t) = \sqrt{t}\mathbf{i} + t^{3/2}\mathbf{j} + t^2\mathbf{k};\qquad t_0 = 1.$

23. A particle moves around a circle of radius r with constant speed v_0. Find its tangential and normal components of acceleration.

    Do everything with respect to arc length.

THEORETICAL EXERCISES

T-1. Prove (a) and (d) of Theorem 4.1.

T-2. Derive Eq. (12).

T-3. Show that the acceleration of a particle moving along a curve with constant speed is in the direction of the normal.

T-4. Suppose a particle moves so that its acceleration is always perpendicular to its velocity. Show that its speed is constant.

T-5. Suppose a particle moves along a curve C given by $\mathbf{X} = \mathbf{X}(t)$ with $z'(t) \equiv 0$. Show that

$$\kappa(t) = \frac{\mid x'(t)y''(t) - y'(t)x''(t) \mid}{[(x'(t))^2 + (y'(t))^2]^{3/2}} \, .$$

T-6. Use the result of Exercise T-5 to show that the curvature of a curve defined by $y = f(x)$ is

$$\kappa(x) = \frac{\mid f''(x) \mid}{(1 + (f'(x))^2)^{3/2}}$$

at the point $(x, f(x))$.

T-7. *Prove:* If a particle moves so that $\mathbf{A}(t) \times \mathbf{V}(t)$ is constant, then the motion is in a plane.

3.4 Parametrically Defined Curves

In this section we generalize some of the ideas of Section 3.3. We consider *parametric functions*; that is, vector valued functions of a single real variable. A parametric function $\mathbf{\Phi}$ is a vector valued function of the form

$$\mathbf{\Phi}(t) = \begin{bmatrix} \Phi_1(t) \\ \vdots \\ \Phi_n(t) \end{bmatrix}$$

where $\Phi_1, \ldots, \Phi_n$ are continuous real valued functions defined on a closed interval $[a, b]$. As t varies over $[a, b]$, $\Phi(t)$ traverses a curve C in R^n. When we wish to specify $[a, b]$ and the dimension of the space containing C, we shall write

$$\Phi : [a, b] \to R^n.$$

Example 4.1   Let $\Phi : [0, \pi] \to R^2$ be defined by

$$\Phi(t) = \begin{bmatrix} \cos t \\ \sin t \end{bmatrix} \qquad (0 \le t \le \pi).$$

Since $\cos^2 t + \sin^2 t = 1$, $-1 \le \cos t \le 1$, and $0 \le \sin t \le 1$ for $0 \le t \le \pi$, it follows that C is the upper half of the unit circle (Fig. 4.1).

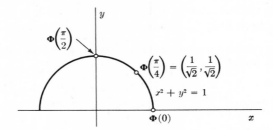

FIGURE 4.1

The parametric functions

$$\Psi(s) = \begin{bmatrix} -\sin 2s \\ \cos 2s \end{bmatrix} \qquad \left(-\frac{\pi}{4} \le s \le \frac{\pi}{4} \right)$$

and

$$\eta(r) = \begin{bmatrix} \sin r \\ \cos r \end{bmatrix} \qquad \left(-\frac{\pi}{2} \le r \le \frac{\pi}{2} \right)$$

define the same curve as Φ. As s and t increase over their respective intervals, $\Phi(t)$ and $\Psi(s)$ traverse the semicircle from $(1, 0)$ to $(-1, 0)$; however, as r varies from $-\frac{\pi}{2}$ to $\frac{\pi}{2}$, $\eta(r)$ moves in the opposite direction.

In connection with the line integral (Section 6.5), the manner in which $X = \Phi(t)$ traverses the curve C is important. Two parametric functions are *equivalent* if they not only define the same curve C, but prescribe the same order for traversing its points. Thus in Example 4.1, Φ and Ψ are equivalent, but Φ and η are not.

Definition 4.1   Let C be defined by $\mathbf{X} = \boldsymbol{\Phi}(t)$ $(a \leq t \leq b)$. Then $\boldsymbol{\Phi}(a)$ and $\boldsymbol{\Phi}(b)$ are respectively the *initial* and *final endpoints* of C; if $\boldsymbol{\Phi}(a) = \boldsymbol{\Phi}(b)$ then C is a *closed curve*. If $\boldsymbol{\Phi}(t_1)$ and $\boldsymbol{\Phi}(t_2)$ are distinct whenever $a \leq t_1 < t_2 < b$ or $a < t_1 < t_2 \leq b$, then C is *simple*. A *simple closed curve* is both closed and simple.

Example 4.2   The curve defined in Example 4.1 is simple. The curve defined by

$$\mathbf{X}(t) = \begin{bmatrix} \cos t \\ \sin t \end{bmatrix} \qquad (0 \leq t \leq 2\pi)$$

is the unit circle, traversed in the counterclockwise direction; it is a simple closed curve (Fig. 4.2).

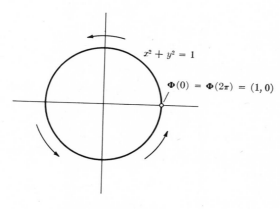

FIGURE 4.2

Example 4.3   The curve defined by

$$\mathbf{X}(t) = \begin{bmatrix} \cos t \\ \sin t \end{bmatrix} \qquad (0 \leq t \leq 4\pi)$$

is closed but not simple, and the curve defined by

$$\mathbf{X}(t) = \begin{bmatrix} \cos t \\ \sin t \end{bmatrix} \qquad (0 \leq t \leq 3\pi)$$

is neither closed nor simple. The curve shown in Fig. 4.3 is closed but not simple.

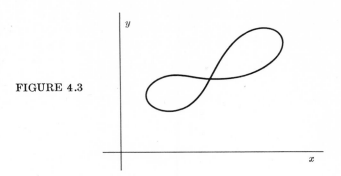

FIGURE 4.3

Definition 4.2   The *derivative* of a parametric function $\mathbf{\Phi}:[a,\ b] \to R^n$ is defined by

$$\mathbf{\Phi}' = \begin{bmatrix} \Phi'_1 \\ \Phi'_2 \\ \vdots \\ \Phi'_n \end{bmatrix},$$

whenever the vector on the right exists.

By $\Phi'_i(a)$ we mean a right hand derivative; thus

$$\Phi'_i(a) = \lim_{t \to a^+} \frac{\Phi_i(t) - \Phi_i(a)}{t - a}$$

where $\lim_{t \to a^+}$ means that t approaches a from above. Similarly $\Phi'_i(b)$ is a left hand derivative:

$$\Phi'_1(b) = \lim_{t \to b^-} \frac{\Phi_i(t) - \Phi_i(b)}{t - b}.$$

Definition 4.3   The curve C defined by $\mathbf{X} = \mathbf{\Phi}(t)$ $(a \le t \le b)$ is *continuously differentiable* if Φ'_i exists and is continuous in $[a,\ b]$ for $i = 1$, $2,\ldots, n$; if in addition $\mathbf{\Phi}'$ does not vanish in $[a,\ b]$, then C is *smooth*. The *unit tangent vector* to C at a point $\mathbf{X}_0 = \mathbf{\Phi}(t_0)$ where $\mathbf{\Phi}'(t_0) \ne \mathbf{0}$ is defined by

$$\mathbf{T}(t_0) = \frac{\mathbf{\Phi}'(t_0)}{|\ \mathbf{\Phi}'(t_0)\ |};$$

thus a smooth curve has a continuously turning unit tangent vector.

This definition of the unit tangent is analogous to that given for $n = 3$ in Section 3.3

Example 4.4 Let C be defined by

$$\mathbf{X} = \mathbf{\Phi}(t) = \begin{bmatrix} \cos t \\ \sin t \\ e^t \\ e^{-t} \end{bmatrix} \qquad (-1 < t < 1).$$

Then

$$\mathbf{\Phi}'(t) = \begin{bmatrix} -\sin t \\ \cos t \\ e^t \\ -e^{-t} \end{bmatrix}$$

and

$$|\mathbf{\Phi}'(t)| = \sqrt{(-\sin t)^2 + (\cos t)^2 + (e^t)^2 + (-e^{-t})^2}$$
$$= \sqrt{1 + 2 \cosh 2t} \neq 0;$$

thus $\mathbf{\Phi}'$ is continuous and nonvanishing in $[-1, 1]$ and therefore C is smooth. The unit tangent vector to C at $\mathbf{X} = \mathbf{\Phi}(t)$ is

$$\mathbf{T}(t) = \frac{1}{\sqrt{1 + 2 \cosh 2t}} \begin{bmatrix} -\sin t \\ \cos t \\ e^t \\ -e^{-t} \end{bmatrix}.$$

The requirement that a curve be smooth is quite stringent; many common and useful curves do not satisfy it, but are made up of smooth sections joined together.

Definition 4.4 A curve C defined by $\mathbf{X} = \mathbf{\Phi}(t)$ $(a \leq t \leq b)$ is *piecewise smooth* if there exist points $a = t_0 < t_1 < \cdots < t_n = b$ such that the curves C_i defined by $\mathbf{X} = \mathbf{\Phi}(t)$ $(t_{i-1} \leq t \leq t_i)$ are smooth $(i = 1, \ldots, n)$.

Example 4.5 Let C consist of the line segment from $(0, 0)$ to $(1, 1)$, followed by the line segment from $(1, 1)$ to $(1, 0)$ (Fig. 4.4). C is defined by

(1) $$\mathbf{X}(t) = \begin{bmatrix} t \\ t \end{bmatrix} \qquad (0 \le t \le 1),$$

and

(2) $$\mathbf{X}(t) = \begin{bmatrix} 1 \\ 2-t \end{bmatrix} \qquad (1 \le t \le 2).$$

Since C has no tangent at $(1, 1)$ it is not smooth. However C_1 and C_2, defined by (1) and (2), respectively, are smooth; hence C is piecewise smooth.

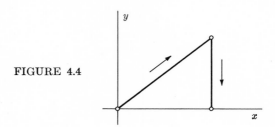

FIGURE 4.4

The following theorem, which we state without proof, gives a method for constructing parametric functions equivalent to a given one.

Theorem 4.1   If $\mathbf{\Psi}:[c, d] \to R^n$ defines a curve C and α is a strictly increasing continuous function which maps $[a, b]$ onto $[c, d]$, then

$$\mathbf{X} = \mathbf{\Phi}(t) = \mathbf{\Psi}(\alpha(t)) \qquad (a \le t \le b)$$

is an equivalent parametric representation of C. We say that $\mathbf{\Phi}$ is obtained from $\mathbf{\Psi}$ by the *change of parameter $s = \alpha(t)$*.

Example 4.6   Consider the change of parameter

$$s = \alpha(t) = \frac{t-a}{b-a} d + \frac{t-b}{a-b} c,$$

which maps $[a, b]$ onto $[c, d]$. If $\mathbf{\Psi}:[c, d] \to R^n$ defines a curve C then

$$\mathbf{X} = \mathbf{\Psi}(\alpha(t)) = \mathbf{\Psi}\left(\frac{t-a}{b-a} d + \frac{t-b}{a-b} c\right) \qquad (a \le t \le b)$$

is an equivalent representation of C. Setting $a = 0$ and $b = 1$ we find that any curve has a representation with parameter interval $[0, 1]$.

Example 4.7   In Example 4.1, $\mathbf{\Phi}$ can be obtained from $\mathbf{\Psi}$ by the change
of parameter $s = \alpha(t) = (t/2) - (\pi/4)$, since

$$\mathbf{\Psi}\left(\frac{t}{2} - \frac{\pi}{4}\right) = \begin{bmatrix} -\sin\left(t - \dfrac{\pi}{2}\right) \\ \\ \cos\left(t - \dfrac{\pi}{2}\right) \end{bmatrix}$$

$$= \begin{bmatrix} \cos t \\ \sin t \end{bmatrix} = \mathbf{\Phi}(t)$$

and α is strictly increasing and maps $[0, \pi]$ onto $\left[-\dfrac{\pi}{4}, \dfrac{\pi}{4}\right]$.

Definition 4.5   The change of parameter $s = \alpha(t)$ $(a \le t \le b)$ is *smooth*
if α is continuously differentiable and $\alpha'(t) > 0$ in $[a, b]$; it is *piecewise
smooth* if there exist points $a = t_0 < t_1 < \cdots < t_n = b$ such that $s = \alpha(t)$
is smooth in $[t_{j-1}, t_j]$ $(j = 1, \ldots, n)$.

Example 4.8   The linear change of parameter discussed in Example 4.7
is smooth; so are

$$s = \alpha(t) = \sin t \qquad \left(-\frac{\pi}{4} < t < \frac{\pi}{4}\right)$$

and

$$s = \alpha(t) = e^t \qquad (a \le t \le b; \quad a, b \text{ arbitrary}).$$

However,

$$s = \alpha(t) = t^3 \qquad (-1 \le t \le 1)$$

is not smooth, because $\alpha'(0) = 0$;

$$s = \alpha(t) = \sin t \qquad \left(0 \le t \le \frac{\pi}{2}\right)$$

is not smooth, because $\alpha'(\pi/2) = 0$. Finally

$$s = \alpha(t) = \begin{cases} t & (0 \le t \le \frac{1}{2}), \\ \\ 2t - \frac{1}{2} & (\frac{1}{2} \le t \le 1) \end{cases}$$

is not smooth, because $\alpha'(\frac{1}{2})$ does not exist (Fig. 4.5); however, it is piecewise smooth.

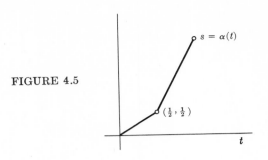

FIGURE 4.5

If a parametric function Φ defines a smooth curve C, then we shall also say that Φ is smooth; similarly if C is piecewise smooth, we shall say that Φ is piecewise smooth. We then have the following theorem.

Theorem 4.2 If $\Psi:[c,\,d] \to R^n$ is a smooth parametric function and $s = \alpha(t)$ is a smooth change of parameter which maps $[a,\,b]$ onto $[c,\,d]$ then

$$(3) \qquad\qquad \Phi(t) = \Psi(\alpha(t)) \qquad (a \le t \le b)$$

is also smooth, and equivalent to Ψ. Conversely, if Φ and Ψ are equivalent smooth representations of a curve C then there is a smooth change of parameter $s = \alpha(t)$ which satisfies (3). These statements remain true if "smooth" is replaced by "piecewise smooth."

The proof of the first statement leans heavily on the relationship

$$(4) \qquad\qquad \Phi'(t) = \Psi'(\alpha(t))\alpha'(t),$$

which follows from the chain rule (Exercise T-1), and holds wherever the indicated derivatives exist.

The next definition generalizes the definition of arc length given in Section 3.3.

Definition 4.6 Let C be defined by a smooth parametric function $\mathbf{X} = \Phi(t)$ $(a \le t \le b)$. Then the *arc length* of C is the integral

$$L(C) = \int_a^b |\, \Phi'(t)\,|\; dt$$

$$= \int_a^b \sqrt{(\Phi_1'(t))^2 + \cdots + (\Phi_n'(t))^2}\; dt.$$

Example 4.9 If C is a curve in R^2 defined by

$$\mathbf{X}(x) = \begin{bmatrix} x \\ f(x) \end{bmatrix} \qquad (a \le x \le b),$$

then

$$L(C) = \int_a^b \sqrt{1 + (f'(x))^2}\, dx,$$

which should be familiar to the student.

Example 4.10 If C is

$$\mathbf{X}(t) = \begin{bmatrix} \cos t \\ \sin t \end{bmatrix} \qquad (0 \le t \le 2\pi),$$

the unit circle traversed counterclockwise, then

$$L(C) = \int_0^{2\pi} \sqrt{(-\sin t)^2 + (\cos t)^2}\, dt$$

$$= \int_0^{2\pi} dt = 2\pi,$$

again a familiar result.

Example 4.11 Let C be the line segment in R^n from $\mathbf{X}_1 = (a_1, \ldots, a_n)$ to $\mathbf{X}_2 = (b_1, \ldots, b_n)$. It can be represented by

$$\mathbf{X} = \mathbf{\Phi}(t) = \begin{bmatrix} a_1 + t(b_1 - a_1) \\ \vdots \\ a_n + t(b_n - a_n) \end{bmatrix} \qquad (0 \le t \le 1);$$

hence

$$\mathbf{\Phi}'(t) = \begin{bmatrix} b_1 - a_1 \\ \vdots \\ b_n - a_n \end{bmatrix}$$

and

$$L(C) = \int_0^1 \sqrt{(b_1 - a_1)^2 + \cdots + (b_n - a_n)^2}\, dt$$

$$= \sqrt{(b_1 - a_1)^2 + \cdots + (b_n - a_n)^2} \int_0^1 dt$$

$$= \sqrt{(b_1 - a_1)^2 + \cdots + (b_n - a_n)^2} = |X_2 - X_1|,$$

which agrees with our earlier definition of length of a line segment (Section 1.1).

The following theorem shows that the arc length of a curve C does not depend upon the particular parametric representation chosen to represent C.

Theorem 4.3 If $\mathbf{X} = \mathbf{\Phi}(t)$ $(a \leq t \leq b)$ and $\mathbf{X} = \mathbf{\Psi}(s)$ $(c \leq s \leq d)$ are equivalent piecewise smooth representations of the same curve C, then

(5)
$$\int_a^b |\, \mathbf{\Phi}'(t)\,|\, dt = \int_c^d |\, \mathbf{\Psi}'(s)\,|\, ds;$$

that is, $L(C)$ may be computed by means of any piecewise smooth representation of C.

Proof. Let $s = \alpha(t)$ be the change of variable relating $\mathbf{\Phi}$ and $\mathbf{\Psi}$ according to (3). The rule for change of variable of integration implies that

$$\int_c^d |\, \mathbf{\Psi}'(s)\,|\, ds = \int_a^b |\, \mathbf{\Psi}'(\alpha(t))\,|\, \alpha'(t)\, dt$$

$$= \int_a^b |\, \mathbf{\Psi}'(\alpha(t))\alpha'(t)\,|\, dt$$

since $|\, \alpha'(t)\,| = \alpha'(t) > 0$. Now (4) yields (5).

EXERCISES 3.4   # 2-10 Even , 18

In Exercises 1, 2, and 3 show that $\mathbf{X} = \mathbf{\Phi}(t)$ and $\mathbf{X} = \mathbf{\Psi}(s)$ define the same curve.

1. $\mathbf{\Phi}(t) = \begin{bmatrix} t \\ t^2 \\ t^3 \end{bmatrix}$ $(1 \leq t \leq 2)$; $\mathbf{\Psi}(s) = \begin{bmatrix} s^2 \\ s^4 \\ s^6 \end{bmatrix}$ $(1 \leq s \leq \sqrt{2})$.

2. $\mathbf{\Phi}(t) = \begin{bmatrix} e^t \\ e^{-t} \\ 2\cosh t \end{bmatrix}$ $(0 \leq t \leq 1)$; $\mathbf{\Psi}(s) = \begin{bmatrix} s \\ \dfrac{1}{s} \\ s + \dfrac{1}{s} \end{bmatrix}$ $(1 \leq s \leq e)$.

Describe the curves in Exercises 3 through 7 as closed, simple, both, or neither.

3.

4. $\quad \mathbf{X} = \boldsymbol{\Phi}(t) = \begin{bmatrix} \cos t \\ \sin t \\ -\sin t \\ -\cos t \end{bmatrix} \qquad (0 \le t \le \pi).$

5. $\quad \mathbf{X} = \boldsymbol{\Phi}(t) = \begin{bmatrix} \cos t \\ \sin t \\ -\sin t \\ -\cos t \end{bmatrix} \qquad (0 \le t \le 2\pi).$

6.

7. $\quad \mathbf{X} = \boldsymbol{\Phi}(t) = \begin{bmatrix} \cos t \\ \sin t \\ -\sin t \\ -\cos t \end{bmatrix} \qquad (0 \le t \le 3\pi).$

8. Show that the curve defined by

$$\mathbf{X} = \boldsymbol{\Phi}(t) = \begin{bmatrix} e^t \cos t \\ e^t \sin t \\ \sqrt{2}\, e^t \end{bmatrix} \qquad (0 \le t \le \pi)$$

is continuously differentiable.

9.   Show that the curve defined by

$$\mathbf{X} = \mathbf{\Phi}(t) = \begin{bmatrix} \dfrac{t^2}{2} \\[2mm] \dfrac{\sqrt{2}}{3} t^3 \\[2mm] \dfrac{t^4}{4} \end{bmatrix} \qquad (0 \le t \le 1)$$

is continuously differentiable.

10.   Show that the curve C defined by

$$\mathbf{X} = \mathbf{\Phi}(t) = \begin{bmatrix} e^t \cos t \\[1mm] e^{-t} \sin t \\[1mm] e^t \\[1mm] e^{-t} \end{bmatrix} \qquad (-\pi \le t \le \pi)$$

is smooth and find the tangent vector to C at $\mathbf{X}_0 = \mathbf{\Phi}(0)$.

11.   Show that the curve C defined by

$$\mathbf{X} = \mathbf{\Phi}(t) = \begin{bmatrix} t^2 + 1 \\[1mm] t^3 + t \\[1mm] t^4 \end{bmatrix} \qquad (0 \le t \le 2)$$

is smooth and find the tangent vector to C at $\mathbf{X}_0 = \mathbf{\Phi}(1)$.

12.   Show that the curve defined by

$$\mathbf{X} = \mathbf{\Phi}(t) = \begin{bmatrix} |t| \\[1mm] |t - 1| \\[1mm] t^2 \end{bmatrix} \qquad (-2 \le t \le 2)$$

is piecewise smooth, but not smooth.

13. Show that the curve defined by

$$\mathbf{X} = \boldsymbol{\Phi}(t) = \begin{bmatrix} e^t \\ \sin t \\ |\cos t| \end{bmatrix} \qquad (-\pi \le t \le 3\pi)$$

is piecewise smooth, but not smooth.

In Exercises 14, 15, and 16 find two parametric functions equivalent to $\boldsymbol{\Phi}$.

14. $\boldsymbol{\Phi}(t) = \begin{bmatrix} t \\ t^2 \\ t^3 \\ t^4 \end{bmatrix} \qquad (0 \le t \le 1).$

15. $\boldsymbol{\Phi}(t) = \begin{bmatrix} -\sin t \\ \cos t \\ \sin t \end{bmatrix} \qquad \left(0 \le t \le \dfrac{\pi}{2}\right).$

16. $\boldsymbol{\Phi}(t) = \begin{bmatrix} e^t \sin t \\ e^t \cos t \\ e^{-t} \cos t \\ e^{-t} \sin t \end{bmatrix} \qquad (0 \le t \le \pi).$

17. Find a smooth change of parameter mapping $[-2, -1]$ onto $[1, 2]$. (See Example 4.6).

18. Find a parametric representation for the circle $x^2 + y^2 = 1$ traversed once in the counterclockwise direction, starting from $\left(\dfrac{1}{\sqrt{2}}, \dfrac{1}{\sqrt{2}}\right)$.

19. Find a parametric equation for the ellipse $(x^2/4) + (y^2/9) = 1$ traversed in the clockwise direction, starting from $(-2, 0)$.

20. Let C consist of the portion of the parabola $y = x^2$ from $(0, 0)$ to $(1, 1)$, followed by the line segment from $(1, 1)$ to $(1, 0)$, followed

by the line segment from $(1, 0)$ to $(0, 0)$. Show that C is piecewise smooth and compute its arc length.

21. Compute the arc length of the curve defined in Exercise 8.

22. Compute the arc length of the curve defined in Exercise 9.

23. Let C be the curve defined in Example 4.1. Verify Theorem 4.3 by computing the arc length of C using the equivalent parametric functions $\mathbf{\Phi}$ and $\mathbf{\Psi}$ defined in Examples 4.1 and 4.7.

24. (a) Show that the function

$$\mathbf{\Psi}(s) = \begin{bmatrix} \dfrac{s}{2} \\[2mm] \dfrac{\sqrt{2}}{3} s^{3/2} \\[2mm] \dfrac{s^2}{4} \end{bmatrix} \qquad (0 \leq s \leq 1)$$

is equivalent to the function $\mathbf{\Phi}$ defined in Exercise 9.

(b) Verify Theorem 4.3 by computing the arc length of C using the representation $\mathbf{X} = \mathbf{\Psi}(s)$, and comparing the result with that of Exercise 9.

THEORETICAL EXERCISES

T-1. *Prove:* If $\mathbf{\Psi}$ and α (as defined in Theorem 4.2) are smooth, then $\mathbf{\Phi} = \mathbf{\Psi} \circ \alpha$ is smooth. (*Hint:* Use the chain rule.)

T-2. *Prove:* If $\mathbf{\Phi}:[a, b] \to R^n$ and $\mathbf{\Psi}:[c, d] \to R^n$, represent same smooth curve, then they are related by a smooth change of parameter. (*Hint:* Since $\mathbf{\Phi}$ and $\mathbf{\Psi}$ are equivalent,

$$\mathbf{\Phi}(t) = \mathbf{\Psi}(\alpha(t)) \qquad (a \leq t \leq b)$$

where α is strictly increasing and continuous in $[a, b]$. By considering difference quotients, show that

$$\alpha'(t_0) = \frac{|\mathbf{\Phi}'(t_0)|}{|\mathbf{\Psi}'(\alpha(t_0))|}$$

for each t_0 in $[a, b]$.

3.5 Coordinate Systems in R^3

We have considered R^3 simply as a set of ordered triples on which certain operations such as addition and scalar multiplication are defined. However, we also have a physical conception of a three-dimensional universe which exists on its own, independent of our mathematical constructions. We can combine these two ideas by regarding the triples (x, y, z) as being assigned as addresses to points which already exist on their own. Thus, if P_0 is an arbitrary point in space we choose three mutually perpendicular directions through P_0 and associate the triple (x, y, z) with the point P that we reach by walking x units along the first direction, then y units along the second, and then z units along the third (Fig. 5.1).

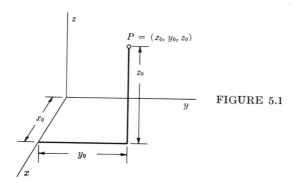

FIGURE 5.1

Thus we do not have to regard the point as being the same as the triple (x, y, z), although we have previously found it convenient to do so, and will again. Rather, we can think of the triple simply as the address of the point, in accordance with a particular kind of addressing method, called a *cartesian*, or *rectangular coordinate* system.

There are other coordinate systems for three-dimensional space. We shall consider two of the most important ones.

Cylindrical Coordinates

Let a cartesian coordinate system have its origin at a point P_0, and suppose P is a point with coordinates (x, y, z) with respect to this system. If ρ and θ satisfy

(1)
$$x = \rho \cos \theta$$
$$y = \rho \sin \theta,$$

then ρ, θ, and z are said to be *cylindrical coordinates* of P. There are infinitely many values of θ which satisfy (1), but we can specify θ uniquely, if $(x, y) \neq (0, 0)$, by stipulating that $\rho > 0$ and $0 \leq \theta < 2\pi$. Then (1) can be solved to obtain θ and ρ:

$$\rho = \sqrt{x^2 + y^2},$$

$$\theta = \cos^{-1} \frac{x}{\rho} = \sin^{-1} \frac{y}{\rho}$$

$$= \tan^{-1} \frac{y}{x} \qquad (0 \leq \theta < 2\pi).$$

If $x^2 + y^2 = 0$, then θ is arbitrary.

The student should recognize that ρ and θ are simply the polar coordinates of the projection P' of P on the xy-plane (Fig. 5.2).

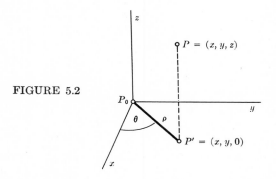

FIGURE 5.2

Example 5.1   If P has cylindrical coordinates $(\rho, \theta, z) = (3, \pi/6, -2)$ then

$$x = 3 \cos \frac{\pi}{6} = 3 \frac{\sqrt{3}}{2},$$

$$y = 3 \sin \frac{\pi}{6} = \frac{3}{2},$$

$$z = -2;$$

hence the position vector of P with respect to P_0 is

$$\overrightarrow{P_0 P} = \frac{3\sqrt{3}}{2} \mathbf{i} + \frac{3}{2} \mathbf{j} - 2\mathbf{k}.$$

Example 5.2 If P has rectangular coordinates $(2, 2\sqrt{3}, -3)$ then

$$\rho = \sqrt{2^2 + (2\sqrt{3})^2 + (-3)^2} = 5,$$

(2) $$\theta = \tan^{-1}\sqrt{3},$$

$$z = -3;$$

hence its cylindrical coordinates are

$$(\rho, \theta, z) = \left(5, \frac{\pi}{3}, -3\right).$$

Notice that we have used more than (2) to determine θ: x and y are both positive, which implies that P is in the first quadrant. Without this information we could not decide whether $\theta = \pi/3$ or $\theta = 4\pi/3$.

Spherical Coordinates

Again let a rectangular coordinate system have its origin at P_0, and let P be an arbitrary point. Then the *spherical coordinates* of P are r, θ and ϕ, where r is the distance from P_0 to P, θ is the angle from the positive x-axis to the vector from P_0 to P', the projection of P on the xy-plane, and ϕ is the angle between the positive z-axis and the vector from P_0 to P. From Fig. 6.3,

$$x = r \sin \phi \cos \theta,$$

(3) $$y = r \sin \phi \sin \theta,$$

$$z = r \cos \phi.$$

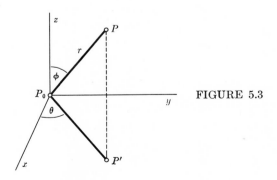

FIGURE 5.3

If $x^2 + y^2 \neq 0$, there is a unique triple (r, θ, ϕ) which satisfies (3) and the inequalities $r > 0$, $0 \leq \phi \leq \pi$, and $0 \leq \theta < 2\pi$:

$$r = \sqrt{x^2 + y^2 + z^2},$$

$$\theta = \cos^{-1} \frac{x}{\sqrt{x^2 + y^2}} = \sin^{-1} \frac{y}{\sqrt{x^2 + y^2}}$$

$$= \tan^{-1} \frac{y}{x} \qquad (0 \leq \theta < 2\pi),$$

$$\phi = \cos^{-1} \frac{z}{\sqrt{x^2 + y^2 + z^2}} \qquad (0 \leq \phi \leq \pi).$$

If $x^2 + y^2 = 0$, then θ can be chosen arbitrarily; if $x^2 + y^2 + z^2 = 0$, then both θ and ϕ can be chosen arbitrarily.

Example 5.3   If P has spherical coordinates $(r, \theta, \phi) = (3, \pi/3, \pi/4)$ then

$$x = 3 \sin \frac{\pi}{4} \cos \frac{\pi}{3}$$

$$= 3 \left(\frac{\sqrt{2}}{2}\right)\left(\frac{1}{2}\right) = \frac{3\sqrt{2}}{4},$$

$$y = 3 \sin \frac{\pi}{4} \sin \frac{\pi}{3}$$

$$= 3 \left(\frac{\sqrt{2}}{2}\right)\left(\frac{\sqrt{3}}{2}\right) = \frac{3\sqrt{6}}{4},$$

$$z = 3 \cos \frac{\pi}{4} = \frac{3\sqrt{2}}{2};$$

hence the position vector of P with respect to P_0 is

$$\overrightarrow{P_0P} = \frac{3}{\sqrt{2}} \left(\frac{1}{2}\mathbf{i} + \frac{\sqrt{3}}{2}\mathbf{j} + \mathbf{k}\right).$$

Example 5.4 If P has rectangular coordinates $(-9/2, -3\sqrt{3}/2, 3)$ then

$$r = \sqrt{\left(-\frac{9}{2}\right)^2 + \left(-\frac{3\sqrt{3}}{2}\right)^2 + 3^2} = 6,$$

$$\theta = \tan^{-1}\left(\frac{-\frac{3\sqrt{3}}{2}}{-\frac{9}{2}}\right) = \tan^{-1}\frac{1}{\sqrt{3}} = \frac{7\pi}{6},$$

$$\phi = \cos^{-1}\frac{3}{6} = \frac{\pi}{3}.$$

In determining θ we have used the fact that x and y are both negative, which implies that θ is in the third quadrant.

Translation and Rotation of Rectangular Coordinate Systems

Suppose two rectangular coordinate systems have origins at P_0 and P_0' and let $(\mathbf{i}, \mathbf{j}, \mathbf{k})$ and $(\mathbf{i}', \mathbf{j}', \mathbf{k}')$ be unit vectors along their coordinate axes. Let P be a point with coordinates (x, y, z) and (x', y', z') with respect to the two systems; thus

(4) $$\overrightarrow{P_0P} = x\mathbf{i} + y\mathbf{j} + z\mathbf{k}$$

and

(5) $$\overrightarrow{P_0'P} = x'\mathbf{i}' + y'\mathbf{j}' + z'\mathbf{k}'.$$

We wish to find the relationship between (x, y, z) and (x', y', z').
From Fig. 5.4,

(6) $$\overrightarrow{P_0'P} = -\overrightarrow{P_0P_0'} + \overrightarrow{P_0P}.$$

Suppose P_0' has coordinates (x_0, y_0, z_0) with respect to the (x, y, z) system; thus

$$\overrightarrow{P_0P_0'} = x_0\mathbf{i} + y_0\mathbf{j} + z_0\mathbf{k}.$$

Then from (4), (5) and (6)

(7) $$x'\mathbf{i}' + y'\mathbf{j}' + z'\mathbf{k}' = (x - x_0)\mathbf{i} + (y - y_0)\mathbf{j} + (z - z_0)\mathbf{k}.$$

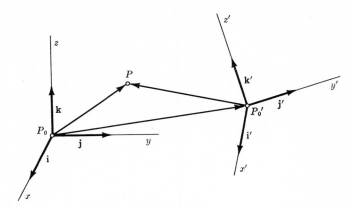

FIGURE 5.4

Since $\mathbf{i'}$, $\mathbf{j'}$, and $\mathbf{k'}$ form a basis for R^3, we can write

$$\mathbf{i} = l_1\mathbf{i'} + m_1\mathbf{j'} + n_1\mathbf{k'},$$

(8)
$$\mathbf{j} = l_2\mathbf{i'} + m_2\mathbf{j'} + n_2\mathbf{k'},$$

$$\mathbf{k} = l_3\mathbf{i'} + m_3\mathbf{j'} + n_3\mathbf{k'}.$$

Substituting this in (7) and collecting coefficients of $\mathbf{i'}$, $\mathbf{j'}$, and $\mathbf{k'}$ yields

$$x'\mathbf{i'} + y'\mathbf{j'} + z'\mathbf{k'} = [l_1(x - x_0) + l_2(y - y_0) + l_3(z - z_0)]\mathbf{i'}$$
$$+ [m_1(x - x_0) + m_2(y - y_0) + m_3(z - z_0)]\mathbf{j'}$$
$$+ [n_1(x - x_0) + n_2(y - y_0) + n_3(z - z_0)]\mathbf{k'}.$$

Thus, from the linear independence of $\mathbf{i'}$, $\mathbf{j}$, and $\mathbf{k'}$,

$$x' = l_1(x - x_0) + l_2(y - y_0) + l_3(z - z_0),$$

$$y' = m_1(x - x_0) + m_2(y - y_0) + m_3(z - z_0),$$

$$z' = n_1(x - x_0) + n_2(y - y_0) + n_3(z - z_0).$$

This can be rewritten in vector–matrix form as

(9)
$$\begin{bmatrix} x' \\ y' \\ z' \end{bmatrix} = \mathbf{M} \begin{bmatrix} x - x_0 \\ y - y_0 \\ z - z_0 \end{bmatrix},$$

where

$$\mathbf{M} = \begin{bmatrix} l_1 & l_2 & l_3 \\ m_1 & m_2 & m_3 \\ n_1 & n_2 & n_3 \end{bmatrix}.$$

Since $\mathbf{i'}$, $\mathbf{j'}$, and $\mathbf{k'}$ are orthonormal it follows from (8) that

$$l_r l_s + m_r m_s + n_r n_s = \begin{cases} 0 & \text{if } r \ne s; \\ 1 & \text{if } r = s. \end{cases}$$

Using this and the definition of matrix multiplication it is easy to verify that

(10) $$\mathbf{M^T M} = \mathbf{I},$$

where

$$\mathbf{I} = \begin{bmatrix} 1 & 0 & 0 \\ 0 & 1 & 0 \\ 0 & 0 & 1 \end{bmatrix}$$

and $\mathbf{M^T}$ is the transpose of $\mathbf{M}$:

$$\mathbf{M^T} = \begin{bmatrix} l_1 & m_1 & n_1 \\ l_2 & m_2 & n_2 \\ l_3 & m_3 & n_3 \end{bmatrix}.$$

Thus $\mathbf{M}$ is an *orthogonal* matrix (Section 2.6).

We can visualize the change of coordinates (9) as occurring in two steps. Let

(11) $$\begin{bmatrix} \bar{x} \\ \bar{y} \\ \bar{z} \end{bmatrix} = \begin{bmatrix} x - x_0 \\ y - y_0 \\ z - z_0 \end{bmatrix};$$

then

(12)
$$\begin{bmatrix} x' \\ y' \\ z' \end{bmatrix} = \mathbf{M} \begin{bmatrix} \bar{x} \\ \bar{y} \\ \bar{z} \end{bmatrix}.$$

Clearly $(\bar{x}, \bar{y}, \bar{z})$ are the coordinates of P with respect to a coordinate system with origin at P_0' and axes parallel to those of the unprimed system (Fig. 6.4). Thus the axes of the $(\bar{x}, \bar{y}, \bar{z})$ system can be obtained by translating those of the (x, y, z) system parallel to themselves from P_0 to P_0'. For this reason, (1) is called a *translation* of coordinate systems. The axes of the $(\bar{x}, \bar{y}, \bar{z})$ system can be brought into coincidence with those of the (x', y', z') system by a rotation; thus (12) is called a *rotation* of coordinates. (More precisely (12) is called a rotation if $\mathbf{M}$ is an orthogonal matrix.)

We have shown that the change of coordinates from one cartesian coordinate system to another can be considered as a translation followed by a rotation.

Example 5.5   Consider the change of coordinates

(13)
$$x' = \tfrac{2}{3}x + \tfrac{2}{3}y + \tfrac{1}{3}z - \tfrac{5}{3},$$
$$y' = -\tfrac{2}{3}x + \tfrac{1}{3}y + \tfrac{2}{3}z + \tfrac{5}{3},$$
$$z' = \tfrac{1}{3}x - \tfrac{2}{3}y + \tfrac{2}{3}z + \tfrac{2}{3}.$$

We wish to verify that this change of coordinates can be viewed as a translation followed by a rotation. First, we determine the coordinates (x_0, y_0, z_0) of P_0', the origin of the (x', y', z') system with respect to P_0, the origin of the (x, y, z) system. The coordinates of P_0' with respect to the (x', y', z') system are $(0, 0, 0)$. Thus, from (13),

$$0 = \tfrac{2}{3}x_0 + \tfrac{2}{3}y_0 + \tfrac{1}{3}z_0 - \tfrac{5}{3}$$

$$0 = -\tfrac{2}{3}x_0 + \tfrac{1}{3}y_0 + \tfrac{2}{3}z_0 + \tfrac{5}{3}$$

$$0 = \tfrac{1}{3}x_0 - \tfrac{2}{3}y_0 + \tfrac{2}{3}x_0 + \tfrac{2}{3}z_0 + \tfrac{2}{3},$$

which must be solved for x_0, y_0, z_0. This system of equations can be written as

$$\mathbf{M} \begin{bmatrix} x_0 \\ y_0 \\ z_0 \end{bmatrix} = \begin{bmatrix} \tfrac{5}{3} \\ -\tfrac{5}{3} \\ -\tfrac{2}{3} \end{bmatrix}$$

where

$$\mathbf{M} = \begin{bmatrix} \frac{2}{3} & \frac{2}{3} & \frac{1}{3} \\ -\frac{2}{3} & \frac{1}{3} & \frac{2}{3} \\ \frac{1}{3} & -\frac{2}{3} & \frac{2}{3} \end{bmatrix}.$$

Hence

$$\begin{bmatrix} x_0 \\ y_0 \\ z_0 \end{bmatrix} = \mathbf{M}^{-1} \begin{bmatrix} \frac{5}{3} \\ -\frac{5}{3} \\ \frac{2}{3} \end{bmatrix}$$

$$= \mathbf{M}^{\mathrm{T}} \begin{bmatrix} \frac{5}{3} \\ -\frac{5}{3} \\ -\frac{2}{3} \end{bmatrix} = \begin{bmatrix} 2 \\ 1 \\ -1 \end{bmatrix}.$$

Thus, the change of coordinates consists of the translation

$$\begin{bmatrix} \bar{x} \\ \bar{y} \\ \bar{z} \end{bmatrix} = \begin{bmatrix} x - 2 \\ y - 1 \\ z + 1 \end{bmatrix}$$

followed by the rotation

$$\begin{bmatrix} x' \\ y' \\ z' \end{bmatrix} = \mathbf{M} \begin{bmatrix} \bar{x} \\ \bar{y} \\ \bar{z} \end{bmatrix}.$$

EXERCISES 3.5

1. Find cylindrical coordinates of the points with the following rectangular coordinates.

(a) $(5\sqrt{3},\ -5,\ 2)$;   (b) $(-2,\ -2,\ 5)$;

(c) $(0,\ 1,\ 0)$;   (d) $(3, 3\sqrt{3}, -3)$.

2. Find rectangular coordinates of the points with the following cylindrical coordinates.

(a) $\left(2, \dfrac{3\pi}{4}, 3\right)$;   (b) $\left(4, \dfrac{5\pi}{3}, 4\right)$;

(c) $(2, 0, -3)$;   (d) $\left(5, \dfrac{\pi}{6}, 2\right)$.

3. Find spherical coordinates for the points with the following rectangular coordinates.

(a) $(\sqrt{2}, \sqrt{2}, 2)$;   (b) $(-3, 3\sqrt{3}, 0)$;

(c) $\left(\dfrac{1}{2}, \dfrac{-\sqrt{3}}{2}, -\sqrt{3}\right)$;   (d) $\left(\dfrac{-\sqrt{3}}{2}, \dfrac{1}{2}, -\sqrt{3}\right)$.

4. Find rectangular coordinates for the points with the following spherical coordinates.

(a) $\left(2, \dfrac{5\pi}{6}, \dfrac{\pi}{3}\right)$;   (b) $\left(4, \dfrac{3\pi}{4}, \dfrac{\pi}{2}\right)$;

(c) $\left(2, \dfrac{5\pi}{3}, \dfrac{\pi}{4}\right)$;   (d) $\left(3, \dfrac{\pi}{6}, \dfrac{3\pi}{4}\right)$.

5. Find spherical coordinates for the points with the following cylindrical coordinates.

(a) $\left(3, \dfrac{\pi}{3}, 2\right)$;   (b) $\left(4, \dfrac{2\pi}{3}, -3\right)$;

(c) $\left(2, \dfrac{5\pi}{6}, 4\right)$;   (d) $\left(2, \dfrac{3\pi}{4}, 3\right)$.

6. Find cylindrical coordinates for the points with the following spherical coordinates.

(a) $\left(3, \dfrac{2\pi}{3}, \dfrac{\pi}{4}\right)$;   (b) $\left(5, \dfrac{3\pi}{4}, \dfrac{\pi}{2}\right)$;

(c) $\left(4, \dfrac{5\pi}{4}, 0\right)$;   (d) $\left(2, \dfrac{7\pi}{6}, \dfrac{\pi}{3}\right)$.

7. In Eq. (9) let $(x_0, y_0, z_0) = (1, -2, 3)$ and

$$\mathbf{M} = \begin{bmatrix} 1 & 0 & 0 \\ 0 & \dfrac{1}{\sqrt{2}} & \dfrac{-1}{\sqrt{2}} \\ 0 & \dfrac{-1}{\sqrt{2}} & \dfrac{-1}{\sqrt{2}} \end{bmatrix}.$$

Find (x', y', z') for the points with the following (x, y, z) coordinates.

(a) $(1, 0, 2)$;   (b) $(-2, -3, 4)$;   (c) $(0, 0, 1)$;
(d) $(2, 3, 5)$.

8. Verify that

$$\mathbf{M} = \begin{bmatrix} \frac{2}{3} & -\frac{2}{3} & \frac{1}{3} \\ \frac{2}{3} & \frac{1}{3} & -\frac{2}{3} \\ \frac{1}{3} & \frac{2}{3} & \frac{2}{3} \end{bmatrix}$$

is an orthogonal matrix.

THEORETICAL EXERCISES

T-1. Verify Eq. (10).

T-2. Let P_0, P_1, and P_2 have coordinates (x_i, y_i, z_i), $i = 1, 2, 3$, with respect to a rectangular coordinate system, and let them have coordinates (x_i', y_i', z_i') with respect to a second system, related to the first by a translation and a rotation. Then there are two ways to compute the length of the vector $\overrightarrow{P_0P_1}$:

$$|\overrightarrow{P_0P_1}| = \sqrt{(x_1 - x_0)^2 + (y_1 - y_0)^2 + (z_1 - z_0)^2}$$

or

$$|\overrightarrow{P_0P_1}| = \sqrt{(x_1' - x_0')^2 + (y_1' - y_0')^2 + (z_1' - z_0')^2}.$$

Show that the two results are the same. Show also that the corresponding ways for computing the cosine of the angle between P_0P_1 and P_0P_2 yield the same result.

(As a consequence of this problem, we say that distance and angle are *invariant under translation and rotation.*)

3.6 Surfaces in R^3

A *surface* S in R^3 is the set of points that satisfy

$$f(x, y, z) = 0,$$

where f is a function such as

$$f(x, y, z) = x + y + z$$

or

$$f(x, y, z) = x^2 + y^2 + z^2 - 1.$$

(In Section 4.1 we shall say more precisely what we mean by a function.) Perhaps the simplest surfaces in R^3 are the planes, which we have already studied; thus

$$Ax + By + Cz + D = 0$$

defines a plane if $A^2 + B^2 + C^2 \neq 0$.

In this section we present examples of surfaces which will be used for illustrative purposes throughout the rest of the book.

When sketching a surface S it is often useful to find the points, if any, where S intersects the coordinate axes and the coordinate planes. The x-coordinates of the points of intersection of S with the x-axis are called the *x-intercepts of S*; *y-intercepts* and *z-intercepts* are defined similarly. The curve in which S intersects a plane Γ is called *the trace of S in Γ*.

Cylinders

Let C be a curve in a plane Γ and L be a line, not in Γ, through a point of C. The set of points on lines through C parallel to L comprise a surface S, which is called a *cylinder* (Fig. 6.1). The curve C is the *directrix* of S and L is a *generator*; we can think of S as being obtained by moving L parallel to itself along C.

Example 6.1   Let Γ be the xy-plane, C the circle

(1)
$$x^2 + y^2 = 1,$$

and L the line through $(0, 1, 0)$ parallel to the z-axis. As L is moved around C parallel to itself, it sweeps out the cylinder shown in Fig. 6.2. This is called a *right circular cylinder*, because its directrix is a circle and its generator is perpendicular to the plane of the directrix. In this case, (1) is also

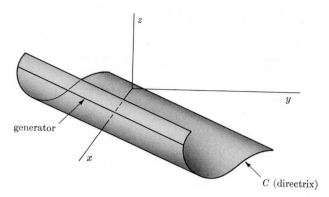

FIGURE 6.1

the equation of S, since a point (x, y, z) is on S if and only if x and y satisfy (1).

Example 6.2 The cylinders defined by

$$\frac{x^2}{a^2} + \frac{y^2}{b^2} = 1$$
$$(a, b > 0)$$
$$\frac{x^2}{a^2} - \frac{y^2}{b^2} = 1$$

and

$$y = ax^2 \qquad (a > 0)$$

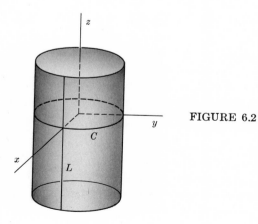

FIGURE 6.2

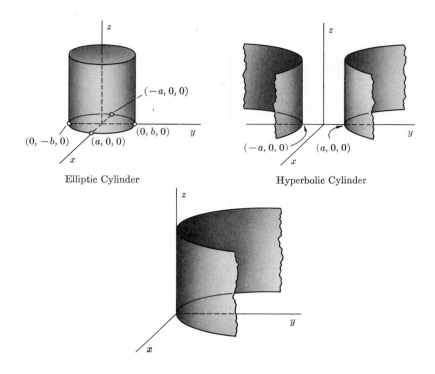

FIGURE 6.3

are named after their traces in the xy-plane; thus they are *elliptic, hyperbolic,* and *parabolic* cylinders, respectively. Portions of these surfaces above the xy-plane and below a plane parallel to it are shown in Fig. 6.3.

Quadric Surfaces

We devote the rest of this section to *quadric surfaces*; that is, surfaces defined by equations of the form

(2) $Ax^2 + By^2 + Cz^2 + 2Dxy + 2Exz + 2Fyz + Gx + Hy + Iz + J = 0,$

where at least one of the coefficients A, B, C, D, E, and F is nonzero. The cylinders of Example 6.2 are surfaces of this kind. There are basically six other kinds of quadric surfaces, examples of which are given below. There are also degenerate cases, as in the following example.

Example 6.3 (Degenerate Quadric Surfaces) The surface defined by (2) may be a plane; thus

$$x + y + z = 0,$$

which defines a plane, can be written as

$$(x + y + z)^2 = x^2 + y^2 + z^2 + 2xy + 2xz + 2yz = 0,$$

which is of the form (2). It is clearly preferable to study planes on their own, rather than as special cases of quadric surfaces; hence they are considered to be trivial, or *degenerate*, quadric surfaces. Other examples of degenerate quadric surfaces are as follows.

(a) Pairs of planes: The locus of points satisfying

$$x^2 - y^2 = 0$$

comprises the two planes defined by

$$x - y = 0$$

and

$$x + y = 0.$$

(b) Lines: The points satisfying

$$(x - x_0)^2 + (y - y_0)^2 = 0$$

comprise the line through $(x_0, y_0, 0)$ parallel to the z-axis.

(c) Single points: Only the point (x_0, y_0, z_0) satisfies

$$(x - x_0)^2 + (y - y_0)^2 + (z - z_0)^2 = 0.$$

(d) No locus: There are no points in R^3 that satisfy

$$x^2 + y^2 + z^2 + 1 = 0.$$

We now give simple examples of the six kinds of nondegenerate, non-cylindrical quadric surfaces. In all cases we assume that a, b, and c are positive.

Example 6.4 (The Ellipsoid) The surface S defined by

$$\frac{x^2}{a^2} + \frac{y^2}{b^2} + \frac{z^2}{c^2} = 1$$

is an *ellipsoid*; if $a = b = c$, then S is a sphere of radius a with center at the origin. The intercepts of S are $(\pm a, 0, 0)$, $(0, \pm b, 0)$, and $(0, 0, \pm c)$. The trace of S in any plane parallel to (and sufficiently close to) a coordinate plane is an ellipse. In particular the traces in the xy-, yz- and xz-planes are defined by

$$\frac{x^2}{a^2} + \frac{y^2}{b^2} = 1,$$

$$\frac{y^2}{b^2} + \frac{z^2}{c^2} = 1,$$

and

$$\frac{x^2}{a^2} + \frac{z^2}{c^2} = 1.$$

The ellipsoid and its traces in the coordinate planes are shown in Fig. 6.4.

FIGURE 6.4

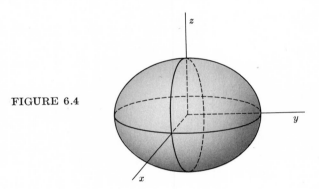

Example 6.5 (The Hyperboloid of One Sheet)   The x- and y-intercepts of the surface defined by

$$\frac{x^2}{a^2} + \frac{y^2}{b^2} - \frac{z^2}{c^2} = 1$$

are $(\pm a, 0, 0)$ and $(0, \pm b, 0)$; it has no z-intercepts. Its trace in the plane $z = k$ is the ellipse

$$\frac{x^2}{a^2} + \frac{y^2}{b^2} = 1 + \frac{k^2}{c^2},$$

while its trace in the plane $y = k$ is the hyperbola

$$\frac{x^2}{a^2} - \frac{z^2}{c^2} = 1 - \frac{k^2}{b^2},$$

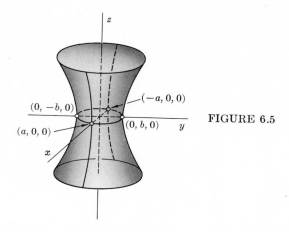

FIGURE 6.5

if $|k| < b$. Similarly, its trace in the plane $x = k$ is the hyperbola

$$\frac{y^2}{b^2} - \frac{z^2}{c^2} = 1 - \frac{k^2}{a^2}$$

if $|k| < a$. This surface, which is known as a *hyperboloid of one sheet*, is shown in Fig. 6.5.

Example 6.6 (The Hyperboloid of Two Sheets) The surface defined by

$$\frac{x^2}{a^2} - \frac{y^2}{b^2} - \frac{z^2}{c^2} = 1$$

has no y- or z-intercepts; its x-intercepts are $(\pm a, 0, 0)$. Its traces in planes parallel to the xy- and xz-planes are hyperbolas. It has no trace in the plane $x = k$ unless $|k| > a$, in which case the trace is an ellipse. This surface has two halves which are mirror images of each other (Fig. 6.6). It is called a *hyperboloid of two sheets*.

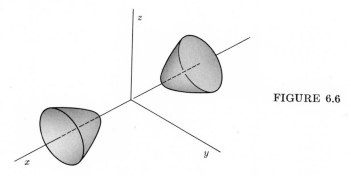

FIGURE 6.6

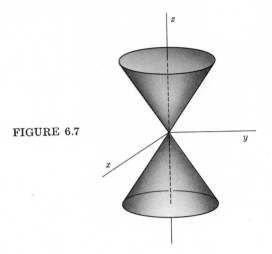

FIGURE 6.7

Example 6.7 (The Elliptic Cone)   The surface defined by

$$\frac{x^2}{a^2} + \frac{y^2}{b^2} = \frac{z^2}{c^2}$$

intersects the coordinate axes only at $(0, 0, 0)$, which is also its trace in the xy-plane. The trace in the plane $z = k \neq 0$ is an ellipse, while its traces in the planes $x = k$ and $y = k$ are hyperbolas which reduce to pairs of inter-secting lines if $k = 0$. This surface is an *elliptic cone*; it is shown in Fig. 6.7.

Example 6.8 (The Elliptic Paraboloid)   The *elliptic paraboloid* defined by

$$\frac{x^2}{a^2} + \frac{y^2}{b^2} = \frac{z}{c},$$

passes through $(0, 0, 0)$, but does not intersect the coordinate axes else-where. Its traces in planes parallel to the yz- and xz-planes are parabolas, and its trace in the plane $z = k > 0$ is an ellipse; it does not intersect any plane $z = k < 0$. It is shown in Fig. 6.8.

Example 6.9 (The Hyperbolic Paraboloid)   The surface defined by

$$\frac{x^2}{a^2} - \frac{y^2}{b^2} = \frac{z}{c}$$

intersects the coordinate axes only at the origin. It intersects the plane $z = k$ in a hyperbola which reduces to the pair of lines $y = \pm(b/a)x$ if

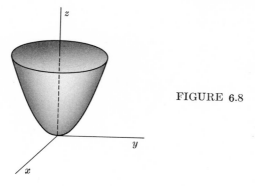

FIGURE 6.8

$k = 0$. Its trace in the plane $y = k$ is a parabola opening upward, and its trace in plane $x = k$ is a parabola opening downward. Near the origin the surface looks like a saddle (Fig. 6.9); it is called a *hyperbolic paraboloid*.

We have said that Examples 6.1 and 6.3 through 6.9 contain all of the nondegenerate quadric surfaces. This does not mean that the equations given in those examples are the only ones of the form (2); this is obviously false, since none of the former contain xy, xz, or yz terms, while the latter may. Rather, it means that, if (2) defines a nondegenerate quadric surface S with respect to some rectangular coordinate system, then there is a rectangular coordinate system (x', y', z'), which can be obtained from the (x, y, z) system by a translation and a rotation, relative to which the equation of S takes one of the *standard forms* discussed in the examples.

Example 6.10 Consider the second-degree equation

$$36x^2 - 72x + 16y^2 + 64y + 9z^2 - 18z = 42,$$

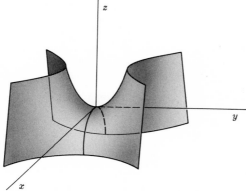

FIGURE 6.9

which can be rewritten as

$$36(x^2 - 2x) + 16(y^2 + 4y) + 9(z^2 - 2z) = 42.$$

Completing the squares yields

$$36(x^2 - 2x + 1) + 16(y^2 + 4y + 4)$$
$$+ 9(z^2 - 2z + 1) = 42 + 36 + 64 + 9 = 151$$

or

$$\frac{(x-1)^2}{6} + \frac{(y+2)^2}{9} + \frac{(z-1)^2}{16} = 1.$$

By setting $x' = x - 1$, $y' = y + 2$, and $z' = z - 1$, we obtain

$$\frac{(x')^2}{6} + \frac{(y')^2}{9} + \frac{(z')^2}{16} = 1,$$

which is the equation of an ellipsoid (Example 6.4).

Reduction to standard form in this example was very easy, requiring only a translation which was found by completing the squares. If (2) has mixed terms with nonzero coefficients, one must also find an appropriate rotation. We shall not go into this problem here.

EXERCISES 3.6

In Exercises 1 through 20 identify and sketch each surface.

1. $9x^2 - 4y^2 - 36z = 0$.

2. $y^2 - 16z^2 = 0$.

3. $x^2 - 25y^2 + 9z^2 = 225$.

4. $x^2 + 9y^2 + 16z^2 + 4 = 0$.

5. $25y^2 - 100z^2 = 400$.

6. $4x^2 + 9y^2 - 36z^2 = 0$.

7. $x^2 + 4z^2 = 0$.

8. $-4x^2 - 9y^2 + z^2 = 36$.

9. $y - 3z^2 = 0$.

10. $9x^2 + 4y^2 - 36z = 0$.

11. $x^2 + y^2 + z^2 = 36$.

12. $4x^2 + y^2 = 4$.

13. $4x^2 + 9y^2 - 16z^2 = 144$.

14. $x - 3z^2 = 0$.

15. $4x^2 + 9y^2 + 16z^2 = 144$.

16. $4x^2 - 10y^2 - 25z^2 = 100$.

17. $-4x^2 - 100y + 25z^2 = 0$.

18. $4x^2 - 100y + 25z^2 = 0$.

19. $16x^2 - 9z^2 = 144$.

20. $4x^2 - 16y^2 + z^2 = 0$.

In Exercises 21 through 30 translate axes and identify and sketch the surface.

21. $4x^2 + 9y^2 + z^2 + 8x - 18y - 4z - 19 = 0.$

22. $-5y^2 + 20y + z - 23 = 0.$

23. $-x^2 + 16y^2 - 4z^2 - 4x - 96y - 16z + 62 = 0.$

24. $y^2 - z^2 - 9x - 4y + 8z - 39 = 0.$

25. $x^2 - 4z^2 - 4x + 8z = 0.$

26. $x^2 + 3y^2 + 2z^2 - 6x - 6y + 4z + 14 = 0.$

27. $x^2 + 4y^2 + 4x + 16y - 16z + 4 = 0.$

28. $4x^2 - y^2 + z^2 - 16x + 8y - 6z + 9 = 0.$

29. $x^2 + y^2 + z^2 - 2x - 4y + 6z + 5 = 0.$

30. $4x^2 - 9y^2 + z^2 + 8x + 18y - 6z + 4 = 0.$

Chapter 4

Differential Calculus of Real-Valued Functions

4.1 Functions, Limits, and Continuity

Functions

Let D be a subset of R^n. A rule f which assigns a unique real number to each point of D will be called a *real valued function on* D. The value that f assigns to

$$\mathbf{X} = (x_1, \ldots, x_n)$$

will be denoted by

$$f(\mathbf{X}) = f(x_1, \ldots, x_n),$$

except that for $n = 1$, 2, and 3 we shall often write $f(x), f(x, y)$, and $f(x, y, z)$; $f(\mathbf{X})$ is a real number which we call the *value of f at* $\mathbf{X}$.

We distinguish between the function, which is f, and its *value* at a point $\mathbf{X}$, which is the real number $f(\mathbf{X})$, except where it is too clumsy to do so. For instance, if c is a fixed real number and f is the function whose value for every $\mathbf{X}$ in R^n is c, we shall simply write $f = c$; in particular, $f = 0$ is the function whose value is zero for all $\mathbf{X}$. Notice that $f(\mathbf{X}) = c$ and $f = c$ do not mean the same thing; $f(\mathbf{X}) = c$ means that c is the value of f at a particular $\mathbf{X}$, while $f = c$ means that c is the value of f at every $\mathbf{X}$ in R^n.

The set D on which f is defined is called the *domain* of f; the set of values assumed by $f(\mathbf{X})$ as $\mathbf{X}$ varies over D is called the *range*, or *image* of f. The specification of the domain is an essential part of the definition of the function.

Example 1.1 Let $n = 2$ and

$$f(\mathbf{X}) = \begin{cases} 1 & (|\mathbf{X}| > 1), \\ -1 & (|\mathbf{X}| < 1). \end{cases}$$

Here D comprises all points of R^2 except those for which $|\mathbf{X}| = 1$, and the range consists of the two points ± 1.

Example 1.2 Let $D = R^2$ and

(1)
$$f(x, y) = xy.$$

The range of f is R^1.

Example 1.3 Let D be all $\mathbf{X}$ in R^2 except those for which $x = y$ and

(2)
$$f(x, y) = \frac{1}{x - y}.$$

The range of f is the set of nonzero real numbers.

Usually a function will be defined by a single formula valid over its entire domain D. For brevity we adopt the convention that if f is given by a formula, and D is unspecified, then D should be understood to be the largest subset of R^n for which the formula makes sense. Thus in Example 1.2 we would simply write (1); since the right side is defined for all $\mathbf{X}$ in R^2, it would be understood that $D = R^2$. The right side of (2) is undefined for $x = y$; hence the domain of f in Example 1.3 would be understood to be as given there, unless specifically stated to the contrary. Also, if f is defined simply by

$$f(x, y) = \sqrt{1 - x^2 - y^2}$$

then D is the set of all $\mathbf{X}$ such that $|\mathbf{X}| \le 1$, since the right side defines a real number if and only if $\mathbf{X}$ is in this set.

Definition 1.1 Suppose f_1 and f_2 have domains D_1 and D_2 which have points in common, and let D be the set of common points. We define $f_1 + f_2$, $f_1 - f_2$, and $f_1 f_2$ for $\mathbf{X}$ in D by

$$(f_1 + f_2)(\mathbf{X}) = f_1(\mathbf{X}) + f_2(\mathbf{X}),$$

$$(f_1 - f_2)(\mathbf{X}) = f_1(\mathbf{X}) - f_2(\mathbf{X}),$$

and

$$(f_1 f_2)(\mathbf{X}) = f_1(\mathbf{X}) f_2(\mathbf{X}).$$

We also define f_1/f_2 by

$$(f_1/f_2)(\mathbf{X}) = \frac{f_1(\mathbf{X})}{f_2(\mathbf{X})},$$

but the domain of f_1/f_2 is the set D' of points in D for which $f_2(\mathbf{X}) \ne 0$ (f_1/f_2 is not defined if $f_2(\mathbf{X}) = 0$ for every $\mathbf{X}$ in D).

Example 1.4 Let

$$f_1(x, y) = \sqrt{1 - x^2 - y^2}$$

and

$$f_2(x, y) = x\sqrt{1 - 2x^2 - 2y^2}.$$

Then D_1 is the set of (x, y) such that $x^2 + y^2 \leq 1$, and D_2 is the set of (x, y) such that $x^2 + y^2 \leq \frac{1}{2}$; hence $D = D_2$,

$$(f_1 + f_2)(x, y) = \sqrt{1 - x^2 - y^2} + x\sqrt{1 - 2x^2 - 2y^2},$$

$$(f_1 - f_2)(x, y) = \sqrt{1 - x^2 - y^2} - x\sqrt{1 - 2x^2 - 2y^2}$$

and

$$(f_1 f_2)(x, y) = x\sqrt{(1 - 2x^2 - 2y^2)(1 - x^2 - y^2)}$$

for all $\mathbf{X}$ in D_2. Since $f_2(0, y) = 0$, D' is the set of (x, y) such that $x^2 + y^2 \leq \frac{1}{2}$ and $x \neq 0$. If (x, y) is in D', then

$$(f_1/f_2)(x, y) = \frac{1}{x}\sqrt{\frac{1 - x^2 - y^2}{1 - 2x^2 - 2y^2}}.$$

Definition 1.2 Let g be a function of a single variable t with domain S on the real axis, and let f be a real valued function with domain D in R^n. If D_0 is the set of all $\mathbf{X}$ in D such that $f(\mathbf{X})$ is in S, then the *composite function* $g \circ f$ is defined by

$$(g \circ f)(\mathbf{X}) = g(f(\mathbf{X}))$$

for $\mathbf{X}$ in D_0. (Of course, we must assume that D_0 actually contains some points—that is, it is *nonempty*.) For $n = 2$ the situation is depicted in Fig. 1.1.

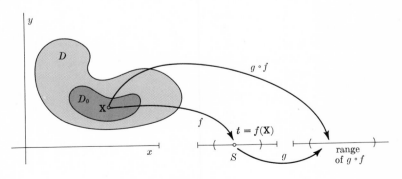

FIGURE 1.1

Example 1.5 Let

$$g(t) = \frac{1}{\sqrt{1 - t^2}}$$

and

$$f(x, y) = \sin (x^2 + y^2);$$

then S is the unit interval $-1 < t < 1$, and D is R^2. Since $|\sin \theta| < 1$ unless $\theta = (2k + 1)\pi/2$ $(k = 0, \pm 1, \dots)$,

$$(g \circ f)(x, y) = \frac{1}{\sqrt{1 - \sin^2 (x^2 + y^2)}},$$

provided $x^2 + y^2 \neq (2k + 1)\pi/2$ $(k = 0, 1, \dots)$.

Limits

The student should recall the concepts of limit and continuity, which are fundamental to the calculus of functions of one variable. We now generalize these ideas to real valued functions of several variables; in Section 5.1 we shall extend them further to vector valued functions.

Definition 1.3 If X_0 is a point in R^n and $\rho > 0$, the set of points X that satisfy $|X - X_0| < \rho$ will be called the *n-ball of radius ρ about* X_0. The set obtained by excluding X_0 from this n-ball will be called the *deleted n-ball of radius ρ about* X_0; X is in this deleted n-ball if and only if $0 < |X - X_0| < \rho$.

When $n = 1$, 2, or 3, we shall also refer to the corresponding n-balls by their more familiar names: *interval*, *disk*, and *ball*.

Example 1.6 If $n = 3$, the set of points whose coordinates satisfy

$$\sqrt{x^2 + (y - 1)^2 + (z - 2)^2} < 5$$

is the 3-ball (or simply ball) of radius 5 about $X_0 = (0, 1, 2)$. The deleted ball of radius 5 about X_0 is obtained from this set by excluding X_0; it consists of all (x, y) such that

$$0 < \sqrt{x^2 + (y - 1)^2 + (z - 2)^2} < 5.$$

A function f is said to *approach the limit L as X approaches* X_0,

$$\lim_{X \to X_0} f(X) = L,$$

if for any $\epsilon > 0$ all values of $f(X)$ for X in some deleted n-ball about X_0 are in the interval $(L - \epsilon, L + \epsilon)$. More formally, for each $\epsilon > 0$ there exists a positive number δ such that

$$|f(\mathbf{X}) - L| < \epsilon$$

for all $\mathbf{X}$ satisfying

$$0 < |\mathbf{X} - \mathbf{X}_0| < \delta.$$

For $n = 1$ this definition reduces to the familiar definition studied in the first calculus course.

The essential idea here is that f has a limit L as $\mathbf{X}$ approaches $\mathbf{X}_0$ if it is possible, by restricting $\mathbf{X}$ to a sufficiently small deleted n-ball about $\mathbf{X}_0$, to restrict the values $f(\mathbf{X})$ to within any previously given open interval about the real number L. Notice that the value $f(\mathbf{X}_0)$ does not enter into the definition; in fact, f need not be defined at $\mathbf{X}_0$.

Example 1.7 Let $n = 1$ and

$$f(x) = x^2;$$

then $\lim\limits_{x \to 2} f(x) = 4$, because

$$
\begin{aligned}
|f(x) - 4| &= |x^2 - 4| \\
&= |x - 2||x + 2| \\
&= |x - 2||(x - 2) + 4| \\
&\leq |x - 2|(|x - 2| + 4).
\end{aligned}
$$

Given $\epsilon > 0$, take $\delta = \min(1, \epsilon/5)$. Then if $|x - 2| < \delta$,

$$|f(x) - 4| < \frac{\epsilon}{5}(1 + 4) = \epsilon.$$

Example 1.8 Let $n = 3$, $\mathbf{X}_0 = (2, 1, 1)$, and $f(x, y, z) = x + 3y + 2z$; then $\lim\limits_{\mathbf{X} \to \mathbf{X}_0} f(\mathbf{X}) = 7$. To verify this, we calculate as follows:

$$
\begin{aligned}
|f(\mathbf{X}) - 7| &= |x + 3y + 2z - 7| \\
&= |(x - 2) + 3(y - 1) + 2(z - 1)| \\
&\leq |x - 2| + 3|y - 1| + 2|z - 1| \\
&\leq |\mathbf{X} - \mathbf{X}_0| + 3|\mathbf{X} - \mathbf{X}_0| + 2|\mathbf{X} - \mathbf{X}_0| \\
&= 6|\mathbf{X} - \mathbf{X}_0|;
\end{aligned}
$$

given $\epsilon > 0$, take $\delta = \epsilon/6$.

The definition of limit implicitly assumes that f is defined for all points of a sufficiently small deleted n-ball about $\mathbf{X}_0$. We now reformulate the definition to avoid this restriction, which is too stringent for our purposes.

Definition 1.4 A point $\mathbf{X}_0$ is a *limit point* of a subset D of R^n if every deleted n-ball about $\mathbf{X}_0$, no matter how small, contains points of D.

A limit point of D need not be in D, nor is a point of D necessarily a limit point (see Example 1.9).

Definition 1.5 A point $\mathbf{X}_0$ in a subset D of R^n is an *interior* point of D if there is an n-ball about $\mathbf{X}_0$ which is contained entirely in D.

Notice that an interior point is necessarily a limit point, but a limit point need not be an interior point.

Example 1.9 Let $n = 2$ and D be those points (x, y) such that $0 \leq x < y$ (Fig. 1.2), together with $(x, y) = (1, 0)$.

Every point in D except $(1, 0)$ and those for which $x = 0$ is an interior point. Points on the positive y-axis $(x = 0, y > 0)$ are limit points of D which are contained in D, but are not interior points; points on the half line $x = y \geq 0$ are limit points which are not in D; $(1, 0)$ is not a limit point of D.

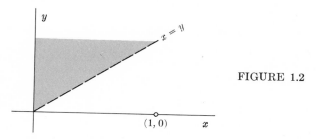

FIGURE 1.2

Definition 1.6 A set S is *open* if all of its points are interior points.

Example 1.10 The n-ball B of radius ρ about any point $\mathbf{X}_0$ is open. Let $\mathbf{X}_1$ be any point in B and $|\mathbf{X}_1 - \mathbf{X}_0| = \rho_1 < \rho$. If $\mathbf{X}$ is any point such that $|\mathbf{X} - \mathbf{X}_1| < \rho - \rho_1$, then the triangle inequality implies that

$$|\mathbf{X} - \mathbf{X}_0| = |(\mathbf{X} - \mathbf{X}_1) + (\mathbf{X}_1 - \mathbf{X}_0)|$$
$$\leq |\mathbf{X} - \mathbf{X}_1| + |\mathbf{X}_1 - \mathbf{X}_0|$$
$$< \rho - \rho_1 + \rho_1 = \rho;$$

hence the n-ball about $\mathbf{X}_1$ of radius $\rho - \rho_1$ is in B (Fig. 1.3). We have shown

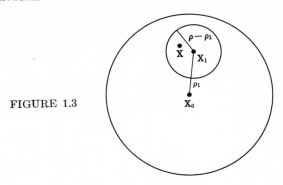

FIGURE 1.3

that $\mathbf{X}_1$ is an interior point (Definition 1.5) of the n-ball B. Since $\mathbf{X}_1$ was arbitrarily chosen, every point of B is an interior point. Hence, B is open, by Definition 1.6.

Example 1.11   The set D defined in Example 1.9 is not open, since $(1, 0)$ and points on the positive y-axis are in D, but are not interior points. However, the set D° consisting of all (x, y) such that $0 < x < y$ is open; it is called the *interior* of D. In general, the *interior* of a set S is the set of interior points of S; it is an open set. (Exercise T-13).

Definition 1.7   Let $\mathbf{X}_0$ be a limit point of the domain D of a real valued function f. Then we say that $\lim_{\mathbf{X} \to \mathbf{X}_0} f(\mathbf{X}) = L$ if for every $\epsilon > 0$ there exists a $\delta > 0$, depending upon ϵ, such that

$$|f(\mathbf{X}) - L| < \epsilon$$

for all $\mathbf{X}$ in D that satisfy $0 < |\mathbf{X} - \mathbf{X}_0| < \delta$. We also say that f *has limit* L at $\mathbf{X}_0$.

Example 1.12   Let $n = 2$ and

$$f(x, y) = \frac{x^2 - y^2}{x - y} \, ;$$

then f is not defined on the line $x = y$. However, these points are limit points of the domain of f. Let $\mathbf{X}_0 = (c, c)$. If $x \neq y$, we can write

$$f(\mathbf{X}) = x + y$$

and the student can verify that

$$\lim_{\mathbf{X} \to \mathbf{X}_0} f(\mathbf{X}) = 2c.$$

It is essential to recognize that Definition 1.7 allows no restriction on the manner in which $\mathbf{X}$ approaches $\mathbf{X}_0$, except that $\mathbf{X}$ be in the domain of f.

Example 1.13 Let $n = 1$ and

$$f(x) = \begin{cases} -1 + x & (x < 0), \\ 1 + x & (x > 0); \end{cases}$$

then f does not have a limit at $x = 0$, since in any open interval containing the origin there are points x_1 and x_2 such that $f(x_1) < -\frac{1}{2}$ and $f(x_2) > \frac{1}{2}$ (Fig. 1.4). If we were careless and examined only positive values of x, we might incorrectly conclude that $\lim_{x \to 0} f(x) = 1$; similarly, examining only negative values of x might lead to the conclusion that $\lim_{x \to 0} f(x) = -1$, also invalid. In fact, f has no limit at $x = 0$.

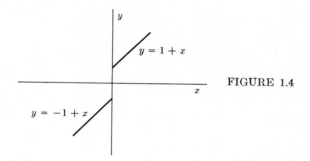

FIGURE 1.4

Example 1.14 Let $n = 2$, $\mathbf{X}_0 = (0, 0)$, and

$$f(x, y) = 2xy \left(1 + \frac{1}{x^2 + y^2} \right), \qquad (x, y) \neq (0, 0).$$

Suppose $\mathbf{X}$ is constrained to approach the origin along the ray $x = \rho \cos \psi$, $y = \rho \sin \psi$, where $\rho > 0$ and ψ is fixed but arbitrary (Fig. 1.5). Then

$$f(\rho \cos \psi, \rho \sin \psi) = 2(1 + \rho^2) \sin \psi \cos \psi$$

$$= (1 + \rho^2) \sin 2\psi,$$

which approaches $\sin 2\psi$ as ρ approaches zero. Can we conclude that $\lim_{\mathbf{X} \to 0} f(\mathbf{X}) = \sin 2\psi$? Clearly not, since the value of the right side depends upon the direction of approach.

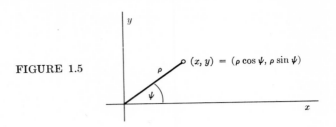

FIGURE 1.5

These two examples should convince the student that it is not safe to conclude that $\lim\limits_{X \to X_0} f(X) = L$ after restricting the way in which X approaches X_0. One can, however, conclude that a limit does not exist if different limiting values are obtained for different paths of approach, as in the last two examples.

Example 1.15 Let

$$f(x, y) = \frac{1}{x^2 + y^2} \,;$$

f has no limit at $(0, 0)$, because it grows beyond all bounds as (x, y) approaches $(0, 0)$. That is, given any M, no matter how large, $|f(x, y)| > M$ if $0 < x^2 + y^2 < 1/M$. We say in this case that

$$\lim_{(x,y) \to (0,0)} \frac{1}{x^2 + y^2} = \infty.$$

In general we say that

$$\lim_{X \to X_0} f(X) = \infty$$

if for every $M > 0$, no matter how large, there is a $\delta > 0$ such that $|f(X)| > M$ if $0 < |X - X_0| < \delta$. Thus

$$\lim_{(x,y) \to (0,0)} \frac{1}{\sin xy} = \infty$$

and

$$\lim_{(x,y,z) \to (0,0,1)} \frac{1}{x + y + z - 1} = \infty.$$

Theorem 1.1 If $\lim\limits_{X \to X_0} f(X) = L_1$, and $\lim\limits_{X \to X_0} f(X) = L_2$, then $L_1 = L_2$; that is, if a limit exists it is unique.

Proof. Suppose $L_1 \neq L_2$. From the triangle inequality,

$$| L_1 - L_2 | = | (L_1 - f(\mathbf{X})) + (f(\mathbf{X}) - L_2) |$$

$$\leq | L_1 - f(\mathbf{X}) | + | f(\mathbf{X}) - L_2 |.$$

From the definition of limit, each of the last two terms is less than

$$\frac{| L_1 - L_2 |}{4}$$

for $\mathbf{X}$ sufficiently close to $\mathbf{X}_0$ (take $\epsilon = | L_1 - L_2 |/4$ in Definition 1.7). This implies that

$$| L_1 - L_2 | < \frac{| L_1 - L_2 |}{2},$$

a contradiction; hence $L_1 = L_2$.

For more complicated functions the "epsilon–delta" argument required in the definition of limit becomes cumbersome, and it is often difficult to guess what the limit is in the first place. The following two theorems are useful for finding $\lim_{\mathbf{X} \to \mathbf{X}_0} f(\mathbf{X})$ if f can be obtained from simpler functions by a finite sequence of additions, subtractions, multiplications, divisions, and compositions.

Theorem 1.2 Let $\mathbf{X}_0$ be a limit point of the set of points common to the domains of f_1 and f_2. If $\lim_{\mathbf{X} \to \mathbf{X}_0} f_1(\mathbf{X}) = L_1$, and $\lim_{\mathbf{X} \to \mathbf{X}_0} f_2(\mathbf{X}) = L_2$, then

(a) $\lim_{\mathbf{X} \to \mathbf{X}_0} (f_1 + f_2)(\mathbf{X}) = L_1 + L_2, \lim_{\mathbf{X} \to \mathbf{X}_0} (f_1 - f_2)(\mathbf{X}) = L_1 - L_2;$

(b) $\lim_{\mathbf{X} \to \mathbf{X}_0} (f_1 f_2)(\mathbf{X}) = L_1 L_2;$ and

(c) if $L_2 \neq 0$, $\lim_{\mathbf{X} \to \mathbf{X}_0} (f_1/f_2)(\mathbf{X}) = L_1/L_2.$

Proof. The proof is analogous to that given in the first year calculus course; hence we shall prove only (b), leaving the rest for the exercises. Write

$$f_1(\mathbf{X})f_2(\mathbf{X}) - L_1 L_2$$

$$= (f_1(\mathbf{X}) - L_1)(f_2(\mathbf{X}) - L_2) + L_2(f_1(\mathbf{X}) - L_1) + L_1(f_2(\mathbf{X}) - L_2);$$

then

$$|f_1(\mathbf{X})f_2(\mathbf{X}) - L_1L_2|$$
$$\leq |f_1(\mathbf{X}) - L_1||f_2(\mathbf{X}) - L_2| + |L_2||f_1(\mathbf{X}) - L_1| + |L_1||f_2(\mathbf{X}) - L_2|.$$

The terms on the right can be made as small as we please by taking $\mathbf{X}$ sufficiently close to $\mathbf{X}_0$, which proves (b).

Example 1.16 Let $\mathbf{X}_0 = (1, 2)$ and find $\lim_{\mathbf{X} \to \mathbf{X}_0} f(x, y)$, where $f(x, y) = xy/(x + y)$.

Solution. Let $g_1(x, y) = x$, and $g_2(x, y) = y$; then $\lim_{\mathbf{X} \to \mathbf{X}_0} g_1(x, y) = \lim_{x \to 1} x = 1$ and $\lim_{\mathbf{X} \to \mathbf{X}_0} g_2(x, y) = 2$. (We assume that the student can evaluate one dimensional limits, using techniques learned in his first calculus course). Now let $f_1 = g_1 g_2$, and $f_2 = g_1 + g_2$; then $f = f_1/f_2$. From Theorem 1.2

$$\lim_{\mathbf{X} \to \mathbf{X}_0} f_2(x, y) = \lim_{\mathbf{X} \to \mathbf{X}_0} g_1(x, y) + \lim_{\mathbf{X} \to \mathbf{X}_0} g_2(x, y) = 1 + 2 = 3,$$

$$\lim_{\mathbf{X} \to \mathbf{X}_0} f_1(x, y) = (\lim_{\mathbf{X} \to \mathbf{X}_0} g_1(x, y))(\lim_{\mathbf{X} \to \mathbf{X}_0} g_2(x, y)) = 1 \cdot 2 = 2,$$

and

$$\lim_{\mathbf{X} \to \mathbf{X}_0} f(x, y) = \frac{\lim_{\mathbf{X} \to \mathbf{X}_0} f_1(x, y)}{\lim_{\mathbf{X} \to \mathbf{X}_0} f_2(x, y)} = \frac{2}{3}.$$

Continuity

Definition 1.8 A real valued function f with domain D is said to be *continuous at a point* $\mathbf{X}_0$ if

(a) $\mathbf{X}_0$ is in D (thus $f(\mathbf{X}_0)$ is defined);
(b) $\mathbf{X}_0$ is a limit point of D; and
(c) $\lim_{\mathbf{X} \to \mathbf{X}_0} f(\mathbf{X}) = f(\mathbf{X}_0)$.

We shall also agree, as a matter of convenience, that f is continuous at any point of D which is not a limit point of D. (Such points are *isolated points*; for example, $(1, 0)$ in Example 1.9.)

If f is continuous at every point of a set S, we say that f *is continuous on* S.

The student should convince himself that the conditions on $\mathbf{X}_0$ in (a) and (b) are merely those required so that (c) should make sense.

Often our understanding of a definition is aided by examining ways in which it can fail to be satisfied.

Example 1.17   Let $n = 2$ and

$$f(x, y) = \frac{1}{1 - x^2 - y^2} \qquad (0 < x^2 + y^2 < 1),$$

$$f(0, 0) = 2, \qquad f(0, 1) = 1, \qquad f(0, 2) = 5.$$

Then $\lim_{\mathbf{X} \to (0,0)} f(\mathbf{X}) = 1 \neq 2$; hence f is not continuous at $(0, 0)$ even though f is defined and has a limit there. At $(0, 1)$, f is defined but has no limit, and again is not continuous. The function f is continuous at $(0, 2)$, which is an isolated point of its domain. If $0 < |\mathbf{X}_0| < 1$, then f is continuous at $\mathbf{X}_0$ (Exercise 26).

Example 1.18   Let g be a function of the single variable x with domain D, and let

$$f(x, y) = g(x)$$

on D^*, the set of points $\mathbf{X} = (x, y)$ such that x is in D. If g is continuous at x_0, and y_0 is arbitrary, then f is continuous at $\mathbf{X}_0 = (x_0, y_0)$.

Proof.   Given $\epsilon > 0$, there is a $\delta > 0$ such that

$$|g(x) - g(x_0)| < \epsilon$$

if $|x - x_0| < \delta$ and x is in D. Since $|x - x_0| \leq |\mathbf{X} - \mathbf{X}_0|$, it follows that

$$|f(x, y) - f(x_0, y_0)| = |g(x) - g(x_0)| < \epsilon$$

if $|\mathbf{X} - \mathbf{X}_0| < \delta$ and $\mathbf{X}$ is in D^*.

This result has a natural generalization. Suppose f is defined on a subset of R^n $(n > 1)$, but its values are independent of some of the variables, say the last $n - k$; that is,

$$f(x_1, \ldots, x_n) = g(x_1, x_2, \ldots, x_k).$$

Then if g is continuous at $(c_1, \ldots, c_k)$, under the definition of continuity for functions on R^k, it follows that f is continuous at $(c_1, \ldots, c_k, x_{k+1}, \ldots, x_n)$ for any $x_{k+1}, \ldots, x_n$, under the definition of continuity for functions on R^n.

Theorem 1.3   Let D be the set of points common to the domains of f_1 and f_2. Let $\mathbf{X}_0$ be in D and be a limit point of D. Then if f_1 and f_2 are continuous at

$\mathbf{X}_0$, so are $f_1 + f_2, f_1 - f_2$, and $f_1 f_2$. If $f_2(\mathbf{X}_0) \neq 0$, then f_1/f_2 is also continuous at $\mathbf{X}_0$.

Proof.   Again we consider only the product, the other cases being left as an exercise (Exercise T-2). By assumption, the limits

$$\lim_{\mathbf{X} \to \mathbf{X}_0} f_1(\mathbf{X}) = L_1$$

and

$$\lim_{\mathbf{X} \to \mathbf{X}_0} f_2(\mathbf{X}) = L_2$$

exist. By Theorem 1.2,

(3) $$\lim_{\mathbf{X} \to \mathbf{X}_0} (f_1 f_2)(\mathbf{X}) = L_1 L_2.$$

Since f_1 and f_2 are continuous at $\mathbf{X}_0$, $L_1 = f(\mathbf{X}_0)$ and $L_2 = f(\mathbf{X}_0)$. Therefore (3) can be rewritten

$$\lim_{\mathbf{X} \to \mathbf{X}_0} (f_1 f_2)(\mathbf{X}) = f_1(\mathbf{X}_0) f_2(\mathbf{X}_0)$$
$$= (f_1 f_2)(\mathbf{X}_0);$$

hence $f_1 f_2$ is continuous at $\mathbf{X}_0$.

Example 1.19   By the same sequence of applications of the assertions of Theorem 1.3 that we performed with those of Theorem 1.2 in Example 1.16

$$f(x, y) = \frac{xy}{x + y}$$

is continuous at $\mathbf{X}_0 = (1, 2)$.

The following theorem states that "a continuous function of a continuous function is continuous."

Theorem 1.4   Let f and g be as in Definition 1.2. Assume that f is continuous at $\mathbf{X}_0$, that $t_0 = f(\mathbf{X}_0)$ is an interior point of the domain S of g, and that g is continuous at t_0. Then the composite function $g \circ f$ is continuous at $\mathbf{X}_0$.

Proof.   Let $\epsilon > 0$ be given. Since t_0 is an interior point of S and g is continuous at t_0, there is a $\delta_1 > 0$ such that $g(t)$ is defined and $|g(t) - g(t_0)| < \epsilon$ if $|t - t_0| < \delta_1$. As $\lim_{\mathbf{X} \to \mathbf{X}_0} f(\mathbf{X}) = f(\mathbf{X}_0) = t_0$, there is a $\delta > 0$ such that $|f(\mathbf{X}) - t_0| < \delta_1$ if $|\mathbf{X} - \mathbf{X}_0| < \delta$. Hence, if $|\mathbf{X} - \mathbf{X}_0| < \delta$, then $|g(f(\mathbf{X})) - g(t_0)| < \epsilon$, which completes the proof. The situation is depicted in Fig. 1.6: if $\mathbf{X}$ is in the n-ball of radius δ about $\mathbf{X}_0$, then $t = f(\mathbf{X})$ is in the interval $(t_0 - \delta_1, t_0 + \delta_1)$ and therefore $g(f(\mathbf{X}))$ is in the interval $(g(t_0) - \epsilon, g(t_0) + \epsilon)$.

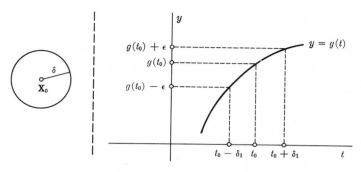

FIGURE 1.6

Example 1.20 Let $\mathbf{X}_0 = (1, 2)$,

$$f(x, y) = \frac{xy}{x + y}$$

and

$$g(t) = \sqrt{t}.$$

From Example 1.19, f is continuous at $\mathbf{X}_0$. Furthermore $t_0 = f(\mathbf{X}_0) = \frac{2}{3}$ is an interior point of the domain of g, and g is continuous at t_0. Consequently $g \circ f$ is continuous at $\mathbf{X}_0$. It follows that

$$\lim_{\mathbf{X} \to \mathbf{X}_0} \sqrt{\frac{xy}{x + y}} = g(t_0) = \sqrt{\frac{2}{3}}.$$

Example 1.18 and Theorems 1.3 and 1.4 allow us to build up an extensive collection of continuous functions by addition, subtraction, multiplication, division, and composition of continuous functions.

Example 1.21 Let

$$f(x, y, z) = z + \log [2 + \sin (x^2 + y^2)].$$

This is continuous on all of R^3 by the following argument:

(a) Let $f_1(x, y, z) = x$, $f_2(x, y, z) = y$, and $f_3(x, y, z) = z$; then f_1, f_2, and f_3 are continuous on R^3 (Example 1.18).

(b) f_1^2 ($=f_1 f_1$), f_2^2, and $f_4 = f_1^2 + f_2^2$ are continuous on R^3 (Theorem 1.3):

$$f_4(x, y, z) = x^2 + y^2.$$

(c) Let $g_1(t) = \sin t$; then $f_5 = g_1 \circ f_4$ is continuous on R^3 (Theorem 1.4):

$$f_5(x, y, z) = \sin (x^2 + y^2).$$

(d) $f_6 = 2$ is continuous on R^3, hence $f_7 = f_5 + f_6$ is also:

$$f_7(x, y, z) = 2 + \sin(x^2 + y^2).$$

(e) $g_2(t) = \log t$ is continuous for $t \geq 0$; since the values of f_7 are positive, $f_8 = g_1 \cdot f_7$ is continuous on R^3 (Theorem 1.4).

(f) $f = f_3 + f_8$ is continuous on R^3 (Theorem 1.3).

October 20

EXERCISES 4.1   1, 2, 12a

1. Find the domains of the following functions.

(a) $f(x, y, z) = \sqrt{x^2 + y^2 - 1}.$ (b) $f(x, y) = (xy)^{3/2}.$

(c) $f(x, y) = \dfrac{x^2 + y^2}{xy}.$ (d) $f(x, y) = \dfrac{1}{\sin(x^2 - y^2)}.$

(e) $f(x, y) = \dfrac{1}{\cos(x^2 - y^2)}.$ (f) $f(x, y) = \cos xy.$

2. Find the domains of the following functions.

(a) $f(x, y, z) = x^2 + y^2 + z.$ (b) $f(x, y) = \log(x - y).$

(c) $f(x, y, z) = x^2 - \dfrac{y}{z - 1}.$ (d) $f(x, y) = \sqrt{x^2 - 1}.$

(e) $f(x, y) = \dfrac{x}{y^2 - 1}.$ (f) $f(x, y) = e^{x^2 + v^2}.$

3. Let $f(x, y) = \sqrt{1 - x^2 - y^2}$ and $g(x, y) = \log(x - y)$. What are the domains of $f + g, f - g, fg,$ and f/g?

4. Let $g(t) = e^{t^2}$ and $f(x, y) = \sqrt{x^2 - y^2}.$

(a) Find $(g \circ f)(x, y)$. What is the domain of $g \circ f$?

(b) Find $(g \circ f)(3, 2).$

5. Let $g(t) = 1/(1 - t)$ and $f(x, y) = 1 + \cos(x + y).$

(a) Find $(g \circ f)(x, y)$. What is the domain of $g \circ f$?

(b) Find $(g \circ f)(\pi/2, 3\pi/2).$

6. (a) Sketch the disk of radius $\rho = 2$ about $\mathbf{X}_0 = (1, 2)$.

   (b) Repeat (a) for $\rho = 1$ and $\mathbf{X}_0 = (3, 4)$.

7. Let S be an n-ball of radius ρ about $\mathbf{X}_0$ and let $\mathbf{X}_1$ be in S. Find the radius of the largest n-ball about $\mathbf{X}_1$ that is contained entirely within S.

8. Find the limit points of the subsets of R^1 defined by

   (a) $0 \le x \le 1$;   (b) $0 < x \le 1$;

   (c) $|x| = 1$;   (d) $x = 1, \dfrac{1}{2}, \ldots, \dfrac{1}{n}, \ldots$

9. Find the limit points of the subsets of R^2 defined by

   (a) $x^2 + y^2 \le 1$;

   (b) $(x, y) = (0, 0)$ or $(x, y) = (1, 1)$;

   (c) $x^2 + y^2 \le 1$, $x \ne 0$, $y \ne 0$;

   (d) $0 < x \le y$;

   (e) $(x, y) \ne (0, 0)$;

   (f) $x^2 + y^2 = 1$ or $(x, y) = (1, 2)$.

10. For the sets in Exercise 8, state which points are (a) interior points, (b) limit points but not interior points, (c) *isolated points* (not interior or limit points).

11. Repeat Exercise 10 for the sets in Exercise 9.

12. For each of the following functions find $\lim\limits_{\mathbf{X} \to \mathbf{X}_0} f(x, y)$ if it exists. When there is a limit, prove it by an "epsilon–delta" argument.

    (a) $f(\mathbf{X}) = x + 2y$, $\mathbf{X}_0 = (1, 2)$.

    (b) $f(\mathbf{X}) = \dfrac{x^2 - y^2}{x + y}$, $\mathbf{X}_0 = (1, -1)$.

    (c) $f(\mathbf{X}) = \dfrac{x^2 y}{x^2 + y^2}$, $\mathbf{X}_0 = (0, 0)$.

    (d) $f(\mathbf{X}) = \sin \dfrac{1}{x} - y$, $\mathbf{X}_0 = (0, 3)$.

13.   Repeat Exercise 12 for the following functions.

(a) $f(\mathbf{X}) = 3x + 2y,$ $\mathbf{X}_0 = (2, 1).$

(b) $f(\mathbf{X}) = \dfrac{x^2 + y^2}{x - y},$ $\mathbf{X}_0 = (1, 1).$

(c) $f(\mathbf{X}) = \dfrac{xy}{\sqrt{x^2 + y^2}},$ $\mathbf{X}_0 = (0, 0).$

(d) $f(x, y) = x^2 + y^2 + z^2,$ $\mathbf{X}_0 = (-1, 0, 1).$

14.   Using Theorems 1.2 and 1.4, find $\lim\limits_{\mathbf{X} \to \mathbf{X}_0} f(\mathbf{X})$, when it exists, for each of the following functions.

(a) $f(\mathbf{X}) = \log \sqrt{1 - x^2 - y^2 - z^2},$ $\mathbf{X}_0 = (0, 0, 0).$

(b) $f(\mathbf{X}) = \exp\left(\dfrac{1}{\sqrt{1 - \sin(1 + x - y)(\pi/2)}}\right),$ $\mathbf{X}_0 = (0, 0).$

(c) $f(\mathbf{X}) = \cos^2(1 + 2x^2 + y^2) + \sin^2 \dfrac{1}{1 - x^2},$ $\mathbf{X}_0 = (0, 0).$

15.   Repeat Exercise 14 for the following functions.

(a) $f(\mathbf{X}) = \cos \dfrac{\pi}{1 - 3xyz},$ $\mathbf{X}_0 = (1, 1, 1).$

(b) $f(\mathbf{X}) = \log \exp(\sqrt{x^2 + 2xy + y^2}),$ $\mathbf{X}_0 = (1, 1).$

(c) $f(\mathbf{X}) = \sin \log(|x| + |y|),$ $\mathbf{X}_0 = (0, 0).$

16.   Let $g(t) = t^3$ and $f(x, y) = (x^2 - y^2)/(x + y)$. Find
$$\lim_{(x, y) \to (2, -2)} g(f(x, y)).$$

17.   Let $g(t) = \sqrt{t}$ and $f(x, y) = \cos(x + y)$. Find
$$\lim_{(x, y) \to (2\pi, 4\pi)} g(f(x, y)).$$

18.   Prove that f is continuous at $\mathbf{X}_0$:

(a) $f(x, y, z) = x^2 + 2y^2 - z,$ $\mathbf{X}_0 = (a, b, c).$

(b) $f(x, y) = \cos(x^2 + y^2),$ $\mathbf{X}_0 = \left(\sqrt{\dfrac{\pi}{2}}, \sqrt{\pi}\right).$

(c) $f(x, y, z) = \exp(x^2 + y + z^2)$, $\mathbf{X}_0 = (1, 0, 1)$.

(d) $f(x, y) = \log(x^2 + y^2)$, $\mathbf{X}_0 = (1, -1)$.

19. Where are the functions of Exercise 1 continuous?

20. Where are the functions of Exercise 2 continuous?

21. Where is $g \circ f$ of Exercise 4 continuous?

22. Where are the functions of Exercise 14 continuous?

23. Where are the functions of Exercise 15 continuous?

24. Where are the following functions continuous?

(a) $f(x, y) = \begin{cases} \dfrac{x + y}{x - y} & \text{if } x \neq y, \\[2ex] 0 & \text{if } x = y. \end{cases}$

(b) $f(x, y) = \begin{cases} x \sin \dfrac{1}{x} - y & \text{if } x \neq 0, \\[2ex] -y & \text{if } x = 0. \end{cases}$

(c) $f(x, y) = \begin{cases} \dfrac{x^2 - y^2}{x - y} & \text{if } x \neq y, \\[2ex] x + y & \text{if } x = y. \end{cases}$

25. How would you define f at $(0, 0)$ to make f, with your extended definition, continuous at $\mathbf{X}_0$?

(a) $f(x, y) = \dfrac{xy}{\sqrt{x^2 + y^2}}$.   (b) $f(x, y) = \dfrac{\sin xy}{xy}$.

(c) $f(x, y) = \dfrac{x^2 + y^2 - 2xy}{x - y}$.

THEORETICAL EXERCISES

T-1.   Complete the proof of Theorem 1.2.

T-2.   Complete the proof of Theorem 1.3.

T-3.   A function f is said to be *bounded* near a limit point $\mathbf{X}_0$ of its domain D if there are positive numbers M and ϵ such that $|f(\mathbf{X})| < M$ whenever $\mathbf{X}$ is in D and $0 < |\mathbf{X} - \mathbf{X}_0| < \epsilon$. (If this is not the case, f is said to be *unbounded* near $\mathbf{X}_0$.) *Prove*: If $\lim_{\mathbf{X} \to \mathbf{X}_0} f(\mathbf{X}) = L$, then f is bounded near $\mathbf{X}_0$, and if $L \neq 0$, $1/f$ is also.

T-4.   *Prove*: If f is continuous at $\mathbf{X}_0$, then it is bounded near $\mathbf{X}_0$.

T-5.   If f is unbounded near $\mathbf{X}_0$, does it follow that $\lim_{\mathbf{X} \to \mathbf{X}_0} f(\mathbf{X}) = \infty$? (*Hint*: Consider $f(x) = (1/x)\sin(1/x)$.)

T-6.   Prove that every point of an n-ball is an interior point.

T-7.   A point $\mathbf{X}_0$ is a *boundary point* of a set S in R^n if every n-ball about $\mathbf{X}_0$ contains points in S and points not in S. Find all boundary points of the sets in Exercises 8 and 9.

T-8.   Show that a boundary point of a set S is either a limit point or an isolated point of S.

T-9.   Let f and g be continuous at $\mathbf{X}_0$; let c and d be real numbers. Show that $cf + dg$ is continuous at $\mathbf{X}_0$.

T-10. A function f is said to have a *removable discontinuity* at $\mathbf{X}_0$ if it is not continuous at $\mathbf{X}_0$, but can be defined (or possibly redefined) there by $f(\mathbf{X}_0) = y_0$ so that, with the new definition, f is continuous at $\mathbf{X}_0$. *Prove*: If f has a removable discontinuity, then y_0 is unique.

T-11. If $f_1, \ldots, f_n$ are functions, define the *sum* function $(f_1 + \cdots + f_n)$ and the *product* function $f_1 \cdots f_n$.

T-12. The *line segment connecting two points* $\mathbf{X}_1$ *and* $\mathbf{X}_2$ in R^n is defined to be the set of points of the form

$$\mathbf{X} = \mathbf{X}_1 + t(\mathbf{X}_2 - \mathbf{X}_1) \qquad (0 \leq t \leq 1).$$

A set D in R^n is *convex* if whenever it contains a pair of points it also contains the line segment connecting them. Show that an n-ball is convex.

T-13. Prove that the interior of a set is open.

4.2 Directional and Partial Derivatives

To prepare the way for the new ideas in this section, let us first recall the definition of the derivative of a function of one variable. If g is such a function and t_0 a point of its domain, consider the function G defined by

$$G(t) = \frac{g(t) - g(t_0)}{t - t_0}$$

for t in the domain of g and distinct from t_0. If $\lim_{t \to t_0} G(t)$ exists, we call its value the *derivative* of g at t_0, and denote it by $g'(t_0)$; thus

$$g'(t_0) = \lim_{t \to t_0} \frac{g(t) - g(t_0)}{t - t_0}.$$

An equivalent definition is

$$g'(t_0) = \lim_{h \to 0} \frac{g(t_0 + h) - g(t_0)}{h}.$$

Here, t_0 can be any point of the domain of g where the limit exists; hence we can omit the subscript and define the derivative g' at such points by

$$(1) \qquad g'(t) = \lim_{h \to 0} \frac{g(t + h) - g(t)}{h}.$$

Example 2.1 Let $t_0 = 2$ and $g(t) = 2t^2 + 3$; for $t \neq 2$ the difference quotient is

$$G(t) = \frac{g(t) - g(2)}{t - 2}$$

$$= \frac{2t^2 + 3 - 11}{t - 2} = \frac{2(t^2 - 4)}{t - 2}$$

$$= 2(t + 2).$$

The student should verify that

$$\lim_{t \to 2} G(t) = 8;$$

hence $g'(2) = 8$.

To compute the value of g' at any arbitrary point of its domain, we form the difference quotient on the right side of (1):

$$\frac{g(t + h) - g(t)}{h} = \frac{2(t + h)^2 + 3 - 2t^2 - 3}{h}$$

$$= 4t + 2h;$$

hence

$$g'(t) = \lim_{h \to 0} \frac{g(t + h) - g(t)}{h}$$

$$= 4t$$

for all t.

Example 2.2   Let $t_0 = 0$ and $g(t) = \sqrt{|t|}$. The difference quotient is

$$G(t) = \frac{\sqrt{|t|} - 0}{t - 0} = \pm \frac{1}{\sqrt{|t|}} \qquad (t \neq 0),$$

which does not approach a limit as t approaches zero; hence g does not have a derivative at $t_0 = 0$.

Directional Derivative

The following definition generalizes the idea of the derivative to functions whose domains are in R^n.

Definition 2.1   Let f be a real valued function defined on a subset D of R^n, let $\mathbf{X}_0$ be an interior point of D, and $\mathbf{U}$ be a unit vector in R^n. Then the *directional derivative of f at $\mathbf{X}_0$ in the direction of $\mathbf{U}$* is defined by

$$(2) \qquad \frac{\partial f}{\partial \mathbf{U}} (\mathbf{X}_0) = \lim_{t \to 0} \frac{f(\mathbf{X}_0 + t\mathbf{U}) - f(\mathbf{X}_0)}{t},$$

if the limit exists.

The problem of calculating directional derivatives quickly reduces to a problem in the ordinary calculus. Since $\mathbf{X}_0$ is an interior point of D, the

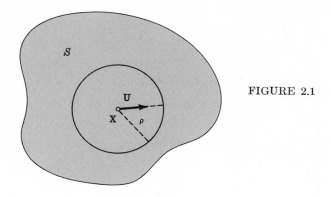

FIGURE 2.1

line segment S defined by

$$\mathbf{X} = \mathbf{X}_0 + t\mathbf{U}, \qquad -\rho < t < \rho,$$

is in D for ρ sufficiently small (Fig. 2.1), and therefore

$$(3) \qquad\qquad g(t) = f(\mathbf{X}_0 + t\mathbf{U})$$

is defined for $|t| < \rho$. The difference quotient on the right side of (2) can be rewritten as

$$\frac{f(\mathbf{X}_0 + t\mathbf{U}) - f(\mathbf{X}_0)}{t} = \frac{g(t) - g(0)}{t} \; ;$$

letting t approach zero yields

$$\frac{\partial f}{\partial \mathbf{U}} (\mathbf{X}_0) = g'(0).$$

Example 2.3   Let

$$\mathbf{U} = \begin{bmatrix} \dfrac{1}{\sqrt{2}} \\[3ex] -\dfrac{1}{\sqrt{2}} \end{bmatrix}$$

and

$$f(x, y) = x^2 + y^2 + xy.$$

In (3) take

$$g(t) = f\left(x + \frac{t}{\sqrt{2}}, y - \frac{t}{\sqrt{2}}\right)$$

$$= \left(x + \frac{t}{\sqrt{2}}\right)^2 + \left(y - \frac{t}{\sqrt{2}}\right)^2 + \left(x + \frac{t}{\sqrt{2}}\right)\left(y - \frac{t}{\sqrt{2}}\right);$$

then

$$g'(t) = \frac{2}{\sqrt{2}}\left(x + \frac{t}{\sqrt{2}}\right) - \frac{2}{\sqrt{2}}\left(y - \frac{t}{\sqrt{2}}\right) + \frac{1}{\sqrt{2}}\left(y - \frac{t}{\sqrt{2}}\right) - \frac{1}{\sqrt{2}}\left(x + \frac{t}{\sqrt{2}}\right)$$

$$= \frac{1}{\sqrt{2}}\left(x + \frac{t}{\sqrt{2}}\right) - \frac{1}{\sqrt{2}}\left(y - \frac{t}{\sqrt{2}}\right),$$

and

$$\frac{\partial f}{\partial \mathbf{U}}(x, y) = g'(0)$$

$$= \frac{x - y}{\sqrt{2}}.$$

At

$$\mathbf{X}_0 = (1, 2),$$

$$\frac{\partial f}{\partial \mathbf{U}}(1, 2) = -\frac{1}{\sqrt{2}}.$$

Geometrical Interpretation of Directional Derivatives

If g is a function of a single variable which has a derivative at $t = t_0$, then $g'(t_0)$ is the slope of the tangent to the curve defined by $y = g(t)$ at the point $(t_0, g(t_0))$ (Fig. 2.2); in fact, the equation of the tangent line is

$$y = g(t_0) + g'(t_0)(x - x_0),$$

or, in symmetric form,

(4)
$$x - x_0 = \frac{y - g(t_0)}{g'(t_0)}.$$

Partial derivatives can also be interpreted geometrically. Suppose S is a

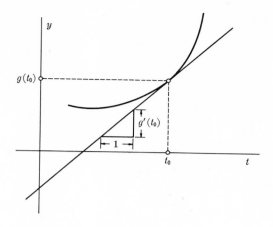

FIGURE 2.2

surface in R^3 defined by

$$z - f(x, y) = 0,$$

where f has domain D in R^2. Let (x_0, y_0) be an interior point of D, let

$$\mathbf{U} = \begin{bmatrix} u \\ v \end{bmatrix}$$

be a unit vector, and let Γ be the plane through $(x_0, y_0, 0)$ parallel to

$$\begin{bmatrix} u \\ v \\ 0 \end{bmatrix}$$

and the z-axis. Now suppose C is the trace of S on Γ (Fig. 2.3). A point (x, y, z) is on Γ if and only if $x = x_0 + tu$ and $y = y_0 + tv$; it is on S if and only if $z = f(x, y)$. Since C is the set of points which are on both Γ and S, it follows that (x, y, z) is on C if and only if

$$\begin{bmatrix} x \\ y \\ z \end{bmatrix} = \mathbf{\Phi}(t) = \begin{bmatrix} x_0 + tu \\ y_0 + tv \\ f(x_0 + tu, y_0 + tv) \end{bmatrix}.$$

This is a parametric equation for C. A tangent vector to C at

$P_0 = (x_0, y_0, z(x_0, y_0))$ is given by

$$\Phi'(0) = \begin{bmatrix} u \\ v \\ \dfrac{\partial f}{\partial \mathbf{U}}(x_0, y_0) \end{bmatrix};$$

consequently the tangent line T to C at P_0 (which is defined to be the line through P_0 in the direction of $\Phi'(0)$) is given by

$$\frac{x - x_0}{u} = \frac{y - y_0}{v} = \frac{z - f(x_0, y_0)}{\dfrac{\partial f}{\partial \mathbf{U}}(x_0, y_0)},$$

which is analogous to (4).

The tangent line is in Γ; if α is the angle between it and $\mathbf{U}$ (Fig. 2.3), then $\tan \alpha$ is the *slope of T in the plane* Γ, which is analogous to the familiar

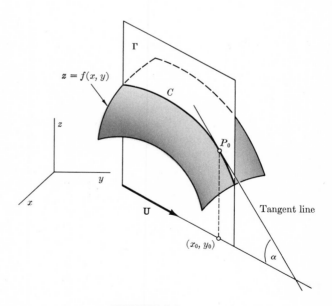

FIGURE 2.3

definition of slope for a line in R^2. Since

$$\cos \alpha = \frac{\mathbf{U} \cdot \mathbf{\Phi}'(0)}{\mid \mathbf{\Phi}'(0) \mid}$$

$$= \frac{1}{\sqrt{1 + \left(\dfrac{\partial f}{\partial \mathbf{U}}\,(x_0,\, y_0)\right)^2}}\,,$$

it follows that

$$\sin \alpha = \frac{\dfrac{\partial f}{\partial \mathbf{U}}\,(x_0,\, y_0)}{\sqrt{1 + \left(\dfrac{\partial f}{\partial \mathbf{U}}\,(x_0,\, y_0)\right)^2}}\,,$$

and the slope is given by

$$\tan \alpha = \frac{\partial f}{\partial \mathbf{U}}\,(x_0,\, y_0).$$

Partial Derivatives

Directional derivatives in the directions of the basis vectors

$$\mathbf{E}_1 = \begin{bmatrix} 1 \\ 0 \\ 0 \\ \vdots \\ 0 \end{bmatrix},\quad \mathbf{E}_2 = \begin{bmatrix} 0 \\ 1 \\ 0 \\ \vdots \\ 0 \end{bmatrix},\dots,\quad \mathbf{E}_{n-1} = \begin{bmatrix} 0 \\ \vdots \\ 0 \\ 1 \\ 0 \end{bmatrix},\quad \mathbf{E}_n = \begin{bmatrix} 0 \\ \vdots \\ 0 \\ 0 \\ 1 \end{bmatrix}$$

occur so often that they deserve special attention. If $\mathbf{U} = \mathbf{E}_i$ we can write (2), with $t = h_i$, as

$$(5) \qquad \frac{\partial f}{\partial \mathbf{E}_i}\,(\mathbf{X}) = \lim_{h_i \to 0} \frac{f(\mathbf{X} + h_i \mathbf{E}_i) - f(\mathbf{X})}{h_i}$$

$$= \lim_{h_i \to 0} \frac{f(x_1,\dots,x_{i-1},\,x_i + h_i,\,x_{i+1},\dots,x_n) - f(x_1,\dots,x_{i-1},\,x_i,\,x_{i+1},\dots,x_n)}{h_i}$$

Definition 2.2   The directional derivative $(\partial f/\partial \mathbf{E}_i)\,(\mathbf{X})$ is called *the partial derivative of f with respect to x_i*. It is also denoted by $f_{x_i}(\mathbf{X})$ or $(\partial f/\partial x_i)\,(\mathbf{X})$.

If we ignore the other variables, the last member of (5) resembles the difference quotient that arises in the definition of an ordinary derivative. This leads to the following rule for calculating $f_{x_i}(\mathbf{X})$:

Treat all variables other than x_i as constants and differentiate f as if it were a function of x_i only, according to the rules of the one dimensional calculus.

Example 2.4 Let

$$f(x, y) = x^2 + y^2 + xy.$$

Considering y as a constant and differentiating with respect to x yields

$$f_x(x, y) = 2x + y;$$

considering x as a constant and differentiating with respect to y yields

$$f_y(x, y) = 2y + x.$$

The partial derivative f_{x_i} defines a new function of $\mathbf{X}$. Since it is obtained by differentiating a function of the single variable x_i according to the rules of the one-dimensional calculus, we have the following theorem.

Theorem 2.1 If $f_{x_i}(\mathbf{X})$ and $g_{x_i}(\mathbf{X})$ exist, then

$$\frac{\partial(f + g)}{\partial x_i}(\mathbf{X}) = \frac{\partial f}{\partial x_i}(\mathbf{X}) + \frac{\partial g}{\partial x_i}(\mathbf{X}),$$

$$\frac{\partial(f - g)}{\partial x_i}(\mathbf{X}) = \frac{\partial f}{\partial x_i}(\mathbf{X}) - \frac{\partial g}{\partial x_i}(\mathbf{X}),$$

$$\frac{\partial(fg)}{\partial x_i}(\mathbf{X}) = g(\mathbf{X})\frac{\partial f}{\partial x_i} + f(\mathbf{X})\frac{\partial g}{\partial x_i}(\mathbf{X})$$

and, if $g(\mathbf{X}) \neq 0$,

$$\frac{\partial}{\partial x_i}\left(\frac{f}{g}\right)(\mathbf{X}) = \frac{g(\mathbf{X})\dfrac{\partial f}{\partial x_i}(\mathbf{X}) - f(\mathbf{X})\dfrac{\partial g}{\partial x_i}(\mathbf{X})}{[g(\mathbf{X})]^2}.$$

When $n = 1$ the partial derivative f_x and the ordinary derivative $f' = df/dx$ are the same; hence in this case we shall use one of the latter symbols rather than f_x or $\partial f/\partial x$.

Higher Order Partial Derivatives

The partial derivative f_{x_i} (or, for that matter, any directional derivative) defines a function whose domain is the set of points at which it exists. We can ask the same questions about it as we did about f: Is it continuous? Does it have partial derivatives? If the partial derivative of f_{x_i} with respect to x_j exists, we denote it by

$$\frac{\partial}{\partial x_j}\left(\frac{\partial f}{\partial x_i}\right) = \frac{\partial^2 f}{\partial x_j \, \partial x_i} = f_{x_i x_j};$$

similarly

$$\frac{\partial}{\partial x_k}\left(\frac{\partial^2 f}{\partial x_j \, \partial x_i}\right) = f_{x_i x_j x_k}.$$

In general the function obtained by differentiating f with respect to x_{i_1}, $x_{i_2}, \ldots, x_{i_r}$ (in that order) is denoted by

$$\frac{\partial^r f}{\partial x_{i_r} \cdots \partial x_{i_2} \, \partial x_{i_1}} = f_{x_{i_1} x_{i_2} \cdots x_{i_r}};$$

here the integer r is the *order* of the partial derivative.

Example 2.5   Let $n = 3$ and

$$f(\mathbf{X}) = x^3 + 3xy + yz + 4z^2.$$

There are three first-order partial derivatives,

$$f_x(\mathbf{X}) = 3x^2 + 3y,$$
$$f_y(\mathbf{X}) = 3x + z,$$
$$f_z(\mathbf{X}) = y + 8z,$$

and nine second-order:

$$f_{xx}(\mathbf{X}) = 6x, \qquad f_{xy}(\mathbf{X}) = 3, \qquad f_{xz}(\mathbf{X}) = 0,$$
$$f_{yx}(\mathbf{X}) = 3, \qquad f_{yy}(\mathbf{X}) = 0, \qquad f_{yz}(\mathbf{X}) = 1,$$
$$f_{zx}(\mathbf{X}) = 0, \qquad f_{zy}(\mathbf{X}) = 1, \qquad f_{zz}(\mathbf{X}) = 8.$$

In general there are n^r rth-order partial derivatives of a function of n variables; however, it can be shown that the value of $f_{x_{i_1} x_{i_2} \cdots x_{i_r}}$ at a point $\mathbf{X}_0$ depends only upon which variables occur in the subscript and not on their order, provided all partial derivatives of order up to and including r are continuous at $\mathbf{X}_0$. For instance, $f_{xy} = f_{yx}, f_{xz} = f_{zx}$, and $f_{yz} = f_{zy}$ in Example

2.5. Whenever we speak of rth-order partial derivatives, we shall assume that this condition is satisfied. Subject to this agreement we can define

$$\frac{\partial^s f}{\partial x_1^{s_1} \cdots \partial x_n^{s_n}}$$

where $s = s_1 + \cdots + s_n$, to be the function obtained by differentiating f s_1 times with respect to x_1, s_2 times with respect to $x_2, \ldots, s_n$ times with respect to x_n.

Example 2.6 If f is a function of two variables, we write

$$f_{xx} = \frac{\partial^2 f}{\partial x^2}, \qquad f_{yy} = \frac{\partial^2 f}{\partial y^2},$$

and

$$f_{xy} = f_{yx} = \frac{\partial^2 f}{\partial x \partial y}.$$

For third-order derivatives,

$$f_{xxy} = f_{xyx} = f_{yxx} = \frac{\partial^3 f}{\partial x^2 \partial y}, \quad \text{etc.}$$

EXERCISES 4.2   1, 3, 5

1. Compute $(\partial f / \partial \mathbf{U})(\mathbf{X}_0)$ for the following functions at $\mathbf{X}_0 = (1, 2)$ in the direction

$$\mathbf{U} = \begin{bmatrix} \dfrac{1}{\sqrt{2}} \\[2mm] \dfrac{1}{\sqrt{2}} \end{bmatrix}.$$

(a) $f(x, y) = x^2 + y^2$.   (b) $f(x, y) = x^2 - xy + 2y^2$.

(c) $f(x, y) = x$.   (d) $f(x, y) = \dfrac{x}{x + y}$.

2. Repeat Exercise 1 for $\mathbf{X}_0 = (1, 2, 3)$ and

$$\mathbf{U} = \begin{bmatrix} \dfrac{1}{\sqrt{3}} \\[2ex] -\dfrac{1}{\sqrt{3}} \\[2ex] \dfrac{1}{\sqrt{3}} \end{bmatrix}.$$

(a) $f(x, y, z) = \log(x^2 + y^2 + z^2)$.

(b) $f(x, y, z) = x + y + z$.

(c) $f(x, y, z) = e^{x+y+z}$.

(d) $f(x, y, z) = \dfrac{1}{x^2 + y^2 + z^2}$.

3. Find $\partial f/\partial x$, $\partial f/\partial y$, and $\partial f/\partial z$ for each of the following; also, calculate $f_x(\mathbf{X}_0), f_y(\mathbf{X}_0)$, and $f_z(\mathbf{X}_0)$.

(a) $f(\mathbf{X}) = 2x^2 + y^2 + z$, $\mathbf{X}_0 = (1, 1, 2)$.

(b) $f(\mathbf{X}) = e^x \cos(x^2 + y^2 + z^2)$, $\mathbf{X}_0 = \left(\sqrt{\dfrac{\pi}{4}}, \sqrt{\dfrac{\pi}{4}}, 0\right)$.

(c) $f(\mathbf{X}) = \log(1 + x^2 + y + z)$, $\mathbf{X}_0 = (0, 0, 0)$.

(d) $f(\mathbf{X}) = \exp(\log\sqrt{x^2 + y^2 + 2xy})$, $\mathbf{X}_0 = (1, 1)$.

4. Compute f_x, f_y, and f_z for each of the following.

(a) $f(x, y, z) = xe^{yz}$.   (b) $f(x, y, z) = \dfrac{e^{xyz}}{1 + z}$.

(c) $f(x, y, z) = \log(x + y)$.   (d) $f(x, y, z) = z + \cos x^2 y$.

5. Compute $f_{xy}(-1, 2)$ and $f_{yx}(-1, 2)$ for

(a) $f(x, y) = 3x^2 y - 2xy^2$;

(b) $f(x, y) = 2xy - y^2$.

6. Compute all second partial derivatives for the functions in Exercises 3(a) and 4(a).

7. Compute all first partial derivatives for each of the following.

   (a) $f(x, y, z) = xye^z + 2y^2z - 3xz \cos y$.

   (b) $f(x, y) = \dfrac{2x^2 + 3y^2 - 5x^3y}{2x + y^2}$.

   (c) $f(x) = 3x + 4$.

8. Let
$$f(x, y, z) = 2x^2yz + \cos(2x - 3y + 4z).$$

   Compute

   (a) $\dfrac{\partial^2 f}{\partial x\,\partial y}$,   (b) $\dfrac{\partial^3 f}{\partial x\,\partial y^2}$,   (c) $\dfrac{\partial^3 f}{\partial x\,\partial y\,\partial z}$,

   (d) $\dfrac{\partial^4 f}{\partial x^2\,\partial y^2}$,   (e) $\dfrac{\partial^4 f}{\partial x^2\,\partial y\,\partial z}$.

9. If $f(x, y) = e^x \cos y$, show that $f_{xx} + f_{yy} = 0$.

10. If $f(x, y) = \cos xy$, show that $xf_x(x, y) - yf_y(x, y) = 0$.

11. Where do the following functions fail to have first partial derivatives?

    (a) $f(x, y) = \sqrt{x + y}$.

    (b) $f(x, y) = \sqrt{1 - x^2 - y^2}$.

    (c) $f(x, y) = \log(1 + y \sin x)$.

THEORETICAL EXERCISES

T-1. In R^1 there are only two unit vectors: $\mathbf{U}_1 = [1]$ and $\mathbf{U}_2 = [-1]$. Show that
$$\frac{\partial f}{\partial \mathbf{U}_1}(x) = -\frac{\partial f}{\partial \mathbf{U}_2}(x) = f'(x).$$

T-2. If $\mathbf{V} = -\mathbf{U}$, show that
$$\frac{\partial f}{\partial \mathbf{V}}(\mathbf{X}) = -\frac{\partial f}{\partial \mathbf{U}}(\mathbf{X}).$$

T-3. If $\mathbf{X}_0$ is an interior point of the domain of f, we say that f is *continuous at $\mathbf{X}_0$ in the direction* $\mathbf{U}$ if $g(t) = f(\mathbf{X}_0 + t\mathbf{U})$ is continuous at $t = 0$.

Prove: (a) If f is continuous at $\mathbf{X}_0$, then it is continuous at $\mathbf{X}_0$ in every direction. (b) If $(\partial f/\partial \mathbf{U})(\mathbf{X}_0)$ exists, then f is continuous at $\mathbf{X}_0$ in the direction $\mathbf{U}$.

T-4.   Prove that

$$\frac{\partial(f+g)}{\partial \mathbf{U}}(\mathbf{X}) = \frac{\partial f}{\partial \mathbf{U}}(\mathbf{X}) + \frac{\partial g}{\partial \mathbf{U}}(\mathbf{X}).$$

T-5.   Let (x_0, y_0) be an interior point of the domain of f. Show that

$$g(y) = \frac{f(x, y_0) - f(x_0, y_0)}{x - x_0}$$

has a removable discontinuity (Exercise T-10, Section 4.1) at y_0 if and only if $f_x(x_0, y_0)$ exists.

4.3 Differentiable Functions

A function g of one variable is said to be differentiable at a point $t = t_0$ if it has a derivative there; if this is the case then

$$\lim_{t \to t_0} \frac{g(t) - g(t_0)}{t - t_0} = g'(t_0).$$

We can rewrite this as

$$\lim_{t \to t_0} \frac{g(t) - g(t_0) - g'(t_0)(t - t_0)}{t - t_0} = 0,$$

or better yet,

$$g(t) = g(t_0) + g'(t_0)(t - t_0) + \epsilon(t, t_0)(t - t_0),$$

where

$$\lim_{t \to t_0} \epsilon(t, t_0) = 0.$$

Thus, differentiability at t_0 implies that, near t_0, g can be approximated so well by

$$h(t) = g(t_0) + g'(t_0)(t - t_0)$$

that the error $g(t) - h(t)$ approaches zero faster than $t - t_0$ (Fig. 3.1). Since h is simple function of t, while g may be quite complicated, this is a desirable situation.

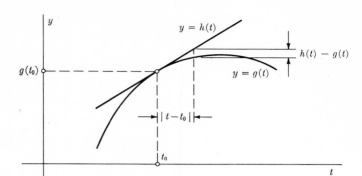

FIGURE 3.1

This interpretation of differentiability motivates the following definition for functions of n variables.

Definition 3.1 Let f be a real valued function defined on a domain D in R^n and let $\mathbf{X}_0$ be an interior point of D. We say that f is *differentiable at* $\mathbf{X}_0$ if there exist constants $a_1, a_2, \ldots, a_n$ such that

$$(1) \qquad \lim_{\mathbf{X} \to \mathbf{X}_0} \frac{f(\mathbf{X}) - f(\mathbf{X}_0) - L(\mathbf{X} - \mathbf{X}_0)}{|\mathbf{X} - \mathbf{X}_0|} = 0,$$

where L is the function from R^n to R whose value at $\mathbf{Y} = (y_1, \ldots, y_n)$ is

$$L(\mathbf{Y}) = a_1 y_1 + a_2 y_2 + \cdots + a_n y_n.$$

Thus, L is a linear transformation from R^n to R.

If f is differentiable at every point of a set S, we say that f is *differentiable on* S.

Example 3.1 Let $\mathbf{X}_0 = (x_0, y_0)$ and $f(x, y) = 2x^2 + y^2$. To show that f is differentiable at $\mathbf{X}_0$, we must find constants a_1 and a_2 such that

$$\lim_{\mathbf{X} \to \mathbf{X}_0} \frac{2x^2 + y^2 - 2x_0^2 - y_0^2 - a_1(x - x_0) - a_2(y - y_0)}{|\mathbf{X} - \mathbf{X}_0|} = 0.$$

Letting $\mathbf{X}$ approach $\mathbf{X}_0$ along the line $x = x_0$, we find that

$$0 = \lim_{y \to y_0} \frac{y^2 - y_0^2 - a_2(y - y_0)}{y - y_0}$$

$$= \lim_{y \to y_0} (y + y_0 - a_2);$$

thus $a_2 = 2y_0$. Similarly, approaching $\mathbf{X}_0$ along the line $y = y_0$ leads to the conclusion that $a_1 = 4x_0$. Thus if L satisfies the requirement of Definition 3.1, then

$$L(\mathbf{X} - \mathbf{X}_0) = 4x_0(x - x_0) + 2y_0(y - y_0).$$

We have not, however, shown that f is differentiable at $\mathbf{X}_0$; to do this we must prove that

$$E(\mathbf{X}) = \frac{2x^2 + y^2 - 2x_0^2 - y_0^2 - 4x_0(x - x_0) - 2y_0(y - y_0)}{|\mathbf{X} - \mathbf{X}_0|}$$

approaches zero as $\mathbf{X}$ approaches $\mathbf{X}_0$, not only along lines parallel to the coordinate axes, but along arbitrary paths. This is accomplished by re-writing,

$$E(\mathbf{X}) = \frac{2(x - x_0)^2 + (y - y_0)^2}{|\mathbf{X} - \mathbf{X}_0|}$$

$$\leq \frac{2|\mathbf{X} - \mathbf{X}_0|^2}{|\mathbf{X} - \mathbf{X}_0|} = 2|\mathbf{X} - \mathbf{X}_0|;$$

thus, the conclusion follows.

The way in which we obtained a_1 and a_2 in this example made it clear that they are unique. This is true in general, as is shown in the following theorem.

Theorem 3.1 If f is differentiable at $\mathbf{X}_0$, then the partial derivatives $f_{x_1}(\mathbf{X}_0), \ldots, f_{x_n}(\mathbf{X}_0)$ exist, and in Definition 3.1,

$$(2) \qquad a_i = f_{x_i}(\mathbf{X}_0) \qquad (i = 1, 2, \ldots, n);$$

thus if there is a linear transformation that satisfies (1), it is unique.

Proof. In (1) let $\mathbf{X}$ approach $\mathbf{X}_0 = (c_1, \ldots, c_n)$ along the line $x_j = c_j$ $(i \neq j)$; then (1) reduces to

$$\lim_{x_i \to c_i} \frac{f(c_1, \ldots, c_{i-1}, x_i, c_{i+1}, \ldots, c_n) - f(c_1, \ldots, c_{i-1}, c_i, c_{i+1}, \ldots, c_n) - a_i(x_i - c_i)}{x_i - c_i}$$

$$= 0$$

so that (2) follows from the definition of $f_{x_i}(\mathbf{X}_0)$.

Thus, in Example 3.1, $a_1 = 4x_0 = f_x(\mathbf{X}_0)$, and $a_2 = 2y_0 = f_y(\mathbf{X}_0)$.

The Differential

Definition 3.2   If f is differentiable at $\mathbf{X}_0$, then $d_{\mathbf{X}_0} f$, the *differential of f at $\mathbf{X}_0$*, is the linear transformation from R^n to R whose value at $\mathbf{Y} = (y_1, \ldots, y_n)$ is

(3) $(d_{\mathbf{X}_0} f)(\mathbf{Y}) = f_{x_1}(\mathbf{X}_0) y_1 + f_{x_2}(\mathbf{X}_0) y_2 + \cdots + f_{x_n}(\mathbf{X}_0) y_n.$

When using the differential, we shall almost always write $\mathbf{Y} = \mathbf{X} - \mathbf{X}_0$ in (3); thus if $\mathbf{X}_0 = (c_1, \ldots, c_n)$,

$$(d_{\mathbf{X}_0} f)(\mathbf{X} - \mathbf{X}_0) = f_{x_1}(\mathbf{X}_0)(x_1 - c_1) + \cdots + f_{x_n}(\mathbf{X}_0)(x_n - c_n).$$

If $n = 2$, we shall usually write

$$(d_{\mathbf{X}_0} f)(\mathbf{X} - \mathbf{X}_0) = f_x(\mathbf{X}_0)(x - x_0) + f_y(\mathbf{X}_0)(y - y_0),$$

and if $n = 3$,

$$(d_{\mathbf{X}_0} f)(\mathbf{X} - \mathbf{X}_0) = f_x(\mathbf{X}_0)(x - x_0) + f_y(\mathbf{X}_0)(y - y_0) + f_z(\mathbf{X}_0)(z - z_0).$$

Example 3.2   Let $\mathbf{X}_0 = (1, 2)$ and

$$f(x, y) = 2x^2 + y^2 + xy;$$

then

$$f_x(x, y) = 4x + y,$$
$$f_y(x, y) = 2y + x,$$

and

$$(d_{\mathbf{X}_0} f)(\mathbf{X} - \mathbf{X}_0) = 6(x - 1) + 5(y - 2).$$

Also, if $\mathbf{X}_1 = (3, 5)$, then

$$(d_{\mathbf{X}_0} f)(\mathbf{X}_1 - \mathbf{X}_0) = 6(3 - 1) + 5(5 - 2) = 27.$$

Example 3.3   Let $\mathbf{X}_0 = \left(\dfrac{\pi}{8}, \dfrac{\pi}{8}, 1\right)$ and $f(x, y, z) = z \cos(x + y)$; then

$$f_x(x, y, z) = f_y(x, y, z) = -z \sin(x + y),$$
$$f_z(x, y, z) = \cos(x + y),$$

and

$$(d_{\mathbf{X}_0} f)(\mathbf{X} - \mathbf{X}_0) = -\frac{1}{\sqrt{2}}\left(x - \frac{\pi}{8}\right) - \frac{1}{\sqrt{2}}\left(y - \frac{\pi}{8}\right) + \frac{1}{\sqrt{2}}(z - 1).$$

From (1) it follows that, if f is differentiable at $\mathbf{X}_0$, the difference $f(\mathbf{X}) - f(\mathbf{X}_0)$ can be approximated so well near $\mathbf{X}_0$ by $(d_{\mathbf{X}_0} f)(\mathbf{X} - \mathbf{X}_0)$

that the error

$$f(\mathbf{X}) - f(\mathbf{X}_0) - (d_{\mathbf{X}_0}f)(\mathbf{X} - \mathbf{X}_0)$$

approaches zero faster than $|\mathbf{X} - \mathbf{X}_0|$. When we wish to use this fact in applications, we shall write simply

$$f(\mathbf{X}) - f(\mathbf{X}_0) \cong (d_{\mathbf{X}_0}f)(\mathbf{X} - \mathbf{X}_0);$$

the precise meaning of this relationship will be as given by (1); that is,

$$\lim_{\mathbf{X} \to \mathbf{X}_0} \frac{f(\mathbf{X}) - f(\mathbf{X}_0) - (d_{\mathbf{X}_0}f)(\mathbf{X} - \mathbf{X}_0)}{|\mathbf{X} - \mathbf{X}_0|} = 0.$$

Example 3.4 Suppose a box is designed to have length $l_0 = 10$ in., width $w_0 = 8$ in. and height $h_0 = 4$ in., and that the manufacturing process is such that the lengths are in error by at most one percent. To estimate an upper bound for the error in the volume V of the box we write $V = lwh$, and let $\mathbf{X} = (l, w, h)$ and $\mathbf{X}_0 = (l_0, w_0, h_0)$. Then

$$(d_{\mathbf{X}_0}V)(\mathbf{X} - \mathbf{X}_0) = \frac{\partial V}{\partial l}(l_0, w_0, h_0)(l - l_0) + \frac{\partial V}{\partial w}(l_0, w_0, h_0)(w - w_0)$$

$$+ \frac{\partial V}{\partial h}(l_0, w_0, h_0)(h - h_0)$$

$$= w_0 h_0 l_0 \left(\frac{l - l_0}{l_0} + \frac{w - w_0}{w_0} + \frac{h - h_0}{h_0} \right).$$

Since $l_0 w_0 h_0 = (10 \text{ in.})(8 \text{ in.})(4 \text{ in.}) = 320 \text{ in.}^3$ and

$$\left| \frac{l - l_0}{l_0} \right| \le .01, \qquad \left| \frac{w - w_0}{w_0} \right| \le .01, \qquad \text{and} \qquad \left| \frac{h - h_0}{h_0} \right| \le .01,$$

it follows that

$$|(d_{\mathbf{X}_0}V)(\mathbf{X} - \mathbf{X}_0)| \le (320 \text{ in.}^3)(0.01 + 0.01 + 0.01) = 9.6 \text{ in.}^3.$$

Thus the error in the volume, which is given by

$$\Delta V = V(\mathbf{X}) - V(\mathbf{X}_0),$$

satisfies

$$|\Delta V| \le 9.6 \text{ in.}^3 + |E|,$$

where E is the error committed in approximating ΔV by the differential. However, $|E|$ is small compared to the terms in the differential; hence we take 9.6 in.³ as a reasonable bound for the error.

For functions of one variable, differentiability and existence of a differential are equivalent, as we pointed out in the beginning of this section; however, if $n \geq 2$, the function may fail to be differentiable at a point even though its partial derivatives exist there.

Example 3.5   Let $\mathbf{X}_0 = (0, 0)$ and $f(x, y) = \sqrt{|xy|}$. Since $f(0, y) = f(x, 0) = 0$, it follows that $f_x(0, 0) = f_y(0, 0) = 0$, and $d_{\mathbf{X}_0} f = 0$. Hence,

$$\frac{f(x, y) - f(0, 0) - (d_{\mathbf{X}_0} f)(\mathbf{X})}{|\mathbf{X}|} = \sqrt{\frac{|xy|}{x^2 + y^2}},$$

which does not approach a limit as $\mathbf{X}$ approaches $(0, 0)$ (see Example 1.14.); thus f is not differentiable at $(0, 0)$.

Example 3.6   Let $\mathbf{X}_0 = (c_1, \ldots, c_n)$, $\mathbf{Y} = (y_1, \ldots, y_n)$, and $f_i(\mathbf{X}) = x_i$; then since

$$(d_{\mathbf{X}_0} f_i)(\mathbf{Y}) = \frac{\partial f_i}{\partial x_1}(\mathbf{X}_0) y_1 + \frac{\partial f_i}{\partial x_2}(\mathbf{X}_0) y_2 + \cdots + \frac{\partial f_i}{\partial x_n}(\mathbf{X}_0) y_n$$

and

$$\frac{\partial f_i}{\partial x_k}(\mathbf{X}_0) = \begin{cases} 0 & \text{if } k \neq i, \\ 1 & \text{if } k = i, \end{cases}$$

it follows that

$$(d_{\mathbf{X}_0} f_i)(\mathbf{Y}) = y_i$$

and

$$\frac{f_i(\mathbf{X}) - f_i(\mathbf{X}_0) - (d_{\mathbf{X}_0} f_i)(\mathbf{X} - \mathbf{X}_0)}{|\mathbf{X} - \mathbf{X}_0|} = \frac{x_i - c_i - (x_i - c_i)}{|\mathbf{X} - \mathbf{X}_0|}$$

$$= 0.$$

Thus f_i is differentiable at $\mathbf{X}_0$. Since the argument holds for all $\mathbf{X}_0$, f_i is differentiable on all of R^n.

The differential of f_i is so important that we introduce a special, simpler notation for it.

Definition 3.3   The linear transformation from R^n to R that has the value y_i at $\mathbf{Y} = (y_1, \ldots, y_n)$ is called the *differential of* x_i and is denoted by dx_i. Thus

$$dx_i(\mathbf{Y}) = y_i.$$

The linear transformation dx_i is really the differential of f_i as defined in Example 3.6, and one could correctly argue that the proper notation for it would be $d_{\mathbf{X}_0} f_i$. We adopt the special notation because it is convenient and traditional. The student should observe that although the symbol dx_i does not identify the dimension of the space on which it is defined, this will always be clear from the context.

If $n = 2$ we shall usually write dx and dy; thus

$$dx(x - x_0, y - y_0) = x - x_0,$$
$$dy(x - x_0, y - y_0) = y - y_0;$$

hence the differential in Example 3.2 is

$$d_{\mathbf{X}_0} f = 6\, dx + 4\, dy.$$

If $n = 3$,

$$dx(x - x_0, y - y_0, z - z_0) = x - x_0,$$
$$dy(x - x_0, y - y_0, z - z_0) = y - y_0,$$
$$dz(x - x_0, y - y_0, z - z_0) = z - z_0;$$

hence the differential in Example 3.3 is

$$d_{\mathbf{X}_0} f = -\frac{1}{\sqrt{2}}\,(dx + dy - dz).$$

From (3), the differential of an arbitrary function can be written as

$$d_{\mathbf{X}_0} f = f_{x_1}(\mathbf{X}_0)\, dx_1 + \cdots + f_{x_n}(\mathbf{X}_0)\, dx_n.$$

When we do not need to emphasize the particular point $\mathbf{X}_0$, we shall write this more simply as

(4) $$df = f_{x_1}\, dx_1 + \cdots + f_{x_n}\, dx_n,$$

which is a convenient symbolic shorthand for what we really mean:

$$d_{\mathbf{X}} f = f_{x_1}(\mathbf{X})\, dx_1 + \cdots + f_{x_n}(\mathbf{X})\, dx_n.$$

When we are dealing with the differential of a specific function, say

$$f(x, y) = x^2 + 3y^3,$$

at a general point, we shall write

$$df = 2x\, dx + 9y^2\, dy;$$

this is not entirely consistent with (4), but the inconsistency is harmless.

Example 3.7 Let $\mathbf{X}_0 = (c_1, \ldots, c_n)$ and $f(\mathbf{X}) = \alpha + \beta_1 x_1 + \cdots + \beta_n x_n$, where $\alpha, \beta_1, \ldots, \beta_n$ are constants. Let $L = \beta_1\, dx_1 + \cdots + \beta_n\, dx_n$; then L is a linear transformation from R^n to R and, if $\mathbf{X}_0$ is an arbitrary point in R^n,

$$f(\mathbf{X}) - f(\mathbf{X}_0) - L(\mathbf{X} - \mathbf{X}_0) = 0.$$

Thus f is differentiable on R^n, and

$$df = \beta_1 \, dx_1 + \cdots + \beta_n \, dx_n.$$

In particular, if $\beta_1 = \cdots = \beta_n = 0$, then $df = 0$; that is, the differential of a constant function is zero.

Example 3.8 If f is a function of a single variable t, then

$$df = f' \, dt,$$

where dt is the linear transformation on R defined by $dt(y) = y$. The function df/dt is a constant with respect to y:

$$\frac{df}{dt} = f'.$$

Theorem 3.2 If f is differentiable at $\mathbf{X}_0 = (c_1, \ldots, c_n)$, then it is continuous at $\mathbf{X}_0$.

Proof. Write

$$f(\mathbf{X}) - f(\mathbf{X}_0) = (f(\mathbf{X}) - f(\mathbf{X}_0) - (d_{\mathbf{X}_0} f)(\mathbf{X} - \mathbf{X}_0)) + (d_{\mathbf{X}_0} f)(\mathbf{X} - \mathbf{X}_0).$$

The first term approaches zero as $\mathbf{X}$ approaches $\mathbf{X}_0$, by hypothesis; the second does also, since, by Schwarz's inequality,

$$| (d_{\mathbf{X}_0} f)(\mathbf{X} - \mathbf{X}_0) | = | f_{x_1}(\mathbf{X}_0)(x_1 - c_1) + \cdots + f_{x_n}(\mathbf{X}_0)(x_n - c_n) |$$

$$\leq [f_{x_1}^2(\mathbf{X}_0) + \cdots + f_{x_n}^2(\mathbf{X}_0)]^{1/2}$$

$$\times [(x_1 - c_1)^2 + \cdots + (x_n - c_n)^2]^{1/2},$$

or

$$| (d_{\mathbf{X}_0} f)(\mathbf{X} - \mathbf{X}_0) | \leq [f_{x_1}^2(\mathbf{X}_0) + \cdots + f_{x_n}^2(\mathbf{X}_0)]^{1/2} | \mathbf{X} - \mathbf{X}_0 |.$$

We saw in Theorem 3.1 that if f is differentiable at $\mathbf{X}_0$, then $f_{x_1}(\mathbf{X}_0), \ldots, f_{x_n}(\mathbf{X}_0)$ exist. We also saw in Example 3.5 that f may fail to be differentiable at $\mathbf{X}_0$ even though the partial derivatives exist there. The next theorem gives sufficient (although not necessary) conditions for differentiability.

Theorem 3.3 Let f be a real valued function defined on a domain D of R^n, and let $\mathbf{X}_0$ be an interior point of D. If the partial derivatives f_{x_i} $(i = 1, 2, \ldots, n)$ exist in some n-ball about $\mathbf{X}_0$ and are continuous at $\mathbf{X}_0$, then f is differentiable at $\mathbf{X}_0$.

Before proving this theorem we recall, without proof, a useful result from the calculus of one variable.

Lemma 3.1 (Mean Value Theorem for Functions of One Variable) Let g be continuous in the closed interval $[a, b]$ and have a derivative at each point of the open interval (a, b). Then there is a point c, $a < c < b$, such that

$$f(b) - f(a) = f'(c)(b - a).$$

Proof of Theorem 3.3. We confine our attention to $n = 3$; the proof is easily extended to any n.

Assume that $|\mathbf{X} - \mathbf{X}_0| < \rho$ and that the ball B_ρ of radius ρ about $\mathbf{X}_0$ is contained in D. Let

$$\mathbf{X} = (x, y, z), \qquad \mathbf{X}_0 = (x_0, y_0, z_0)$$

and define

$$\mathbf{X}_1 = (x_0, y, z), \qquad \mathbf{X}_2 = (x_0, y_0, z);$$

then $\mathbf{X}_1$ and $\mathbf{X}_2$ are in B_ρ and, since B_ρ is convex (Exercise T-12, Section 2.1), so are the line segments L_1 from $\mathbf{X}$ to $\mathbf{X}_1$, L_2 from $\mathbf{X}_1$ to $\mathbf{X}_2$, and L_3 from $\mathbf{X}_2$ to $\mathbf{X}_0$ (Fig. 3.2).

Now

$$f(\mathbf{X}) - f(\mathbf{X}_0) = (f(\mathbf{X}) - f(\mathbf{X}_1)) + (f(\mathbf{X}_1) - f(\mathbf{X}_2))$$
$$+ (f(\mathbf{X}_2) - f(\mathbf{X}_0)).$$

The arguments in each of the three parentheses differ only in their first, second and third coordinates, respectively. Applying Lemma 3.1 to $f(\mathbf{X}) - f(\mathbf{X}_1)$, $f(\mathbf{X}_1) - f(\mathbf{X}_2)$, and $f(\mathbf{X}_2) - f(\mathbf{X}_0)$ as functions of single variables yields

$$(5) \quad f(\mathbf{X}) - f(\mathbf{X}_0) = f_x(\hat{\mathbf{X}}_1)(x - x_0) + f_y(\hat{\mathbf{X}}_2)(y - y_0) + f_z(\hat{\mathbf{X}}_3)(z - z_0),$$

where

$$\hat{\mathbf{X}}_1 = (\hat{x}, y, z), \qquad \hat{\mathbf{X}}_2 = (x_0, \hat{y}, z), \qquad \hat{\mathbf{X}}_3 = (x_0, y_0, \hat{z});$$

$\hat{x}$ is between x and x_0, $\hat{y}$ is between y and y_0, and $\hat{z}$ is between z and z_0. Thus $\hat{\mathbf{X}}_1$ is on L_1, $\hat{\mathbf{X}}_2$ is on L_2, and $\hat{\mathbf{X}}_3$ is on L_3.

Now subtract $(d_{\mathbf{X}_0} f)(\mathbf{X} - \mathbf{X}_0)$ from both sides of (5):

$$(6) \quad f(\mathbf{X}) - f(\mathbf{X}_0) - (d_{\mathbf{X}_0} f)(\mathbf{X} - \mathbf{X}_0)$$
$$= (f_x(\hat{\mathbf{X}}_1) - f_x(\mathbf{X}_0))(x - x_0) + (f_y(\hat{\mathbf{X}}_2) - f_y(\mathbf{X}_0))(y - y_0)$$
$$+ (f_z(\hat{\mathbf{X}}_3) - f_z(\mathbf{X}_0))(z - z_0).$$

Given an arbitrary $\epsilon > 0$, there is a $\delta > 0$ such that $|f_x(\mathbf{X}) - f_x(\mathbf{X}_0)| < \epsilon/3$, $|f_y(\mathbf{X}) - f_y(\mathbf{X}_0)| < \epsilon/3$, and $|f_z(\mathbf{X}) - f_z(\mathbf{X}_0)| < \epsilon/3$ whenever $|\mathbf{X} - \mathbf{X}_0| < \delta$, because the partial derivatives are continuous at $\mathbf{X}_0$. Since

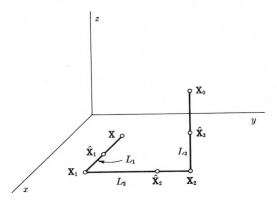

FIGURE 3.2

$|\hat{\mathbf{X}}_i - \mathbf{X}_0| < |\mathbf{X} - \mathbf{X}_0|$ $(i = 1,\ 2,\ 3)$, it follows from (6) that, if $|\mathbf{X} - \mathbf{X}_0| < \delta$, then

$$\frac{|f(\mathbf{X}) - f(\mathbf{X}_0) - (d_{\mathbf{X}_0}f)(\mathbf{X} - \mathbf{X}_0)|}{|\mathbf{X} - \mathbf{X}_0|}$$

$$< \frac{\epsilon}{3}\left(\frac{|x - x_0| + |y - y_0| + |z - z_0|}{|\mathbf{X} - \mathbf{X}_0|}\right) \le \epsilon,$$

since $|x - x_0|$, $|y - y_0|$, and $|z - z_0|$ are each $\le |\mathbf{X} - \mathbf{X}_0|$; thus f is differentiable at $\mathbf{X}_0$.

Theorem 3.4   If f and g are differentiable at $\mathbf{X}_0$ and α is a constant, then αf, $f + g$, and fg are differentiable at $\mathbf{X}_0$; the same is true of f/g provided $g(\mathbf{X}_0) \ne 0$. The differentials are given by

$$d_{\mathbf{X}_0}(\alpha f) = \alpha\, d_{\mathbf{X}_0}f,$$

$$d_{\mathbf{X}_0}(f + g) = d_{\mathbf{X}_0}f + d_{\mathbf{X}_0}g,$$

$$d_{\mathbf{X}_0}(fg) = f(\mathbf{X}_0)\, d_{\mathbf{X}_0}g + g(\mathbf{X}_0)\, d_{\mathbf{X}_0}f,$$

$$d_{\mathbf{X}_0}(f/g) = \frac{g(\mathbf{X}_0)\, d_{\mathbf{X}_0}f - f(\mathbf{X}_0)\, d_{\mathbf{X}_0}g}{[g(\mathbf{X}_0)]^2}.$$

Proof.   Because it is much easier, we give the proof under the additional assumption that f and g have continuous first partial derivatives near $\mathbf{X}_0$. Then αf, $f + g$, fg, and f/g (assuming $g(\mathbf{X}_0) \ne 0$) have continuous partial derivatives near $\mathbf{X}_0$, because of Theorem 2.1, and Theorem 1.3, (applied to the partial derivatives of f and g). From Theorem 3.3 it now follows that

αf, $f + g$, fg, and f/g are differentiable at $\mathbf{X}_0$ and, for example,

$$
\begin{aligned}
d_{\mathbf{X}_0}(fg) &= (fg)_{x_1}(\mathbf{X}_0)\ dx_1 + \cdots + (fg)_{x_n}(\mathbf{X}_0)\ dx_n \\
&= (f_{x_1}(\mathbf{X}_0)g(\mathbf{X}_0) + f(\mathbf{X}_0)g_{x_1}(\mathbf{X}_0))\ dx_1 + \cdots \\
&\quad + (f_{x_n}(\mathbf{X}_0)g(\mathbf{X}_0) + f(\mathbf{X}_0)g_{x_n}(\mathbf{X}_0))\ dx_n \\
&= g(\mathbf{X}_0)\,(f_{x_1}(\mathbf{X}_0)\ dx_1 + \cdots + f_{x_n}(\mathbf{X}_0)\ dx_n) \\
&\quad + f(\mathbf{X}_0)\,(g_{x_1}(\mathbf{X}_0)\ dx_1 + \cdots + g_{x_n}(\mathbf{X}_0)\ dx_n) \\
&= g(\mathbf{X}_0)\ d_{\mathbf{X}_0}f + f(\mathbf{X}_0)\ d_{\mathbf{X}_0}g.
\end{aligned}
$$

The other assertions of the theorem can be obtained by similar manipulations.

When there is no need to emphasize $\mathbf{X}_0$ the equations of Theorem 3.4 can be written more simply as

$$d(\alpha f) = \alpha\ df,$$

$$d(f + g) = df + dg,$$

$$d(fg) = f\ dg + g\ df,$$

$$d(f/g) = \frac{g\ df - f\ dg}{g^2}.$$

Directional Derivatives from the Differential

If f is differentiable at $\mathbf{X}_0$, $\mathbf{U}$ is a unit vector, and t a real parameter, then setting $\mathbf{X} = \mathbf{X}_0 + t\mathbf{U}$ in (1) yields

$$\lim_{t \to 0} \frac{f(\mathbf{X}_0 + t\mathbf{U}) - f(\mathbf{X}_0) - (d\mathbf{X}_0 f)(t\mathbf{U})}{t} = 0.$$

Since $d_{\mathbf{X}_0}f$ is a linear transformation, $(d_{\mathbf{X}_0}f)(t\mathbf{U}) = t(d_{\mathbf{X}_0}f)(\mathbf{U})$; substituting this in the last equation yields

$$\lim_{t \to 0} \frac{f(\mathbf{X}_0 + t\mathbf{U}) - f(\mathbf{X}_0)}{t} = (d_{\mathbf{X}_0}f)(\mathbf{U}),$$

which proves the following theorem.

Theorem 3.5   If f is differentiable at a point $\mathbf{X}_0$ and

$$
\mathbf{U} = \begin{bmatrix} u_1 \\ u_2 \\ \vdots \\ u_n \end{bmatrix}
$$

is an arbitrary unit vector, then the directional derivative of f at $\mathbf{X}_0$ in the direction of $\mathbf{U}$ exists, and is given by

$$\frac{\partial f}{\partial \mathbf{U}}(\mathbf{X}_0) = (d_{\mathbf{X}_0} f)(\mathbf{U})$$

$$= f_{x_1}(\mathbf{X}_0) u_1 + f_{x_2}(\mathbf{X}_0) u_2 + \cdots + f_{x_n}(\mathbf{X}_0) u_n.$$

Example 3.9 Let $\mathbf{X}_0 = (1, 2)$,

$$\mathbf{U} = \begin{bmatrix} \dfrac{1}{\sqrt{2}} \\[2ex] -\dfrac{1}{\sqrt{2}} \end{bmatrix},$$

and

$$f(x, y) = x^2 + y^2 + xy.$$

The partial derivatives,

$$f_x(x, y) = 2x + y, \qquad f_y(x, y) = 2y + x,$$

are continuous everywhere. In particular f is differentiable at $\mathbf{X}_0$ and

$$d_{\mathbf{X}_0} f = 4\, dx + 5\, dy;$$

thus

$$\frac{\partial f}{\partial \mathbf{U}}(\mathbf{X}_0) = 4(dx)\left(\frac{1}{\sqrt{2}}, -\frac{1}{\sqrt{2}}\right) + 5(dy)\left(\frac{1}{\sqrt{2}}, -\frac{1}{\sqrt{2}}\right)$$

$$= 4\left(\frac{1}{\sqrt{2}}\right) + 5\left(\frac{-1}{\sqrt{2}}\right) = -\frac{1}{\sqrt{2}},$$

which agrees with the result obtained in Example 2.3.

EXERCISES 4.3   1, 3 a.b., 5

1. Using the method of Example 3.1, find constants a and b such that

$$\lim_{(x,y)\to(x_0,y_0)} \frac{f(x, y) - f(x_0, y_0) - a(x - x_0) - b(y - y_0)}{\sqrt{(x - x_0)^2 + (y - y_0)^2}} = 0,$$

where

(a) $f(x, y) = x^2 + y^2 + 2xy + x + 1$, $(x_0, y_0) = (0, 0)$.

(b) $f(x, y) = 3x + 4y - 6$, $(x_0, y_0) = (2, 5)$.

2. Find $d_{X_0} f$ and $(d_{X_0} f)(X_1 - X_0)$:

    (a) $f(X) = x^2 + 3xy + y^2 + 2z$, $X_0 = (1, 0, 2)$, $X_1 = (2, 1, 3)$.

    (b) $f(X) = \log(x + y + z)$, $X_0 = (1, 2, -2)$, $X_1 = (3, 1, 1)$.

    (c) $f(X) = \sqrt{x^2 + y^2 + z^2}$, $X_0 = \left(\dfrac{1}{\sqrt{2}}, \dfrac{1}{2}, -\dfrac{1}{2}\right)$, $X_1 = \left(\dfrac{1}{\sqrt{2}}, 1, -1\right)$

    (d) $f(X) = e^x \cos(y + z)$, $X_0 = \left(0, \dfrac{\pi}{2}, 0\right)$, $X_1 = \left(0, \dfrac{\pi}{2}, \dfrac{\pi}{2}\right)$.

3. Find $d_{X_0} f$ and $(d_{X_0} f)(X_1 - X_0)$: ~~Theo. 349 (3.5)~~

    (a) $f(X) = 3x^3 + x^2 y + x + y$, $X_0 = (-1, 0, 1)$, $X_1 = (3, 2, 1)$.

    (b) $f(X) = \log(x^2 + y^2 + z^2)$, $X_0 = \left(\dfrac{1}{\sqrt{3}}, -\dfrac{1}{\sqrt{3}}, \dfrac{1}{\sqrt{3}}\right)$,

$$X_1 = \left(\frac{2}{\sqrt{3}}, \frac{1}{\sqrt{3}}, -\frac{1}{\sqrt{3}}\right).$$

    (c) $f(X) = e^{\cos xyz}$, $X_0 = \left(1, 1, \dfrac{\pi}{2}\right)$, $X_1 = \left(2, 2, \dfrac{3\pi}{2}\right)$.

    (d) $f(X) = \dfrac{x^2 + y^2}{xy}$, $X_0 = (1, 1, 3)$, $X_1 = (2, 3, 17)$.

4. Verify (4) directly for f and X_0 as given in Exercise 1(a), Exercise 2(a), and Exercise 3(a).

5. The area of a triangle with sides of length a and b and included angle ψ is $A = \frac{1}{2}ab \sin \psi$. Let $a = 5$, $b = 10$, and $\psi = 30$ degrees, to within errors of at most 1 percent in the lengths and 1 degree in the angle. Find an upper bound for the error in A.

6. A satellite is above a point P on a plane Γ. A radar at a point Q in Γ, at a distance $d = 300$ miles from P, measures the line of sight range from Q to the satellite and finds it to be $r = 500$ miles. From this information the height of the satellite above P is calculated. If the error in d is at most 2 miles and the error in the range is at most $\frac{1}{2}$ percent, what is the maximum error in h?

P 352 ±7≤,(14)← P. 337 Se. 4.2
P365   #9
P 3&3 #1(b)
P 38 5 #9,13

7. Find:

(a) $dx(0, 1, 2)$,   (b) $dy(0, 1, 2)$,   (c) $dz(0, 1, 2)$,

(d) $dx_i(2, 5, 6, 7, 9)$ $(1 \leq i \leq 5)$.

8. Express $d_{\mathbf{X}_0} f$, as calculated in Exercise 2, in terms of dx, dy, and dz.

9. Express $d_{\mathbf{X}_0} f$, as calculated in Exercise 3, in terms of dx, dy, and dz.

10. Write the general expression for df for each function in Exercise 3.

11. Write the general expression for df for each function in Exercise 3.

12. Let $\mathbf{X}_0 = (-1, \ 0, \ 1)$, $f(\mathbf{X}) = x^2 + 2x^2 yz + x + 3y + 1$, and $g(\mathbf{X}) = x^2 + 3y + z$. Find $d_{\mathbf{X}_0}(2f), d_{\mathbf{X}_0}(f + g), d_{\mathbf{X}_0}(fg)$, and $d_{\mathbf{X}_0}(f/g)$.

13. Repeat Exercise 12 for $\mathbf{X}_0 = (\pi/8, \pi/16, \pi/16), f(\mathbf{X}) = \sin(x + y + z)$, and $g(\mathbf{X}) = \cos(x + y + z)$.

14. Calculate the directional derivatives called for in Exercise 1, Section 2.2, by the method of Theorem 3.5.

15. Calculate the directional derivatives called for in Exercise 2, Section 2.2, by the method of Theorem 3.5.

THEORETICAL EXERCISES

T-1. Give a direct proof, without assuming continuity of the first partial derivatives, that $f + g$ is differentiable at $\mathbf{X}_0$ if f and g are.

T-2. If there are constants $a_1, \ldots, a_n$, not all zero, such that $a_1 f_1(\mathbf{X}) + \cdots + a_n f_n(\mathbf{X}) = 0$ for all $\mathbf{X}$ in a set S in R^n, then we say that $f_1, \ldots, f_n$ are *linearly dependent on* S. Prove: If $f_1, \ldots, f_n$ are linearly dependent on a set S and differentiable at a point $\mathbf{X}_0$ in S, then $d_{\mathbf{X}_0} f_1, \ldots, d_{\mathbf{X}_0} f_n$ are linearly dependent on R^n.

T-3. Prove Theorem 3.3 for $n = 2$.

4.4   The Mean-Value Theorem

We shall now generalize the mean-value theorem (Lemma 3.1) to functions of several variables.

Theorem 4.1 Let f be continuous at distinct points $\mathbf{X}_1$ and $\mathbf{X}_2$, and differentiable on the line segment L joining them. Then there is a point

$\mathbf{X}_0$ on L such that

$$f(\mathbf{X}_2) - f(\mathbf{X}_1) = (d_{\mathbf{X}_0} f)(\mathbf{X}_2 - \mathbf{X}_1).$$

Proof.   Define the unit vector

$$\mathbf{U} = \frac{\mathbf{X}_2 - \mathbf{X}_1}{|\mathbf{X}_2 - \mathbf{X}_1|}$$

and

$$g(t) = f(\mathbf{X}_1 + t\mathbf{U}) \qquad (0 \leq t \leq |\mathbf{X}_2 - \mathbf{X}_1|).$$

Since f is continuous on L (including the endpoints), g is continuous on the closed interval $[0, |\mathbf{X}_2 - \mathbf{X}_1|]$. We now show that g has a derivative in the open interval $(0, |\mathbf{X}_2 - \mathbf{X}_1|)$.

Let t_0 be a fixed but arbitrary point in the open interval; then

$$\frac{g(t) - g(t_0)}{t - t_0} = \frac{f(\mathbf{X}_1 + t\mathbf{U}) - f(\mathbf{X}_1 + t_0\mathbf{U})}{t - t_0}.$$

Setting $t - t_0 = \tau$ and $\mathbf{X}_1 + t_0\mathbf{U} = \hat{\mathbf{X}}$, we rewrite this as

$$\frac{g(t) - g(t_0)}{t - t_0} = \frac{f(\hat{\mathbf{X}} + \tau\mathbf{U}) - f(\hat{\mathbf{X}})}{\tau}.$$

Since f is differentiable at $\hat{\mathbf{X}}$, the right side approaches $(\partial f / \partial \mathbf{U})(\hat{\mathbf{X}})$ as τ approaches zero; thus $g'(t_0)$ exists and

$$g'(t_0) = \lim_{t \to t_0} \frac{g(t) - g(t_0)}{t - t_0}$$

(1)
$$= \lim_{t \to t_0} \frac{f(\hat{\mathbf{X}} + \tau\mathbf{U}) - f(\hat{\mathbf{X}})}{\tau}$$

$$= \frac{\partial f}{\partial \mathbf{U}}(\hat{\mathbf{X}}) = (d_{\hat{\mathbf{x}}} f)(\mathbf{U}),$$

where the last equality follows from Theorem 3.5.

Since g has a derivative at all points of $(0, |\mathbf{X}_2 - \mathbf{X}_1|)$, the mean-value theorem for functions of one variable implies that there is a point s_0 $(0 < s_0 < |\mathbf{X}_2 - \mathbf{X}_1|)$ such that

$$g(|\mathbf{X}_2 - \mathbf{X}_1|) - g(0) = |\mathbf{X}_2 - \mathbf{X}_1| g'(s_0).$$

But $g(|\mathbf{X}_2 - \mathbf{X}_1|) = f(\mathbf{X}_2)$ and $g(0) = f(\mathbf{X}_1)$; using (1) with $t_0 = s_0$ and $\mathbf{X}_0 = \mathbf{X}_1 + s_0\mathbf{U}$, we rewrite the last equation as

$$f(\mathbf{X}_2) - f(\mathbf{X}_1) = |\mathbf{X}_2 - \mathbf{X}_1| (d_{\mathbf{X}_0} f)(\mathbf{U}) = (d_{\mathbf{X}_0} f)(\mathbf{X}_2 - \mathbf{X}_1),$$

which completes the proof, since $\mathbf{X}_0$ is on L.

If $n = 3$ the last equation can be written

$$f(\mathbf{X_2}) - f(\mathbf{X_1}) = f_x(\mathbf{X_0})(x_2 - x_1) + f_y(\mathbf{X_0})(y_2 - y_1) + f_z(\mathbf{X_0})(z_2 - z_1).$$

As an application of the mean-value theorem we now establish an analog of the chain rule of the one dimensional calculus, which says that if f is a differentiable function of x and x is a differentiable function of t, then the derivative of $h(t) = f(x(t))$ is $h'(t) = f'(x(t))x'(t)$.

Example 4.1   Let

$$f(x) = \frac{1}{1 - x^2} \qquad (|x| < 1)$$

and

$$x(t) = \sin t \qquad \left(|t| < \frac{\pi}{2}\right).$$

Then

$$h(t) = f(x(t)) = \frac{1}{(1 - \sin^2 t)^2}$$

is defined if $|t| < \pi/2$, and f and x are differentiable:

$$f'(x) = \frac{2x}{(1 - x^2)^2},$$

$$x'(t) = \cos t.$$

Thus

$$h'(t) = f'(x(t))x'(t) = \frac{2 \sin t}{(1 - \sin^2 t)^2} \cos t$$

$$= \frac{2 \sin t}{\cos^3 t}.$$

To state sufficient conditions under which an analogous rule holds for a function of several variables, we need a preliminary definition.

Definition 4.1   A function f is said to be *continuously differentiable* on a domain D of R^n if all of its first partial derivatives exist and are continuous on D.

The next theorem is an immediate consequence of this definition and Theorem 3.3.

Theorem 4.2 If f is continuously differentiable in a domain D of R^n, then f is differentiable on D.

Theorem 4.3 Let f be continuously differentiable in a domain D of R^n, let $x_1, \ldots, x_n$ be differentiable functions of t for $a < t < b$, and $X(t) = (x_1(t), \ldots, x_n(t))$ be an interior point of D for all t in (a, b). Define $h(t) = f(\mathbf{X}(t))$; then h has a derivative in (a, b), given by

$$(2) \qquad h'(t) = f_{x_1}(\mathbf{X}(t)) x_1'(t) + \cdots + f_{x_n}(\mathbf{X}(t)) x_n'(t)$$

$$= (d_{\mathbf{X}_0} f)(x_1'(t), \ldots, x_n'(t)).$$

Proof. Let t_0 be fixed in (a, b); since

$$|\mathbf{X}(t) - \mathbf{X}(t_0)| = \sqrt{(x_1(t) - x_1(t_0))^2 + \cdots + (x_n(t) - x_n(t_0))^2}$$

and each x_i is continuous at t_0 (being differentiable there), it follows that

$$\lim_{t \to t_0} |\mathbf{X}(t) - \mathbf{X}(t_0)| = 0.$$

Thus, for t sufficiently near t_0, the line segment connecting $\mathbf{X}(t)$ and $\mathbf{X}(t_0)$ lies entirely in D (because of the convexity of the n-ball), and Theorem 4.1 implies that there is a point $\hat{\mathbf{X}}(t)$ on this segment such that

$$f(\mathbf{X}(t)) - f(\mathbf{X}(t_0)) = (d_{\hat{\mathbf{X}}(t)} f)(\mathbf{X}(t) - \mathbf{X}(t_0)).$$

Hence

$$(3) \qquad \frac{h(t) - h(t_0)}{t - t_0} = f_{x_1}(\hat{\mathbf{X}}(t)) \frac{x_1(t) - x_1(t_0)}{t - t_0} + \cdots + f_{x_n}(\hat{\mathbf{X}}(t)) \frac{x_n(t) - x_n(t_0)}{t - t_0}.$$

Now, $|\hat{\mathbf{X}}(t) - \mathbf{X}(t_0)| \le |\mathbf{X}(t) - \mathbf{X}(t_0)|$, which implies that $\hat{\mathbf{X}}(t)$ approaches $\mathbf{X}(t_0)$ as t approaches t_0; hence the continuity of f_{x_i} implies that

$$\lim_{t \to t_0} f_{x_i}(\hat{\mathbf{X}}(t)) = f_{x_i}(\mathbf{X}(t_0)) \qquad (1 \le i \le n).$$

The differentiability of x_i implies that

$$\lim_{t \to t_0} \frac{x_i(t) - x_i(t_0)}{t - t_0} = x_i'(t_0).$$

Now let t approach t_0 in (3) to find that

$$h'(t_0) = f_{x_1}(\mathbf{X}(t_0)) x_1'(t_0) + \cdots + f_{x_n}(\mathbf{X}(t_0)) x_n'(t_0).$$

Since this argument can be carried out for each t_0 in (a, b), we can replace t_0 by t in the last equation and obtain (2). This completes the proof.

Example 4.2 Let

$$f(x, y) = 2x^2 - y^2,$$

$$x(t) = 2t^3 + t,$$

and

$$y(t) = t + 1.$$

We wish to find the derivative of $h(t) = f(x(t), y(t))$ at $t = 0$. Then $f_x(x, y) = 4x$, $f_y(x, y) = -2y$, $x'(t) = 6t^2 + 1$, and $y'(t) = 1$; thus

$$h'(0) = f_x(x(0), y(0))x'(0) + f_y(x(0), y(0))y'(0)$$

$$= f_x(0, 1)x'(0) + f_y(0, 1)y'(0)$$

$$= (0)(1) + (-2)(1)$$

$$= -2.$$

In many applications of the chain rule, $(n - 1)$ of the components of **X** are functions of the remaining one. For example, suppose $n = 3$ and y and z are functions of x, so that $h(x) = f(x, y(x), z(x))$. To apply Theorem 4.3, we introduce an intermediate variable t and write $\mathbf{X}(t) = (t, y(t), z(t))$. Then $h(t) = f(t, y(t), z(t))$ and Theorem 4.3 yields

$$h'(t) = f_x(t, y(t), z(t)) + f_y(t, y(t), z(t))y'(t)$$

$$+ f_z(t, y(t), z(t))z'(t),$$

since $x'(t) = 1$. The name of the independent variable is immaterial; hence the last equation can be rewritten

$$h'(x) = f_x(x, y(x), z(x)) + f_y(x, y(x), z(x))y'(x)$$

$$+ f_z(x, y(x), z(x))z'(x).$$

The student should recognize that the introduction of the intermediate variable t served merely as a bookkeeping device to facilitate the application of Theorem 4.2; it can be omitted when he is thoroughly familiar with this form of the chain rule.

Example 4.3 Let $f(x, y, z) = x^2 + y^2 + z^2$, $y(x) = \cos x$, and $z(x) = e^x$; then $f_x(x, y, z) = 2x$, $f_y(x, y, z) = 2y$, $f_z(x, y, z) = 2z$, $y'(x) = -\sin x$, and $z'(x) = e^x$. Thus, if $h(x) = f(x, y(x), z(x))$, then

$$h'(x) = 2x + 2y(x)(-\sin x) + 2z(x)e^x$$

$$= 2x - 2 \sin x \cos x + 2e^{2x}.$$

The same result can be obtained by first expressing h explicitly in terms of x,

$$h(x) = x^2 + \cos^2 x + e^{2x},$$

and then differentiating with respect to x directly; however, this does not detract from the usefulness of the chain rule, which applies to cases where direct substitution is not convenient.

The chain rule will be further extended in Section 3.3.

EXERCISES 4.4

In the following exercises let $h(t) = f(\mathbf{X}(t))$.

1. Evaluate $h'(t_0)$ by the chain rule and then by substituting $\mathbf{X}(t)$ for $\mathbf{X}$ and differentiating directly with respect to t:

   (a) $f(x, y) = x^2 + y^2$; $x(t) = \cos t$, $y(t) = \sin t$; $t_0 = \dfrac{\pi}{4}$.

   (b) $f(x, y, z) = \cos xyz$; $x(t) = 1$, $y(t) = t^2$, $z(t) = t^3$; $t_0 = \sqrt[5]{\dfrac{\pi}{2}}$

   (c) $f(x, y) = \dfrac{x^2 + y^2}{xy}$; $x(t) = \sqrt{t}$, $y(t) = \sqrt[3]{t}$; $t_0 = 1$.

   (d) $f(x, y, z) = \log(x + y + z)$; $x(t) = e^t$, $y(t) = -e^t$, $z(t) = t^2$; $t_0 = 1$.

   (e) $f(x, y, z) = \tan \dfrac{xy}{z}$; $x(t) = e^t$, $y(t) = e^{-t}$, $z(t) = t$; $t_0 = \dfrac{1}{\pi}$.

2. Let $f(x, y, z) = \log(x^2 + y^2 + z^2)$, $x(t) = 1 + t/\sqrt{3}$, $y(t) = 2 - t/\sqrt{3}$, and $z(t) = 3 + t/\sqrt{3}$. Calculate $h'(0)$ and, by comparing the result with that of Exercise 2(a) of Section 2.2, show that $h'(0) = \dfrac{\partial F}{\partial \mathbf{U}}(\mathbf{X}_0)$, with $\mathbf{U}$ and $\mathbf{X}_0$ as defined there.

3. Repeat Exercise 2 for $f(x, y) = x/(x + y)$, $x(t) = 1 + t/\sqrt{2}$, and $y(t) = 2 + t/\sqrt{2}$, comparing the result with that of Exercise 1(d), Section 2.2.

4. Find $h'(0)$:

   (a) $f(x, y, z) = x^{37}y^{12}z^6$; $x(t) = \cos t$, $y(t) = \sin t$, $z(t) = e^t$.

   (b) $f(x_1, \ldots, x_n) = x_1^2 + \cdots + x_n^2$; $x_i(t) = a_i + u_i t$ $(1 \le i \le n)$.

5. Find $h'(1)$:

   (a) $f(x, y, z) = e^{x+y^2+z}$; $x(t) = 2\log t$, $y(t) = t - 1$, $z(t) = \sin(t - 1)$.

   (b) $f(x, y, z) = \tan^{-1} xyz$; $x(t) = e^t$, $y(t) = e^{-t}$, $z(t) = 1$.

6. Let $f(x, y, z) = \cos^3 x + y^3 e^z$, $y(x) = \sin x$, $z(x) = x^2$, and $h(x) = f(x, y(x), z(x))$; find $h'(x)$.

7. Let $f(x, y, z) = xz \tan y$; $x(y) = \log y$, $z(y) = e^y$, and $h(y) = f(x(y), y, z(y))$; find $h'(y)$.

8. Let $f(\mathbf{X}) = x^2 + y^2 + z^2$, $\mathbf{X}_1 = (2, 0, -1)$, and $\mathbf{X}_2 = (1, -1, 1)$. Verify Theorem 4.1 by finding $\mathbf{X}_0$ on the line connecting $\mathbf{X}_2$ and $\mathbf{X}_1$ such that

$$f(\mathbf{X}_2) - f(\mathbf{X}_1) = (d_{\mathbf{X}_0} f)(\mathbf{X}_2 - \mathbf{X}_1).$$

THEORETICAL EXERCISES

T-1. Let

$$\mathbf{U} = \begin{bmatrix} u_1 \\ \vdots \\ u_n \end{bmatrix}$$

be a unit vector, $\mathbf{X}_0 = (c_1, \ldots, c_n)$, and $x_i(t) = c_i + u_i t$ $(1 \le i \le n)$. Define $h(t) = f(\mathbf{X}(t))$ and show that $h'(0) = (\partial f/\partial \mathbf{U})(\mathbf{X}_0)$.

T-2. Prove Theorem 4.2.

4.5   Graphs and Tangent Planes

The *graph* of a real-valued function f with domain D in R^n is the set of points in R^{n+1} of the form $(\mathbf{X}, f(\mathbf{X})) = (x_1, \ldots, x_n, f(\mathbf{X}))$, where $\mathbf{X}$ is in D. Thus, the graph of f consists of those points $(\mathbf{X}, x_{n+1})$ in R^{n+1} which satisfy

$$x_{n+1} = f(\mathbf{X}).$$

This equation is *the equation of the graph.*

   Example 5.1   Let $n = 1$ and

$$f(x) = \sqrt{1 - x^2}.$$

FIGURE 5.1

The graph of f is a semicircle (Fig. 5.1); its equation is

$$y = \sqrt{1 - x^2}.$$

Example 5.2 Let

$$f(x, y) = \sqrt{1 - x^2 - y^2};$$

the domain of f is the disk of radius 1 about the origin, and the graph of f is a hemisphere (Fig. 5.2); its equation is

$$z = \sqrt{1 - x^2 - y^2}.$$

Example 5.3 If $n > 2$ and

$$f(\mathbf{X}) = \sqrt{1 - x_1^2 - \cdots - x_n^2},$$

then the graph of f consists of all points in R^{n+1} of the form

$$(x_1, \ldots, x_n, \sqrt{1 - x_1^2 - \cdots - x_n^2})$$

such that $x_1^2 + \cdots + x_n^2 \leq 1$; it is impossible, however, to depict it on a two-dimensional page. The equation of the graph is

$$x_{n+1} = \sqrt{1 - x_1^2 - \cdots - x_n^2}.$$

The student should recall that if f is a function of one variable, then $f'(x_0)$ is the slope of the tangent to the graph of f at the point $(x_0, f(x_0))$. We shall discuss this in detail, to motivate the generalization of this idea to functions of n variables.

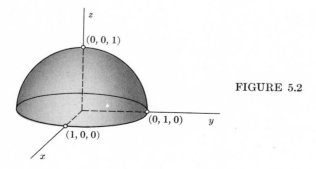

FIGURE 5.2

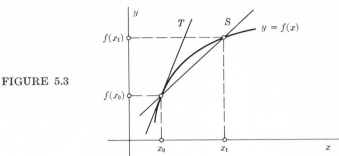

FIGURE 5.3

The point $(x_0, f(x_0))$ and any second point $(x_1, f(x_1))$ on the graph of f determine a secant line S, whose equation is

$$y = f(x_0) + \frac{f(x_1) - f(x_0)}{x_1 - x_0} (x - x_0)$$

(Fig. 5.3). As x_1 approaches x_0 the slope of the secant,

$$m = \frac{f(x_1) - f(x_0)}{x_1 - x_0},$$

approaches $f'(x_0)$, and the secant rotates into coincidence with the line T, given by

$$y = f(x_0) + f'(x_0)(x - x_0),$$

which is tangent to the graph of f at $x = x_0$.

Thus, the tangent to the graph of f at x_0 is the "limit" of the secants connecting $(x_0, f(x_0))$ and $(x_1, f(x_1))$ as x_1 approaches x_0. This limit exists if and only if f has a derivative at x_0.

Example 5.4 Let $x_0 = 1$ and $f(x) = x^2$; then $f(1) = 1$, $f'(1) = 2$, and the equation of the tangent to the graph of f at $(1, 1)$ is

$$y = 1 + 2(x - 1)$$

(Fig. 5.4).

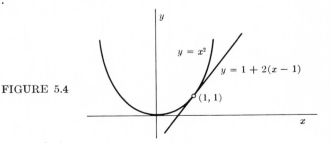

FIGURE 5.4

Example 5.5 Let $x_0 = 1$ and $f(x) = |x - 1|$. This function has no derivative at x_0; neither does its graph have a tangent at $(1, 0)$, since if $x_1 < 1$, the secant connecting $(x_1, |x_1 - 1|)$ and $(1, 0)$ has the equation

$$y = -(x - 1),$$

while if $x_1 > 1$, it has the equation

$$y = (x - 1)$$

(Fig. 5.5).

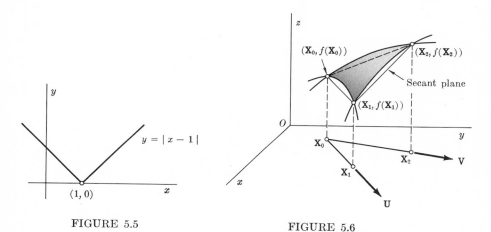

FIGURE 5.5   FIGURE 5.6

If f, a function of n (≥ 2) variables, is differentiable at $\mathbf{X}_0$, then the graph of f has a tangent plane at $(\mathbf{X}_0, f(\mathbf{X}_0))$. Before defining this plane for arbitrary n, we present an intuitive discussion for $n = 2$.

Let f be differentiable at $\mathbf{X}_0 = (x_0, y_0)$. Let $\mathbf{X}_1 = (x_1, y_1)$ and $\mathbf{X}_2 = (x_2, y_2)$ be distinct points such that the line segments from $\mathbf{X}_0$ to $\mathbf{X}_1$ and from $\mathbf{X}_0$ to $\mathbf{X}_2$ are in the domain of f, and not collinear. Then there is a unique secant plane in R^3 through $(\mathbf{X}_0, f(\mathbf{X}_0))$, $(\mathbf{X}_1, f(\mathbf{X}_1))$, and $(\mathbf{X}_2, f(\mathbf{X}_2))$ (Fig. 5.6); its equation is

$$(1) \quad z = f(\mathbf{X}_0) + \frac{(y - y_0)(x_2 - x_0) - (y_2 - y_0)(x_1 - x_0)}{(y_1 - y_0)(x_2 - x_0) - (y_2 - y_0)(x_1 - x_0)} (f(\mathbf{X}_1) - f(\mathbf{X}_0))$$

$$+ \frac{(y_1 - y_0)(x - x_0) - (y - y_0)(x_1 - x_0)}{(y_1 - y_0)(x_2 - x_0) - (y_2 - y_0)(x_1 - x_0)} (f(\mathbf{X}_2) - f(\mathbf{X}_0)).$$

(See Section 3.2, in particular, Example 2.5 and the discussion preceding it.) In allowing $\mathbf{X}_1$ and $\mathbf{X}_2$ to approach $\mathbf{X}_0$, we must restrict the method of approach so that $(\mathbf{X}_0, f(\mathbf{X}_0))$, $(\mathbf{X}_1, f(\mathbf{X}_1))$, and $(\mathbf{X}_2, f(\mathbf{X}_2))$ determine a plane; we do this by setting

$$\text{(2)} \qquad \mathbf{X_1} = \mathbf{X_0} + t\mathbf{U}, \qquad \mathbf{X_2} = \mathbf{X_0} + t\mathbf{V},$$

where $\mathbf{U}$ and $\mathbf{V}$ are nonparallel unit vectors:

$$\mathbf{U} = \begin{bmatrix} u_1 \\ u_2 \end{bmatrix}, \qquad \mathbf{V} = \begin{bmatrix} v_1 \\ v_2 \end{bmatrix} \qquad (u_2 v_1 - u_1 v_2) \neq 0.$$

We shall study the behavior of the secant plane as t approaches zero; although this is not the most general way in which the secant plane can approach its limiting position, it is sufficiently general to motivate our definition.

With $\mathbf{X_1}$ and $\mathbf{X_2}$ replaced by (2), the equation of the secant plane becomes

$$\text{(3)} \quad z = f(\mathbf{X_0}) + \frac{(y - y_0)v_1 - (x - x_0)v_2}{u_2 v_1 - u_1 v_2} \frac{f(\mathbf{X_0} + t\mathbf{U}) - f(\mathbf{X_0})}{t}$$

$$+ \frac{(x - x_0)u_2 - (y - y_0)u_1}{u_2 v_1 - u_1 v_2} \frac{f(\mathbf{X_0} + t\mathbf{V}) - f(\mathbf{X_0})}{t}.$$

In Section 2.3 we showed that

$$\lim_{t \to t_0} \frac{f(\mathbf{X_0} + t\mathbf{U}) - f(\mathbf{X_0})}{t} = \frac{\partial f}{\partial \mathbf{U}}(\mathbf{X_0}) = f_x(\mathbf{X_0})u_1 + f_y(\mathbf{X_0})u_2;$$

similarly

$$\lim_{t \to t_0} \frac{f(\mathbf{X_0} + t\mathbf{V}) - f(\mathbf{X_0})}{t} = f_x(\mathbf{X_0})v_1 + f_y(\mathbf{X_0})v_2.$$

Thus, as t approaches zero in (3), the secant plane "approaches" the *tangent plane,*

$$z = f(\mathbf{X_0}) + \frac{(y - y_0)v_1 - (x - x_0)v_2}{u_2 v_1 - u_1 v_2} [f_x(\mathbf{X_0})u_1 + f_y(\mathbf{X_0})u_2]$$

$$+ \frac{(x - x_0)u_2 - (y - y_0)u_1}{u_2 v_1 - u_1 v_2} [f_x(\mathbf{X_0})v_1 + f_y(\mathbf{X_0})v_2].$$

This is the equation of a plane through $(\mathbf{X_0}, f(\mathbf{X_0}))$. A remarkable thing about it is that it does not depend on $\mathbf{U}$ and $\mathbf{V}$, as can be seen by collecting coefficients of $f_x(\mathbf{X_0})$ and $f_y(\mathbf{X_0})$:

$$z = f(\mathbf{X_0}) + f_x(\mathbf{X_0})(x - x_0) + f_y(\mathbf{X_0})(y - y_0)$$

$$= f(\mathbf{X_0}) + (d_{\mathbf{X_0}} f)(\mathbf{X} - \mathbf{X_0})$$

(Fig. 5.7).

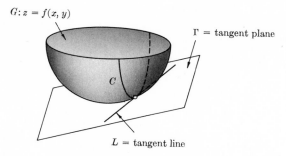

$G: z = f(x, y)$

Γ = tangent plane

C

L = tangent line

FIGURE 5.7

Definition 5.1 Let $n \geq 1$ and f be differentiable at $\mathbf{X}_0$. The *tangent plane to the graph of* f *at* $(\mathbf{X}_0, f(\mathbf{X}_0))$ is the graph defined by

$$x_{n+1} = f(\mathbf{X}_0) + (d_{\mathbf{X}_0} f)(\mathbf{X} - \mathbf{X}_0).$$

This plane is the "limit" of secant planes to the graph of f, just as we have seen above for $n = 2$.

Thus, the tangent plane to the graph of f at $(\mathbf{X}_0, f(\mathbf{X}_0))$ is itself the graph of the function T defined by

$$T(\mathbf{X}) = f(\mathbf{X}_0) + (d_{\mathbf{X}_0} f)(\mathbf{X} - \mathbf{X}_0).$$

From the differentiability of f at $\mathbf{X}_0$ it follows that

$$\lim_{\mathbf{X} \to \mathbf{X}_0} \frac{f(\mathbf{X}) - T(\mathbf{X})}{\mathbf{X} - \mathbf{X}_0} = 0.$$

The tangent plane is the only plane with this property; that is, if $\mathbf{X}_0 = (c_1, \ldots, c_n)$ and

$$\lim_{\mathbf{X} \to \mathbf{X}_0} \frac{f(\mathbf{X}) - a - b_1(x_1 - c_1) - \cdots - b_n(x_n - c_n)}{|\mathbf{X} - \mathbf{X}_0|} = 0,$$

then the plane defined by

$$x_{n+1} = a + b_1(x_1 - c_1) + \cdots + b_n(x_n - c_n)$$

is, in fact, the tangent plane.

Example 5.6 To find the tangent plane to the graph of

$$f(x, y) = \sqrt{1 - x^2 - y^2}$$

at $(x_0, y_0, f(x_0, y_0)) = (1/2, -1/2, 1/\sqrt{2})$, compute

$$f_x\left(\frac{1}{2}, -\frac{1}{2}\right) = -f_y\left(\frac{1}{2}, -\frac{1}{2}\right) = -\frac{1}{\sqrt{2}};$$

thus the equation of the tangent plane is

$$z - \frac{1}{\sqrt{2}} = -\frac{1}{\sqrt{2}}\left(x - \frac{1}{2}\right) + \frac{1}{\sqrt{2}}\left(y + \frac{1}{2}\right),$$

or

$$z = \frac{1}{\sqrt{2}}(-x + y + 2).$$

Let G be the graph of f and Γ the tangent plane to G at $(\mathbf{X}_0, f(\mathbf{X}_0))$. If C is a curve in G passing through $(\mathbf{X}_0, f(\mathbf{X}_0))$ and L is the tangent line to C at $(\mathbf{X}_0, f(\mathbf{X}_0))$ (Fig. 5.7), then L is in Γ. To see this for $n = 2$, let C be given parametrically by $x = x(t)$, $y = y(t)$, and $z = z(t)$, where $(x(t_0), y(t_0), z(t_0)) = (x_0, y_0, f(x_0, y_0))$. Since C is in G,

$$z(t) = f(x(t), y(t));$$

from the chain rule,

(4) $z'(t_0) = f_x(x(t_0), y(t_0))x'(t_0) + f_y(x(t_0), y(t_0))y'(t_0)$

$$= f_x(x_0, y_0)x'(t_0) + f_y(x_0, y_0)y'(t_0).$$

If (x_1, y_1, z_1) is on L then

$$x'(t_0) = \lambda(x_1 - x_0), \qquad y'(t_0) = \lambda(y_1 - y_0),$$

$$z'(t_0) = \lambda(z_1 - f(x_0, y_0)),$$

where λ is a nonzero constant. Substituting these in (4) yields

$$z_1 - f(x_0, y_0) = f_x(x_0, y_0)(x_1 - x_0) + f_y(x_0, y_0)(y_1 - y_0);$$

hence (x_1, y_1, z_1) is in Γ, from Definition 5.1.

A similar proof holds for $n > 2$.

EXERCISES 4.5

In Exercises 1 through 7 sketch the graph of the given function.

1. $f(x) = x^2 + 1$.

2. $f(x, y) = \sqrt{36 - 9x^2 - 4y^2}$.

3. $f(x, y) = 4x^2 + 9y^2$.

4. $f(x, y) = \sqrt{36x^2 - 9y^2}$.

5. $f(x, y) = \sqrt{9x^2 - 4y^2 - 36}$.

6. $f(x, y) = \sqrt{16x^2 + 4y^2}$.

7. $f(x, y) = \sqrt{9x^2 + 4y^2 - 36}$.

8. Let $f(x, y) = x^2 + y^2$; find the equation of the secant plane to the graph of f through $(0, 2, f(0, 2))$, $(1, 1, f(1, 1))$, and $(-1, 3, f(-1, 3))$.

In Exercises 9 through 14 find the equation of the tangent plane to the graph of f at $(\mathbf{X}_0, f(\mathbf{X}_0))$.

9. $f(x, y) = \sqrt{36 - 9x^2 - 4y^2}$, $\mathbf{X}_0 = (1, 2)$.

10. $f(x, y, z) = 1 + 4x^2 - y^2 + 2z^2$, $\mathbf{X}_0 = (1, 1, 1)$.

11. $f(x, y) = \sqrt{4x^2 + 9y^2 - 36}$, $\mathbf{X}_0 = (3, 2)$.

12. $f(x, y) = 9 - x^2$, $\mathbf{X}_0 = (2, 4)$.

13. $f(x, y) = x^2 + 2y^2$, $\mathbf{X}_0 = (1, -2)$.

14. Let f and $\mathbf{X}_0$ be as in Exercise 13. Find the equation of the line through $\mathbf{X}_0$ perpendicular to the tangent plane to the graph of f at $(\mathbf{X}_0, f(\mathbf{X}_0))$.

THEORETICAL EXERCISES

T-1. If f is differentiable at $\mathbf{X}_0 = (c_1, \ldots, c_n)$, show that the line through $\mathbf{X}_0$ perpendicular to the tangent plane to the graph of f at $(\mathbf{X}_0, f(\mathbf{X}_0))$ is given by

$$x_i = c_i + tf_{x_i}(\mathbf{X}_0) \qquad (1 \le i \le n),$$
$$x_{n+1} = f(\mathbf{X}_0) - t.$$

T-2. Find the equation of the tangent plane to the graph of $f(\mathbf{X}) = \sin(x_1 + \cdots + x_n)$ at $(\mathbf{X}_0, f(\mathbf{X}_0))$, where

(a) $\mathbf{X}_0 = \left(\dfrac{\pi}{n}, \cdots, \dfrac{\pi}{n}\right)$;   (b) $\mathbf{X}_0 = \left(\dfrac{\pi}{2n}, \cdots, \dfrac{\pi}{2n}\right)$.

T-3. Repeat Exercise T-2 for $f(\mathbf{X}) = \log(x_1^2 + \cdots + x_n^2)$ and

(a) $\mathbf{X}_0 = \left(\dfrac{1}{\sqrt{n}}, \ldots, \dfrac{1}{\sqrt{n}}\right)$;   (b) $\mathbf{X}_0 = (e, 0, \ldots, 0)$.

T-4. Verify Eq. (1) for the secant plane through $(\mathbf{X}_0, f(\mathbf{X}_0))$, $(\mathbf{X}_1, f(\mathbf{X}_1))$, and $(\mathbf{X}_2, f(\mathbf{X}_2))$.

4.6 Implicit Functions

If f is defined on a domain D in R^n, then its graph G, as defined in Section 2.5, is the set of points in R^{n+1} of the form $(\mathbf{X}, f(\mathbf{X}))$, where $\mathbf{X}$ is in D. No two points in G can differ only in their $(n+1)$th coordinates; if $(\mathbf{X}, x_{n+1})$ and $(\mathbf{X}, \hat{x}_{n+1})$ are both in G, then

$$x_{n+1} = \hat{x}_{n+1} = f(\mathbf{X}).$$

Now let us change our point of view. Suppose G is a subset of R^{n+1} with the property that no two points of G differ only in their $(n+1)$th coordinates, and let D be the set of all $\mathbf{X}$ in R^n that appear as the first n coordinates of points of G; thus for each $\mathbf{X}$ in D there is exactly one real number x_{n+1} such that $(\mathbf{X}, x_{n+1})$ is in G. Define $x_{n+1} = f(\mathbf{X})$; that is, $f(\mathbf{X})$ is the unique real number such that $(\mathbf{X}, f(\mathbf{X}))$ is in G. Then f is a function on D, and G is its graph; rather than starting with the function and arriving at the graph, we have used the graph to define the function. We say that f in this case is defined *implicitly* by G; we also say that G *defines* x_{n+1} *as a function of* $(x_1, \ldots, x_n)$.

It is convenient at this point to introduce the notation

$$S = \{\mathbf{X} \mid \cdots\}$$

to mean that S is the set of points which satisfy the conditions listed to the right of the vertical line.

Example 6.1 Let $n = 2$ and

$$(1) \qquad G = \{(x, y, z) \mid x^2 + y^2 + z^2 = 1, \quad z \geq 0\};$$

thus G is a hemisphere (Fig. 6.1), and it defines a function f with domain

$$D = \{(x, y) \mid x^2 + y^2 \leq 1\},$$

since for each (x, y) in D there is exactly one nonnegative z such that (x, y, z) is in G:

$$(2) \qquad z = f(x, y) = \sqrt{1 - x^2 - y^2}.$$

The hemisphere

$$G_1 = \{(x, y, z) \mid x^2 + y^2 + z^2 = 1, \quad z \leq 0\}$$

(Fig. 6.2) defines a function f_1 with domain D:

$$z = f_1(x, y) = -\sqrt{1 - x^2 - y^2}.$$

However, the sphere

$$S = \{(x, y, z) \mid x^2 + y^2 + z^2 = 1\}$$

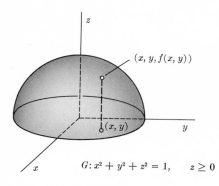

$G: x^2 + y^2 + z^2 = 1, \quad z \geq 0$

FIGURE 6.1

$G_1: x^2 + y^2 + z^2 = 1, \quad z \leq 0$

FIGURE 6.2

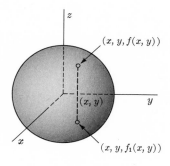

FIGURE 6.3

$S: x^2 + y^2 + z^2 = 1$

is not a graph, because for every $(x, y) \neq (0, 0)$ in D, the distinct points $(x, y, f(x, y))$ and $(x, y, f_1(x, y))$ are in S. (Fig. 6.3).

Sometimes it is possible to obtain an explicit expression such as (2) for an implicitly defined function, but this is not essential; one could argue that (2) does no more than provide a convenient notation for the function defined by (1). This is precisely the way in which many functions are introduced into analysis.

Example 6.2 Let $n = 1$ and

$$G = \{(x, y) \mid e^y - x = 0, \quad 0 < x < \infty\};$$

then G defines y as a function of x on $(0, \infty)$, because $e^{y_1} = e^{y_2}$ if and only if $y_1 = y_2$. This function is the natural logarithm of x;

$$y = f(x) = \log x \qquad (-\infty < x < \infty)$$

(Fig. 6.4).

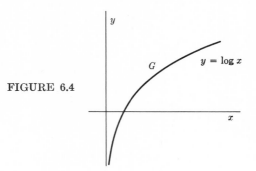

FIGURE 6.4

Example 6.3   Let $n = 1$ and

$$G = \left\{ (x, y) \mid x - \sin y = 0, \quad -\frac{\pi}{2} \leq y \leq \frac{\pi}{2} \right\}$$

(Fig. 6.5); again, G defines y as a function of x, since there is only one value of y in $[-\pi/2, \pi/2]$ whose sine equals a given x in $[-1, 1]$. This function is the inverse sine function; it is denoted by

$$y = f(x) = \text{Arcsin } x.$$

Placing other restrictions on y in the definition of G produces other inverse sine functions; thus

$$G_1 = \left\{ (x, y) \mid x - \sin y = 0, \quad \frac{\pi}{2} \leq y \leq \frac{3\pi}{2} \right\}$$

(Fig. 6.6) also defines y as a function of x.

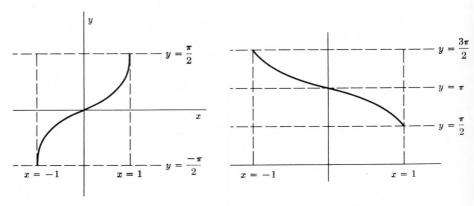

FIGURE 6.5   FIGURE 6.6

A set G in R^{n+1} may define some coordinate other than the $(n+1)$th in terms of the remaining ones.

Example 6.4 Let

$$G = \{(1, 1, 1), (1, 1, 2), (1, 2, 3)\};$$

then G does not define z as a function of x and y because the first two points differ only in their z coordinates. However, G defines y as a function f of (x, z) with domain

$$D = \{(1, 1), (1, 2), (1, 3)\}$$

and values

$$f(1, 1) = 1, \qquad f(1, 2) = 1, \qquad f(1, 3) = 2.$$

Even though y is not the last coordinate in this example, let us agree to call G the *graph of f*, and say that G is a *graph with respect to y*.

A set G can be a graph with respect to more than one coordinate.

Example 6.5 In Example 6.4, G also defines x as a function g of (y, z) with domain

$$D_1 = \{(1, 1), \quad (1, 2), \quad (2, 3)\}$$

and values

$$g(1, 1) = g(1, 2) = g(2, 3) = 1.$$

Example 6.6 Let

$$G = \{(x_1, \ldots, x_{n+1}) \mid x_1^2 + \cdots + x_{n+1}^2 = 1, \quad x_i \geq 0 \quad (1 \leq i \leq n+1)\};$$

then G defines each x_i as a function of the other coordinates, and is therefore a graph with respect to each coordinate.

Let F be defined on a subset of R^{n+1} and let N be the set of points at which F vanishes. We next discuss conditions under which N or certain of its subsets define x_{n+1} as a function of $\mathbf{X} = (x_1, \ldots, x_n)$. (The choice of x_{n+1} as the preferred coordinate is one of convenience; our discussion could be conducted equally well in terms of any other coordinate.)

Example 6.7 If

$$F(\mathbf{X}, x_{n+1}) = f(\mathbf{X}) - x_{n+1},$$

then N defines x_{n+1} in terms of $\mathbf{X}$ and the functional relationship is quite explicit:

$$x_{n+1} = f(\mathbf{X}).$$

In this case N is the graph of f.

Example 6.8 If

$$F(\mathbf{X}, x_{n+1}) = x_1^2 + \cdots + x_{n+1}^2 - 1,$$

then N is not a graph (because, for example $(0, \ldots, 0, 1)$ and $(0, \ldots, 0, -1)$ are both in N), but some of its subsets are. For example, the subset of N for which $x_{n+1} \geq 0$ defines x_{n+1} as a function of $\mathbf{X}$:

$$x_{n+1} = f(\mathbf{X}) = \sqrt{1 - x_1^2 - \cdots - x_n^2}.$$

The following theorem, which we state without proof, gives useful sufficient conditions which ensure that certain subsets of N define functions implicitly.

Implicit Function Theorem

Theorem 6.1 (Implicit Function Theorem) Let F be a function of $(x_1, \ldots, x_{n+1})$. Suppose that $(\mathbf{X}_0, c)$ is an interior point of the domain of F and let $F_{x_1}, \ldots, F_{x_{n+1}}$ be continuous in an $(n + 1)$-ball about $(\mathbf{X}_0, c)$. Suppose that $F(\mathbf{X}_0, c) = 0$ and $F_{x_{n+1}}(\mathbf{X}_0, c) \neq 0$. Then there is a set D in R^n with $\mathbf{X}_0$ as an interior point and a unique continuous function f with domain D, such that $f(\mathbf{X}_0) = c$ and $F(\mathbf{X}, f(\mathbf{X})) = 0$ for all $\mathbf{X}$ in D. Furthermore, f has continuous first partial derivatives on D given by

$$(3) \qquad f_{x_i}(\mathbf{X}) = \frac{-F_{x_i}(\mathbf{X}, f(\mathbf{X}))}{F_{x_{n+1}}(\mathbf{X}, f(\mathbf{X}))} \qquad (1 \leq i \leq n).$$

We say that $F(\mathbf{X}, x_{n+1}) = 0$ *defines f implicitly near* $\mathbf{X}_0$, or that *it defines x_{n+1} as a function of* $\mathbf{X}$ *near* $\mathbf{X}_0$. We write

$$x_{n+1} = f(\mathbf{X}).$$

Example 6.9 Let $(\mathbf{X}_0, c) = (0, \ldots, 0)$ and

$$F(\mathbf{X}, x_{n+1}) = a_1 x_1 + \cdots + a_{n+1} x_{n+1}.$$

Then $F_{x_i} = a_i$ $(1 \leq i \leq n + 1)$ and Theorem 6.1 implies that $F(\mathbf{X}, x_{n+1}) = 0$ defines x_{n+1} as a function of $\mathbf{X}$ near $\mathbf{X}_0$, provided $a_{n+1} \neq 0$. Elementary algebra bears this out, for then

$$x_{n+1} = f(\mathbf{X}) = -\frac{1}{a_{n+1}}(a_1 x_1 + \cdots + a_n x_n)$$

satisfies $F(\mathbf{X}, f(\mathbf{X})) = 0$ for all $\mathbf{X}$, and

$$f_{x_i} = -\frac{a_i}{a_{n+1}} = \frac{-F_{x_i}}{F_{x_{n+1}}},$$

which agrees with (3).

Example 6.10   Let

$$F(x, y, z) = x^2 + y^2 + z^2 - 1$$

and

$$(x_0, y_0, z_0) = \left(\frac{1}{\sqrt{2}}, 0, -\frac{1}{\sqrt{2}}\right);$$

then

(4) $F_x(x, y, z) = 2x,$ $F_y(x, y, z) = 2y,$ and $F_z(x, y, z) = 2z.$

Since

$$F_z\left(\frac{1}{\sqrt{2}}, 0, \frac{-1}{\sqrt{2}}\right) = -F_z\left(\frac{1}{\sqrt{2}}, 0, -\frac{1}{\sqrt{2}}\right)$$

$$= \sqrt{2},$$

it follows that $F(x, y, z) = 0$ defines $z = f(x, y)$ such that $f(1/\sqrt{2}, 0) = -1/\sqrt{2}$ and $x = g(y, z)$ such that $g(0, 1/\sqrt{2}) = 1/\sqrt{2}$. These functions can also be written explicitly as

(5) $$f(x, y) = -\sqrt{1 - x^2 - y^2}$$

and

$$g(y, z) = \sqrt{1 - y^2 - z^2},$$

but this is not true for the general implicit functions, so we shall not make use of it in the following. It is possible to compute the partial derivatives of f at $(1/\sqrt{2}, 0)$ without using the explicit formula (5). From (3) and (4)

(6) $$f_x(x, y) = -\frac{x}{z}, \qquad f_y(x, y) = -\frac{y}{z};$$

hence

(7) $$f_x\left(\frac{1}{\sqrt{2}}, 0\right) = 1, \qquad f_y\left(\frac{1}{\sqrt{2}}, 0\right) = 0$$

(note that $z = -1/\sqrt{2}$ when $(x, y) = (1/\sqrt{2}, 0)$).
  To obtain $f_{xx}(1/\sqrt{2}, 0)$, replace z by $f(x, y)$ in (6):

$$f_x(x, y) = -\frac{x}{f(x, y)}.$$

Then

(8)
$$f_{xx}(x, y) = \frac{\partial}{\partial x}\left(\frac{-x}{f(x, y)}\right)$$

$$= \frac{-f(x, y) + xf_x(x, y)}{[f(x, y)]^2}.$$

Setting $(x, y) = (1/\sqrt{2}, 0)$ and using (7) yields

$$f_{xx}\left(\frac{1}{\sqrt{2}}, 0\right) = \frac{4}{\sqrt{2}}.$$

Similar computations yield $f_{xy}(1/\sqrt{2}, 0)$ and $f_{yy}(1/\sqrt{2}, 0)$; then by differentiating (8) and using the known values for the first and second derivatives, we could calculate the third derivatives at $(1/\sqrt{2}, 0)$, etc.

Example 6.11   Let $F(x, y, z) = x^2yz + 2xy^2z^3 - 3x^3y^3z^5$ and $(x_0, y_0, z_0) = (1, 1, 1)$, so that $F(x_0, y_0, z_0) = 0$. Then

$$F_x(x, y, z) = 2xyz + 2y^2z^3 - 9x^2y^3z^5,$$

$$F_y(x, y, z) = x^2z + 4xyz^3 - 9x^3y^2z^5,$$

and

$$F_z(x, y, z) = x^2y + 6xy^2z^2 - 15x^2y^3z^4.$$

Since $F_z(x_0, y_0, z_0) = -8$, $z = f(x, y)$ on some domain D containing $(1, 1)$ as an interior point. However, it is extremely difficult to exhibit z explicitly as a function of x and y. Nevertheless, $f_x(1, 1)$ and $f_y(1, 1)$ can be computed:

$$f_x(1, 1) = \frac{-F_x(1, 1, 1)}{F_z(1, 1, 1)} = -\frac{(-5)}{(-8)} = \frac{-5}{8}$$

and

$$f_y(1, 1) = \frac{-F_y(1, 1, 1)}{F_z(1, 1, 1)} = -\frac{(-4)}{(-8)} = -\frac{1}{2}$$

Higher partial derivatives at $(1, 1)$ can be computed as in Example 6.10.

For $n = 1$, (3) can be obtained by formally applying the chain rule. Thus if $F(x, f(x)) = 0$ for all x in some interval, differentiation with respect to x yields

$$F_x(x, f(x)) + F_y(x, f(x))f'(x) = 0;$$

hence,

(9)
$$f'(x) = -\frac{F_x(x, f(x))}{F_y(x, f(x))}$$

provided $F_y(x, f(x)) \neq 0$.

The conditions of Theorem 6.1 are sufficient, but not necessary. That is, $F(\mathbf{X}, x_{n+1}) = 0$ *may* define a function f near $\mathbf{X}_0$ even though F fails to satisfy the hypotheses of Theorem 6.1 at $(\mathbf{X}_0, c)$. (See Exercises 9 and 10).
By writing

$$y = f(x), \qquad z = f(x, y), \qquad \text{and} \qquad x_{n+1} = f(x_1, \ldots, x_n)$$

in this section, we have introduced a new convention, since we are now denoting the *value* of a function at a point in R^n by the *name* of a coordinate in R^{n+1}. This is not consistent with our previous notation and can sometimes lead to ambiguity; nevertheless it is convenient for simplifying what might otherwise be extremely ugly expressions. For $n = 1$ the student should not be shocked by this notation; he must have written $y = f(x)$ hundreds of times in his first calculus course! He has also written

$$y' = \frac{dy}{dx} = f'(x), \qquad y'' = \frac{d^2y}{dx^2} = f''(x),$$

etc. We shall extend this idea and write

$$z = f(x, y),$$

$$z_x = \frac{\partial z}{\partial x} = \frac{\partial f}{\partial x}(x, y),$$

$$z_{xy} = \frac{\partial^2 z}{\partial x \, \partial y} = \frac{\partial^2 f}{\partial x \, \partial y}(x, y),$$

and so forth, when adherence to our earlier (still preferred) notation is cumbersome. Thus, we write (9) as

$$y' = \frac{-F_x(x, y)}{F_y(x, y)},$$

or, if we want an even more abbreviated expression,

$$y' = \frac{-F_x}{F_y}.$$

The student manipulates such an expression at his own risk, and should always bear in mind what it really stands for.

EXERCISES 4.6

1. Which of the following sets define $x_4 = f(x_1, x_2, x_3)$? For those which do, determine the domain and range of f.

    (a) $G_1 = \{(1, 5, 1, 1), \quad (1, 1, 2, 2), \quad (1, 1, 1, 3), \quad (1, 1, 3, 2)\}.$

(b) $G_2 = \{(1, 1, 1, 1), \quad (1, 1, 2, 2), \quad (1, 1, 1, 2), \quad (1, 1, 3, 2)\}$.

(c) $G_3 = \{(2, 7, 5, 3), \quad (1, 7, 5, 3), \quad (3, 6, 2, 3), \quad (7, 5, 3, 1)\}$.

(d) $G_4 = \{(1, 2, 3, 4), \quad (2, 3, 4, 1), \quad (3, 4, 1, 2), \quad (4, 1, 2, 3)\}$.

2. $G = \{(x, y) \mid \log y - x = 0\}$ defines $y = f(x)$. What is $f(x)$ explicitly?

3. Which of the following sets defines $y = f(x)$? For those which do, determine the domain and range of f.

(a) $G_1 = \{(x, y) \mid \cos y - x = 0\}$.

(b) $G_2 = \{(x, y) \mid \cos y - x = 0, \quad |y| < \pi/2\}$.

(c) $G_3 = \{(x, y) \mid \cos y - x = 0, \quad 0 \le y \le \pi\}$.

(d) $G_4 = \{(x, y) \mid \cos y - x = 0, \quad \pi \le y \le 2\pi\}$.

4. Which of the following sets define $z = f(x, y)$? For those which do, determine the domain and range of f.

(a) $G_1 = \{(x, y, z) \mid x^2 + y^2 - z^2 - 1 = 0, \quad z \ge 0\}$.

(b) $G_2 = \{(x, y, z) \mid x^2 + y^2 - z^2 - 1 = 0, \quad |z| < 1\}$.

(c) $G_3 = \{(x, y, z) \mid x^2 - y^2 + z^2 - 1 = 0, \quad y \ge 0\}$.

(d) $G_4 = \{(x, y, z) \mid x^2 - y^2 + z^2 - 1 = 0, \quad z \ge 0\}$.

5. Which of the sets G_1, G_2, G_3, and G_4 in Exercise 1 define $x_1 = g(x_2, x_3, x_4)$? For those which do, specify the domains and ranges of the functions that they define.

6. Repeat Exercise 5 for $x = g(y)$ in Exercise 3.

7. Repeat Exercise 5 for $y = h(x, z)$ in Exercise 4.

8. Show that $1 - e^z \cos (x - y) = 0$ defines $z = f(x, y)$ near $(\pi/2, \pi/2)$. Find $f(\pi/2, \pi/2)$ and all first and second partial derivatives of f at $(\pi/2, \pi/2)$. Also, obtain explicit expressions for f_x and f_y.

9. Show that $1 - e^z \cos(x - y) = 0$ does not define $z = f(x, y)$ near $(\pi/2, 0)$. (It is *not* enough to simply state that the hypotheses of the implicit function theorem are not satisfied, since this theorem states sufficient, not necessary, conditions for the existence of an implicit function.)

10. $F(x, y) = x - y^3 = 0$ defines $y = f(x)$ near $x = 0$ (with $f(0) = 0$) even though F does not satisfy the hypothesis of Theorem 6.1 at $(0, 0)$. Does this violate Theorem 6.1?

11. Let $F(x, y) = y^2(1 - xy) - 3y[1 + \log(x + 1)] + 2e^{xy}$; then $F(x, y) = 0$ defines two functions, $y = g(x)$ and $y = h(x)$, near $x = 0$. (Does this contradict Theorem 6.1?) Find $g(0)$, $g'(0)$, $h(0)$, and $h'(0)$.

12. The equation

$$x^2y^5z^2w^5 + 2xy^2w^3 - 3x^3z^2w = 0$$

defines $w = f(x, y, z)$ with $f(1, 1, 1) = 1$. Find $f_x(1, 1, 1)$, $f_y(1, 1, 1)$ and $f_z(1, 1, 1)$.

13. The equation of Exercise (12) also defines $y = g(x, z, w)$ with $g(1, 1, 1) = 1$. Find $g_x(1, 1, 1)$, $g_z(1, 1, 1)$, and $g_w(1, 1, 1)$.

14. Suppose $y = f(x)$ is defined by

$$x^4 - x^3y^2 + y^5 + 1 = 0;$$

find $f'(x)$ in terms of x and y.

THEORETICAL EXERCISES

T-1. Let $F(x, y) = 0$ define $y = f(x)$; by formally applying the chain rule to (9), show that

$$y'' = -\frac{1}{F_y} (F_{xx} + 2F_{xy}y' + F_{yy}(y')^2).$$

Compare this with the expression for $f''(x)$, completely written out in analogy with (9).

T-2. Show that the result of Exercise T-1 is consistent with the expected result if $F(x, y) = y - f(x)$ (that is, if y is defined explicitly as a function of x).

T-3. Give assumptions on F and f that make the derivation of (9) by the chain rule legitimate.

4.7 The Gradient

In Theorem 3.5, we found a useful representation for directional derivatives: if f is differentiable at $\mathbf{X}_0$ and

$$\mathbf{U} = \begin{bmatrix} u_1 \\ \vdots \\ u_n \end{bmatrix}$$

is a unit vector, then

$$(1) \qquad \frac{\partial f}{\partial \mathbf{U}}(\mathbf{X}_0) = f_{x_1}(\mathbf{X}_0)u_1 + \cdots + f_{x_n}(\mathbf{X}_0)u_n.$$

We have also seen that $(\partial f/\partial \mathbf{U})(\mathbf{X}_0)$ is the rate of change of f, at $\mathbf{X}_0$, in the direction of $\mathbf{U}$. Question: What is the maximum rate of change of f at $\mathbf{X}_0$, and in what direction is it attained?

The answer to this question is provided by Schwarz's inequality (Theorem 5.2, Section 2.5).

Definition 7.1 Let f be a function with domain D in R^n which is differentiable at $\mathbf{X}_0$. The vector

$$\nabla_{\mathbf{X}_0} f = \begin{bmatrix} f_{x_1}(\mathbf{X}_0) \\ \vdots \\ f_{x_n}(\mathbf{X}_0) \end{bmatrix}$$

is called the *gradient of f at $\mathbf{X}_0$*. If $n = 3$ we may also write

$$\nabla_{\mathbf{X}_0} f = f_x(\mathbf{X}_0)\mathbf{i} + f_y(\mathbf{X}_0)\mathbf{j} + f_z(\mathbf{X}_0)\mathbf{k}.$$

Example 7.1 Let $\mathbf{X}_0 = (1, 1, \ldots, 1)$ and

$$f(\mathbf{X}) = x_1^2 + x_2^2 + \cdots + x_n^2;$$

then

$$f_{x_i}(\mathbf{X}) = 2x_i, \qquad f_{x_i}(\mathbf{X}_0) = 2 \qquad (1 \le i \le n);$$

thus

$$\nabla_{\mathbf{X}_0} f = \begin{bmatrix} 2 \\ 2 \\ \vdots \\ 2 \end{bmatrix}.$$

Example 7.2 Let $\mathbf{X}_0 = (-1, 2, 1)$ and $f(x, y, z) = xy^2z^2$; then

$$f_x(x, y, z) = y^2z^2, \qquad f_y(x, y, z) = 2xyz^2, \qquad f_z(x, y, z) = 2xy^2z$$

and

$$f_x(\mathbf{X}_0) = 4, \qquad f_y(\mathbf{X}_0) = -4, \qquad f_z(\mathbf{X}_0) = -8.$$

Now (1) can be rewritten as an inner product:

$$\frac{\partial f}{\partial \mathbf{U}}(\mathbf{X}_0) = \mathbf{U} \cdot \nabla_{\mathbf{X}_0} f.$$

From Schwarz's inequality, since $|\mathbf{U}| = 1$,

$$\left| \frac{\partial f}{\partial \mathbf{U}}(\mathbf{X}_0) \right| \le |\nabla_{\mathbf{X}_0} f|$$

and equality is attained if and only if $\mathbf{U}$ and $\nabla_{\mathbf{X}_0} f$ have the same direction. (This statement includes the case where $f_{x_1}(\mathbf{X}_0) = \cdots = f_{x_n}(\mathbf{X}_0) = 0$, if we agree that the zero vector has every direction.) This proves the following theorem.

Theorem 7.1   If f is differentiable at $\mathbf{X}_0$ and $\mathbf{U}$ is a unit vector, then

$$\frac{\partial f}{\partial \mathbf{U}} (\mathbf{X}_0) = \mathbf{U} \cdot \nabla_{\mathbf{X}_0} f.$$

The maximum value of the directional derivative of f at $\mathbf{X}_0$ is $|\nabla_{\mathbf{X}_0} f|$; it is attained in the direction of

$$\mathbf{U}_0 = \frac{\nabla_{\mathbf{X}_0} f}{|\nabla_{\mathbf{X}_0} f|},$$

and only in this direction, unless $\nabla_{\mathbf{X}_0} f = 0$, in which case $\mathbf{U}_0$ is undefined and $(\partial f/\partial \mathbf{U})(\mathbf{X}_0) = 0$ in all directions.

Example 7.3   Let $\mathbf{X}_0 = \left(\dfrac{1}{\sqrt{2}}, -\dfrac{1}{\sqrt{2}} \right)$ and $f(x, y) = x^2 + y^2$; then

$$\nabla_{\mathbf{X}_0} f = \begin{bmatrix} \sqrt{2} \\ -\sqrt{2} \end{bmatrix}$$

and, if

$$\mathbf{U} = \begin{bmatrix} u_1 \\ u_2 \end{bmatrix}$$

is an arbitrary unit vector,

$$\frac{\partial f}{\partial \mathbf{U}} (\mathbf{X}_0) = \mathbf{U} \cdot \nabla_{\mathbf{X}_0} f$$

$$= \sqrt{2}(u_1 - u_2).$$

The direction of the maximum rate of change is

$$\mathbf{U}_0 = \frac{\nabla_{\mathbf{X}_0} f}{|\nabla_{\mathbf{X}_0} f|}$$

$$= \frac{1}{2} \begin{bmatrix} \sqrt{2} \\ -\sqrt{2} \end{bmatrix} = \begin{bmatrix} \dfrac{1}{\sqrt{2}} \\ -\dfrac{1}{\sqrt{2}} \end{bmatrix}$$

and

$$\frac{\partial f}{\partial \mathbf{U}_0} (\mathbf{X}_0) = | \nabla_{\mathbf{X}_0} f | = 2.$$

Example 7.4 Let f be as in Example 7.1 and $\mathbf{X}_0 = (0, 0)$; then

$$\nabla_{\mathbf{X}_0} f = \begin{bmatrix} 0 \\ 0 \end{bmatrix} \quad \text{and} \quad \frac{\partial f}{\partial \mathbf{U}} (\mathbf{X}_0) = 0$$

for all $\mathbf{U}$.

The gradient provides a convenient representation for the tangent plane to the graph of an implicit function. Suppose $F(x_0, y_0, z_0) = 0$, $F_z(x_0, y_0, z_0) \neq 0$, and F satisfies the hypotheses of the implicit function theorem near $\mathbf{X}_0$. Then $F(x, y, z) = 0$ defines $z = f(x, y)$, where f is differentiable and satisfies $F(x, y, f(x, y)) = 0$ for all (x, y) in some domain D containing (x_0, y_0) in its interior; hence the graph of f has a tangent plane at $(x_0, y_0, f(x_0, y_0)) = (x_0, y_0, z_0)$ with the equation

(2) $\qquad z = z_0 + (x - x_0)f_x(x_0, y_0) + (y - y_0)f_y(x_0, y_0).$

From the implicit function theorem,

$$f_x(x_0, y_0) = \frac{-F_x(x_0, y_0, z_0)}{F_z(x_0, y_0, z_0)},$$

$$f_y(x_0, y_0) = \frac{-F_y(x_0, y_0, z_0)}{F_z(x_0, y_0, z_0)};$$

substituting these in (2) and rearranging terms yields

(3) $\qquad (x - x_0)F_x(x_0, y_0, z_0) + (y - y_0)F_y(x_0, y_0, z_0)$

$$+ (z - z_0)F_z(x_0, y_0, z_0) = 0,$$

which can be written more briefly as

$$(\mathbf{X} - \mathbf{X}_0) \cdot \nabla_{\mathbf{X}_0} F = 0.$$

Thus the gradient is normal to the tangent plane (Fig. 7.1).

Example 7.5 Let

$$F(x, y, z) = x^2 + y^2 + z^2 - 1,$$

and

$$\mathbf{X}_0 = \left(\frac{1}{\sqrt{3}}, -\frac{1}{\sqrt{6}}, \frac{1}{\sqrt{2}} \right).$$

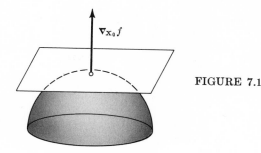

FIGURE 7.1

Then $F(x, y, z) = 0$ defines $z = f(x, y) = \sqrt{1 - x^2 - y^2}$ near $\mathbf{X_0}$ and

$$f_x(x, y) = \frac{-x}{\sqrt{1 - x^2 - y^2}},$$

$$f_y(x, y) = \frac{-y}{\sqrt{1 - x^2 - y^2}}.$$

Hence $f_x\left(\dfrac{1}{\sqrt{3}}, -\dfrac{1}{\sqrt{6}}\right) = -\sqrt{\dfrac{2}{3}}$, $f_y\left(\dfrac{1}{\sqrt{3}}, -\dfrac{1}{\sqrt{6}}\right) = \dfrac{1}{\sqrt{3}}$; and, from (2), the equation of the tangent plane to the graph of f at $\mathbf{X_0}$ is

(4)
$$z = \frac{1}{\sqrt{2}} - \sqrt{\frac{2}{3}}\left(x - \frac{1}{\sqrt{3}}\right) + \frac{1}{\sqrt{3}}\left(y + \frac{1}{\sqrt{6}}\right).$$

On the other hand,

$$F_x(x, y, z) = 2x, \qquad F_y(x, y, z) = 2y, \qquad F_z(x, y, z) = 2z,$$

so that

$$F_x\left(\frac{1}{\sqrt{3}}, -\frac{1}{\sqrt{6}}, \frac{1}{\sqrt{2}}\right) = \frac{2}{\sqrt{3}}, \qquad F_y\left(\frac{1}{\sqrt{3}}, -\frac{1}{\sqrt{6}}, \frac{1}{\sqrt{2}}\right) = -\frac{2}{\sqrt{6}},$$

$$F_z\left(\frac{1}{\sqrt{3}}, -\frac{1}{\sqrt{6}}, \frac{1}{\sqrt{2}}\right) = \frac{2}{\sqrt{2}},$$

and the equation of the tangent plane in the form (3) is

(5)
$$\frac{1}{\sqrt{3}}\left(x - \frac{1}{\sqrt{3}}\right) - \frac{1}{\sqrt{6}}\left(y + \frac{1}{\sqrt{6}}\right) + \frac{1}{\sqrt{2}}\left(z - \frac{1}{\sqrt{2}}\right) = 0,$$

which is easily shown to be equivalent to (4).

In (3) the equation of the tangent plane does not explicitly involve f;

this is an obvious advantage if it is difficult or impossible to obtain f explicitly.

Suppose we make the additional assumption that $F_x(x_0, y_0, z_0) \neq 0$; then

$$F(x, y, z) = 0$$

also defines $x = g(y, z)$ such that $g(y_0, z_0) = x_0$ and $F(g(y, z), y, z)) = 0$ at all (y, z) in some domain. The equation of the tangent plane to the graph of g at $(x_0, y_0, z_0) = (g(y_0, z_0), y_0, z_0)$ is

(6) $x = x_0 + (y - y_0)g_y(y_0, z_0) + (z - z_0)g_z(y_0, z_0).$

From the implicit function theorem,

$$g_y(y_0, z_0) = \frac{-F_y(x_0, y_0, z_0)}{F_x(x_0, y_0, z_0)},$$

$$g_z(y_0, z_0) = \frac{-F_z(x_0, y_0, z_0)}{F_x(x_0, y_0, z_0)},$$

which, substituted into (6), yields (3). Thus the graphs of f and g have the same tangent plane at (x_0, y_0, z_0). If $F(x, y, z) = 0$ also defined $y = h(x, z)$ near (x_0, z_0), the equation of the tangent plane to the graph of h at (x_0, y_0, z_0) would also be (3).

Example 7.6   In Example 7.5, we could have considered that

$$F(x, y, z) = 0$$

defines

$$x = g(y, z) = \sqrt{1 - y^2 - z^2};$$

then

$$g_y(y, z) = \frac{-y}{\sqrt{1 - y^2 - z^2}}$$

and

$$g_z(y, z) = \frac{-z}{\sqrt{1 - y^2 - z^2}}.$$

Hence

$$g_y\left(-\frac{1}{\sqrt{6}}, \frac{1}{\sqrt{2}}\right) = \frac{1}{\sqrt{2}},$$

$$g_z\left(-\frac{1}{\sqrt{6}}, \frac{1}{\sqrt{2}}\right) = -\sqrt{\frac{3}{2}}$$

and, from (6), the equation of the tangent plane to the graph of g at $\mathbf{X}_0$ is

$$x = \frac{1}{\sqrt{3}} + \frac{1}{\sqrt{2}}\left(y + \frac{1}{\sqrt{6}}\right) - \sqrt{\frac{3}{2}}\left(z + \frac{1}{\sqrt{2}}\right),$$

which is equivalent to (5).

In view of this discussion it makes good sense to associate the plane (3) with F rather than with any particular implicit function defined by F. Before pursuing this further, we need two definitions.

Definition 7.2   Let $\mathbf{X}_0$ be in the domain of F. Then the *level set of F through* $\mathbf{X}_0$ is the set of points $\mathbf{X}$ such that $F(\mathbf{X}) = F(\mathbf{X}_0)$; we shall also describe it as the *level set defined by* $F(\mathbf{X}) = F(\mathbf{X}_0)$, or simply as the level set defined by $F(\mathbf{X}) = c$, where $c = F(\mathbf{X}_0)$. If $n = 2$ we shall refer to level sets as *level curves*; if $n = 3$ we shall speak of *level surfaces*.

Example 7.7   Let

$$F(x, y) = x^2 + y^2;$$

the level curve through (x_0, y_0) is the circle

$$x^2 + y^2 = x_0^2 + y_0^2.$$

The level curves through $(3, 4)$ and $(0, 4)$ are the circles defined by $F(x, y) = 5$ and $F(x, y) = 4$, respectively (Fig. 7.2).

Example 7.8   Let

$$F(x, y, z) = x + y + z;$$

the level surface through (x_0, y_0, z_0) is the plane

$$(x - x_0) + (y - y_0) + (z - z_0) = 0.$$

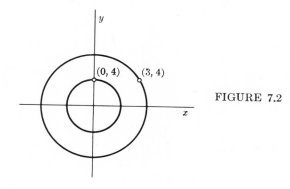

FIGURE 7.2

If $(x_0, y_0, z_0) = (1, 0, 2)$ then the level surface is the plane

$$x + y + z = 3.$$

Definition 7.3 Let F have continuous first partial derivatives in an n-ball about $\mathbf{X}_0 = (c_1, \ldots, c_n)$, and suppose $F_{x_i}(\mathbf{X}_0) \neq 0$ for some i. Let L be the level set of F through $\mathbf{X}_0$. Then the *tangent plane* (*line, if* $n = 2$) *to* L *at* $\mathbf{X}_0$ is the plane whose equation is

$$(7) \qquad (x_1 - c_1) F_{x_1}(\mathbf{X}_0) + \cdots + (x_n - c_n) F_{x_n}(\mathbf{X}_0) = 0.$$

For each i such that $F_{x_i}(\mathbf{X}_0) \neq 0$, $F(\mathbf{X}) - F(\mathbf{X}_0) = 0$ defines x_i as a function f_i of the other coordinates (to see this take $F = F - F(\mathbf{X}_0)$; then F satisfies the hypotheses of the implicit function theorem). By the argument given earlier for $n = 3$, (7) is the tangent plane to the graph of f_i at $\mathbf{X}_0$; hence calling (7) a tangent plane in this definition is consistent with the usage in Section 2.5.

Example 7.9 To find the equation of the tangent plane to the sphere

$$x^2 + y^2 + z^2 = 1$$

at

$$\mathbf{X}_0 = \left(\frac{1}{\sqrt{3}}, \frac{-1}{\sqrt{6}}, \frac{1}{\sqrt{2}} \right),$$

take

$$F(x, y, z) = x^2 + y^2 + z^2 - 1.$$

The sphere is the level surface of F through $\mathbf{X}_0$. Since

$$F_x(\mathbf{X}_0) = \frac{2}{\sqrt{3}}, \qquad F_y(\mathbf{X}_0) = \frac{-2}{\sqrt{6}}, \qquad F_z(\mathbf{X}_0) = \frac{2}{\sqrt{2}},$$

the equation of the tangent plane is

$$\frac{1}{\sqrt{3}} \left(x - \frac{1}{\sqrt{3}} \right) - \frac{1}{\sqrt{6}} \left(y + \frac{1}{\sqrt{6}} \right) + \frac{1}{\sqrt{2}} \left(z - \frac{1}{\sqrt{2}} \right) = 0.$$

We obtained the same result in Example 7.5; the difference here is that we have dispensed with the intermediate step of introducing an implicit function.

If a level set $F(\mathbf{X}) = F(\mathbf{X}_0)$ has a tangent plane at $\mathbf{X}_0$, we define the *normal to the level set at* $\mathbf{X}_0$ to be the normal to the tangent plane. Since (7) can be rewritten as

$$(\mathbf{X} - \mathbf{X}_0) \cdot \nabla_{\mathbf{X}_0} F = 0,$$

it follows that this normal is in the direction of $\nabla_{X_0}F$. More specifically, the unit vector **N** in the direction of the gradient is called the *outward normal*; $-\mathbf{N}$ is the *inward normal*.

Example 7.10 In Example 7.9

$$\nabla_{X_0}F = \begin{bmatrix} \dfrac{2}{\sqrt{3}} \\[2mm] -\dfrac{2}{\sqrt{6}} \\[2mm] \dfrac{2}{\sqrt{2}} \end{bmatrix} = 2\mathbf{X}_0$$

that is, the normal to the sphere at $\mathbf{X}_0$ is along the radius vector from the origin to $\mathbf{X}_0$, which is geometrically evident from Fig. 7.3.

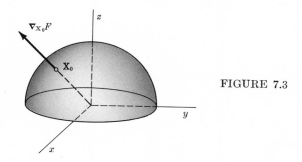

FIGURE 7.3

Example 7.11 Let $T(x, y, z)$ be the temperature at (x, y, z) in some domain D. Then the level surfaces of T are called *isothermals*. The normal to the isothermal through (x_0, y_0, z_0) is $\nabla_{X_0}T$; thus the direction of the greatest rate of temperature increase is normal to the isothermal through $\mathbf{X}_0$.

EXERCISES 4.7

1. Find $\nabla_{X_0}f$:

    (a) $f(\mathbf{X}) = a_1x_1 + \cdots + a_nx_n,$ $\mathbf{X}_0$ arbitrary.

    (b) $f(\mathbf{X}) = x^2 - 2xy + z^2,$ $\mathbf{X}_0 = (1, 1, 2).$

(c) $f(\mathbf{X}) = \log(x + y + z)$, $\quad \mathbf{X}_0 = (1, 3, 2)$.

(d) $f(\mathbf{X}) = e^{x^2 + y^2 + z^2}$, $\quad \mathbf{X}_0 = (1, 0, 0)$.

2. In each of the following $F(\mathbf{X}, w) = 0$ defines $w = f(\mathbf{X})$ near $\mathbf{X}_0$. Find $\nabla_{\mathbf{X}_0} f$.

(a) $F(\mathbf{X}, w) = x^2 + y^2 + z^2 - w^2 + 1$, $\quad f(\mathbf{X}_0) = 2$,
$\mathbf{X}_0 = (1, -1, 1)$.

(b) $F(\mathbf{X}, w) = (2x + 3y + z + 2)^2 - e^w$, $\quad \mathbf{X}_0 = (1, -2, 3)$.

(c) $F(\mathbf{X}, w) = \tan xyz - \log w$, $\quad \mathbf{X}_0 = (0, 1, 2)$.

(d) $F(\mathbf{X}, w) = x + 3xy + z^2 + e^{\sin w}$, $\quad f(\mathbf{X}_0) = \pi$,
$\mathbf{X}_0 = (-2, 0, 1)$.

3. For each function in Exercise 1 find the direction and magnitude of the greatest rate of increase of f at $\mathbf{X}_0$.

4. Find the direction in which each function in Exercise 2 decreases most rapidly at $\mathbf{X}_0$.

5. Go back to Exercise 1, (c) and (d), and find $\dfrac{\partial f}{\partial \mathbf{U}}(\mathbf{X}_0)$ for

(a) $\mathbf{U} = \begin{bmatrix} \dfrac{1}{\sqrt{2}} \\ 0 \\ \dfrac{-1}{\sqrt{2}} \end{bmatrix}$,
(b) $\mathbf{U} = \begin{bmatrix} \dfrac{1}{\sqrt{3}} \\ \dfrac{-1}{\sqrt{3}} \\ \dfrac{1}{\sqrt{3}} \end{bmatrix}$.

6. Go back to Exercise 2, (a) and (b), and find $(\partial f / \partial \mathbf{U})(\mathbf{X}_0)$ for

(a) $\mathbf{U} = \begin{bmatrix} \dfrac{1}{\sqrt{2}} \\ 0 \\ \dfrac{-1}{\sqrt{2}} \end{bmatrix}$,
(b) $\mathbf{U} = \begin{bmatrix} \dfrac{1}{\sqrt{3}} \\ \dfrac{-1}{\sqrt{3}} \\ \dfrac{1}{\sqrt{3}} \end{bmatrix}$.

7. Let $F(x, y, z) = x^2 + y^2 + e^{z-2} - 3$; then $F(x, y, z) = 0$ defines $z = f(x, y)$, $y = g(x, z)$, and $x = h(y, z)$ such that the graphs of

f, g, and h all contain $(1, -1, 2)$. Show by direct computation that the three graphs have the same tangent plane at $(1, -1, 2)$.

8. Repeat Exercise 7 for $F(x, y, z) = x + y + \log(1 + x^2 + y^2) + z^2 + 2z$ and $(0, 0, 0)$.

9. What is the level surface of $g(x, y, z) = 2x + 4y - 3z$ through $\mathbf{X}_0 = (0, 1, 2)$?

10. What is the level curve of $f(x, y) = \sin xy$ through $(0, 0)$?

11. What is the level surface of $f(x, y, z) = \sin(x^2 + y^2 + z^2)$ through $(\sqrt{\pi}, 0, 0)$?

12. What is the level curve of $f(x, y) = x^2 - y^2$ through $(1, 1)$?

In Problems 13 through 17 find the tangent plane (line if $n = 2$) to the level set of F through $\mathbf{X}_0$.

13. $F(\mathbf{X}) = x^2 + 3xy + y^2 + zx$, $\quad \mathbf{X}_0 = (1, 2, -1)$.

14. $F(\mathbf{X}) = \sin(x^2 + y^2 + z^2)$, $\quad \mathbf{X}_0 = \left(\sqrt{\dfrac{\pi}{3}}, \sqrt{\dfrac{\pi}{3}}, \sqrt{\dfrac{\pi}{3}} \right)$.

15. $F(\mathbf{X}) = \sin(x^2 + y^2 + z^2)$, $\quad \mathbf{X}_0 = (\sqrt{\pi}, 0, 0)$.

16. $F(\mathbf{X}) = x^2 - y^2$, $\quad \mathbf{X}_0 = (2, 1)$.

17. $F(\mathbf{X}) = e^x + \log(1 + xy)$, $\quad \mathbf{X}_0 = (0, 0)$.

In Problems 18 through 21 find the outward normal to the given level set at $\mathbf{X}_0$.

18. $x^2 + 4y^2 = 5$, $\quad \mathbf{X}_0 = (1, 1)$.

19. $\sin(x^2 - y^2) = 0$, $\quad \mathbf{X}_0 = (1, 1)$.

20. $\sin xyz - \log(1 + x^2 + y^2 + z^2) = \log 2$; $\quad \mathbf{X}_0 = (1, 0, 0)$.

21. $\sin^2 x + \sin^2 y + \sin^2 z = 1$, $\quad \mathbf{X}_0 = \left(\dfrac{\pi}{4}, \dfrac{\pi}{4}, 0 \right)$.

THEORETICAL EXERCISES

T-1. Find $(\partial F / \partial \mathbf{n})(\mathbf{X}_0)$, where $\mathbf{n}$ is a unit vector along the outward normal to the level set $F(\mathbf{X}) = F(\mathbf{X}_0)$ at $\mathbf{X}_0$.

T-2. Let f and g be differentiable at $\mathbf{X}_0$. Show that

(a) $\nabla_{\mathbf{X}_0}(\alpha f) = \alpha \nabla_{\mathbf{X}_0} f \quad (\alpha = \text{constant})$,

(b) $\nabla_{X_0}(f + g) = \nabla_{X_0} f + \nabla_{X_0} g$,

(c) $\nabla_{X_0}(fg) = f(X_0) \nabla_{X_0} g + g(X_0) \nabla_{X_0} f$,

and, if $g(X_0) \neq 0$,

(d) $\nabla_{X_0}(f/g) = \dfrac{g(X_0) \nabla_{X_0} f - f(X_0) \nabla_{X_0} g}{[g(X_0)]^2}$.

4.8   Taylor's Theorem

By Definition 3.1 (Section 2.3), a function f which is differentiable at $X_0 = (c_1, \ldots, c_n)$ can be so well approximated near X_0 by

$$T_1(X) = f(X_0) + (d_{X_0} f)(X - X_0)$$

$$= f(X_0) + \sum_{r=1}^{n} \frac{\partial f}{\partial x_r}(X_0)(x_r - c_r)$$

that

$$\lim_{X \to X_0} \frac{f(X) - T_1(X)}{|X - X_0|} = 0.$$

The function T_1, or more generally, any function of the form

$$P(X) = a + b_1(x_1 - c_1) + \cdots + b_n(x_n - c_n),$$

where $a, b_1, \ldots, b_n$ are constants, is a *polynomial of first degree in* $X - X_0$ provided at least one of $b_1, \ldots, b_n$ is nonzero. If $b_1 = \cdots = b_n = 0$ and $a \neq 0$, then P_1 is a polynomial of degree zero. The *zero polynomial*, which is the polynomial whose coefficients all vanish, is arbitrarily assigned the degree minus infinity.

In this section we consider the problem of approximating functions near a point X_0 by means of polynomials of higher degree. Since the situation is simplest for functions of one variable, we consider that case first and in more detail.

Definition 8.1   Let g be a function of a single variable t which possesses k derivatives at $t = t_0$. Then the *kth Taylor polynomial of g about t_0* is defined by

$$T_k(t) = \sum_{r=0}^{k} \frac{g^{(r)}(t_0)}{r!}(t - t_0)^r$$

$$= g(t_0) + \frac{g'(t_0)}{1!}(t - t_0) + \cdots + \frac{g^{(k)}(t_0)}{k!}(t - t_0)^k.$$

Clearly the *degree* of T_k, which is defined to be the highest power of $(t - t_0)$ that appears in T_k with a nonzero coefficient, or minus infinity if all the coefficients vanish, does not exceed k.

Example 8.1 Let $g(t) = \log t$ and $t_0 = 1$. Then $g(1) = 0$ and $g^{(r)}(1) = (-1)^{r-1}(r - 1)!$ if $r \geq 1$. Hence

$$T_0(t) = 0$$

and

$$T_k(t) = \sum_{r=1}^{k} \frac{(-1)^r}{r} (t - 1)^r \qquad (k \geq 1).$$

Example 8.2 Let $g(t) = e^t$ and $t_0 = 0$. Then $g^{(r)}(0) = 1$ for all $r \geq 0$ and

$$T_k(t) = \sum_{r=0}^{k} \frac{t^r}{r!}.$$

Example 8.3 Let $g(t) = \cos t$; then $g'(t) = -\sin t$, $g''(t) = -\cos t$, $g'''(t) = \sin t$, and the cycle repeats. The derivatives of odd order vanish at $t = 0$, while those of even order alternate between $+1$ and -1. Thus the Taylor polynomials about $t = 0$ are given by

$$T_0(t) = T_1(t) \quad = 1,$$

$$T_2(t) = T_3(t) \quad = 1 - \frac{t^2}{2!},$$

$$T_4(t) = T_5(t) \quad = 1 - \frac{t^2}{2!} + \frac{t^4}{4!},$$

$$\vdots$$

$$T_{2k}(t) = T_{2k+1}(t) = \sum_{r=0}^{k} (-1)^r \frac{t^{2r}}{(2r)!}.$$

Taylor polynomials have the following remarkable property:

Theorem 8.1 (Taylor's Theorem) Let g have k continuous derivatives in an open interval J about t_0 and let T_k be its kth Taylor polynomial about t_0:

$$T_k(t) = g(t_0) + \frac{g'(t_0)}{1!} (t - t_0) + \cdots + \frac{g^{(k)}(t_0)}{k!} (t - t_0)^k.$$

Then

(1)
$$\lim_{t \to t_0} \frac{g(t) - T_k(t)}{(t - t_0)^k} = 0,$$

and T_k is the only polynomial of degree $\leq k$ with this property.

For $k = 1$ this theorem reduces to the situation discussed in the beginning of Section 2.3. We shall give a partial proof for arbitrary k, leaving the proof of uniqueness to the student (Exercise T-2).

Proof. Let $\epsilon > 0$ be given. From the definition of derivative, there is a $\delta > 0$ such that if $0 < |t - t_0| < \delta$ then

$$\left| \frac{g^{(k-1)}(t) - g^{(k-1)}(t_0)}{t - t_0} - g^{(k)}(t_0) \right| < \epsilon,$$

which can be rewritten as

$$|g^{(k-1)}(t) - g^{(k-1)}(t_0) - g^{(k)}(t_0)(t - t_0)| < \epsilon |t - t_0|.$$

Therefore, if $0 < |t - t_0| < \delta$,

$$\left| \int_{t_0}^{t} [g^{(k-1)}(\tau) - g^{(k-1)}(t_0) - g^{(k)}(t_0)(\tau - t_0)] \, d\tau \right| < \epsilon \left| \int_{t_0}^{t} (\tau - t_0) \, d\tau \right|,$$

or

$$\left| g^{(k-2)}(t) - g^{(k-2)}(t_0) - \frac{g^{(k-1)}(t_0)}{1!}(t - t_0) - \frac{g^{(k)}(t_0)}{2!}(t - t_0)^2 \right| < \epsilon \frac{|t - t_0|^2}{2!}.$$

If $k = 2$ we are finished, since ϵ is arbitrary; if $k > 2$, we integrate $k - 2$ more times to obtain

$$|g(t) - T_k(t)| < \epsilon \frac{|t - t_0|^k}{k!},$$

which completes the proof.

Example 8.4 Let $t_0 = 0$ and

$$g(t) = \frac{1}{1 - t};$$

then $g^{(r)}(0) = r!$ and

$$T_k(t) = 1 + t + \cdots + t^k.$$

Theorem 8.1 implies that

$$\lim_{t\to 0} t^{-k}\left(\frac{1}{1-t} - T_k(t)\right) = 0,$$

which in this case can be verified directly, since

$$\frac{1}{1-t} = 1 + t + \cdots + t^k + \frac{t^{k+1}}{1-t};$$

thus

$$t^{-k}\left(\frac{1}{1-t} - T_k(t)\right) = \frac{t}{1-t}.$$

Since T_k is the only polynomial of degree not greater than k for which (1) holds, it follows that $T_k = g$ if g is a polynomial of degree $\leq k$.

A polynomial in one variable written as a linear combination of powers of $t - t_0$ is said to be *expanded about* $t = t_0$. Theorem 8.1 provides a convenient way to find such an expansion.

Example 8.5 The polynomial

$$p(t) = 3 + 4(t - 1) + (t - 1)^3$$

is expanded about $t = 1$. To obtain its expansion about $t = -1$ we use the fact that $p = T_3$, where

$$T_3(t) = \sum_{r=0}^{3} \frac{p^{(r)}(-1)}{r!}(t + 1)^r.$$

Thus,

$$p(-1) = -13$$

$$p'(t) = 4 + 3(t - 1)^2, \qquad p'(-1) = 16;$$

$$p''(t) = 6(t - 1), \qquad p''(-1) = -12;$$

$$p''' = 6,$$

and

$$p(t) = p(-1) + p'(-1)(t + 1) + \frac{p''(-1)}{2!}(t + 1)^2 + \frac{p'''(-1)}{3!}(t + 1)^3$$

$$= -13 + 16(t + 1) - 6(t + 1)^2 + (t + 1)^3.$$

Although Theorem 8.1 says that the error $g(t) - T_k(t)$ goes to zero faster than $(t - t_0)^k$, it does not tell us how small $|t - t_0|$ must be to ensure that the error is below some given bound. The following theorem is useful in this connection, provided $g^{(k+1)}$ exists and is continuous in some interval about t_0.

Theorem 8.2   Let g be as in Theorem 8.1 and also let $g^{(k+1)}$ be continuous in J. For t in J let $M(t, t_0)$ be a nonnegative number such that

$$(2) \qquad\qquad | g^{(k+1)}(x) | \leq M(t, t_0)$$

if x is in the closed interval with endpoints t and t_0. Then

$$(3) \qquad\qquad | g(t) - T_k(t) | \leq \frac{M(t, t_0)}{(k+1)!} | t - t_0 |^{k+1}.$$

Proof.   From the fundamental theorem of the calculus,

$$g(t) = g(t_0) + \int_{t_0}^{t} g'(x) \, dx$$

$$= T_0(t) + \int_{t_0}^{t} g'(x) \, dx$$

if t is in J. Integration by parts yields

$$g(t) = g(t_0) + \frac{g'(x)}{1!} (x - t) \, \Big|_{t_0}^{t} - \int_{t_0}^{t} g''(x) (x - t) \, dx$$

$$= g(t_0) + \frac{g'(t_0)}{1!} (t - t_0) - \int_{t_0}^{t} g''(x) (x - t) \, dx$$

$$= T_1(t) - \int_{t_0}^{t} g''(x) (x - t) \, dx.$$

If $k > 1$, we integrate by parts again to obtain

$$f(t) = T_1(t) - g''(x) \frac{(x - t)^2}{2!} \, \Big|_{t_0}^{t} + \frac{1}{2!} \int_{t_0}^{t} g'''(x) (x - t)^2 \, dx$$

$$= T_1(t) + \frac{g''(t_0)}{2!} (t - t_0)^2 + \frac{1}{2!} \int_{t_0}^{t} g'''(x) (x - t)^2 \, dx$$

$$= T_2(t) + \frac{1}{2!} \int_{t_0}^{t} g'''(x) (x - t)^2 \, dx.$$

Continuing in this way we obtain

$$g(t) = T_k(t) + \frac{(-1)^k}{k!} \int_{t_0}^{t} g^{(k+1)}(x)(x-t)^k \, dx.$$

From this and (2),

$$| g(t) - T_k(t) | \leq \frac{1}{k!} \left| \int_{t_0}^{t} g^{(k+1)}(x)(x-t)^k \, dx \right|$$

$$\leq \frac{M(t, t_0)}{k!} \left| \int_{t_0}^{t} (x-t)^k \, dx \right|$$

$$= \frac{M(t, t_0)}{(k+1)!} | t - t_0 |^{k+1},$$

which completes the proof.

Example 8.6   Let $t_0 = 0$ and $g(t) = e^t$; from Example 8.2,

$$T_k(t) = 1 + \frac{t}{1!} + \cdots + \frac{t^k}{k!}.$$

If $t \geq 0$, then $M(t, 0) = e^t$ and

$$| e^t - T_k(t) | \leq \frac{e^t}{(k+1)!} t^{k+1};$$

if $t \leq 0$, then $M(t, 0) = 1$ and

$$| e^t - T_k(t) | \leq \frac{| t |^{k+1}}{(k+1)!}.$$

To choose k so that the error is less than a given $\epsilon > 0$ for all t in $[-1, 1]$ we combine these estimates to obtain

$$| e^t - T_k(t) | \leq \frac{e}{(k+1)!}.$$

Thus, any k such that

$$\frac{e}{(k+1)!} < \epsilon$$

will be satisfactory.

Example 8.7   To compute $\sqrt{5}$ accurate to within an error less than 0.0005, we can use an appropriate Taylor polynomial about $t = 0$ for

$$g(t) = \sqrt{t+4},$$

since we know $g(0) = \sqrt{4} = 2$. Thus we argue that $\sqrt{5}$ is approximately equal to $T_k(1)$ for some k, and Theorem 8.2 tells us how to choose k: we have only to compute the error bound (3) at $t = 1$ for $k = 0, 1, \ldots$, until we arrive at a bound less than 0.0005. The following calculation shows that the choice $k = 3$ yields an acceptable error:

$$(4) \quad \begin{cases} g(t) = (t + 4)^{1/2}, & g'(t) = \tfrac{1}{2}(t + 4)^{-1/2}, & g''(t) = -\tfrac{1}{4}(t + 4)^{-3/2}, \\ g'''(t) = \tfrac{3}{8}(t + 4)^{-5/2}, & g^{(4)}(t) = -\tfrac{15}{16}(t + 4)^{-7/2}. \end{cases}$$

The maximum of $\mid g^{(4)}(t) \mid$ in $[0, 1]$ is

$$\mid g^4(0) \mid = \frac{15}{16}\left(\frac{1}{2^7}\right),$$

and therefore, from (3), since $g(1) = \sqrt{5}$,

$$\mid \sqrt{5} - T_3(1) \mid \leq \left(\frac{15}{16}\right)\left(\frac{1}{2^7}\right)\frac{(1)^4}{4!} < 0.0004.$$

From (4), $g(0) = 2$, $g'(0) = 1/4$, $g''(0) = -1/32$, and $g'''(0) = 3/256$: thus

$$T_3(t) = 2 + \frac{1}{4}t - \frac{1}{64}t^2 + \frac{1}{512}t^3$$

and

$$T_3(1) = \frac{1145}{512} \cong 2.2363 \cong \sqrt{5}.$$

From a table of square roots,

$$\sqrt{5} \cong 2.2361,$$

rounded off to four decimal places.

The following theorem is a useful by-product of the proof of Theorem 8.2.

Theorem 8.3 (Taylor's Theorem with Integral Remainder) Let $g, g', \ldots, g^{(k+1)}$ be continuous in some open interval J about t_0. Then

$$g(t) = T_k(t) + R_k(t)$$

for all t in J, where the *remainder* $R_k(t)$ is given by

$$R_k(t) = \frac{1}{k!}\int_{t_0}^{t} g^{(k+1)}(x)(t - x)^k \, dx.$$

Taylor Polynomials in n Variables

Taylor's theorem for functions of n variables at first appears to be quite formidable. However, when expressed in terms of a sensible notation, it can be recognized as a natural generalization of Theorem 8.1.

Definition 8.2 If all partial derivatives of the function f through order r (≥ 1) exist and are continuous at a point $\mathbf{X}_0$ in R^n, then the polynomial

$$(5) \qquad (d_{\mathbf{X}_0}^r f)(\mathbf{Y}) = \sum_r \frac{r!}{r_1! \cdots r_n!} \frac{\partial^r f(\mathbf{X}_0)}{\partial x_1^{r_1} \cdots \partial x_n^{r_n}} y_1^{r_1} \cdots y_n^{r_n},$$

is called the *rth differential* of f at $\mathbf{X}_0$. Here $\sum_r$ indicates the sum over all ordered n-tuples $(r_1, \ldots, r_n)$ of nonnegative integers such that $r_1 + \cdots + r_n = r$. Since the first differential is the differential previously defined in Section 2.3, we write $d_{\mathbf{X}_0}^1 f = d_{\mathbf{X}_0} f$; by convention, $d_{\mathbf{X}_0}^0 f = f(\mathbf{X}_0)$ (a constant).

We observe that $d_{\mathbf{X}_0}^r f$ is a *homogeneous* polynomial of degree r (provided at least one of its coefficients is nonzero); that is, the sum of the powers of y_i ($1 \leq i \leq n$) appearing in each term of (5) is r.

Example 8.8 If $n = 1$ then

$$(d_{x_0}^r f)(y) = f^{(r)}(x_0) y^r.$$

Example 8.9 If $n = 2$ then

$$(d_{\mathbf{X}_0}^r f)(\mathbf{Y}) = \sum_r \frac{r!}{r_1! r_2!} \frac{\partial^r f(\mathbf{X}_0)}{\partial x_1^{r_1} \partial x_2^{r_2}} y_1^{r_1} y_2^{r_2}.$$

Setting $r_1 = m$ and $r_2 = r - m$, we can write this more conveniently as

$$(d_{\mathbf{X}_0}^r f)(\mathbf{Y}) = \sum_{m=0}^{r} \binom{r}{m} \frac{\partial^r f(\mathbf{X}_0)}{\partial x_1^m \partial x_2^{r-m}} y_1^m y_2^{r-m},$$

where $\binom{r}{m}$ is the binomial coefficient:

$$\binom{r}{m} = \frac{r!}{m!(r-m)!}.$$

Most often in the applications we shall be interested in $(d_{\mathbf{X}_0}^r f)(\mathbf{X} - \mathbf{X}_0)$;

for $n = 2$ it is convenient to write this as

$$(d^r_{X_0} f)\,(X - X_0) = \sum_{m=0}^{r} \binom{r}{m} \frac{\partial^r f(X_0)}{\partial x^m \partial y^{r-m}}\,(x - x_0)^m (y - y_0)^{r-m};$$

thus,

$$(d^0_{X_0} f)\,(X - X_0) = f(X_0),$$

$$(d_{X_0} f)\,(X - X_0) = \frac{\partial f}{\partial x}\,(X_0)\,(x - x_0) + \frac{\partial f}{\partial y}\,(X_0)\,(y - y_0),$$

$$(d^2_{X_0} f)\,(X - X_0) = \frac{\partial^2 f}{\partial x^2}\,(X_0)\,(x - x_0)^2 + 2\,\frac{\partial^2 f}{\partial x\,\partial y}\,(X_0)\,(x - x_0)\,(y - y_0)$$

$$+ \frac{\partial^2 f}{\partial y^2}\,(X_0)\,(y - y_0)^2,$$

$$(d^3_{X_0} f)\,(X - X_0) = \frac{\partial^3 f}{\partial x^3}\,(X_0)\,(x - x_0)^3 + 3\,\frac{\partial^3 f}{\partial x^2\,\partial y}\,(X_0)\,(x - x_0)^2 (y - y_0)$$

$$+ 3\,\frac{\partial^3 f}{\partial x\,\partial y^2}\,(X_0)\,(x - x_0)\,(y - y_0)^2$$

$$+ \frac{\partial^3 f}{\partial y^3}\,(X_0)\,(y - y_0)^3, \quad \text{etc.}$$

Example 8.10 Suppose $X_0 = (c_1, \ldots, c_n)$ and p is a homogeneous polynomial of degree r in $X - X_0$; that is

$$p(X) = \sum_r a_{r_1 \ldots r_n} (x_1 - c_1)^{r_1} \cdots (x_n - c_n)^{r_n}.$$

Then all differentials of p except the rth vanish, and

(6) $$p(X) = \frac{1}{r!}\,(d^r_{X_0} p)\,(X - X_0).$$

(To see this, use the result of Exercise T-3.) For example, the polynomial

$$p(X) = (x_1 + x_2 + \cdots + x_n)^r$$

is homogeneous of degree r in X; its partial derivatives vanish at $X = 0$, except those of order r, all of which equal $r!$ (Prove.) Substituting this result into (5) and using (6) yields

$$(x_1 + \cdots + x_n)^r = \sum_r \frac{r!}{r_1! \cdots r_n!}\,x_1^{r_1} \cdots x_n^{r_n};$$

for this reason the coefficients $\dfrac{r!}{r_1! \cdots r_n!}$, where $r_1 + \cdots + r_n = r$, are called

the *multinomial coefficients*. They are generalizations of the familiar binomial coefficients.

We can now define Taylor polynomials for functions of several variables.

Definition 8.3 If all partial derivatives of f through the rth order exist and are continuous at $\mathbf{X}_0 = (c_1, \ldots, c_n)$ then the *Taylor polynomial of f about $\mathbf{X}_0$* is defined by

$$T_k(\mathbf{X}) = \sum_{r=0}^{k} \frac{1}{r!} \, (d^r_{\mathbf{X}_0} f)\, (\mathbf{X} - \mathbf{X}_0),$$

$$= \sum_{r=0}^{k} \sum_r \frac{1}{r_1! \cdots r_n!} \frac{\partial^r f(\mathbf{X}_0)}{\partial x_1^{r_1} \cdots \partial x_n^{r_n}} \, (x_1 - c_1)^{r_1} \cdots (x_n - c_n)^{r_n}.$$

We point out that Example 8.8 shows that this definition is consistent with Definition 8.1 when $n = 1$.

Example 8.11 Let

$$f(\mathbf{X}) = e^{x_1 + x_2 + \ldots + x_n};$$

then all partial derivatives of f at $\mathbf{X}_0 = 0$ are equal to 1. Hence

$$T_k(\mathbf{X}) = \sum_{r=0}^{k} \sum_r \frac{x_1^{r_1} \cdots x_n^{r_n}}{r_1! \cdots r_n!}$$

Example 8.12 Let $f(x, y) = \log(1 + 2x + y)$ and $\mathbf{X}_0 = (0, 0)$. Then

$$\frac{\partial f}{\partial x}(x, y) = \frac{2}{1 + 2x + y}, \qquad \frac{\partial f}{\partial y}(x, y) = \frac{1}{1 + 2x + y};$$

$$\frac{\partial^2 f}{\partial x^2}(x, y) = \frac{-4}{(1 + 2x + y)^2}, \qquad \frac{\partial^2 f}{\partial x\,\partial y}(x, y) = \frac{-2}{(1 + 2x + y)^2},$$

$$\frac{\partial^2 f}{\partial y^2}(x, y) = \frac{-1}{(1 + 2x + y)^2};$$

in general,

$$\frac{\partial^r f(x, y)}{\partial x^m \, \partial y^{r-m}} = \frac{(-1)^{r-1}(r - 1)!\, 2^m}{(1 + 2x + y)^r}.$$

Therefore,

$$T_0(x, y) = 0,$$

$$T_1(x, y) = 2x + y,$$

$$T_2(x, y) = 2x + y - 2x^2 - xy - \tfrac{1}{2}y^2,$$

$$\vdots$$

$$T_k(x, y) = \sum_{r=0}^{k} \frac{(-1)^{r-1}}{r} \sum_{m=0}^{r} \binom{r}{m} 2^m x^m y^{r-m}.$$

The next theorem plays a role in the approximation of functions of several variables similar to that of Theorem 8.1 for functions of a single variable. However, a detailed study of this question is beyond the scope of this book. Hence we shall merely state the theorem here and apply it, in Section 2.9, to the problem of finding maxima and minima of functions of several variables.

Theorem 8.4 (Taylor's Theorem for Functions of n Variables)   Let f and all its partial derivatives through order k be continuous in an n-ball about $\mathbf{X}_0 = (c_1, \ldots, c_n)$, and let T_k be its kth Taylor polynomial about $\mathbf{X}_0$. Then

$$\lim_{\mathbf{X} \to \mathbf{X}_0} \frac{f(\mathbf{X}) - T_k(\mathbf{X})}{|\mathbf{X} - \mathbf{X}_0|^k} = 0,$$

and T_k is the only polynomial in $\mathbf{X} - \mathbf{X}_0$, of degree $\leq k$, with this property.

EXERCISES 4.8

1.   Find the Taylor polynomial T_2 about t_0 for each of the following functions.

   (a) $f(t) = e^t;$ $t_0 = 1.$

   (b) $f(t) = \sqrt{t};$ $t_0 = 1.$

   (c) $f(t) = \sqrt{1 + t^2};$ $t_0 = 0.$

2.   Find the Taylor polynomial T_3 about t_0 for each of the following functions.

   (a) $f(t) = \tan \pi t;$ $t_0 = \tfrac{1}{4}.$

    (b) $f(t) = \arcsin t;$ $t_0 = 0.$

    (c) $f(t) = e^{3t+1};$ $t_0 = 0.$

3. Expand $p(t) = 2 + 3t + 2t^2 + t^3$ in powers of $t - 1$.

4. Expand $p(t) = 2 - 5(t + 2)^2 + 3(t + 2)^3$ about $t = 3$.

5. Using Theorem 8.2 and the method of Example 8.6, find the smallest value of k such that $\mid T_k(t) - f(t) \mid \leq \epsilon$ whenever $\mid x - x_0 \mid \leq \delta$:

    (a) $f(t) = \sin t,$ $t_0 = 0,$ $\epsilon = 0.001,$ $\delta = \dfrac{\pi}{4}.$

    (b) $f(t) = \sqrt{t},$ $t_0 = 1,$ $\epsilon = 0.0001,$ $\delta = \frac{1}{2}.$

    (c) $f(t) = e^{t^2},$ $t_0 = 0,$ $\epsilon = 0.005,$ $\delta = 1.$

    (d) $f(t) = \log t,$ $t_0 = 2,$ $\epsilon = 0.005,$ $\delta = 1.$

6. Use a Taylor polynomial for $\sqrt{1 + t}$ to compute $\sqrt{1.2}$ to two decimal places.

7. Use a Taylor polynomial for e^t to compute $\sqrt{e}$ to within 0.001.

8. Find $(d_{X_0}^r f)(X - X_0)$ for $r = 0, 1, 2$:

    (a) $f(x, y) = x + y + 2x^2 + 3xy + 4y^2 + 6x^3,$ $X_0 = (0, 0).$

    (b) $f(x, y) = \sin(x + y)\pi,$ $X_0 = (\frac{1}{8}, \frac{1}{8}).$

    (c) $f(x, y) = e^{x^2} + y^2,$ $X_0 = (0, 0).$

9. Find $(d_{X_0}^r f)(X - X_0)$ for $r = 0, 1, 2$:

    (a) $f(x, y) = \log(2x + y),$ $X_0 = (0, 1).$

    (b) $f(x, y) = \sin(x^2 + y^2),$ $X_0 = (0, 0).$

10. Find $(d_{X_0}^r f)(X - X_0)$ for $r = 0, 1, 2$:

    (a) $f(x, y, z) = xyz + x^3 + y,$ $X_0 = (1, 2, -1).$

    (b) $f(x, y, z) = \log(x + y + z),$ $X_0 = (1, 1, -1).$

11. Find $(d_{X_0}^r f)(X - X_0)$ for $r = 0, 1, 2$:

    (a) $f(x, y, z) = \sin(x + y + z)\pi,$ $X_0 = (\frac{1}{6}, \frac{1}{6}, \frac{1}{6}).$

    (b) $f(x, y, z) = \sqrt{x^2 + y^2 + z^2},$ $X_0 = (1, 0, 0).$

12. Find the Taylor polynomial T_2 about X_0 for each function in Exercise 8.

13. Find the Taylor polynomial T_2 about X_0 for each function in Exercise 9.

14. Find the Taylor polynomial T_2 about $\mathbf{X}_0$ for each function in Exercise 10.

15. Find the Taylor polynomial T_2 about $\mathbf{X}_0$ for each function in Exercise 11.

16. Using Theorem 8.4, expand

$$p(x, y) = 3 + x + 2(y - 3) + x^2 + 4x(y - 3) + 2(y - 3)^2$$

about $\mathbf{X}_0 = (1, 2)$.

17. Using Theorem 8.4, expand

$$p(x, y) = 2 + (x - 1) + 3(y - 2) + (x - 1)^2 - (x - 1)(y - 2)$$

about $\mathbf{X}_0 = (2, 3)$.

18. Using Theorem 8.4, expand

$$p(x, y, z) = 4 + x + y + 2z + xz + xyz$$

about $\mathbf{X}_0 = (0, 1, -1)$.

THEORETICAL EXERCISES

T-1. Let p be a polynomial of degree $\leq k$ such that

$$\lim_{t \to t_0} \frac{p(t)}{(t - t_0)^k} = 0.$$

Show that $p = 0$ (that is, $p(t) = 0$ for all t).

T-2. Use Exercise T-1 to establish the uniqueness of the Taylor polynomial in Theorem 8.1.

T-3. If $\phi(\mathbf{X}) = (x_1 - c_1)^{r_1} \cdots (x_n - c_n)^{r_n}$ show that

$$\frac{\partial^{s_1 + \cdots + s_n}}{\partial x_1^{s_1} \cdots \partial x_x^{s_n}} \phi(c_1, \ldots, c_n) = \begin{cases} r_1! \cdots r_n! & \text{if} \quad s_i = r_i \quad (1 \leq i \leq n) \\ 0 & \text{if} \quad s_j \neq r_j \quad \text{for some} \quad j. \end{cases}$$

T-4. If F_k and G_k are the kth Taylor polynomials about $\mathbf{X}_0$ of f and g, respectively, show that $F_k + G_k$ is the kth Taylor polynomial of $f + g$ about $\mathbf{X}_0$.

T-5. Let $n = 1$ and f, g, F_k, and G_k be as in Exercise T-2;

$$F_k(t) = a_0 + a_1(t - t_0) + \cdots + a_k(t - t_0)^k,$$
$$G_k(t) = b_0 + b_1(t - t_0) + \cdots + b_k(t - t_0)^k.$$

Let H_k be the kth Taylor polynomial of $fg = h$ about t_0:

$$H_k = c_0 + c_1(t - t_0) + \cdots + c_k(t - t_0)^k.$$

Show that

$$\begin{aligned}
c_0 &= a_0 b_0, \\
c_1 &= a_0 b_1 + a_1 b_0, \\
c_2 &= a_0 b_2 + a_1 b_1 + a_2 b_0, \\
&\vdots \\
c_k &= a_0 b_k + a_1 b_{k-1} + \cdots + a_k b_0,
\end{aligned}$$

which implies that H_k can be obtained by multiplying $F_k(t)$ by $G_k(t)$, collecting coefficients of like powers of $(t - t_0)$ through degree k, and ignoring higher powers of $(t - t_0)$. (*Hint*: Use Leibniz's rule for differentiating a product.)

T-6. As an application of Exercise T-5 find H_4 for $h(t) = e^{3t}$, taking $f(t) = e^t$ and $g(t) = e^{2t}$. Also, obtain H_4 directly.

T-7. Suppose f and g are functions of one variable and $h = f \circ g$ (composite function). Let $y_0 = g(t_0)$ and G_k be the kth Taylor polynomial of g about t_0:

$$G_k(t) = y_0 + b_1(t - t_0) + \cdots + b_k(t - t_0)^k.$$

Let F_k be the kth Taylor polynomial of f about y_0:

$$F_k(y) = a_0 + a_1(y - y_0) + \cdots + a_k(y - y_0)^k.$$

Then H_k, the kth Taylor polynomial of h about t_0, can be obtained by substituting $y = G_k(t)$ into F_k, collecting coefficients of like powers of $(t - t_0)$, and retaining only those powers which do not exceed k. Thus, if

$$F_k(G_k(t)) = c_0 + c_1(t - t_0) + \cdots + c_k(t - t_0)^k + \cdots,$$

then

$$H_k(t) = c_0 + c_1(t - t_0) + \cdots + c_k(t - t_0)^k.$$

We shall not prove this result. However, the student can use it to find H_4, with $t_0 = 0$, for the following functions. Also, he should find H_4 by some other means to check his result.

(a) $h(t) = e^{t^2}$.

(b) $h(t) = \log e^t$.

(c) $h(t) = e^{\sin t}$.

(d) $h(t) = e^{\cos t}$.

4.9 Maxima and Minima

The problem of locating extreme values of a function of one variable is an important application of the one-dimensional calculus. We now consider this problem for functions of several variables.

Let f be defined on a domain D in R^n. Then f is said to have a *relative maximum* at $\mathbf{X}_0$ if there is an n-ball S about $\mathbf{X}_0$ such that

$$(1) \qquad\qquad f(\mathbf{X}) \leq f(\mathbf{X}_0)$$

for all $\mathbf{X}$ belonging to both S and D; if (1) holds for all $\mathbf{X}$ in D, then f is said to have an *absolute maximum* at $\mathbf{X}_0$. Thus, an absolute maximum is a relative maximum, but a relative maximum need not be an absolute maximum.

Definitions of *relative minimum* and *absolute minimum* are obtained by replacing "maximum" by "minimum" in the last paragraph and replacing (1) by

$$f(\mathbf{X}) \geq f(\mathbf{X}_0).$$

An *extreme value* of f is a relative maximum or minimum, and a point where an extreme value is attained is an *extreme point*.

Example 9.1 Let

$$f(x, y) = \begin{cases} 1 - x^2 - y^2 & (x^2 + y^2 < 1), \\[2mm] 0 & (1 \leq x^2 + y^2 \leq 2), \\[2mm] \tfrac{1}{2} & ((x, y) = (3, 0)). \end{cases}$$

This function has an absolute maximum at $(0, 0)$; at every (x, y) satisfying $1 \leq x^2 + y^2 \leq 2$, f has an absolute minimum, and, at those for which $1 < x^2 + y^2 \leq 2$, a relative maximum as well; f has both a relative maximum and minimum at $(3, 0)$, simply because there is a circle about $(3, 0)$ which contains no other point of its domain.

In the one dimensional calculus it is shown that if f is differentiable on the open interval (a, b) and has an extreme point x_0 in (a, b), then $f'(x_0) = 0$. Since $d_{x_0} f = f'(x_0) \, dx$, this can be rewritten as $d_{x_0} f = 0$; in this form the result is true for arbitrary n.

Theorem 9.1 If $\mathbf{X}_0$ is both an interior point of the domain of f and an extreme point of f, and if f is differentiable at $\mathbf{X}_0$, then $d_{\mathbf{X}_0} f = 0$.

Proof. If $d_{\mathbf{X}_0}f \neq 0$ there is a unit vector $\mathbf{U}$ such that $(d_{\mathbf{X}_0}f)(\mathbf{U}) = \sigma > 0$. From Theorem 3.5 (Section 2.3),

$$\lim_{t \to 0} \frac{f(\mathbf{X}_0 + t\mathbf{U}) - f(\mathbf{X}_0)}{t} = \sigma;$$

hence there is a $\delta > 0$ such that

$$\frac{f(\mathbf{X}_0 + t\mathbf{U}) - f(\mathbf{X}_0)}{t} > 0$$

if $0 < |t| < \delta$. Therefore

$$f(\mathbf{X}_0 + t\mathbf{U}) > f(\mathbf{X}_0) \qquad (0 < t < \delta)$$

and

$$f(\mathbf{X}_0 + t\mathbf{U}) < f(\mathbf{X}_0) \qquad (-\delta < t < 0),$$

which is impossible if $\mathbf{X}_0$ is an extreme point. Thus $d_{\mathbf{X}_0}f = 0$ if $\mathbf{X}_0$ is an extreme point.

Example 9.2 Let

$$f(x, y) = 4 - 4x^2 - 9y^2;$$

then f has an absolute maximum at $(0, 0)$ since $4x^2 + 9y^2 \geq 0$, with equality only if $x = y = 0$. The theorem asserts that $(\partial f/\partial x)(0, 0) = (\partial f/\partial y)(0, 0) = 0$, which is easily verified, since $(\partial f/\partial x)(x, y) = -8x$ and $(\partial f/\partial y)(x, y) = -18y$. Since $(0, 0)$ is the only point where the differential vanishes, f has no other extreme points.

The tangent plane to the graph of f at $(0, 0)$ is horizontal (Fig. 9.1). Theorem 9.1 says that this is always true at an extreme point where f is differentiable.

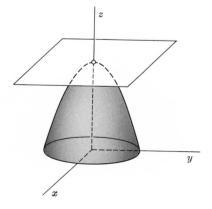

FIGURE 9.1

Differentiability of f at $\mathbf{X}_0$ requires at least that $\mathbf{X}_0$ be an interior point of the domain D of f; hence Theorem 9.1 does not apply to extreme values at boundary points (limit points that are not interior points) of D, or at isolated points such as $(3, 0)$ in Example 9.1. Obviously, Theorem 9.1 cannot be used to test for an extreme value at an interior point where f is not differentiable.

Example 9.3   Let $n = 1$ and

$$f(x) = |\, x \,| \qquad (-1 \le x \le 1);$$

then f has an absolute minimum at $x = 0$ and absolute maxima at $x = \pm 1$. Theorem 9.1 does not apply to them since ± 1 are not interior points of the domain of f and, although $x = 0$ is an interior point, f is not differentiable there. Since $f'(x) = 1$ if $0 < x < 1$ and $f'(x) = -1$ if $-1 < x < 0$, Theorem 9.1 implies that f has no other extreme points; this is also evident from the graph of f (Fig. 9.2).

FIGURE 9.2

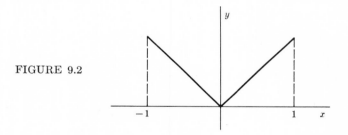

In the rest of this section we shall restrict our attention to extreme points where f is differentiable.

The converse of Theorem 9.1 is false, as is shown by the next two examples.

Example 9.4   Let $\mathbf{X}_0 = (0, 0)$ and

$$f(x, y) = x^2 - y^2;$$

then $d_{\mathbf{X}_0} f = 0$ since $f_x(0, 0) = f_y(0, 0) = 0$. Nevertheless, $(0, 0)$ is not an extreme point of f. Near $(0, 0)$ the graph of f looks like a saddle (Fig. 9.3).

Example 9.5   Let n be arbitrary, $\mathbf{X}_0 = (0, 0, \ldots, 0)$ and

$$f(\mathbf{X}) = x_1^3;$$

then $d_{\mathbf{X}_0} f = 0$ but $\mathbf{X}_0$ is not an extreme point, because $f(\mathbf{X}) < 0$ if $x_1 \le 0$ and $f(\mathbf{X}) > 0$ if $x_1 \ge 0$. Figure 9.4 shows the graph of $f(x) = x^3$ $(n = 1)$; in this case $x = 0$ is a point of inflection.

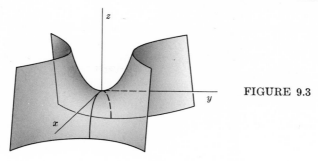

FIGURE 9.3

Definition 9.1 A point $\mathbf{X}_0$ at which f is differentiable is called a *critical point of f* if $d_{\mathbf{X}_0} f = 0$.

We have seen that an extreme point must be a critical point, but that a critical point need not be an extreme point. Although this may seem to be only half of a result, it is nevertheless quite useful. In many problems the nature of a critical point can be deduced from the problem itself, and it is only necessary to locate it.

Example 9.6 Find the distance from the point $\mathbf{X}_1 = (0, 1, -1)$ to the plane $\Gamma: 2x + 3y + z = 1$.

Solution. The distance between $\mathbf{X}_1$ and any point $\mathbf{X}$ is

$$(2) \qquad |\mathbf{X} - \mathbf{X}_1| = \sqrt{x^2 + (y - 1)^2 + (z + 1)^2}$$

and the distance from $\mathbf{X}_1$ to Γ is the minimum value attained by $|\mathbf{X} - \mathbf{X}_1|$ as $\mathbf{X}$ varies over Γ. To be sure that we consider only points on Γ we substitute

$$z + 1 = 2 - 2x - 3y$$

in (2) to obtain

$$|\mathbf{X} - \mathbf{X}_1| = \sqrt{x^2 + (y - 1)^2 + (2 - 2x - 3y)^2}$$

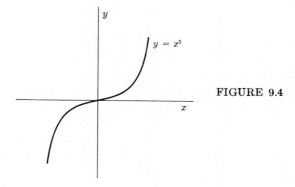

$y = x^3$

FIGURE 9.4

for $\mathbf{X}$ on P. Minimizing $|\mathbf{X} - \mathbf{X}_1|$ is equivalent to minimizing $|\mathbf{X} - \mathbf{X}_1|^2$, so our problem reduces to finding the minimum of

$$\rho(x, y) = x^2 + (y - 1)^2 + (2 - 2x - 3y)^2.$$

The partial derivative are

$$\frac{\partial \rho}{\partial x}(x, y) = 10x + 12y - 8$$

and

$$\frac{\partial \rho}{\partial y}(x, y) = 12x + 20y - 14.$$

The only point at which both derivatives vanish is

$$(x_0, y_0) = \left(\frac{-2}{14}, \frac{11}{14}\right).$$

Theorem 9.1 does not, by itself, tell us that ρ attains its maximum at (x_0, y_0), but the following argument does.

(a)   It is geometrically evident that the problem has a solution; hence ρ must have an absolute minimum at some point $(\hat{x}_0, \hat{y}_0)$.

(b)   Since ρ is differentiable everywhere, the differential must vanish at $(\hat{x}_0, \hat{y}_0)$, from Theorem 9.1.

(c)   The differential vanishes only at (x_0, y_0); hence $(\hat{x}_0, \hat{y}_0) = (x_0, y_0)$.

The minimum value of ρ is

$$\rho(x_0, y_0) = \tfrac{1}{14}$$

and the distance from $\mathbf{X}_1$ to P is

$$d = \sqrt{\rho(x_0, y_0)} = \frac{1}{\sqrt{14}}.$$

Example 9.7 (Least Squares Curve Fitting)   Suppose $t_1, \ldots, t_m$ are distinct and $y_1, \ldots, y_m$ are arbitrary real numbers $(m > 2)$. For each first degree polynomial

$$p(t) = a + bt,$$

define

$$Q(a, b) = \sum_{i=1}^{m} (y_i - a - bt_i)^2.$$

The polynomial

$$p_0(t) = a_0 + b_0 t$$

for which Q is minimized is called the *least squares linear polynomial fit* to the points $(t_1, y_1), \ldots, (t_m, y_m)$. To locate the critical points of Q we equate the first partial derivatives to zero:

$$\frac{\partial Q}{\partial a}(a_0, b_0) = -2 \sum_{i=1}^{m} (y_i - a_0 - b_0 t_i) = 0,$$

$$\frac{\partial Q}{\partial b}(a_0, b_0) = -2 \sum_{i=1}^{m} (y_i - a_0 - b_0 t_i) t_i = 0.$$

Hence (a_0, b_0) satisfies the system

(3)
$$\sum_{i=1}^{m} y_i = a_0 m + b_0 \sum_{i=1}^{m} t_i$$

$$\sum_{i=1}^{m} t_i y_i = a_0 \sum_{i=1}^{m} t_i + b_0 \sum_{i=1}^{m} t_i^2,$$

whose determinant can be shown to be

$$m \sum_{i=1}^{m} t_i^2 - (\sum_{i=1}^{m} t_i)^2 = \tfrac{1}{2} \sum_{i,j=1}^{m} (t_i - t_j)^2 > 0.$$

Therefore (3) has exactly one solution (a_0, b_0); if we assume that the curve fitting problem has a solution, then a_0 and b_0 must be its coefficients, by the reasoning used in Example 9.6.

Taylor's theorem yields a test which is sometimes useful for deciding whether a critical point is an extreme point. To establish this test, we first need a definition.

Definition 9.2 A homogeneous polynomial p in $\mathbf{X} - \mathbf{X}_0$ is said to be *positive (negative) semidefinite* if $p(\mathbf{X}) \geq 0$ $(p(\mathbf{X}) \leq 0)$ for all $\mathbf{X}$; if, in addition, $p(\mathbf{X}) = 0$ only when $\mathbf{X} = \mathbf{X}_0$, p is said to be *positive (negative) definite*. A *semidefinite (definite)* homogeneous polynomial is one that is either positive or negative semidefinite (definite).

Example 9.8 Let $n = 2$. Then

$$p_1(x, y) = (x - 1)^2 + 2(x - 1)y + 2y^2,$$

which is homogeneous in $(x - 1, y)$, is positive definite, since it can be rewritten as

$$p_1(x, y) = (x + y - 1)^2 + y^2.$$

The polynomial

$$p_2(x, y) = x^2 - 2xy + y^2 = (x - y)^2$$

is positive semidefinite. Clearly $-p_1$ and $-p_2$ are negative definite and negative semidefinite, respectively, while

$$p_3(x, y) = x^2 - y^2$$

is not semidefinite.

Theorem 9.2 Let the partial derivatives of f through order $k \geq 2$ be continuous in an n-ball about the critical point $\mathbf{X}_0$, and

(4)
$$d^r_{\mathbf{X}_0} f = 0 \qquad (1 \leq r \leq k - 1),$$
$$d^k_{\mathbf{X}_0} f \neq 0;$$

thus $d^k_{\mathbf{X}_0} f$ is the first nonvanishing higher differential of f at $\mathbf{X}_0$. Then

(a) $\mathbf{X}_0$ is not an extreme point unless $d^k_{\mathbf{X}_0} f$ is semidefinite (in particular, it is not an extreme point if k is odd);

(b) $\mathbf{X}_0$ is a relative minimum (maximum) if $d^k_{\mathbf{X}_0} f$ is positive (negative) definite;

(c) if $d^k_{\mathbf{X}_0} f$ is semidefinite but not definite, no general conclusion can be drawn (that is, $\mathbf{X}_0$ is not necessarily an extreme point, although it may be).

Proof. From (4) and Taylor's theorem,

(5)
$$\lim_{\mathbf{X} \to \mathbf{X}_0} \frac{f(\mathbf{X}) - f(\mathbf{X}_0) - \dfrac{(d^k_{\mathbf{X}_0} f)(\mathbf{X} - \mathbf{X}_0)}{k!}}{|\mathbf{X} - \mathbf{X}_0|^k} = 0,$$

which implies that

(6)
$$\lim_{t \to 0} \frac{f(\mathbf{X}_0 + t\mathbf{U}) - f(\mathbf{X}_0) - \dfrac{t^k}{k!}(d^k_{\mathbf{X}_0} f)(\mathbf{U})}{t^k} = 0,$$

where $\mathbf{U}$ is any unit vector and t is a real variable. For (a), if $d^k_{\mathbf{X}_0} f$ is not semidefinite, there are unit vectors $\mathbf{U}_1$ and $\mathbf{U}_2$ such that

$$(d^k_{\mathbf{X}_0} f)(\mathbf{U}_1) = \sigma_1 > 0 \qquad \text{and} \qquad (d^k_{\mathbf{X}_0} f)(\mathbf{U}_2) = -\sigma_2 < 0,$$

and from (6) there are positive numbers δ_1 and δ_2 such that

$$\frac{f(\mathbf{X}_0 + t\mathbf{U}_1) - f(\mathbf{X}_0)}{t^k} > 0$$

if $0 < |t| < \delta_1$, and

$$\frac{f(\mathbf{X}_0 + t\mathbf{U}_2) - f(\mathbf{X}_0)}{t^k} < 0$$

if $0 < |t| < \delta_2$. Thus if $0 < t < \min(\delta_1, \delta_2)$,

$$f(\mathbf{X}_0 + t\mathbf{U}_1) > f(\mathbf{X}_0)$$

and

$$f(\mathbf{X}_0 + t\mathbf{U}_2) < f(\mathbf{X}_0);$$

hence $\mathbf{X}_0$ is not an extreme point of f.

If k is odd and $d_{\mathbf{X}_0}^k f \neq 0$, it cannot be semidefinite (Exercise T-1). This completes the proof of (a).

For (b), assume that $d_{\mathbf{X}_0}^k f$ is positive definite; then it can be shown that there is a number $\rho > 0$ such that

$$(7) \qquad \frac{(d_{\mathbf{X}_0}^k f)(\mathbf{X} - \mathbf{X}_0)}{k!} > \rho \, |\, \mathbf{X} - \mathbf{X}_0 \,|^k$$

for all $\mathbf{X}$. From (5) and the definition of limit, the length of

$$\frac{f(\mathbf{X}) - f(\mathbf{X}_0) - \dfrac{(d_{\mathbf{X}_0}^k f)(\mathbf{X} - \mathbf{X}_0)}{k!}}{|\, \mathbf{X} - \mathbf{X}_0 \,|^k}$$

can be made as small as we wish by choosing $|\, \mathbf{X} - \mathbf{X}_0 \,|$ sufficiently small (but not zero); in particular, it can be made less than $\rho/2$. Thus

$$f(\mathbf{X}) - f(\mathbf{X}_0) - \frac{(d_{\mathbf{X}_0}^k f)(\mathbf{X} - \mathbf{X}_0)}{k!} > -\frac{\rho}{2} \, |\, \mathbf{X} - \mathbf{X}_0 \,|^k$$

if $\mathbf{X}$ is sufficiently close to $\mathbf{X}_0$; from this and (7),

$$f(\mathbf{X}) - f(\mathbf{X}_0) > \frac{\rho}{2} \, |\, \mathbf{X} - \mathbf{X}_0 \,|^k$$

for such $\mathbf{X}$. Therefore f has a relative minimum at $\mathbf{X}_0$.

The proof of the other half of (b) is similar to this and we leave it to the student (Exercise T-2). Part (c) is proved in Example 9.9.

Corollary 1 Let f, a function of a single variable, have $k \geq 2$ continuous derivatives in an open interval containing a critical point x_0 and suppose that

$$f^{(r)}{}_{(x_0)} = 0 \qquad (1 \leq r \leq k - 1)$$

while

$$f^{(k)}{}_{(x_0)} \neq 0.$$

Then x_0 is a point of inflection if k is odd; if k is even and $f^{(k)}{}_{(x_0)} > 0$ $(f^{(k)}{}_{(x_0)} < 0)$ then x_0 is a relative minimum (maximum).

For the definition of a point of inflection, see Example 9.5. If $k = 2$ this corollary reduces to the second derivative test.

Example 9.9 The following polynomials are homogeneous in $\mathbf{X}$; $\mathbf{X}_0 = (0, 0)$ is a critical point, $p(0, 0) = 0$, and $k = 2$ in all cases.

(a) $p(x, y) = x^2 - y^2 + x^4 + y^4$; then $(d^2_{\mathbf{X}_0}p)(x, y) = 2(x^2 - y^2)$ is not semidefinite. As Theorem 9.2 asserts, $(0, 0)$ is not an extreme point, since $p(t, 2t) = -t^2(3 - 17t^2)$, which is negative if $0 < t\sqrt{3/17}$, while $p(2t, t) = t^2(3 + 17t^2)$, which is positive for all $t \neq 0$.

(b) $p(x, y) = x^2 + y^2 - x^4 - y^4$; then $(d^2_{\mathbf{X}_0}p)(x, y) = 2(x^2 + y^2)$, which is positive definite. Since $p(x, y) = x^2(1 - x^2) + y^2(1 - y^2)$, it is clear that $p(x, y) \geq 0$ if $-1 < x, y < 1$; hence $\mathbf{X}_0$ is a relative minimum, as asserted by Theorem 9.2.

(c) $p(x, y) = x^2 - 2xy + y^2 + x^4 + y^4$; then $(d^2_{\mathbf{X}_0}p)(x, y) = 2(x^2 - 2xy + y^2) = -2(x - y)^2$ is positive semidefinite, but not positive definite; however $\mathbf{X}_0$ is relative minimum because for all (x, y), $p(x, y) = (x - y)^2 + x^4 + y^4 \geq 0$.

(d) $p(x, y) = x^2 - 2xy + y^2 - x^4 - y^4$; again $(d^2_{\mathbf{X}_0}p)(x, y) = 2(x - y)^2$ is positive semidefinite, but not positive definite. In this case, however, $\mathbf{X}_0$ is not an extreme point, since $p(t, t) = -2t^4$ is negative for all $t \neq 0$, while $p(t, -t) = 2t^2(2 - t^2)$ is positive if $0 < |t| < \sqrt{2}$.

The following consequence of Theorem 9.2 is of special interest.

Theorem 9.3 Let f be a function of (x, y) with continuous second partial derivatives f_{xx}, f_{xy}, and f_{yy} in a disk about a critical point $\mathbf{X}_0 = (x_0, y_0)$, and define

(8)
$$D = f_{xx}(\mathbf{X}_0)f_{yy}(\mathbf{X}_0) - f^2_{xy}(\mathbf{X}_0).$$

(a) If $D > 0$, $\mathbf{X}_0$ is an extreme point. It is a relative minimum if $f_{xx}(\mathbf{X}_0) > 0$, a relative maximum if $f_{xx}(\mathbf{X}_0) < 0$.

(b) If $D < 0$, $\mathbf{X}_0$ is not an extreme point.

Proof. Let $\mathbf{X} - \mathbf{X}_0 = (u, v)$ and write

$$(d^2_{\mathbf{X}_0}f)(u, v) = p(u, v) = Au^2 + 2Buv + Cv^2$$

where $A = f_{xx}(\mathbf{X}_0)$, $B = f_{xy}(\mathbf{X}_0)$, $C = f_{yy}(\mathbf{X}_0)$, and $D = AC - B^2$. If $D > 0$, then $A \neq 0$, and we can write

$$p(u, v) = A\left(u^2 + \frac{2B}{A}uv + \frac{B^2}{A^2}v^2\right) + \left(C - \frac{B^2}{A}\right)v^2$$

$$= A\left(u + \frac{B}{A}v\right)^2 + \frac{D}{A}v^2,$$

which vanishes only if $(u, v) = (0, 0)$. Hence $d^2_{X_0} f$ is positive definite if $A > 0$ or negative definite if $A < 0$, and (a) follows from (b) of Theorem 9.2.

If $D < 0$ there are three possibilities:

(i) $A \neq 0$; then $p(1, 0) = A$ and $p\left(\dfrac{-B}{A}, 1\right) = \dfrac{D}{A}$;

(ii) $C \neq 0$; then $p(0, 1) = C$ and $p\left(1, \dfrac{-B}{C}\right) = \dfrac{D}{C}$;

(iii) $A = C = 0$; then $B \neq 0$ and $p(1, 1) = -p(1, -1) = 2B$.

The two values of p in each of these cases are of opposite sign; hence if $D < 0$, $d^2_{X_0} f$ is not semidefinite and X_0 is not an extreme point, from (a) of Theorem 9.2.

Example 9.10   Consider the polynomials of Example 9.9, taking $X_0 = (0, 0)$ in all cases.

(a) $f_{xx}(0, 0) = -f_{yy}(0, 0) = 2$, and $f_{xy}(0, 0) = 0$; hence $D = -4$, and $(0, 0)$ is not an extreme point.

(b) $f_{xx}(0, 0) = f_{yy}(0, 0) = 2$, and $f_{xy}(0, 0) = 0$; hence $D = 4$, and $(0, 0)$ is a relative minimum.

(c) and (d) $f_{xx}(0, 0) = f_{yy}(0, 0) = -f_{xy}(0, 0) = 2$; hence $D = 0$ and the test yields no information.

EXERCISES 4.9

1.   Find the extreme values of f, with domain D:

(a) $f(x) = 2x^3 - 9x^2 + 12x + 6$; $D = \{x \mid |x| \leq \frac{3}{2}\}$.

(b) $f(x) = x$; $D = \{x \mid 0 < x < 1\}$.

(c) $f(x) = \sqrt{1 - x^2}$; $D = \{x \mid |x| \leq 1\}$.

(d) $f(x) = |x^2 - 1|$; $D = R^1$.

(e) $f(x, y) = e^{|x+y|}$; $D = R^2$.

(f) $f(x, y) = \sqrt{|\sin xy|}$; $D = R^2$.

2.   Find all critical points:

(a) $f(x, y) = x^3 + y^3 - x - y$.

(b) $f(x, y, z) = \cos(x^2 + y^2 + z^2)$.

(c) $f(x, y) = 2x^2 + 2xy - 2x + 3y^2 + 2y + 2$.

(d) $f(x, y) = (x^2 + y^2)e^{-2xy}$.

(e) $f(x, y) = (xy)^2 e^{-(x^2 + y^2)}$

3. Find all critical points:

(a) $f(x, y) = \dfrac{x^3}{3} + x^2 y - xy^2 + \dfrac{y^3}{3}$.

(b) $f(x, y) = x^2 e^{-x^2} + y^2 e^{-y^2}$.

(c) $f(x, y) = x \log x + y \log y$.

(d) $f(x, y) = e^{x^2 - y^2}$.

4. Show that among all rectangles with a given perimeter P, the square has the largest area.

5. Find the distance from $(2, -1, 3)$ to the plane $2x + y - 2z = 4$.

6. Find the least squares linear polynomial fit to the points $(2, 1)$, $(1, 3)$, and $(3, 4)$.

7. A rectangular box with no top is to have a given volume V. Find dimensions for the box so as to minimize its surface area.

8. A rectangular box with no top is to have a given surface area S. Find dimensions for the box so as to maximize its volume.

9. Find the distance from the point $(1, 2, -1)$ to the line $x = 2 - t$, $y = 3 + t, z = 1 - 2t$.

10. Find the distance between the lines $x = 2 - t, y = 3 + t, z = 1 - 2t$ and $x = 1 - s, y = 2 - s, z = 3 + s$.

11. Find the dimensions of the largest rectangular box with three faces in the coordinate planes and a vertex in the plane

$$\frac{x}{a} + \frac{y}{b} + \frac{z}{c} = 1.$$

12. Find the distance between the line $x = 1 - 2t, y = 3 + t, z = 1 - t$ and the plane $x + 2y - z = 3$.

13. Classify the critical points of the functions in Exercise 3; that is, state whether they are maxima, minima, or neither.

14. Repeat Exercise 13 for the functions in Exercise 2.

15. Find the volume of the largest rectangular parallelepiped that can be inscribed in the ellipsoid

$$\frac{x^2}{a^2} + \frac{y^2}{b^2} + \frac{z^2}{c^2} = 1.$$

16. Classify the critical points:

    (a) $f(x, y) = \cos(x^2 + y^2)$.   (b) $f(x, y, z) = e^{x^3 + y^3 + z^3}$.

    (c) $f(x, y) = e^{x^2 + y^2}$.   (d) $f(x, y, z) = e^{x^4 + y^3 + z^3}$.

17. Classify the critical points:

    (a) $f(x, y) = \sin(x^2 + y^2)$.   (b) $f(x, y, z) = e^{x^2 - y^2 + z^2}$.

    (c) $f(x, y, z) = e^{x^4 + y^4 + z^4}$.   (d) $f(x, y) = e^{x^4 + y^3}$.

    (e) $f(x, y) = e^{x^4 + y^6}$.

THEORETICAL EXERCISES

T-1. Show that the only semidefinite homogeneous polynomial of odd degree is $p = 0$.

T-2. Prove (b) of Theorem 9.2 if $d^k_{X_0} f$ is negative definite.

T-3. Minimize $f(x_1, \ldots, x_n) = x_1^2 + \cdots + x_n^2$ subject to $x_1 + \cdots + x_n = 1$.

T-4. Let $t_1, \ldots, t_n$ be distinct, $y_1, \ldots, y_n$ be arbitrary, and $m \geq k + 1$. For each polynomial of degree not exceeding k,

$$p(t) = a_0 + a_1 t + \cdots + a_k t^k,$$

define

$$Q(a_0, \ldots, a_k) = \sum_{i=1}^{m} (y_i - a_0 - a_1 t_i - \cdots - a_k t_i^k)^2.$$

The polynomial

$$p_0(t) = \alpha_0 + \alpha_1 t + \cdots + \alpha_k t^k$$

whose coefficients minimize Q is called the *least squares polynomial fit of degree k to the points* $(t_1, y_1), \ldots, (t_n, y_n)$. Show that $\alpha_0, \ldots, \alpha_k$

satisfy the *normal equations*:

$$\sum_{j=0}^{k} s_{i+j}\alpha_j = d_i \qquad (0 \le i \le k),$$

where

$$s_r = \sum_{j=1}^{m} t_j^r \qquad \text{and} \qquad d_r = \sum_{j=1}^{m} y_j t_j^r.$$

T-5. Using Theorem 9.3 show that $f(x, y) = ax^2 + 2bxy + cy^2$ is positive definite if $a > 0$ and

$$\begin{vmatrix} a & b \\ b & c \end{vmatrix} > 0.$$

T-6. If f is differentiable on R^n, how are the critical points of f related to those of f^2?

4.10   The Method of Lagrange Multipliers

Suppose f and $g_1, \ldots, g_k$ are defined on a subset D of R^n and that

(1) $$g_1(\mathbf{X}) = \cdots = g_m(\mathbf{X}) = 0$$

for all $\mathbf{X}$ in some nonempty subset D_1 of D. Then $\mathbf{X}_0$ is said to be a *relative maximum point of f subject to the constraints (1)* if there is an n-ball S about $\mathbf{X}_0$ such that

$$f(\mathbf{X}) \le f(\mathbf{X}_0)$$

for every $\mathbf{X}$ which is in both S and D_1. If

$$f(\mathbf{X}) \ge f(\mathbf{X}_0)$$

under the same assumptions on $\mathbf{X}$ and $\mathbf{X}_0$, then $\mathbf{X}_0$ is a *relative minimum point of f subject to the constraints* (1). In either case we also say that $\mathbf{X}_0$ is a *relative extreme point of f subject to the constraints* (1). We also use the terms *constrained minimum point, constrained maximum point,* and *constrained extreme point.*

Example 9.6, Section 2.9 involved a constrained minimum point, since we wished to find the point in the plane

$$2x + 3y + z = 1$$

that is closest to $(0, 1, -1)$. To solve this problem we must minimize the function

$$f(x, y, z) = x^2 + (y - 1)^2 + (z + 1)^2$$

subject to the constraint

(2) $$g(x, y, z) = 2x + 3y + z - 1 = 0.$$

We accomplished this by solving (2) for z and minimizing

$$h(x, y) = x^2 + (y - 1)^2 + (2 - 2x - 3y)^2.$$

In many applications it is inconvenient to solve the constraint equations as we did there. Fortunately the method of Lagrange multipliers makes this unnecessary.

We shall derive the method only for the simplest case, and state it for the general case. We assume that all functions in this section are continuously differentiable on their domains.

Two Variables Subject to One Constraint

Suppose we wish to extremize a function f of two variables (x, y) subject to the constraint

(3) $$g(x, y) = 0.$$

Let $\mathbf{X}_0 = (x_0, y_0)$ be a constrained extreme point and suppose that $g_y(\mathbf{X}_0) \neq 0$. (An argument similar to the following could be given under the assumption that $g_x(\mathbf{X}_0) \neq 0$.) Then the implicit function theorem implies that (3) defines y as a differentiable function of x is an open interval J about x_0; that is, there is a function $y = y(x)$ such that $y(x_0) = y_0$ and

(4) $$g(x, y(x)) = 0$$

for all x in J. Now define

$$h(x) = f(x, y(x));$$

since $(x, y(x))$ satisfies the constraint equation for all x in J and $\mathbf{X}_0$ is a relative extreme point of f subject to the constraint, it follows that x_0 must be a relative extreme point of h with respect to x. Therefore

(5) $$h'(x_0) = f_x(\mathbf{X}_0) + f_y(\mathbf{X}_0) y'(x_0) = 0.$$

Differentiating both sides of (4) and setting $x = x_0$ yields

(6) $$g_x(\mathbf{X}_0) + g_y(\mathbf{X}_0) y'(x_0) = 0.$$

From (5) and (6), the gradients

$$\nabla_{\mathbf{X}_0} f = \begin{bmatrix} f_x(\mathbf{X}_0) \\ f_y(\mathbf{X}_0) \end{bmatrix} \quad \text{and} \quad \nabla_{\mathbf{X}_0} g = \begin{bmatrix} g_x(\mathbf{X}_0) \\ g_y(\mathbf{X}_0) \end{bmatrix}$$

are both perpendicular to the vector $\begin{bmatrix} 1 \\ y'(x_0) \end{bmatrix}$. Hence they are parallel, and there is a constant λ such that

$$(7) \qquad\qquad\qquad \nabla_{\mathbf{X}_0} f = \lambda\, \nabla_{\mathbf{X}_0} g;$$

therefore

$$(8) \qquad\qquad f_x(\mathbf{X}_0) - \lambda g_x(\mathbf{X}_0) = f_y(\mathbf{X}_0) - \lambda g_y(\mathbf{X}_0) = 0.$$

Geometrically, (7) means that at $\mathbf{X}_0$ the normals to the level curves $g(x, y) = 0$ and $f(x, y) = f(\mathbf{X}_0)$ are parallel; consequently, the two curves have a common tangent. We say in this case that the curves are tangent (Fig. 10.1). In general, two surfaces in R^n are *tangent at a point* $\mathbf{X}_0$ if they have the same tangent plane at $\mathbf{X}_0$.

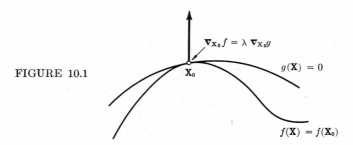

FIGURE 10.1

Analytically (8) means that any extreme point of f subject to the constraint (3) is a critical point of

$$(9) \qquad\qquad\qquad F = f - \lambda g$$

for some value of λ. This is the basis of the method of Lagrange multipliers for finding constrained extreme points, which we now state for a function of two variables subject to one constraint:

(a)   Find all critical points of the *auxiliary function* (9), treating λ as a fixed but unspecified constant.

(b)   Determine which of the points obtained in (a) satisfy the constraint (3). Any such point is called a *critical point of f subject to the constraint*, or simply a *constrained critical point of f*.

(c)   Determine which of the constrained critical points are actually constrained extreme points.

In general, not all of the constrained critical points discovered in part (b) are actually constrained extreme points; however, no point which is not a constrained critical point can be a constrained extreme point.

The parameter λ introduced in the auxiliary function is called a *Lagrange multiplier*. The power of the method of Lagrange multipliers is due to the fact that it does not require solving the constraint equation for one of the variables as a function of the other.

Example 10.1   Suppose we wish to choose the dimensions of a rectangle with given perimeter p so that its area A is a maximum. We must maximize

$$A(x, y) = xy$$

subject to the constraint

(10)
$$2x + 2y - p = 0.$$

The auxiliary function is

$$F(x, y) = xy - \lambda(2x + 2y - p),$$

which has critical points satisfying

$$\frac{\partial F}{\partial x}(x_0, y_0) = y_0 - 2\lambda = 0,$$

$$\frac{\partial F}{\partial y}(x_0, y_0) = x_0 - 2\lambda = 0.$$

Thus any constrained critical point of A must satisfy $x_0 = y_0$ and (10), which implies that $x_0 = y_0 = p/4$. Since we know on geometrical grounds that the problem has a solution, we conclude that, of all rectangles with given perimeter p, the one with the largest area is the square with sides of length $p/4$.

In this example we could just as well have taken

$$F_1(x, y) = xy - \lambda(2x + 2y)$$

to be the auxiliary function, since it differs from F only by a constant, and therefore has the same critical points. We could also have replaced λ by $\lambda/2$, and considered the auxiliary function

$$F_2(x, y) = xy - \lambda(x + y).$$

In general it is legitimate to add an arbitrary constant to an auxiliary function and replace λ by $c\lambda$, where c is any convenient constant. This allows us to dispense with "nuisance constants" in our calculations.

Example 10.2   To find the points on the circle $x^2 + y^2 = 80$ which are closest to and farthest from $(1, 2)$, we must extremize

$$f(x, y) = (x - 1)^2 + (y - 2)^2$$

subject to

(11) $$g(x, y) = x^2 + y^2 - 80 = 0.$$

Define the auxiliary function

$$F(x, y) = (x - 1)^2 + (y - 2)^2 - \lambda(x^2 + y^2);$$

its critical points satisfy

$$\frac{\partial F}{\partial x}(x_0, y_0) = 2(x_0 - 1) - 2\lambda x_0 = 0,$$

$$\frac{\partial F}{\partial y}(x_0, y_0) = 2(y_0 - 2) - 2\lambda y_0 = 0.$$

Therefore

$$\frac{x_0 - 1}{x_0} = \frac{y_0 - 2}{y_0} = \lambda,$$

which implies that $y_0 = 2x_0$. Substituting this in (11) yields

$$5x_0^2 = 80$$

or

$$x_0 = \pm 4.$$

Therefore there are two constrained critical points: $(4, 8)$ and $(-4, -8)$. Since $f(4, 8) = 45$ and $f(-4, -8) = 125$, it follows that $(4, 8)$ is the point on the circle closest to $(1, 2)$ and $(-4, -8)$ is the farthest (Fig. 10.2).

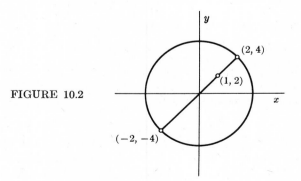

FIGURE 10.2

The General Case

The method of Lagrange multipliers for extremizing a function of n variables subject to m constraints is based on the following theorem, which we state without proof.

Theorem 10.1 Let f and $g_1, \ldots, g_m$ be continuously differentiable on a domain D of R^n, where $n > m \geq 1$, and suppose that at every $\mathbf{X}$ in D the matrix

$$\begin{bmatrix} \dfrac{\partial g_1}{\partial x_1}(\mathbf{X}) & \dfrac{\partial g_1}{\partial x_2}(\mathbf{X}) & \cdots & \dfrac{\partial g_1}{\partial x_n}(\mathbf{X}) \\[2ex] \dfrac{\partial g_2}{\partial x_1}(\mathbf{X}) & \dfrac{\partial g_2}{\partial x_2}(\mathbf{X}) & \cdots & \dfrac{\partial g_2}{\partial x_n}(\mathbf{X}) \\[1ex] \vdots & \vdots & & \vdots \\[1ex] \dfrac{\partial g_m}{\partial x_1}(\mathbf{X}) & \dfrac{\partial g_m}{\partial x_2}(\mathbf{X}) & \cdots & \dfrac{\partial g_m}{\partial x_n}(\mathbf{X}) \end{bmatrix}.$$

has rank m (Section 2.4). Then at any point $\mathbf{X}_0$ in D where f attains an extreme value subject to the constraints

$$(12) \qquad g_1(\mathbf{X}) = \cdots = g_m(\mathbf{X}) = 0,$$

the gradient of f is a linear combination of the gradients of $g_1, \ldots, g_m$; that is,

$$(13) \qquad \nabla_{\mathbf{X}_0} f = \lambda_1 \, \nabla_{\mathbf{X}_0} g_1 + \cdots + \lambda_m \, \nabla_{\mathbf{X}_0} g_m.$$

A proof of this theorem is sketched in Exercises T-2 and T-3, Section 3.5.

We can rewrite (13) in terms of components as

$$\frac{\partial f}{\partial x_i}(\mathbf{X}_0) - \lambda_1 \frac{\partial g_1}{\partial x_i}(\mathbf{X}_0) - \cdots - \lambda_m \frac{\partial g_m}{\partial x_i}(\mathbf{X}_0) = 0 \qquad (1 \leq i \leq m);$$

thus, if we define the auxiliary function

$$(14) \qquad F = f - \lambda_1 g_1 - \cdots - \lambda_m g_m,$$

it follows that

$$F_{x_1}(\mathbf{X}_0) = F_{x_2}(\mathbf{X}_0) = \cdots = F_{x_n}(\mathbf{X}_0) = 0.$$

We therefore conclude that any extremum of f subject to (12) is a critical point of F, for some choice of the constraints $\lambda_1, \ldots, \lambda_m$. This is the basis for the general method of Lagrange multipliers for locating constrained extreme points, which consists of the following three steps.

(a) Find all critical points of the auxiliary function (14), treating the *Lagrange multipliers* $\lambda_1, \ldots, \lambda_m$ as fixed but arbitrary constants.

(b) Determine which of the points obtained in (a) satisfy the constraints (12). These are the *constrained critical points* of f.

(c) Determine which of the constrained critical points are actually constrained extreme points.

We shall now consider several examples of the method of Lagrange multipliers. We observe again that, in forming the auxiliary function (14), it is permissible to omit additive constants and replace $\lambda_1, \ldots, \lambda_m$ by $c_1\lambda_1, \ldots, c_m\lambda_m$, where $c_1, \ldots, c_m$ are constants.

Example 10.3 Suppose we wish to design a rectangular box, open at the top, with maximum volume V and given surface area A. Let x and y be the dimensions of the base and let z be the height (Fig. 10.3). We must maximize

$$V(x, y, z) = xyz$$

subject to the constraint

(15)
$$xy + 2yz + 2xz = A.$$

Define the auxiliary function

$$F(x, y, z) = xyz - \lambda(xy + 2yz + 2xz);$$

its critical points satisfy

(16)
$$y_0 z_0 - \lambda(y_0 + 2z_0) = 0,$$

(17)
$$x_0 z_0 - \lambda(x_0 + 2z_0) = 0,$$

(18)
$$x_0 y_0 - \lambda(2y_0 + 2x_0) = 0.$$

We need only consider critical points of F with positive coordinates. For such points (16) and (17) imply that

$$\frac{y_0 + 2z_0}{y_0 z_0} = \frac{x_0 + 2z_0}{x_0 z_0},$$

from which it follows that $x_0 = y_0$. Now (18) yields $\lambda = x_0/4$. Substituting this in (17) and solving for z_0 yields $z_0 = x_0/2$.

Thus the length, width, and height of the box are x_0, x_0, and $x_0/2$, respectively.

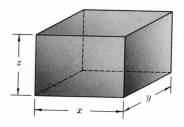

FIGURE 10.3

From (15),

$$3x_0^2 = A;$$

thus

$$x_0 = \sqrt{\frac{A}{3}}$$

and the maximum volume is

$$V = \frac{x_0^3}{2} = \frac{A^{3/2}}{6\sqrt{3}}.$$

In the remaining examples in this section we shall, for simplicity of notation, omit the zero subscript which we have previously attached to the coordinates of constrained critical points.

Example 10.4 By the distance d between the sphere

$$x^2 + y^2 + z^2 = 1$$

and the plane

$$x + y + z = 2\sqrt{3}$$

we mean the minimum value of

$$f(\mathbf{X}_1, \mathbf{X}_2) = \sqrt{(x_1 - x_2)^2 + (y_1 - y_2)^2 + (z_1 - z_2)^2}$$

as $\mathbf{X}_1 = (x_1, y_1, z_1)$ and $\mathbf{X}_2 = (x_2, y_2, z_2)$ vary over the sphere and plane, respectively. To find d we must minimize f subject to the constraints

(19)
$$x_1^2 + y_1^2 + z_1^2 = 1$$

and

(20)
$$x_2 + y_2 + z_2 = 2\sqrt{3}.$$

Clearly we can minimize f^2 to simplify the calculations. Consider the auxiliary function of six variables,

$$
\begin{aligned}
F(\mathbf{X}_1, \mathbf{X}_2) &= f^2(\mathbf{X}_1, \mathbf{X}_2) - \lambda_1(x_1^2 + y_1^2 + z_1^2) - 2\lambda_2(x_2 + y_2 + z_2) \\
&= (x_1 - x_2)^2 + (y_1 - y_2)^2 + (z_1 - z_2)^2 \\
&\quad - \lambda_1(x_1^2 + y_1^2 + z_1^2) - 2\lambda_2(x_2 + y_2 + z_2).
\end{aligned}
$$

Equating the partial derivatives of F to zero yields

(21) $\qquad x_1 - x_2 - \lambda_1 x_1 = y_1 - y_2 - \lambda_1 y_1 = z_1 - z_2 - \lambda_1 z_1 = 0$

and

(22) $\qquad x_1 - x_2 + \lambda_2 = y_1 - y_2 + \lambda_2 = z_1 - z_2 + \lambda_2 = 0.$

From (22),

(23) $\qquad\qquad\qquad x_1 - x_2 = y_1 - y_2 = z_1 - z_2;$

since their common value cannot be zero (the plane and the sphere do not intersect), it follows that $\lambda_1 \neq 0$ in (21), and therefore

$$x_1 = y_1 = z_1.$$

Therefore, from (19),

(24) $\qquad\qquad\qquad x_1 = y_1 = z_1 = \pm \dfrac{1}{\sqrt{3}}.$

Now it follows from (20) and (23) that

$$x_2 = y_2 = z_2 = \frac{2}{\sqrt{3}}.$$

Clearly we must choose the plus sign in the last member of (24), since

$$(1/\sqrt{3}, \quad 1/\sqrt{3}, \quad 1/\sqrt{3})$$

is closer to

$$(2/\sqrt{3}, \quad 2/\sqrt{3}, \quad 2/\sqrt{3})$$

than is

$$(-1/\sqrt{3}, \quad -1/\sqrt{3}, \quad -1/\sqrt{3}).$$

Thus the distance from the plane to the sphere is

$$d = \sqrt{\left(\frac{2}{\sqrt{3}} - \frac{1}{\sqrt{3}}\right)^2 + \left(\frac{2}{\sqrt{3}} - \frac{1}{\sqrt{3}}\right)^2 + \left(\frac{2}{\sqrt{3}} - \frac{1}{\sqrt{3}}\right)^2} = 1.$$

Example 10.5   Let

$$\mathbf{A} = \begin{bmatrix} a_{11} & a_{12} & \cdots & a_{1n} \\ a_{21} & a_{22} & \cdots & a_{2n} \\ \vdots & \vdots & & \vdots \\ a_{n1} & a_{n2} & \cdots & a_{nn} \end{bmatrix}$$

be a symmetric matrix; that is,

(25) $\qquad\qquad\qquad\qquad a_{ij} = a_{ji}.$

Let Q be the associated quadratic form:

(26)
$$Q(\mathbf{X}) = \mathbf{X} \cdot \mathbf{AX} = \sum_{i,j=1}^{n} a_{ij} x_i x_j.$$

(Here we think of $\mathbf{X}$ as a column vector.) Suppose we wish to maximize Q subject to the requirement

(27)
$$|\mathbf{X}| = 1.$$

Using (25) and equating the partial derivatives of

(28)
$$F(\mathbf{X}) = Q(\mathbf{X}) - \lambda |\mathbf{X}|^2.$$

to zero yields

$$\sum_{j=1}^{n} a_{ij} x_j = \lambda x_i \qquad (1 \le i \le n)$$

(verify!), or, in matrix notation,

$$\mathbf{AX} = \lambda \mathbf{X},$$

which means that $\mathbf{X}$ is an eigenvector and λ an eigenvalue of $\mathbf{A}$. If $\mathbf{X}$ satisfies this equation then, from (26) and (27),

$$Q(\mathbf{X}) = \mathbf{X} \cdot (\lambda \mathbf{X}) = \lambda |\mathbf{X}|^2 = \lambda.$$

Therefore, the maximum value of Q subject to (27) is equal to the largest eigenvalue of $\mathbf{A}$, and it is attained when $\mathbf{X}$ is an associated eigenvector. Similarly, the minimum value of Q subject to (27) is the smallest eigenvalue of Q.

Example 10.6   Let us minimize

$$Q(\mathbf{X}) = \sum_{r=1}^{n} x_r^2$$

subject to the constraints

(29)
$$\sum_{r=1}^{n} x_r = 1, \qquad \sum_{r=1}^{n} r x_r = 0.$$

Define the auxiliary function

$$F(\mathbf{X}) = \sum_{r=1}^{n} x_r^2 - 2\lambda_0 \sum_{r=1}^{n} x_r - 2\lambda_1 \sum_{r=1}^{n} r x_r;$$

equating its partial derivatives to zero yields

$$x_r = \lambda_0 + \lambda_1 r \qquad (1 \le r \le n).$$

To determine λ_0 and λ_1 we substitute in (29) to find that

$$\sum_{r=1}^{n} (\lambda_0 + \lambda_1 r) = 1,$$

$$\sum_{r=1}^{n} (\lambda_0 + \lambda_1 r) r = 0;$$

thus λ_0 and λ_1 satisfy the simultaneous equations

(30)
$$s_0 \lambda_0 + s_1 \lambda_1 = 1,$$
$$s_1 \lambda_0 + s_2 \lambda_1 = 0,$$

where

$$s_0 = \sum_{1}^{n} 1,$$

$$s_1 = \sum_{1}^{n} r = \frac{n(n+1)}{2},$$

$$s_2 = \sum_{1}^{n} r^2 = \frac{n(n+1)(2n+1)}{6}.$$

Solving (30) yields

$$\lambda_0 = \frac{2(2n+1)}{n(n-1)},$$

$$\lambda_1 = -\frac{6}{n(n-1)}.$$

Thus the only extreme point of Q subject to (29) is $\hat{\mathbf{X}} = (\hat{x}_1, \ldots, \hat{x}_n)$ where

$$\hat{x}_r = \frac{2(2n+1) - 6r}{n(n-1)} \qquad (1 \le r \le n).$$

To see that $\hat{\mathbf{X}}$ is a constrained minimum, suppose $\mathbf{X} = (x_1, \ldots, x_n)$ is any point which satisfies the constraints; then

$$\sum_{r=1}^{n} (x_r - \hat{x}_r) = \sum_{r=1}^{n} (x_r - \hat{x}_r) r = 0,$$

and therefore

$$\sum_{r=1}^{n} \hat{x}_r (x_r - \hat{x}_r) = 0.$$

Hence

$$\sum_{r=1}^{n} x_r^2 = \sum_{r=1}^{n} [\hat{x}_r + (x_r - \hat{x}_r)]^2$$

$$= \sum_{r=1}^{n} \hat{x}_r^2 + \sum_{r=1}^{n} (x_r - \hat{x}_r)^2 \geq \sum_{r=1}^{n} \hat{x}_r^2,$$

and equality occurs only if $x_r - \hat{x}_r = 0 \ (1 \leq r \leq n)$.

EXERCISES 4.10

Use the method of Lagrange multipliers in the following problems.

1. Find the dimensions of the rectangle with minimum perimeter subject to the requirement that its area be A.

2. A rectangular box with no top is to have given volume V. Find dimensions for the box so as to minimize its surface area.

3. Find the extreme points of $f(x, y) = 2x + y$ subject to $x^2 + y^2 = 4$.

4. Find the distance from $(1, 2, -3)$ to the plane $x + 2y - z = 4$.

5. Find the extreme points of $f(x, y, z) = 2x + y + 2z$ subject to $x^2 + y^2 = 4$ and $x + z = 2$.

6. Find the maximum and minimum of $f(x, y, z) = x + y^2 + 2z$ subject to $4x^2 + 9y^2 - 36z^2 = 36$.

7. Find the dimensions of the largest rectangular box with three faces in the coordinate planes and a vertex in the plane

$$\frac{x}{a} + \frac{y}{b} + \frac{z}{c} = 1.$$

8. Find the distance from the point $(-1, 2, 3)$ to the line of intersection of the planes $x + 2y - 3z = 4$ and $2x - y + 2z = 5$.

9. Find the points on the ellipse $4x^2 + 9y^2 = 36$ which are closest to and farthest from $(0, 0)$.

10. Find the extreme value of xy on the line $3x + 2y = 6$.

11. Find the volume of the largest rectangular parallelepiped that can be inscribed in the ellipsoid

$$\frac{x^2}{a^2} + \frac{y^2}{b^2} + \frac{z^2}{c^2} = 1.$$

12. Find the points on the ellipsoid

$$4x^2 + 9y^2 + 36z^2 = 1.$$

whose distances from the origin are relative extremes.

13. Extremize $f(x, y, z) = x^2 + y^2 + z^2$ subject to $2x + 3y - 4z = 5$.

14. Find the points on the curve $x^2 + 4xy + y^2 = 36$ which are closest to the origin.

15. Find the distance between the circle $x^2 + y^2 = 1$ and the part of the hyperbola $xy = 1$ which is in the first quadrant.

In Exercises 16, 17, and 18 let

$$\mathbf{X} = \begin{bmatrix} x_1 \\ x_2 \\ x_3 \\ x_4 \end{bmatrix}, \qquad \mathbf{A} = \begin{bmatrix} 0 & 0 & -2 & 0 \\ 0 & -2 & 0 & 0 \\ -2 & 0 & 3 & 0 \\ 0 & 0 & 0 & 1 \end{bmatrix}$$

and $Q(\mathbf{X}) = \mathbf{X}^{\mathrm{T}} \cdot \mathbf{A} \mathbf{X}$.

16. Find the maximum value of $Q(\mathbf{X})$ subject to the constraint $x_1^2 + x_2^2 + x_3^2 + x_4^2 = 1$.

17. Find the maximum value of $Q(\mathbf{X})$ subject to the constraint in Exercise 16, and the additional constraint $x_1 - 2x_3 = 0$.

18. Find the maximum value of $Q(\mathbf{X})$ subject to the constraints of Exercise 17 and the additional constraint $x_4 = 0$.

THEORETICAL EXERCISES

T-1. Prove the Schwarz inequality by extremizing $F(\mathbf{U}, \mathbf{V}) = \mathbf{U} \cdot \mathbf{V}$ where $\mathbf{U}$ and $\mathbf{V}$ are n-vectors) subject to appropriate constraints.

T-2. What is wrong with defining the auxiliary function in Example 10.5 as $F(\mathbf{X}) = Q(\mathbf{X}) - \lambda |\mathbf{X}|$?

T-3. Let α be an eigenvalue of the symmetric matrix $\mathbf{A}$ of Example 10.5 and let $\mathbf{U}$ be the associated eigenvector. Show that the extreme values of $Q(\mathbf{X})$ subject to the constraints $|\mathbf{X}| = 1$ and $\mathbf{X} \cdot \mathbf{U} = 0$ are eigenvalues of $\mathbf{A}$, and that they are attained when $\mathbf{X}$, considered as a column vector, is an associated eigenvector.

T-4. Let $n \geq k + 1$ and $c_0, \ldots, c_k$ be given constants. Show that the minimum value of

$$Q(\mathbf{X}) = \sum_{r=1}^{n} x_r^2$$

subject to the constraints

$$\sum_{r=1}^{n} x_r r^i = c_i \qquad (0 \leq i \leq k)$$

is attained when

$$x_r = \lambda_0 + \lambda_1 r + \cdots + \lambda_k r^k,$$

where

$$\sum_{j=0}^{k} \sigma_{i+j} \lambda_j = c_i \qquad (0 \leq i \leq k)$$

and

$$\sigma_i = \sum_{r=1}^{n} r^i.$$

Chapter 5

Differential Calculus of Vector-Valued Functions

5.1 Functions, Limits, and Continuity

If $f_1, \ldots, f_m$ are real valued functions defined on a set D in R^n, then

$$\mathbf{f} = \begin{bmatrix} f_1 \\ \vdots \\ f_m \end{bmatrix}$$

is the function that assigns to every $\mathbf{X}$ in D the m-vector

$$(1) \qquad\qquad \mathbf{f}(\mathbf{X}) = \begin{bmatrix} f_1(\mathbf{X}) \\ \vdots \\ f_m(\mathbf{X}) \end{bmatrix}.$$

We say that $\mathbf{f}$ is *vector valued*, and $f_1, \ldots, f_m$ are called the *coordinate functions* or *components* of $\mathbf{f}$. The *range* (or *image*) of $\mathbf{f}$ is the set of m-vectors obtained from (1) as $\mathbf{X}$ varies over D, which is the *domain* of $\mathbf{f}$.

As a convenient method for stating the dimensions of the spaces containing the domain and range of $\mathbf{f}$, we shall write

$$(2) \qquad\qquad \mathbf{f}: \quad R^n \to R^m$$

when we mean that D is a subset of R^n, and that the range of $\mathbf{f}$ is a subset of R^m. Many authors use this notation to mean that the domain of $\mathbf{f}$ is *all* of R^n. It should be emphasized that when we use this notation we mean only that the domain of $\mathbf{f}$ is a subset D of R^n; of course this includes the possibility that $D = R^n$.

If f is a real valued function such as those studied in Chapter 4, we shall write

$$f: \quad R^n \to R^1$$

or simply

$$f: \quad R^n \to R;$$

when $m = 1$, it is to be understood that $\mathbf{f} = f_1$ in (1).

When we write (2) without further definition of the domain of $\mathbf{f}$, we shall mean that D is the largest subset of R^n common to the domains of $f_1, \ldots, f_m$.

Example 1.1   Let $\mathbf{f}: R^2 \to R^3$ have components

$$f_1(x, y) = x + y,$$

$$f_2(x, y) = x - y,$$

$$f_3(x, y) = x^2 + y^2.$$

Thus

$$\mathbf{f} = \begin{bmatrix} f_1 \\ f_2 \\ f_3 \end{bmatrix} \quad \text{and} \quad \mathbf{f}(x, y) = \begin{bmatrix} x + y \\ x - y \\ x^2 + y^2 \end{bmatrix}.$$

The domain of $\mathbf{f}$ is all of R^2 and its range is a subset of R^3.

Example 1.2   Let $\mathbf{f}: R^2 \to R^2$ be defined by

$$\mathbf{f}(x, y) = \begin{bmatrix} \dfrac{1}{x - y} \\ x^2 + y^2 \end{bmatrix};$$

the components of $\mathbf{f}$ are

$$f_1(x, y) = \frac{1}{x - y},$$

$$f_2(x, y) = x^2 + y^2.$$

The domain of $\mathbf{f}$ is all of R^2 except points of the form (x, x), which are not in the domain of f_1. The range of $\mathbf{f}$ is a subset of R^2.

Example 1.3 Let

$$
\mathbf{f}(x, y) = \begin{bmatrix} x \\ y \\ \sqrt{1 - x^2 - y^2} \end{bmatrix}.
$$

The domain of $\mathbf{f}$ is the set of (x, y) such that $x^2 + y^2 \leq 1$; its range is a subset of R^3.

If D is the largest subset of R^n on which both

$$
\mathbf{f} = \begin{bmatrix} f_1 \\ \vdots \\ f_m \end{bmatrix} \qquad \text{and} \qquad \mathbf{g} = \begin{bmatrix} g_1 \\ \vdots \\ g_m \end{bmatrix}
$$

are defined, then we define

$$
\mathbf{f} + \mathbf{g} = \begin{bmatrix} f_1 + g_1 \\ \vdots \\ f_m + g_m \end{bmatrix} \qquad \text{and} \qquad \mathbf{f} - \mathbf{g} = \begin{bmatrix} f_1 - g_1 \\ \vdots \\ f_m - g_m \end{bmatrix}
$$

on D. If $\mathbf{f}: R^n \to R^m$ and $\mathbf{g}: R^{n_1} \to R^{m_1}$, then $\mathbf{f} + \mathbf{g}$ and $\mathbf{f} - \mathbf{g}$ are not defined unless $n = n_1$ and $m = m_1$.

If c is a constant then

$$
c\mathbf{f} = \begin{bmatrix} cf_1 \\ \vdots \\ cf_m \end{bmatrix}.
$$

Example 1.4 Let

$$
\mathbf{f}(x, y) = \begin{bmatrix} x \\ \sqrt{1 - x^2 - y^2} \\ y \end{bmatrix}
$$

and

$$
\mathbf{g}(x, y) = \begin{bmatrix} x \\ x\sqrt{1 - 2x^2 - 2y^2} \\ y \end{bmatrix};
$$

then

$$(\mathbf{f} + \mathbf{g})\,(x, y) \;=\; \begin{bmatrix} 2x \\ \sqrt{1 - x^2 - y^2} + x\sqrt{1 - 2x^2 - 2y^2} \\ 2y \end{bmatrix}$$

$$(x^2 + y^2 \leq 1/2).$$

Example 1.5 Let

$$\mathbf{f}(x, y) \;=\; \begin{bmatrix} x \\ y \\ x^2 - y^2 \end{bmatrix}$$

and $c = -3$. Then $-3\mathbf{f}$ has values

$$(-3\mathbf{f})\,(x, y) \;=\; \begin{bmatrix} -3x \\ -3y \\ -3x^2 + 3y^2 \end{bmatrix}.$$

Definition 1.1 Let $\mathbf{f}: R^n \to R^m$, $\mathbf{g}: R^m \to R^p$, and let D be the subset of the domain of $\mathbf{f}$ consisting of those n-vectors $\mathbf{X}$ such that the m-vector $\mathbf{f}(\mathbf{X})$ is in the domain of $\mathbf{g}$. Then the *composite function* $\mathbf{g} \circ \mathbf{f}$ is defined for $\mathbf{X}$ in D by

$$(\mathbf{g} \circ \mathbf{f})\,(\mathbf{X}) \;=\; \mathbf{g}(\mathbf{f}(\mathbf{X})).$$

Thus $\mathbf{g} \circ \mathbf{f}: R^n \to R^p$.

If $m = p = 1$, this reduces to Definition 1.2, Section 4.1.

Composite functions are so important that we consider them in more detail. Let

$$\mathbf{f}(\mathbf{X}) \;=\; \begin{bmatrix} f_1(\mathbf{X}) \\ \vdots \\ f_m(\mathbf{X}) \end{bmatrix}$$

and

(3)
$$\mathbf{g}(u_1, \ldots, u_m) \;=\; \begin{bmatrix} g_1(u_1, \ldots, u_m) \\ \vdots \\ g_p(u_1, \ldots, u_m) \end{bmatrix};$$

then $(\mathbf{g} \circ \mathbf{f})\,(\mathbf{X})$ is obtained by setting $u_1 = f_1(\mathbf{X}), \ldots, u_m = f_m(\mathbf{X})$ in (3):

$$(4) \qquad\qquad (\mathbf{g} \circ \mathbf{f})\,(\mathbf{X}) = \begin{bmatrix} g_1(\,f_1(\mathbf{X}), \ldots, f_m(\mathbf{X})\,) \\ \vdots \\ g_p(\,f_1(\mathbf{X}), \ldots, f_m(\mathbf{X})\,) \end{bmatrix}.$$

However, according to Definition 1.1 with $\mathbf{g}$ replaced by g_i and $p = 1$,

$$g_i(\,f_1(\mathbf{X}), \ldots, f_m(\mathbf{X})\,) = (g_i \circ f)\,(\mathbf{X});$$

thus, from (4),

$$\mathbf{g} \circ \mathbf{f} = \begin{bmatrix} g_1 \circ \mathbf{f} \\ \vdots \\ g_p \circ \mathbf{f} \end{bmatrix};$$

that is, $(g \circ \mathbf{f})_i = g_i \circ \mathbf{f}$.

Example 1.6 Suppose $\mathbf{f}\colon R^n \to R^m$ is a linear transformation (Section 2.3). Thus

$$\mathbf{f} = \begin{bmatrix} f_1 \\ \vdots \\ f_m \end{bmatrix},$$

where

$$f_r(\mathbf{X}) = \sum_{j=1}^{n} a_{rj} x_j \qquad (1 \le r \le m)$$

and the a_{rj} are constants.

Now suppose $\mathbf{g}$ is a linear transformation from R^m to R^p; thus if $\mathbf{U} = (u_1, \ldots, u_m)$ is a point in R^m then

$$\mathbf{g} = \begin{bmatrix} g_1 \\ \vdots \\ g_p \end{bmatrix}$$

where

$$g_i(\mathbf{U}) = \sum_{s=1}^{m} b_{is} u_s \qquad (1 \le i \le p).$$

Since the range of $\mathbf{f}$ is contained in the domain of $\mathbf{g}$, we can define the composite function $\mathbf{g} \circ \mathbf{f}$. To obtain $(\mathbf{g} \circ \mathbf{f})_i$, we take

$$u_s = f_s(\mathbf{X}) = \sum_{j=1}^{n} a_{sj} x_j;$$

then

$$(\mathbf{g} \circ \mathbf{f})_i(\mathbf{X}) = (g_i \circ \mathbf{f})(\mathbf{X})$$

$$= \sum_{s=1}^{m} b_{is} \sum_{j=1}^{n} a_{sj}x_j$$

$$= \sum_{j=1}^{n} \left(\sum_{s=1}^{m} b_{is}a_{sj} \right) x_j.$$

Thus $\mathbf{g} \circ \mathbf{f}$ is a linear transformation from R^n to R^p:

$$(\mathbf{g} \circ \mathbf{f})_i(\mathbf{X}) = \sum_{j=1}^{n} c_{ij}x_j \qquad (1 \leq i \leq p),$$

where

$$c_{ij} = \sum_{s=1}^{m} b_{is}a_{sj} \qquad (1 \leq i \leq p, \quad 1 \leq j \leq m).$$

In terms of the matrices $\mathbf{M_f} = (a_{ij})$, $\mathbf{M_g} = (b_{ij})$, and $\mathbf{M_{g \circ f}} = (c_{ij})$ we have

$$\mathbf{M_{g \circ f}} = \mathbf{M_g}\mathbf{M_f}.$$

Example 1.7 Let

$$\mathbf{f}(x, y) = \begin{bmatrix} x \\ x \cos y \\ x \sin y \end{bmatrix} \qquad (\mathbf{f} \colon R^2 \to R^3)$$

and

$$\mathbf{g}(u, v, w) = \begin{bmatrix} uv \\ vw \\ uw \\ \sqrt{1 - u^2 - v^2 - w^2} \end{bmatrix} \qquad (\mathbf{g} \colon R^3 \to R^4);$$

then

$$(5) \qquad (\mathbf{g} \circ \mathbf{f})(x, y) = \begin{bmatrix} x^2 \cos y \\ x^2 \sin y \cos y \\ x^2 \sin y \\ \sqrt{1 - 2x^2} \end{bmatrix} \qquad (\mathbf{g} \circ \mathbf{f}: \ R^2 \to R^4),$$

and the domain of $\mathbf{g} \circ \mathbf{f}$ is the set of points (x, y) such that $|x| \leq 1/\sqrt{2}$.

Definition 1.2   If $\mathbf{X}_0$ is a limit point of the domain of

$$\mathbf{f} = \begin{bmatrix} f_1 \\ \vdots \\ f_m \end{bmatrix} \qquad \text{and} \qquad \mathbf{U}_0 = \begin{bmatrix} b_1 \\ \vdots \\ b_m \end{bmatrix},$$

we say that $\lim_{\mathbf{X} \to \mathbf{X}_0} \mathbf{f}(\mathbf{X}) = \mathbf{U}_0$ if $\lim_{\mathbf{X} \to \mathbf{X}_0} f_i(\mathbf{X}) = b_i$ $(1 \leq i \leq m)$.

Example 1.8   Let $\mathbf{X}_0 = (-1, 2)$ and

$$\mathbf{f}(x, y) = \begin{bmatrix} 2x + 3y \\ x + y \\ x - y \end{bmatrix};$$

then $\lim_{\mathbf{X} \to \mathbf{X}_0} f_1(x, y) = 4$, $\lim_{\mathbf{X} \to \mathbf{X}_0} f_2(x, y) = 1$, and $\lim_{\mathbf{X} \to \mathbf{X}_0} f_3(x, y) = -3$. Hence

$$\lim_{\mathbf{X} \to \mathbf{X}_0} \mathbf{f}(x, y) = \begin{bmatrix} 4 \\ 1 \\ -3 \end{bmatrix}.$$

Example 1.9   Let $\mathbf{X}_0 = (0, 0)$ and

$$\mathbf{f}(x, y) = \begin{bmatrix} x + y \\ \dfrac{2xy}{x^2 + y^2} \end{bmatrix}.$$

Then $\lim\limits_{X \to X_0} f_1(x, y) = 0$, but f_2 has no limit as (x, y) approaches $(0, 0)$. Hence $\lim\limits_{X \to X_0} f(x, y)$ does not exist.

Theorem 1.1 $\lim\limits_{X \to X_0} \mathbf{f}(\mathbf{X}) = \mathbf{U}_0$ if and only if $\lim\limits_{X \to X_0} |\mathbf{f}(X) - \mathbf{U}_0| = 0$.

Proof. Write

$$|\mathbf{f}(X) - \mathbf{U}_0| = \sqrt{(f_1(\mathbf{X}) - b_1)^2 + \cdots + (f_m(\mathbf{X}) - b_m)^2}.$$

The sum of squares on the right approaches zero as $\mathbf{X}$ approaches $\mathbf{X}_0$ if and only if the individual terms do; that is, if and only if

$$\lim_{X \to X_0} f_i(\mathbf{X}) = b_i \qquad (1 \leq i \leq m).$$

The following corollary gives a useful interpretation of Theorem 1.1.

Corollary 1.1 If $\mathbf{f}: R^n \to R^m$, then $\lim\limits_{X \to X_0} \mathbf{f}(\mathbf{X}) = \mathbf{U}_0$ if and only if, for every $\epsilon > 0$, there is a $\delta > 0$ such that $|\mathbf{f}(X) - \mathbf{U}_0| < \epsilon$ whenever $\mathbf{X}$ is in the domain of $\mathbf{f}$ and $0 < |\mathbf{X} - \mathbf{X}_0| < \delta$.

We leave the proof to the student.

From Definition 1.2 it follows that all questions about limits of a vector valued function can be answered by examining its components. Thus, limits are unique, and the following theorems can be proved by applying Theorem 1.2 of Section 4.1 to the components of $\mathbf{f}$ and $\mathbf{g}$.

Theorem 1.2 Let $\mathbf{X}_0$ be a limit point of the set of points common to the domains of $\mathbf{f}$ and $\mathbf{g}$. If $\lim\limits_{X \to X_0} \mathbf{f}(\mathbf{X}) = \mathbf{U}_0$ and $\lim\limits_{X \to X_0} \mathbf{g}(\mathbf{X}) = \mathbf{U}_1$, then

$$\lim_{X \to X_0} (\mathbf{f} + \mathbf{g})(\mathbf{X}) = \mathbf{U}_0 + \mathbf{U}_1$$

and

$$\lim_{X \to X_0} (\mathbf{f} - \mathbf{g})(\mathbf{X}) = \mathbf{U}_0 - \mathbf{U}_1.$$

Theorem 1.3 Let c be a constant and $\lim\limits_{X \to X_0} \mathbf{f}(\mathbf{X}) = \mathbf{U}_0$. Then

$$\lim_{X \to X_0} (c\mathbf{f})(\mathbf{X}) = c\mathbf{U}_0.$$

Definition 1.3 A vector valued function is said to be *continuous* at a point $\mathbf{X}_0$ if each of its components is continuous at $\mathbf{X}_0$.

Example 1.10 Let $\mathbf{X}_0 = (0, 0)$ and

$$\mathbf{f}(x, y) = \begin{bmatrix} \cos(x + y) \\ e^{x+y} \\ 2x^2 - y^2 + 1 \end{bmatrix};$$

then $\mathbf{f}$ is continuous at $\mathbf{X}_0$, since each of its components is continuous at $\mathbf{X}_0$. Thus

$$\lim_{\mathbf{X} \to \mathbf{X}_0} \mathbf{f}(\mathbf{X}) = \begin{bmatrix} 1 \\ 1 \\ 1 \end{bmatrix} = \mathbf{f}(\mathbf{X}_0).$$

Example 1.11 Let $\mathbf{X}_0 = (1, 1)$ and

$$\mathbf{f}(x, y) = \begin{cases} \begin{bmatrix} x - y \\ \dfrac{x^2 - y^2}{x - y} \end{bmatrix} & \text{if } x \neq y, \\ \begin{bmatrix} 0 \\ 0 \end{bmatrix} & \text{if } x = y. \end{cases}$$

Then f_1 is continuous at $\mathbf{X}_0$, but f_2 is not, because

$$\lim_{\mathbf{X} \to \mathbf{X}_0} f_2(x, y) = \lim_{\mathbf{X} \to \mathbf{X}_0} (x + y)$$

$$= 2 \neq 0 = f_2(\mathbf{X}_0);$$

hence $\mathbf{f}$ is not continuous at $\mathbf{X}_0$.

Example 1.12 A linear transformation $\mathbf{L}: R^n \to R^m$

$$\mathbf{L} = \begin{pmatrix} L_1 \\ \vdots \\ L_m \end{pmatrix}$$

is continuous on R^n, since the real valued linear transformations $L_1, \ldots, L_m$ are continuous on R^n.

The following theorem can be obtained by applying Theorem 1.3 of Section 4.1 to the component functions of **f** and **g**.

Theorem 1.4 (a) If a vector valued function **f** is continuous at $\mathbf{X}_0$ and c is a constant, then $c\mathbf{f}$ is continuous at $\mathbf{X}_0$. (b) Let D be the set of points common to the domains of **f** and **g** (both $R^n \to R^m$). Let $\mathbf{X}_0$ be in D and be a limit point of D. Then, if **f** and **g** are continuous at $\mathbf{X}_0$, so are $\mathbf{f} + \mathbf{g}$ and $\mathbf{f} - \mathbf{g}$.

A continuous function of a continuous function is continuous; this is stated more precisely in the following theorem, which generalizes Theorem 1.4 of Section 4.1.

Theorem 1.5 Let **f** be continuous at $\mathbf{X}_0$, and $\mathbf{U}_0 = \mathbf{f}(\mathbf{X}_0)$ be an interior point of the domain of **g**, at which **g** is continuous. Then the composite function $\mathbf{g} \circ \mathbf{f}$ is continuous at $\mathbf{X}_0$.

Example 1.13 Let **f** and **g** be as in Example 1.7, and let $\mathbf{X}_0 = (1/2, \pi/2)$. Then **f** is continuous at $\mathbf{X}_0$ and

$$\mathbf{U}_0 = \mathbf{f}(\mathbf{X}_0) = \begin{bmatrix} \frac{1}{2} \\ 0 \\ \frac{1}{2} \end{bmatrix}$$

is an interior point of the domain of **g**; furthermore, **g** is continuous at $\mathbf{U}_0$. Theorem 1.5 implies that $\mathbf{g} \circ \mathbf{f}$ is continuous at $\mathbf{X}_0$, which can be verified from the explicit expression (5) for $(\mathbf{g} \circ \mathbf{f})(\mathbf{X})$.

EXERCISES 5.1

1. Find the coordinate functions and domains of the following functions.

(a) $\mathbf{f}(x, y) = \begin{bmatrix} \cos xy \\ \sqrt{x^2 + y^2 - 4} \\ \dfrac{1}{xy} \end{bmatrix}$. (b) $\mathbf{f}(x, y) = \begin{bmatrix} e^{x^2 + y^2} \\ \dfrac{x}{y^2 - 1} \end{bmatrix}$.

(c) $\mathbf{f}(x, y, z) = \begin{bmatrix} x^2 + y^2 + z^2 \\ x^2 \\ -y^2 \end{bmatrix}.$ (d) $\mathbf{f}(x, y) = \begin{bmatrix} \sqrt{4 - x^2} \\ \sqrt{4 - y^2} \\ x + y \end{bmatrix}.$

2. Repeat Exercise 1 for the following functions.

(a) $\mathbf{f}(x, y, z) = \begin{bmatrix} x + \dfrac{1}{z} \\ z \\ y^2 - 9 \end{bmatrix}.$ (b) $\mathbf{f}(x, y) = \begin{bmatrix} \log(x - y) \\ e^{x-y} \end{bmatrix}.$

(c) $\mathbf{f}(x, y, z) = \begin{bmatrix} \dfrac{1}{x - y} \\ \dfrac{1}{y^2 - z^2} \\ \dfrac{1}{x - z} \end{bmatrix}.$ (d) $\mathbf{f}(x, y) = \begin{bmatrix} \dfrac{1}{\cos(x^2 - y^2)} \\ e^{x^2 - y^2} \\ \sqrt{x^2 - y^2} \end{bmatrix}.$

3. For each function in Exercise 1, fill in m and n in "$\mathbf{f}: R^n \to R^m$."

4. For each function in Exercise 2, fill in m and n in "$\mathbf{f}: R^n \to R^m$."

5. Let

$$\mathbf{f}(x, y) = \begin{bmatrix} x^2 + y^2 \\ \dfrac{1}{xy} \\ y \end{bmatrix}, \qquad \mathbf{g}(x, y) = \begin{bmatrix} \dfrac{1}{xy} \\ \sqrt{2 - x^2 - y^2} \\ x \end{bmatrix}.$$

(a) What are the domains of $\mathbf{f} + \mathbf{g}$ and $\mathbf{f} - \mathbf{g}$?

(b) Find $(\mathbf{f} + \mathbf{g})(\tfrac{1}{2}, \tfrac{1}{2})$ and $(\mathbf{f} - \mathbf{g})(\tfrac{1}{2}, -\tfrac{1}{2})$.

(c) Find $(3\mathbf{f} + 2\mathbf{g})(1, 1)$.

6. Let

$$\mathbf{f}(x, y, z) = \begin{bmatrix} x^2 + y^2 - z^2 \\ \dfrac{1}{x^2 - y^2} \end{bmatrix}, \qquad \mathbf{g}(x, y, z) = \begin{bmatrix} x + y - z \\ \sqrt{x^2 + y^2} \end{bmatrix}.$$

(a) What are the domains of $\mathbf{f} + \mathbf{g}$ and $\mathbf{f} - \mathbf{g}$?

(b) Find $(\mathbf{f} + \mathbf{g})(1, 0, 1)$ and $(\mathbf{f} - \mathbf{g})(1, 0, -1)$.

(c) Find $(2\mathbf{f} + 4\mathbf{g})(0, 1, 0)$.

7. Let

$$\mathbf{f}(x, y) = \begin{bmatrix} 2x + y \\ 3x - 2y \\ -2x + 3y \end{bmatrix}, \qquad \mathbf{g}(u, z, w) = \begin{bmatrix} u - v + w \\ 2u + v + 2w \\ 3u - v \end{bmatrix}.$$

(a) Find $(\mathbf{g} \circ \mathbf{f})(2, -1)$.

(b) Find $(\mathbf{g} \circ \mathbf{f})(x, y)$.

(c) Verify that the matrices of $\mathbf{f}$, $\mathbf{g}$, and $\mathbf{g} \circ \mathbf{f}$ satisfy

$$\mathbf{M}_{\mathbf{g} \circ \mathbf{f}} = \mathbf{M}_{\mathbf{g}} \mathbf{M}_{\mathbf{f}}.$$

(d) Is $\mathbf{f} \circ \mathbf{g}$ defined?

8. Let

$$\mathbf{f}(x, y) = \begin{bmatrix} x + 2y \\ 2x - y \end{bmatrix}, \qquad \mathbf{g}(u, v) = \begin{bmatrix} 3u + v \\ 2u - v \end{bmatrix}.$$

(a) Find $(\mathbf{g} \circ \mathbf{f})(x, y)$ and $(\mathbf{f} \circ \mathbf{g})(u, v)$.

(b) Find $\mathbf{M}_{\mathbf{g} \circ \mathbf{f}}$ and $\mathbf{M}_{\mathbf{f} \circ \mathbf{g}}$.

9. Let

$$\mathbf{f}(x, y, z) = \begin{bmatrix} \log(x - y) \\ \log z \end{bmatrix}, \qquad \mathbf{g}(u, v) = \begin{bmatrix} e^{u-v} \\ e^{v/2} \end{bmatrix}.$$

(a) Find $(\mathbf{g} \circ \mathbf{f})(2, 0, 4)$.

(b)   Find $(\mathbf{g} \circ \mathbf{f})(x, y, z)$.

(c)   What is the domain of $\mathbf{g} \circ \mathbf{f}$?

10.   Let

$$
\mathbf{f}(x, y, z) = \begin{bmatrix} x + y + z \\ \\ x - y \end{bmatrix}, \qquad \mathbf{g}(u, v) = \begin{bmatrix} u \\ u + v \\ v \end{bmatrix}.
$$

(a)   Find $(\mathbf{g} \circ \mathbf{f})(x, y, z)$.

(b)   Find $(\mathbf{f} \circ \mathbf{g})(u, v, w)$.

11.   Let

$$
\mathbf{f}(x, y) = \begin{bmatrix} x \tan y \\ \\ x \end{bmatrix}, \qquad \mathbf{g}(u, z) = \begin{bmatrix} 1 \\ u^2 + v^2 \\ v \end{bmatrix}.
$$

(a)   Find $(\mathbf{g} \circ \mathbf{f})(1, \pi/4)$.

(b)   Write $(\mathbf{g} \circ \mathbf{f})(x, y)$ in its simplest form.

12.   Find $\lim\limits_{\mathbf{X} \to \mathbf{X}_0} \mathbf{f}(\mathbf{X})$, if it exists:

(a) $\mathbf{f}(\mathbf{X}) = \begin{bmatrix} \dfrac{x^2 - y^2}{x + y} \\ \\ x - y \end{bmatrix}$, $\mathbf{X}_0 = (1, -1)$.

(b) $\mathbf{f}(\mathbf{X}) = \begin{bmatrix} \dfrac{x^2 - y^2}{x + y} \\ \\ \dfrac{x - y}{x^2 - y^2} \end{bmatrix}$, $\mathbf{X}_0 = (1, -1)$.

(c) $\mathbf{f}(\mathbf{X}) = \begin{bmatrix} \log \sqrt{1 - x^2 - y^2 - z^2} \\ \\ e^{x^2 + y^2 + z^2} \end{bmatrix}$, $\mathbf{X}_0 = (0, 0, 0)$.

13. Find $\lim_{X \to X_0} f(X)$, if it exists:

(a) $f(X) = \begin{bmatrix} \dfrac{x}{y^2 - 1} \\ 2x - 3y \\ \dfrac{xy}{\sqrt{x^2 + y^2}} \end{bmatrix}$, $X_0 = (3, 4)$.

(b) $f(X) = \begin{bmatrix} \cos \dfrac{\pi}{4 - 2xyz} \\ \sin \dfrac{2\pi}{2x + y + z} \\ \dfrac{x}{2x - y - z} \end{bmatrix}$, $X_0 = (1, 1, 1)$.

(c) $f(X) = \begin{bmatrix} \dfrac{xy}{\sqrt{x^2 + y^2}} \\ \dfrac{x^2 + y^2 - 2xy}{x - y} \\ \sqrt{x^2 + y^2} \end{bmatrix}$, $X_0 = (0, 0)$.

14. Let $X_0 = (1, 2)$ and

$$f(X) = \begin{bmatrix} 2x + 3 \\ x + y \\ y + 1 \end{bmatrix}.$$

(a) Find $L = \lim_{X \to X_0} f(X)$.

(b) Find $|f(X) - L|$.

(c) Show that $\lim_{X \to X_0} |f(X) - L| = 0$.

15. Repeat Exercise 14 for $\mathbf{X}_0 = (1, 0, -1)$ and

$$\mathbf{f(X)} = \begin{bmatrix} x + y + z + 2 \\ x - y + z - 3 \end{bmatrix}.$$

16. Let $\mathbf{f}$ and $\mathbf{g}$ be as in Exercise 10, $\mathbf{h} = \mathbf{g} \circ \mathbf{f}$ and $\mathbf{X}_0 = (1, 0, -1)$.

    (a) Find $\mathbf{h}$ explicitly in terms of (x, y, z) and evaluate

    $$\lim_{\mathbf{X} \to \mathbf{X}_0} \mathbf{h}(x, y, z).$$

    (b) Find $\lim_{\mathbf{X} \to \mathbf{X}_0} \mathbf{h}(x, y, z)$ by means of Theorem 1.4.

17. Repeat Exercise 16 for $\mathbf{f}$ and $\mathbf{g}$ as defined in Exercise 9, and $\mathbf{X}_0 = (4, 2, 1)$.

18. Where are the functions of Exercise 1 continuous?

19. Where are the functions of Exercise 2 continuous?

20. Where are the following functions continuous?

    (a) $\mathbf{f}(x, y) = \begin{cases} \begin{bmatrix} \dfrac{x^2 - y^2}{x + y} \\ x - y \end{bmatrix} & \text{if } x \neq -y \\ \begin{bmatrix} x - y \\ x - y \end{bmatrix} & \text{if } x = -y. \end{cases}$

    (b) $\mathbf{f}(x, y) = \begin{cases} \begin{bmatrix} \dfrac{x^2 - y^2}{x + y} \\ x - y \end{bmatrix} & \text{if } x \neq -y \\ \begin{bmatrix} 0 \\ x - y \end{bmatrix} & \text{if } x = -y. \end{cases}$

$$(c) \quad \mathbf{f}(x, y) = \begin{cases} \begin{bmatrix} x + y \\ xy \end{bmatrix} & \text{if } x \neq y \\ \\ \begin{bmatrix} 2x \\ 4 \end{bmatrix} & \text{if } x = y. \end{cases}$$

21. Let **f** and **g** be as defined in Exercise 5. Where are **f** + **g** and **f** − **g** continuous?

22. Let **f** and **g** be as defined in Exercise 6. Where are **f** + **g** and **f** − **g** continuous?

23. Where is **g** ∘ **f** continuous, if **f** and **g** are as defined in Exercise 7?

24. Where is **g** ∘ **f** continuous, if **f** and **g** are as defined in Exercise 9?

25. Where are **g** ∘ **f** and **f** ∘ **g** continuous, if **f** and **g** are as defined in Exercise 11?

THEORETICAL EXERCISES

T-1. Let $\mathbf{f}: R^n \to R^m$ and $\mathbf{g}: R^p \to R^q$. If both **f** ∘ **g** and **g** ∘ **f** are defined, what relations must hold between m, n, p, and q?

T-2. Show that Corollary 1.1 follows from Theorem 1.1.

T-3. Using Theorem 1.1, prove Theorems 1.2 and 1.3.

T-4. Prove Theorem 1.5.

T-5. Let $\mathbf{f}: R^n \to R^m$, $h: R^n \to R$, and define $\mathbf{F} = h\mathbf{f}$ by

$$\mathbf{F}(\mathbf{X}) = \begin{bmatrix} h(\mathbf{X})f_1(\mathbf{X}) \\ \vdots \\ h(\mathbf{X})f_m(\mathbf{X}) \end{bmatrix}.$$

Prove: If h and **f** are continuous at $\mathbf{X}_0$, so is **F**.

5.2 Differentiable Functions

In this section we generalize the idea of differentiability, previously defined for real valued functions, to vector valued functions. Looking back at the corresponding generalization of continuity, the student might guess that a differentiable vector valued function is nothing more than an ordered m-tuple of differentiable real valued functions; this is exactly the case, as we shall now see.

Definition 2.1   A vector valued function

$$\mathbf{f} = \begin{bmatrix} f_1 \\ \vdots \\ f_m \end{bmatrix}$$

is said to be *differentiable* at $\mathbf{X}_0$ if its components are differentiable at $\mathbf{X}_0$. If $\mathbf{f}$ is differentiable at every point of a set S, we say that $\mathbf{f}$ is *differentiable on S*.

Example 2.1   Let

$$\mathbf{f}(x, y) = \begin{bmatrix} 2x^2y + y^2x \\ x^2 - y^2 \end{bmatrix}.$$

The components of $\mathbf{f}$ have continuous partial derivatives everywhere; hence according to Theorem 3.3, Section 4.3, they are differentiable on R^2 and so is $\mathbf{f}$, by definition.

If $\mathbf{f}$ is differentiable at $\mathbf{X}_0$ then

$$(1) \qquad \lim_{\mathbf{X} \to \mathbf{X}_0} \frac{f_i(\mathbf{X}) - f_i(\mathbf{X}_0) - (d_{\mathbf{X}_0} f_i)(\mathbf{X} - \mathbf{X}_0)}{|\mathbf{X} - \mathbf{X}_0|} = 0$$

for $i = 1, \ldots, m$, and the definition of limit for vector valued functions implies that

$$(2) \qquad \lim_{\mathbf{X} \to \mathbf{X}_0} \frac{\mathbf{f}(\mathbf{X}) - \mathbf{f}(\mathbf{X}_0) - \mathbf{L}(\mathbf{X} - \mathbf{X}_0)}{|\mathbf{X} - \mathbf{X}_0|} = \mathbf{0},$$

where $\mathbf{L}$ is the linear transformation

$$\mathbf{L} = \begin{bmatrix} d_{\mathbf{X}_0} f_1 \\ \vdots \\ d_{\mathbf{X}_0} f_m \end{bmatrix}.$$

This motivates the following definition.

Definition 2.2 If $f: R^n \to R^m$ is differentiable at $\mathbf{X}_0$, then the *differential of f at* $\mathbf{X}_0$ is the linear transformation from R^n to R^m given by

$$(3) \qquad \mathbf{d}_{\mathbf{X}_0}\mathbf{f} = \begin{bmatrix} d_{\mathbf{X}_0} f_1 \\ \vdots \\ d_{\mathbf{X}_0} f_m \end{bmatrix}.$$

Let $\mathbf{Y} = (y_1, \ldots, y_n)$ and $\mathbf{U} = (\mathbf{d}_{\mathbf{X}_0}\mathbf{f})(\mathbf{Y})$; then the components of $\mathbf{U}$ are

$$u_1 = \frac{\partial f_1}{\partial x_1}(\mathbf{X}_0)\, y_1 + \frac{\partial f_1}{\partial x_2}(\mathbf{X}_0)\, y_2 + \cdots + \frac{\partial f_1}{\partial x_n}(\mathbf{X}_0)\, y_n,$$

$$u_2 = \frac{\partial f_2}{\partial x_1}(\mathbf{X}_0)\, y_1 + \frac{\partial f_2}{\partial x_2}(\mathbf{X}_0)\, y_2 + \cdots + \frac{\partial f_2}{\partial x_n}(\mathbf{X}_0)\, y_n,$$

$$\vdots$$

$$u_m = \frac{\partial f_m}{\partial x_1}(\mathbf{X}_0)\, y_1 + \frac{\partial f_m}{\partial x_2}(\mathbf{X}_0)\, y_2 + \cdots + \frac{\partial f_m}{\partial x_n}(\mathbf{X}_0)\, y_n.$$

Definition 2.3 If $\mathbf{f}$ is as in Definition 2.2, then the matrix of $\mathbf{d}_{\mathbf{X}_0}\mathbf{f}$ is called the *Jacobian matrix of* $\mathbf{f}$ *at* $\mathbf{X}_0$ and denoted by $\mathbf{J}_{\mathbf{X}_0}\mathbf{f}$; thus

$$\mathbf{J}_{\mathbf{X}_0}\mathbf{f} = \begin{bmatrix} \dfrac{\partial f_1}{\partial x_1}(\mathbf{X}_0) & \dfrac{\partial f_1}{\partial x_2}(\mathbf{X}_0) & \cdots & \dfrac{\partial f_1}{\partial x_n}(\mathbf{X}_0) \\[2ex] \dfrac{\partial f_2}{\partial x_1}(\mathbf{X}_0) & \dfrac{\partial f_2}{\partial x_2}(\mathbf{X}_0) & \cdots & \dfrac{\partial f_2}{\partial x_n}(\mathbf{X}_0) \\[2ex] \vdots & \vdots & & \vdots \\[1ex] \dfrac{\partial f_m}{\partial x_1}(\mathbf{X}_0) & \dfrac{\partial f_m}{\partial x_2}(\mathbf{X}_0) & \cdots & \dfrac{\partial f_m}{\partial x_n}(\mathbf{X}_0) \end{bmatrix}.$$

In matrix notation, $(\mathbf{d}_{\mathbf{X}_0}\mathbf{f})(\mathbf{Y}) = (\mathbf{J}_{\mathbf{X}_0}\mathbf{f})\mathbf{Y}$, where we consider $\mathbf{Y}$ as a column vector on the right side.

Just as we did when discussing real valued functions, we shall write simply

$$\mathbf{df} = \begin{bmatrix} df_1 \\ df_2 \\ \vdots \\ df_m \end{bmatrix}$$

when it is not important to emphasize $\mathbf{X}_0$; in the same situation we shall write

$$\mathbf{Jf} = \begin{bmatrix} \dfrac{\partial f_1}{\partial x_1} & \cdots & \dfrac{\partial f_1}{\partial x_n} \\ \vdots & & \vdots \\ \dfrac{\partial f_m}{\partial x_1} & \cdots & \dfrac{\partial f_m}{\partial x_n} \end{bmatrix}.$$

Example 2.2 Let $\mathbf{X}_0 = (0, 1, 2)$ and

$$\mathbf{f}(x, y, z) = \begin{bmatrix} 2x^2 + y^2 + z \\ x^2 - y^2 \end{bmatrix};$$

then

$$\mathbf{Jf} = \begin{bmatrix} \dfrac{\partial f_1}{\partial x} & \dfrac{\partial f_1}{\partial y} & \dfrac{\partial f_1}{\partial z} \\ \dfrac{\partial f_2}{\partial x} & \dfrac{\partial f_2}{\partial y} & \dfrac{\partial f_2}{\partial z} \end{bmatrix} = \begin{bmatrix} 4x & 2y & 1 \\ 2x & -2y & 0 \end{bmatrix}.$$

In this equation we have a harmless inconsistency: the entries in the first matrix are functions, while those in the second are their values at the point (x, y, z). We shall not hesitate to write the Jacobian matrices of particular functions in this way when it does not lead to confusion.

Now

$$\mathbf{df} = \begin{bmatrix} df_1 \\ df_2 \end{bmatrix} = \begin{bmatrix} 4x\,dx + 2y\,dy + dz \\ 2x\,dx - 2y\,dy \end{bmatrix},$$

$$\mathbf{d}_{\mathbf{X}_0}\mathbf{f} = \begin{bmatrix} d_{\mathbf{X}_0} f_1 \\ d_{\mathbf{X}_0} f_2 \end{bmatrix} = \begin{bmatrix} 2\,dy + dz \\ -2\,dy \end{bmatrix}$$

and

(4) $$(\mathbf{d}_{\mathbf{X}_0}\mathbf{f})\,(\mathbf{X} - \mathbf{X}_0) = (\mathbf{d}_{\mathbf{X}_0}\mathbf{f})\,(x, y - 1, z - 2)$$

$$= \begin{bmatrix} 2y + z - 4 \\ -2y + 2 \end{bmatrix}.$$

Example 2.3 If **f** is a vector valued function of one variable,

$$\mathbf{f}(t) = \begin{bmatrix} f_1(t) \\ \vdots \\ f_m(t) \end{bmatrix},$$

then the Jacobian matrix is a column vector,

$$\mathbf{Jf} = \begin{bmatrix} f_1' \\ \vdots \\ f_m' \end{bmatrix},$$

and

$$\mathbf{df} = \begin{bmatrix} f_1' \, dt \\ \vdots \\ f_m' \, dt \end{bmatrix}.$$

For instance, let $t_0 = 0$ and

$$\mathbf{f}(t) = \begin{bmatrix} e^t \\ \sin t \\ \cos t \end{bmatrix};$$

then

$$\mathbf{Jf} = \begin{bmatrix} e^t \\ \cos t \\ -\sin t \end{bmatrix}, \qquad \mathbf{df} = \begin{bmatrix} e^t \, dt \\ \cos t \, dt \\ -\sin t \, dt \end{bmatrix},$$

$$\mathbf{d}_{t_0}\mathbf{f} = \begin{bmatrix} dt \\ dt \\ 0 \end{bmatrix}, \qquad (\mathbf{d}_{t_0}\mathbf{f})(t - t_0) = \begin{bmatrix} t \\ t \\ 0 \end{bmatrix}.$$

If a vector valued function **f** is differentiable at $\mathbf{X}_0$ then (2) can be rewritten as

$$(5) \qquad \lim_{\mathbf{X} \to \mathbf{X}_0} \frac{\mathbf{f}(\mathbf{X}) - \mathbf{f}(\mathbf{X}_0) - (\mathbf{d}_{\mathbf{X}_0}\mathbf{f})(\mathbf{X} - \mathbf{X}_0)}{|\,\mathbf{X} - \mathbf{X}_0\,|} = \mathbf{0},$$

which is identical in appearance with the definition of differentiability for real valued functions; the difference is that (5) is a statement about vectors rather than real numbers. It says that

$$(6) \qquad \mathbf{f}(\mathbf{X}) \cong \mathbf{f}(\mathbf{X}_0) + (\mathbf{d}_{\mathbf{X}_0}\mathbf{f})(\mathbf{X} - \mathbf{X}_0),$$

where the magnitude of the error vector approaches zero "very fast" as $\mathbf{X}$ approaches $\mathbf{X}_0$; that is, the ratio of the length of the error vector to $|\mathbf{X} - \mathbf{X}_0|$ approaches zero as $|\mathbf{X} - \mathbf{X}_0|$ approaches zero.

The function $\mathbf{g}$ defined by

$$\mathbf{g}(\mathbf{X}) = \mathbf{f}(\mathbf{X}_0) + (\mathbf{d}_{\mathbf{X}_0}\mathbf{f})(\mathbf{X} - \mathbf{X}_0)$$

is the sum of a constant vector and a linear transformation. A function of this kind is called an *affine function*. That is, an affine function $\mathbf{g}: R^n \to R^m$ is defined by

$$\mathbf{g} = \begin{bmatrix} g_1 \\ \vdots \\ g_m \end{bmatrix},$$

where

$$g_1(\mathbf{X}) = b_1 + a_{11}(x_1 - c_1) + \cdots + a_{1n}(x_n - c_n),$$

(7) $$g_2(\mathbf{X}) = b_2 + a_{21}(x_1 - c_1) + \cdots + a_{2n}(x_n - c_n),$$

$$\vdots$$

$$g_m(\mathbf{X}) = b_m + a_{m1}(x_1 - c_1) + \cdots + a_{mn}(x_n - c_n),$$

and all quantities on the right side except $(x_1, \ldots, x_n)$ are constants. We can write (7) in matrix form as

$$\mathbf{g}(\mathbf{X}) = \mathbf{B} + \mathbf{A}(\mathbf{X} - \mathbf{X}_0)$$

where

$$\mathbf{B} = \begin{bmatrix} b_1 \\ \vdots \\ b_m \end{bmatrix},$$

$$\mathbf{A} = \begin{bmatrix} a_{11} & a_{12} & \cdots & a_{1n} \\ a_{21} & a_{22} & \cdots & a_{2n} \\ \vdots & \vdots & & \vdots \\ a_{m1} & a_{m2} & & a_{mn} \end{bmatrix}$$

and

$$\mathbf{X} - \mathbf{X}_0 = \begin{bmatrix} x_1 - c_1 \\ \vdots \\ x_n - c_n \end{bmatrix}.$$

Equation (6) says that a differentiable vector valued function can be closely approximated by an affine function. This is analogous to the statement that a differentiable real valued function can be closely approximated by a polynomial of degree ≤ 1, to which (5) reduces when $m = 1$.

Example 2.4 Let $\mathbf{f}$ and $\mathbf{X}_0$ be as defined in Example 2.2. Since

$$\mathbf{f}(\mathbf{X}_0) = \begin{bmatrix} 3 \\ -1 \end{bmatrix},$$

it follows from (4) that $\mathbf{f}$ is closely approximated near $\mathbf{X}_0$ by the affine function

$$\mathbf{g}(\mathbf{X}) = \begin{bmatrix} 3 \\ -1 \end{bmatrix} + \begin{bmatrix} 2y + z - 4 \\ -2y + 2 \end{bmatrix}$$

$$= \begin{bmatrix} 2y + z - 1 \\ -2y + 1 \end{bmatrix}.$$

The next theorem is obtained by applying the corresponding results for real valued functions to the components of $\mathbf{f}$ and $\mathbf{g}$.

Theorem 2.1 If $\mathbf{f}$ is differentiable at $\mathbf{X}_0$, then it is continuous there. Also $c\mathbf{f}$, where c is a constant, is differentiable at $\mathbf{X}_0$; furthermore

$$\mathbf{d}_{\mathbf{X}_0}(c\mathbf{f}) = c\,\mathbf{d}_{\mathbf{X}_0}\mathbf{f}.$$

If $\mathbf{g}$ is also differentiable at $\mathbf{X}_0$, the same is true of $\mathbf{f} + \mathbf{g}$ and $\mathbf{f} - \mathbf{g}$, and

$$\mathbf{d}_{\mathbf{X}_0}(\mathbf{f} + \mathbf{g}) = \mathbf{d}_{\mathbf{X}_0}\mathbf{f} + \mathbf{d}_{\mathbf{X}_0}\mathbf{g} \qquad \text{and} \qquad \mathbf{d}_{\mathbf{X}_0}(\mathbf{f} - \mathbf{g}) = \mathbf{d}_{\mathbf{X}_0}\mathbf{f} - \mathbf{d}_{\mathbf{X}_0}\mathbf{g}.$$

A vector valued function may fail to be differentiable at a point $\mathbf{X}_0$ even though the components have partial derivatives there, as is shown by the following example.

Example 2.5 Let $\mathbf{X}_0 = (0, 0)$ and

$$f(x, y) = \begin{bmatrix} \sqrt{|xy|} \\ x \end{bmatrix};$$

then

$$\frac{\partial f_1}{\partial x}(\mathbf{X}_0) = \frac{\partial f_1}{\partial y}(\mathbf{X}_0) = \frac{\partial f_2}{\partial y}(\mathbf{X}_0) = 0$$

and

$$\frac{\partial f_2}{\partial x}(\mathbf{X}_0) = 1.$$

However, f_1 is not differentiable at $\mathbf{X}_0$ (Example 3.5, Section 4.3). Hence neither is $\mathbf{f}$.

The following theorem gives sufficient conditions for differentiability at a point.

Theorem 2.2 A vector valued function is differentiable at $\mathbf{X}_0$ if all first partial derivatives of its components exist in some n-ball about $\mathbf{X}_0$, and are continuous at $\mathbf{X}_0$.

Proof. The hypothesis implies that the components are differentiable at $\mathbf{X}_0$ (Theorem 3.3, Section 4.3); hence $\mathbf{f}$ is differentiable by definition.

Example 2.6 As previously noted in Example 2.1,

$$\mathbf{f}(x, y) = \begin{bmatrix} 2x^2y + y^2x \\ \\ x^2 - y^2 \end{bmatrix}$$

is differentiable everywhere, and

$$\mathbf{df} = \begin{bmatrix} (4xy + y^2)\,dx + (2x^2 + 2xy)\,dy \\ \\ 2x\,dx - 2y\,dy \end{bmatrix}.$$

Example 2.7 If $\mathbf{g}$ is the affine function (7), then $\mathbf{g}$ is differentiable on R^n, because the partial derivatives of $g_1, \cdots, g_m$, being constants, are continuous. By evaluating the partial derivatives, we find that

$$\mathbf{Jg} = \begin{bmatrix} a_{11} & a_{12} & \cdots & a_{1n} \\ a_{21} & a_{22} & \cdots & a_{2n} \\ \vdots & \vdots & & \vdots \\ a_{m1} & a_{m2} & \cdots & a_{mn} \end{bmatrix}.$$

For any $\mathbf{X}_0$,

$$\mathbf{g}(\mathbf{X}) - \mathbf{g}(\mathbf{X}_0) - (\mathbf{d}_{\mathbf{X}_0}\mathbf{g})(\mathbf{X} - \mathbf{X}_0) = 0$$

(verify!), a stronger result than (5).

Our view has been that a differentiable vector valued function is simply an ordered m-tuple of differentiable real valued functions; thus we started from (1) and deduced (2). An alternate approach would be to take (2) as the starting point; that is, define $\mathbf{f}: R^n \to R^m$ to be differentiable if there is a linear transformation

$$\mathbf{L} = \begin{bmatrix} L_1 \\ \vdots \\ L_m \end{bmatrix}$$

that satisfies (2). This definition, however, is equivalent to ours, since (2) and the definition of limit for vector valued functions imply that

$$\lim_{X \to X_0} \frac{f_i(X) - f_i(X_0) - L_i(X - X_0)}{|X - X_0|} = 0 \qquad (1 \leq i \leq m),$$

which in turn implies that f_i is differentiable at X_0 and $L_i = d_{X_0} f_i$. Hence

$$L = \begin{bmatrix} d_{X_0} f_1 \\ \vdots \\ d_{X_0} f_m \end{bmatrix} = d_{X_0} f,$$

as defined by (3).

EXERCISES 5.2 # 1b, 9a, 10a   Page 462 # 1a

1. Find $d_{X_0}f$ and $(d_{X_0}f)(X_1 - X_0)$:

   (a) $f(X) = \begin{bmatrix} \sqrt{x^2 + y^2 + z^2} \\ 3x + 2y + 4z + 1 \end{bmatrix}$; $\begin{aligned} X_0 &= (1, 2, -2), \\ X_1 &= (1, 1, 1). \end{aligned}$

   (b) $f(X) = \begin{bmatrix} \log(x - y) \\ x^3 + y^2 \\ 2x^2 + y \end{bmatrix}$; $X_0 = (2, 1)$, $X_1 = (3, 2)$.

   (c) $f(X) = \begin{bmatrix} ze^{x+y} \\ \sqrt{x + y} \\ (xyz)^{3/2} \end{bmatrix}$; $X_0 = (1, 1, 1)$, $X_1 = (1, 1, 2)$.

2. Find $J_{X_0}f$ for each f in Exercise 1.

3. Find $d_{X_0}f$ and $(d_{X_0}f)(X_1 - X_0)$:

   (a) $f(X) = \begin{bmatrix} x + y + z \\ x^2 + y^2 + z^2 \end{bmatrix}$; $X_0 = (1, 1, 2)$, $X_1 = (1, 2, 3)$.

(b) $\mathbf{f}(\mathbf{X}) = \begin{bmatrix} \sin(x+y) \\ \tan(x+y) \\ \cos(x+y) \end{bmatrix}$; $\mathbf{X}_0 = \left(\dfrac{\pi}{8}, \dfrac{\pi}{8} \right)$, $\mathbf{X}_1 = \left(\dfrac{\pi}{4}, -\dfrac{\pi}{4} \right)$.

(c) $\mathbf{f}(\mathbf{X}) = \begin{bmatrix} e^x \\ x^2 \\ x+1 \end{bmatrix}$; $x_0 = 0$, $x_1 = 1$.

4. Find $\mathbf{J}_{\mathbf{X}_0}\mathbf{f}$ for each $\mathbf{f}$ in Exercise 3.

5. Find $\mathbf{df}$:

(a) $\mathbf{f}(x, y) = \begin{bmatrix} \cos(x+y) \\ \sin(x+y) \end{bmatrix}$.

(b) $\mathbf{f}(x) = \begin{bmatrix} x \\ x^2 \\ x^3 \end{bmatrix}$.

(c) $\mathbf{f}(x, y, z) = \begin{bmatrix} e^x \sin yz \\ e^y \sin xz \\ e^z \sin xy \end{bmatrix}$.

6. Find $\mathbf{Jf}$ for each $\mathbf{f}$ in Exercise 5.

7. Find $\mathbf{df}$:

(a) $\mathbf{f}(x, y, z, w) = \begin{bmatrix} e^x(y+z+w) \\ \sin(x+y+z+w) \\ xyz \end{bmatrix}$.

(b) $\mathbf{f}(x, y) = \begin{bmatrix} \sin(1-x-y) \\ 2x+xy \\ x^3+y^2 \end{bmatrix}$.

(c)
$$\mathbf{f}(x) = \begin{bmatrix} e^x \sin x \\ e^x \cos x \end{bmatrix}.$$

8. Find $\mathbf{Jf}$ for each $\mathbf{f}$ in Exercise 7.

9. Let $\mathbf{X}_0 = (1, 2)$,

$$\mathbf{f}(x, y) = \begin{bmatrix} x^2 + 2y^2 \\ x^2 + y \end{bmatrix} \quad \text{and} \quad \mathbf{g}(x, y) = \begin{bmatrix} 2x - y \\ x^3 + y^2 \end{bmatrix}.$$

Find

(a) $\mathbf{d}_{\mathbf{X}_0}(3\mathbf{f})$, (b) $\mathbf{d}_{\mathbf{X}_0}(\mathbf{f} + \mathbf{g})$, (c) $\mathbf{d}_{\mathbf{X}_0}(2\mathbf{f} - 3\mathbf{g})$.

10. For $\mathbf{X}_0$, $\mathbf{f}$, and $\mathbf{g}$ as defined in Exercise 9, compute

(a) $\mathbf{J}_{\mathbf{X}_0}\mathbf{f}$ and $\mathbf{J}_{\mathbf{X}_0}\mathbf{g}$;

(b) $\mathbf{J}_{\mathbf{X}_0}(3\mathbf{f})$ and $3\,\mathbf{J}_{\mathbf{X}_0}\mathbf{f}$;

(c) $\mathbf{J}_{\mathbf{X}_0}(\mathbf{f} + \mathbf{g})$ and $\mathbf{J}_{\mathbf{X}_0}\mathbf{f} + \mathbf{J}_{\mathbf{X}_0}\mathbf{g}$.

11. For each $\mathbf{f}$ and corresponding $\mathbf{X}_0$ in Exercise 1 find an affine transformation $\mathbf{g}$ such that

$$\lim_{\mathbf{X} \to \mathbf{X}_0} \frac{\mathbf{f}(\mathbf{X}) - \mathbf{g}(\mathbf{X})}{|\mathbf{X} - \mathbf{X}_0|} = \mathbf{0}.$$

12. Repeat Exercise 11 for each $\mathbf{f}$ in Exercise 3.

THEORETICAL EXERCISES

T-1. If f is a real valued function, show that $\nabla_{\mathbf{X}_0} f$ is the transpose of $\mathbf{J}_{\mathbf{X}_0} f$.

T-2. Prove Theorem 2.1.

T-3. Prove that $\mathbf{J}(c\mathbf{f}) = c\,\mathbf{Jf}$.

T-4. Prove that $\mathbf{J}(\mathbf{f} + \mathbf{g}) = \mathbf{Jf} + \mathbf{Jg}$.

T-5. What is $\mathbf{dg}$ in Example 2.7?

5.3   The Chain Rule

In this section we generalize the chain rule of Section 4.4, Chapter 4.

Theorem 3.1   Let $\mathbf{f}: R^n \to R^m$ be differentiable at $\mathbf{X}_0$ and $\mathbf{g}: R^m \to R^p$ be differentiable at $\mathbf{U}_0 = \mathbf{f}(\mathbf{X}_0)$. Then $\mathbf{h} = \mathbf{g} \circ \mathbf{f}$ is differentiable at $\mathbf{X}_0$ and

$$(1) \qquad\qquad d_{\mathbf{X}_0}\mathbf{h} = d_{\mathbf{U}_0}\mathbf{g} \circ d_{\mathbf{X}_0}\mathbf{f};$$

thus, the differential of the composite function is the composition of the differentials.

Proof.   We first observe that $\mathbf{X}_0$ is an interior point of the domain of $\mathbf{h}$ (Exercise T-1), so that it is legitimate to ask if $\mathbf{h}$ is differentiable at $\mathbf{X}_0$.

From the differentiability of $\mathbf{f}$ and $\mathbf{g}$,

$$(2) \qquad \mathbf{g}(\mathbf{U}) - \mathbf{g}(\mathbf{U}_0) = (d_{\mathbf{U}_0}\mathbf{g})(\mathbf{U} - \mathbf{U}_0) + |\,\mathbf{U} - \mathbf{U}_0\,|\,\mathbf{E}_1(\mathbf{U})$$

and

$$(3) \qquad \mathbf{f}(\mathbf{X}) - \mathbf{f}(\mathbf{X}_0) = (d_{\mathbf{X}_0}\mathbf{f})(\mathbf{X} - \mathbf{X}_0) + |\,\mathbf{X} - \mathbf{X}_0\,|\,\mathbf{E}_2(\mathbf{X}),$$

where

$$(4) \qquad\qquad \begin{aligned} &\lim_{\mathbf{U} \to \mathbf{U}_0} \mathbf{E}_1(\mathbf{U}) = \mathbf{0}, \\[4pt] &\lim_{\mathbf{X} \to \mathbf{X}_0} \mathbf{E}_2(\mathbf{X}) = \mathbf{0}. \end{aligned}$$

(Here the zeroes stand for zero vectors of dimensions p and m, respectively.) Substituting $\mathbf{U} = \mathbf{f}(\mathbf{X})$ and $\mathbf{U}_0 = \mathbf{f}(\mathbf{X}_0)$ in (2) gives

$$\mathbf{g}(\mathbf{f}(\mathbf{X})) - \mathbf{g}(\mathbf{f}(\mathbf{X}_0)) = (d_{\mathbf{U}_0}\mathbf{g})(\mathbf{f}(\mathbf{X}) - \mathbf{f}(\mathbf{X}_0)) + |\,\mathbf{f}(\mathbf{X}) - \mathbf{f}(\mathbf{X}_0)\,|\,\mathbf{E}_1(\mathbf{f}(\mathbf{X})),$$

and substituting (3) into this yields

$$\mathbf{g}(\mathbf{f}(\mathbf{X})) - \mathbf{g}(\mathbf{f}(\mathbf{X}_0)) = (d_{\mathbf{U}_0}\mathbf{g})((d_{\mathbf{X}_0}\mathbf{f})(\mathbf{X} - \mathbf{X}_0) + |\,\mathbf{X} - \mathbf{X}_0\,|\,\mathbf{E}_2(\mathbf{X}))$$
$$+\,|\,(d_{\mathbf{X}_0}\mathbf{f})(\mathbf{X} - \mathbf{X}_0) + |\,(\mathbf{X} - \mathbf{X}_0)\,|\,\mathbf{E}_2(\mathbf{X})\,|\,\mathbf{E}_1(\mathbf{f}(\mathbf{X})).$$

Defining

$$\mathbf{E}_3(\mathbf{X}) = \mathbf{g}(\mathbf{f}(\mathbf{X})) - \mathbf{g}(\mathbf{f}(\mathbf{X}_0)) - (d_{\mathbf{U}_0}\mathbf{g})((d_{\mathbf{X}_0}\mathbf{f})(\mathbf{X} - \mathbf{X}_0)),$$

and using the linearity of $d_{\mathbf{U}_0}\mathbf{g}$, we find that

$$(5) \qquad \mathbf{E}_3(\mathbf{X}) = |\,\mathbf{X} - \mathbf{X}_0\,|\,(d_{\mathbf{U}_0}\mathbf{g})(\mathbf{E}_2(\mathbf{X}))$$
$$+\,|\,(d_{\mathbf{X}_0}\mathbf{f})(\mathbf{X} - \mathbf{X}_0) + |\,\mathbf{X} - \mathbf{X}_0\,|\,\mathbf{E}_2(\mathbf{X})\,|\,\mathbf{E}_1(\mathbf{f}(\mathbf{X})).$$

There is a constant M such that

$$| (\mathbf{d}_{\mathbf{X}_0}\mathbf{f}) (\mathbf{X} - \mathbf{X}_0) | < M | \mathbf{X} - \mathbf{X}_0 |$$

(Exercise T-2); hence from (5),

(6) $\quad \dfrac{| \mathbf{E}_3(\mathbf{X}) |}{| \mathbf{X} - \mathbf{X}_0 |} \leq | (\mathbf{d}_{\mathbf{U}_0}\mathbf{g}) (\mathbf{E}_2(\mathbf{X})) | + \left| M + | \mathbf{E}_2(\mathbf{X}) | \right| \, | \mathbf{E}_1(\mathbf{f}(\mathbf{X})) |.$

The right side approaches zero as $\mathbf{X}$ approaches $\mathbf{X}_0$ (Exercise T-3); hence

$$\lim_{\mathbf{X} \to \mathbf{X}_0} \frac{\mathbf{g} \circ \mathbf{f}(\mathbf{X}) - \mathbf{g} \circ \mathbf{f}(\mathbf{X}_0) - (\mathbf{d}_{\mathbf{U}_0}\mathbf{g} \circ \mathbf{d}_{\mathbf{X}_0}\mathbf{f}) (\mathbf{X} - \mathbf{X}_0)}{| \mathbf{X} - \mathbf{X}_0 |} = 0.$$

Since $\mathbf{d}_{\mathbf{U}_0}\mathbf{g} \circ \mathbf{d}_{\mathbf{X}_0}\mathbf{f}$ is a linear transformation, $\mathbf{h}$ is differentiable at $\mathbf{X}_0$ and (1) follows. This completes the proof.

In Example 1.6, we saw that the matrix of the composition of two linear transformations is the product of their matrices; hence from (1)

(7) $\qquad\qquad\qquad \mathbf{J}_{\mathbf{X}_0}\mathbf{h} = (\mathbf{J}_{\mathbf{U}_0}\mathbf{g}) (\mathbf{J}_{\mathbf{X}_0}\mathbf{f}),$

so that, from the definition of matrix multiplication,

(8) $\qquad\qquad \dfrac{\partial h_i}{\partial x_j} (\mathbf{X}_0) = \sum_{r=1}^{m} \dfrac{\partial g_i}{\partial u_r} (\mathbf{f}(\mathbf{X}_0)) \, \dfrac{\partial f_r}{\partial x_j} (\mathbf{X}_0).$

When it is not important to emphasize a particular point $\mathbf{X}_0$, we shall simplify the notation by rewriting the last two equations as

$$\mathbf{Jh} = (\mathbf{Jg}) (\mathbf{Jf})$$

and

$$\frac{\partial h_i}{\partial x_j} = \sum_{r=1}^{m} \frac{\partial g_i}{\partial u_r} \frac{\partial f_r}{\partial x_j};$$

however, it must be remembered, when calculating $(\partial h_i/\partial x_j) (\mathbf{X})$, that the argument of $\partial g_i/\partial u_r$ is the vector $\mathbf{f}(\mathbf{X})$. To further simplify the formula for $\partial h_i/\partial x_j$, we shall often dispense with the symbol $\mathbf{f}$ entirely and write $\mathbf{f} = \mathbf{U} = \mathbf{U}(\mathbf{X})$ (a natural extension of the informal notation introduced in Section 2.6); we then have

$$\frac{\partial h_i}{\partial x_j} (\mathbf{X}) = \sum_{r=1}^{m} \frac{\partial g_i}{\partial u_r} (\mathbf{U}(\mathbf{X})) \frac{\partial u_r}{\partial x_j} (\mathbf{X}),$$

or simply

$$\frac{\partial h_i}{\partial x_j} = \sum_{r=1}^{m} \frac{\partial g_i}{\partial u_r} \frac{\partial u_r}{\partial x_j}.$$

Then (1) becomes

$$d_{X_0}h = d_{U_0}g \circ d_{X_0}U,$$

(where $U_0 = U(X_0)$), or simply

$$dh = dg \circ dU.$$

The simplified form of (7) is

$$Jh = (Jg)(JU).$$

Example 3.1 Let $X_0 = (1, 0, -1)$,

$$U = \begin{bmatrix} u \\ v \end{bmatrix} = \begin{bmatrix} \sqrt{x^2 + y^2 + z^2} \\ \sqrt{x^2 - y^2} \end{bmatrix}$$

and

$$g(u, v) = \begin{bmatrix} u^2 \\ u^2 + v^2 \\ v^2 \end{bmatrix}.$$

In this case **h** can be conveniently expressed directly in terms of (x, y, z):

$$h(x, y, z) = g(u(x, y, z), v(x, y, z))$$

$$= \begin{bmatrix} x^2 + y^2 + z^2 \\ 2x^2 + z^2 \\ x^2 - y^2 \end{bmatrix},$$

and

$$Jh = \begin{bmatrix} \dfrac{\partial h_1}{\partial x} & \dfrac{\partial h_1}{\partial y} & \dfrac{\partial h_1}{\partial z} \\ \dfrac{\partial h_2}{\partial x} & \dfrac{\partial h_2}{\partial y} & \dfrac{\partial h_2}{\partial z} \\ \dfrac{\partial h_3}{\partial x} & \dfrac{\partial h_3}{\partial y} & \dfrac{\partial h_3}{\partial z} \end{bmatrix} = \begin{bmatrix} 2x & 2y & 2z \\ 4x & 0 & 2z \\ 2x & -2y & 0 \end{bmatrix}.$$

Hence

$$\mathbf{J}_{x_0}\mathbf{h} = \begin{bmatrix} 2 & 0 & -2 \\ 4 & 0 & -2 \\ 2 & 0 & 0 \end{bmatrix}.$$

Now let us compute $\mathbf{Jh}$ by the chain rule:

$$\mathbf{JU} = \begin{bmatrix} \dfrac{\partial u}{\partial x} & \dfrac{\partial u}{\partial y} & \dfrac{\partial u}{\partial z} \\[2mm] \dfrac{\partial v}{\partial x} & \dfrac{\partial v}{\partial y} & \dfrac{\partial v}{\partial z} \end{bmatrix}$$

$$= \begin{bmatrix} \dfrac{x}{\sqrt{x^2 + y^2 + z^2}} & \dfrac{y}{\sqrt{x^2 + y^2 + z^2}} & \dfrac{z}{\sqrt{x^2 + y^2 + z^2}} \\[4mm] \dfrac{x}{\sqrt{x^2 + y^2}} & \dfrac{-y}{\sqrt{x^2 - y^2}} & 0 \end{bmatrix}$$

and

$$\mathbf{Jg} = \begin{bmatrix} \dfrac{\partial g_1}{\partial u} & \dfrac{\partial g_1}{\partial v} \\[2mm] \dfrac{\partial g_2}{\partial u} & \dfrac{\partial g_2}{\partial v} \\[2mm] \dfrac{\partial g_3}{\partial u} & \dfrac{\partial g_3}{\partial v} \end{bmatrix} = \begin{bmatrix} 2u & 0 \\ 2u & 2v \\ 0 & 2v \end{bmatrix}.$$

Now, $\mathbf{U}_0 = (\sqrt{2}, 1)$ and

$$\mathbf{J}_{x_0}\mathbf{h} = \mathbf{J}_{U_0}\mathbf{g}\, \mathbf{J}_{x_0}\mathbf{U} = \begin{bmatrix} 2\sqrt{2} & 0 \\ 2\sqrt{2} & 2 \\ 0 & 2 \end{bmatrix} \begin{bmatrix} \dfrac{1}{\sqrt{2}} & 0 & -\dfrac{1}{\sqrt{2}} \\ 1 & 0 & 0 \end{bmatrix} = \begin{bmatrix} 2 & 0 & -2 \\ 4 & 0 & -2 \\ 2 & 0 & 0 \end{bmatrix},$$

which agrees with the previous result.

The differential of $\mathbf{h}$ at $\mathbf{X}_0$ is

$$\mathbf{d}_{\mathbf{X}_0}\mathbf{h} = \begin{bmatrix} 2\,dx - 2\,dz \\[2mm] 4\,dx - 2\,dz \\[2mm] 2\,dx \end{bmatrix}.$$

For simplicity we take g to be real valued in the following examples. The results are easily generalized to vector valued functions.

Example 3.2 Let g be a real valued function of $\mathbf{U} = (u_1, \ldots, u_m)$, which is in turn a function of a single variable x, and

$$h(x) = g(u_1(x), \ldots, u_m(x)).$$

Then

$$\mathbf{JU} = \begin{bmatrix} \dfrac{du_1}{dx} \\[3mm] \dfrac{du_2}{dx} \\[2mm] \vdots \\[2mm] \dfrac{du_m}{dx} \end{bmatrix} \quad \text{and} \quad J\mathbf{g} = \begin{bmatrix} \dfrac{\partial g}{\partial u_1} & \dfrac{\partial g}{\partial u_2} & \cdots & \dfrac{\partial g}{\partial u_m} \end{bmatrix};$$

hence

$$\mathbf{Jh} = \frac{dh}{dx} = \begin{bmatrix} \dfrac{\partial g}{\partial u_1} & \dfrac{\partial g}{\partial u_2} & \cdots & \dfrac{\partial g}{\partial u_m} \end{bmatrix} \begin{bmatrix} \dfrac{du_1}{dx} \\[3mm] \dfrac{du_2}{dx} \\[2mm] \vdots \\[2mm] \dfrac{du_m}{dx} \end{bmatrix}$$

$$= \frac{\partial g_1}{\partial u_1}\frac{du_1}{dx} + \cdots + \frac{\partial g_m}{\partial u_m}\frac{du_m}{dx},$$

which, aside from differences in notation, agrees with the result of Theorem 4.3, Section 2.4.

Example 3.3   Let g be a real valued function of (x, y).

$$
\begin{bmatrix} x \\ y \end{bmatrix} = \begin{bmatrix} x(u, v) \\ y(u, v) \end{bmatrix}
$$

and

$$
h(u, v) = g(x(u, v), y(u, v)).
$$

(Note that x and y are now the intermediate variables, rather than u and v). Then

$$
\begin{bmatrix} \dfrac{\partial h}{\partial u} \cdot \dfrac{\partial h}{\partial v} \end{bmatrix} = \begin{bmatrix} \dfrac{\partial g}{\partial x} & \dfrac{\partial g}{\partial y} \end{bmatrix} \begin{bmatrix} \dfrac{\partial x}{\partial u} & \dfrac{\partial x}{\partial v} \\[2mm] \dfrac{\partial y}{\partial u} & \dfrac{\partial y}{\partial v} \end{bmatrix} ;
$$

hence

$$
\frac{\partial h}{\partial u} = \frac{\partial g}{\partial x} \frac{\partial x}{\partial u} + \frac{\partial g}{\partial y} \frac{\partial y}{\partial u} ,
$$

$$
\frac{\partial h}{\partial v} = \frac{\partial g}{\partial x} \frac{\partial x}{\partial v} + \frac{\partial g}{\partial y} \frac{\partial y}{\partial v} .
$$

Example 3.4   Let (x, y) and (r, θ) be cartesian and polar coordinates in the plane;

$$
\begin{bmatrix} x \\ y \end{bmatrix} = \begin{bmatrix} r \cos \theta \\ r \sin \theta \end{bmatrix} .
$$

If g is a real valued function of (x, y) and $h(r, \theta) = g(r \cos \theta, r \sin \theta)$, then

$$
\begin{bmatrix} \dfrac{\partial h}{\partial r} & \dfrac{\partial h}{\partial \theta} \end{bmatrix} = \begin{bmatrix} \dfrac{\partial g}{\partial x} & \dfrac{\partial g}{\partial y} \end{bmatrix} \begin{bmatrix} \dfrac{\partial x}{\partial r} & \dfrac{\partial x}{\partial \theta} \\[2mm] \dfrac{\partial y}{\partial r} & \dfrac{\partial y}{\partial \theta} \end{bmatrix}
$$

$$
= \begin{bmatrix} \dfrac{\partial g}{\partial x} & \dfrac{\partial g}{\partial y} \end{bmatrix} \begin{bmatrix} \cos \theta & -r \sin \theta \\ \sin \theta & r \cos \theta \end{bmatrix} .
$$

Hence

$$\frac{\partial h}{\partial r} = \cos \vartheta \frac{\partial g}{\partial x} + \sin \theta \frac{\partial g}{\partial y} ,$$

$$\frac{\partial h}{\partial \theta} = -r \sin \theta \frac{\partial g}{\partial x} + r \cos \theta \frac{\partial g}{\partial y} .$$

Example 3.5   Let (x, y, z) and (ρ, θ, z) be cartesian and cylindrical coordinates in R^3:

$$\begin{bmatrix} x \\ y \\ z \end{bmatrix} = \begin{bmatrix} \rho \cos \theta \\ \rho \sin \theta \\ z \end{bmatrix} .$$

If g is a real valued function of (x, y, z) and

$$h(\rho, \theta, z) = g(\rho \cos \theta, \rho \sin \theta, z),$$

then

$$\begin{bmatrix} \dfrac{\partial h}{\partial \rho} & \dfrac{\partial h}{\partial \theta} & \dfrac{\partial h}{\partial z} \end{bmatrix} = \begin{bmatrix} \dfrac{\partial g}{\partial x} & \dfrac{\partial g}{\partial y} & \dfrac{\partial g}{\partial z} \end{bmatrix} \begin{bmatrix} \dfrac{\partial x}{\partial \rho} & \dfrac{\partial x}{\partial \theta} & \dfrac{\partial x}{\partial z} \\ \dfrac{\partial y}{\partial \rho} & \dfrac{\partial y}{\partial \theta} & \dfrac{\partial y}{\partial z} \\ \dfrac{\partial z}{\partial \rho} & \dfrac{\partial z}{\partial \theta} & \dfrac{\partial z}{\partial z} \end{bmatrix}$$

$$= \begin{bmatrix} \dfrac{\partial g}{\partial x} & \dfrac{\partial g}{\partial y} & \dfrac{\partial g}{\partial z} \end{bmatrix} \begin{bmatrix} \cos \theta & -\rho \sin \theta & 0 \\ \sin \theta & \rho \cos \theta & 0 \\ 0 & 0 & 1 \end{bmatrix} ;$$

hence

$$\frac{\partial h}{\partial r} = \cos \theta \frac{\partial g}{\partial x} + \sin \theta \frac{\partial g}{\partial y} ,$$

$$\frac{\partial h}{\partial \theta} = -\rho \sin \theta \frac{\partial g}{\partial x} + \rho \cos \theta \frac{\partial g}{\partial y} ,$$

$$\frac{\partial h}{\partial z} = \frac{\partial g}{\partial z} .$$

Example 3.6   Let (r, θ, ϕ) be spherical coordinates in R^3; then

$$
\begin{bmatrix} x \\ y \\ z \end{bmatrix} = \begin{bmatrix} r \cos \theta \sin \phi \\ r \sin \theta \sin \phi \\ r \cos \phi \end{bmatrix}
$$

If g is a real valued function of (x, y, z) and $h(r, \theta, \phi) = g(r \cos \theta \sin \phi,$ $r \sin \theta \sin \phi, r \cos \phi)$, then

$$
\begin{bmatrix} \dfrac{\partial h}{\partial r} & \dfrac{\partial h}{\partial \theta} & \dfrac{\partial h}{\partial \phi} \end{bmatrix} = \begin{bmatrix} \dfrac{\partial g}{\partial x} & \dfrac{\partial g}{\partial y} & \dfrac{\partial g}{\partial z} \end{bmatrix} \begin{bmatrix} \cos \theta \sin \phi & -r \sin \theta \sin \phi & r \cos \theta \cos \phi \\ \sin \theta \sin \phi & r \cos \theta \sin \phi & r \cos \theta \cos \phi \\ \cos \phi & 0 & -r \sin \phi \end{bmatrix}.
$$

It may happen that some of the variables $x_1, \ldots, x_n$ are functions of the others, as in Example 4.3 of Section 2.4 and the following example.

Example 3.7   Let g be a real valued function of (x, y, z, w),

$$
\begin{bmatrix} x \\ y \end{bmatrix} = \begin{bmatrix} x(z, w) \\ y(z, w) \end{bmatrix}
$$

and

$$
h(z, w) = g(x(z, w), y(z, w), z, w);
$$

then

$$
\mathbf{U} = \begin{bmatrix} x(z, w) \\ y(z, w) \\ z \\ w \end{bmatrix}
$$

and

$$
\begin{bmatrix} \dfrac{\partial h}{\partial z} & \dfrac{\partial h}{\partial w} \end{bmatrix} = \begin{bmatrix} \dfrac{\partial g}{\partial x} & \dfrac{\partial g}{\partial y} & \dfrac{\partial g}{\partial z} & \dfrac{\partial g}{\partial w} \end{bmatrix} \begin{bmatrix} \dfrac{\partial x}{\partial z} & \dfrac{\partial x}{\partial w} \\ \dfrac{\partial y}{\partial z} & \dfrac{\partial y}{\partial w} \\ 1 & 0 \\ 0 & 1 \end{bmatrix}.
$$

Thus

$$\frac{\partial h}{\partial z} = \frac{\partial g}{\partial x}\frac{\partial x}{\partial z} + \frac{\partial g}{\partial y}\frac{\partial y}{\partial z} + \frac{\partial g}{\partial z},$$

$$\frac{\partial h}{\partial w} = \frac{\partial g}{\partial x}\frac{\partial x}{\partial w} + \frac{\partial g}{\partial y}\frac{\partial y}{\partial w} + \frac{\partial g}{\partial w}.$$

Second derivatives of composite functions can be computed by repetition of the chain rule, as in the following examples.

Example 3.8   In Example 3.3 we found that

$$\frac{\partial h}{\partial u} = \frac{\partial g}{\partial x}\frac{\partial x}{\partial u} + \frac{\partial g}{\partial y}\frac{\partial y}{\partial u};$$

hence

$$\frac{\partial^2 h}{\partial u^2} = \frac{\partial}{\partial u}\left(\frac{\partial h}{\partial u}\right) = \frac{\partial}{\partial u}\left(\frac{\partial g}{\partial x}\frac{\partial x}{\partial u}\right) + \frac{\partial}{\partial u}\left(\frac{\partial g}{\partial y}\frac{\partial y}{\partial u}\right)$$

$$= \frac{\partial x}{\partial u}\frac{\partial}{\partial u}\left(\frac{\partial g}{\partial x}\right) + \frac{\partial g}{\partial x}\frac{\partial^2 x}{\partial u^2} + \frac{\partial y}{\partial u}\frac{\partial}{\partial u}\left(\frac{\partial g}{\partial y}\right) + \frac{\partial g}{\partial y}\frac{\partial^2 y}{\partial u^2}.$$

In calculating $(\partial/\partial u)\,(\partial g/\partial x)$ we must recall that we are actually after

$$\frac{\partial}{\partial u}\left(\frac{\partial g}{\partial x}\left[(x(u,v),\,y(u,z))\right]\right);$$

the safest procedure is to define

$$g_1(x,\,y) = \frac{\partial g}{\partial x}\,(x,\,y) \text{ and } h_1(u,\,v) = g_1(x(u,\,v),\,y(u,\,v)),$$

and apply the chain rule to h_1:

$$\frac{\partial h_1}{\partial u} = \frac{\partial g_1}{\partial x}\frac{\partial x}{\partial u} + \frac{\partial g_1}{\partial y}\frac{\partial y}{\partial u}.$$

But

$$\frac{\partial h_1}{\partial u} = \frac{\partial}{\partial u}\left(\frac{\partial g}{\partial x}\right), \qquad \frac{\partial g_1}{\partial x} = \frac{\partial^2 g}{\partial x^2}, \qquad \text{and} \qquad \frac{\partial g_1}{\partial y} = \frac{\partial^2 g}{\partial x\partial y};$$

hence

$$\frac{\partial}{\partial u}\left(\frac{\partial g}{\partial x}\right) = \frac{\partial^2 g}{\partial x^2}\frac{\partial x}{\partial u} + \frac{\partial^2 g}{\partial x\,\partial y}\frac{\partial y}{\partial u}.$$

Similarly,

$$\frac{\partial}{\partial u}\left(\frac{\partial g}{\partial y}\right) = \frac{\partial^2 g}{\partial x \, \partial y}\frac{\partial x}{\partial u} + \frac{\partial^2 g}{\partial y^2}\frac{\partial y}{\partial u} ;$$

hence

$$\frac{\partial^2 h}{\partial u^2} = \frac{\partial^2 g}{\partial x^2}\left(\frac{\partial x}{\partial u}\right)^2 + 2\frac{\partial^2 g}{\partial x \, \partial y}\frac{\partial x}{\partial u}\frac{\partial x}{\partial v} + \frac{\partial^2 g}{\partial y^2}\left(\frac{\partial y}{\partial u}\right)^2 + \frac{\partial g}{\partial x}\frac{\partial^2 x}{\partial u^2} + \frac{\partial g}{\partial y}\frac{\partial^2 y}{\partial u^2}.$$

Again we emphasize that all partial derivatives of g are to be evaluated at $(x, y) = (x(u, v), y(u, v))$.

Example 3.9 In Example 3.4,

$$(9) \quad \frac{\partial^2 h}{\partial \theta \, \partial r} = \frac{\partial}{\partial \theta}\left(\frac{\partial h}{\partial r}\right) = \frac{\partial}{\partial \theta}\left(\cos \theta \frac{\partial g}{\partial x} + \sin \theta \frac{\partial g}{\partial y}\right)$$

$$= -\sin \theta \frac{\partial g}{\partial x} + \cos \theta \frac{\partial g}{\partial y} + \cos \theta \frac{\partial}{\partial \theta}\left(\frac{\partial g}{\partial x}\right) + \sin \theta \frac{\partial}{\partial \theta}\left(\frac{\partial g}{\partial y}\right).$$

But

$$\frac{\partial}{\partial \theta}\left(\frac{\partial g}{\partial x}\right) = \frac{\partial^2 g}{\partial x^2}\frac{\partial x}{\partial \theta} + \frac{\partial^2 g}{\partial x \, \partial y}\frac{\partial y}{\partial \theta}$$

$$= -r \sin \theta \frac{\partial^2 g}{\partial x^2} + r \cos \theta \frac{\partial^2 g}{\partial x \, \partial y}$$

and

$$\frac{\partial}{\partial \theta}\left(\frac{\partial g}{\partial y}\right) = \frac{\partial^2 g}{\partial x \, \partial y}\frac{\partial x}{\partial \theta} + \frac{\partial^2 g}{\partial y^2}\frac{\partial y}{\partial \theta}$$

$$= -r \sin \theta \frac{\partial^2 g}{\partial x \, \partial y} + r \cos \theta \frac{\partial^2 g}{\partial y^2}.$$

Substituting these in (9) yields

$$\frac{\partial^2 h}{\partial \theta \, \partial r} = -\sin \theta \frac{\partial g}{\partial x} + \cos \theta \frac{\partial g}{\partial y} + r(\cos^2 \theta - \sin^2 \theta)\frac{\partial^2 g}{\partial x \, \partial y}$$

$$+ r \sin \theta \cos \theta \left(\frac{\partial^2 g}{\partial y^2} - \frac{\partial^2 g}{\partial x^2}\right).$$

Example 3.10   In Example 3.7,

$$(10) \quad \frac{\partial^2 h}{\partial z \, \partial w} = \frac{\partial}{\partial z}\left(\frac{\partial h}{\partial w}\right) = \frac{\partial}{\partial z}\left(\frac{\partial g}{\partial x}\frac{\partial x}{\partial w} + \frac{\partial g}{\partial y}\frac{\partial y}{\partial w} + \frac{\partial g}{\partial w}\right)$$

$$= \frac{\partial x}{\partial w}\frac{\partial}{\partial z}\left(\frac{\partial g}{\partial x}\right) + \frac{\partial g}{\partial x}\frac{\partial^2 x}{\partial z \, \partial w}$$

$$+ \frac{\partial y}{\partial w}\frac{\partial}{\partial z}\left(\frac{\partial g}{\partial y}\right) + \frac{\partial g}{\partial y}\frac{\partial^2 y}{\partial z \, \partial w} + \frac{\partial}{\partial z}\left(\frac{\partial g}{\partial w}\right).$$

To obtain $(\partial/\partial z)(\partial g/\partial x)$, we first define $g_1(x, y) = (\partial g/\partial x)(x, y)$ and $h_1(z, w) = g_1(x(z, w), y(z, w), z, w)$; then

$$\frac{\partial}{\partial z}\left(\frac{\partial g}{\partial x}\right) = \frac{\partial h_1}{\partial z} = \frac{\partial g_1}{\partial x}\frac{\partial x}{\partial z} + \frac{\partial g_1}{\partial y}\frac{\partial y}{\partial z} + \frac{\partial g_1}{\partial z}$$

$$= \frac{\partial^2 g}{\partial x^2}\frac{\partial x}{\partial z} + \frac{\partial^2 g}{\partial x \, \partial y}\frac{\partial y}{\partial z} + \frac{\partial^2 g}{\partial x \, \partial z}.$$

Similarly,

$$\frac{\partial}{\partial z}\left(\frac{\partial g}{\partial y}\right) = \frac{\partial^2 g}{\partial x \, \partial y}\frac{\partial x}{\partial z} + \frac{\partial^2 g}{\partial y^2}\frac{\partial y}{\partial z} + \frac{\partial^2 g}{\partial y \, \partial z}$$

and

$$\frac{\partial}{\partial z}\left(\frac{\partial g}{\partial w}\right) = \frac{\partial^2 g}{\partial x \, \partial w}\frac{\partial x}{\partial z} + \frac{\partial^2 g}{\partial y \, \partial w}\frac{\partial y}{\partial z} + \frac{\partial^2 g}{\partial z \, \partial w}.$$

Substitution of these last three expressions into (10) yields $\partial^2 h/\partial z \, \partial w$.

After attaining proficiency with the chain rule, the student will probably prefer to dispense with the explicit introduction of intermediate functions such as g_1 in the last example.

EXERCISES   5.3   #1a.

1. Find $\mathbf{h} = \mathbf{g} \circ \mathbf{f}$ explicitly in terms of (x, y, z) and compute $\mathbf{d}_{\mathbf{X}_0}\mathbf{h}$; then find $\mathbf{d}_{\mathbf{X}_0}\mathbf{h}$ by the chain rule.

(a) $\mathbf{f}(x, y, z) = \begin{bmatrix} 2x - y + z \\ e^{x^2 - y^2} \end{bmatrix}$, $\mathbf{g}(u, v) = \begin{bmatrix} u + v \\ u - v \end{bmatrix}$;

$$\mathbf{X}_0 = (1, 1, -2).$$

(b) $\mathbf{f}(x, y) = \begin{bmatrix} e^x \cos y \\ e^x \sin y \end{bmatrix}$, $\mathbf{g}(u, v) = \begin{bmatrix} u^2 + v^2 \\ u^2 - v^2 \end{bmatrix}$;

$$\mathbf{X}_0 = (0, 0).$$

2.   Repeat Exercise 1 for the following functions.

(a) $\mathbf{f}(x, y) = \begin{bmatrix} x + 2y \\ 2x - y^2 \end{bmatrix}$, $\mathbf{g}(u, v) = \begin{bmatrix} u + 2v \\ u - v^2 \\ u^2 + v \end{bmatrix}$;

$$\mathbf{X}_0 = (1, -2).$$

(b) $\mathbf{f}(x, y) = \begin{bmatrix} \log(x + y) \\ \log(x - y) \end{bmatrix}$, $\mathbf{g}(u, v) = \begin{bmatrix} e^{2v} \\ e^{u+v} \end{bmatrix}$;

$$\mathbf{X}_0 = (2, 1).$$

3.   Find $\mathbf{J}_{\mathbf{X}_0}\mathbf{f}$, $\mathbf{J}_{\mathbf{f}(\mathbf{X}_0)}\mathbf{g}$, $\mathbf{J}_{\mathbf{f}(\mathbf{X}_0)}\mathbf{g} \cdot \mathbf{J}_{\mathbf{X}_0}\mathbf{f}$ and $\mathbf{J}_{\mathbf{X}_0}\mathbf{h}$ in Exercise 1.

4.   Repeat Exercise 3 for the functions in Exercise 2.

5.   Let

$$\mathbf{U} = \begin{bmatrix} u \\ v \\ w \end{bmatrix} = \begin{bmatrix} \sqrt{x^2 + y^2 + z^2} \\ \tan^{-1}\dfrac{y}{x} \\ \tan^{-1}\dfrac{z}{\sqrt{x^2 + y^2}} \end{bmatrix} \quad \left(|z| < \frac{\pi}{2}, |w| < \frac{\pi}{2} \right),$$

$$\mathbf{g}(u, v, w) = \begin{bmatrix} uv \\ vw \\ uw \end{bmatrix},$$

and $\mathbf{h} = \mathbf{g} \circ \mathbf{U}$.

(a)   Find $\mathbf{J}\mathbf{U}$ and $\mathbf{J}\mathbf{g}$.

(b)   Find $\mathbf{J}_{\mathbf{X}_0}\mathbf{U}$, $\mathbf{J}_{\mathbf{U}_0}\mathbf{g}$, and $\mathbf{J}_{\mathbf{X}_0}\mathbf{h}$, with $\mathbf{X}_0 = (1/\sqrt{2}, -1/\sqrt{2}, 0)$.

(c)   Find $\mathbf{d}_{\mathbf{X}_0}\mathbf{U}$, $\mathbf{d}_{\mathbf{U}_0}\mathbf{g}$ and $\mathbf{d}_{\mathbf{X}_0}\mathbf{h}$.

6. Repeat Exercise 5 with $\mathbf{X}_0 = (2, 1, 2)$,

$$\mathbf{U} = \begin{bmatrix} x^2 - y^2 \\ 2xy \\ z \end{bmatrix}, \quad \text{and} \quad \mathbf{g}(u, v, w) = \begin{bmatrix} u + w^2 \\ \dfrac{u}{w} \end{bmatrix}.$$

7. Let $h(t) = f(\mathbf{X}(t))$ and find $h'(t_0)$ and $d_{t_0}h$:

(a) $f(x, y, z) = \sqrt{x^2 + y + z^4}, \qquad \mathbf{X}(t) = \begin{bmatrix} t \\ t^2 \\ \sqrt{t} \end{bmatrix} \qquad (t_0 = 1).$

(b) $f(x, y, z) = xye^z + 2y^2z - 3xz \cos y, \qquad \mathbf{X}(t) = \begin{bmatrix} e^t \\ \sin t \\ 1 - \cos t \end{bmatrix}$

$$(t_0 = 0).$$

(c) $f(x, y, z) = \log(x + y + z), \qquad \mathbf{X}(t) = \begin{bmatrix} t + 2 \\ t^2 \\ -2 \sin \pi t \end{bmatrix}$

$$(t_0 = 1).$$

8. If $h(r, \theta) = g(r \cos \theta, r \sin \theta)$, show that

$$\left(\frac{\partial h}{\partial r}\right)^2 + \frac{1}{r^2}\left(\frac{\partial h}{\partial \theta}\right)^2 = \left(\frac{\partial g}{\partial x}\right)^2 + \left(\frac{\partial g}{\partial y}\right)^2.$$

9. Let $\mathbf{h}(r, \theta) = \mathbf{g}(r \cos \theta, r \sin \theta)$ and $\mathbf{U}_0 = (r_0, \theta_0) = (2, \pi/4)$. Find $d_{\mathbf{U}_0}\mathbf{h}$:

(a) $g(x, y) = x^2 + y.$ (b) $\mathbf{g}(x, y) = \begin{bmatrix} x - y^2 \\ 2x + y \end{bmatrix}.$

(c) $\mathbf{g}(x, y) = \begin{bmatrix} (x + y)^2 \\ x^2 + y^2 \end{bmatrix}.$

10. Let f and g be functions of a single variable and $F(x, y) = f(y + ax) + g(y - ax)$. Show that $F_{xx} = a^2 F_{yy}$.

11. Let $\mathbf{h}(r, \theta, z) = \mathbf{g}(r \cos \theta, r \sin \theta, z)$ and $\mathbf{U}_0 = (r_0, \theta_0, z_0) = (1, \pi/2, -1)$. Find $\mathbf{d}_{\mathbf{U}_0}\mathbf{h}$:

   (a) $g(x, y, z) = z e^{xy}$.   (b) $\mathbf{g}(x, y, z) = \begin{bmatrix} x^3 + y^3 + 2yz \\ x^2 + y^2 + z^2 \end{bmatrix}$.

   (c) $\mathbf{g}(x, y, z) = \begin{bmatrix} \dfrac{xy}{z} \\ \dfrac{xz}{y} \end{bmatrix}$.

12. If f is a function of one variable and $h(u, v) = f(u/v)$, show that
$$v \frac{\partial h}{\partial v} + u \frac{\partial h}{\partial u} = 0.$$

13. Let $\mathbf{h}(r, \theta, \phi) = \mathbf{g}(r \cos \theta \sin \phi, r \sin \theta \sin \phi, r \cos \phi)$ and $\mathbf{U}_0 = (r_0, \theta_0, \phi_0) = (1, \pi/4, -\pi/4)$. Find $\mathbf{d}_{\mathbf{U}_0}\mathbf{h}$:

   (a) $\mathbf{g}(x, y, z) = \begin{bmatrix} \dfrac{x^2 - y^2}{z^2} \\ x^2 + y^2 + z^2 \end{bmatrix}$.   (b) $\mathbf{g}(x, y, z) = \begin{bmatrix} x^2 \\ y^2 \\ z^2 \end{bmatrix}$.

   (c) $\mathbf{g}(x, y, z) = \dfrac{x}{x + y}$.

14. If $h(x, y) = f(ax + by)$, show that
$$a \frac{\partial h}{\partial y} - b \frac{\partial h}{\partial x} = 0.$$

15. Let $g(x, y, z, w) = \log(1 + x + y + z + w)$,
$$\begin{bmatrix} z \\ w \end{bmatrix} = \begin{bmatrix} x^2 - y^2 \\ 2xy \end{bmatrix}$$
and $h(x, y) = g(x, y, z(x, y), w(x, y))$. Find $\mathbf{d}_{(1,1)}h$.

16. Let $g(x, y, z, w) = e^{x+y+z+w}$, $z = z(x, y)$, $w = w(x)$, and $h(x, y) = g(x, y, z(x, y), w(x))$. Find dg.

17. Find $(\partial^2 h/\partial x^2)(1, 1)$, $(\partial^2 h/\partial x \, \partial y)(1, 1)$, and $(\partial^2 h/\partial y^2)(1, 1)$ in Exercise 15.

18. If f is a function of (u, v) and $h(x, y) = f(x + y, x - y)$, show that

$$\frac{\partial h}{\partial x}\frac{\partial h}{\partial y} = \left(\frac{\partial f}{\partial u}\right)^2 - \left(\frac{\partial f}{\partial v}\right)^2 .$$

19. Calculate $\partial^2 h/\partial u \, \partial v$ and $\partial^2 h/\partial v^2$ in Example 3.3. (*Hint*: See Example 3.8.)

20. If U is a real valued function of (x, y) and $h(r, \theta) = U(r \cos \theta, r \sin \theta)$, show that

$$U_{xx} + U_{yy} = h_{rr} + \frac{1}{r} h_r + \frac{1}{r^2} h_{\theta\theta} .$$

THEORETICAL EXERCISES

T-1. Prove that $\mathbf{X}_0$ is an interior point of the domain of $\mathbf{g} \circ \mathbf{f}$ in Theorem 3.1. (*Hint*: From the definition of differentiability, $\mathbf{X}_0$ and $\mathbf{U}_0$ are interior points of the domains of $\mathbf{f}$ and $\mathbf{g}$, respectively; also $\mathbf{f}$, being differentiable at $\mathbf{X}_0$, is continuous there.)

T-2. If $\mathbf{L}$ is a linear transformation,

$$\mathbf{L}(\mathbf{X}) = \begin{bmatrix} a_{11}x_1 + \cdots + a_{1n}x_n \\ \vdots \\ a_{m1}x_1 + \cdots + a_{mn}x_n \end{bmatrix},$$

show that $|\mathbf{L}(\mathbf{X})| < mn \, K \, |\mathbf{X}|$, where $K = \max |a_{ij}|$.

T-3. Show that the right side of (6) approaches zero as $\mathbf{X}$ approaches $\mathbf{X}_0$. *Hint*: Use (4), the continuity of $\mathbf{f}$ at $\mathbf{X}_0$ and Theorem 1.5.

T-4. Prove (8) directly, under the assumption that $\mathbf{f}$ has first partial derivatives at $\mathbf{X}_0$ and that $\mathbf{g}$ has continuous first partial derivatives near $\mathbf{U}_0 = \mathbf{f}(\mathbf{X}_0)$. (*Hint*: See the proof of Theorem 4.3, Section 4.4.)

5.4 Vector and Scalar Fields

Vector functions of the form $\mathbf{F} \colon R^n \to R^n$ occur in so many applications that we give them a special name.

Definition 4.1   A *vector field* is a vector valued function **F** whose domain and range are both subsets of R^n.

Example 4.1   Consider a fluid flowing in a channel in R^3 with velocity at each point **X** depending on **X** alone, and not on time. (We say that the flow is *steady*.) With each point **X** of the channel we associate the velocity **V(X)** of the fluid at that point; the vector function so defined is a vector field. We can depict the vector field by drawing the velocity vector **V(X)** at each **X** in the channel (Fig. 4.1).

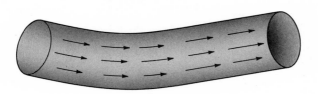

FIGURE 4.1

Example 4.2   Let a particle P_0 of mass m_0 be fixed at the origin of R^3. According to Newton's law of gravitation the force exerted by P_0 on a unit mass at an arbitrary point $\mathbf{X} \neq \mathbf{0}$ is given by

$$\mathbf{F(X)} = -\frac{Gm_0}{|\,\mathbf{X}\,|^3}\,\mathbf{X},$$

where G is a universal constant. Thus **F** is a continuous vector field for all $\mathbf{X} \neq \mathbf{0}$. The direction of **F(X)** is toward the origin and its magnitude varies inversely with $|\,\mathbf{X}\,|^2$; **F** is an *inverse square law* force field.

The following definition introduces a convenient terminology for a familiar concept.

Definition 4.2   A real valued function $f\colon R^n \to R$ is called a *scalar field*.

Example 4.3   In connection with the inverse square law force field of Example 4.2, we may associate with each $\mathbf{X} \neq \mathbf{0}$ the real number

$$f(\mathbf{X}) = |\,\mathbf{F(X)}\,| = \frac{Gm_0}{|\,\mathbf{X}\,|^2}\,;$$

then f is a scalar field.

The Gradient as a Vector Field

In Section 2.5 we defined the gradient of a real valued function $f \colon R^n \to R$ at a point $\mathbf{X}_0$ to be the vector

$$\nabla_{\mathbf{X}_0} f = \begin{bmatrix} \dfrac{\partial f}{\partial x_1}(\mathbf{X}_0) \\[2mm] \dfrac{\partial f}{\partial x_2}(\mathbf{X}_0) \\[2mm] \vdots \\[2mm] \dfrac{\partial f}{\partial x_n}(\mathbf{X}_0) \end{bmatrix},$$

or, if $n = 3$,

$$\nabla_{\mathbf{X}_0} f = \frac{\partial f}{\partial x}(\mathbf{X}_0)\,\mathbf{i} + \frac{\partial f}{\partial y}(\mathbf{X}_0)\,\mathbf{j} + \frac{\partial f}{\partial z}(\mathbf{X}_0)\,\mathbf{k},$$

provided the indicated partial derivatives exist. If f is a scalar field which has partial derivatives at every point $\mathbf{X}$ of a domain D we can define the vector field ∇f, called the *gradient of f*, to be the vector field whose value at each $\mathbf{X}$ in D is $\nabla_{\mathbf{X}} f$; thus

$$\nabla f = \begin{bmatrix} \dfrac{\partial f}{\partial x_1} \\[2mm] \dfrac{\partial f}{\partial x_2} \\[2mm] \vdots \\[2mm] \dfrac{\partial f}{\partial x_n} \end{bmatrix}$$

or, if $n = 3$,

$$(1) \qquad \nabla f = \frac{\partial f}{\partial x}\mathbf{i} + \frac{\partial f}{\partial y}\mathbf{j} + \frac{\partial f}{\partial z}\mathbf{k}.$$

Example 4.4   Let f be the scalar field defined by $f(x, y, z) = 3x^2yz$. Then

$$\nabla f = 6xyz\mathbf{i} + 3x^2z\mathbf{j} + 3x^2y\mathbf{k}.$$

Example 4.5 The scalar field f defined by

$$f(\mathbf{X}) = \frac{Gm_0}{|\mathbf{X}|} = \frac{Gm_0}{\sqrt{x^2 + y^2 + z^2}}$$

is defined for all $\mathbf{X} \neq \mathbf{0}$ in R^3. Since

$$f_x(\mathbf{X}) = -\frac{Gm_0 x}{|\mathbf{X}|^3}, \qquad f_y(\mathbf{X}) = -\frac{Gm_0 y}{|\mathbf{X}|^3}, \qquad f_z(\mathbf{X}) = -\frac{Gm_0 z}{|\mathbf{X}|^3},$$

it follows that the inverse square law force field of Example 4.2 is the gradient of f.

If a vector field $\mathbf{F}$ can be expressed as the gradient of a scalar field f,

$$\mathbf{F} = \nabla f,$$

then $\mathbf{F}$ is said to be *conservative*, and $-f$ is called the *potential function* of $\mathbf{F}$. We shall encounter these ideas again in Section 4.5.

The next theorem follows from Exercise T-2, Section 4.7.

Theorem 4.1 Let f and g be differentiable scalar fields in a domain D in R^n and let a be a real number. Then

$$\text{(a)} \qquad \nabla(af) = a\,\nabla f,$$

$$\text{(b)} \quad \nabla(f + g) = \nabla f + \nabla g,$$

and

$$\text{(c)} \qquad \nabla(fg) = f\,\nabla g + g\,\nabla f$$

on D. Moreover $\nabla(f/g)$ is defined on the subset D' of D where $g(\mathbf{X}) \neq 0$, and

$$\text{(d)} \qquad \nabla(f/g) = \frac{g\,\nabla f - f\,\nabla g}{g^2}.$$

If $n = 3$ we may view ∇f as the vector field obtained by applying the operator

$$\nabla = \frac{\partial}{\partial x}\mathbf{i} + \frac{\partial}{\partial y}\mathbf{j} + \frac{\partial}{\partial z}\mathbf{k}$$

to f. Thus we write (1) as

$$\nabla f = \left(\frac{\partial}{\partial x}\mathbf{i} + \frac{\partial}{\partial y}\mathbf{j} + \frac{\partial}{\partial z}\mathbf{k}\right)f,$$

which suggests that we are multiplying the "vector" ∇ by a scalar f. Of course this is not actually the case, but it is sometimes useful to think of ∇ as a vector (which it is not!), especially in connection with the divergence and curl, defined below. The operator ∇ is called the "del" operator.

Divergence of a Vector Field

Throughout the rest of this section we consider only scalar and vector fields in R^3, except where it is specifically stated otherwise.

Definition 4.3 Let $\mathbf{F} = F_1 \mathbf{i} + F_2 \mathbf{j} + F_3 \mathbf{k}$ be a differentiable vector field defined on a domain D in R^3. The *divergence of* $\mathbf{F}$, denoted by div $\mathbf{F}$, is the scalar field defined by

$$(2) \qquad \text{div } \mathbf{F} = \frac{\partial F_1}{\partial x} + \frac{\partial F_2}{\partial y} + \frac{\partial F_3}{\partial z} ;$$

its value at a point $\mathbf{X}_0$ in D is

$$\text{div } \mathbf{F}(\mathbf{X}_0) = \frac{\partial F_1}{\partial x}(\mathbf{X}_0) + \frac{\partial F_2}{\partial y}(\mathbf{X}_0) + \frac{\partial F_3}{\partial z}(\mathbf{X}_0).$$

Example 4.6 If

$$\mathbf{F}(\mathbf{X}) = x\mathbf{i} + (xz + 1)\mathbf{j} + (yz + x)\mathbf{k}$$

then

$$\text{div } \mathbf{F}(\mathbf{X}) = (1) + (0) + (y) = 1 + y;$$

thus if $\mathbf{X}_0 = (2, -2, 3)$, then

$$\text{div } \mathbf{F}(\mathbf{X}_0) = 1 - 2 = -1.$$

It is sometimes useful to view the right side of (2) as the "inner product" of ∇ and $\mathbf{F}$; thus

$$\nabla \cdot \mathbf{F} = \left(\frac{\partial}{\partial x}\mathbf{i} + \frac{\partial}{\partial y}\mathbf{j} + \frac{\partial}{\partial z}\mathbf{k} \right) \cdot (F_1\mathbf{i} + F_2\mathbf{j} + F_3\mathbf{k}) = \text{div } \mathbf{F}.$$

However, caution is in order here. The inner product of two vectors is commutative ($\mathbf{U} \cdot \mathbf{V} = \mathbf{V} \cdot \mathbf{U}$), but $\nabla \cdot \mathbf{F}$ and $\mathbf{F} \cdot \nabla$ are different: $\nabla \cdot \mathbf{F}$ is a scalar field, while $\mathbf{F} \cdot \nabla$ is the operator

$$\mathbf{F} \cdot \nabla = F_1 \frac{\partial}{\partial x} + F_2 \frac{\partial}{\partial y} + F_3 \frac{\partial}{\partial z} .$$

Definition 4.3 appears to be quite arbitrary. Nevertheless, the divergence of a vector field has important geometrical and physical applications which we shall explore further in Chapter 6. For now we shall simply show that although div $\mathbf{F}$ is defined in terms of a particular coordinate system, its value at a given point is determined by the point itself, and not by the particular rectangular coordinate system chosen to identify points.

To see this, let us recall the viewpoint discussed in Section 3.5, where we distinguished between a point P in R^3 and its coordinates (x, y, z) with respect to a particular rectangular coordinate system. Suppose (x_1, x_2, x_3) and (y_1, y_2, y_3) are the coordinates of the same point P with respect to two rectangular systems. Denote the unit vectors along the (x_1, x_2, x_3) axes by $\mathbf{i}_1, \mathbf{i}_2$, and $\mathbf{i}_3$, and those along the (y_1, y_2, y_3) axes by $\mathbf{j}_1, \mathbf{j}_2$, and $\mathbf{j}_3$. Then we can write

(3)
$$\mathbf{i}_r = \sum_{s=1}^{3} l_{sr} \mathbf{j}_s,$$

where the matrix

$$\mathbf{M} = \begin{bmatrix} l_{11} & l_{12} & l_{13} \\ l_{21} & l_{22} & l_{23} \\ l_{31} & l_{32} & l_{33} \end{bmatrix}$$

is orthogonal; that is

(4)
$$\mathbf{M}^{\mathrm{T}} = \mathbf{M}^{-1}.$$

This result was derived, in different notation, in Section 3.5. It was also shown there that the coordinates of a given point P with respect to the two systems are related by

(5)
$$\begin{bmatrix} y_1 \\ y_2 \\ y_3 \end{bmatrix} = \mathbf{M} \begin{bmatrix} x_1 - x_1^0 \\ x_2 - x_2^0 \\ x_3 - x_3^0 \end{bmatrix},$$

where (x_1^0, x_2^0, x_3^0) are the coordinates of the origin of the (y_1, y_2, y_3) system with respect to the (x_1, x_2, x_3) system.

To emphasize that $\mathbf{F}$ is a function of P let us temporarily denote its values by $\mathbf{F}(P)$ instead of the usual $\mathbf{F}(\mathbf{X})$. Then we can express $\mathbf{F}(P)$ as a linear combination of $\mathbf{i}_1, \mathbf{i}_2$, and $\mathbf{i}_3$, and also as a linear combination of $\mathbf{j}_1, \mathbf{j}_2$, and $\mathbf{j}_3$:

(6)
$$\mathbf{F}(P) = F_1(x_1, x_2, x_3)\mathbf{i}_1 + F_2(x_1, x_2, x_3)\mathbf{i}_2 + F_3(x_1, x_2, x_3)\mathbf{i}_3$$
$$= G_1(y_1, y_2, y_3)\mathbf{j}_1 + G_2(y_1, y_2, y_3)\mathbf{j}_2 + G_3(y_1, y_2, y_3)\mathbf{j}_3 .$$

By substituting (3) into (6) and equating coefficients of j_1, j_2, and j_3 on both sides of the second equality, we obtain

$$\begin{bmatrix} G_1(y_1, y_2, y_3) \\ G_2(y_1, y_2, y_3) \\ G_3(y_1, y_2, y_3) \end{bmatrix} = \mathbf{M} \begin{bmatrix} F_1(x_1, x_2, x_3) \\ F_2(x_1, x_2, x_3) \\ F_3(x_1, x_2, x_3) \end{bmatrix},$$

or, in terms of components,

$$(7) \qquad G_r(y_1, y_2, y_3) = \sum_{s=1}^{3} l_{rs} F_s(x_1, x_2, x_3) \qquad (r = 1, 2, 3)$$

(Exercise T-1). Differentiating both sides with respect to y_r yields

$$(8) \qquad \frac{\partial G_r}{\partial y_r} = \sum_{s=1}^{3} l_{rs} \sum_{q=1}^{3} \frac{\partial F_s}{\partial x_q} \frac{\partial x_q}{\partial y_r} \qquad (r = 1, 2, 3),$$

where the partial derivatives of G_r are to be evaluated at (y_1, y_2, y_3) and those of F_s at (x_1, x_2, x_3).

Now (4) and (5) imply that

$$\begin{bmatrix} x_1 - x_1^0 \\ x_2 - x_2^0 \\ x_3 - x_3^0 \end{bmatrix} = M^{\mathrm{T}} \begin{bmatrix} y_1 \\ y_2 \\ y_3 \end{bmatrix},$$

from which it follows that

$$\frac{\partial x_q}{\partial y_r} = l_{rq}$$

(Exercise T-2). Substituting this in (8) yields

$$\frac{\partial G_r}{\partial y_r} = \sum_{s=1}^{3} l_{rs} \sum_{q=1}^{3} l_{rq} \frac{\partial F_s}{\partial x_q} \qquad (r = 1, 2, 3).$$

Adding these equations for $r = 1$, 2, and 3 yields

$$\sum_{r=1}^{3} \frac{\partial G_r}{\partial y_r} = \sum_{r=1}^{3} \sum_{s=1}^{3} l_{rs} \sum_{q=1}^{3} l_{rq} \frac{\partial F_s}{\partial x_q}.$$

It is convenient to rewrite the sum on the right so that the summation on r is performed first; thus

$$(9) \qquad \sum_{r=1}^{3} \frac{\partial G_r}{\partial x_r} = \sum_{s=1}^{3} \sum_{q=1}^{3} \left(\sum_{r=1}^{3} l_{rs} l_{rq} \right) \frac{\partial F_s}{\partial x_q}.$$

According to (4), $\mathbf{M}\mathbf{M}^{\mathrm{T}} = \mathbf{I}$; thus

$$\sum_{r=1}^{3} l_{rs}l_{rq} = \begin{cases} 0 & \text{if} \quad q \neq s, \\ 1 & \text{if} \quad q = s \end{cases}$$

(Exercise T-3). Therefore (9) reduces to

(10) $$\sum_{r=1}^{3} \frac{\partial G_r}{\partial x_r} (y_1, y_2, y_3) = \sum_{s=1}^{3} \frac{\partial F_s}{\partial x_s} (x_1, x_2, x_3).$$

We have therefore proved the following theorem.

Theorem 4.2 Let $\mathbf{F}$ be a vector field defined at all points P in a domain D, which is expressed by

(11) $\quad \mathbf{F}(P) = F_1(x_1, x_2, x_3)\mathbf{i}_1 + F_2(x_1, x_2, x_3)\mathbf{i}_2 + F_3(x_1, x_2, x_3)\mathbf{i}_3$

with respect to a given rectangular coordinate system in R^3, where F_1, F_2, and F_3 are differentiable functions of (x_1, x_2, x_3). Furthermore suppose

(12) $\quad \mathbf{F}(P) = G_1(y_1, y_2, y_3)\mathbf{j}_i + G_2(y_1, y_2, y_3)\mathbf{j}_2 + G_3(y_1, y_2, y_3)\mathbf{j}_3$

is a representation of the same vector field with respect to a second rectangular coordinate system. Then div $\mathbf{F}(P)$ can be computed in terms of either coordinate system, since (10) holds whenever (x_1, x_2, x_3) and (y_1, y_2, y_3) are the coordinates of the same point P with respect to the two coordinate systems.

We leave the verification of the following theorem to the student (Exercise T-4).

Theorem 4.3 If a and b are constants, $\mathbf{F}$ and $\mathbf{G}$ differentiable vector fields, and g a differentiable scalar field, then

(a) div$(a\mathbf{F} + b\mathbf{G}) = a$ div $\mathbf{F} + b$ div $\mathbf{G}$;
(b) div$(g\mathbf{F}) = g$ div $\mathbf{F} + (\nabla g) \cdot \mathbf{F}$.

The divergence of a vector field

$$\mathbf{F} = \begin{bmatrix} F_1 \\ F_2 \\ \vdots \\ F_n \end{bmatrix} \qquad (R^n \to R^n)$$

can be defined for arbitrary $n > 1$ by

$$\text{div } \mathbf{F} = \sum_{i=1}^{n} \frac{\partial F_i}{\partial x_i}.$$

The Laplacian of a Scalar Field

If f is a scalar field with second partial derivatives f_{xx}, f_{yy}, and f_{zz} in a domain D in R^3, then the divergence of $\boldsymbol{\nabla} f$ is given by

$$\boldsymbol{\nabla} \cdot \boldsymbol{\nabla} f = \left(\frac{\partial}{\partial x} \mathbf{i} + \frac{\partial}{\partial y} \mathbf{j} + \frac{\partial}{\partial z} \mathbf{k} \right) \cdot \left(\frac{\partial f}{\partial x} \mathbf{i} + \frac{\partial f}{\partial y} \mathbf{j} + \frac{\partial f}{\partial z} \mathbf{k} \right)$$

$$= \frac{\partial^2 f}{\partial x^2} + \frac{\partial^2 f}{\partial y^2} + \frac{\partial^2 f}{\partial z^2}.$$

The expression on the right is called the *Laplacian of f* and is denoted more briefly by $\nabla^2 f$. The operator

$$\nabla^2 = \frac{\partial^2}{\partial x^2} + \frac{\partial^2}{\partial y^2} + \frac{\partial^2}{\partial z^2}$$

is called the *Laplacian operator*.

Example 4.7 If

$$f(x, y, z) = x^3 + 2xy^2 + yz^3$$

then

$$\nabla^2 f = 6x + 4x + 6yz.$$

A scalar field f is said to be *harmonic* in a domain D if

(13) $$\nabla^2 f = 0$$

in D; (13) is called *Laplace's equation*. Harmonic functions have many applications in the natural sciences. We shall study them further in Chapter 6.

Example 4.8 If

$$f(x, y, z) = x^2 + y^2 - 2z^2$$

then

$$\nabla^2 f = 2 + 2 - 2(2) = 0;$$

hence f is harmonic.

The Laplacian $\nabla^2 f$ can be shown to be independent of the particular rectangular coordinate system chosen for R^3 (Exercise T-5). It can also be defined for scalar fields in R^n for any $n > 1$ by

$$\nabla^2 f = \sum_{r=1}^{n} \frac{\partial^2 f}{\partial x_r^2}.$$

Curl of a Vector Field

The *curl* of a vector field $\mathbf{F} = F_1\mathbf{i} + F_2\mathbf{j} + F_3\mathbf{k}$ is defined by

$$(14) \quad \operatorname{curl} \mathbf{F} = \left(\frac{\partial F_3}{\partial y} - \frac{\partial F_2}{\partial z}\right)\mathbf{i} + \left(\frac{\partial F_1}{\partial z} - \frac{\partial F_3}{\partial x}\right)\mathbf{j} + \left(\frac{\partial F_2}{\partial x} - \frac{\partial F_1}{\partial y}\right)\mathbf{k}.$$

The value of **curl F** at a point $\mathbf{X}_0$ is denoted by $\operatorname{curl} \mathbf{F}(\mathbf{X}_0)$. This definition is easily remembered if we write

$$\operatorname{curl} \mathbf{F} = \nabla \times \mathbf{F} = \begin{vmatrix} \mathbf{i} & \mathbf{j} & \mathbf{k} \\ \dfrac{\partial}{\partial x} & \dfrac{\partial}{\partial y} & \dfrac{\partial}{\partial z} \\ F_1 & F_2 & F_3 \end{vmatrix}.$$

Formal expansion of this determinant yields (14).

Example 4.9   If

$$\mathbf{F}(\mathbf{X}) = 3x^2\mathbf{i} + xy\mathbf{j} + z\mathbf{k}$$

then

$$\operatorname{curl} \mathbf{F} = \begin{vmatrix} \mathbf{i} & \mathbf{j} & \mathbf{k} \\ \dfrac{\partial}{\partial x} & \dfrac{\partial}{\partial y} & \dfrac{\partial}{\partial z} \\ 3x^2 & xy & z \end{vmatrix}$$

$$= \left(\frac{\partial}{\partial y}(z) - \frac{\partial}{\partial z}(xy)\right)\mathbf{i} + \left(\frac{\partial}{\partial z}(3x^2) - \frac{\partial}{\partial x}(z)\right)\mathbf{j}$$

$$+ \left(\frac{\partial}{\partial x}(xy) - \frac{\partial}{\partial y}(3x^2)\right)\mathbf{k} = y\mathbf{k}.$$

Thus if $\mathbf{X}_0 = (2, 3, -4)$ then

$$\text{curl } \mathbf{F}(\mathbf{X}_0) = -4\mathbf{k}.$$

Example 4.10 If f is a real valued function with continuous second partial derivatives, then

$$(15) \quad \text{curl } (\nabla f) = \begin{vmatrix} \mathbf{i} & \mathbf{j} & \mathbf{k} \\ \dfrac{\partial}{\partial x} & \dfrac{\partial}{\partial y} & \dfrac{\partial}{\partial z} \\ f_x & f_y & f_z \end{vmatrix}$$

$$= (f_{zy} - f_{yz})\mathbf{i} + (f_{xz} - f_{zx})\mathbf{j} + (f_{yx} - f_{xy})\mathbf{k}$$

$$= \mathbf{0},$$

since the mixed derivatives are equal. A vector field $\mathbf{F}$ for which $\text{curl } \mathbf{F} = \mathbf{0}$ is said to be *irrotational*. Thus we have shown that any vector field which is the gradient of a scalar field with continuous second partial derivatives is irrotational. Notice that (15) can be written formally as

$$\nabla \times (\nabla f) = \mathbf{0}.$$

Example 4.11 If $\mathbf{F}$ has continuous second partial derivatives then

$$(16) \quad \text{div}(\text{curl } \mathbf{F}) = \frac{\partial}{\partial x}\left(\frac{\partial F_3}{\partial y} - \frac{\partial F_2}{\partial z}\right) + \frac{\partial}{\partial y}\left(\frac{\partial F_1}{\partial z} - \frac{\partial F_3}{\partial x}\right)$$

$$+ \frac{\partial}{\partial z}\left(\frac{\partial F_2}{\partial x} - \frac{\partial F_1}{\partial y}\right)$$

$$= \frac{\partial^2 F_3}{\partial x \partial y} - \frac{\partial^2 F_2}{\partial x \partial z} + \frac{\partial^2 F_1}{\partial y \partial z} - \frac{\partial^2 F_3}{\partial y \partial z} + \frac{\partial^2 F_2}{\partial z \partial x} - \frac{\partial^2 F_1}{\partial z \partial y}$$

$$= 0.$$

Notice that (16) can be written as

$$\nabla \cdot (\nabla \times \mathbf{F}) = 0.$$

The curl of a vector field is independent of the rectangular coordinate system used in its definition (Exercise T-7).

Theorem 4.4 If a and b are constants, $\mathbf{F}$ and $\mathbf{G}$ differentiable vector fields, and g a differentiable scalar field, then

(a) $\mathbf{curl}(a\mathbf{F} + b\mathbf{G}) = a\,\mathbf{curl}\,\mathbf{F} + b\,\mathbf{curl}\,\mathbf{G}$;

(b) $\mathbf{curl}(g\mathbf{F}) = \mathbf{curl}\,\mathbf{F} + \nabla g \times \mathbf{F}$.

We leave the proof of this theorem to the student (Exercise T-8).

EXERCISES 5.4

1. Compute ∇f.

(a) $f(\mathbf{X}) = e^{xy} \sin z$.

(b) $f(\mathbf{X}) = \log |\,\mathbf{X}\,|$.

(c) $f(\mathbf{X}) = \dfrac{x^2 z}{2} + xz$.

2. Compute ∇f.

(a) $f(\mathbf{X}) = 4x - 2y + z$.

(b) $f(\mathbf{X}) = x \log x + y \log y + z \log z$.

(c) $f(\mathbf{X}) = \cos xyz$.

3. Verify Theorem 4.1 for $a = 2, f(x, y, z) = xyz^2$, and $g(x, y, z) = yz^2$.

4. Compute $(\operatorname{div}\mathbf{F})(\mathbf{X})$:

(a) $\mathbf{F}(\mathbf{X}) = x^2 y\mathbf{i} + xyz\mathbf{j} + yz^2\mathbf{k}$;

(b) $\mathbf{F}(\mathbf{X}) = e^{xy} \sin z\,\mathbf{i} + e^{xz} \cos y\,\mathbf{j} + e^{yz} \sin x\,\mathbf{k}$;

(c) $\mathbf{F}(\mathbf{X}) = (y + z)x\mathbf{i} + (x - z^2)y^2\mathbf{j} + (y^2 - z^2)\mathbf{k}$;

(d) $\mathbf{F}(\mathbf{X}) = \nabla f$, where $f(x, y, z) = ye^x - xe^y$;

(e) $\mathbf{F}(\mathbf{X}) = \nabla f$, where $f(x, y, z) = 4x - 2y + z$.

5. Compute $(\operatorname{div}\mathbf{F})(\mathbf{X})$:

(a) $\mathbf{F}(\mathbf{X}) = y \log x\mathbf{i} + x \log y\mathbf{j} + xy \log z\mathbf{k}$;

(b) $\mathbf{F}(\mathbf{X}) = 2\mathbf{i} + xy\mathbf{j} + 3\mathbf{k}$;

(c) $\mathbf{F}(\mathbf{X}) = x^3\mathbf{i} + y^2\mathbf{j} + z^3\mathbf{k}$;

(d) $\mathbf{F}(\mathbf{X}) = \nabla f$,   where $f(x, y, z) = x^3 + y^3 + z^3$;

(e) $\mathbf{F}(\mathbf{X}) = \nabla f$,   where $f(x, y, z) = \cos xyz$.

6. Verify Theorem 4.3 for $a = 2$, $b = -3$,

$$\mathbf{F}(\mathbf{X}) = x^2\mathbf{i} + y^2z\mathbf{j} + (y^2 - z^2)\mathbf{k},$$

and

$$\mathbf{G}(\mathbf{X}) = x^3y\mathbf{i} + (y^2 + x)\mathbf{j} + (y^2 - z^2)\mathbf{k}.$$

7. Compute $\nabla^2 f$.

  (a) $f(\mathbf{X}) = ze^{xy}$.

  (b) $f(\mathbf{X}) = e^{yz} \cos x$.

  (c) $f(\mathbf{X}) = \log(1 + x^2 + y^2 + z^2)$.

  (d) $f(\mathbf{X}) = x^2 - y^2 + 2z^2$.

  (e) $f(\mathbf{X}) = \sin xyz$.

8. Compute $\nabla^2 f$.

  (a) $f(\mathbf{X}) = (x^2 + y^2)e^{xyz}$.

  (b) $f(\mathbf{X}) = e^{-(x^2+y^2+z^2)}$.

  (c) $f(\mathbf{X}) = \log(\cos x + \sin y)$.

  (d) $f(\mathbf{X}) = x^2 - 3xyz + 2xz$.

  (e) $f(\mathbf{X}) = y \log x + z \log y$.

9. Show that f is harmonic.

  (a) $f(\mathbf{X}) = \dfrac{1}{\sqrt{x^2 + y^2 + z^2}}$.

  (b) $f(\mathbf{X}) = x^2 - y^2 + 2z$.

  (c) $f(\mathbf{X}) = \sinh 2x \cos y \sin \sqrt{3}\, z$.

(d) $f(\mathbf{X}) = \sin \dfrac{x}{\sqrt{2}} \cos \dfrac{y}{\sqrt{2}} \cosh z.$

(e) $f(\mathbf{X}) = \sinh x \cos \sqrt{2y} \cosh y.$

10.   Compute **curl F**.

(a) $\mathbf{F}(\mathbf{X}) = (x^2 + y^3)\mathbf{i} + (xz^3 + x^2y)\mathbf{j} + (x^2 + xyz)\mathbf{k}.$

(b) $\mathbf{F}(\mathbf{X}) = \left(x^2 + \dfrac{y^2}{2}\right)\mathbf{i} + \left(xy + \dfrac{z^2}{2}\right)\mathbf{j} + (yz)\mathbf{k}.$

(c) $\mathbf{F}(\mathbf{X}) = e^x y \mathbf{i} + e^y z \mathbf{j} + e^z xy \mathbf{k}.$

(d) $\mathbf{F}(\mathbf{X}) = y \sin x \, \mathbf{i} + z \cos y \, \mathbf{j} + xy \sin z \, \mathbf{k}.$

11.   Compute **curl F**.

(a) $\mathbf{F}(\mathbf{X}) = (xyz + x)\mathbf{i} + \left(\dfrac{x^2z}{2} + y\right)\mathbf{j} + (xz)\mathbf{k}.$

(b) $\mathbf{F}(\mathbf{X}) = (2x + y + z)\mathbf{i} + (x - y - z)\mathbf{j} + (y - z)\mathbf{k}.$

(c) $\mathbf{F}(\mathbf{X}) = x^2 y e^z \mathbf{i} + x e^y z^2 \mathbf{j} + e^x y^2 z \mathbf{k}.$

(d) $\mathbf{F}(\mathbf{X}) = e^x \sin y \, \mathbf{i} + e^y \cos x \, \mathbf{j} + e^z \cos z \, \mathbf{k}.$

12.   Verify Theorem 4.3 for a, b, **F**, **G**, and g as defined in Example 4.6.

THEORETICAL EXERCISES

T-1.   Verify Eq. (7).

T-2.   If (x_1, x_2, x_3) and (y_1, y_2, y_3) are related by (5), show that

$$\frac{\partial x_q}{\partial y_r} = l_{rq}.$$

T-3.   If

$$\mathbf{M} = \begin{bmatrix} l_{11} & l_{12} & l_{13} \\ l_{21} & l_{22} & l_{23} \\ l_{31} & l_{32} & l_{33} \end{bmatrix}$$

is an orthogonal matrix $(\mathbf{M}^{-1} = \mathbf{M}^{\mathbf{T}})$ show that

$$\sum_{j=1}^{3} l_{jr} l_{js} = \sum_{j=1}^{3} l_{rj} l_{sj} = \begin{cases} 0 & \text{if} \quad r \neq s, \\ 1 & \text{if} \quad r = s. \end{cases}$$

T-4. Prove Theorem 4.3.

T-5. Let f and g be real valued functions of three variables which satisfy

$$f(x_1, x_2, x_3) = g(y_1, y_2, y_3)$$

whenever (x_1, x_2, x_3) and (y_1, y_2, y_3) are related by (5), where $\mathbf{M}$ is an orthogonal matrix. Let f and g have continuous second partial derivatives. Show that

$$\sum_{r=1}^{3} \frac{\partial^2 f}{\partial x_r^2} (x_1, x_2, x_3) = \sum_{s=1}^{3} \frac{\partial^2 g}{\partial y_s^2} (y_1, y_2, y_3).$$

T-6. Under the assumption of Exercise T-5 (except that we need only assume that f and g are differentiable), show that

$$\sum_{r=1}^{3} \frac{\partial f}{\partial x_r} (x_1, x_2, x_3) \, \mathbf{i}_r = \sum_{s=1}^{3} \frac{\partial g}{\partial y_s} (y_1, y_2, y_3) \, \mathbf{j}_s.$$

(*Hint*: Use Eq. (3) of Section 3.4). Use this result to give a precise meaning to the statement: The gradient of a vector field is independent of the particular rectangular coordinate system chosen for R^3.

T-7. If the vector field $\mathbf{F} = \mathbf{F}(P)$ is defined with respect to two rectangular coordinate systems by (11) and (12), show that

$$\begin{vmatrix} \mathbf{i}_1 & \mathbf{i}_2 & \mathbf{i}_3 \\[2mm] \dfrac{\partial}{\partial x_1} & \dfrac{\partial}{\partial x_2} & \dfrac{\partial}{\partial x_3} \\[4mm] F_1 & F_2 & F_3 \end{vmatrix} = \begin{vmatrix} \mathbf{j}_1 & \mathbf{j}_2 & \mathbf{j}_3 \\[2mm] \dfrac{\partial}{\partial y_1} & \dfrac{\partial}{\partial y_2} & \dfrac{\partial}{\partial y_3} \\[4mm] G_1 & G_2 & G_3 \end{vmatrix}.$$

(*Hint*: Refer to the proof of Theorem 4.2.) Use this result to give a precise meaning to the statement: The curl of a vector field is independent of the particular rectangular coordinate system chosen for R^3.

T-8. Prove Theorem 4.4.

T-9. Let $\mathbf{F}$ be defined by

$$\mathbf{F}(\mathbf{X}) = -\frac{Gm_0}{|\mathbf{X}|^3}\,\mathbf{X}.$$

Show that div $\mathbf{F} = 0$ and curl $\mathbf{F} = 0$.

T-10. Show that $\mathrm{div}(\mathbf{F} \times \mathbf{G}) = \mathbf{G}\cdot\mathrm{curl}\,\mathbf{F} - \mathbf{F}\cdot\mathrm{curl}\,\mathbf{G}$.

T-11. Show that $\mathrm{div}(\nabla f \times \nabla g) = 0$.

T-12. Show that $\mathrm{div}\,u\nabla v = \nabla u \cdot \nabla v + u\nabla^2 v$.

T-13. The *Laplacian of a vector field* $\mathbf{F}$ is defined by

$$\nabla^2\mathbf{F} = \nabla^2 F_1\,\mathbf{i} + \nabla^2 F_2\,\mathbf{j} + \nabla^2 F_3\,\mathbf{k}.$$

Show that

$$\mathrm{curl}(\mathrm{curl}\,\mathbf{F}) = \nabla(\mathrm{div}\,\mathbf{F}) - \nabla^2\mathbf{F}.$$

5.5 Implicit Functions

In this section we study the implicit functions theorem for vector valued functions, a generalization of Theorem 6.1 of Section 4.6.

It will be convenient to denote points in R^{n+m} by $(\mathbf{X}, \mathbf{U})$, where $\mathbf{X} = (x_1, \ldots, x_n)$ and $\mathbf{U} = (u_1, \ldots, u_m)$.

Let $\mathbf{F}: R^{n+m} \to R^m$. For a given $\mathbf{X}$ the vector equation

(1) $$\mathbf{F}(\mathbf{X}, \mathbf{U}) = \mathbf{0},$$

which is equivalent to the system

$$F_1(\mathbf{X}, u_1, \ldots, u_m) = 0,$$
$$\vdots$$
$$F_m(\mathbf{X}, u_1, \ldots, u_m) = 0,$$

may have no solution, one solution, or more than one solution for the unknown $\mathbf{U}$.

Definition 5.1 A rule which assigns to each $\mathbf{X}$ in a domain D exactly one solution $\mathbf{U}$ of (1) defines a function $\mathbf{f}: R^n \to R^m$ on D such that $\mathbf{F}(\mathbf{X}, \mathbf{f}(\mathbf{X})) = \mathbf{0}$ for all $\mathbf{X}$ in D. We say that $\mathbf{F}(\mathbf{X}, \mathbf{U}) = \mathbf{0}$ *defines* $\mathbf{f}$ *implicitly*, and that it defines $\mathbf{U} = \mathbf{f}(\mathbf{X})$ on D. If $\mathbf{X}_0$ is an interior point of D, we say that $\mathbf{F}(\mathbf{X}, \mathbf{U}) = \mathbf{0}$ *defines* $\mathbf{f}$ *implicitly near* $\mathbf{X}_0$.

Example 5.1 If F is real valued and satisfies the hypotheses of Theorem 6.1, Section 4.6, at $(\mathbf{X}_0, c)$, then $F(\mathbf{X}_0, u) = 0$ defines near $\mathbf{X}_0$ the unique continuous function $\mathbf{U} = f(\mathbf{X})$ such that $f(\mathbf{X}_0) = c$.

Example 5.2 If $\mathbf{F}(\mathbf{X}, \mathbf{U}) = \mathbf{U} - \mathbf{f}(\mathbf{X})$ where $\mathbf{f}: R^n \to R^m$ has domain D, then $\mathbf{F}(\mathbf{X}, \mathbf{U}) = \mathbf{0}$ defines $\mathbf{f}$ on D.

Example 5.3 Let $\mathbf{F}$ be the linear transformation

$$\mathbf{F}(\mathbf{X}, \mathbf{U}) = \begin{bmatrix} a_{11}x_1 + \cdots + a_{1n}x_n + b_{11}u_1 + \cdots + b_{1m}u_m \\ a_{21}x_1 + \cdots + a_{2n}x_n + b_{21}u_1 + \cdots + b_{2m}u_m \\ \vdots \\ a_{m1}x_1 + \cdots + a_{mn}x_n + b_{m1}u_1 + \cdots + b_{mm}u_m \end{bmatrix}$$

$$= \mathbf{AX} + \mathbf{BU}.$$

If $\mathbf{B}$ is nonsingular (that is, there exists a matrix $\mathbf{B}^{-1}$ such that $\mathbf{BB}^{-1} = \mathbf{B}^{-1}\mathbf{B} = \mathbf{I}$, the $m \times m$ identity matrix), then $\mathbf{F}(\mathbf{X}, \mathbf{U}) = \mathbf{0}$ defines

$$\mathbf{U} = \mathbf{f}(\mathbf{X}) = -\mathbf{B}^{-1}\mathbf{AX}$$

on R^n.

The general problem of existence of implicit functions defined by $\mathbf{F}(\mathbf{X}, \mathbf{U}) = \mathbf{0}$ is too broad, so we shall consider a more specific question: If $\mathbf{F}(\mathbf{X}_0, \mathbf{U}_0) = \mathbf{0}$, what conditions on $\mathbf{F}$ are sufficient to guarantee that there is a unique *differentiable* function $\mathbf{f}$, defined on a domain D including $\mathbf{X}_0$, such that $\mathbf{f}(\mathbf{X}_0) = \mathbf{U}_0$ and $\mathbf{F}(\mathbf{X}, \mathbf{f}(\mathbf{X})) = \mathbf{0}$ for all $\mathbf{X}$ in D?

In Example 5.3 we saw that a linear transformation from R^{n+m} to R^m defines an implicit function on R^n if a certain $m \times m$ submatrix of its matrix is nonsingular. The fact that the matrix of a linear transformation is its Jacobian matrix is the clue to a generalization of this result.

Let $\mathbf{F}: R^{n+m} \to R^m$ (not necessarily a linear transformation) and divide its Jacobian matrix into two parts:

$$\mathbf{JF} = \begin{bmatrix} \dfrac{\partial F_1}{\partial x_1} & \cdots & \dfrac{\partial F_1}{\partial x_n} & \vdots & \dfrac{\partial F_1}{\partial u_1} & \cdots & \dfrac{\partial F_1}{\partial u_m} \\ \vdots & & \vdots & \vdots & \vdots & & \vdots \\ \dfrac{\partial F_m}{\partial x_1} & \cdots & \dfrac{\partial F_m}{\partial x_n} & \vdots & \dfrac{\partial F_m}{\partial u_1} & \cdots & \dfrac{\partial F_m}{\partial u_m} \end{bmatrix}.$$

Generalizing the symbol for partial derivatives, we denote the matrices to the left and right of the dashed line by $\mathbf{F}_\mathbf{X}$ and $\mathbf{F}_\mathbf{U}$, respectively. Then in Example 5.3, $\mathbf{F}_X = \mathbf{A}$ and $\mathbf{F}_U = \mathbf{B}$ are constant matrices, and $\mathbf{F}(\mathbf{X}, \mathbf{U}) = \mathbf{0}$ defines $\mathbf{U} = \mathbf{f}(\mathbf{X})$ if $\mathbf{F}_U$ is nonsingular. For more general functions the result is not as simple as this, but the following generalization holds.

Theorem 5.1 (The Implicit Function Theorem) Let $\mathbf{F} \colon R^{n+m} \to R^m$ be a function of $(x_1, \ldots, x_n, u_1, \ldots, u_m)$ with continuous partial derivatives $\mathbf{F}_{x_1}, \ldots, \mathbf{F}_{x_n}, \mathbf{F}_{u_1}, \ldots, \mathbf{F}_{u_m}$ in an $(n+m)$-ball about $(\mathbf{X}_0, \mathbf{U}_0)$. Suppose that $\mathbf{F}(\mathbf{X}_0, \mathbf{U}_0) = \mathbf{0}$ and $\mathbf{F}_U(\mathbf{X}_0, \mathbf{U}_0)$ is nonsingular. Then there is a set D in R^n with $\mathbf{X}_0$ as an interior point, and a unique continuous function $\mathbf{f} \colon R^n \to R^m$ with domain D, such that $\mathbf{f}(\mathbf{X}_0) = \mathbf{U}_0$ and $\mathbf{F}(\mathbf{X}, \mathbf{f}(\mathbf{X})) = \mathbf{0}$. Furthermore, $\mathbf{F}_U(\mathbf{X}, \mathbf{f}(\mathbf{X}))$ is nonsingular, $\mathbf{f}$ has continuous partial derivatives, and

$$(2) \qquad \mathbf{J}_X \mathbf{f} = -[\mathbf{F}_U(\mathbf{X}, \mathbf{f}(\mathbf{X}))]^{-1} \mathbf{F}_X(\mathbf{X}, \mathbf{f}(\mathbf{X}))$$

for all $\mathbf{X}$ in D.

This theorem reduces to Theorem 6.1, Section 4.6, when $m = 1$. Its proof is beyond the scope of this book; however, given that $\mathbf{f}$ is differentiable and $\mathbf{F}(\mathbf{X}, \mathbf{f}(\mathbf{X})) = \mathbf{0}$, (2) follows from the chain rule. In terms of the components of $\mathbf{F}$,

$$F_i(\mathbf{X}, f_i(\mathbf{X}), \ldots, f_m(\mathbf{X})) = 0 \qquad (1 \le i \le m);$$

differentiating both sides with respect to x_j yields

$$\frac{\partial F_i}{\partial x_j} + \frac{\partial F_i}{\partial u_1} \frac{\partial f_1}{\partial x_j} + \frac{\partial F_i}{\partial u_2} \frac{\partial f_2}{\partial x_j} + \cdots + \frac{\partial F_i}{\partial u_m} \frac{\partial f_m}{\partial x_j} = 0 \qquad (1 \le i, j \le m)$$

for all $\mathbf{X}$ in D, where the partial derivatives of $\mathbf{f}$ and $\mathbf{F}$ are evaluated at $\mathbf{X}$ and $(\mathbf{X}, \mathbf{f}(\mathbf{X}))$, respectively. In vector form,

$$\begin{bmatrix} \dfrac{\partial F_1}{\partial x_j} \\ \vdots \\ \dfrac{\partial F_m}{\partial x_j} \end{bmatrix} + \mathbf{F}_U \begin{bmatrix} \dfrac{\partial f_1}{\partial x_j} \\ \vdots \\ \dfrac{\partial f_m}{\partial x_j} \end{bmatrix} = \mathbf{0} \qquad (1 \le j \le m).$$

The vectors here are the columns of $\mathbf{F}_X$ and $\mathbf{J}\mathbf{f}$; hence these m equations can be combined to yield

$$\mathbf{F}_X + \mathbf{F}_U \mathbf{J}\mathbf{f} = \mathbf{0},$$

from which (2) follows.

In the informal notation, where we write $\mathbf{U} = \mathbf{f}$, (2) becomes

$$
(3) \quad
\begin{bmatrix}
\dfrac{\partial u_1}{\partial x_1} & \dfrac{\partial u_1}{\partial x_2} & \cdots & \dfrac{\partial u_1}{\partial x_n} \\[2ex]
\dfrac{\partial u_2}{\partial x_1} & \dfrac{\partial u_2}{\partial x_2} & \cdots & \dfrac{\partial u_2}{\partial x_n} \\[2ex]
\vdots & \vdots & & \vdots \\[2ex]
\dfrac{\partial u_m}{\partial x_1} & \dfrac{\partial u_m}{\partial x_2} & \cdots & \dfrac{\partial u_m}{\partial x_n}
\end{bmatrix}
$$

$$
= -
\begin{bmatrix}
\dfrac{\partial F_1}{\partial u_1} & \dfrac{\partial F_1}{\partial u_2} & \cdots & \dfrac{\partial F_1}{\partial u_m} \\[2ex]
\dfrac{\partial F_2}{\partial u_1} & \dfrac{\partial F_2}{\partial u_2} & \cdots & \dfrac{\partial F_2}{\partial u_m} \\[2ex]
\vdots & \vdots & & \vdots \\[2ex]
\dfrac{\partial F_m}{\partial u_1} & \dfrac{\partial F_m}{\partial u_2} & \cdots & \dfrac{\partial F_m}{\partial u_m}
\end{bmatrix}^{-1}
\begin{bmatrix}
\dfrac{\partial F_1}{\partial x_1} & \dfrac{\partial F_1}{\partial x_2} & \cdots & \dfrac{\partial F_1}{\partial x_n} \\[2ex]
\dfrac{\partial F_2}{\partial x_1} & \dfrac{\partial F_2}{\partial x_2} & \cdots & \dfrac{\partial F_2}{\partial x_n} \\[2ex]
\vdots & \vdots & & \vdots \\[2ex]
\dfrac{\partial F_m}{\partial x_1} & \dfrac{\partial F_m}{\partial x_2} & \cdots & \dfrac{\partial F_m}{\partial x_n}
\end{bmatrix}.
$$

Theorem 5.1 is an existence theorem; it says that under certain conditions (1) has a differentiable solution for all $\mathbf{X}$ in some domain D, but does not tell how to determine D or the solution. It is nevertheless a useful theorem, for often it is enough just to know that a solution exists in some domain. Furthermore, even if $\mathbf{f}$ cannot be obtained explicitly, its partial derivatives can be evaluated at any $\mathbf{X}$ in D where $\mathbf{f}(\mathbf{X})$ is known, by means of (2).

Example 5.4 Let

$$
\mathbf{F}(\mathbf{X}, \mathbf{U}) =
\begin{bmatrix}
x^2 + y^2 + z^2 - u^2 - v^2 \\[1ex]
(x - y)^2 + u + v
\end{bmatrix};
$$

thus $m = 2$ and $n = 3$. Then

$$
\mathbf{F}_{\mathbf{X}}(\mathbf{U}) =
\begin{bmatrix}
2x & 2y & 2z \\[1ex]
2x - 2y & 2y - 2x & 0
\end{bmatrix}
$$

and

$$
\mathbf{F}_{\mathbf{U}}(\mathbf{X}, \mathbf{U}) =
\begin{bmatrix}
-2u & -2v \\[1ex]
1 & 1
\end{bmatrix}.
$$

If

$$\mathbf{X}_0 = \left(\frac{1}{\sqrt{3}}, \frac{1}{\sqrt{3}}, -\frac{1}{\sqrt{3}} \right) \quad \text{and} \quad \mathbf{U}_0 = \left(\frac{1}{\sqrt{2}}, -\frac{1}{\sqrt{2}} \right),$$

then $\mathbf{F}(\mathbf{X}_0, \mathbf{U}_0) = \mathbf{0}$ and

$$[\mathbf{F}_U(\mathbf{X}_0, \mathbf{U}_0)]^{-1} = \begin{bmatrix} -\sqrt{2} & \sqrt{2} \\ 1 & 1 \end{bmatrix}^{-1}$$

$$= \begin{bmatrix} -\dfrac{1}{2\sqrt{2}} & \dfrac{1}{2} \\ \dfrac{1}{2\sqrt{2}} & \dfrac{1}{2} \end{bmatrix};$$

hence $\mathbf{F}(\mathbf{X}, \mathbf{U}) = \mathbf{0}$ defines $\mathbf{U} = \mathbf{U}(\mathbf{X})$ near $\mathbf{X}_0$ and

$$\begin{bmatrix} u_x(\mathbf{X}_0) & u_y(\mathbf{X}_0) & u_z(\mathbf{X}_0) \\ v_x(\mathbf{X}_0) & v_y(\mathbf{X}_0) & v_z(\mathbf{X}_0) \end{bmatrix} = -(\mathbf{F}_U(\mathbf{X}_0, \mathbf{U}_0))^{-1}\, \mathbf{F}_X(\mathbf{X}_0, \mathbf{U}_0)$$

$$= -\begin{bmatrix} -\dfrac{1}{2\sqrt{2}} & \dfrac{1}{2} \\ \dfrac{1}{2\sqrt{2}} & \dfrac{1}{2} \end{bmatrix} \begin{bmatrix} \dfrac{2}{\sqrt{3}} & \dfrac{2}{\sqrt{3}} & -\dfrac{2}{\sqrt{3}} \\ 0 & 0 & 0 \end{bmatrix}$$

$$= \begin{bmatrix} \dfrac{1}{\sqrt{6}} & \dfrac{1}{\sqrt{6}} & -\dfrac{1}{\sqrt{6}} \\ -\dfrac{1}{\sqrt{6}} & -\dfrac{1}{\sqrt{6}} & \dfrac{1}{\sqrt{6}} \end{bmatrix}.$$

It is not necessary to memorize (2) or (3), since the partial derivatives of an implicit function can be calculated from the chain rule and Cramer's rule, as in the next example.

Example 5.5 Suppose we wish to find only $u_x(\mathbf{X}_0)$ in Example 5.4. Since

$$x^2 + y^2 + z^2 - u^2 - v^2 = 0,$$

$$(x - y)^2 + u + v = 0,$$

differentiation with respect to x yields

$$2x - 2uu_x - 2vv_x = 0,$$

$$2(x - y) + u_x + v_x = 0.$$

From Cramer's rule,

$$u_x = \frac{\begin{vmatrix} 2x & 2v \\ 2(x - y) & -1 \end{vmatrix}}{\begin{vmatrix} 2u & 2v \\ -1 & -1 \end{vmatrix}}$$

and

$$u_x(\mathbf{X}_0) = \frac{\begin{vmatrix} \dfrac{2}{\sqrt{3}} & -\sqrt{2} \\ 0 & -1 \end{vmatrix}}{\begin{vmatrix} \sqrt{2} & -\sqrt{2} \\ -1 & -1 \end{vmatrix}} = \frac{1}{\sqrt{6}}.$$

Example 5.6 Let $(x_0, y_0, z_0) = (1, 0, 0)$ and

$$\mathbf{F}(x, y, z) = \begin{bmatrix} x^2 + y^2 + z^2 + y - z \\ xyz + e^x \sin(y + z) \end{bmatrix};$$

then

$$\mathbf{F}_{(y, z)} = \begin{bmatrix} \dfrac{\partial F_1}{\partial y} & \dfrac{\partial F_1}{\partial z} \\ \dfrac{\partial F_2}{\partial y} & \dfrac{\partial F_2}{\partial z} \end{bmatrix} = \begin{bmatrix} 2y + 1 & 2z - 1 \\ xz + e^x \cos(y + z) & xy + e^x \cos(y + z) \end{bmatrix},$$

and

$$\mathbf{F}_{(y, z)}(1, 0, 0) = \begin{bmatrix} 1 & -1 \\ e & e \end{bmatrix}$$

is nonsingular. Therefore $\mathbf{F}(x, y, z) = \mathbf{0}$ defines

$$\begin{bmatrix} y \\ z \end{bmatrix} = \begin{bmatrix} y(x) \\ z(x) \end{bmatrix}$$

near $x_0 = 1$, with

$$\begin{bmatrix} y(0) \\ z(0) \end{bmatrix} = \begin{bmatrix} 0 \\ 0 \end{bmatrix}.$$

The derivatives satisfy

$$2x + (2y + 1)y' + (2z - 1)z' = 0$$

$$yz + e^x \sin(y + z) + [xz + e^x \cos(y + z)]y'$$

$$+ [xy + e^x \cos(y + z)]z' = 0;$$

hence

$$2 + y'(0) - z'(0) = 0,$$

$$y'(0) + z'(0) = 0,$$

which has the solution $y'(0) = -1, z'(0) = 1$.

The positions of $u_1, \ldots, u_m$ among the $(n + m)$ coordinates of a point in R^{n+m} are not important; they do not have to be last as we have assumed until now. Any m variables for which the inverse matrix on the right of (3) exists are defined implicitly by $\mathbf{F} = \mathbf{0}$, with derivatives given by (3).

Example 5.7 In Example 5.4,

$$\mathbf{F}_{(x, u)} = \begin{bmatrix} \dfrac{\partial F_1}{\partial x} & \dfrac{\partial F_1}{\partial u} \\ \\ \dfrac{\partial F_2}{\partial x} & \dfrac{\partial F_2}{\partial u} \end{bmatrix} = \begin{bmatrix} 2x & -2u \\ 2(x - y) & 1 \end{bmatrix},$$

and

$$\mathbf{F}_{(x, u)}(x_0, u_0) = \begin{bmatrix} \dfrac{2}{\sqrt{3}} & -\sqrt{2} \\ \\ 0 & 1 \end{bmatrix}$$

is nonsingular. Hence $\mathbf{F}(x, y, z, u, v) = \mathbf{0}$ also defines

$$\begin{bmatrix} x \\ u \end{bmatrix} = \begin{bmatrix} x(y, z, v) \\ u(y, z, v) \end{bmatrix}$$

such that

$$\begin{bmatrix} x(\mathbf{V}_0) \\ u(\mathbf{V}_0) \end{bmatrix} = \begin{bmatrix} \dfrac{1}{\sqrt{3}} \\[2ex] \dfrac{1}{\sqrt{2}} \end{bmatrix},$$

where $\mathbf{V}_0 = (1/\sqrt{3}, -1/\sqrt{3}, 1/\sqrt{2})$. Furthermore, for $\mathbf{V} = (y, z, v)$ near $\mathbf{V}_0$,

$$\begin{bmatrix} \dfrac{\partial x}{\partial y} & \dfrac{\partial x}{\partial z} & \dfrac{\partial x}{\partial v} \\[3ex] \dfrac{\partial u}{\partial y} & \dfrac{\partial u}{\partial z} & \dfrac{\partial u}{\partial v} \end{bmatrix} = - \begin{bmatrix} \dfrac{\partial F_1}{\partial x} & \dfrac{\partial F_1}{\partial u} \\[3ex] \dfrac{\partial F_2}{\partial x} & \dfrac{\partial F_2}{\partial u} \end{bmatrix}^{-1} \begin{bmatrix} \dfrac{\partial F_1}{\partial y} & \dfrac{\partial F_1}{\partial z} & \dfrac{\partial F_1}{\partial v} \\[3ex] \dfrac{\partial F_2}{\partial y} & \dfrac{\partial F_2}{\partial z} & \dfrac{\partial F_2}{\partial v} \end{bmatrix}$$

$$= - \begin{bmatrix} 2x & -2u \\ 2(x-y) & 1 \end{bmatrix}^{-1} \begin{bmatrix} 2y & 2z & -2v \\ 2(y-x) & 0 & 1 \end{bmatrix}.$$

EXERCISES 5.5

1. Let

$$\mathbf{F}(x, y, u, v) = \begin{bmatrix} x + 2y - u + v \\ -2x + y + 2u + 2v \end{bmatrix};$$

then $\mathbf{F}(x, y, u, v) = \mathbf{0}$ defines $\begin{bmatrix} u \\ v \end{bmatrix} = \mathbf{f}(x, y)$ on R^2. Find $\mathbf{f}$ explicitly;

also find $\mathbf{Jf}$.

2. Repeat Exercise 1 for

$$\mathbf{F}(x, y, u, v) = \begin{bmatrix} 3x + 2y - 3u + 2v \\ 5x - 2y + u - 2v \end{bmatrix}.$$

3. Let $\mathbf{X} = (x, y)$ and $\mathbf{U} = (u, v)$ in Eqs. (1) and (2). Find $\mathbf{F_X}$ and $\mathbf{F_U}$ for $\mathbf{F}$ in Exercise 1, and calculate $\mathbf{Jf}$ by means of (2).

4. Repeat Exercise 3 for $\mathbf{F}$ as defined in Exercise 2.

5. Let $\mathbf{X}_0 = (x_0, y_0) = (1, 1)$ and

$$\mathbf{F}(x, y, z, w) = \begin{bmatrix} xyz - 2yz + xw \\ 2xy - y^2 - w^2 \end{bmatrix}.$$

Then $\mathbf{F}(x, y, z, w) = \mathbf{0}$ defines near $\mathbf{X}_0$ two differentiable functions,

$$\begin{bmatrix} z \\ w \end{bmatrix} = \mathbf{f}(x, y) \qquad \text{and} \qquad \begin{bmatrix} z \\ w \end{bmatrix} = \mathbf{g}(x, y).$$

(Does this contradict Theorem 5.1?) Find $\mathbf{f}(\mathbf{X}_0)$, $\mathbf{g}(\mathbf{X}_0)$, $\mathbf{J}_{\mathbf{X}_0}\mathbf{f}$, $\mathbf{J}_{\mathbf{X}_0}\mathbf{g}$, $\mathbf{d}_{\mathbf{X}_0}\mathbf{f}$, and $\mathbf{d}_{\mathbf{X}_0}\mathbf{g}$.

6. Let

$$\mathbf{F}(x, y, z) = \begin{bmatrix} e^x \cos z - y \\ e^x \sin z - y + 1 \end{bmatrix};$$

then $\mathbf{F}(x, y, z) = \mathbf{0}$ defines exactly one continuous function

$$\begin{bmatrix} x \\ y \end{bmatrix} = \begin{bmatrix} x(z) \\ y(z) \end{bmatrix}$$

near $z_0 = 0$. Find $x(0)$, $y(0)$, $x'(0)$, and $y'(0)$.

7. Let

$$\mathbf{F}(x, y, z, w) = \begin{bmatrix} x^2y + xy^2 + z^2 - w^2 \\ e^{x+y} - w \end{bmatrix};$$

then $\mathbf{F}(x, y, z, w) = \mathbf{0}$ defines

$$\begin{bmatrix} z \\ w \end{bmatrix} = \begin{bmatrix} z(x, y) \\ w(x, y) \end{bmatrix}$$

near $(x_0, y_0) = (0, 0)$ such that

$$\begin{bmatrix} z(0, 0) \\ w(0, 0) \end{bmatrix} = \begin{bmatrix} 1 \\ 1 \end{bmatrix}.$$

Using the chain rule, find $\dfrac{\partial z}{\partial y}(0, 0)$.

8. Let $\mathbf{F}$ be as in Exercise 7; then $\mathbf{F}(x, y, z, w) = \mathbf{0}$ defines

$$\begin{bmatrix} x \\ z \end{bmatrix} = \begin{bmatrix} x(y, w) \\ z(y, w) \end{bmatrix}$$

near $(y_0, w_0) = (0, 1)$ such that

$$\begin{bmatrix} x(0, 1) \\ z(0, 1) \end{bmatrix} = \begin{bmatrix} 0 \\ 1 \end{bmatrix}.$$

Using the chain rule, find $(\partial z/\partial y)(0, 1)$.

9. Let

$$\mathbf{F}(x, y, z, u, v, w) = \begin{bmatrix} e^x \cos y + e^z \cos u + e^v \cos w + x - 3 \\ e^x \sin y + e^z \sin u + e^v \cos w - 1 \\ e^x \tan y + e^z \tan u + e^v \tan w + z \end{bmatrix}.$$

(a) Find $\mathbf{F}_{(x,y,z)}(0, 0, 0)$ and $\mathbf{F}_{(u,v,w)}(0, 0, 0)$.

(b) Let

$$\begin{bmatrix} u \\ v \\ w \end{bmatrix} = \mathbf{f}(x, y, z)$$

be defined by $\mathbf{F}(x, y, z, u, v, w) = \mathbf{0}$ near $(x_0, y_0, z_0) = (0, 0, 0)$ such that $\mathbf{f}(0, 0, 0) = \mathbf{0}$. Find $\mathbf{J}_{(0,0,0)}\mathbf{f}$.

10. In Exercise 1, which of the pairs (x, y), (x, u), (x, v), (y, u), (y, v), and (u, v) are defined as functions of the remaining variables by $\mathbf{F}(x, y, u, v) = \mathbf{0}$?

11. Repeat Exercise 10 for $\mathbf{F}$ as defined in Exercise 2.

12. Let $\mathbf{F}$ be as in Exercise 5. Which of the pairs (x, y), (x, z), (x, w), (y, z), (y, w), and (z, w) are, according to the implicit function theorem, defined as functions of the remaining variables by $\mathbf{F}(x, y, z, w) = \mathbf{0}$, such that $(x_0, y_0, z_0, w_0) = (1, 1, 1, 1)$?

THEORETICAL EXERCISES

T-1. Suppose $n = m$ in Theorem 5.1 and $\mathbf{F_X}(X_0, \mathbf{U}_0)$ is also nonsingular. Let $\mathbf{g}$ be the differentiable function defined near $\mathbf{U}_0$ such that $\mathbf{g}(\mathbf{U}_0) = X_0$ and $\mathbf{F}(\mathbf{g}(\mathbf{U}), \mathbf{U}) = 0$. Show that $\mathbf{J_X f} = (\mathbf{J_U g})^{-1}$.

T-2. Let $1 \leq m < n$ and suppose f and $g_1, \ldots, g_m$ are continuously differentiable real valued functions on a domain D in R^n. Let $X_0 = (c_1, c_2, \ldots, c_n)$ be an interior point of D at which f assumes an extreme value subject to the constraints

(a)
$$g_1(\mathbf{X}) = g_2(\mathbf{X}) = \cdots = g_m(\mathbf{X}) = 0,$$

and suppose that the matrix

$$\mathbf{A} = \begin{bmatrix} \dfrac{\partial g_1}{\partial x_1}(\mathbf{X}_0) & \dfrac{\partial g_1}{\partial x_2}(\mathbf{X}_0) & \cdots & \dfrac{\partial g_1}{\partial x_m}(\mathbf{X}_0) \\[2ex] \dfrac{\partial g_2}{\partial x_1}(\mathbf{X}_0) & \dfrac{\partial g_2}{\partial x_2}(\mathbf{X}_0) & \cdots & \dfrac{\partial g_2}{\partial x_m}(\mathbf{X}_0) \\[2ex] \vdots & \vdots & & \vdots \\[2ex] \dfrac{\partial g_m}{\partial x_1}(\mathbf{X}_0) & \dfrac{\partial g_m}{\partial x_2}(\mathbf{X}_0) & \cdots & \dfrac{\partial g_m}{\partial x_m}(\mathbf{X}_0) \end{bmatrix}$$

is nonsingular. Show that, for $m + 1 \leq i \leq n$,

(b)
$$\begin{bmatrix} \dfrac{\partial f}{\partial x_1}(\mathbf{X}_0) \\[2ex] \dfrac{\partial f}{\partial x_2}(\mathbf{X}_0) \\[2ex] \vdots \\[2ex] \dfrac{\partial f}{\partial x_m}(\mathbf{X}_0) \\[2ex] \dfrac{\partial f}{\partial x_i}(\mathbf{X}_0) \end{bmatrix} = \sum_{j=1}^{m} \lambda_{ij} \begin{bmatrix} \dfrac{\partial g_j}{\partial x_1}(\mathbf{X}_0) \\[2ex] \dfrac{\partial g_j}{\partial x_2}(\mathbf{X}_0) \\[2ex] \vdots \\[2ex] \dfrac{\partial g_j}{\partial x_m}(\mathbf{X}_0) \\[2ex] \dfrac{\partial g_j}{\partial x_i}(\mathbf{X}_0) \end{bmatrix}.$$

Hint: From the implicit function theorem, (a) defines $x_1, \ldots, x_m$ as functions of $x_{m+1}, \ldots, x_n$ near $(x_{m+1}, \ldots, x_n) = (c_{m+1}, \ldots, c_n)$, with $x_i(c_{m+1}, \ldots, c_n) = c_i \qquad (1 \leq i \leq n)$.

Therefore the function h defined by

$$h(x_{m+1}, \ldots, x_n) = f(x_1(x_{m+1}, \ldots, x_n), \ldots, x_m(x_{m+1}, \ldots, x_n), x_{m+1}, \ldots, x_n)$$

assumes a relative extreme value at $(c_{m+1}, \ldots, c_n)$. Show that this implies that the vectors on both sides of (b) are all orthogonal to

$$
\mathbf{N}_i =
\begin{bmatrix}
\dfrac{\partial x_1}{\partial x_i}(c_{m+1}, \ldots, c_n) \\[2ex]
\dfrac{\partial x_2}{\partial x_i}(c_{m+1}, \ldots, c_n) \\[2ex]
\vdots \\[2ex]
\dfrac{\partial x_m}{\partial x_i}(c_{m+1}, \ldots, c_n) \\[2ex]
1
\end{bmatrix}
$$

and are therefore linearly dependent. Since the vectors on the right side of (b) are linearly independent (verify this from the fact that the columns of a nonsingular matrix are linearly independent), (b) now follows.

T-3. Show that the nonsingularity of the matrix $\mathbf{A}$ in Exercise T-2 implies that $\lambda_{i1}, \ldots, \lambda_{im}$ are in fact independent of i, and conclude that $\nabla_{\mathbf{X}_0} f$ is a linear combination of $\nabla_{\mathbf{X}_0} g_1, \ldots, \nabla_{\mathbf{X}_0} g_m$. This result is the basis for the method of Lagrange multipliers for locating extreme values of a function of n variables subject to m constraints (Section 4.10).

5.6 Inverse Functions and Coordinate Transformations

In Example 5.2 of the last section we observed that

(1) $$\mathbf{F}(\mathbf{X}, \mathbf{U}) = \mathbf{U} - \mathbf{f}(\mathbf{X}) = 0$$

defines $\mathbf{U} = \mathbf{f}(\mathbf{X})$ on the domain of $\mathbf{f}$. We now consider a less trivial question: Under what conditions on $\mathbf{f}$ does (1) define $\mathbf{X}$ as a function of $\mathbf{U}$? This involves the idea of an inverse function, which we now develop.

In the following D is the domain of $\mathbf{f}$ and S is a subset of D.

Definition 6.1 A function f is said to be *one-to-one on* S if $f(X_1)$ and $f(X_2)$ are distinct whenever X_1 and X_2 are distinct points of S; thus, $f(X_1) = f(X_2)$ implies $X_1 = X_2$. If $S = D$ we say simply that f *is one-to-one.*

Example 6.1 Let

$$f(x, y) = \begin{bmatrix} x + y \\ x - y \end{bmatrix};$$

then f is one-to-one, because if

$$\begin{bmatrix} x_1 + y_1 \\ x_1 - y_1 \end{bmatrix} = \begin{bmatrix} x_2 + y_2 \\ x_2 - y_2 \end{bmatrix},$$

then

$$(x_1 - x_2) + (y_1 - y_2) = 0,$$
$$(x_1 - x_2) - (y_1 - y_2) = 0,$$

from which it follows that $x_1 - x_2 = y_1 - y_2 = 0$.

Definition 6.2 *The restriction of* f *to a subset* S *of its domain* D is denoted by f_S and defined by $f_S(X) = f(X)$ for X in S; it is undefined if X is not in S. If $S = D$ we write simply $f_S = f$.

From Definitions 6.1 and 6.2, f is one-to-one on S if and only f_S is one-to-one.

Example 6.2 Let

$$f(x, y) = \begin{bmatrix} x^2 + y^2 \\ x^2 - y^2 \end{bmatrix};$$

then f is not one-to-one on any set which contains, along with a point $(x_0, y_0) \neq (0, 0)$, one of the points $(-x_0, y_0)$, $(x_0, -y_0)$, and $(-x_0, -y_0)$ which is distinct from (x_0, y_0). In particular, f is not one-to-one on any set containing points on the coordinate axes in its interior. It is one-to-one on any set S contained within one of the quadrants of R^2 (Exercise T-1). In particular, if S is the first quadrant,

$$S = \{(x, y) \mid x \geq 0, \quad y \geq 0\},$$

then f_S is one-to-one.

Definition 6.3 Let $f: R^n \to R^m$ be one-to-one on S and let $f(S)$ be the set of values attained by f as X varies over S. For each U in $f(S)$ let $f_S^{-1}(U)$ be the unique X in S such that $U = f(X)$ (Fig. 6.1); then f_S^{-1} is a function

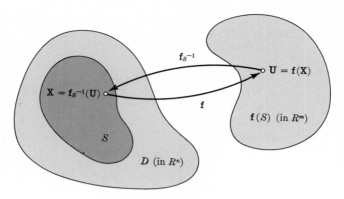

FIGURE 6.1

with domain $f(S)$, which we call the *inverse of f on S*, or simply the *inverse of f* if $S = D$. We say that $f_{\bar{S}}^{-1}$ is the inverse of f near X_0 if X_0 is in the interior of S.

This terminology is somewhat misleading. It does not mean that X_0 is in the domain of $f_{\bar{S}}^{-1}$ (in fact, it is in the range); it means that the restriction of f to some set S containing X_0 in its interior is one-to-one.

Example 6.3 We saw in Example 6.1 that if

$$f(x, y) = \begin{bmatrix} x + y \\ x - y \end{bmatrix}$$

then $S = R^2$. The image T is also R^2, since for (u, v) in R^2 there is a unique (x, y) such that

$$\begin{bmatrix} u \\ v \end{bmatrix} = \begin{bmatrix} x + y \\ x - y \end{bmatrix};$$

explicitly,

$$\begin{bmatrix} x \\ y \end{bmatrix} = f^{-1}(u, v) = \begin{bmatrix} \dfrac{u + v}{2} \\ \dfrac{u - v}{2} \end{bmatrix}.$$

Example 6.4 Let

$$f(x, y) = \begin{bmatrix} x^2 + y^2 \\ x^2 - y^2 \end{bmatrix}$$

as in Example 6.2, and let S be the first quadrant of R^2. Then $\mathbf{f}_S$ is one-to-one (Exercise T-1); to find $T = \mathbf{f}(S)$, we observe that if

$$\begin{bmatrix} u \\ v \end{bmatrix} = \begin{bmatrix} x^2 + y^2 \\ x^2 - y^2 \end{bmatrix}$$

then

$$u + v = 2x^2,$$

$$u - v = 2y^2;$$

hence $u + v \geq 0$ and $u - v \geq 0$ if (u, v) is in T. Conversely, any point which satisfies these inequalities is the image of

$$(2) \qquad \begin{bmatrix} x \\ y \end{bmatrix} = \begin{bmatrix} \sqrt{\dfrac{u+v}{2}} \\ \sqrt{\dfrac{u-v}{2}} \end{bmatrix},$$

which is in S. Therefore

$$\mathbf{f}(S) = \{(u, v) \mid u - v \geq 0, \quad u + v \geq 0\}$$

is the shaded area shown in Fig. 6.2. The inverse of $\mathbf{f}_S$ is given by

$$\mathbf{f}_S^{-1}(u, v) = \begin{bmatrix} \sqrt{\dfrac{u+v}{2}} \\ \sqrt{\dfrac{u-v}{2}} \end{bmatrix}.$$

A similar argument shows that if S_1 is the second quadrant then $\mathbf{f}(S_1) = \mathbf{f}(S)$ and

$$\mathbf{f}_{S_1}^{-1}(u, v) = \begin{bmatrix} -\sqrt{\dfrac{u+v}{2}} \\ \sqrt{\dfrac{u-v}{2}} \end{bmatrix}.$$

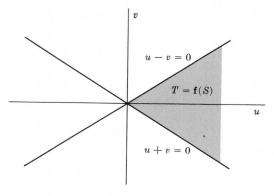

FIGURE 6.2

Theorem 6.1 Let S be a subset of the domain D of $\mathbf{f}$, and suppose $\mathbf{f}_S$ is one-to-one. Then $\mathbf{f}_{\bar{S}}^1$ is one-to-one on $T = \mathbf{f}(S)$ and $\mathbf{f}_S$ is the inverse of $\mathbf{f}_{\bar{S}}^1$. Furthermore $\mathbf{f}_{\bar{S}}^1(\mathbf{f}_S(\mathbf{X})) = \mathbf{X}$ for all $\mathbf{X}$ in S and $\mathbf{f}_S(\mathbf{f}_{\bar{S}}^1(\mathbf{U})) = \mathbf{U}$ for all $\mathbf{U}$ in T.

We leave the proof of this theorem to the student (Exercise T-2).

We are primarily interested in the existence of inverse functions when the range and domain of $\mathbf{f}$ are both in R^n. The following theorem gives sufficient conditions under which such a function has a differentiable inverse near $\mathbf{X}_0$.

Theorem 6.2 (The Inverse Function Theorem) Let $\mathbf{f} \colon R^n \to R^n$ have continuous first partial derivatives on D and let $\mathbf{X}_0$ be an interior point of D. If $\mathbf{J}_{\mathbf{X}_0}\mathbf{f}$ is nonsingular then there is an n-ball C about $\mathbf{X}_0$ such that

(a) $\mathbf{J}_{\mathbf{X}}\mathbf{f}$ is nonsingular for all $\mathbf{X}$ in C;

(b) $\mathbf{f}$ is one-to-one on C and $\mathbf{f}_{\bar{C}}^1$ exists on $\mathbf{f}(C)$.

(c)   every point of $\mathbf{f}(C)$ is an interior point and $\mathbf{f}_C^{-1}$ has continuous first partial derivatives on $\mathbf{f}(C)$;

(d)   if $\mathbf{U}$ is in $\mathbf{f}(C)$ then

(3)
$$\mathbf{J}_{\mathbf{U}}\mathbf{f}_{\bar{C}}^1 = (\mathbf{J}_{\mathbf{X}}\mathbf{f})^{-1},$$

where $\mathbf{X}$ is the unique point in C such that $\mathbf{U} = \mathbf{f}(\mathbf{X})$.

We omit the proof of this theorem, which is quite formidable. However, (3) can be obtained formally from Eq. (2) of Section 5.4 (Exercise T-3).

Example 6.5   Suppose f is a linear transformation,

$$f(\mathbf{X}) = \begin{bmatrix} a_{11}x_1 + a_{12}x_2 + \cdots + a_{1n}x_n \\ \vdots \\ a_{n1}x_1 + a_{n2}x_2 + \cdots + a_{nn}x_n \end{bmatrix} = \mathbf{AX},$$

where $\mathbf{A}^{-1}$ exists. Since $\mathbf{J}_{\mathbf{X}_0}f = \mathbf{A}$ for all $\mathbf{X}$, Theorem 6.2 implies that f has an inverse near every point $\mathbf{X}_0$. However, in this case we already know from linear algebra that f has an inverse on all of R^n, and that

$$\mathbf{f}^{-1}(\mathbf{U}) = \mathbf{A}^{-1}\mathbf{U}.$$

Furthermore, if $\mathbf{U} = \mathbf{f}(\mathbf{X})$ then

$$\mathbf{J}_{\mathbf{U}}\mathbf{f}^{-1} = \mathbf{A}^{-1} = (\mathbf{J}_{\mathbf{X}}\mathbf{f})^{-1},$$

which is consistent with (3).

Example 6.6   Let $\mathbf{X}_0 = (1, 2)$ and

$$\mathbf{f}(x, y) = \begin{bmatrix} x^2 + y^2 \\ x^2 - y^2 \end{bmatrix};$$

then

$$\mathbf{Jf} = \begin{bmatrix} 2x & 2y \\ 2x & -2y \end{bmatrix},$$

and

$$\mathbf{J}_{\mathbf{X}_0}\mathbf{f} = \begin{bmatrix} 2 & 2 \\ 2 & -4 \end{bmatrix}$$

is nonsingular. Theorem 6.1 implies that f has an inverse near $\mathbf{X}_0$, and that

$$(4) \qquad \mathbf{J}_{\mathbf{U}}\mathbf{f}_S^{-1} = (\mathbf{J}_{\mathbf{X}}\mathbf{f})^{-1} = \begin{bmatrix} \dfrac{1}{4x} & \dfrac{1}{4x} \\[2mm] \dfrac{1}{4y} & -\dfrac{1}{4y} \end{bmatrix}$$

for all $\mathbf{X}$ in some set S containing $\mathbf{X}_0$, but the theorem does not tell us exactly what S is. However, we know from Example 6.4 that S is the first quadrant and that $\mathbf{f}^{-1}$ is given explicitly by (2), from which (4) can be verified directly.

Example 6.7 Let $\mathbf{X}_0 = (1, -1)$ and

$$\mathbf{f}(x, y) = \begin{bmatrix} x^4 y^5 - 4x \\ x^3 y^2 - 3y \end{bmatrix} ;$$

then

$$\mathbf{J}_{\mathbf{X}_0}\mathbf{f} = \begin{bmatrix} -8 & 5 \\ 3 & -5 \end{bmatrix}$$

is nonsingular and Theorem 6.2 implies that $\mathbf{f}$ has an inverse near $\mathbf{X}_0$. This time it is difficult to obtain $\mathbf{f}^{-1}$ and $\mathbf{J}_{\mathbf{U}}\mathbf{f}$ explicitly; however, (3) yields

$$(\mathbf{J}_{\mathbf{U}_0}\mathbf{f}_C^{-1}) = (\mathbf{J}_{\mathbf{X}_0}\mathbf{f})^{-1} = \begin{bmatrix} -\dfrac{5}{25} & -\dfrac{5}{25} \\[2ex] -\dfrac{3}{25} & -\dfrac{8}{25} \end{bmatrix},$$

where $\mathbf{U}_0 = (-5, 4)$. Moreover,

$$\mathbf{d}_{\mathbf{U}_0}\mathbf{f}^{-1} = \begin{bmatrix} -\dfrac{5}{25}\,du - \dfrac{5}{25}\,dv \\[2ex] -\dfrac{3}{25}\,du - \dfrac{8}{25}\,dv \end{bmatrix}.$$

If $\mathbf{f}: R^n \to R^n$ is one-to-one on S then we can identify a point $\mathbf{X}_0$ in S by the n-vector $\mathbf{U}_0 = \mathbf{f}(\mathbf{X}_0)$ that $\mathbf{f}$ assigns to $\mathbf{X}_0$. We say that $\mathbf{U} = \mathbf{f}(\mathbf{X})$ is a *coordinate transformation near* $\mathbf{X}_0$, thus considering $\mathbf{f}$ not as a mapping from S to $\mathbf{f}(S)$, but as a way of giving new names to the points of S.

Example 6.8 We saw in Example 6.4 that

$$\begin{bmatrix} u \\ v \end{bmatrix} = \mathbf{f}(x, y) = \begin{bmatrix} x^2 + y^2 \\ x^2 - y^2 \end{bmatrix}$$

defines a coordinate transformation in the first quadrant of R^2. The point whose new coordinates are u_0 and v_0 is the intersection of the circle $x^2 + y^2 = u_0$ and the branch of the hyperbola $x^2 - y^2 = v_0$ that lies in the first quadrant (Fig. 6.3).

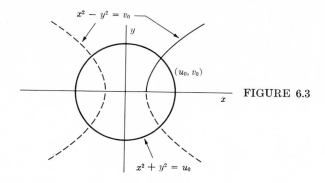

FIGURE 6.3

Definition 6.4   Let $\mathbf{f}: R^n \to R^n$ be differentiable on a set S. The determinant of $\mathbf{Jf}$ is called the *Jacobian of* $\mathbf{f}$, and is denoted by

$$(5) \qquad \frac{\partial(f_1, \ldots, f_n)}{\partial(x_1, \ldots, x_n)} ;$$

its value at a particular point $\mathbf{X}_0$ is denoted by

$$(6) \qquad \frac{\partial(f_1, \ldots, f_n)}{\partial(x_1, \ldots, x_n)} \bigg|_{\mathbf{X}_0}.$$

Consistent with the informal notation $\mathbf{U} = \mathbf{f}(\mathbf{X})$, we also write (5) and (6) as

$$\frac{\partial(u_1, \ldots, u_n)}{\partial(x_1, \ldots, x_n)}$$

and

$$\frac{\partial(u_1, \ldots, u_n)}{\partial(x_1, \ldots, x_n)} \bigg|_{\mathbf{X}_0}.$$

Theorem 6.3   If $\mathbf{f}: R^n \to R^n$ has continuous first derivatives near $\mathbf{X}_0$ and

$$\frac{\partial(u_1, \ldots, u_n)}{\partial(x_1, \ldots, x_n)} \bigg|_{\mathbf{X}_0} \neq 0,$$

then $\mathbf{U} = \mathbf{f}(\mathbf{X})$ defines a coordinate transformation on a set S containing $\mathbf{X}_0$ in its interior and $\mathbf{X} = \mathbf{f}_S^{-1}(\mathbf{U})$ defines a coordinate transformation near $\mathbf{U}_0 = \mathbf{f}(\mathbf{X}_0)$. Furthermore,

$$(7) \qquad \frac{\partial(u_1, \ldots, u_n)}{\partial(x_1, \ldots, x_n)} \frac{\partial(x_1, \ldots, x_n)}{\partial(u_1, \ldots, u_n)} = 1$$

for $\mathbf{X}$ sufficiently near $\mathbf{X}_0$.

We leave the proof, which follows quickly from Theorem 6.2, to the student (Exercise T-4).

Example 6.9 Let $\mathbf{f}$ be the transformation from rectangular to polar coordinates in R^2:

$$\mathbf{f}(x, y) = \begin{bmatrix} r(x, y) \\ \theta(x, y) \end{bmatrix},$$

where

$$r = \sqrt{x^2 + y^2},$$

$$\theta = \cos^{-1} \frac{x}{\sqrt{x^2 + y^2}}$$

$$= \sin^{-1} \frac{y}{\sqrt{x^2 + y^2}} \qquad (0 \leq \theta < 2\pi).$$

(Thus, θ is the unique angle in $(0, 2\pi)$ such that $\cos \theta = x/\sqrt{x^2 + y^2}$ and $\sin \theta = y/\sqrt{x^2 + y^2}$.) The domain of $\mathbf{f}$ is all of R^2 except $(0, 0)$, where θ is undefined, and

$$\begin{bmatrix} \dfrac{\partial r}{\partial x} & \dfrac{\partial r}{\partial y} \\[2mm] \dfrac{\partial \theta}{\partial x} & \dfrac{\partial \theta}{\partial y} \end{bmatrix} = \begin{bmatrix} \dfrac{x}{\sqrt{x^2 + y^2}} & \dfrac{y}{\sqrt{x^2 + y^2}} \\[2mm] -\dfrac{y}{x^2 + y^2} & \dfrac{x}{x^2 + y^2} \end{bmatrix};$$

hence

(8)
$$\frac{\partial(r, \theta)}{\partial(x, y)} = \frac{1}{\sqrt{x^2 + y^2}} = \frac{1}{r}.$$

The inverse function theorem implies that

$$\begin{bmatrix} r \\ \theta \end{bmatrix} = \mathbf{f}(x, y)$$

defines a coordinate transformation near every $\mathbf{X}_0 \neq (0, 0)$. However, we know more: $\mathbf{f}$ is a coordinate transformation on its entire domain, with inverse

$$\begin{bmatrix} x \\ y \end{bmatrix} = \mathbf{f}^{-1}(r, \theta) = \begin{bmatrix} r \cos \theta \\ r \sin \theta \end{bmatrix}.$$

By direct calculation,

$$\frac{\partial(x, y)}{\partial(r, \theta)} = \begin{bmatrix} \cos\theta & -r\sin\theta \\ \sin\theta & r\cos\theta \end{bmatrix} = r,$$

so that

$$\frac{\partial(r, \theta)}{\partial(x, y)} \frac{\partial(x, y)}{\partial(r, \theta)} = 1,$$

as is implied by (7) and (8).

In this example (and in Example 6.5), $\mathbf{f}$ has an inverse on its entire domain D. A hasty reading of Theorem 6.2 might lead the student to conclude that this is so because $\partial(u, v)/\partial(x, y)$ does not vanish in D. However, this would be faulty reasoning. The inverse function theorem is a "local" theorem. It says only that if $\mathbf{f}$ satisfies certain hypotheses near $\mathbf{X}_0$, then $\mathbf{f}$ has an inverse near $\mathbf{X}_0$; however, $\mathbf{f}$ may have an inverse near every point of D, but fail to have an inverse on D, as in the following example.

Example 6.10   Let

$$\mathbf{f}(x, y) = \begin{bmatrix} e^x \cos y \\ e^x \sin y \end{bmatrix};$$

then

$$\frac{\partial(u, v)}{\partial(x, y)} = \begin{vmatrix} e^x \cos y & -e^x \sin y \\ e^x \sin y & e^x \cos y \end{vmatrix} = e^{2x}$$

never vanishes, and Theorem 6.2 implies that $\mathbf{f}$ has inverse near every point of R^2. Nevertheless, $\mathbf{f}$ is not one-to-one on R^2, since

$$\mathbf{f}(x, y + 2\pi) = f(x, y),$$

and therefore $\mathbf{f}$ does not have an inverse on R^2.

We shall meet coordinate transformations and Jacobians again when we discuss change of variables in multiple integrals (Section 6.3).

EXERCISES 5.6

1.   Show that

$$\mathbf{f}(x, y) = \begin{bmatrix} x + y \\ 2x + 3y \end{bmatrix}$$

is one-to-one and that its range is R^2. Find $\mathbf{f}^{-1}$ and $\mathbf{Jf}^{-1}$.

2.   Repeat Exercise 1 for

$$f(x, y) = \begin{bmatrix} x - y \\ x + 4y \end{bmatrix}.$$

3.   Let

$$f(x, y) = \begin{bmatrix} x^2 + 2y^2 \\ 2x^2 + y^2 \end{bmatrix}.$$

   (a)   Show that f is one-to-one in any quadrant of R^2.

   (b)   Find f_S^{-1}, where S is the first quadrant. What is the domain
         of f_S^{-1}?

   (c)   Find f_T^{-1}, where T is the second quadrant. What is the domain
         of f_T^{-1}?

   (d)   Does f have an inverse near $(0, 1)$?

4.   Repeat Exercise 3 for

$$f(x, y) = \begin{bmatrix} 3x^2 - 2y^2 \\ x^2 + y^2 \end{bmatrix}.$$

5.   Let f be as in Exercise 3, $X_0 = (1, 2)$ and $U_0 = f(X_0)$. Using the in-
     verse function theorem, find $J_{U_0}f_S^{-1}$, where S is a small disk about X_0.

6.   Repeat Exercise 5 with f as defined in Exercise 4.

7.   Let

$$f(x, y) = \begin{bmatrix} x^2y + yx \\ 2xy + xy^2 \end{bmatrix}.$$

   (a)   Show that f satisfies the hypotheses of the inverse function
         theorem at $X_0 = (1, 1)$.

   (b)   Find an affine transformation L that approximates f_S^{-1} for U
         near $U_0 = f(X_0)$, where S is a small disk about X_0.

8.   Repeat Exercise 7 for

$$f(x, y) = \begin{bmatrix} x^2 - y^2 \\ 2xy \end{bmatrix}$$

     and $X_0 = (1, -1)$.

9. Find $\partial(r, \theta, \phi)/\partial(x, y, z)$ and $\partial(x, y, z)/\partial(r, \theta, \phi)$ if

$$\begin{bmatrix} x \\ y \\ z \end{bmatrix} = \begin{bmatrix} r \cos \theta \sin \phi \\ r \sin \theta \sin \phi \\ r \cos \phi \end{bmatrix}.$$

10. Find $\partial(r, \theta, z)/\partial(x, y, z)$ and $\partial(x, y, z)/\partial(r, \theta, z)$ if

$$\begin{bmatrix} x \\ y \\ z \end{bmatrix} = \begin{bmatrix} r \cos \theta \\ r \sin \theta \\ z \end{bmatrix}.$$

11. Find:

(a) $\dfrac{\partial(x, y, z)}{\partial(x, y, z)}$,   (b) $\dfrac{\partial(x, z, y)}{\partial(x, y, z)}$,   (c) $\dfrac{\partial(x, y, z)}{\partial(z, y, x)}$.

THEORETICAL EXERCISES

T-1. Prove the assertions made about

$$\mathbf{f}(x, y) = \begin{bmatrix} x^2 + y^2 \\ x^2 - y^2 \end{bmatrix}$$

in Example 6.2.

T-2. Prove Theorem 6.1.

T-3. Formally derive Eq. (4) from Eq. (2), Section 3.5.

T-4. Use Theorem 6.2 to prove Eq. (7).

T-5. Set $n = 1$ in Theorem 6.2 and show that if f has a continuous derivative near x_0 and $f'(x_0) \neq 0$, then f has an inverse near x_0. Find an equation relating the derivatives of f and f^{-1}.

Chapter 6

Integration

6.1 Multiple Integrals

In this section we define the integral of a real valued function of n variables. First we recall the definition for $n = 1$. Let f be defined on $[a, b]$ and Δ be a partition of $[a, b]$, defined by points $x_0, x_1, \ldots, x_m$ such that

$$a = x_0 < x_1 < \cdots < x_m = b.$$

The largest of the lengths of the subintervals so defined is the *norm* of Δ; it is denoted by

$$\| \Delta \| = \max_{1 \leq i \leq m} (x_i - x_{i-1}).$$

A sum of the form

$$\sigma = \sum_{i=1}^{m} f(\xi_i)(x_i - x_{i-1}),$$

where ξ_i is a fixed but arbitrary point in $[x_i, x_{i-1}]$, is called a *Riemann sum*. If there is a number I such that $| \sigma - I |$ can be made as small as we wish by taking $\| \Delta \|$ sufficiently small—regardless of how $\xi_1, \ldots, \xi_n$ are chosen in their respective intervals—then we say that f is *integrable* on $[a, b]$ and define I to be its *integral*; we write

$$(1) \qquad\qquad I = \int_a^b f(x) \, dx.$$

Thus if f is integrable on $[a, b]$ there corresponds to each $\epsilon > 0$ a $\delta > 0$ such that

$$\left| \sigma - \int_a^b f(x) \, dx \right| < \epsilon$$

provided $|| \Delta || < \delta$ and $x_{i-1} \leq \xi_i \leq x_i$ $(1 \leq i \leq m)$. We shall call an integral of this kind an *ordinary* integral to distinguish it from the other integrals defined in this chapter.

The student should recall that if f is integrable and nonnegative on $[a, b]$, then the integral (1) is the area of the two-dimensional region T bounded by the curve $y = f(x)$, the x-axis and the lines $x = a$ and $x = b$:

$$T = \{ (x, y) \mid 0 \leq y \leq f(x), \quad a \leq x \leq b \}$$

(Fig. 1.1).

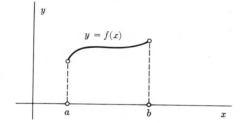

FIGURE 1.1

The most general conditions under which f is integrable on $[a, b]$ cannot be given here. A useful sufficient condition is that f be continuous on $[a, b]$, or that there exist points $a = \alpha_0 < \alpha_1 < \cdots < \alpha_k = b$ such that f is continuous in (α_{i-1}, α_i) and the right-hand limit $f(\alpha_{i-1}+)$ and the left-hand limit $f(\alpha_i-)$ exist for $1 \leq i \leq k$. (For definitions of right- and left-hand limits see Exercise T-5). In this case f is said to be *piecewise continuous* in $[a, b]$. A continuous function is piecewise continuous, but the converse is false. The graph of a piecewise continuous function is shown in Fig. 1.2.

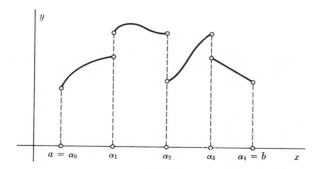

FIGURE 1.2

Double Integrals

Now suppose f is defined on a rectangle

$$S = \{(x, y) \mid a \leq x \leq b, \quad c \leq y \leq d\}.$$

Let

$$a = x_0 < x_1 < \cdots < x_m = b,$$
$$c = y_0 < y_1 < \cdots < y_n = d,$$

and construct a grid on S by drawing vertical line segments through $x = x_i$ from $y = c$ to $y = d$ $(1 \leq i \leq m)$ and horizontal lines through $y = y_j$ from $x = a$ to $x = b$ $(1 \leq j \leq n)$. Let Δ be the resulting partition of S into mn subrectangles

$$S_{ij} = \{(x, y) \mid x_{i-1} \leq x \leq x_i, \, y_{j-1} \leq y \leq y_j\} \quad (1 \leq i \leq m, \, 1 \leq j \leq n)$$

(Fig. 1.3), and let $\| \Delta \|$, the *norm* of Δ, be the length of the longest side appearing in any subrectangle.

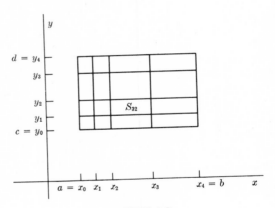

FIGURE 1.3

The sum

$$(2) \qquad \sigma = \sum_{i=1}^{m} \sum_{j=1}^{n} f(\mathbf{X}_{ij}) \, (x_i - x_{i-1}) \, (y_j - y_{j-1}),$$

where $\mathbf{X}_{ij}$ is an arbitrary point in S_{ij}, is called a *Riemann sum*. If there is a number I such that $| \sigma - I |$ can be made as small as we wish by taking $\| \Delta \|$ sufficiently small, no matter how $\mathbf{X}_{ij}$ is chosen in S_{ij}, then we say that f is *integrable on* S. We call I the *double integral of f over* S and write

$$(3) \qquad I = \int_{S} f \, dx \, dy;$$

when dealing with a specific function, say $f(x, y) = x^2 + 2xy$, we write

$$I = \int_S (x^2 + 2xy) \, dx \, dy.$$

Another common notation for the double integral is

$$I = \iint_S f \, dx \, dy;$$

however we shall use (3).

Example 1.1 Suppose $f(x, y) = 1$; then

$$\sigma = \sum_{i=1}^{m} \sum_{j=1}^{n} (x_i - x_{i-1})(y_j - y_{j-1})$$

$$= \left(\sum_{i=1}^{m} (x_i - x_{i-1}) \right) \left(\sum_{j=1}^{n} (y_j - y_{j-1}) \right)$$

$$= (x_m - x_0)(y_n - y_0)$$

$$= (b - a)(d - c).$$

Thus

$$\int_S 1 \, dx \, dy = \int_S dx \, dy$$

is the area of S. We can also interpret the integral as the volume of a rectangular box with base S and unit height.

The following theorem, whose proof we omit, gives a useful sufficient condition for the existence of the double integral.

Theorem 1.1 If $f: R^2 \to R$ is continuous on a rectangle S then

$$\int_S f \, dx \, dy$$

exists.

The next example illustrates how the continuity of f leads to the existence of the integral.

Example 1.2 Let

$$S = \{(x, y) \mid -1 \le x \le 1, \quad 0 \le y \le 2\}$$

(Fig. 1.4) and consider

$$\int_S (x + 3y) \, dx \, dy.$$

A typical Riemann sum is

$$\sigma = \sum_{i=1}^{m} \sum_{j=1}^{n} (\xi_i + 3\eta_j)(x_i - x_{i-1})(y_j - y_{j-1}),$$

which can be rewritten

$$\sigma = \sigma_0 + (\sigma - \sigma_0)$$

where

$$\sigma_0 = \sum_{i=1}^{m} \sum_{j=1}^{n} \left(\frac{x_i + x_{i-1}}{2} + 3\frac{y_j + y_{j-1}}{2} \right)(x_i - x_{i-1})(y_j - y_{j-1})$$

and

$$(4) \quad \sigma - \sigma_0 = \sum_{i=1}^{m} \sum_{j=1}^{n} \left[\xi_i + 3\eta_j - \frac{x_i + x_{i-1}}{2} - 3\frac{y_j + y_{j-1}}{2} \right]$$

$$\times (x_i - x_{i-1})(y_j - y_{j-1}).$$

Since $x_{i-1} < \xi_i < x_i$,

$$\left| \xi_i - \frac{x_i + x_{i-1}}{2} \right| \leq \frac{x_i - x_{i-1}}{2};$$

similarly

$$\left| \eta_j - \frac{y_j + y_{j-1}}{2} \right| \leq \frac{y_j - y_{j-1}}{2}.$$

Hence

$$\left| \xi_i + 3\eta_j - \frac{x_i + x_{i-1}}{2} - 3\frac{y_j + y_{j-1}}{2} \right| \leq 2 \, \| \Delta \|$$

and

$$| \sigma - \sigma_0 | \leq 2 \, \| \Delta \| \sum_{i=1}^{m} \sum_{j=1}^{n} (x_i - x_{i-1})(y_j - y_{j-1})$$

$$= 2 \, \| \Delta \| \, (2)(2) = 8 \, \| \Delta \|,$$

from (4) and Example 1.1, with $a = x_0 = -1$, $b = x_m = 1$, $c = y_0 = 0$,

$d = y_n = 2.$ Now

$$\sigma_0 = \tfrac{1}{2} \sum_{i=1}^{m} \sum_{j=1}^{n} (x_i^2 - x_{i-1}^2)(y_j - y_{j-1}) + \tfrac{3}{2} \sum_{i=1}^{m} \sum_{j=1}^{n} (x_i - x_{i-1})(y_j^2 - y_{j-1}^2)$$

$$= \tfrac{1}{2} \left(\sum_{i=1}^{m} (x_i^2 - x_{i-1}^2) \right) \left(\sum_{j=1}^{n} (y_j - y_{j-1}) \right)$$

$$+ \tfrac{3}{2} \left(\sum_{i=1}^{n} (x_i - x_{i-1}) \right) \left(\sum_{j=1}^{m} (y_j^2 - y_{j-1}^2) \right)$$

$$= \tfrac{1}{2}(x_n^2 - x_0^2)(y_m - y_0) + \tfrac{3}{2}(x_n - x_0)(y_m^2 - y_0^2)$$

$$= 12.$$

Hence

$$| \sigma - 12 | \leq 8 \, \| \, \Delta \, \|$$

and therefore

$$\int_S (x + 3y) \, dx \, dy = 12.$$

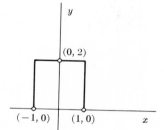

FIGURE 1.4

If f is integrable over S and $f(x, y) \geq 0$ in S, then the volume of the three dimensional region

$$T = \{ (x, y, z) \mid a \leq x \leq b, \quad c \leq y \leq d, \quad 0 \leq z \leq f(x, y) \}$$

(Fig. 1.5) is defined by

$$V(T) = \int_S f \, dx \, dy.$$

This definition is motivated by the interpretation of a typical term in the Riemann sum (2),

$$f(\mathbf{X}_{ij})(x_i - x_{i-1})(y_j - y_{j-1}),$$

as the volume of a rectangular parallelepiped with base S_{ij} and height $f(\mathbf{X}_{ij})$. Geometrically, the collection of these parallelepipeds approximates T increasingly well as $\| \Delta \|$ is made small; therefore it is reasonable to define $V(T)$ so that it is closely approximated by the sum of the volumes of the parallelepipeds.

If $f = 1$, then T is the rectangular box of Example 1.1, where we saw that

$$\int_S dx\, dy = (b - a)(d - c),$$

which is the volume of the box; thus the new definition of volume agrees with the old in this case. This is also true for other three-dimensional geometrical shapes whose volumes have been previously defined without recourse to the double integral, as we shall see in later sections after developing more practical ways to evaluate double integrals.

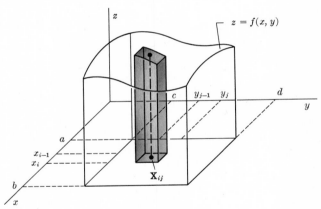

FIGURE 1.5

If f is integrable and nonpositive on S, then we define the volume of

$$T = \{ (x, y, z) \mid a \leq x \leq b, \quad c \leq y \leq d, \quad f(x, y) \leq z \leq 0 \}$$

to be

$$V(T) = - \int_S f\, dx\, dy.$$

If f is an arbitrary integrable function on S then we define

$$f^+(x, y) = \begin{cases} f(x, y) & \text{if} \quad f(x, y) \geq 0, \\ 0 & \text{if} \quad f(x, y) < 0; \end{cases}$$

and

$$f^-(x, y) = \begin{cases} f(x, y) & \text{if} \quad f(x, y) \leq 0, \\ 0 & \text{if} \quad f(x, y) > 0. \end{cases}$$

It can be shown that f^+ and f^- are integrable on S, and the volume of

$$T = \{ (x, y, z) \mid a \leq x \leq b, c \leq y \leq d, \min(0, f(x, y)) \leq z \leq \max(0, f(x, y)) \}$$

is defined to be

$$V(T) = \int_S f^+ \, dx \, dy - \int_S f^- \, dx \, dy.$$

Example 1.3 Let $S = \{0 \leq x \leq 1, \ 0 \leq y \leq 1\}$ and $f(x, y) = x - y$;
then T is shown in Fig. 1.6 and

$$f^+(x, y) = \begin{cases} x - y & (0 \leq y \leq x \leq 1), \\ 0 & (0 \leq x < y \leq 1); \end{cases}$$

$$f^-(x, y) = \begin{cases} x - y & (0 \leq x \leq y \leq 1), \\ 0 & (0 \leq y < x \leq 1). \end{cases}$$

The volume of T is given by

$$V(T) = \int_S f^+ \, dx \, dy - \int_S f^- \, dx \, dy,$$

which is calculated in Exercise 38, Section 6.2.

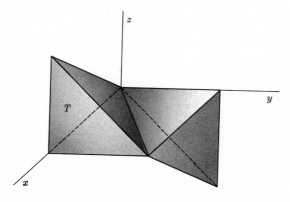

FIGURE 1.6

The assumption that the domain of f is rectangular was introduced merely for convenience. If f is defined on any bounded set R in the plane, then let S be a rectangle which contains R and define F on S by

$$F(x, y) = \begin{cases} f(x, y) & ((x, y) \text{ in } R), \\ 0 & ((x, y) \text{ in } S \text{ but not in } R). \end{cases}$$

Then define

$$\int_R f \, dx \, dy = \int_S F \, dx \, dy$$

if the integral on the right exists.

It can be shown that this definition is independent of the particular rectangle S chosen to enclose R; in fact the definition can be recast so that it does not involve an enclosing rectangle at all.

Example 1.4 Let g_1 and g_2 be continuous and $g_1(x) \leq g_2(x)$ in $[a, b]$. Then

$$R = \{ (x, y) \mid a \leq x \leq b, \quad g_1(x) \leq y \leq g_2(x) \}$$

(Fig. 1.7) is an example of a more general type of domain which occurs in applications of the double integral.

FIGURE 1.7

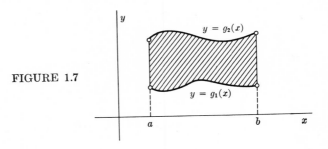

Example 1.5 Let h_1 and h_2 be continuous and $h_1(y) \leq h_2(y)$ in $[c, d]$. Then

$$R = \{ (x, y) \mid h_1(y) \leq x \leq h_2(y), \quad c \leq y \leq d \}$$

(Fig. 1.8) is an example of another type of domain which occurs in applications of the double integral.

The following theorem, whose proof we omit, gives sufficient conditions for the existence of the double integral which are more general than those given in Theorem 1.1.

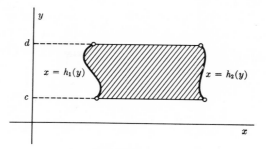

FIGURE 1.8

Theorem 1.2   If R consists of a piecewise smooth closed curve (Definition 5.4, Section 3.4) and the bounded region enclosed by it, and if f is continuous in R except possibly on a finite number of segments of smooth curves, then

$$\int_R f \, dx \, dy$$

exists.

The regions depicted in Figs. 1.7 and 1.8 have the properties required in this theorem, provided g_1, g_2, h_1, and h_2 are piecewise smooth.

Example 1.6   Let R be the shaded region bounded by the curves $y = -x^2 + 9$ and $y = x^2 - 1$ (Fig. 1.9), and let

$$f(x, y) = \begin{cases} x^2 + y^2 & (x \le 0), \\ x^2 + y^2 + 1 & (x > 0). \end{cases}$$

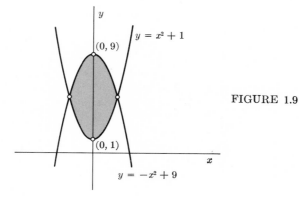

FIGURE 1.9

Then f is continuous in R except on the smooth curve defined by $x = 0$, $1 \leq y \leq 9$; hence f is integrable over R.

Theorem 1.2 tells us only that the integral exists under certain conditions; it does not provide a way to find its value (except, of course, by considering the Riemann sums). We shall consider this important matter in the next section.

If f is nonnegative and integrable over R then $\int_R f \, dx \, dy$ is the *volume* of

the three-dimensional region

$$T = \{(x, y, z) \mid 0 \leq z \leq f(x, y), \quad (x, y) \text{ in } R\};$$

the motivation for this definition is the same as that given earlier for the case where R is a rectangle. If $f = 1$, then

$$\int_R f \, dx \, dy = \int_R dx \, dy$$

is the area of R. We have already seen (Example 1.1) that this is consistent with the usual definition if R is a rectangle. The theory of ordinary integrals also provides a definition of area for a more general two dimensional region.

We leave it to the student to define the volume of

(5) $\qquad T = \{(x, y, z) \mid \min(f(x, y), 0) \leq z \leq \max(f(x, y), 0)\}$

if f has negative values in R (Exercise T-1).

The following theorem, which we state without proof, shows that the double integral has properties already familiar to the student from his study of ordinary integrals.

Theorem 1.3 Let f and g be integrable over a domain R. Then

(a) $f + g$ is integrable over R and

$$\int_R (f + g) \, dx \, dy = \int_R f \, dx \, dy + \int_R g \, dx \, dy;$$

(b) if k is a constant then kf is integrable over R and

$$\int_R (kf) \, dx \, dy = k \int_R f \, dx \, dy;$$

(c) if $f(x, y) \geq 0$ for all (x, y) in R, then

$$\int_R f \, dx \, dy \geq 0,$$

and

(d)   If R consists of regions R_1 and R_2 with no points except possibly on their bounding curves in common, then

$$\int_R f \, dx \, dy = \int_{R_1} f \, dx \, dy + \int_{R_2} f \, dx \, dy.$$

Parts (a), (b), and (c) can be established by considering the Riemann sums that approximate the various integrals (Exercise T-2). Figure 1.10 illustrates two kinds of domains to which (d) applies; in both cases R comprises R_1 and R_2.

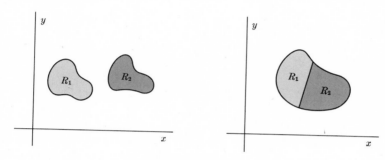

FIGURE 1.10

Corollary 1.1   If in addition to the assumptions of Theorem 1.2, $g(x, y) \leq f(x, y)$ in R, then

$$\int_R g \, dx \, dy \leq \int_R f \, dx \, dy.$$

Proof.   From (a) and (b), $F = f - g$ is integrable over R and

$$\int_R F \, dx \, dy = \int_R f \, dx \, dy - \int_R g \, dx \, dy;$$

from (c)

$$\int_R F \, dx \, dy \geq 0,$$

which yields the corollary.

Corollary 1.2   If g and $|\, g \,|$ are integrable over R then

(6) $$\left| \int_R g \, dx \, dy \right| \leq \int_R |\, g \,| \, dx \, dy.$$

Proof.   Since $g \leq |g|$ and $-g \leq |g|$, it follows from Corollary 1.1 that

$$\int_R g \, dx \, dy \leq \int_R |g| \, dx \, dy$$

and

$$-\int_R g \, dx \, dy \leq \int_R |g| \, dx \, dy.$$

These two inequalities together imply (6).

Multiple Integrals

We now give the general definition for the integral of a real valued function of n variables. A *rectangular parallelepiped* in R^n is a set of the form

$$(7) \qquad S = \{\mathbf{X} \mid a_i \leq x_i \leq b_i, \quad i = 1, \ldots, n\};$$

the sides of S have length $b_1 - a_1, \ldots, b_n - a_n$ and its *volume* is

$$V(S) = (b_1 - a_1) \cdots (b_n - a_n).$$

Example 1.7   If $n = 1$, S is a line segment, and $V(S)$ is its length; if $n = 2$, S is a rectangle and $V(S)$ its area, and if $n = 3$, S is a rectangular box and $V(S)$ its familiar volume.

If for each $i = 1, \ldots, n$ we cut through S with planes

$$x_i = x_{i0}, \qquad x_i = x_{i1}, \ldots, \qquad x_i = x_{in_i},$$

where

$$a_i = x_{i0} < x_{i1} < \cdots < x_{in_i} = b_i,$$

then we obtain a partition Δ of S into subparallelepipeds $S_1, \ldots, S_k$; furthermore

$$V(S) = V(S_1) + \cdots + V(S_k).$$

(Fig. 1.11 depicts the situation for $n = 3$.) The order in which the subparallelepipeds are numbered is immaterial.

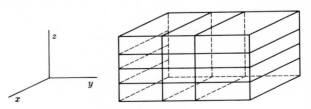

FIGURE 1.11

The *norm* of Δ, denoted by $\|\Delta\|$, is the length of the largest side appearing on $S_1, \ldots, S_k$. Now suppose f is defined on S; in each S_i choose an arbitrary point $\mathbf{X}_i$ and form the Riemann sum,

$$\sigma = \sum_{i=1}^{k} f(\mathbf{X}_i) V(S_i).$$

We say that f *is integrable over* S if there is a number I such that $|\sigma - I|$ can be made as small as we wish by taking $\|\Delta\|$ sufficiently small, independently of how $\mathbf{X}_i$ is chosen in S_i; I is called the *n-fold multiple integral of f over* S, and we write

$$I = \int_S f \, dx_1 \cdots dx_n.$$

If f is defined on an arbitrary bounded set R in R^n, then let S be a rectangular parallelepiped of the form (7) which contains S, and define

$$F(\mathbf{X}) = \begin{cases} f(\mathbf{X}) & (\mathbf{X} \text{ in } R), \\ 0 & (\mathbf{X} \text{ in } S \text{ but not in } R) \end{cases}$$

and

$$\int_R f \, dx_1 \cdots dx_n = \int_S F \, dx_1 \cdots dx_n$$

if the integral on the right exists. If $f = 1$, then

$$\int_R dx_1 \cdots dx_n$$

is the *n-dimensional volume* of T.

We shall not state general conditions under which n-fold multiple integrals exist. In succeeding sections there are many examples of multiple integrals whose existence will be demonstrated by direct calculation.

Theorem 1.1 and its corollaries, with $dx\, dy$ replaced by $dx_1 \cdots dx_n$, also hold for n-fold multiple integrals.

The student has no doubt noticed that $dx_1, \ldots, dx_n$ as used in this section do not stand for linear transformations, as they did in Chapter 4. They are nothing more than bookkeeping devices to identify the variables of integration; indeed many writers simply omit them and write, for instance,

$$\int_R f \, dx \, dy \, dz = \int_R f.$$

It is possible, however, to ascribe a fictitious geometric meaning to these symbols which aids in understanding multiple integrals. Consider $dx_1, \ldots, dx_n$ to be small, or *differential*, lengths measured parallel to the $x_1, \ldots, x_n$ axes, respectively. Then

$$dV_n = dx_1 \cdots dx_n$$

can be viewed as a *differential* volume (strictly an intuitive notion, which we shall not attempt to define precisely), and we can write an n-fold multiple integral as

$$(8) \qquad \int_R f(\mathbf{X}) \, dx_1 \cdots dx_n = \int_R f(\mathbf{X}) \, dV_n.$$

If $n = 1, 2,$ or 3, we write simply

$$dV_1 = dx \qquad \text{(differential length)},$$
$$dV_2 = dA \qquad \text{(differential area)},$$

and

$$dV_3 = dV.$$

We now view the multiple integral (8) as an integral with respect to n-dimensional volume. This idea will be exploited in the following sections.

EXERCISES 6.1

In Exercises 1 through 12 let

$$S = \{(x, y) \mid -1 \leq x \leq 1, \quad 0 \leq y \leq 2\}$$

and $f(x, y) = x + 3y$. Compute the Riemann sum associated with the given partition Δ and the given choice of $\mathbf{X}_{ij}$.

1. $\Delta: \ x_0 = -1 < 0 < 1 = x_2; \ y_0 = 0 < 1 < 2 = y_2.$

$$\mathbf{X}_{ij} = \left(\frac{x_{i-1} + x_i}{2}, \frac{y_{j-1} + y_j}{2} \right).$$

2. $\Delta: \ x_0 = -1 < 0 < 1 = x_2, \ y_0 = 0 < \frac{1}{2} < 1 < \frac{3}{2} < 2 = y_4.$

$$\mathbf{X}_{ij} = \left(\frac{x_{i-1} - x_i}{2}, \frac{y_{j-1} + y_j}{2} \right).$$

3. $\Delta: \ x_0 = -1 < -\frac{1}{2} < 0 < \frac{1}{2} < 1 = x_4, \ y_0 = 0 < \frac{1}{2} < 1 < \frac{3}{2} < 2 = y_4.$

$$\mathbf{X}_{ij} = \left(\frac{x_{i-1} + x_i}{2}, \frac{y_{j-1} + y_j}{2} \right).$$

4.   Let Δ be as in Exercise 3 and $\mathbf{X}_{ij} = (x_{i-1}, y_{j-1})$.

5.   Let Δ be as in Exercise 3 and $\mathbf{X}_{ij} = (x_i, y_{j-1})$.

6.   Let Δ be as in Exercise 3 and $\mathbf{X}_{ij} = (x_{i-1}, y_j)$.

7.   Let Δ be as in Exercise 3 and $\mathbf{X}_{ij} = (x_i, y_j)$.

In Exercises 8 through 12 show that the Riemann sum can be made to be as close to 12 (see Example 1.2) as we wish, by taking m and n sufficiently large. Use the formulas

$$\sum_{r=1}^{k} r = \frac{k(k+1)}{2}, \qquad \sum_{r=1}^{k} r^2 = \frac{k(k+1)(2k+1)}{6}.$$

8. Δ: $x_i = -1 + 2i/n \ (0 \le i \le n), \ y_j = 2j/m \ (0 \le j \le m)$;

$$\mathbf{X}_{ij} = \left(\frac{x_i + x_{i-1}}{2}, \frac{y_j + y_{j-1}}{2} \right).$$

9.   Let Δ be as in Exercise 8 and $\mathbf{X}_{ij} = (x_{i-1}, y_{j-1})$.

10.   Let Δ be as in Exercise 8 and $\mathbf{X}_{ij} = (x_i, y_{j-1})$.

11.   Let Δ be as in Exercise 8 and $\mathbf{X}_{ij} = (x_{i-1}, y_j)$.

12.   Let Δ be as in Exercise 8 and $\mathbf{X}_{ij} = (x_i, y_j)$.

13.   Use the method of Example 1.2 to compute

$$\int_S (2x + 3y) \, dx \, dy, \quad \text{where } S = \{(x, y) \mid 2 \le x \le 4, \ 1 \le y \le 3\}.$$

14.   Use the method of Example 1.2 to compute

$$\int_S (3x + y) \, dx \, dy, \quad \text{where } S = \{(x, y) \mid 1 \le x \le 4, \ 2 \le y \le 6\}.$$

15.   Use the result of Exercise 13 to obtain the volume of

$$T = \{(x, y, z) \mid 2 \le x \le 4, \ 1 \le y \le 3, \ 0 \le z \le 2x + 3y\}.$$

16.   Use the result of Exercise 14 to obtain the volume of

$$T = \{(x, y, z) \mid 1 \le x \le 4, \ 2 \le y \le 6, \ 0 \le z \le 3x + y\}.$$

17.   Find the volume of

$$T = \{(x, y, z) \mid -0 \le x \le 1, \ 0 \le y \le 1, \ x + y \le z \le 0\}.$$

THEORETICAL EXERCISES

T-1. Define the volume of
$$T = \{ (x, y, z) \mid (x, y) \text{ in } R, \ \min(f(x,y), 0) \le z \le \max(f(x,y), 0) \},$$
where R is a region in R^2.

T-2. Prove (a), (b), and (c) of Theorem 1.3.

T-3. Suppose f is integrable and $m \le f(x, y) \le M$ on a bounded set S.
Show that
$$mA(S) \le \int_S f(x, y) \, dx \, dy \le MA(S).$$

T-4. Let S be the subset of R^n defined by
$$S = \{ \mathbf{X} \mid a_i \le x_i \le b_i \quad (i = 1, \ldots, n) \}.$$

Suppose f_i is a continuous real valued function of the single variable
x_i on $[a_i, b_i]$. Show directly from the definition of the integral that
$$\int_S f_1 f_2 \cdots f_n \, dV = \left(\int_{a_1}^{b_1} f_1(x_1) \, dx_1 \right) \left(\int_{a_2}^{b_2} f_2(x_2) \, dx_2 \right) \cdots \left(\int_{a_n}^{b_n} f_n(x_n) \, dx_n \right).$$

T-5. Suppose f is defined on (a, b) and $a < \alpha < b$. Then we define the
left-hand limit $f(\alpha-)$ to be $\lim_{x \to \alpha} f_1(x)$, where f_1 is defined by $f_1(x) =$
$f(x)$ if $a \le x < \alpha$, and undefined elsewhere (See also Definition 1.7,
Section 4.1).

    (a)   Give an analogous definition for the *right-hand limit $f(\alpha+)$*.

    (b)   Show that $\lim_{x \to \alpha} f(x)$ exists if and only if $f(\alpha+) = f(\alpha-)$. In
this case, what is $\lim_{x \to \alpha} f(x)$?

6.2 Iterated Integrals

It is theoretically possible to evaluate a multiple integral from its Riemann
sums, as we did in Example 1.2. However, this is impractical for all but the
simplest integrals. In this section we show how to evaluate an n-fold
multiple integral by considering n successive integrals, each with respect
to a single variable.

If f is continuous on the rectangle

$$S = \{(x, y) \mid a \leq x \leq b, \quad c \leq y \leq d\}$$

we can define a function F on $[c, d]$ by

$$F(y) = \int_a^b f(x, y) \, dx,$$

where y is considered to be constant as we integrate with respect to x. It can be shown that F is continuous, and therefore integrable, on $[c, d]$; then

$$J_1 = \int_c^d F(y) \, dy = \int_c^d \left(\int_a^b f(x, y) \, dx \right) dy$$

is an *iterated integral* of f over S. Another iterated integral can be obtained by integrating with respect to y first,

$$G(x) = \int_c^d f(x, y) \, dy,$$

and then with respect to x:

$$J_2 = \int_a^b G(x) \, dx = \int_a^b \left(\int_c^d f(x, y) \, dy \right) dx.$$

Example 2.1   Let $f(x, y) = x + 3y$ and

$$S = \{(x, y) \mid -1 \leq x \leq 1, \quad 0 \leq y \leq 2\};$$

then

$$F(y) = \int_{-1}^1 (x + 3y) \, dx$$

$$= \left(\frac{x^2}{2} + 3xy \right) \Big|_{x=-1}^1 = (\tfrac{1}{2} + 3y) - (\tfrac{1}{2} - 3y) = 6y.$$

Hence

$$J_1 = \int_0^2 F(y) \, dy$$

$$= \int_0^2 6y \, dy = 3y^2 \Big|_0^2 = 12.$$

Also,

$$G(x) = \int_0^2 (x + 3y)\, dx$$

$$= (xy + \tfrac{3}{2}y^2)\,\Big|_{y=0}^2 = 2x + 6$$

and

$$J_2 = \int_{-1}^1 (2x + 6)\, dx$$

$$= (x^2 + 6x)\,\Big|_{-1}^1 = 12.$$

We have shown previously (Example 1.2) that

$$\int_S (x + 3y)\, dx\, dy = 12;$$

thus, in this example the two iterated integrals are equal and their common value is the double integral. That this is no accident is indicated by the following theorem.

Theorem 2.1 If the double integral of f over the rectangle S exists, and if either of the iterated integrals J_1 and J_2 exist, then both exist and

$$\int_S f(x, y)\, dA = \int_a^b \left(\int_c^d f(x, y)\, dy \right) dx$$

$$= \int_c^d \left(\int_a^b f(x, y)\, dx \right) dy.$$

We shall not prove this theorem, nor other similar results that are given below. Rather, we ask the student to accept it on the basis of the following geometric evidence.

If $f(x, y) \geq 0$ and T is the three-dimensional region bounded above by $z = f(x, y)$ and below by S, then

$$F(\eta) = \int_a^b f(x, \eta)\, dx$$

is the area of the cross section cut out of T by the plane $y = \eta$ (Fig. 2.1).

If $y_j \leq \eta_j \leq y_{j-1}$, then

$$F(\eta_j)(y_j - y_{j-1})$$

approximates the volume of that part of T between the planes $y = y_{j-1}$ and $y = y_j$; hence the Riemann sum

$$\sum_{j=1}^{n} F(\eta_j)(y_j - y_{j-1}),$$

which approximates

$$\int_c^d \left(\int_a^b f(x, y)\ dx \right) dy,$$

also approximates the volume of T, which is given by the double integral. Hence it is to be expected that $J_1 = I$.

By considering cross sections of T cut out by planes $x = $ constant, we also find that $J_2 = I$.

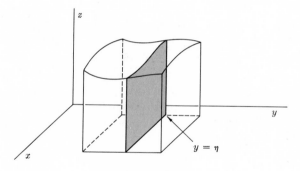

FIGURE 2.1

We shall adopt the notation

$$\int_c^d \left(\int_a^b f(x, y)\ dx \right) dy = \int_c^d dy \int_a^b f(x, y)\ dx,$$

and

$$\int_a^b \left(\int_c^d f(x, y)\ dy \right) dx = \int_a^b dx \int_c^d f(x, y)\ dy,$$

which exhibits the order of integration without cumbersome parentheses.

Suppose g_1 and g_2 are continuous and $g_1(x) \le g_2(x)$ in $[a, b]$, and let f be defined on

$$R = \{ (x, y) \mid a \le x \le b, \quad g_1(x) \le y \le g_2(x) \}$$

(Fig. 2.2). For each fixed x in $[a, b]$ let

$$G(x) = \int_{g_1(x)}^{g_2(x)} f(x, y) \, dy;$$

then

$$\int_a^b G(x) \, dx = \int_a^b \left(\int_{g_1(x)}^{g_2(x)} f(x, y) \, dy \right) dx = \int_a^b dx \int_{g_1(x)}^{g_2(x)} f(x, y) \, dy$$

is also an iterated integral. It can be shown that if this iterated integral and the double integral of f over R both exist, then they are equal:

$$(1) \qquad \int_R f(x, y) \, dA = \int_a^b dx \int_{g_1(x)}^{g_2(x)} f(x, y) \, dy.$$

Example 2.2   Let $f = 1$, then

$$\int_a^b dx \int_{g_1(x)}^{g_2(x)} dy = \int_a^b (g_2(x) - g_1(x)) \, dx = \int_a^b g_2(x) \, dx - \int_a^b g_1(x) \, dx.$$

The first integral is the area under the curve $y = g_2(x)$ and the second is the area under the curve $y = g_1(x)$ $(a \le x \le b)$; the difference is the area of R, which is $\int_R dA$ (Fig. 2.2).

To convert a double integral into an iterated integral, one must specify the limits of integration in the latter. This problem can be solved intuitively by considering the multiple integral as a sum of symbolic terms of the form

$$f(x, y) \, dx \, dy,$$

and viewing the iterated integral as a way of computing this sum by means of a specific order of summation. Thus,

$$(2) \qquad dx \int_{g_1(x)}^{g_2(x)} f(x, y) \, dy$$

is the sum over the shaded strip in Fig. 2.2, and the sum of all terms of the form (2) is the double integral, which leads to (1).

Example 2.3   Let it be required to find

$$I = \int_R (x + y) \, dA,$$

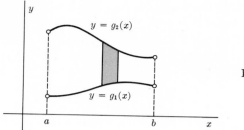

FIGURE 2.2

where R is the domain bounded by the parabola $y = x^2$ and the line $y = 2x$. We first sum $(x + y) \, dx \, dy$ over the vertical strip shown in Fig. 2.3 to obtain

$$dx \int_{x^2}^{2x} (x + y) \, dy,$$

and then over all such strips $(0 \leq x \leq 2)$ to obtain

$$I = \int_0^2 dx \int_{x^2}^{2x} (x + y) \, dy$$

$$= \int_0^2 \left[\left(xy + \frac{y^2}{2} \right) \Big|_{y=x^2}^{2x} \right] dx$$

$$= \int_0^2 \left(4x^2 - x^3 - \frac{x^4}{2} \right) dx$$

$$= \left(\frac{4}{3} x^3 - \frac{x^4}{4} - \frac{x^5}{10} \right) \Big|_0^2 = \frac{52}{15}.$$

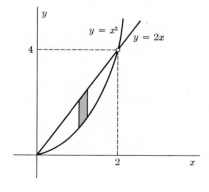

FIGURE 2.3

Example 2.4   The double integral of Example 2.3 can also be calculated by means of a second iterated integral, since the equations of the boundaries of R can also be written as $x = y/2$ and $x = \sqrt{y}$ $(0 \le y \le 4)$. Summing $(x + y)\, dx\, dy$ over the horizontal strip in Fig. 2.4 yields

$$dy \int_{y/2}^{\sqrt{y}} (x + y)\, dx$$

and then summing over all such horizontal strips yields

$$I = \int_0^4 dy \int_{y/2}^{\sqrt{y}} (x + y)\, dx$$

$$= \int_0^4 \left[\left(\frac{x^2}{2} + xy \right) \bigg|_{x=y/2}^{\sqrt{y}} \right] dy$$

$$= \int_0^4 \left(\frac{y}{2} + y^{3/2} - \frac{5}{8} y^2 \right) dy$$

$$= \left(\frac{y^2}{4} + \frac{2}{5} y^{5/2} - \frac{5}{24} y^3 \right) \bigg|_0^4 = \frac{52}{15}.$$

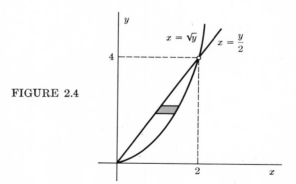

FIGURE 2.4

Examples 2.3 and 2.4 show that I can be evaluated by either of the iterated integrals, and, since they are roughly of the same difficulty, neither has an advantage over the other. The next example depicts a situation in which one of the iterated integrals is definitely simpler than the other.

Example 2.5   Find the area A of the region R bounded by $y = x^2 + 1$ and $y = 9 - x^2$ (Fig. 2.5).

Summing over the vertical strip in Fig. 2.5 yields

$$dx \int_{x^2+1}^{9-x^2} dy;$$

hence

$$A = \int_{-2}^{2} dx \int_{x^2+1}^{9-x^2} dy$$

$$= \int_{-2}^{2} \left[(9 - x^2) - (x^2 + 1) \right] dx$$

$$= 2 \int_{0}^{2} (8 - 2x^2) \, dx$$

$$= 2 \left(8x - \frac{2}{3} x^3 \right) \Big|_{0}^{2} = \frac{64}{3}.$$

To find A by means of an iterated integral with the opposite order of integration requires extra care. Summing over a horizontal strip yields

$$dy \int_{-\sqrt{9-y}}^{\sqrt{9-y}} dx$$

if $1 \leq y \leq 5$ and

$$dy \int_{-\sqrt{y-1}}^{\sqrt{y-1}} dx$$

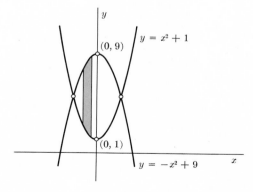

FIGURE 2.5

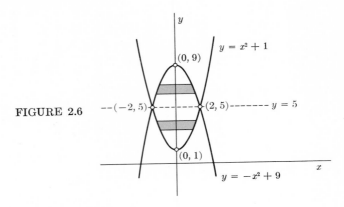

FIGURE 2.6

if $5 \leq y \leq 9$ (Fig. 2.6); thus

$$A = \int_1^5 dy \int_{-\sqrt{9-y}}^{\sqrt{9-y}} dx + \int_5^9 dy \int_{-\sqrt{y-1}}^{\sqrt{y-1}} dx$$

$$= 2 \int_1^5 \sqrt{9-y} \, dx + 2 \int_5^9 \sqrt{y-1} \, dy$$

$$= -\frac{4}{3}(9-y)^{3/2} \Big|_1^5 + \frac{4}{3}(y-1)^{3/2} \Big|_5^9 = \frac{64}{3}.$$

It is sometimes easier to evaluate an iterated integral by replacing it with an equivalent iterated integral with the opposite order of integration. The determination of the limits of integration of the latter is a geometrical problem.

Example 2.6 Let

$$J = \int_{1/2}^1 dy \int_{-\sqrt{1-y^2}}^{\sqrt{1-y^2}} f(x, y) \, dx,$$

and suppose we wish to find an equivalent iterated integral with the order of integration reversed. For each y in $[\frac{1}{2}, 1]$,

$$dy \int_{-\sqrt{1-y^2}}^{\sqrt{1-y^2}} f(x, y) \, dx$$

is a symbolic sum over a horizontal strip from $(-\sqrt{1 - y^2}, y)$ to $(\sqrt{1 - y^2}, y)$ (Fig. 2.7a). As y varies from $\frac{1}{2}$ to 1, these strips cover the part of the unit circle for which $\frac{1}{2} \leq y \leq 1$ (Fig. 2.7b). The same area can be covered first be summing over vertical strips from $(x, \frac{1}{2})$ to $(x, \sqrt{1 - x^2})$

to obtain

$$dx \int_{1/2}^{\sqrt{1-x^2}} f(x, y) \; dy \qquad \left(-\frac{\sqrt{3}}{2} \le x \le \frac{\sqrt{3}}{2} \right)$$

(Fig. 2.7c), and then summing over all such strips to obtain

$$J = \int_{-\sqrt{3}/2}^{\sqrt{3}/2} dx \int_{1/2}^{\sqrt{1-x^2}} f(x, y) \; dy.$$

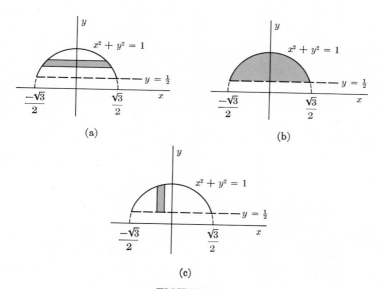

FIGURE 2.7

Iterated Integrals in n *Dimensions*

An iterated integral in n dimensions is of the form

$$J = \int_a^b dx_1 \int_{\eta_1(x_1)}^{\eta_2(x_1)} dx_2 \cdots \int_{\psi_1(x_1,\ldots,x_{n-2})}^{\psi_2(x_1,\ldots,x_{n-2})} dx_{n-1} \int_{\varphi_1(x_1,\ldots,x_{n-1})}^{\varphi_2(x_1,\ldots,x_{n-1})} f(x_1, \ldots, x_n) \; dx_n;$$

here the integration is with respect to x_n first, then $x_{n-1}, \ldots,$ then x_1. Iterated integrals in n variables can be written with $n!$ different orders of integration.

Example 2.7   Let $n = 3$ and

$$R = \{ (x, y, z) \mid x^2 + y^2 + z^2 \le 1, \quad x \ge 0, \quad y \ge 0, \quad z \ge 0 \};$$

thus R is the part of the unit sphere that lies in the first octant (Fig. 2.8a). Suppose we wish to express

$$I = \int_R f(x, y, z)\ dx\ dy\ dz$$

as an iterated integral where the integrations are with respect to x, y, and z, in that order. First, summing from $x = 0$ to $x = \sqrt{1 - y^2 - z^2}$ gives

$$dz\ dy \int_0^{\sqrt{1-y^2-z^2}} f(x, y, z)\ dx,$$

the sum over the parallelepiped shown in Fig. 2.8b; summing this from $y = 0$ to $y = \sqrt{1 - z^2}$ yields

(3) $$dz \int_0^{\sqrt{1-z^2}} dy \int_0^{\sqrt{1-y^2-z^2}} f(x, y, z)\ dx,$$

the sum over the slab in Fig. 2.8c. Finally, summing (3) from $z = 0$ to

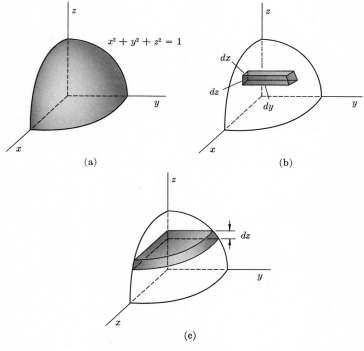

FIGURE 2.8

$z = 1$ yields

$$I = \int_0^1 dz \int_0^{\sqrt{1-z^2}} dy \int_0^{\sqrt{1-y^2-z^2}} f(x, y, z) \ dx.$$

Example 2.8 There are six iterated integrals by which

$$I = \int_R f(x, y, z) \ dx \ dy \ dz$$

can be computed. We leave it to the student to show that if R is the region bounded by the coordinate planes and the plane

$$x + 2y + 3z = 1$$

(Fig. 2.9), then

$$I = \int_0^{1/3} dz \int_0^{(1-3z)/2} dy \int_0^{1-2y-3z} f(x, y, z) \ dx$$

$$= \int_0^{1/2} dy \int_0^{(1-2y)/3} dz \int_0^{1-2y-3z} f(x, y, z) \ dx$$

$$= \int_0^1 dx \int_0^{(1-x)/3} dz \int_0^{(1-x-3z)/2} f(x, y, z) \ dy$$

$$= \int_0^{1/3} dz \int_0^{1-3z} dx \int_0^{(1-x-3z)/2} f(x, y, z) \ dy$$

$$= \int_0^1 dx \int_0^{(1-x)/2} dy \int_0^{(1-x-2y)/3} f(x, y, z) \ dz$$

$$= \int_0^{1/2} dy \int_0^{1-2y} dx \int_0^{(1-x-2y)/3} f(x, y, z) \ dz.$$

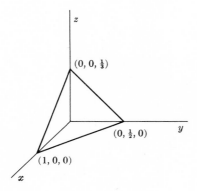

FIGURE 2.9

It is convenient and legitimate to rearrange the integrand of an iterated integral so that factors which do not depend upon the earlier variables of integration appear as far to the left as possible. Thus, we would write

$$\int_a^b dy \int_{\phi_1(y)}^{\phi_2(y)} y^2 \sin xy \, dx = \int_a^b y^2 \, dy \int_{\phi_1(y)}^{\phi_2(y)} \sin xy \, dx,$$

since y^2 can be treated as a constant while integrating with respect to x. As another example,

$$\int_a^b dz \int_{\phi_1(z)}^{\phi_2(z)} dy \int_{\psi_1(y,z)}^{\psi_2(y,z)} z^2 e^{yz} x^2 \, dx = \int_a^b z^2 \, dz \int_{\phi_1(z)}^{\phi_2(z)} e^{yz} \, dy \int_{\psi_1(y,z)}^{\psi_2(y,z)} x^2 \, dx \; ,$$

since z^2 can be treated as constant while integrating with respect to x and y, and e^{yz} can be treated as a constant while integrating with respect to x.

EXERCISES 6.2   # 20, # 31 $P_g. 554$ # 11a.

In Exercises 1 through 9 evaluate the given iterated integrals and sketch the region of integration.

1. $\displaystyle\int_0^3 dy \int_{-1}^2 (2xy^2 + 3x^2 y) \, dx.$ 2. $\displaystyle\int_{-2}^5 dx \int_2^3 (3x^2 + y^4) \, dy.$

3. $\displaystyle\int_{-1}^3 dx \int_{-2x+1}^{-x^2+4} xy \, dy.$ 4. $\displaystyle\int_{-\sqrt{2}}^{\sqrt{2}} dx \int_{x^2}^{-x^2+4} (x + 2y) \, dy.$

5. $\displaystyle\int_0^1 dx \int_{x+1}^{-x+3} (x - y) \, dy.$ 6. $\displaystyle\int_1^3 dy \int_0^{\log y} 2y \, dx.$

7. $\displaystyle\int_{-2}^0 dx \int_2^5 dy \int_{-3}^2 (x - 2y + 3z) \, dz.$

8. $\displaystyle\int_{-2}^2 dx \int_{-\sqrt{4-x^2}}^{\sqrt{4-x^2}} dy \int_0^{2-y} 2x \, dz.$

9. $\displaystyle\int_0^2 dx \int_0^{2+3x} dy \int_0^{2-y+3x} x \, dz.$

In Exercises 10 through 14 evaluate the given iterated integral and sketch the region of integration. Then compute the corresponding double integral by means of an iterated integral with the opposite order of integration.

10. $\displaystyle\int_0^2 dx \int_{x^2-4}^{4-x^2} x^2\,dy.$

11. $\displaystyle\int_{-3}^3 dy \int_0^{9-y^2} (2x+y^2)\,dx.$

12. $\displaystyle\int_{-1}^1 dx \int_0^{|x|} (2x^2y+y^3)\,dy.$

13. $\displaystyle\int_{-2}^3 dx \int_0^{e^x} xy^2\,dy.$

14. $\displaystyle\int_0^1 dy \int_y^{\sin^{-1}y} dx.$

In Exercises 15 through 19 sketch S and evaluate $\displaystyle\int_S f(x, y)\,dx\,dy.$

15. $f(x, y) = x$; S is the triangle with vertices $(1, 2)$, $(4, -1)$, and $(4, 5)$.

16. $f(x, y) = 4x$; S is bounded by $y = 3 - x^2$ and $y = 2x$.

17. $f(x, y) = 2x$; S is bounded by $y = x^2$ and $x = y^2$.

18. $f(x, y) = xy$; S is bounded by $y = |x|$ and $y = 6 - x^2$.

19. $f(x, y) = x + y$; S is bounded by $x = 0$, $x = \pi/6$, $y = \sin x$, and $y = \cos x$.

In Exercises 20 through 23 sketch S and evaluate $\displaystyle\int_S f(x, y, z)\,dx\,dy\,dz.$

20. $f(x, y, z) = yz$, S is the solid in the first octant bounded by $z = \sqrt{x^2 + y^2}$, $x^2 + y^2 = 9$, $x = 0$, $y = 0$, and $z = 0$.

21. $f(x, y, z) = x + y + z$; S is bounded by $x + y + z = 1$ and the coordinate planes.

22. $f(x, y, z) = x^2$; S is bounded by $4x^2 + y^2 = 4$, $z + x = 2$, and $z = 0$.

23. $f(x, y, z) = 2xz$; S is in the first octant, bounded by $z = y$ and $y = 4 - x^2$.

In Exercises 24 through 30 find the area of the region bounded by the given curves.

24. $y = x^2 + 4$, $y = x^2 - 4$, $x = -1$, $x = 1$.

25. $y = x + 1$, $y = 3 - x$, $x = 0$.

26. $y = 4 - x^2,$ $y = 2 - x.$

27. $y = x,$ $y = \dfrac{1}{x},$ $y = 0,$ $x = 2.$

28. $y = \sqrt{x},$ $y = x,$ $y = \tfrac{1}{2}x.$

29. $x = y^2 - 9,$ $x = 9 - y^2.$

$y = e^x,$ $y = -x,$ $x = 3.$

In Exercises 31 through 38 find the volume of S.

31. $S = \{(x, y, z) \mid 0 \le y \le x^2,\ \ 0 \le x \le 2,\ \ 0 \le z \le y^2\}.$

32. $S = \{(x, y, z) \mid x \ge 0,\ y \ge 0,\ x^2 + y^2 \le 16,\ 0 \le z \le x + y + 5\}.$

33. $S = \{(x, y, z) \mid x \ge 0,\ \ y \ge 0,\ \ 0 \le z \le 4 - 4x^2 - y^2\}.$

34. $S = \{(x, y, z) \mid x \ge 0,\ y \ge 0,\ z \ge 0,\ x^2 + y^2 \le 4,\ y^2 + z^2 \le 4\}.$

35. $S = \{(x, y, z) \mid 0 \le z \le x^2 + y^2,$ and $(x, y, 0)$ is in the triangle with vertices $(0, 1, 0),$ $(0, 0, 0),$ and $(1, 0, 0)\}.$

36. $S = \{(x, y, z) \mid y^2 \le z \le 16 - 4x^2 - y^2\}.$

37. $S = \{(x, y, z) \mid x^2 + y^2 \le z \le 8 - x^2 - y^2\}.$

38. Find the volume of the solid T defined in Example 1.3.

THEORETICAL EXERCISES

T-1. Show that

$$\int_0^1 dx_1 \int_0^{x_1} dx_2 \cdots \int_0^{x_{n-1}} dx_n = \frac{1}{n!}$$

T-2. Show that

$$\int_0^x dx_1 \int_0^{x_1} dx_2 \cdots \int_0^{x_{n-1}} f(x_n)\, dx_n = \frac{1}{(n-1)!} \int_0^x (x - t)^{n-1} f(t)\, dt.$$

T-3. Show that

$$\int_{a_1}^{b_1} dx_1 \int_{a_2}^{b_2} dx_2 \cdots \int_{a_n}^{b_n} f_1(x_1) f_2(x_2) \cdots f_n(x_n)\, dx_n$$

$$= \left(\int_{a_1}^{b_1} f_1(x_1)\, dx_1 \right) \left(\int_{a_2}^{b_2} f_2(x_2)\, dx_2 \right) \cdots \left(\int_{a_n}^{b_n} f_n(x_2)\, dx_2 \right).$$

(Compare this result with the result of Exercise T-4, Section 4.1.)

6.3 Change of Variables

In evaluating ordinary integrals it is often convenient to introduce a new variable of integration. The justification for this is provided by the following theorem.

Theorem 3.1 Let g be continuously differentiable in $[\alpha, \beta]$ and suppose that

$$(1) \qquad a = g(\alpha) \le g(u) \le g(\beta) = b$$

for $\alpha \le u \le \beta$. Let f be defined on $[a, b]$. Then f is integrable on $[a, b]$ if and only if $(f \circ g)g'$ is integrable on $[\alpha, \beta]$, and

$$(2) \qquad \int_a^b f(x)\ dx = \int_\alpha^\beta f(g(u))g'(u)\ du.$$

If (1) is replaced by

$$(3) \qquad a = g(\beta) \le g(u) \le g(\alpha) = b,$$

then (2) must be replaced by

$$(4) \qquad \int_a^b f(x)\ dx = -\int_\alpha^\beta f(g(u))g'(u)\ du.$$

In either case we say that the integral on the right is obtained from the one on the left by the *change of variable* $x = g(u)$.

Example 3.1 Let

$$I = \int_0^1 \sqrt{1 - x^2}\ dx$$

and

$$g_1(u) = \sin u;$$

then, if $0 \le u \le \pi/2$,

$$0 = g_1(0) \le g_1(u) \le g_1\left(\frac{\pi}{2}\right) = 1.$$

Thus (1) is satisfied and, since

$$g_1'(u) = \cos u,$$

the change of variable $x = \sin u$ yields

$$I = \int_0^{\pi/2} \sqrt{1 - \sin^2 u}\, \cos u \; du$$

$$= \int_0^{\pi/2} \cos^2 u \; du$$

$$= \frac{1}{2} \int_0^{\pi/2} (1 + \cos 2u) \; du$$

$$= \frac{1}{2} \left(u + \frac{\sin 2u}{2} \right) \Bigg|_0^{\pi/2} = \frac{\pi}{4}.$$

A second change of variable is given by

$$x = g_2(u) = \cos u,$$

for which

$$0 = g_1 \left(\frac{\pi}{2} \right) \leq g_1(u) \leq g_1(0) = 1$$

if $0 \leq u \leq \pi/2$ and

$$g_2'(u) = -\sin u;$$

hence (3) is satisfied and, from (4),

$$I = -\int_0^{\pi/2} \sqrt{1 - \cos^2 u} (-\sin u) \; du$$

$$= \int_0^{\pi/2} \sin^2 u \; du$$

$$= \frac{1}{2} \int_0^{\pi/2} (1 - \cos 2u) \; du$$

$$= \frac{1}{2} \left(u - \frac{\sin 2u}{2} \right) \Bigg|_0^{\pi/2} = \frac{\pi}{4}.$$

If we require that $x = g(u)$ be one-to-one (Definition 6.1, Section 5.6), then it must be strictly increasing if it satisfies (1) or strictly decreasing if it satisfies (3); in either case g' cannot change sign in $[\alpha, \beta]$ and both (2) and (4) can be replaced by

(5) $$\int_a^b f(x)\ dx = \int_\alpha^\beta f(g(u))\ |\ g'(u)\ |\ du.$$

We now generalize this form of the rule for change of variable to multiple integrals.

Theorem 3.2 Let T be a domain in R^n and $\mathbf{g}\colon R^n \to R^n$ be a continuously differentiable function which is one-to-one on T and maps T onto a domain S in R^n. Let

$$D_{\mathbf{g}}(\mathbf{U}) = \frac{\partial(g_1, \ldots, g_n)}{\partial(u_1, \ldots, u_n)}\bigg|_{\mathbf{U}}$$

be the Jacobian of $\mathbf{g}$ (Definition 5.4, Section 5.5). Then $f\colon R^n \to R$ is integrable on S if and only if $(f \circ \mathbf{g})\ |\ D_{\mathbf{g}}\ |$ is integrable on T, and

(6) $$\int_S f(\mathbf{X})\ dx_1 \cdots dx_n = \int_T f(\mathbf{g}(\mathbf{U}))\ |\ D_{\mathbf{g}}(\mathbf{U})\ |\ du_1 \cdots du_n.$$

(Note that the *absolute value* $|\ D_{\mathbf{g}}(\mathbf{U})\ |$ appears in the integral on the right.) We say that the integral on the right is obtained from the one on the left by the *change of variable* $\mathbf{X} = \mathbf{g}(\mathbf{U})$.

The proof of this theorem is quite difficult. Hence, we shall simply illustrate it and make is plausible by means of examples.

Example 3.2 If $n = 1$ and we take $S = [a, b]$ and $T = [\alpha, \beta]$, then $D_g = g'$ and (6) reduces to (5).

Example 3.3 Let $f = 1$, T be a rectangle

$$T = \{(u, v)\ |\ u_0 \le u \le u_1, \quad v_0 \le v \le v_1\}$$

(Fig. 3.1), and $\mathbf{g}$ be the affine function

(7) $$\begin{bmatrix} x \\ y \end{bmatrix} = \begin{bmatrix} x_0 \\ y_0 \end{bmatrix} + \begin{bmatrix} a & b \\ c & d \end{bmatrix} \begin{bmatrix} u \\ v \end{bmatrix};$$

if $\mathbf{g}$ is one-to-one, then its Jacobian matrix,

$$\mathbf{J_g} = \begin{bmatrix} a & b \\ c & d \end{bmatrix},$$

is nonsingular and

$$D_{\mathbf{g}} = ad - bc \ne 0.$$

To find S, we solve (7) for $\begin{bmatrix} u \\ v \end{bmatrix}$:

$$\begin{bmatrix} u \\ v \end{bmatrix} = \frac{1}{D_g} \begin{bmatrix} d & -b \\ -c & a \end{bmatrix} \begin{bmatrix} x - x_0 \\ y - y_0 \end{bmatrix}.$$

The lines $u = u_0$ and $u = u_1$ are transformed into the parallel lines

$$L_1: \quad d(x - x_0) - b(y - y_0) = D_g u_0$$

and

$$L_2: \quad d(x - x_0) - b(y - y_0) = D_g u_1;$$

also, $v = v_0$ and $v = v_1$ are transformed into the parallel lines

$$M_1: \quad -c(x - x_0) + a(y - y_0) = D_g v_0$$

and

$$M_2: \quad -c(x - x_0) + a(y - y_0) = D_g v_1.$$

Thus S is a parallelogram. The vertices P_1, P_2, P_3, and P_4 of T are mapped onto the vertices P_1', P_2', P_3' and P_4', respectively, of S (Fig. 3.1).

According to (6) the area of S is

$$(8) \quad A(S) = \int_S dx\,dy = \int_T |\,ad - bc\,|\,du\,dv$$

$$= |\,ad - bc\,| \int_S du\,dv$$

$$= |\,ad - bc\,|\,(u_1 - u_0)(v_1 - v_0),$$

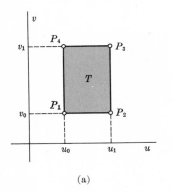

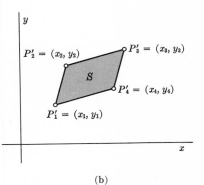

(a)   (b)

FIGURE 3.1

which can be verified independently as follows. Let $P_i' = (x_i, y_i)$ $(1 \leq i \leq 4)$; then

(9) $\qquad A(S) = |(x_3 - x_2)(y_2 - y_1) - (x_2 - x_1)(y_3 - y_2)|$

(See Section 3.2). From (7),

$$
\begin{bmatrix} x_3 - x_2 \\ y_3 - y_2 \end{bmatrix} = \begin{bmatrix} a & b \\ c & d \end{bmatrix} \begin{bmatrix} u_1 - u_0 \\ v_1 - v_1 \end{bmatrix}
$$

$$
= (u_1 - u_0) \begin{bmatrix} a \\ c \end{bmatrix},
$$

$$
\begin{bmatrix} x_2 - x_1 \\ y_2 - y_1 \end{bmatrix} = \begin{bmatrix} a & b \\ c & d \end{bmatrix} \begin{bmatrix} u_0 - u_0 \\ v_1 - v_0 \end{bmatrix}
$$

$$
= (v_1 - v_0) \begin{bmatrix} b \\ d \end{bmatrix},
$$

and substitution into (9) yields

$$
A(S) = |ad - bc| (u_1 - u_0)(v_1 - v_0),
$$

in agreement with (8).

Example 3.4 Again let $f = 1$ and let T be a rectangle,

$$
T = \{(u, v) \mid \alpha \leq u \leq \beta, \quad \gamma \leq v \leq \delta\},
$$

and suppose **g** (not necessarily an affine function) maps T one-to-one on a region S in R^2. According to (6),

(10) $\qquad \displaystyle\int_S dx\, dy = \int_T \left| \frac{\partial(g_1, g_2)}{\partial(u, v)} \right| du\, dv.$

Thus the integral on the right must equal $A(S)$, the area of S. To see that this is so, let Δ be a partition of T defined by

$$
\alpha = u_0 < u_1 < \cdots < u_m = \beta
$$

and

$$
\gamma = v_0 < v_1 < \cdots < v_m = \delta,
$$

and let T_{ij} be a typical subrectangle of T:

$$
T_{ij} = \{(u, v) \mid u_{i-1} \leq u \leq u_i, \quad v_{j-1} \leq v \leq v_j\}.
$$

Denote by S_{ij} the image of T_{ij} under $\mathbf{g}$ and let its area be $A(S_{ij})$ (Fig. 3.2a); then

$$A(S) = \sum_{i=1}^{m} \sum_{j=1}^{n} A(S_{ij}).$$

Now let us obtain an approximate value for $A(S_{ij})$. Let $\mathbf{U}_{ij}$ be a point in T_{ij}; since $\mathbf{g}$ is differentiable the transformation $\mathbf{X} = \mathbf{g}(\mathbf{U})$ can be approximated near $\mathbf{U}_{ij}$ (in particular, on T_{ij} if the latter is sufficiently small) by the affine function

(11) $$\mathbf{X} = \mathbf{h}(\mathbf{U}) = \mathbf{g}(\mathbf{U}_{ij}) + (d_{\mathbf{U}_{ij}}\mathbf{g})(\mathbf{U} - \mathbf{U}_{ij}),$$

which has Jacobian determinant

$$D_{\mathbf{g}}(\mathbf{U}_{ij}) = \left. \frac{\partial(g_1, g_2)}{\partial(u_1, u_2)} \right|_{\mathbf{U}_{ij}}.$$

According to Example 3.3 the transformation (11) maps T_{ij} onto a parallelogram S'_{ij} with area

(12) $$A(S'_{ij}) = |D_{\mathbf{g}}(\mathbf{U}_{ij})| (u_i - u_{i-1})(v_j - v_{j-1}).$$

If T_{ij} is sufficiently small, then S_{ij} and S'_{ij} are approximately equal (Fig. 3.2b) and therefore have approximately the same area; thus from (12)

$$A(S_{ij}) \cong |D_g(\mathbf{U}_{ij})| (u_i - u_{i-1})(v_j - v_{j-1}),$$

and

$$\sum_{i=1}^{m} \sum_{j=1}^{n} A(S_{ij}) \cong \sum_{i=1}^{m} \sum_{j=1}^{n} |D_{\mathbf{g}}(\mathbf{U}_{ij})| (u_i - u_{i-1})(v_j - v_{j-1}).$$

The sum on the left is the area of S, which is equal to

$$\int_S dx\, dy,$$

while the sum on the right is a Riemann sum associated with

$$\int_T |D_g(\mathbf{U})|\, du\, dv.$$

Hence (10) follows by letting $\|\Delta\|$ approach zero.

Example 3.5 Let $(u, v) = (r, \theta)$ and $\mathbf{g}$ be the transformation from polar to rectangular coordinates:

$$\begin{bmatrix} x \\ y \end{bmatrix} = \mathbf{g}(r, \theta) = \begin{bmatrix} r\cos\theta \\ r\sin\theta \end{bmatrix}.$$

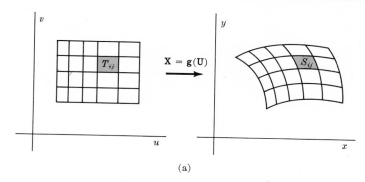

(a)

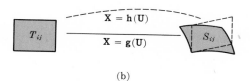

(b)

FIGURE 3.2

If

$$T = \{(r, \theta) \mid 0 < r_0 \leq r \leq r_1, \quad \theta_0 \leq \theta \leq \theta_1\},$$

(Fig. 3.3a), then **g** maps T one to one on the region S shown in Fig. 3.3b. Since

(13)
$$\frac{\partial(x, y)}{\partial(r, \theta)} = \begin{vmatrix} \cos \theta & \sin \theta \\ -r \sin \theta & r \cos \theta \end{vmatrix} = r,$$

the area of S is given, according to (6), by

$$A(S) = \int_T r \, dr \, d\theta$$

$$= \int_{\theta_0}^{\theta_1} d\theta \int_{r_0}^{r} r \, dr$$

$$= (\theta_1 - \theta_0) \frac{r_1^2 - r_0^2}{2}$$

$$= (\theta_1 - \theta_0)(r_1 - r_0) \frac{r_1 + r_0}{2}.$$

Now let $\theta_1 - \theta_0 = \Delta\theta$ and $r_1 - r_0 = \Delta r$; if Δr is small compared to r_0,

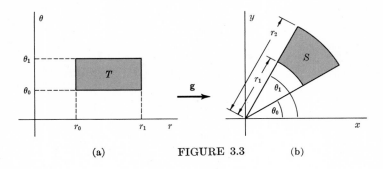

(a)   FIGURE 3.3   (b)

we can write

$$\Delta(S) \cong r \, \Delta r \, \Delta \theta,$$

where r is any number satisfying $r_1 \leq r \leq r_2$. For this reason we call the symbol $r \, dr \, d\theta$ the *differential element of area, in polar coordinates, at the point with polar coordinates* (r, θ).

Example 3.6 Let Q be the part of the closed unit ball that lies in the first octant; thus

$$Q = \{(x, y, z) \mid 0 \leq x^2 + y^2 + z^2 \leq 1, \quad x \geq 0, \quad y \geq 0, \quad z \geq 0\}.$$

(Fig. 3.4). The volume of Q is

$$(14) \qquad V(Q) = \int_S \sqrt{1 - x^2 - y^2} \, dx \, dy,$$

where S is the quarter of the unit circle in the first quadrant of the (x, y) plane. The integral (14) can be evaluated by means of the iterated integral

$$V(S) = \int_0^1 dx \int_0^{\sqrt{1-x^2}} \sqrt{1 - x^2 - y^2} \, dy,$$

but this requires the evaluation of two rather formidable ordinary integrals. This complication can be avoided by thinking of S in terms of polar coordinates; thus S is the image of the rectangle

$$T = \left\{(r, \theta) \mid 0 \leq r \leq 1, \quad 0 \leq \theta \leq \frac{\pi}{2}\right\}$$

under the transformation

$$\begin{bmatrix} x \\ y \end{bmatrix} = \mathbf{g}(r, \theta) = \begin{bmatrix} r \cos \theta \\ r \sin \theta \end{bmatrix}.$$

According to Theorem 3.2,

$$V(Q) = \int_S \sqrt{1 - x^2 - y^2}\, dx\, dy = \int_T \sqrt{1 - r^2 \cos^2 \theta - r^2 \sin^2 \theta}\ r\, dr\, d\theta.$$

The integral in polar coordinates can be evaluated by means of a simple iterated integral:

$$V(Q) = \int_T \sqrt{1 - r^2}\ r\, dr\, d\theta$$

$$= \int_0^{\pi/2} d\theta \int_0^1 \sqrt{1 - r^2}\ r\, dr$$

$$= \frac{\pi}{2}\left[-\frac{1}{3}(1 - r^2)^{3/2}\right]\Bigg|_0^1 = \frac{\pi}{6}.$$

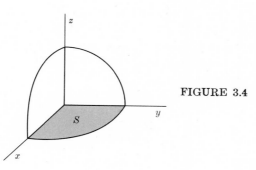

FIGURE 3.4

In this example we applied Theorem 3.2 and obtained a correct result even though the hypotheses of the theorem are not satisfied: **g** is not one-to-one on T, since it maps every point of the form $(0, \theta)$ onto $(x, y) = (0, 0)$. The fact is that (6) holds even if the hypotheses of Theorem 3.2 are relaxed so that **g** fails to be one-to-one on a finite number of smooth surfaces (also curves or points) of dimension less than n in T.

The rule for the change of variables from rectangular to polar coordinates is so important that we state it separately. The next theorem follows from Theorem 3.2 and (13).

Theorem 3.3   Let S be a region in R^2 and T be the set of polar coordinates of points in S:

$$T = \{(r, \theta) \mid (x, y) = (r \cos \theta, r \sin \theta)\quad \text{is in } S\}.$$

Then

$$\int_S f(x, y)\ dx\ dy = \int_T f(r \cos \theta, r \sin \theta) r\ dr\ d\theta,$$

provided f is integrable over S. In particular, the area of S is given by

$$A(S) = \int_T r\ dr\ d\theta.$$

Example 3.7   Find the area of the region S in R^2 bounded by the curve whose polar coordinates satisfy

$$r = 1 - \cos \theta \qquad (0 \le \theta \le 2\pi).$$

(This heart-shaped curve, shown in Fig. 3.5a, is called a *cardioid*.) According to Theorem 3.3,

$$A(S) = \int_T r\ dr\ d\theta$$

where T is the region of the (θ, r) plane shown in Fig. 3.5b. The area can be calculated by the iterated integral

$$A(S) = \int_0^{2\pi} d\theta \int_0^{1-\cos\theta} r\ dr$$

$$= \int_0^{2\pi} \left(\frac{r^2}{2} \Big|_0^{1-\cos\theta} \right) d\theta$$

$$= \frac{1}{2} \int_0^{2\pi} (1 - \cos \theta)^2\ d\theta$$

$$= \frac{1}{2} \int_0^{2\pi} (1 - 2 \cos \theta + \cos^2 \theta)\ d\theta$$

$$= \frac{1}{2} \int_0^{2\pi} \left(\frac{3}{2} - 2 \cos \theta + \frac{\cos 2\theta}{2} \right) d\theta = \frac{3}{4} (2\pi) = \frac{3}{2} \pi.$$

The next example illustrates Theorem 3.2 for another set of curvilinear coordinates in the plane.

Example 3.8   Let

$$I = \int_S (x^4 - y^4) e^{xy}\ dx\ dy$$

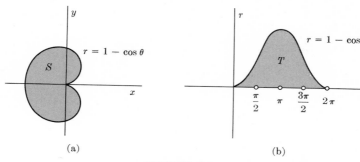

FIGURE 3.5

where S is the region in the first quadrant bounded by the hyperbolas $xy = 1$, $xy = 2$, $x^2 - y^2 = 2$, and $x^2 - y^2 = 3$ (Fig. 3.6a). Through each point in S there is exactly one hyperbola of the form

(15) $$x^2 - y^2 = u \qquad (2 \leq u \leq 3)$$

and exactly one of the form

(16) $$xy = v \qquad (1 \leq v \leq 2);$$

hence u and v define a coordinate system in S. The change of variable

$$\begin{bmatrix} x \\ y \end{bmatrix} = \mathbf{g}(u, v)$$

maps the rectangle

$$T = \{(u, v) \mid 2 \leq u \leq 3, \ 1 \leq v \leq 2\}$$

in the (u, v) plane one-to-one onto S (Fig. 3.6b); hence

(17) $$I = \int_T (x^4 - y^4)e^{xy} \left| \frac{\partial(x, y)}{\partial(u, v)} \right| \, du \, dv,$$

where, of course, the integrand must be expressed in terms of u and v. From (15) and (16) we know $\mathbf{g}^{-1}$, not $\mathbf{g}$; however, this suits our purpose, since

$$\frac{\partial(x, y)}{\partial(u, v)} = \frac{1}{\dfrac{\partial(u, v)}{\partial(x, y)}} = \frac{1}{\begin{vmatrix} 2x & -2y \\ y & x \end{vmatrix}} = \frac{1}{2(x^2 + y^2)}.$$

Substituting this into (17) yields

$$I = \frac{1}{2} \int_T u e^v \, du \, dv = \frac{1}{2} \int_2^3 u \, du \int_1^2 e^v \, dv = \frac{1}{2} \left(\frac{u^2}{2} \right) \Big|_2^3 \left(e^v \right) \Big|_1^2 = \frac{5}{4} e(e - 1).$$

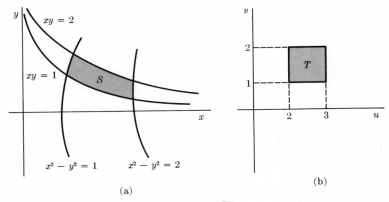

(a)

(b)

FIGURE 3.6

Example 3.9   Let

$$I = \int_S x e^{v^2 - x^2} \, dx \, dy$$

where S is the rectangle bounded by the lines $x + y = 0$, $x + y = 2$, $x - y = -4$, and $x - y = -2$ (Fig. 3.7a). Let

$$\begin{bmatrix} u \\ v \end{bmatrix} = \mathbf{g}^{-1}(x, y) = \begin{bmatrix} x + y \\ x - y \end{bmatrix};$$

then S is the image of the rectangle

$$T = \{(u, v) \mid 0 \le u \le 2, \quad -4 \le v \le -2\}$$

(Fig. 3.7b) under the transformation

$$\begin{bmatrix} x \\ y \end{bmatrix} = \mathbf{g}(u, v) = \begin{bmatrix} \dfrac{u + v}{2} \\ \dfrac{u - v}{2} \end{bmatrix}.$$

Since

$$\frac{\partial(x, y)}{\partial(u, v)} = \begin{vmatrix} \frac{1}{2} & \frac{1}{2} \\ \frac{1}{2} & -\frac{1}{2} \end{vmatrix} = -\frac{1}{2},$$

we have

$$I = \int_T \frac{u + v}{2} e^{uv} \left(\frac{1}{2}\right) du \, dv$$

$$= \frac{1}{4} \int_{-4}^{-2} dv \int_0^2 (u + v) e^{uv} \, du.$$

We leave the evaluation of the iterated integral for the student.

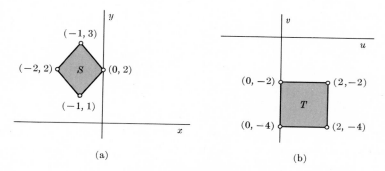

(a)   (b)

FIGURE 3.7

The two most common changes of variables for triple integrals are from rectangular to cylindrical coordinates and from rectangular to spherical coordinates. We first consider the former.

Theorem 3.4 Let S be a region in R^3 and T be the set of cylindrical coordinates of points in S:

$$T = \{(r, \theta, z) \mid (x, y, z) = (r \cos \theta, r \sin \theta, z) \quad \text{is in } S\}.$$

Then

$$\int_S f(x, y, z) \, dx \, dy \, dz = \int_T f(r \cos \theta, r \sin \theta, z) r \, dr \, d\theta \, dz$$

provided f is integrable over S. In particular, the volume of S is given by

$$V(R) = \int_T r \, dr \, d\theta \, dz.$$

Proof. The transformation

(18)
$$\begin{bmatrix} x \\ y \\ z \end{bmatrix} = \begin{bmatrix} r\cos\theta \\ r\sin\theta \\ z \end{bmatrix}$$

maps T onto S. Since

$$\frac{\partial(x, y, z)}{\partial(r, \theta, z)} = \begin{vmatrix} \cos\theta & -r\sin\theta & 0 \\ \sin\theta & r\cos\theta & 0 \\ 0 & 0 & 1 \end{vmatrix} = r,$$

the result follows from Theorem 3.2. (The fact that (18) fails to be one-to-one when $r = 0$ is unimportant; see the remark following Example 3.6.)

Example 3.10 Let S be the region in R^3 whose cylindrical coordinates satisfy $r_0 \leq r \leq r_1$, $\theta_0 \leq \theta \leq \theta_1$, and $z_0 \leq z \leq z_1$ (Fig. 3.8a). Thus, S is the image of a rectangular parallelepiped T in R^3, shown in Fig. 3.8b, under the transformation (18); hence the volume of S is

$$V(S) = \int_S dx\,dy\,dz = \int_S r\,dr\,d\theta\,dz$$

$$= \int_{z_0}^{z_1} dz \int_{\theta_0}^{\theta_1} d\theta \int_{r_0}^{r_1} r\,dr$$

$$= \frac{r_0 + r_1}{2}\,(r_1 - r_0)\,(\theta_1 - \theta_0)\,(z_1 - z_0).$$

Now let $r_1 - r_0 = \Delta r$, $\theta_1 - \theta_0 = \Delta\theta$, and $z_1 - z_0 = \Delta z$; if Δr is small compared to r_0, we have

$$V(S) \approx r\,\Delta r\,\Delta\theta\,\Delta z,$$

provided $r_0 \leq r \leq r_1$. For this reason we call the symbol $r\,dr\,d\theta\,dz$ the *differential element of volume, in cylindrical coordinates, at the point with cylindrical coordinates* (r, θ, z).

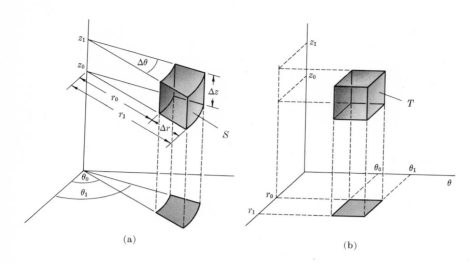

(a)

(b)

FIGURE 3.8

Example 3.11   Let

$$I = \int_S z \, dx \, dy \, dz,$$

where S is the hemisphere defined by

$$x^2 + y^2 + z^2 \le a^2, \qquad z \ge 0$$

(Fig. 3.9a). The transformation (18) maps

$$T = \{(r, \theta, z) \mid 0 \le \theta \le 2\pi, \quad 0 \le z \le \sqrt{a^2 - r^2}, \quad 0 \le r \le a\}$$

(Fig. 3.9b) onto S; hence

$$I = \int_T zr \, dr \, d\theta \, dz = \int_0^{2\pi} d\theta \int_0^a r \, dr \int_0^{\sqrt{a^2 - r^2}} z \, dz$$

$$= 2\pi \int_0^a r \, dr \int_0^{\sqrt{a^2 - r^2}} z \, dz = 2\pi \int_0^a r \left(\frac{z^2}{2} \Big|_0^{\sqrt{a^2 - r^2}} \right) dr$$

$$= \pi \int_0^a (a^2 r - r^3) \, dr = \pi \left(\frac{a^2 r^2}{2} - \frac{r^4}{4} \right) \Big|_0^a = \frac{\pi a^4}{4}.$$

Now let us consider the rule for change of variable from rectangular to spherical coordinates.

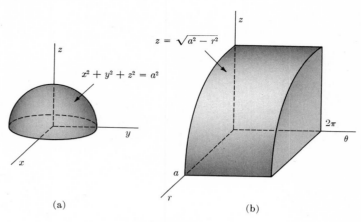

(a)

(b)

FIGURE 3.9

Theorem 3.5 Let S be a region in R^3 and T be the set of spherical co-ordinates of points in S:

$$T = \{(r, \theta, \phi) \mid (x, y, z) = (r \cos \theta \sin \phi, r \sin \theta \sin \phi, r \cos \phi) \quad \text{is in } S\}.$$

Then

$$\int_S f(x, y, z) \, dx \, dy \, dz = \int_T f(r \cos \theta \sin \phi, r \sin \theta \sin \phi, r \cos \phi) r^2 \sin \phi \, dr \, d\theta \, d\phi$$

provided f is integrable over S. In particular, the volume of S is given by

$$(19) \qquad\qquad V(S) = \int_T r^2 \sin \phi \, dr \, d\theta \, d\phi.$$

   Proof.   The transformation

$$(20) \qquad\qquad \begin{bmatrix} x \\ y \\ z \end{bmatrix} = \begin{bmatrix} r \cos \theta \sin \phi \\ r \sin \theta \sin \phi \\ r \cos \phi \end{bmatrix}$$

maps T onto S. Since

$$\frac{\partial(x, y, z)}{\partial(r, \theta, \phi)} = \begin{vmatrix} \cos \theta \sin \phi & -r \sin \theta \sin \phi & r \cos \theta \cos \phi \\ \sin \theta \sin \phi & r \cos \theta \sin \phi & r \sin \theta \cos \phi \\ \cos \phi & 0 & -r \sin \phi \end{vmatrix}$$

$$= -r^2 \sin \phi,$$

the result follows from Theorem 3.2.

Because of (19) we call the symbol $r^2 \sin \phi \, dr \, d\theta \, d\phi$ the *differential element of volume in, spherical coordinates, at the point with spherical coordinates* (r, θ, ϕ).

By a computation similar to that of Example 3.10, it can be shown that if S is the region in R^3 defined by $r_0 \leq r \leq r_0 + \Delta r,\ \theta_0 \leq \theta \leq \theta_0 + \Delta\theta,$ $\phi_0 \leq \phi \leq \phi_0 + \Delta\phi$, then

$$V(S) = r_0^2 \sin \phi_0 \, \Delta r \, \Delta\theta \, \Delta\phi$$

for small Δr, $\Delta\theta$, and $\Delta\phi$ (Fig. 3.10).

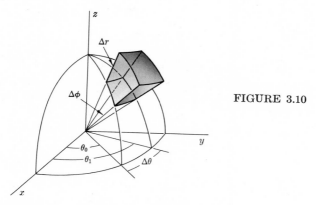

FIGURE 3.10

Example 3.12 Let S be the sphere defined by

$$x^2 + y^2 + z^2 \leq a^2$$

(Fig. 3.11a). The polar coordinates of points in S fill the rectangular parallelepiped

$$T = \{(r, \theta, \phi) \mid 0 \leq r \leq a,\ \ 0 \leq \theta \leq 2\pi,\ \ 0 \leq \phi \leq \pi\}$$

(Fig. 3.11b); hence

$$V(S) = \int_T r^2 \sin \phi \, dr \, d\theta \, d\phi$$

$$= \int_0^{2\pi} d\theta \int_0^{\pi} \sin \phi \, d\phi \int_0^a r^2 \, dr$$

$$= \left(\theta \,\Big|_0^{2\pi}\right)\left(-\cos \phi \,\Big|_0^{\pi}\right)\left(\frac{r^3}{3}\,\Big|_0^a\right)$$

$$= (2\pi)\,(2)\,\frac{a^3}{3} = \frac{4}{3}\pi a^3.$$

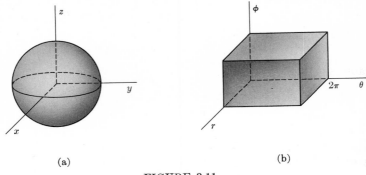

(a)

(b)

FIGURE 3.11

Example 3.13 To evaluate the integral I of Example 3.11 by transforming it to spherical coordinates, observe that the transformation (20) maps

$$T = \left\{(r, \theta, \phi) \mid 0 \le r \le a, \quad 0 \le \theta \le 2\pi, \quad 0 \le \phi \le \frac{\pi}{2}\right\}$$

onto S. Hence

$$\int_S z \, dx \, dy \, dz = \int_T (r \cos \phi) r^2 \sin \phi \, dr \, d\theta \, d\phi$$

$$= \int_0^{2\pi} d\theta \int_0^{\pi/2} \cos \phi \sin \phi \, d\phi \int_0^a r^3 \, dr$$

$$= \left(\theta \Big|_0^{2\pi}\right)\left(\frac{\sin^2 \phi}{2}\Big|_0^{\pi/2}\right)\left(\frac{r^4}{4}\Big|_0^a\right)$$

$$= (2\pi)\left(\frac{1}{2}\right)\left(\frac{a^4}{4}\right) = \frac{\pi a^4}{4}.$$

EXERCISES 6.3

1. Evaluate:

(a) $\displaystyle\int_0^1 \frac{dx}{4 + x^2}$;

(b) $\displaystyle\int_0^1 \sqrt{9 - x^2} \, dx$;

(c) $\displaystyle\int_2^4 \frac{\sqrt{x^2 - 4}}{x^2} \, dx$;

(d) $\displaystyle\int_0^1 \frac{e^{\sqrt{1+2x}}}{\sqrt{1 + 2x}} \, dx.$

2. Find S, the image of

$$T = \{(u, v) \mid 0 \leq u \leq 1, \quad 2 \leq v \leq 3\}$$

under the transformation

$$\begin{bmatrix} x \\ y \end{bmatrix} = \begin{bmatrix} 3 \\ -2 \end{bmatrix} + \begin{bmatrix} 1 & 2 \\ -1 & 3 \end{bmatrix} \begin{bmatrix} u \\ v \end{bmatrix}.$$

Also, find the area of S.

3. Repeat Exercise 2 for

$$T = \{(u, v) \mid -2 \leq u \leq 2, \quad -3 \leq v \leq 0\}$$

and

$$\begin{bmatrix} x \\ y \end{bmatrix} = \begin{bmatrix} -1 \\ 3 \end{bmatrix} + \begin{bmatrix} 0 & 2 \\ -1 & 1 \end{bmatrix} \begin{bmatrix} u \\ v \end{bmatrix}.$$

In Exercises 4 through 6 find the image of T under the given change of variables, and, find the area of the image.

4. $T = \left\{(r, \theta) \mid 1 \leq r \leq 3, \ \dfrac{\pi}{6} \leq \theta \leq \dfrac{\pi}{4}\right\}$; $\begin{bmatrix} x \\ y \end{bmatrix} = \begin{bmatrix} r \cos \theta \\ r \sin \theta \end{bmatrix}.$

5. $T = \left\{(u, v) \mid 0 \leq u \leq 1, \ 0 \leq v \leq \dfrac{\pi}{2}\right\}$; $\begin{bmatrix} x \\ y \end{bmatrix} = \begin{bmatrix} e^u \cos v \\ e^u \sin v \end{bmatrix}.$

6. $T = \{(u, v) \mid 0 \leq u \leq 1, \ 0 \leq v \leq 1\};$ $\begin{bmatrix} x \\ y \end{bmatrix} = \begin{bmatrix} 2uv \\ u^2 - v^2 \end{bmatrix}.$

In Exercises 7 through 9 find the image of T under the given change of variables, and find the volume of the image.

7. $T = \left\{(r, \theta, z) \mid 1 \leq r \leq 3, \ \dfrac{\pi}{4} \leq \theta \leq \dfrac{3\pi}{2}, \ -2 \leq z \leq 3\right\}$;

$$\begin{bmatrix} x \\ y \\ z \end{bmatrix} = \begin{bmatrix} r \cos \theta \\ r \sin \theta \\ z \end{bmatrix}.$$

8. $T = \left\{ (r, \theta, \phi) \,\middle|\, 2 \le r \le 5, \ \dfrac{\pi}{3} \le \theta \le \dfrac{3\pi}{4}, \ \dfrac{\pi}{6} \le \phi \le \dfrac{3\pi}{4} \right\};$

$$\begin{bmatrix} x \\ y \\ z \end{bmatrix} = \begin{bmatrix} r \sin \phi \cos \theta \\ r \sin \phi \sin \theta \\ r \cos \phi \end{bmatrix}.$$

9. $T = \{ (u, v, w) \mid 0 \le u \le 1, \ \ 0 \le v \le \pi, \ \ 0 \le w \le 2\pi \};$

$$\begin{bmatrix} x \\ y \\ z \end{bmatrix} = \begin{bmatrix} 2u \sin v \cos w \\ u \sin v \sin w \\ 3u \cos v \end{bmatrix}.$$

10. Evaluate by changing variables to polar coordinates in the equivalent double integrals:

   (a) $\displaystyle \int_0^2 dx \int_{-\sqrt{4-x^2}}^{\sqrt{4-x^2}} x^2 y^2 \, dy;$ (b) $\displaystyle \int_0^2 dy \int_{-\sqrt{2y-y^2}}^{\sqrt{2y-y^2}} \sqrt{x^2+y^2} \, dx.$

11. Evaluate by changing variables to polar coordinates in the equivalent double integrals:

   (a) $\displaystyle \int_0^2 dx \int_0^{\sqrt{4-x^2}} e^{(x^2+y^2)} \, dy;$ (b) $\displaystyle \int_0^{\sqrt{2}} dy \int_y^{\sqrt{4-y^2}} \dfrac{dx}{1+x^2+y^2}.$

12. Evaluate

$$\int_S [(y - x^2)^4 + 6xy](y + 2x^2) \, dx \, dy,$$

   where S is the region in the first quadrant bounded by $xy = 1$, $xy = 2$, $y = x^2$, and $y = x^2 + 1$. (*Hint*: See Example 3.8.)

13. Evaluate

$$\int_S x^2 y^6 \, dx \, dy,$$

   where S is the region in the first quadrant bounded by $xy = 1$, $xy = 2$, $y = x$, and $y = 2x$. (*Hint*: See Example 3.8.)

14. Find the area of the region inside the circle $x^2 + y^2 - 2y = 0$ and outside the circle $x^2 + y^2 = 2$.

15. Find the area of the region in the first quadrant outside $r = 2(1 - \cos\theta)$ and inside $r = 2\sin\theta$.

16. Find the area of the region bounded by $r = 3\sin\theta$ and $r = 4\cos\theta$.

17. Find the area of the region bounded by $x^2 + y^2 = 2$ and $y = x^2$.

18. Find the volume of the solid bounded by $z^2 = x^2 + y^2$ and $x^2 + y^2 = 4$.

19. Find the volume of the solid bounded by $z = x^2 + y^2$ and $z^2 = 4x^2 + 4y^2$.

20. Find the volume of the part of the cylinder $x^2 + y^2 = 4$ which is contained in the sphere $x^2 + y^2 + z^2 = 16$.

21. Find the volume of the solid bounded above by $x^2 + y^2 + z^2 = 12$ and below by $z = x^2 + y^2$.

22. Evaluate $\displaystyle\int_S (x^2 + y^2)\, dx\, dy\, dz,$ where S is the solid bounded by $x^2 + y^2 = 4$, $z = 0$, and $z = 3$.

23. Evaluate $\displaystyle\int_S (x^2 + y^2 + z^2)\, dx\, dy\, dz,$ where S is the solid bounded by

$$z = \frac{x^2 + y^2}{2}, \qquad z = 0, \qquad \text{and} \qquad x^2 + y^2 = 16.$$

24. Evaluate $\displaystyle\int_S \frac{e^{x^2+y^2+z^2}}{\sqrt{x^2 + y^2 + z^2}}\, dx\, dy\, dz,$ where

$$S = \{(x, y, z) \mid 3 \le x^2 + y^2 + z^2 \le 5\}.$$

25. Evaluate $\displaystyle\int_S (x^2 + z^2)\, dx\, dy\, dz,$ where

$$S = \{(x, y, z) \mid 3 \le x^2 + y^2 + z^2 \le 5\}.$$

26. Find the volume of the ellipsoid

$$\frac{x^2}{a^2} + \frac{y^2}{b^2} + \frac{z^2}{c^2} = 1.$$

27. Find the volume of the cylinder defined by

$$\frac{x^2}{a^2} + \frac{y^2}{b^2} = 1, \qquad |z| \le 1.$$

28. Find the volume of the parallelepiped bounded by the planes $x + y + z = 1$, $x + y + z = 2$, $x - y + z = 0$, $x - y + z = 1$, $x + y - z = 0$, and $x + y - z = 2$.

29. Evaluate $\int_S (3x + 2y + z) \, dx \, dy \, dz$ where S is the solid defined by

$$|x - y| \le 1, \qquad |y - z| \le 1, \qquad \text{and} \qquad |z + x| \le 1.$$

30. Evaluate $\int_S 4xyz(x^4 - y^4) \, dx \, dy \, dz$ where S is the region in the first octant defined by $0 \le x^2 - y^2 \le 1, 0 \le xy \le 1, 0 \le z^2 + y^2 \le 1$.

31. Evaluate $\int_S xy \, dx \, dy \, dz$, where S is defined by $0 \le z \le 4 - x^2 - 2y^2$.

32. Find the volume of the right circular cone of radius r and height h.

In Exercises 33 and 34, (r, θ, z) are cylindrical coordinates.

33. Find the volume of the solid defined by $0 \le z \le 4 - r^2$.

34. Find the volume of the solid bounded by $z = 0$, $z = r$, and $r = 2$.

THEORETICAL EXERCISES

T-1. Let $\mathbf{V}_1, \ldots, \mathbf{V}_n$ be a basis for R^n. Find the volume of
$$S = \{\mathbf{X} \mid a_i \le \mathbf{X} \cdot \mathbf{V}_i \le b_i \qquad (1 \le i \le n)\}.$$

T-2. Let $u_1, \ldots, u_n$ be constants. Describe how you would evaluate

$$\int_S (u_1 x_1 + u_2 x_2 + \cdots + u_n x_n) \, dV,$$

where S is defined in Exercise T-1.

T-3. Suppose V_n is the volume of the n-ball defined by
$$x_1^2 + x_2^2 + \cdots + x_n^2 \le 1.$$
Find the volume of the region defined by $a_1^2 x_1^2 + a_2^2 x_2^2 + \cdots + a_n^2 x_n^2 \le \rho^2$.

6.4 Physical Applications

Much of elementary mechanics is devoted to the study of systems of point masses; that is, physical systems in which the dimensions of the various parts are of no importance, so that they can be viewed as points to which nonzero masses are assigned. The multiple integral allows us to study mechanical systems comprised of parts with nonzero dimensions.

Mass Distributions

Definition 4.1   Let S be a region in R^3. A real valued function ρ is said to define a *mass distribution in* S if $\rho(\mathbf{X}) \geq 0$ for all $\mathbf{X}$ in S and

$$m(S) = \int_S \rho(\mathbf{X}) \, dx \, dy \, dz$$

exists. The integral $m(S)$ is the *mass of* S; more generally, if S_0 is a subset of S, the mass of S_0 is defined by

$$m(S_0) = \int_{S_0} \rho(\mathbf{X}) \, dx \, dy \, dz.$$

The function ρ is called the *mass density function of* S, and its value, $\rho(\mathbf{X})$, at a point $\mathbf{X}$ in S is *the density of* S at $\mathbf{X}$. If ρ is constant throughout S, then S is said to be *homogeneous*.

Throughout this section we shall assume that ρ is continuous. Suppose $\mathbf{X}_0$ is an interior point of S and

$$S_\delta = \{(x, y, z) \mid |x - x_0| \leq \delta, \quad |y - y_0| \leq \delta, \quad |z - z_0| \leq \delta\},$$

where δ is so small that the cube S_δ is contained in S (Fig. 4.1). The mass of S_δ is

$$m(S_\delta) = \int_{S_\delta} \rho(\mathbf{X}) \, dx \, dy \, dz.$$

For any $\epsilon > 0$ there is a $\delta > 0$ such that

$$|\rho(\mathbf{X}) - \rho(\mathbf{X}_0)| < \epsilon$$

if $\mathbf{X}$ is in S_δ; this follows from the continuity of ρ. Now,

$$m(S_\delta) = \rho(\mathbf{X}_0) \int_{S_\delta} dx \, dy \, dz + \int_{S_\delta} [\rho(\mathbf{X}) - \rho(\mathbf{X}_0)] \, dx \, dy \, dz$$

$$= \rho(\mathbf{X}_0) V(S_\delta) + \int_{S_\delta} [\rho(\mathbf{X}) - \rho(\mathbf{X}_0)] \, dx \, dy \, dz,$$

where $V(S_\delta)$ is the volume of S_δ. Hence

$$\left| \frac{m(S_\delta)}{V(S_\delta)} - \rho(X_0) \right| = \left| \frac{1}{V(S_\delta)} \int_{S_\delta} [\rho(\mathbf{X}) - \rho(\mathbf{X}_0)] \, dx \, dy \, dz \right|$$

$$\leq \frac{1}{V(S_\delta)} \int_{S_\delta} | \rho(\mathbf{X}) - \rho(\mathbf{X}_0) | \, dx \, dy \, dz$$

$$< \frac{\epsilon}{V(S_\delta)} \int_{S_\delta} dx \, dy \, dz = \epsilon.$$

Thus, $\rho(\mathbf{X}_0)$ is approximately equal to the ratio of the mass of S_δ to its volume, and the error of approximation approaches zero as δ approaches zero; that is, the mass $m(S_\delta)$ is approximated by the product $\rho(\mathbf{X}_0) V(S_\delta)$ when δ is small.

FIGURE 4.1

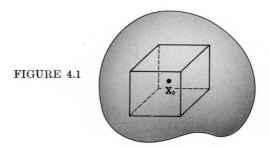

Example 4.1 Let S be the rectangular parallelepiped

$$S = \{ (x, y, z) \mid a \leq x \leq b, \quad c \leq x \leq d, \quad e \leq z \leq f \}$$

and let Δ be a partition of S into subparallelepipeds $S_1, \ldots, S_n$. The Riemann sum

$$\sigma = \sum_{i=1}^{n} \rho(\mathbf{X}_i) V(S_i),$$

where $V(S_i)$ is the volume of S_i and $\mathbf{X}_i$ is a point in S_i, is on the one hand an approximation to $m(S)$ and, on the other, the mass of a system of particles of mass $\rho(\mathbf{X}_1) V(S_1), \ldots, \rho(\mathbf{X}_n) V(S_n)$, situated at $\mathbf{X}_1, \ldots, \mathbf{X}_n$, respectively. We shall use this observation to motivate the definitions of first moment, center of mass, and moment of inertia of a region with a continuous mass distribution. We first recall their definitions for systems of point masses.

Center of Mass of a System of Point Masses

Let Γ be the plane through $\mathbf{X}_0$ and normal to

$$N = \begin{bmatrix} A \\ B \\ C \end{bmatrix} \qquad (A^2 + B^2 + C^2 \neq 0);$$

thus Γ has the equation

(1) $$A(x - x_0) + B(y - y_0) + C(z - z_0) = 0.$$

If $\mathbf{X}_1$ is an arbitrary point in R^3, then the directed distance from $\mathbf{X}_1$ to Γ is

$$d(\mathbf{X}_1) = \frac{A(x_1 - x_0) + B(y_1 - y_0) + C(z_1 - z_0)}{\sqrt{A^2 + B^2 + C^2}}$$

(Section 3.2), and the *moment about* Γ of a particle P_1 situated at $\mathbf{X}_1$ is

$$M_\Gamma = m_1 d(\mathbf{X}_1),$$

where m_1 is the mass of the particle. If P is a system of N particles of mass $m_1, \ldots, m_n$ located at $\mathbf{X}_1, \ldots, \mathbf{X}_n$, then the *moment* of P *about* Γ is the sum of the individual moments:

(2) $$M_\Gamma = \frac{A\left(\sum_{i=1}^n m_i x_i - m x_0\right) + B\left(\sum_{i=1}^n m_i y_i - m y_0\right) + C\left(\sum_{i=1}^n m_i z_i - m z_0\right)}{\sqrt{A^2 + B^2 + C^2}}$$

where

$$m = \sum_{i=1}^n m_i$$

is the total mass of the system.

The moment of P about the yz-plane is denoted by M_{yz}; setting $A = 1$ and $x_0 = B = C = 0$ in (2) yields

$$M_{yz} = \sum_{i=1}^n m_i x_i.$$

Similarly, the moments about the zx- and xy-planes are

$$M_{zx} = \sum_{i=1}^n m_i y_i$$

and

$$M_{zy} = \sum_{i=1}^{n} m_i z_i.$$

In terms of these moments, (2) can be rewritten as

$$(3) \qquad M_\Gamma = \frac{A(M_{yz} - mx_0) + B(M_{zx} - my_0) + C(M_{xy} - mz_0)}{\sqrt{A^2 + B^2 + C^2}}.$$

The *center of mass* of P is $\bar{\mathbf{X}} = (\bar{x}, \bar{y}, \bar{z})$, where

$$\bar{x} = \frac{M_{yz}}{m}, \qquad \bar{y} = \frac{M_{zx}}{m}, \qquad \bar{z} = \frac{M_{xy}}{m};$$

from (3), P has zero moment about every plane through $\mathbf{X}_0$ if and only if $\mathbf{X}_0 = \bar{\mathbf{X}}$.

Example 4.2   Let P consist of point masses of 2, 3, and 4 units at $\mathbf{X}_1 = (1, -1, 2)$, $\mathbf{X}_2 = (2, 1, 0)$, and $\mathbf{X}_3 = (-1, 0, 1)$, respectively (Fig. 4.2). Then

$$M_{yz} = 2(1) + 3(2) + 4(-1) = 4,$$
$$M_{zx} = 2(-1) + 3(1) + 4(0) = 1,$$
$$M_{xy} = 2(2) + 3(0) + 4(1) = 8,$$

and the total mass of P is

$$m = 2 + 3 + 4 = 9.$$

The moment of P about the plane Γ_0 defined by

$$2(x - 4) + (y + 1) - 3z = 0$$

is

$$M_{\Gamma_0} = \frac{2(M_{yz} - 4m) + (M_{zx} + m) - 3M_{xy}}{\sqrt{14}}$$

$$= \frac{2[4 - 4(9)] + [1 + 9] + 3[8]}{\sqrt{14}} = -\frac{78}{\sqrt{14}}.$$

The center of mass of P has coordinates

$$\bar{x} = \frac{M_{yz}}{m} = \frac{4}{9}, \qquad \bar{y} = \frac{M_{zx}}{m} = \frac{1}{9},$$

and

$$\bar{z} = \frac{M_{xy}}{m} = \frac{8}{9}.$$

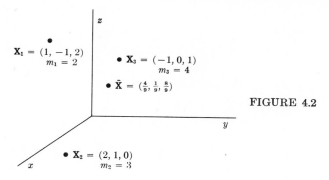

FIGURE 4.2

Center of Mass of a Solid

Now consider the rectangular parallelepiped of Example 4.1, again partitioned into subparallelepipeds $S_1, \ldots, S_n$, and let P_Δ be a system of particles situated at points $\mathbf{X}_1, \ldots, \mathbf{X}_n$ in $S_1, \ldots, S_n$, with masses $\rho(\mathbf{X}_1) V(S_1), \ldots, \rho(\mathbf{X}_n) V(S_n)$. The moment of P_Δ about the yz-plane is

$$M_{yz}(P_\Delta) = \sum_{i=1}^{n} x_i \rho(\mathbf{X}_i) V(S_i),$$

which is a Riemann sum for the integral

$$\int_S x\rho(\mathbf{X}) \, dx \, dy \, dz.$$

As $\|\Delta\|$ decreases the sum approaches the integral and P_Δ approaches S, in an intuitive sense. This and similar reasoning applied to the other moments motivates the following definition.

Definition 4.2 Let S be a region in R^3 with mass density function ρ and total mass

$$m = \int_S \rho(\mathbf{X}) \, dx \, dy \, dz.$$

The *moments of S about the yz-, zx- and xy-planes* are defined by

$$M_{yz} = \int_S x\rho(\mathbf{X}) \, dx \, dy \, dz,$$

$$M_{zx} = \int_S y\rho(\mathbf{X}) \, dx \, dy \, dz$$

and

$$M_{xy} = \int_S z\rho(\mathbf{X}) \; dx \; dy \; dz;$$

the *moment of S about the arbitrary plane* Γ defined by (1) is

$$M_\Gamma = \frac{A(M_{yz} - mx_0) + B(M_{zx} - my_0) + C(M_{xy} - mz_0)}{\sqrt{A^2 + B^2 + C^2}}.$$

The *center of mass of P* is the unique point $\bar{\mathbf{X}}$ with the property that $M_\Gamma = 0$ for every plane Γ which contains $\bar{\mathbf{X}}$; its coordinates are

$$\bar{x} = \frac{M_{yz}}{m}, \qquad \bar{y} = \frac{M_{zx}}{m}, \qquad \bar{z} = \frac{M_{xy}}{m}.$$

The center of mass is called the *centroid* of S if $\rho(\mathbf{X}) = 1$ for all $\mathbf{X}$ in S.

Example 4.3 Let S be the rectangular parallelepiped bounded by the coordinate axes and the three planes defined by $x = 1$, $y = 1$, and $z = 1$ (Fig. 4.3). Thus

$$S = \{(x, y, z) \mid 0 \leq x \leq 1, \quad 0 \leq y \leq 1, \quad 0 \leq z \leq 1\}.$$

Let the mass density function of S be $\rho(x, y, z) = z$. Then the mass of S is

$$m = \int_S z \; dx \; dy \; dz$$

$$= \int_0^1 z \; dz \int_0^1 dy \int_0^1 dx = \tfrac{1}{2}.$$

The moments of S about the coordinate planes are

$$M_{xy} = \int_S z^2 \; dx \; dy \; dz$$

$$= \int_0^1 z^2 \; dz \int_0^1 dy \int_0^1 dx = \tfrac{1}{3},$$

$$M_{yz} = \int_S xz \; dx \; dy \; dz$$

$$= \int_0^1 z \; dz \int_0^1 dy \int_0^1 x \; dx = \tfrac{1}{4},$$

and

$$M_{zx} = \int_S yz \; dx \; dy \; dz$$

$$= \int_0^1 z \; dz \int_0^1 y \; dy \int_0^1 dx = \tfrac{1}{4}.$$

The coordinates of the center of mass are

$$\bar{x} = \frac{M_{yz}}{m} = \frac{\frac{1}{4}}{\frac{1}{2}} = \frac{1}{2},$$

$$\bar{y} = \frac{M_{zx}}{m} = \frac{\frac{1}{4}}{\frac{1}{2}} = \frac{1}{2},$$

and

$$\bar{z} = \frac{M_{xy}}{m} = \frac{\frac{1}{3}}{\frac{1}{2}} = \frac{2}{3}.$$

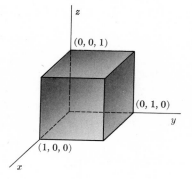

FIGURE 4.3

Example 4.4 To find the centroid of the region S considered in Example 4.3, take $\rho = 1$. Then

$$m = \int_S dx \; dy \; dz = 1,$$

$$M_{xy} = \int_S z \; dx \; dy \; dz = \tfrac{1}{2},$$

$$M_{yz} = \int_S x \; dx \; dy \; dz = \tfrac{1}{2},$$

and

$$M_{zx} = \int_S y \, dx \, dy \, dz = \tfrac{1}{2};$$

hence the centroid is

$$\bar{\mathbf{X}} = (\tfrac{1}{2}, \tfrac{1}{2}, \tfrac{1}{2}).$$

Example 4.5 Let it be required to find the centroid of a homogeneous right circular cone S with height h and base radius a. We first introduce a coordinate system so that the base of the cone is in the xy-plane and its longitudinal axis is along the positive z-axis (Fig. 4.4). The mass of S is

$$m = \rho \int_S dx \, dy \, dz$$

$$= \rho \int_T r \, dr \, d\theta \, dz,$$

where, in cylindrical coordinates,

$$(4) \quad T = \left\{ (r, \theta, z) \,\Big|\, 0 \le r \le a\left(1 - \frac{z}{h}\right), \quad 0 \le \theta \le 2\pi, \quad 0 \le z \le h \right\};$$

thus

$$m = \rho \int_0^{2\pi} d\theta \int_0^h dz \int_0^{a(1-z/h)} r \, dr = \frac{\pi \rho a^2 h}{3}.$$

The symmetry of S about the z-axis suggests that $\bar{x} = \bar{y} = 0$, which is easily verified by observing, for instance, that

$$M_{yz} = \rho \int_S x \, dx \, dy \, dz$$

can be evaluated by means of an interated integral where the first integration is with respect to x, between symmetric limits. To obtain $\bar{z}$, we calculate

$$M_{xy} = \rho \int_S z \, dx \, dy \, dz$$

$$= \rho \int_0^{2\pi} d\theta \int_0^h z \, dz \int_0^{a(1-z/h)} r \, dr = \frac{\pi \rho a^2 h^2}{12},$$

and

$$\bar{z} = \frac{M_{xy}}{m} = \frac{h}{4}.$$

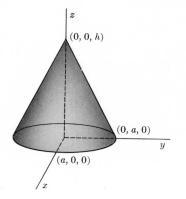

FIGURE 4.4

Example 4.6 Suppose S is the portion of a homogeneous solid sphere of radius 1 that lies in the first octant (Fig. 4.5). Without loss of generality, we can assume $\rho = 1$; then the mass of S is $m = \pi/6$ (Example 3.6). By symmetry, $\bar{x} = \bar{y} = \bar{z}$; hence, to find the centroid of S we need only calculate

$$M_{xy} = \int_S z \, dx \, dy \, dz.$$

This is best accomplished in spherical coordinates:

$$M_{xy} = \int_T r^3 \sin \phi \cos \phi \, dr \, d\theta \, d\phi,$$

where

$$T = \left\{ (r, \theta, \phi) \,\middle|\, 0 \le r \le 1, \quad 0 \le \theta \le \frac{\pi}{2}, \quad 0 \le \phi \le \frac{\pi}{2} \right\}.$$

Thus

$$M_{xy} = \int_0^1 r^3 \, dr \int_0^{\pi/2} d\theta \int_0^{\pi/2} \sin \phi \cos \phi \, d\phi$$

$$= \left(\frac{1}{4} \right) \left(\frac{\pi}{2} \right) \left(\frac{1}{2} \right) = \frac{\pi}{16},$$

and

$$\bar{x} = \bar{y} = \bar{z} = \frac{\dfrac{\pi}{16}}{\dfrac{\pi}{6}} = \frac{3}{8}.$$

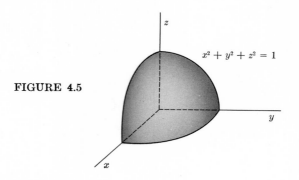

FIGURE 4.5

$$x^2 + y^2 + z^2 = 1$$

Moments of Inertia

Again let P be a system of N particles with masses $m_1, \ldots, m_n$, situated at $\mathbf{X}_1, \ldots, \mathbf{X}_n$, respectively. If the entire system rotates about a line L with instantaneous angular velocity ω then the velocity of the ith particle is

$$v_i = \omega d_i,$$

where d_i is the distance from $\mathbf{X}_i = (x_i, y_i, z_i)$ to L. The kinetic energy of the system is

$$(5) \qquad \text{K.E.} = \frac{1}{2} \sum_{i=1}^{n} m_i v_i^2$$

$$= \frac{\omega^2}{2} \sum_{i=1}^{n} m_i d_i^2.$$

Suppose we have chosen the coordinate system so that L is a line through the origin, defined by

$$(6) \qquad \frac{x}{A} = \frac{y}{B} = \frac{z}{C} \qquad (A^2 + B^2 + C^2 \neq 0);$$

then

$$d_i^2 = \frac{A^2(y_i^2+z_i^2)+B^2(x_i^2+z_i^2)+C^2(x_i^2+y_i^2)-2ABx_iy_i-2BCy_iz_i-2CAz_ix_i}{A^2 + B^2 + C^2}$$

(Section 3.2). From this and (5)

$$(7) \qquad \text{K.E.} = \tfrac{1}{2} I_L \omega^2,$$

where

$$(8) \qquad I_L = \frac{A^2 I_x + B^2 I_y + C^2 I_z - 2ABI_{xy} - 2BCI_{yz} - 2CAI_{zx}}{A^2 + B^2 + C^2}$$

and

$$I_x = \sum_{i=1}^{n} m_i(y_i^2 + z_i^2),$$

$$I_y = \sum_{i=1}^{n} m_i(z_i^2 + x_i^2),$$

$$I_z = \sum_{i=1}^{n} m_i(x_i^2 + y_i^2),$$

$$I_{xy} = \sum_{i=1}^{n} m_i x_i y_i,$$

$$I_{yz} = \sum_{i=1}^{n} m_i y_i z_i,$$

$$I_{zx} = \sum_{i=1}^{n} m_i z_i x_i.$$

The quantity I_L is the *moment of inertia of P about L*; I_x, I_y, and I_z are *the moments of inertia of P about the x-, y- and z-axes*, respectively, and I_{xy}, I_{yz}, and I_{zx} are the *products of inertia of P*. From (8) it is clear that the moment of inertia about any line through the origin can be calculated from these six quantities.

Example 4.7 For the system P defined in Example 4.2,

$$I_x = 2[(-1)^2 + 2^2] + 3[1^2 + 0^2] + 4[0^2 + 1^2] = 17;$$

$$I_y = 2[2^2 + 1^2] + 3[0^2 + 2^2] + 4[1^2 + (-1)^2] = 30;$$

$$I_z = 2[1^2 + (-1)^2] + 3[2^2 + 1^2] + 4[(-1)^2 + 0^2] = 23;$$

$$I_{xy} = 2(1)(-1) + 3(2)(1) + 4(-1)(0) = 4;$$

$$I_{yz} = 2(-1)(2) + 3(1)(0) + 4(0)(1) = -4;$$

and

$$I_{zx} = 2(2)(1) + 3(0)(2) + 4(1)(-1) = 0.$$

The moment of inertia of P about the line L_0 defined by

$$\frac{x}{2} = \frac{y}{-1} = \frac{z}{0}$$

is

$$I_L = \frac{2^2(17) + (-1)^2(30) + 0^2(23) - 2(2)(-1)(4)}{2^2 + (-1)^2 + 0^2}$$
$$\frac{- 2(-1)(0)(-4) - 2(0)(2)(0)}{}$$

$$= \frac{114}{5}.$$

Definition 4.3 Let S be a region in R^3 with mass density function ρ. The *moments of inertia of S about the x-, y- and z-axes* are defined by

$$I_x = \int_S \rho(\mathbf{X})\,(y^2 + z^2)\,dx\,dy\,dz,$$

$$I_y = \int_S \rho(\mathbf{X})\,(z^2 + x^2)\,dx\,dy\,dz,$$

and

$$I_z = \int_S \rho(\mathbf{X})\,(x^2 + y^2)\,dx\,dy\,dz.$$

The *products of inertia* are defined by

$$I_{xy} = \int_S \rho(\mathbf{X})xy\,dx\,dy\,dz,$$

$$I_{yz} = \int_S \rho(\mathbf{X})yz\,dx\,dy\,dz,$$

and

$$I_{zx} = \int_S \rho(\mathbf{X})zx\,dx\,dy\,dz.$$

In terms of these quantities the *moment of inertia I_L of S about the line L,* defined by (6), is given by (8).

If S rotates about L with instantaneous angular velocity ω, then its kinetic energy is given by (7).

Example 4.8 Let L be the line

$$\frac{x}{1} = \frac{y}{2} = \frac{z}{0}$$

and let S be the solid studied in Example 4.3. Then

$$I_L = \frac{I_x + 4I_y - 4I_{xy}}{5},$$

where

$$I_x = \int_S \rho(\mathbf{X})\,(y^2 + z^2)\ dx\ dy\ dz$$

$$= \int_S z(y^2 + z^2)\ dx\ dy\ dz = \tfrac{5}{12},$$

$$I_y = \int_S \rho(\mathbf{X})\,(x^2 + z^2)\ dx\ dy\ dz$$

$$= \int_S z(x^2 + z^2)\ dx\ dy\ dz = \tfrac{5}{12},$$

and

$$I_{xy} = \int_S \rho(\mathbf{X})\,xy\ dx\ dy\ dz$$

$$= \int_S xyz\ dx\ dy\ dz = \tfrac{1}{8}.$$

Thus

$$I_L = \frac{\tfrac{5}{12} + 4(\tfrac{5}{12}) - 4(\tfrac{1}{8})}{5} = \frac{19}{60}.$$

Example 4.9   Suppose we wish to find the moment of inertia I of a solid sphere S of radius a and uniform density ρ, about a line through its center. By symmetry I is independent of the particular line chosen. Consider a coordinate system with origin at the center of S. Then

$$I = I_z = \rho \int_S (x^2 + y^2)\ dx\ dy\ dz,$$

which is best evaluated in spherical coordinates:

$$I = \rho \int_T (r^2 \sin^2 \phi \cos^2 \theta + r^2 \sin^2 \phi \sin^2 \theta) r^2 \sin \phi\ dr\ d\theta\ d\phi$$

$$= \rho \int_T r^4 \sin^3 \phi\ dr\ d\theta\ d\phi,$$

where

$$T = \{(r, \theta, \phi) \mid 0 \leq r \leq a, \quad 0 \leq \theta \leq 2\pi, \quad 0 \leq \phi \leq \pi\}.$$

Thus

$$I = \rho \int_0^{2\pi} d\theta \int_0^{\pi} \sin^3 \phi \, d\phi \int_0^a r^4 \, dr$$

$$= \rho(2\pi) \left(\frac{4}{3}\right)\left(\frac{a^5}{5}\right) = \frac{8}{15} \pi\rho a^5.$$

Example 4.10. The moment of inertia of the right circular cone of Example 4.5 about its longitudinal axis is given by

$$I = \rho \int_S (x^2 + y^2) \, dx \, dy \, dz,$$

which is best evaluated in cylindrical coordinates over T as defined by (4):

$$I = \rho \int_T (r^2 \cos^2 \theta + r^2 \sin^2 \theta) r \, dr \, d\theta \, dz$$

$$= \rho \int_T r^3 \, dr \, d\theta \, dz$$

$$= \rho \int_0^{2\pi} d\theta \int_0^h dz \int_0^{a(1-z/h)} r^3 \, dr = \frac{\pi a^4 h \rho}{10} .$$

Laminas

A *lamina* is a region S in R^2 on which a mass density function ρ, having the dimensions of mass per unit area, is defined. It can be thought of as the idealization of a thin metal sheet of uniform cross section. In analogy with the corresponding definitions given above, the *total mass of S* is

$$m = \int_S \rho(x, y) \, dx \, dy$$

and its *first moments about the x- and y-axes* are defined by

$$M_x = \int_S \rho(x, y) y \, dx \, dy$$

and

$$M_y = \int_S \rho(x, y) x \, dx \, dy.$$

The *moment of S about a line L* given by

$$A(x - x_0) + B(y - y_0) = 0$$

is defined to be

$$M_L = \frac{A(M_y - mx_0) + B(M_x - my_0)}{\sqrt{A^2 + B^2}} \, ;$$

thus S has zero moment about any line through its *center of mass* $\bar{\mathbf{X}}$, which has coordinates

$$\bar{x} = \frac{M_y}{m}, \qquad \bar{y} = \frac{M_x}{m}.$$

If ρ is constant throughout S, $\mathbf{X}$ is again called the *centroid* of S.

Example 4.11 Let S be the region bounded by the parabola $y = x^2$ and the line $y = 2x$ (Fig. 4.6), with density proportional to the distance from the x-axis; thus

$$\rho(x, y) = ky.$$

Then

$$m = \int_S ky \, dx \, dy$$

$$= k \int_0^2 dx \int_{x^2}^{2x} y \, dy = \frac{32k}{15}.$$

The moments of S with respect to the axes are

$$M_x = \int_S kxy \, dx \, dy$$

$$= k \int_0^2 x \, dx \int_{x^2}^{2x} y \, dy = \tfrac{8}{3}k$$

and

$$M_y = \int_S kx^2 \, dx \, dy$$

$$= k \int_0^2 x^2 \, dx \int_{x^2}^{2x} dy = \tfrac{8}{5}k.$$

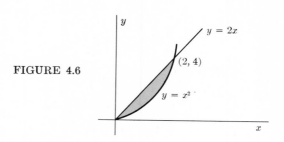

FIGURE 4.6

Thus the center of mass has coordinates

$$\bar{x} = \frac{\frac{8}{5}k}{\frac{32}{15}k} = \frac{3}{4}, \qquad \bar{y} = \frac{\frac{8}{3}k}{\frac{32}{15}k} = \frac{5}{4}.$$

Example 4.12   To find the centroid of the lamina of Example 4.11, take $\rho = 1$. Then

$$m = \int_0^2 dx \int_{x^2}^{2x} dy = \tfrac{4}{3},$$

$$M_x = \int_0^2 dx \int_{x^2}^{2x} y \, dy = \tfrac{32}{15},$$

and

$$M_y = \int_0^2 x \, dx \int_{x^2}^{2x} dy = \tfrac{4}{3};$$

hence

$$\bar{x} = \frac{\frac{4}{3}}{\frac{4}{3}} = 1$$

and

$$\bar{y} = \frac{\frac{32}{15}}{\frac{4}{3}} = \frac{8}{5}.$$

Example 4.13   Let S be the half disk defined by $x^2 + y^2 \leq a^2$, $y \geq 0$ (Fig. 4.7), with density proportional to the distance from the origin; thus

$$\rho(x, y) = k\sqrt{x^2 + y^2}.$$

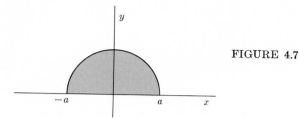

FIGURE 4.7

The mass of S is

$$m = \int_S k\sqrt{x^2 + y^2}\, dx\, dy$$

$$= \int_T kr^2\, dx\, d\theta,$$

where

$$T = \{(r, \theta) \mid 0 \le r \le a, \quad \theta \le 0 \le \pi\};$$

thus

$$m = k\int_0^\pi d\theta \int_0^a r^2\, dr = \frac{k\pi a^3}{3}.$$

The moments of S with respect to the axes are

$$M_x = \int_S ky\sqrt{x^2 + y^2}\, dx\, dy$$

$$= \int_T kr^3 \sin\theta\, dr\, d\theta$$

$$= k\int_0^\pi \sin\theta\, d\theta \int_0^a r^3\, dr = \frac{ka^4}{2}$$

and

$$M_y = \int_S kx\sqrt{x^2 + y^2}\, dx\, dy$$

$$= \int_T kr^3 \cos\theta\, dr\, d\theta$$

$$= k\int_0^\pi \cos\theta\, d\theta \int_0^a r^3\, dr = 0.$$

The coordinates of the center of mass are $\bar{x} = 0$ and

$$\bar{y} = \frac{\dfrac{ka^4}{2}}{\dfrac{k\pi a^3}{3}} = \frac{3a}{2\pi} .$$

Now suppose the entire xy-plane rotates with angular velocity ω about the line L defined by

$$\frac{x}{A} = \frac{y}{B},$$

carrying the lamina S with it (Fig. 4.8). Then it can be shown that the kinetic energy of S is

$$\text{K.E.} = \frac{I_L \omega^2}{2},$$

where

$$I_L = \frac{A^2 I_x + B^2 I_y - 2AB I_{xy}}{A^2 + B^2}$$

is *the moment of inertia of S about L,*

$$I_x = \int_S \rho(x, y) y^2 \, dx \, dy$$

and

$$I_y = \int_S \rho(x, y) x^2 \, dx \, dy$$

are the *moments of inertia of S about the x- and y-axes,* respectively, and

$$I_{xy} = \int_S \rho(x, y) xy \, dx \, dy$$

is its *product of inertia.*

If the xy-plane rotates about a line perpendicular to itself and through the origin, carrying S with it (Fig. 6.9), then the kinetic energy of S is

$$\text{K.E.} = \tfrac{1}{2} I_0 \omega^2,$$

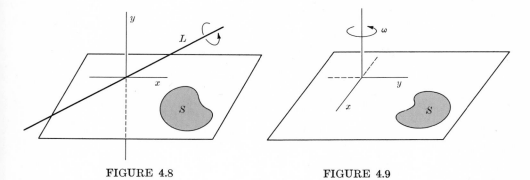

FIGURE 4.8   FIGURE 4.9

where

$$I_0 = \int_S \rho(x, y) (x^2 + y^2) \, dx \, dy$$

$$= I_y + I_x$$

is the *polar moment of inertia of S with respect to the origin.*

Example 4.14   For the lamina of Example 4.11,

$$I_x = \int_S kxy^2 \, dx \, dy$$

$$= k \int_0^2 x \, dx \int_{x^2}^{2x} y^2 \, dy = \frac{32}{5} \, k,$$

$$I_y = \int_S kx^3 \, dx \, dy$$

$$= k \int_0^2 x^3 \, dx \int_{x^2}^{2x} dy = \frac{32}{15} \, k,$$

and

$$I_{xy} = \int_S x^2 y \, dx \, dy$$

$$= k \int_0^2 x^2 \, dx \int_{x^2}^{2x} y \, dy = \frac{128}{35} \, k;$$

hence the moment of inertia of S about the line $x = y/2$ is

$$I_L = \frac{1^2(\frac{32}{5}k) + 2^2(\frac{32}{15}k) - 2(1)(2)\frac{128}{35}k}{1^2 + 2^2} = \frac{32k}{525}.$$

Its polar moment of inertia is

$$I_0 = I_x + I_y$$

$$= \frac{32}{5}k + \frac{32}{15}k = \frac{128}{15}k.$$

EXERCISES 6.4

1. Let P consist of point masses of 4 and 5 units at $(-1, 0, 2)$ and $(0, 1, 2)$. Let Γ_0 be the plane $2x - 3y + z = 0$ and let L_0 be the line

$$\frac{x}{3} = \frac{y}{-1} = \frac{z}{0}.$$

Find: (a) M_{xy}, M_{yz}, and M_{zz}; (b) $M\Gamma_0$; (c) $(\bar{x}, \bar{y}, \bar{z})$; (d) I_x, I_y, and I_z;

(e) I_{xy}, I_{yz}, and I_{zz};   (f) I_{L_0}.

2. Repeat Exercise 1 for the system of point masses of 2, 3, and 5 units at $(-1, 0, 2)$, $(0, 1, 2)$, $(3, 2, -1)$. Let Γ_0 be the plane $3x + 2y + z = -2$ and L_0 be the line

$$\frac{x}{0} = \frac{y}{3} = \frac{z}{0}.$$

In Exercises 3 through 11 let ρ be a mass density function for S. Find: (a) the mass of S; (b) M_{xy}, M_{yz}, and M_{zz}; (c) $M\Gamma_0$; (d) the centroid and center of mass of S; (e) I_x, I_y, and I_z; (f) I_{xy}, I_{yz}, and I_{zz}; (g) I_{L_0}.

3. $S = \{(x, y, z) \mid 0 \leq a \leq \sqrt{x^2 + y^2} \leq b, \mid z \mid < c\}; \rho(x, y, z) = \mid x \mid;$ Γ_0 is defined by $x + y + z = 0$ and L_0 by $x/0 = y/1 = -z/2$.

4. S is bounded by $x + 3y = 3$, $z = 2$ and the coordinate planes; $\rho(x, y, z) = kz; \Gamma_0$ is defined by $x + y - z = 0$ and L_0 by $x = y = z/2$.

5. $S = \{(x, y, z) \mid 0 \leq z \leq 1 - x^2/4 - y^2/4\}; \rho(x, y, z) = kz; \Gamma$ is defined by $x + y + z = 1$ and L_0 by $x = y = (z - 1)/0$.

6. S is a cube in the first octant with sides of length a, with a vertex at the origin and one face in the xy-plane; $\rho(x, y, z) = k(x^2 + y^2 + z^2)$; Γ_0 is the plane $z = a$ and L_0 is the diagonal of S that passes through the origin.

7. $S = \{(x, y, z) \mid x^2 + y^2 + z^2 \leq 4, x \geq 0, y \geq 0, z \geq 0\}$; $\rho(x, y, z) = kxyz$; Γ_0 is defined by $z = 2$ and L_0 by $y = x = z/0$.

8. S is bounded by $x^2 + y^2 + z^2 = 16$ and $x^2 + y^2 = 4$; $\rho(x, y, z) = k$; Γ_0 is defined by $x + y + z = 3$ and L_0 by $y/2 = x/1 = z/0$.

9. S is bounded above by $r = 4$ and below by $\phi = \pi/3$ (spherical coordinates); $\rho(r, \theta, \phi) = kr$; Γ_0 is defined by $x + y + 3z = 3$ and L_0 by $x = y/2 = z/2$.

10. S is bounded by $r = 2$, $r = 4$, $z = r/2$ (cylindrical coordinates) and the xy-plane; $\rho(r, \theta, \phi) = kr$; Γ_0 is defined by $x + y - z = 0$ and L_0 by $x = y/0 = z/0$.

11. $S = \{(x, y, z) \mid x^2/a^2 + y^2/b^2 + z^2/c^2 \leq 1\}$; $\rho(x, y, z) = k|z|$; Γ is defined by $x + y - z = 0$ and L_0 by $x/2 = y/3 = z/0$.

12. A homogeneous cylinder with a mass density ρ_0 has length h and radius a. Find its moment of inertia about its axis.

13. A right circular cone with mass density ρ_0 has height h and base radius a. Find its moment of inertia about a diameter of its base.

14. Find $I_x, I_y, I_z, I_{xy}, I_{yz},$ and I_{zz} for

$$S = \left\{(x, y, z) \ \middle| \ \frac{x^2}{a^2} + \frac{y^2}{b^2} + \frac{z^2}{c^2} \leq 1\right\};$$

take $\rho(x, y, z) = \rho_0$.

In Exercises 15 through 23, S is the given lamina with mass density function ρ. Find: (a) M_x and M_y; (b) M_{L_0}; (c) the centroid and center of mass of S; (d) $I_x, I_y, I_{xy},$ and I_0; (e) I_{L_0}.

15. The rectangle with vertices $(0, 0)$, $(a, 0)$, $(0, b)$, and (a, b), where $a > 0$ and $b > 0$; $\rho(x, y) = kxy$; L_0 is the line $y = (b/a)x$.

16. $\{(x, y) \mid x^2 + y^2 \leq 4, \ y \geq 0\}$; $\rho(x, y) = ky$; L_0 is the line $y = -2x$.

17. $S = \{(x, y) \mid x^2 \leq y \leq x^2 + 4, \ 0 \leq x \leq 2\}$; $\rho(x, y) = kx$; L_0 is the line $2y = x$.

18. S is bounded by the coordinate axes and the line $y = x + 2$; $\rho(x, y) = 2ky$; L_0 is the line $3y = x$.

19. S is bounded by $y = x$, $y = 1/x$, $y = 0$, and $x = 2$; $\rho(x, y) = ky$; L_0 is the line $y = x$.

20. S is bounded by $x = 0$, $y = x + 1$, and $y = 3 - x$; $\rho(x, y) = kx$; L_0 is the line $y = 2x$.

21. S is bounded by $y = x^2$ and $y^2 = x$; $\rho(x, y) = kx$; L_0 is the line $y = x$.

22. S is bounded by $r = 2 \sin \theta$; $\rho(r, \theta) = kr$; L_0 is the line $\theta = 3\pi/4$.

23. S is bounded by $r = 1 + \cos \theta$; $\rho(x, y) = k$; L_0 is the line $\theta = \pi/4$.

THEORETICAL EXERCISES

T-1. Let S_1 and S_2 be two solids with no points in common. *Prove*: The centroid of the system comprised of S_1 and S_2 is on the line segment connecting the centroids of S_1 and S_2.

T-2. Let S be a solid with mass m. Let L_0 be a line through the center of gravity of S and suppose L is a line parallel to L_0 and at a distance h from it. Show that

$$I_L = I_{L_0} + mh^2.$$

In Exercises T-3 and T-4 let $f(x) \geq 0$ in $[a, b]$ and suppose S is the solid swept out when the lamina

$$S_0 = \{(x, y) \mid 0 \leq y \leq f(x), \quad a \leq x \leq b\}$$

is rotated through 2π radians about the x-axis. Let $\rho(x, y, z) = 1$.

T-3. Show that the mass (volume) of S is

$$m = \pi \int_a^b (f(x))^2 \, dx.$$

T-4. Show that $M_{xy} = 0$, $M_{xz} = 0$, and

$$M_{yz} = \pi \int_a^b x(f(x))^2 \, dx.$$

In Exercises T-5 and T-6 let S_0 be the lamina bounded by $y = f(x)$, the x-axis and the lines $x = a$ and $x = b$. Let $\rho(x, y) = 1$.

T-5. Show that the mass (area) of S is

$$m = \int_a^b |f(x)| \, dx.$$

T-6. Show that

$$M_x = \frac{1}{2}\int_a^b f(x) \, |f(x)| \, dx, \qquad M_y = \int_a^b x \, |f(x)| \, dx.$$

6.5 Line Integrals

In this section we consider the line integral, which has its physical origins in the notion of work done in moving a particle along a curve in a force field.

Work

To motivate our definition of the line integral, let us recall the notion of work from elementary physics. The work done by a constant force $\mathbf{F}$ in moving an object along a line segment from $\mathbf{X}_1$ to $\mathbf{X}_2$ in R^2 is

$$W = |\mathbf{F}| \, |\mathbf{X}_2 - \mathbf{X}_1| \cos\theta = \mathbf{F} \cdot (\mathbf{X}_2 - \mathbf{X}_1)$$

where θ is the angle between $\mathbf{F}$ and the line segment (Fig. 5.1).

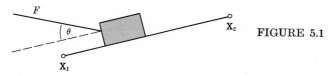

FIGURE 5.1

Now suppose an object moves from $\mathbf{X}_0$ to $\mathbf{X}_f$, along a curve C in R^2, under the influence of a force $\mathbf{F} = \mathbf{F}(\mathbf{X})$ which varies continuously with $\mathbf{X}$ along C. We wish to define the work done on the object by $\mathbf{F}$. Let $\mathbf{X} = \mathbf{X}(t)$ $(a \le t \le b)$ be a smooth parametric representation of C and consider the partition Δ of $[a, b]$:

$$\Delta: \quad a = t_0 < t_1 < \cdots < t_n = b.$$

For the moment replace C by the path C' consisting of line segments $L_1, \ldots, L_n$ connecting $\mathbf{X}(t_{j-1})$ to $\mathbf{X}(t_j), j = 1, \ldots, n$ (Fig. 5.2) and suppose

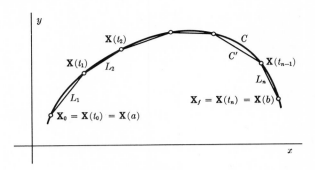

FIGURE 5.2

that the object is moved across C' by a "piecewise constant" force whose value is $\mathbf{F}(\mathbf{X}(\tau_j))$ on L_j, where τ_j is some point in $[t_{j-1}, t_j]$. Then the total work done on the object as it traverses C' is

$$W(\Delta) = \sum_{j=1}^{n} \mathbf{F}(\mathbf{X}(\tau_j)) \cdot (\mathbf{X}(t_j) - \mathbf{X}(t_{j-1})).$$

Applying the mean value theorem to the components of $\mathbf{X}(t)$, we can write

$$\mathbf{X}(t_j) - \mathbf{X}(t_{j-1}) = (t_j - t_{j-1}) \begin{bmatrix} x_1'(\tau_{1j}) \\ x_2'(\tau_{2j}) \\ \vdots \\ x_n'(\tau_{nj}) \end{bmatrix},$$

where $\tau_{1j}, \ldots, \tau_{nj}$ are in the interval (t_{j-1}, t_j). This can be rewritten as

$$\mathbf{X}(t_j) - \mathbf{X}(t_{j-1}) = (t_j - t_{j-1}) \begin{bmatrix} x_1'(\tau_j) \\ x_2'(\tau_j) \\ \vdots \\ x_n'(\tau_j) \end{bmatrix} + (t_j - t_{j-1}) \mathbf{E}_j$$

$$= \mathbf{X}'(\tau_j)(t_j - t_{j-1}) + (t_j - t_{j-1}) \mathbf{E}_j,$$

where

$$\mathbf{E}_j = \begin{bmatrix} x_1'(\tau_{1j}) \\ x_2'(\tau_{2j}) \\ \vdots \\ x_n'(\tau_{nj}) \end{bmatrix} - \begin{bmatrix} x_1'(\tau_j) \\ x_2'(\tau_j) \\ \vdots \\ x_n'(\tau_j) \end{bmatrix}.$$

Thus the work done in moving the object along C' can also be written as

$$W(\Delta) = \sum_{j=1}^{n} \mathbf{F}(\mathbf{X}(\tau_j)) \cdot \mathbf{X}'(\tau_j)(t_j - t_{j-1})$$

$$+ \sum_{j=1}^{n} \mathbf{F}(\mathbf{X}(\tau_j)) \cdot \mathbf{E}_j(t_j - t_{j-1}).$$

From the continuity of $\mathbf{X}'$ on $[a, b]$ it can be shown that the second sum on the right approaches zero as

$$\| \Delta \| = \max_{1 \le v \le n} (t_j - t_{j-1})$$

approaches zero. However, as this occurs, the first sum on right approaches

$$W = \int_a^b \mathbf{F}(\mathbf{X}(t)) \cdot \mathbf{X}'(t) \, dt;$$

consequently we define this integral to be the work done in moving the object from $\mathbf{X}_0 = \mathbf{X}(a)$ to $\mathbf{X}_f = \mathbf{X}(b)$ along C. It is an example of a line integral, defined next.

Henceforth all curves and parametric representations are assumed to be piecewise smooth (Definition 4.5, Section 3.4), and we shall also refer to curves as *paths*. When we say that a path (curve) has endpoints $\mathbf{X}_1$ and $\mathbf{X}_2$, we shall mean that the first named endpoint is the initial point.

Definition 5.1 Let $\mathbf{X} = \mathbf{X}(t)$ $(a \le t \le b)$ define a path C in the domain of a continuous vector field

$$\mathbf{F} = \begin{bmatrix} F_1 \\ \vdots \\ F_n \end{bmatrix}.$$

Then

(1)
$$I = \int_a^b \mathbf{F}(\mathbf{X}(t)) \cdot \mathbf{X}'(t) \, dt$$

$$= \int_a^b \left(\sum_{i=1}^n F_i(\mathbf{X}(t)) x_i'(t) \right) dt$$

is *the line integral of* $\mathbf{F}$ *along* C.

Example 5.1 Let C be given by

$$\mathbf{X} = \mathbf{X}(t) = \begin{bmatrix} t \\ 2t^2 \\ t^3 \end{bmatrix} \qquad (0 \le t \le 1)$$

and

$$\mathbf{F}(\mathbf{X}) = \begin{bmatrix} x \\ xz + 1 \\ yz + x \end{bmatrix};$$

then

$$\mathbf{X}'(t) = \begin{bmatrix} 1 \\ 4t \\ 3t^2 \end{bmatrix}, \qquad \mathbf{F}(\mathbf{X}(t)) = \begin{bmatrix} t \\ t^4 + 1 \\ 2t^5 + t \end{bmatrix},$$

and

$$I = \int_0^1 \left[(t)(1) + (t^4 + 1)(4t) + (2t^5 + t)(3t^2) \right] dt$$

$$= \int_0^1 (6t^7 + 4t^5 + 3t^3 + 5t) \, dt = \tfrac{14}{3}.$$

Properties of Line Integrals

Even though Definition 5.1 appears to depend upon a particular parametric representation of C, the line integral is in fact independent of the choice of parameter. Let $\mathbf{X} = \mathbf{X}(t)$ $(a \le t \le b)$ and $\mathbf{X} = \mathbf{Y}(\tau)$ $(c \le \tau \le d)$ be equivalent representations of C, related by a piecewise smooth change of parameter $t = \alpha(\tau)$; thus

(2) $$\mathbf{Y}(\tau) = \mathbf{X}(\alpha(\tau)) \qquad (c \le \tau \le d).$$

Applying the change of variable $t = \alpha(\tau)$ in (1) yields

(3) $$I = \int_c^d \mathbf{F}(\mathbf{X}(\alpha(\tau))) \cdot \mathbf{X}'(\alpha(\tau)) \alpha'(\tau) \, d\tau;$$

this integral exists because α' is piecewise continuous. From the chain rule and (2),

$$\mathbf{X}'(\alpha(\tau)) \alpha'(\tau) = \mathbf{Y}'(\tau)$$

at all except possibly a finite number of values of τ where one of the derivatives shown may fail to exist; hence (3) can be written as

$$I = \int_c^d \mathbf{F}(\mathbf{Y}(\tau)) \cdot \mathbf{Y}'(\tau) \, d\tau,$$

which is precisely the definition we would have given for I if we had originally represented C by $\mathbf{X} = \mathbf{Y}(\tau)$. This shows that the line integral depends only on **F** and the directed curve C, and not on the parametric representation chosen for C.

When we wish to denote a line integral without reference to a particular parametric function we shall write symbolically

$$I = \int_C \mathbf{F} \cdot d\mathbf{X}$$

$$= \int_C (F_1 \, dx_1 + F_2 \, dx_2 + \cdots + F_n \, dx_n).$$

Example 5.2 The functions

$$\mathbf{X} = \mathbf{X}(t) = \begin{bmatrix} \cos t \\ \sin t \end{bmatrix} \qquad (0 \le t \le \pi)$$

and

$$\mathbf{X} = \mathbf{Y}(\tau) = \begin{bmatrix} -\sin 2\tau \\ \cos 2\tau \end{bmatrix} \qquad \left(-\frac{\pi}{4} \le \tau \le \frac{\pi}{4}\right)$$

are equivalent representations of C, the upper half of the unit circle traversed from $(1, 0)$ to $(0, 1)$ (Fig. 5.3). If

$$\mathbf{F}(x, y) = \begin{bmatrix} \sqrt{x^2 + y^2} \\ \sqrt{1 - x^2} \end{bmatrix}$$

then

$$\mathbf{F}(\mathbf{X}(t)) = \begin{bmatrix} 1 \\ \sin t \end{bmatrix}$$

and

$$\mathbf{F}(\mathbf{Y}(\tau)) = \begin{bmatrix} 1 \\ \cos 2\tau \end{bmatrix}.$$

In terms of $\mathbf{X} = \mathbf{X}(t)$,

$$\int_C \mathbf{F} \cdot d\mathbf{X} = \int_0^\pi [(1)(-\sin t) + (\sin t) \cos t] \, dt$$

$$= (\cos t + \tfrac{1}{2} \sin^2 t) \Big|_0^\pi = -2,$$

while in terms of $\mathbf{X} = \mathbf{Y}(\tau)$

$$\int_C \mathbf{F} \cdot d\mathbf{X} = \int_{-\pi/4}^{\pi/4} \left[(1)(-2\cos 2\tau) + (\cos 2\tau)(-2\sin 2\tau)\right] d\tau$$

$$= -\left(\sin 2\tau + \frac{\sin^2 2\tau}{2}\right)\bigg|_{-\pi/4}^{\pi/4} = -2.$$

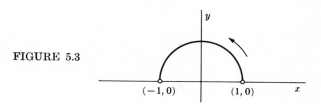

FIGURE 5.3

Theorem 5.1   Let $-C$ be the path obtained by reversing the order in which C is traversed. Then

$$\int_{-C} \mathbf{F} \cdot d\mathbf{X} = -\int_C \mathbf{F} \cdot d\mathbf{X}.$$

Proof.   If C is given by $\mathbf{X} = \mathbf{X}(t)$ $(a \leq t \leq b)$ then $-C$ is given by $\mathbf{X} = \mathbf{Y}(\tau) = \mathbf{X}(-\tau)$ $(-b \leq \tau \leq -a)$; hence

$$\int_{-C} \mathbf{F} \cdot d\mathbf{X} = \int_{-b}^{-a} \mathbf{F}(\mathbf{Y}(\tau)) \cdot \mathbf{Y}'(\tau) \, d\tau$$

$$= \int_{-b}^{-a} \mathbf{F}(\mathbf{X}(-\tau)) \cdot \frac{d}{d\tau} (\mathbf{X}(-\tau)) \, d\tau.$$

Since

$$\frac{d}{d\tau} (\mathbf{X}(-\tau)) = -\mathbf{X}'(-\tau),$$

we have

$$\int_{-C} \mathbf{F} \cdot d\mathbf{X} = -\int_{-b}^{-a} \mathbf{F}(\mathbf{X}(-\tau)) \cdot \mathbf{X}'(-\tau) \, d\tau,$$

and the change of variable $t = -\tau$ yields

$$\int_{-C} \mathbf{F} \cdot d\mathbf{X} = \int_b^a \mathbf{F}(\mathbf{X}(t)) \cdot \mathbf{X}'(t) \ dt$$

$$= -\int_a^b \mathbf{F}(\mathbf{X}(t)) \cdot \mathbf{X}'(t) \ dt$$

$$= -\int_C \mathbf{F} \cdot d\mathbf{X},$$

which completes the proof.

Example 5.3 Let $\mathbf{F}$ and C be as in Example 5.1. Then $-C$ is given by

$$\mathbf{X} = \mathbf{X}(-\tau) = \begin{bmatrix} -\tau \\ 2\tau^2 \\ -\tau^3 \end{bmatrix} \qquad (-1 \le \tau \le 0);$$

hence

$$\frac{d}{d\tau}\mathbf{X}(-\tau) = \begin{bmatrix} -1 \\ 4\tau \\ -3\tau^2 \end{bmatrix},$$

$$F(\mathbf{X}(-\tau)) = \begin{bmatrix} -\tau \\ \tau^4 + 1 \\ -2\tau^5 - \tau \end{bmatrix},$$

and

$$\int_{-C} \mathbf{F} \cdot d\mathbf{X} = \int_{-1}^0 [(-\tau)(-1) + (\tau^4 + 1)(4\tau) + (-2\tau^5 - \tau)(-3\tau^2)] \ d\tau$$

$$= \int_{-1}^0 (6\tau^7 + 4\tau^5 + 3\tau^3 + 5\tau) \ d\tau = -\tfrac{14}{3}$$

$$= -\int_C \mathbf{F} \cdot d\mathbf{X}.$$

The next theorem can be proved by parametrizing C and invoking properties of ordinary integrals (Exercise T-1).

Theorem 5.2 If α and β are real numbers and $\mathbf{F}$ and $\mathbf{G}$ are continuous vector fields whose domains contain C, then

$$\int_C (\alpha\mathbf{F} + \beta\mathbf{G}) \cdot d\mathbf{X} = \alpha \int_C \mathbf{F} \cdot d\mathbf{X} + \beta \int_C \mathbf{G} \cdot d\mathbf{X}.$$

It is sometimes convenient to subdivide a curve C into smaller segments; for example, a piecewise smooth curve can be divided into smooth parts (Fig. 5.4). Let C be given by $\mathbf{X} = \mathbf{X}(t)$ $(a \leq t \leq b)$ and partition $[a, b]$:

$$a = t_0 < t_1 < \cdots < t_n = b.$$

If $\mathbf{F}$ is a continuous vector field on C, then

(4)
$$\int_C \mathbf{F} \cdot d\mathbf{X} = \int_a^b \mathbf{F}(\mathbf{X}(t)) \cdot \mathbf{X}'(t)\, dt$$

$$= \sum_{i=1}^n \int_{t_{i-1}}^{t_i} \mathbf{F}(\mathbf{X}(t)) \cdot \mathbf{X}'(t)\, dt$$

$$= \sum_{i=1}^n \int_{C_i} \mathbf{F} \cdot d\mathbf{X},$$

where, for $i = 1, \ldots, n$, the curve C_i is given by

(5)
$$\mathbf{X} = \mathbf{X}(t) \qquad (t_{i-1} \leq t \leq t_i).$$

The terminal point of C_{i-1} coincides with the initial point of C_i. Symbolically we write

$$C = C_1 + C_2 + \cdots + C_n;$$

then (4) can be rewritten

$$\int_{C_1 + \ldots + C_n} \mathbf{F} \cdot d\mathbf{X} = \sum_{i=1}^n \int_{C_i} \mathbf{F} \cdot d\mathbf{X}.$$

Since the integrals on the right do not depend on the parametric representations chosen for $C_1, \ldots, C_n$, we need not use the representations (5) inherited from C to evaluate them; rather, we may choose any convenient representations for $C_1, \ldots, C_n$.

FIGURE 5.4

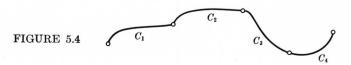

Example 5.4 Let C be the curve from $(0, 0)$ to $(1/\sqrt{2}, 1/\sqrt{2})$ along the parabola $y = \sqrt{2}x^2$, followed by the arc of the unit circle from $(1/\sqrt{2}, 1/\sqrt{2})$ to $(-1, 0)$ (Fig. 5.5). Then $C = C_1 + C_2$, where C_1 is given by

$$\mathbf{X} = \begin{bmatrix} t \\ \sqrt{2}t^2 \end{bmatrix} \qquad \left(0 \leq t \leq \frac{1}{\sqrt{2}}\right)$$

and C_2 by

$$\mathbf{X} = \begin{bmatrix} \cos t \\ \sin t \end{bmatrix} \qquad \left(\frac{\pi}{4} \leq t \leq \pi\right).$$

If

$$\mathbf{F} = \begin{bmatrix} P \\ Q \end{bmatrix}$$

then

$$\int_C \mathbf{F} \cdot d\mathbf{X} = \int_0^{1/\sqrt{2}} [P(t, \sqrt{2}t^2) + 2\sqrt{2}tQ(t, \sqrt{2}t^2)]\, dt$$

$$+ \int_{\pi/4}^{\pi} [-P(\cos t, \sin t) \sin t + Q(\cos t, \sin t) \cos t]\, dt.$$

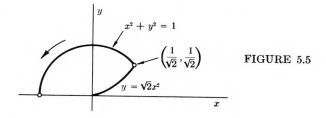

FIGURE 5.5

Independence of Path

Let $\mathbf{F}$ be a continuous vector field in a region D and suppose that C_1 and C_2 are two paths in D with common initial and terminal points $\mathbf{X}_1$ and $\mathbf{X}_2$ (Fig. 5.6). Our next objective is to develop conditions which guarantee that

(6)
$$\int_{C_1} \mathbf{F} \cdot d\mathbf{X} = \int_{C_2} \mathbf{F} \cdot d\mathbf{X}.$$

If this equation holds whenever C_1 and C_2 are two such curves in D, we say that *line integrals of* $\mathbf{F}$ *are independent of their paths* in D. This is not always the case, as is shown by the following example.

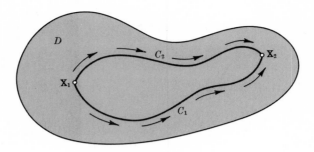

FIGURE 5.6

Example 5.5   Let C_1 be the line segment from $(0, 0)$ to $(1, 1)$ and C_2 be the section of the parabola $y = x^2$ connecting the same two points (Fig. 5.7). Thus C_1 is given by

$$\mathbf{X} = \begin{bmatrix} t \\ t \end{bmatrix} \qquad (0 \le t \le 1)$$

and C_2 by

$$\mathbf{X} = \begin{bmatrix} t \\ t^2 \end{bmatrix} \qquad (0 \le t \le 1).$$

If

$$\mathbf{F}(x, y) = \begin{bmatrix} x + y \\ y \end{bmatrix},$$

then

$$\int_{C_1} \mathbf{F} \cdot d\mathbf{X} = \int_0^1 \left[2t(1) + t(1) \right] dt$$

$$= \int_0^1 3t \, dt = \tfrac{3}{2}$$

and

$$\int_{C_2} \mathbf{F} \cdot d\mathbf{X} = \int_0^1 (t + t^2)(1) + t^2(2t) \,] \, dt$$

$$= \int_0^1 (t + t^2 + 2t^3) \, dt = \tfrac{4}{3};$$

hence (6) is not satisfied.

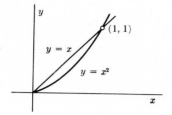

FIGURE 5.7

Example 5.6 Let C_1 and C_2 be as defined in Example 5.5 and

$$\mathbf{G}(x, y) = \begin{bmatrix} y \\ x \end{bmatrix}.$$

Then

$$\int_{C_1} \mathbf{G} \cdot d\mathbf{X} = \int_0^1 [t(1) + t(1)] \, dt$$

$$= \int_0^1 2t \, dt = 1$$

and

$$\int_{C_2} \mathbf{G} \cdot d\mathbf{X} = \int_0^1 [t^2(1) + t(2t)] \, dt$$

$$= \int_0^1 3t^2 \, dt = 1;$$

thus (6) is satisfied. This is not an accident, caused by a lucky choice of C_1 and C_2; if C is any smooth curve connecting $(0, 0)$ and $(1, 1)$ and parametrized by $\mathbf{X} = \mathbf{X}(t)$ $(a \le t \le b)$, then

$$\int_C \mathbf{G} \cdot d\mathbf{X} = \int_a^b [y(t)x'(t) + x(t)y'(t)] \, dt$$

$$= \int_a^b \frac{d}{dt} (y(t)x(t)) \, dt$$

$$= y(b)x(b) - y(a)x(a) = 1,$$

since $x(a) = y(a) = 0$ and $x(b) = y(b) = 1$.

The following definition is necessary for a careful discussion of conditions guaranteeing that line integrals of a vector field are independent of their paths.

Definition 5.2 A set D of points in R^n is said to be *connected* if any two points in D can be joined by a piecewise smooth curve which lies entirely in D. A *region* is an open connected set.

Prior to this we have used "region" in a much broader sense. Henceforth we shall use it in the sense of Definition 5.2.

Example 5.7 In R^2 the disk

$$D = \{(x, y) \mid x^2 + y^2 < 1\}$$

is a region. If we add to D its boundary, we obtain

$$\bar{D} = \{(x, y) \mid x^2 + y^2 \leq 1\},$$

which is still connected, but no longer open; hence $\bar{D}$ is not a region. The set consisting of the two open disks

$$D_1 = \{(x, y) \mid (x - 1)^2 + y^2 < \tfrac{1}{4}\}$$

and

$$D_2 = \{(x, y) \mid x^2 + y^2 < \tfrac{1}{4}\}$$

(Fig. 5.8) is open but not connected, since any curve joining points in D_1 and D_2 must contain at least one point which is in neither. If S_1 is obtained from S by adding the point $(\tfrac{1}{4}, 0)$, then S_1 is connected. However, S_1 is not open and therefore not a region.

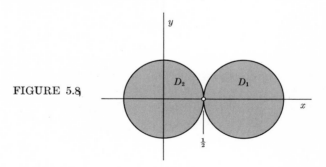

FIGURE 5.8

Theorem 5.3 Let $\mathbf{F}$ be a continuous vector field in a region D. Then the following statements are equivalent:

(a) For any two paths C_1 and C_2 in D with common initial and terminal points,

(7)
$$\int_{C_1} \mathbf{F} \cdot d\mathbf{X} = \int_{C_2} \mathbf{F} \cdot d\mathbf{X}.$$

(b) The vector field $\mathbf{F}$ is the gradient of a continuously differentiable real valued function f in D.

Proof. We first show that (b) implies (a). If $\mathbf{X} = \mathbf{X}(t)$ defines a smooth curve C in D and f is a real valued function such that

$$\mathbf{F} = \nabla f = \begin{bmatrix} \dfrac{\partial f}{\partial x_1} \\ \vdots \\ \dfrac{\partial f}{\partial x_n} \end{bmatrix},$$

then

$$\int_C \mathbf{F} \cdot d\mathbf{X} = \int_a^b \left[F_1(\mathbf{X}(t)) x_1'(t) + \cdots + F_n(\mathbf{X}(t)) x_n'(t) \right] dt$$

$$= \int_a^b \left[\frac{\partial f}{\partial x_1} (\mathbf{X}(t)) x_1'(t) + \cdots + \frac{\partial f}{\partial x_n} (\mathbf{X}(t)) x_n'(t) \right] dt$$

$$= \int_a^b \frac{d}{dt} (f(\mathbf{X}(t))) \, dt = f(\mathbf{X}(b)) - f(\mathbf{X}(a)).$$

If $\mathbf{U}$ and $\mathbf{V}$ are the initial and terminal points of C then $\mathbf{X}(a) = \mathbf{U}$ and $\mathbf{X}(b) = \mathbf{V}$; hence

$$(8) \qquad \int_C \mathbf{F} \cdot d\mathbf{X} = f(\mathbf{V}) - f(\mathbf{U}).$$

This result also holds if C is piecewise smooth but not smooth, for then there are points $a = t_0 < t_1 < \cdots < t_n = b$ such that $\mathbf{X}'$ is continuous in $[t_{i-1}, t_i]$ $(i = 1, \ldots, n)$ (continuous from the interior at the endpoints). Then $(d/dt)f(\mathbf{X}(t))$ is also continuous in these subintervals and

$$\int_{t_{i-1}}^{t_i} \frac{d}{dt} f(\mathbf{X}(t)) \, dt = f(\mathbf{X}(t_i)) - f(\mathbf{X}(t_{i-1}));$$

hence

$$\int_C \mathbf{F} \cdot d\mathbf{X} = \sum_{i=1}^n \int_{t_{i-1}}^{t_i} \frac{d}{dt} f(\mathbf{X}(t)) \, dt$$

$$= \sum_{i=1}^n (f(\mathbf{X}(t_i)) - f(\mathbf{X}(t_{i-1})))$$

$$= f(\mathbf{X}(t_n)) - f(\mathbf{X}(t_0)) = f(\mathbf{V}) - f(\mathbf{U}).$$

Thus (8) holds for all piecewise smooth curves in D. It follows that (7) holds if C_1 and C_2 have common endpoints.

Now suppose (a) holds. Let $\mathbf{X}_0$ be fixed and $\mathbf{X}$ arbitrary in D, and define

$$(9) \qquad\qquad f(\mathbf{X}) = \int_C \mathbf{F} \cdot d\mathbf{X},$$

where C is any piecewise smooth curve in D with initial point $\mathbf{X}_0$ and terminal point $\mathbf{X}$. The assumption is that all such integrals are the same; hence (9) defines a function on D which we denote by

$$f(\mathbf{X}) = \int_{\mathbf{X}_0}^{\mathbf{X}} \mathbf{F} \cdot d\mathbf{X}.$$

Now let $\mathbf{X} = (x_1, \ldots, x_n)$ and $\hat{\mathbf{X}} = (x_1 + \Delta x_1, x_2, \ldots, x_n)$ and suppose that Δx_1 is so small that the line segment L from $\mathbf{X}$ to $\hat{\mathbf{X}}$ is in D. If C is an arbitrary path from $\mathbf{X}_0$ to $\mathbf{X}$ then $C + L$ is a path from $\mathbf{X}_0$ to $\hat{\mathbf{X}}$ (Fig. 5.9 depicts the situation for $n = 2$) and therefore

$$(10) \qquad\qquad f(\hat{\mathbf{X}}) = \int_{\mathbf{X}_0}^{\mathbf{X}} \mathbf{F} \cdot d\mathbf{X} + \int_L \mathbf{F} \cdot d\mathbf{X}$$

$$= f(\mathbf{X}) + \int_L \mathbf{F} \cdot d\mathbf{X}.$$

Since L can be represented by

$$\mathbf{X}(t) = \begin{bmatrix} x_1 + t\,\Delta x_1 \\ x_2 \\ \vdots \\ x_n \end{bmatrix} \qquad (0 \le t \le 1),$$

with

$$\mathbf{X}'(t) = \begin{bmatrix} \Delta x_1 \\ 0 \\ \vdots \\ 0 \end{bmatrix},$$

(10) implies that

$$f(\hat{\mathbf{X}}) - f(\mathbf{X}) = \Delta x_1 \int_0^1 F_1(x_1 + t\,\Delta x_1, x_2, \ldots, x_n)\, dt.$$

From the continuity of F_1, this integral approaches $F_1(\mathbf{X})$ as Δx_1 approaches zero (Exercise T-3); hence

$$\lim_{\Delta x_1 \to 0} \frac{f(\hat{\mathbf{X}}) - f(\mathbf{X})}{\Delta x_1} = F_1(\mathbf{X})$$

and therefore

$$\frac{\partial f}{\partial x_1}(\mathbf{X}) = F_1(\mathbf{X}).$$

The same argument applied to the other variables implies that

$$\frac{\partial f}{\partial x_i}(\mathbf{X}) = F_i(\mathbf{X}) \qquad (i = 1, \ldots, n);$$

therefore (b) follows.

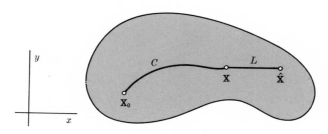

FIGURE 5.9

Conservative Fields; Potential Functions

Let us recall from Section 3.4 that a vector field $\mathbf{F}$ is *conservative* in a region D if $\mathbf{F} = \nabla f$ where f is a continuously differentiable real valued function on D; f is the *scalar field* and $-f$ the *potential function* associated with $\mathbf{F}$.

If $\mathbf{F}$ is conservative, then, formally,

$$\mathbf{F} \cdot d\mathbf{X} = F_1 \, dx_1 + \cdots + F_n \, dx_n$$

$$= \frac{\partial f}{\partial x_1} \, dx_1 + \cdots + \frac{\partial f}{\partial x_n} \, dx_n = df;$$

thus $\mathbf{F} \cdot d\mathbf{X}$ is formally the differential of f. For this reason we say that $\mathbf{F} \cdot d\mathbf{X}$ is an *exact differential* if $\mathbf{F}$ is conservative.

The next theorem follows from these definitions and the proof of Theorem 5.3.

Theorem 5.4 If $\mathbf{F}$ is a conservative vector field on a region D, f the associated scalar field, and C a path in D with endpoints $\mathbf{X}_1$ and $\mathbf{X}_2$, then

$$\int_C \mathbf{F} \cdot d\mathbf{X} = f(\mathbf{X}_2) - f(\mathbf{X}_1);$$

in particular, the integral of $\mathbf{F}$ around any closed path in D is zero.

Example 5.8   We saw in Section 3.4 that the inverse square law force field,

$$\mathbf{F}(\mathbf{X}) = \frac{-Gm_0}{|\mathbf{X}|^3}\mathbf{X},$$

is the gradient of

$$f(\mathbf{X}) = \frac{Gm_0}{|\mathbf{X}|}.$$

Thus $-f(\mathbf{X})$ is the *potential energy* of a unit mass at $\mathbf{X}$. Theorem 5.4 states that the work done by gravity on a unit mass which moves along a path C in this field is equal to the change in the potential energy of the particle.

The following example illustrates the significance of the adjective *conservative*.

Example 5.9   Let $\mathbf{X} = \mathbf{X}(t)$ be the position of a particle of mass m as a function of time, and let $\mathbf{X}'(t)$ be its velocity vector. Suppose that, as t varies from t_0 to t_1, the particle moves from $\mathbf{X}_0$ to $\mathbf{X}_1$ along a path C which lies in a domain D in R^3, in which a conservative force field $\mathbf{F}$ is defined. From Theorem 5.4 the work done on the particle by $\mathbf{F}$ is

$$(11) \qquad W = \int_C \mathbf{F}\cdot d\mathbf{X} = f(\mathbf{X}_1) - f(\mathbf{X}_0),$$

where $\mathbf{F} = \nabla f$. On the other hand, using Newton's second law of motion,

$$\mathbf{F}(\mathbf{X}(t)) = m\mathbf{X}''(t),$$

we can write

$$W = \int_{t_0}^{t_1} \mathbf{F}(\mathbf{X}(t))\cdot\mathbf{X}'(t)\,dt$$

$$= m\int_{t_0}^{t_1} \mathbf{X}''(t)\cdot\mathbf{X}'(t)\,dt$$

$$= \frac{m}{2}\int_{t_0}^{t_1} \frac{d}{dt}(\mathbf{X}'(t)\cdot\mathbf{X}'(t))\,dt$$

$$= \frac{m}{2}\int_{t_0}^{t_1} \frac{d}{dt}|\mathbf{X}'(t)|^2\,dt$$

$$= \frac{m}{2}|\mathbf{X}'(t_1)|^2 - \frac{m}{2}|\mathbf{X}'(t_0)|^2.$$

Comparing this with (11), we see that

$$(12) \qquad \frac{m}{2} \mid \mathbf{X}'(t_1) \mid^2 - f(\mathbf{X}_1) = \frac{m}{2} \mid \mathbf{X}'(t_0) \mid^2 - f(\mathbf{X}_0).$$

The quantity $-f(\mathbf{X})$ is the potential energy of the particle at $\mathbf{X}$, while $(m/2) \mid \mathbf{X}'(t) \mid^2$ is its kinetic energy. From (12) it follows that their sum, the total energy, remains constant as the particle moves through a conservative force field; that is, energy is *conserved*.

Theorem 5.3 says that a vector field $\mathbf{F}$ is conservative if and only if it has a potential function; however, it does not specify how to find a potential function, nor does it provide a practical test for actually deciding whether one exists. The following theorem gives necessary conditions for a continuously differentiable vector field to be conservative.

Theorem 5.5   If

$$\mathbf{F} = \begin{bmatrix} F_1 \\ \vdots \\ F_n \end{bmatrix}$$

is a continuously differentiable, conservative vector field in a region D in R^n, then

$$(13) \qquad \frac{\partial F_i}{\partial x_j} = \frac{\partial F_j}{\partial x_i} \qquad (i, j = 1, \dots, n).$$

In particular, if $n = 2$, and

$$\mathbf{F} = \begin{bmatrix} P \\ Q \end{bmatrix},$$

then

$$(14) \qquad \frac{\partial P}{\partial y} = \frac{\partial Q}{\partial x} \, ;$$

if $n = 3$ and

$$\mathbf{F} = \begin{bmatrix} P \\ Q \\ R \end{bmatrix},$$

then

(15)
$$\frac{\partial P}{\partial y} = \frac{\partial Q}{\partial x}, \quad \frac{\partial Q}{\partial z} = \frac{\partial R}{\partial y}, \quad \frac{\partial R}{\partial x} = \frac{\partial P}{\partial z}.$$

Proof. If $n = 2$ and $\mathbf{F} = \nabla f$ then

$$P = \frac{\partial f}{\partial x}$$

and

$$Q = \frac{\partial f}{\partial y}.$$

Therefore

(16)
$$\frac{\partial P}{\partial y} = \frac{\partial^2 f}{\partial y \, \partial x}$$

and

(17)
$$\frac{\partial Q}{\partial x} = \frac{\partial^2 f}{\partial x \, \partial y}.$$

Since $\partial P / \partial y$ and $\partial Q / \partial x$ are continuous, the mixed second partial derivatives in (16) and (18) are equal (Section 4.2) and (14) follows.

The idea of the proof is the same for $n > 2$. If $\mathbf{F} = \nabla f$ and $1 \leq i < j \leq n$ then

(18)
$$F_i = \frac{\partial f}{\partial x_i}$$

and

(19)
$$F_j = \frac{\partial f}{\partial x_j};$$

differentiating (18) with respect to x_j and (19) with respect to x_i yields

$$\frac{\partial F_i}{\partial x_j} = \frac{\partial^2 f}{\partial x_j \, \partial x_i}$$

and

$$\frac{\partial F_j}{\partial x_i} = \frac{\partial^2 f}{\partial x_i \, \partial x_j},$$

and again the continuity of the $\partial F_i / \partial x_j$ and $\partial F_j / \partial x_i$ implies that the mixed second partial derivatives of f are equal. Thus (13) follows.

Example 5.10   Let

$$\mathbf{F}(x, y) = \begin{bmatrix} y \\ x \end{bmatrix};$$

then $P(x, y) = y$ and $Q(x, y) = x$, so that

$$\frac{\partial P}{\partial y} = \frac{\partial Q}{\partial x} = 1.$$

Thus $\mathbf{F}$ satisfies the necessary conditions for a conservative vector field and therefore *may* have a potential function. It does, since

$$\begin{bmatrix} y \\ x \end{bmatrix} = \boldsymbol{\nabla} f,$$

where

$$f(x, y) = xy.$$

Example 5.11   The inverse square law field of Example 5.8 satisfies (15) if $\mathbf{X} \neq \mathbf{0}$ (Verify) ; we exhibited its scalar field in Example 5.7.

We emphasize that Theorem 5.5 gives necessary, but not sufficient, conditions for a vector field to be conservative. The following example shows that a vector field may satisfy these conditions but fail to be conservative.

Example 5.12   The vector field

$$\mathbf{F}(x, y) = \begin{bmatrix} \dfrac{-y}{x^2 + y^2} \\ \dfrac{x}{x^2 + y^2} \end{bmatrix}$$

satisfies (14) in the domain D consisting of all $(x, y) \neq (0, 0)$; however, if C is the unit circle,

$$\mathbf{X} = \begin{bmatrix} \cos t \\ \sin t \end{bmatrix} \qquad (0 \leq t \leq 2\pi),$$

then

$$\int_C \mathbf{F} \cdot d\mathbf{X} = \int_0^{2\pi} [(-\sin t)(-\sin t) + (\cos t)(\cos t)] \, dt = 2\pi.$$

Therefore $\mathbf{F}$ is not conservative in D, by Theorem 5.4.

The conditions of Theorem 5.5 are also sufficient if the region D is simply connected, which we now define.

Definition 5.3   Consider a closed curve C in a region D to be constructed of a loop of elastic material. If it is possible to deform C to a single point in D without breaking the loop or leaving D, then we say that C *can be contracted to a point within* D. A region D is *simply connected* if any closed curve in D has this property.

This definition is intuitive, but adequate for our purposes; to formulate it in an entirely precise and rigorous fashion would take us beyond the scope of this book.

Example 5.13   In R^2 the interiors of a circle and a half plane are simply connected, as is R^2 itself. However the exterior of the unit disc is not. In fact no domain in R^2 with a "hole" in it is simply connected (Fig. 10); thus a region in R^2 obtained by removing a single point from a simply connected region is not simply connected.

Simply Connected   Not Simply Connected

FIGURE 5.10

Example 5.14   In R^n $(n > 3)$ the region between two concentric n-spheres, say

$$D = \{\mathbf{X} \mid 1 < |\mathbf{X}| < 2\}$$

is simply connected.

Example 5.15   In R^3 the region between two cylinders,

$$D = \{(x, y, z) \mid 1 < x^2 + y^2 < 4, \quad 0 < z < 1\},$$

is not simply connected, since a closed curve C which encloses the axis of D cannot be contracted to a point within D (Fig. 5.11).

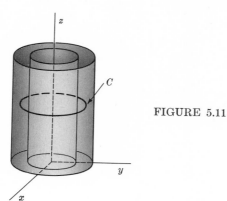

FIGURE 5.11

We state the following useful theorem without proof.

Theorem 5.6 If $\mathbf{F}$ satisfies the hypotheses of Theorem 5.5 in a simply connected domain D, then $\mathbf{F}$ is conservative in D.

Example 5.16 We saw in Example 5.12 that

$$\mathbf{F}(x, y) = \begin{bmatrix} \dfrac{-y}{x^2 + y^2} \\[2ex] \dfrac{x}{x^2 + y^2} \end{bmatrix} \qquad ((x, y) \neq (0, 0))$$

is not conservative in its domain of definition (which is not simple connected). However, Theorem 5.6 guarantees that $\mathbf{F}$ is conservative in any simply connected region which does not contain the origin.

Example 5.17 To see that $\mathbf{F}$ may be conservative even if D is not simply connected, consider

$$\mathbf{F}(x, y) = \begin{bmatrix} \dfrac{-x}{(x^2 + y^2)^{3/2}} \\[2ex] \dfrac{-y}{(x^2 + y^2)^{3/2}} \end{bmatrix} \qquad ((x, y) \neq (0, 0)),$$

whose domain D is again all of R^2 except the origin. It is conservative in D

since $\mathbf{G} = \nabla g$, where

$$g(x, y) = \frac{1}{\sqrt{x^2 + y^2}}.$$

Theorem 5.6 has the following consequence in R^2.

Theorem 5.7   Let a vector field

$$\mathbf{F} = \begin{bmatrix} P \\ Q \end{bmatrix}$$

be continuously differentiable and satisfy

$$\frac{\partial P}{\partial y} = \frac{\partial Q}{\partial x}$$

in a domain D in R^2 which contains two simple closed curves C_1 and C_2, situated as shown in Fig. 5.12, and the region S bounded by them. Then

$$\int_{C_1} P \, dx + Q \, dy = \int_{C_2} P \, dx + Q \, dy,$$

where both curves are traversed counterclockwise.

Proof.   Divide S into simply connected subregions S_1 and S_2 by inserting line segments L_1 and L_2 as shown in Fig. 5.13. By Theorem 5.6,

$$\int_{\Gamma_1} P \, dx + Q \, dy = \int_{\Gamma_2} P \, dx + Q \, dy = 0,$$

where Γ_1 and Γ_2 are the boundaries of S_1 and S_2, traversed counterclock-

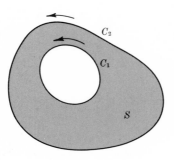

FIGURE 5.12

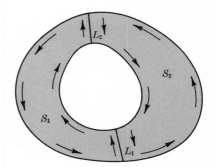

FIGURE 5.13

wise. Hence

(20)
$$0 = \int_{\Gamma_1} P\, dx + Q\, dy + \int_{\Gamma_2} P\, dx + Q\, dy$$

$$= \int_{C_2} P\, dx + Q\, dy - \int_{C_1} P\, dx + Q\, dy$$

(Exercise T-4), which completes the proof.

Example 5.18 Let

$$\mathbf{F}(x, y) = \begin{bmatrix} P(x, y) \\ Q(x, y) \end{bmatrix} = \begin{bmatrix} y^2 \\ 2xy \end{bmatrix};$$

since $\partial P/\partial y = 2y = \partial Q/\partial x$ on all of R^2, which is simply connected, the value of the integral

$$I = \int_{(0,0)}^{(2,1)} y^2\, dx + 2xy\, dy$$

does not depend upon the path of integration. The integral could be evaluated by means of Theorem 5.4, if we wished to find the potential function associated with $\mathbf{F}$. However, this is not necessary; we can choose any convenient path. Let us integrate along the line segment L connecting $(0, 0)$ and $(2, 1)$ (Fig. 5.14). We parametrize L as

$$\mathbf{X}(t) = t \begin{bmatrix} 2 \\ 1 \end{bmatrix} \qquad (0 < t < 1);$$

that is, $x(t) = 2t$ and $y(t) = t$. Thus

$$I = \int_0^1 \left[(t^2)(2) + (2)(2t)(t)(1) \right] dt$$

$$= 6 \int_0^1 t^2\, dt = 2.$$

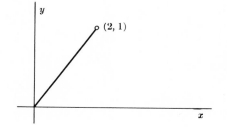

FIGURE 5.14

Finding the Scalar Field of a Conservative Vector Field

We now discuss a formal method for finding the scalar field associated with a conservative vector field.

If P is a real valued function defined on a region D in R^2, then we write

$$P_1(x, y) = \int P(x, y) \, dx$$

to indicate that P_1 is also defined on D and satisfies

$$(21) \qquad\qquad \frac{\partial P_1}{\partial x} = P;$$

thus P_1 is the "partial" indefinite integral of P with respect to x and is determined by holding y constant and integrating P as a function of x. It can be shown that if P_1 and P_2 are both continuously differentiable indefinite integrals (with respect to x) of P, then

$$(22) \qquad\qquad P_2(x, y) = P_1(x, y) + g(y),$$

where g is continuously differentiable with respect to y. Conversely, if P_1 satisfies (21), then the most general indefinite integral of P with respect to x is of the form (22); thus g is analogous to the constant of integration for ordinary indefinite integrals.

If P is a function of $\mathbf{X} = (x_1, \ldots, x_n)$, then P_1 is *an indefinite integral of P with respect to x_i* if

$$\frac{\partial P_1}{\partial x_i} = P;$$

the most general indefinite integral of P with respect to x_i is

$$P_2(\mathbf{X}) = P_1(\mathbf{X}) + g(x_1, \ldots, x_{i-1}, x_{i+1}, \ldots, x_n).$$

Now suppose we wish to express

$$\mathbf{F} = \begin{bmatrix} P \\ Q \end{bmatrix}$$

in the form $\mathbf{F} = \nabla f$ in some region in R^2. Of course, we must first check to see that

$$(23) \qquad\qquad \frac{\partial P}{\partial y} = \frac{\partial Q}{\partial x}.$$

If this is so, let

$$P_1(x, y) = \int P(x, y) \, dx$$

be any indefinite integral of P with respect to x; since

$$\frac{\partial f}{\partial x} = P,$$

it follows that

$$f(x, y) = P_1(x, y) + g(y),$$

where g satisfies

$$\frac{\partial f}{\partial y}(x, y) = \frac{\partial P_1}{\partial y}(x, y) + g'(y) = Q(x, y),$$

or

(24)
$$g'(y) = Q(x, y) - \frac{\partial P_1}{\partial y}(x, y).$$

Assumption (23) enters into the argument at this point, since it guarantees that the right side of (25) which seems to depend on x, is actually independent of it. To see this, consider

$$\frac{\partial}{\partial x}\left(Q - \frac{\partial P_1}{\partial y} \right) = \frac{\partial Q}{\partial x} - \frac{\partial}{\partial x}\left(\frac{\partial P_1}{\partial y} \right)$$

$$= \frac{\partial Q}{\partial x} - \frac{\partial^2 P_1}{\partial x \, \partial y}$$

$$= \frac{\partial Q}{\partial x} - \frac{\partial^2 P_1}{\partial y \, \partial x}$$

$$= \frac{\partial Q}{\partial x} - \frac{\partial}{\partial y}\left(\frac{\partial P_1}{\partial x} \right)$$

$$= \frac{\partial Q}{\partial x} - \frac{\partial P}{\partial y} = 0.$$

Example 5.19   Let

$$\mathbf{F}(x, y) = \begin{bmatrix} P(x, y) \\ Q(x, y) \end{bmatrix} = \begin{bmatrix} xy^2 + x^2 \\ x^2y + y^2 \end{bmatrix};$$

since

$$\frac{\partial P}{\partial y} = 2xy = \frac{\partial Q}{\partial x}$$

for all (x, y), there is a scalar field f such that

(25)
$$\frac{\partial f}{\partial x}(x, y) = xy^2 + x^2$$

and

(26)
$$\frac{\partial f}{\partial y}(x, y) = x^2y + y^2$$

(Theorem 5.6). Integrating both sides of (25) with respect to x (treating y as a constant) yields

$$f(x, y) = \frac{x^2y^2}{2} + \frac{x^3}{3} + g(y).$$

From this,

$$\frac{\partial f}{\partial y}(x, y) = x^2y + g'(y),$$

which, when compared with (26), yields

$$g'(y) = y^2.$$

Therefore

$$g(y) = \frac{y^3}{3} + C,$$

where C is a constant, and

$$f(x, y) = \frac{x^2y^2}{2} + \frac{x^3}{3} + \frac{y^3}{3} + C.$$

It is easy to verify that $\mathbf{F} = \nabla f$.

We emphasize that this is a formal method. It can happen that the indefinite integrals P_1 and g are not defined in a particular region. Indeed, we must expect the method to break down in some cases, since it assumes only the hypotheses of Theorem 5.5 (for $n = 2$), which are not sufficient to guarantee the existence of a potential function in an arbitrary region in R^2.

An analogous procedure is available for $n > 2$. We give an example for $n = 3$.

Example 5.20 Let $n = 3$ and

$$\mathbf{F}(\mathbf{X}) = \begin{bmatrix} P(x, y, z) \\ Q(x, y, z) \\ R(x, y, z) \end{bmatrix} = \begin{bmatrix} x + z \\ y + z \\ x + y \end{bmatrix};$$

then

$$\frac{\partial P}{\partial y} = \frac{\partial Q}{\partial x} = 0, \qquad \frac{\partial Q}{\partial z} = \frac{\partial R}{\partial y} = 1,$$

and

$$\frac{\partial R}{\partial x} = \frac{\partial P}{\partial z} = 1$$

in R^3. Theorem 5.6 implies that there is a function $f \colon R^3 \to R$ such that

(27)
$$\frac{\partial f}{\partial x}(\mathbf{X}) = x + z,$$

(28)
$$\frac{\partial f}{\partial y}(\mathbf{X}) = y + z,$$

and

(29)
$$\frac{\partial f}{\partial z}(\mathbf{X}) = x + y.$$

Integrating (27) with respect to x yields

(30)
$$f(\mathbf{X}) = \frac{x^2}{2} + xz + g(y, z),$$

where g is a "constant of integration"; that is, a function which does not depend upon x. Differentiating (30) with respect to y yields

$$\frac{\partial f}{\partial y}(\mathbf{X}) = \frac{\partial g}{\partial y}(y, z);$$

thus, from (28)

$$\frac{\partial g}{\partial y}(y, z) = y + z$$

and

$$g(y, z) = \frac{y^2}{2} + zy + h(z).$$

Substituting this into (30) yields

$$f(\mathbf{X}) = \frac{x^2}{2} + xz + \frac{y^2}{2} + yz + h(z),$$

and therefore

$$\frac{\partial f(\mathbf{X})}{\partial z} = x + y + h'(z).$$

From (29), $h'(z) = 0$ and h is constant; thus

$$f(\mathbf{X}) = \frac{x^2}{2} + xz + \frac{y^2}{2} + yz + C,$$

where C is an arbitrary constant.

Line Integrals with Respect to Arc Length

These is another type of line integral that we have not yet mentioned.

Definition 5.4   Let C be a smooth curve in R^n, given parametrically by $\mathbf{X} = \mathbf{X}(t)$ $(a \le t \le b)$, and suppose f is a continuous scalar field defined on C. Then the integral

$$\int_a^b f(\mathbf{X}(t)) \, | \, \mathbf{X}'(t) \, | \, dt$$

is called the *line integral of f, with respect to arc length, along* C. It is denoted by

(31) $$\int_C f \, ds.$$

If C is a piecewise smooth curve with smooth segment $C_1, \ldots, C_n$, we define

$$\int_C f \, ds = \int_{C_1} f \, ds + \cdots + \int_{C_n} f \, ds.$$

   In view of the notation (31), the student should not be surprised to learn that the value of this new line integral is independent of the parametrization chosen for C. We leave it to the student to verify this (Exercise T-7), and also that

$$\int_C f \, ds = \int_{-C} f \, ds$$

(Exercise T-8).

If a thin wire is represented as a curve in R^3 with mass density (mass per unit length) function $\rho = \rho(x, y, z)$, then its mass is

$$m = \int_C \rho \, ds.$$

The moments, center of gravity, and moments and products of inertia of the wire can be defined and computed as in Section 4.4.

Example 5.21   Let a wire be in the shape of a helical curve C defined by

$$\mathbf{X}(t) = 2 \cos t \, \mathbf{i} + 2 \sin t \, \mathbf{j} + \mathbf{k} \qquad \left(0 \le t \le \frac{\pi}{2} \right),$$

with mass density function $\rho(x, y, z) = ax$. Then

$$| \mathbf{X}'(t) | = | -2 \sin t \, \mathbf{i} + 2 \cos t \, \mathbf{j} | = 2,$$

and the mass of the wire is

$$m = \int_C \rho \, ds = a \int_C x \, ds$$

$$= a \int_0^{\pi/2} (2 \cos t) 2 \, dt = 4a.$$

The first moments with respect to the coordinate planes are

$$M_{yz} = \int_C x \rho \, ds = a \int_C x^2 \, ds$$

$$= a \int_0^{\pi/2} (2 \cos t)^2 (2) \, dt = 2a\pi,$$

$$M_{zx} = \int_C y \rho \, ds = a \int_C xy \, ds$$

$$= a \int_0^{\pi/2} (2 \cos t) (2 \sin t) (2) \, dt = 4a,$$

and

$$M_{xy} = \int_C z \rho \, ds = a \int_C zx \, ds$$

$$= a \int_0^{\pi/2} (1) (2 \cos t) (2) \, dt = 4a.$$

The coordinates of the center of mass are

$$\bar{x} = \frac{M_{yz}}{m} = \frac{\pi}{2}, \qquad \bar{y} = \frac{M_{zx}}{m} = 1, \qquad \bar{z} = M_{xy} = 1.$$

We leave the computation of the moments and products of inertia to the student.

The line integral of a vector function $\mathbf{F}$ over a curve C can also be viewed as an integral, with respect to arc length, of a certain scalar valued function. If C has the smooth representation $\mathbf{X} = \mathbf{X}(t)$ $(a \leq t \leq b)$, we can assign to every $\mathbf{X}$ on C a unit tangent vector $\mathbf{T}(\mathbf{X})$, defined by

$$\mathbf{T}(\mathbf{X}(t)) = \frac{\mathbf{X}'(t)}{|\mathbf{X}(t)|}.$$

This vector function is independent of the parametric representation chosen for C, except for sign. If $\mathbf{F}$ is a continuous vector valued function on C, then

$$\int_C \mathbf{F} \cdot d\mathbf{X} = \int_a^b \mathbf{F}(\mathbf{X}(t)) \cdot \mathbf{X}'(t) \, dt$$

$$= \int_a^b \mathbf{F}(\mathbf{X}(t)) \cdot \frac{\mathbf{X}'(t)}{|\mathbf{X}'(t)|} \, |\mathbf{X}'(t)| \, dt,$$

which can be rewritten as

$$\int_C \mathbf{F} \cdot d\mathbf{X} = \int_C (\mathbf{F} \cdot \mathbf{T}) \, ds.$$

EXERCISES 6.5

In Exercises 1 through 9 compute the line integral of $\mathbf{F}$ along the given path.

1. $\mathbf{F}(\mathbf{X}) = \begin{bmatrix} x^2 + y \\ y^2 - x \end{bmatrix}$; $\mathbf{X} = \begin{bmatrix} t^2 \\ t^3 \end{bmatrix}$ $(0 \leq t \leq 1)$.

2. $\mathbf{F}(\mathbf{X}) = \begin{bmatrix} xy + z \\ yz - x \\ xyz + x^2 \end{bmatrix}$; $\mathbf{X} = \begin{bmatrix} t \\ t \\ t^2 \end{bmatrix}$ $(0 \leq t \leq 2)$.

3. $\mathbf{F}(\mathbf{X}) = \begin{bmatrix} xy \\ y^2 - x^2 \end{bmatrix}$; $y = x^2 \quad (-1 \le x \le 1)$.

4. $\mathbf{F}(\mathbf{X}) = \begin{bmatrix} x + 1 \\ y^2 - x \end{bmatrix}$ from $(0, 2)$ to $(2, 0)$ along the arc of the circle $x^2 + y^2 = 4$ in the first quadrant.

5. $\mathbf{F}(\mathbf{X}) = \begin{bmatrix} x + y \\ xy \\ z^2 \end{bmatrix}$ along the line segment joining $(0, 0, 0)$ and $(1, 2, 1)$.

6. $\mathbf{F}(\mathbf{X}) = \begin{bmatrix} x^2 y + x \\ x^2 - y \end{bmatrix}$ along the path consisting of $y = x^2$ from $(0, 0)$ to $(1, 1)$ and $x = y^2$ from $(1, 1)$ to $(0, 0)$.

7. $\mathbf{F}(\mathbf{X}) = \begin{bmatrix} x^2 y \\ y + z^2 \\ xz \end{bmatrix}$ along the path consisting of the line segment from $(0, 0, 0)$ to $(1, 2, 1)$, followed by the line segment from $(1, 2, 1)$ to $(2, 3, 4)$, followed by the line segment from $(2, 3, 4)$ to $(4, 2, 1)$.

8. $\mathbf{F}(\mathbf{X}) = \begin{bmatrix} x^2 - 3y \\ 2xy^2 \end{bmatrix}$ along the parabola $y = x^2 + 1$ from $(0, 1)$ to $(2, 5)$.

9. $\mathbf{F}(\mathbf{X}) = \begin{bmatrix} x^2 + y \\ x^2 + y^2 \end{bmatrix}$ counterclockwise around the ellipse $\dfrac{x^2}{a^2} + \dfrac{y^2}{b^2} = 1$.

In Exercises 10 through 15 compute the work done by the force $\mathbf{F}$ in moving a particle along the given path.

10. $\mathbf{F}(\mathbf{X}) = \begin{bmatrix} x + 2y \\ x - y \end{bmatrix}$; $\mathbf{X} = \begin{bmatrix} 2 \cos t \\ 3 \sin t \end{bmatrix}$ $\left(0 \le t \le \dfrac{\pi}{4} \right)$.

11. $\mathbf{F}(\mathbf{X}) = \begin{bmatrix} x - y \\ x + z \\ y - z \end{bmatrix}$; $\mathbf{X} = \begin{bmatrix} a \cos t \\ a \sin t \\ bt \end{bmatrix}$ $\left(0 \le t \le \dfrac{\pi}{3} \right)$.

12. $\mathbf{F}(\mathbf{X}) = \begin{bmatrix} xy + x^2 \\ x^2y - y \end{bmatrix}$; $y = \dfrac{1}{x}$ $(1 \leq x \leq 3)$.

13. $\mathbf{F}(\mathbf{X}) = \begin{bmatrix} xz + y^2 \\ yz - x^2 \\ xy - z^2 \end{bmatrix}$ along the path consisting of the line segment from $(1, 2, 3)$ to $(4, 3, 5)$, followed by the line segment from $(4, 3, 5)$ to $(1, 1, 1)$.

14. $\mathbf{F}(\mathbf{X}) = \begin{bmatrix} x^2 - y^2 + y \\ y^2 + x^2 - x \end{bmatrix}$ along the circular arc $y = \sqrt{4 - x^2}$ from $(2, 0)$ to $(1, \sqrt{3})$, followed by the line segment from $(1, \sqrt{3})$ to $(0, 0)$.

15. $\mathbf{F}(\mathbf{X}) = \begin{bmatrix} x^2y^2 - 2x \\ y + x \end{bmatrix}$ along the parabola $x = 4 - y^2$ from $(0, 2)$ to $(4, 0)$.

16. Show that

$$\mathbf{\Phi}(t) = \begin{bmatrix} t \\ t^2 \\ t^3 \end{bmatrix} \quad (1 \leq t \leq 2)$$

and

$$\mathbf{\Psi}(s) = \begin{bmatrix} s^2 \\ s^4 \\ s^6 \end{bmatrix} \quad (1 \leq s \leq \sqrt{2})$$

are equivalent parametric functions, and let C be the curve that they represent. Evaluate $\displaystyle\int_C \mathbf{F} \cdot \mathbf{dX}$, using both parametrizations, for

$$\mathbf{F}(\mathbf{X}) = \begin{bmatrix} xy + z^2 \\ x^2y - z^2 \\ yz \end{bmatrix}.$$

17.   Repeat Exercise 16 with

$$\mathbf{\Phi}(t) = \begin{bmatrix} e^t \\ e^{-t} \\ 2\cosh t \end{bmatrix} \quad (0 \leq t \leq 1),$$

$$\mathbf{\Psi}(s) = \begin{bmatrix} s \\ \dfrac{1}{s} \\ s + \dfrac{1}{s} \end{bmatrix} \quad (1 \leq s \leq e),$$

and

$$\mathbf{F}(\mathbf{X}) = \begin{bmatrix} x^2 + xy \\ y^2 + xz \\ z^2 + yz \end{bmatrix}.$$

In Exercises 18 through 25 determine whether $\mathbf{F}$ is conservative. If it is conservative find a potential function.

18. $\mathbf{F}(\mathbf{X}) = \begin{bmatrix} e^{xy} + y \\ e^{xy} + x \end{bmatrix}.$

19. $\mathbf{F}(\mathbf{X}) = \begin{bmatrix} x^2 + \dfrac{y^2}{2} \\ xy + \dfrac{z^2}{2} \\ yz \end{bmatrix}.$

20. $\mathbf{F}(\mathbf{X}) = \begin{bmatrix} xyz \\ \dfrac{x^2 z}{2} + y \\ xz \end{bmatrix}.$

21. $\mathbf{F}(\mathbf{X}) = \begin{bmatrix} x \log y + y \\ y \log x + x \end{bmatrix}.$

22. $\mathbf{F(X)} = \begin{bmatrix} 2xe^y - e^xy^2 \\ x^2e^y - 2ye^x \end{bmatrix}$.

23. $\mathbf{F(X)} = \begin{bmatrix} 2xye^z + z^2e^y + y^2ze^x \\ x^2e^z + xz^2e^y + 2yze^x \\ x^2ye^z + 2xze^y + y^2e^x \end{bmatrix}$.

24. $\mathbf{F(X)} = \begin{bmatrix} xy - z \\ \dfrac{x^2}{2} + \dfrac{z^3}{3} \\ z^2y + x \end{bmatrix}$.

25. $\mathbf{F(X)} = \begin{bmatrix} \cos y + y \cos x \\ -x \sin y + \sin x \end{bmatrix}$.

In Exercises 26 through 29 find $\int \mathbf{F \cdot dX}$ for any path C from $\mathbf{X_0}$ to $\mathbf{X_1}$.

26. $\mathbf{F(X)} = \begin{bmatrix} e^x \cos y + e^y \cos x \\ e^y \sin x - e^x \sin y \end{bmatrix}$; $\mathbf{X_0} = \left(0, \dfrac{\pi}{2}\right)$, $\mathbf{X_1} = \left(\dfrac{\pi}{4}, \pi\right)$.

27. $\mathbf{F(X)} = \begin{bmatrix} 2xy + yz \\ x^2 + z^2 + xz \\ 2yz + xy \end{bmatrix}$; $\mathbf{X_0} = (1, 2, 1)$, $\mathbf{X_1} = (-2, 3, 0)$.

28. $\mathbf{F(X)} = \begin{bmatrix} x^2y^2 + \dfrac{y^3}{3} + 1 \\ \frac{2}{3}x^3y + xy^2 + 1 \end{bmatrix}$; $\mathbf{X_0} = (1, 1)$, $\mathbf{X_1} = (-2, 3)$.

29. $\mathbf{F(X)} = \begin{bmatrix} x^2y^3 + xy^2 \\ x^3y^2 + x^2y + yz^2 \\ y^2z \end{bmatrix}$; $\mathbf{X_0} = (0, 1, 2)$, $\mathbf{X_1} = (-1, 3, -4)$.

In Exercises 30 through 34 compute the line integral of f with respect to arc length along the given curve.

30. $f(\mathbf{X}) = x^2 + y^2 + z^2$; $\mathbf{X} = a \cos t\,\mathbf{i} + a \sin t\,\mathbf{j} + bt\mathbf{k}$ $(0 \le t \le 2\pi)$.

31. $f(\mathbf{X}) = 3x^2 - 2xy + y^2$; along the semicircle $x^2 + y^2 = 4$, $y \ge 0$.

32. $f(x, y, z) = z^2$; $\mathbf{X} = \cos t\,\mathbf{i} + \sin t\,\mathbf{j} + e^t\mathbf{k}$ $\left(0 \le t \le \dfrac{\pi}{2}\right)$.

33. $f(x, y, z) = xy;$ $\quad \mathbf{X} = e^t \cos t \, \mathbf{i} + e^t \sin t \, \mathbf{j}$ $\quad \left(0 \leq t \leq \dfrac{\pi}{4} \right).$

34. $f(x, y, z) = xy;$ $\quad \mathbf{X} = \dfrac{t^2}{2} \mathbf{i} + \dfrac{\sqrt{2}}{3} t^3 \, \mathbf{j} + \dfrac{t^0}{4} \mathbf{k}$ $\quad (0 \leq t \leq 1).$

35. Find the moment of inertia of a circular wire of radius R about a diameter. Assume that the mass density is ρ_0, a constant.

36. Find the first moment of the wire of Exercise 35 with respect to a plane perpendicular to the wire and tangent to it.

37. Let a wire be in the form of an equilateral triangle with sides of length L. Suppose the mass density of the base is ρ_0, a constant, and that the density of the other two sides varies linearly with altitude, from ρ_0 at the base to $\rho_0/2$ at the opposite vertex. Find the mass and center of gravity of the triangle, and its moments of inertia with respect to its three sides.

THEORETICAL EXERCISES

T-1. Prove Theorem 5.2.

T-2. Prove Theorem 5.4.

T-3. Let f be a real valued function, continuous in a neighborhood of $\mathbf{X}_0 = (c_1, \ldots, c_n)$. Show that

$$\lim_{y \to 0} \int_0^1 f(c_1 + ty, c_2, \ldots, c_n) \, dt = f(\mathbf{X}_0).$$

T-4. Verify Eq. (20).

T-5. If C is given in R^2 by $y = g(x)$ $(a \leq x \leq b)$, show that

$$\int_C f \, ds = \int_a^b f(x, g(x)) \sqrt{1 + (g'(x))^2} \, dx.$$

T-6. Let C be any piecewise smooth simple closed curve in R^2 and let $\mathbf{X}_0 = (x_0, y_0)$ be any point in the interior of C. Show that

(a) $\displaystyle \int_C \frac{(y - y_0) \, dx - (x - x_0) \, dy}{(x - x_0)^2 + (y - y_0)^2} = -2\pi$

and

(b) $$\int_C \frac{(x - x_0)\, dx + (y - y_0)\, dy}{(x - x_0)^2 + (y - y_0)^2} = 0.$$

(*Hint*: Evaluate the integrals for a small circle centered at $\mathbf{X}_0$, and apply Theorem 5.7.)

T-7. *Prove*: If $\mathbf{X} = \mathbf{\Phi}(t)$ $(a \leq t \leq b)$ and $\mathbf{X} = \mathbf{\Psi}(s)$ $(c \leq s \leq d)$ are equivalent parametrizations of a curve C, then

$$\int_a^b f(\mathbf{\Phi}(t)) \mid \mathbf{\Phi}'(t) \mid dt = \int_c^d f(\mathbf{\Psi}(s)) \mid \mathbf{\Psi}'(s) \mid ds.$$

T-8. *Prove*:

$$\int_C f\, ds = \int_{-C} f\, ds.$$

6.6   Surface Integrals of Scalar Fields

In this section we discuss surface integrals. We consider only integrals over *smooth* surfaces in R^3. Roughly speaking, a smooth surface is one which has a tangent plane at each point.

Definition 6.1   A *smooth surface* S in R^3 is the range of a parametric function

$$\mathbf{\Phi}(u, v) = \phi_1(u, v)\mathbf{i} + \phi_2(u, v)\mathbf{j} + \phi_1(u, v)\mathbf{k}$$

with the following properties:

(a) $\mathbf{\Phi}$ is continuous on a domain D in R^2 consisting of a connected open set D° along with the boundary $B(D)$, which is made up of finitely many piecewise smooth curves.

(b)   If (u, v) is in D° and (u_1, v_1) is a point in D distinct from (u, v), then $\mathbf{\Phi}(u, v) \neq \mathbf{\Phi}(u_1, v_1)$; thus $\mathbf{\Phi}$ is one-to-one on D°.

(c)   The partial derivatives $\partial\mathbf{\Phi}/\partial u$ and $\partial\mathbf{\Phi}/\partial v$ are continuous and

(1) $$\mathbf{N}(u, v) = \frac{\partial\mathbf{\Phi}}{\partial u}(u, v) \times \frac{\partial\mathbf{\Phi}}{\partial v}(u, v)$$

is nonzero in D°. (We shall see that $\mathbf{N}(u, v)$ is perpendicular to the tangent plane to S at the point $\mathbf{X} = \mathbf{\Phi}(u, v)$.)

(d)   If (u_0, v_0) is a point on $B(D)$ then

$$\lim_{(u, v) \to (u_0, v_0)} \frac{\mathbf{N}(u, v)}{|\mathbf{N}(u, v)|}$$

exists.

We say that S is defined by $\mathbf{X} = \boldsymbol{\Phi}(u, v)$, and that $\boldsymbol{\Phi}$ is a *smooth parametric function* of two variables.

Example 6.1   The surface S defined by

$$\mathbf{X} = \boldsymbol{\Phi}(\theta, z) = a \cos \theta \, \mathbf{i} + a \sin \theta \, \mathbf{j} + z \mathbf{k} \qquad (a > 0)$$

for (θ, z) in

$$D = \{(\theta, z) \mid 0 \le \theta \le 2\pi, \ |z| \le 1\}$$

is the part of the cylinder $x^2 + y^2 = a^2$ bounded by $z = 1$ and $z = -1$ (Fig. 6.1a). Clearly $\boldsymbol{\Phi}$ satisfies (a) and (b). Its partial derivatives are continuous and

$$(2) \qquad \mathbf{N}(\theta, z) = \begin{vmatrix} \mathbf{i} & \mathbf{j} & \mathbf{k} \\ -a \sin \theta & a \cos \theta & 0 \\ 0 & 0 & 1 \end{vmatrix}$$

$$= a \cos \theta \, \mathbf{i} + a \sin \theta \, \mathbf{j}$$

for (θ, z) in D°. The boundary $B(D)$ of D consists of points of the forms $(0, z)$ and $(2\pi, z)$ with $|z| < 1$ (Fig. 6.1b). From (2),

$$\frac{\mathbf{N}(\theta, z)}{|\mathbf{N}(\theta, z)|} = \frac{a \cos \theta \, \mathbf{i} + a \sin \theta \, \mathbf{j}}{\sqrt{a^2 \cos^2 \theta + a^2 \sin^2 \theta}} = \cos \theta \, \mathbf{i} + \sin \theta \, \mathbf{j};$$

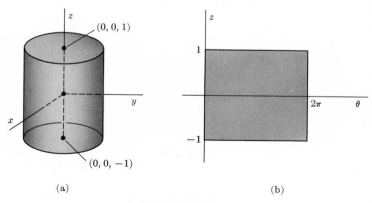

(a)   (b)

FIGURE 6.1

hence

$$\lim_{(\theta, z) \to (0, z_0)} \frac{\mathbf{N}(\theta, z)}{|\mathbf{N}(\theta, z)|} = \lim_{(\theta, z) \to (2\pi, z_0)} \frac{\mathbf{N}(\theta, z)}{|\mathbf{N}(\theta, z)|} = \mathbf{i},$$

and therefore (d) is satisfied. Hence S is smooth.

Example 6.2 The hemisphere S defined by $x^2 + y^2 + z^2 = a^2$, $z > 0$ can be represented by

$$\mathbf{X} = \mathbf{\Phi}(\phi, \theta) = a \sin \phi \cos \theta \, \mathbf{i} + a \sin \phi \sin \theta \, \mathbf{j} + a \cos \phi \mathbf{k}$$

for (ϕ, θ) in

$$D = \left\{ (\phi, \theta) \mid 0 \leq \phi \leq \frac{\pi}{2}, \quad 0 \leq \theta \leq 2\pi \right\};$$

here ϕ and θ are simply the spherical coordinate angles (Fig. 6.2). Now

$$(3) \quad \mathbf{N}(\phi, \theta) = \begin{vmatrix} \mathbf{i} & \mathbf{j} & \mathbf{k} \\ a \cos \phi \cos \theta & a \cos \phi \sin \theta & -a \sin \phi \\ -a \sin \phi \sin \theta & a \sin \phi \cos \theta & 0 \end{vmatrix}$$

$$= a^2 \sin^2 \phi \cos \theta \, \mathbf{i} + a^2 \sin^2 \phi \sin \theta \, \mathbf{j} + a^2 \sin \phi \cos \phi \, \mathbf{k}$$

$$= a \sin \phi \, \mathbf{\Phi}(\phi, \theta),$$

from which it is easily verified that S is smooth.

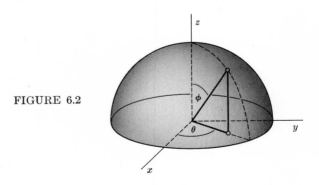

FIGURE 6.2

Example 6.3 If in Example 6.1 we replace the domain of Φ by

$$D_1 = \{(\theta, z) \mid 0 \leq \theta \leq 3\pi, \mid z \mid \leq 1\},$$

then Φ no longer satisfies (b). For example,

$$\Phi\left(\frac{\pi}{2}, \frac{1}{2}\right) = \Phi\left(\frac{5\pi}{2}, \frac{1}{2}\right) = a\mathbf{j} + \tfrac{1}{2}\mathbf{k}.$$

Hence (b) is not satisfied. In fact any point on S for which $y > 0$ is the image of distinct points in D_1°. Some authors would still say that S is smooth, but not *simple*.

For the cylindrical surface of Example 6.1,

$$\mathbf{N}(\theta, z) = a \cos\theta\,\mathbf{i} + a \sin\theta\,\mathbf{j} = x\mathbf{i} + y\mathbf{j};$$

and for the hemisphere of Example 6.2,

$$\mathbf{N}(\phi, \theta) = a \sin\phi\,\Phi(\phi, \theta) = a \sin\phi\,(x\mathbf{i} + y\mathbf{j} + z\mathbf{k}).$$

In both cases, $\mathbf{N}(u, v)$ is normal to the surface in question. (Verify.) This is no accident, as we shall now see. If (u_0, v_0) is in D° then $\mathbf{X} = \Phi(u, v_0)$ defines a curve C_u on S which passes through $\mathbf{X}_0 = \Phi(u_0, v_0)$ (Fig. 6.3). The vector $(\partial\Phi/\partial u)(u_0, v_0)$ is tangent to C_u at $\mathbf{X}_0$ and therefore lies in the plane Γ which is tangent to S at $\mathbf{X}_0$. Similarly, $\mathbf{X} = \Phi(u_0, v)$ defines a curve C_v in S, and $(\partial\Phi/\partial v)(u_0, v_0)$ is in Γ. Thus

$$\mathbf{N}(u_0, v_0) = \frac{\partial\Phi}{\partial u}(u_0, v_0) \times \frac{\partial\Phi}{\partial v}(u_0, v_0)$$

is perpendicular to Γ; that is, $\mathbf{N}(u_0, v_0)$ is normal to S at $\mathbf{X}_0$.

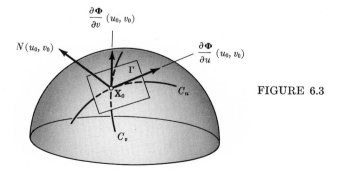

FIGURE 6.3

The following definition of surface area is motivated by the formula for the area of a parallelepiped, derived in Section 3.2.

Definition 6.2 Let the surface S in R^3 be given parametrically by $\mathbf{X} = \mathbf{X}(u, v)$ for (u, v) in a domain D in R^3. Then the *area* of S is defined by

$$(4) \qquad A(S) = \int_D |\mathbf{N}(u, v)| \, du \, dv,$$

provided the integral exists. We shall refer to the symbol

$$dS = |\mathbf{N}(u, v)| \, du \, dv$$

as the *differential element of surface area with respect to the parameters u and v*.

Example 6.4   Suppose S is a parallelogram with vertices $\mathbf{X}_1$, $\mathbf{X}_2$, $\mathbf{X}_3$, and $\mathbf{X}_4$ (Fig. 6.4). In Section 3.2 we showed that the area of S is

$$(5) \qquad A(S) = |(\mathbf{X}_2 - \mathbf{X}_1) \times (\mathbf{X}_3 - \mathbf{X}_1)|.$$

Now we wish to show that Definition 6.2 assigns the same area to S.

The parallelogram can be represented parametrically by

$$\mathbf{X} = \mathbf{X}_1 + u(\mathbf{X}_2 - \mathbf{X}_1) + v(\mathbf{X}_3 - \mathbf{X}_4)$$

for (u, v) in

$$D = \{(u, v) \mid 0 \le u \le 1, \quad 0 \le v \le 1\}.$$

Therefore

$$\frac{\partial \mathbf{X}}{\partial u} = (\mathbf{X}_2 - \mathbf{X}_1), \qquad \frac{\partial \mathbf{X}}{\partial v} = (\mathbf{X}_3 - \mathbf{X}_1),$$

and

$$|\mathbf{N}(u, v)| = |(\mathbf{X}_2 - \mathbf{X}_1) \times (\mathbf{X}_3 - \mathbf{X}_1)| = \text{constant}.$$

From Definition 6.2,

$$A(S) = \int_D |\mathbf{N}(u, v)| \, du \, dv$$

$$= |(\mathbf{X}_2 - \mathbf{X}_1) \times (\mathbf{X}_3 - \mathbf{X}_1)| \int_D du \, dv$$

$$= |(\mathbf{X}_2 - \mathbf{X}_1) \times (\mathbf{X}_3 - \mathbf{X}_1)| \int_0^1 dv \int_0^1 du$$

$$= |(\mathbf{X}_2 - \mathbf{X}_1) \times (\mathbf{X}_3 - \mathbf{X}_1)|,$$

which agrees with (5).

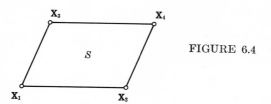

FIGURE 6.4

To motivate Definition 6.2 for more general surfaces, let S be defined by $\mathbf{X} = \mathbf{X}(u, v)$. For simplicity we take the domain of the parametric function to be a rectangle:

$$D = \{(u, v) \mid a \le u \le b, \;\; c \le z \le d\}.$$

Let $a = a_0 < a_1 < \cdots < a_m = b$ and $c = c_0 < c_1 < \cdots < c_n = d$, and let Δ be the partition of D into subrectangles

$$D_{ij} = \{(u, v) \mid a_{i-1} \le u \le a_i, \;\; b_{j-1} \le v \le b_j \;\; (1 \le i \le m, \;\; 1 \le j \le n)\}$$

(Fig. 6.5). Denote by S_{ij} the portion of S given by $\mathbf{X} = \mathbf{X}(u, v)$ $((u, v)$ in $D_{ij})$, suppose $\mathbf{U}_{ij} = (u_{ij}, v_{ij})$ is a point in D_{ij}, and let $\mathbf{X}_{ij} = \mathbf{X}(u_{ij}, v_{ij})$. Finally, let T_{ij} be that portion of the tangent plane to S at $\mathbf{X}_{ij}$ which is given parametrically by

$$\mathbf{X} = \mathbf{X}_{ij} + \frac{\partial \mathbf{X}}{\partial u}(\mathbf{U}_{ij})(u - u_{ij}) + \frac{\partial \mathbf{X}}{\partial v}(\mathbf{U}_{ij})(v - v_{ij}) \qquad ((u, v) \text{ in } D_{ij});$$

then T_{ij} is a parallelogram in R^3 (Fig. 6.6). By an argument similar to that of Example 6.4, it can be shown (Exercise T-5) that the area of T_{ij} is

$$A(T_{ij}) = |\mathbf{N}(u_{ij}, v_{ij})|(a_i - a_{i-1})(b_j - b_{j-1}).$$

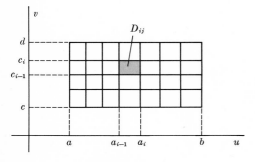

FIGURE 6.5

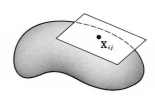

FIGURE 6.6

The total area of all the parallelograms is

$$A_\Delta = \sum_{i=1}^{m} \sum_{j=1}^{n} A(T_{ij}).$$

which approaches the integral on the right side of (4) as $\| \Delta \|$ approaches zero.

Example 6.5 Let S be the cylindrical surface of Example 6.1. From (2),

$$dS = a \, d\theta \, dz;$$

hence

$$A(S) = \int_D a \, d\theta \, dz$$

$$= a \int_0^{2\pi} d\theta \int_{-1}^{1} dz = a(2\pi)(2) = 4\pi a.$$

This result is easily checked: if we cut S along a line parallel to the z-axis and flatten it onto the page, we obtain a rectangle with dimensions 2 and $2\pi a$. (Fig. 6.7). The result is then evident.

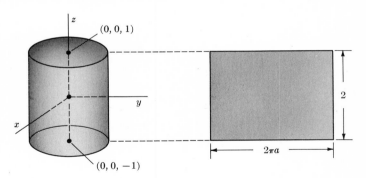

FIGURE 6.7

Example 6.6 For the hemisphere of Example 6.2

$$| \mathbf{N}(\phi, \theta) | = a^2 \sin \phi$$

(see (3)); hence

$$dS = a^2 \sin \phi \, d\theta \, d\phi$$

and

$$A(S) = \int_D a^2 \sin\phi \, d\theta \, d\phi$$

$$= a^2 \int_0^{2\pi} d\theta \int_0^{\pi/2} \sin\phi \, d\phi = 2\pi a^2.$$

By symmetry the surface area of the sphere $x^2 + y^2 + z^2 = a^2$ is $4\pi a^2$.

If S is defined explicitly by

$$z = f(x, y) \qquad ((x, y) \text{ in } D),$$

it can be given the parametric representation

$$\mathbf{X} = \mathbf{X}(x, y) = x\mathbf{i} + y\mathbf{j} + f(x, y)\mathbf{k}.$$

Then

$$\mathbf{N}(x, y) = \begin{vmatrix} \mathbf{i} & \mathbf{j} & \mathbf{k} \\ 1 & 0 & f_x \\ 0 & 1 & f_y \end{vmatrix}$$

$$= -f_x\mathbf{i} - f_y\mathbf{j} + \mathbf{k},$$

and

$$dS = \sqrt{1 + f_x^2 + f_y^2} \, dx \, dy.$$

Example 6.7 Let S be the part of the paraboloid

$$z = f(x, y) = x^2 + y^2$$

which is bounded above by $z = 1$ (Fig. 6.8). It can be represented parametrically by

$$\mathbf{X} = \mathbf{X}(x, y) = x\mathbf{i} + y\mathbf{j} + (x^2 + y^2)\mathbf{k}$$

where (x, y) is in

$$D = \{(x, y) \mid x^2 + y^2 \le 1\}.$$

Since $f_x(x, y) = 2x$ and $f_y(x, y) = 2y$,

$$dS = \sqrt{1 + 4x^2 + 4y^2}$$

and

$$A(S) = \int_D \sqrt{1 + 4x^2 + 4y^2} \, dx \, dy.$$

To evaluate this integral it is natural to change to polar coordinates, which yields

$$A(S) = \int_0^{2\pi} d\theta \int_0^1 \sqrt{1 + 4r^2}\, r\, dr$$

$$= 2\pi \left[\frac{1}{12} (1 + 4r^2)^{3/2} \right] \Big|_0^1 = \frac{\pi}{6} (5^{3/2} - 1).$$

FIGURE 6.8

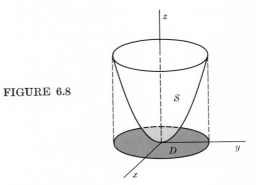

Equivalent Parametrizations

It can be shown that any two parametric representations of an orientable surface S which meet the requirements of Definition 6.1 are equivalent in the following sense.

Definition 6.3 Two parametric functions $\mathbf{\Phi}_1 = \mathbf{\Phi}_1(u, v)$ $((u, v)$ in $D_1)$ and $\mathbf{\Phi}_2 = \mathbf{\Phi}_2(\xi, \eta)$ $((\xi, \eta)$ in $D_2)$ are said to be *equivalent* if they are related by

(6) $$\mathbf{\Phi}_2(g_1(u, v), g_2(u, v)) = \mathbf{\Phi}_1(u, v),$$

where

$$\begin{bmatrix} \xi \\ \eta \end{bmatrix} = \begin{bmatrix} g_1(u, v) \\ g_2(u, v) \end{bmatrix}$$

is a continuously differentiable one to one mapping of D_1 onto D_2 for which

$$\frac{\partial(g_1, g_2)}{\partial(u, v)} \neq 0$$

if (u, v) is in D_1^o.

Example 6.8 Let

$$\mathbf{\Phi}_1(\phi, \theta) = a \sin \phi \cos \theta \, \mathbf{i} + a \sin \phi \sin \theta \, \mathbf{j} + a \cos \phi \, \mathbf{k},$$

$$D_1 = \left\{ (\phi, \theta) \mid 0 \le \phi \le \frac{\pi}{2}, \quad 0 \le \theta \le 2\pi \right\},$$

$$\mathbf{\Phi}_2(x, y) = x\mathbf{i} + y\mathbf{j} + \sqrt{a^2 - x^2 - y^2}\,\mathbf{k}$$

and

$$D_2 = \{ (x, y) \mid 0 \le x^2 + y^2 \le a^2 \}.$$

The transformation

$$\begin{bmatrix} x \\ y \end{bmatrix} = \begin{bmatrix} a \sin \phi \cos \theta \\ a \sin \phi \sin \theta \end{bmatrix}$$

is one-to-one on D_1° and

$$\frac{\partial(x, y)}{\partial(\phi, \theta)} = a^2 \sin \phi \cos \phi$$

is positive on D_1°. Furthermore

$$\mathbf{\Phi}_2(a \sin \phi \cos \theta, a \sin \phi \sin \theta) = \mathbf{\Phi}_1(\phi, \theta);$$

hence $\mathbf{X} = \mathbf{\Phi}_1(\phi, \theta)$ and $\mathbf{X} = \mathbf{\Phi}_2(x, y)$ are equivalent parametric representations of the sphere.

The following theorem states an expected result; namely, that the area of a surface S does not depend upon the particular parametric representation chosen for S.

Theorem 6.1 Let $\mathbf{X} = \mathbf{\Phi}_2(u, v)$ and $\mathbf{X} = \mathbf{\Phi}_2(\xi, \eta)$ be equivalent smooth parametric representations of a surface S, in accordance with Definition 6.3. Then

$$(7) \qquad \int_{D_1} \left| \frac{\partial \mathbf{\Phi}_1}{\partial u} \times \frac{\partial \mathbf{\Phi}_1}{\partial v} \right| du \, dv = \int_{D_2} \left| \frac{\partial \mathbf{\Phi}_2}{\partial \xi} \times \frac{\partial \mathbf{\Phi}_2}{\partial \eta} \right| d\xi \, d\eta.$$

Proof. First we observe that the Jacobian determinant does not change sign since it is continuous and nonvanishing. From (6) and the chain rule,

$$\frac{\partial \mathbf{\Phi}_1}{\partial u} = \frac{\partial g_1}{\partial u} \frac{\partial \mathbf{\Phi}_2}{\partial \xi} + \frac{\partial g_2}{\partial u} \frac{\partial \mathbf{\Phi}_2}{\partial \eta}$$

and

$$\frac{\partial \mathbf{\Phi}_1}{\partial v} = \frac{\partial g_1}{\partial v} \frac{\partial \mathbf{\Phi}_2}{\partial \xi} + \frac{\partial g_2}{\partial v} \frac{\partial \mathbf{\Phi}_2}{\partial \eta},$$

where the partial derivatives of $\mathbf{\Phi}_2$ are evaluated at $(g_1(u, v), g_2(u, v))$, and those of g_1, g_2, and $\mathbf{\Phi}_1$ are evaluated at (u, v). Therefore

$$(8) \quad \frac{\partial \mathbf{\Phi}_1}{\partial u} \times \frac{\partial \mathbf{\Phi}_1}{\partial v} = \frac{\partial g_1}{\partial u} \frac{\partial g_1}{\partial v} \left(\frac{\partial \mathbf{\Phi}_2}{\partial \xi} \times \frac{\partial \mathbf{\Phi}_2}{\partial \xi} \right) + \frac{\partial g_1}{\partial u} \frac{\partial g_2}{\partial v} \left(\frac{\partial \mathbf{\Phi}_2}{\partial \xi} \times \frac{\partial \mathbf{\Phi}_2}{\partial \eta} \right)$$

$$+ \frac{\partial g_2}{\partial u} \frac{\partial g_1}{\partial v} \left(\frac{\partial \mathbf{\Phi}_2}{\partial \eta} \times \frac{\partial \mathbf{\Phi}_2}{\partial \xi} \right) + \frac{\partial g_2}{\partial u} \frac{\partial g_2}{\partial v} \left(\frac{\partial \mathbf{\Phi}_2}{\partial \eta} \times \frac{\partial \mathbf{\Phi}_2}{\partial \eta} \right).$$

Since

$$\frac{\partial \mathbf{\Phi}_2}{\partial \xi} \times \frac{\partial \mathbf{\Phi}_2}{\partial \xi} = \frac{\partial \mathbf{\Phi}_2}{\partial \eta} \times \frac{\partial \mathbf{\Phi}_2}{\partial \eta} = 0$$

and

$$\frac{\partial \mathbf{\Phi}_2}{\partial \eta} \times \frac{\partial \mathbf{\Phi}_2}{\partial \xi} = -\left(\frac{\partial \mathbf{\Phi}_2}{\partial \xi} \times \frac{\partial \mathbf{\Phi}_2}{\partial \eta} \right),$$

it follows from (8) that

$$\frac{\partial \mathbf{\Phi}_1}{\partial u} \times \frac{\partial \mathbf{\Phi}_2}{\partial v} = \left(\frac{\partial g_1}{\partial u} \frac{\partial g_2}{\partial v} - \frac{\partial g_1}{\partial v} \frac{\partial g_2}{\partial u} \right) \left(\frac{\partial \mathbf{\Phi}_2}{\partial \xi} \times \frac{\partial \mathbf{\Phi}_2}{\partial \eta} \right)$$

$$= \frac{\partial (g_1, g_2)}{\partial (u, v)} \left(\frac{\partial \mathbf{\Phi}_2}{\partial \xi} \times \frac{\partial \mathbf{\Phi}_2}{\partial \eta} \right).$$

Therefore

$$\int_{D_1} \left| \frac{\partial \mathbf{\Phi}_1}{\partial u} \times \frac{\partial \mathbf{\Phi}_1}{\partial v} \right| du \, dv = \int_{D_1} \left| \frac{\partial \mathbf{\Phi}_2}{\partial \xi} \times \frac{\partial \mathbf{\Phi}_2}{\partial \eta} \right| \left| \frac{\partial (g_1, g_2)}{\partial (u, v)} \right| du \, dv,$$

and (7) follows from the rule for change of variables in multiple integrals (Theorem 3.2, Section 6.3).

Surface Integrals of Scalar Functions

Let S be a smooth surface defined by $\mathbf{X} = \mathbf{\Phi}(u, v)$ for (u, v) in a domain D, and suppose f is a real valued function defined on S. Then the *surface integral of f over S* is defined by

(9)
$$\int_D f(\mathbf{\Phi}(u, v)) \left| \frac{\partial \mathbf{\Phi}}{\partial u}(u, v) \times \frac{\partial \mathbf{\Phi}}{\partial v}(u, v) \right| du \, dv$$

if the latter exists. It can be shown (Exercise T-1) that if $\mathbf{X} = \mathbf{\Phi}_1(\xi, \eta)$ $((\xi, \eta)$ in $D_1)$ is an equivalent representation of S, then (9) equals

$$\int_{D_1} f(\mathbf{\Phi}_1(\xi, \eta)) \left| \frac{\partial \mathbf{\Phi}_1}{\partial \xi}(\xi, \eta) \times \frac{\partial \mathbf{\Phi}_2}{\partial \eta}(\xi, \eta) \right| d\xi \, d\eta;$$

hence the integral of f over S depends only on f and S, and not on the particular parametrization chosen for S. We indicate this by writing the integral as

(10)
$$\int_S f \, dS.$$

If $f = 1$, then (10) is simply the area of S.

If ρ is defined and nonnegative for all $\mathbf{X}$ on S, we can regard it as defining a mass distribution. This is analogous to the corresponding definition of mass distribution for a solid (Definition 4.1, Section 6.4), except that here $\rho(\mathbf{X})$ has the dimensions of mass per unit area. The mass of S is given by

(11)
$$m = \int_S \rho \, dS.$$

The moments, center of gravity, and moments and products of inertia of S, as defined in Section 6.4, can be computed by means of surface integrals.

Example 6.9 Let S be the hemisphere

$$x^2 + y^2 + z^2 = a^2 \qquad (z > 0),$$

with mass density function

$$\rho(x, y, z) = (x^2 + y^2)z.$$

Again we represent S by

$$\mathbf{X} = a \sin \phi \cos \theta \, \mathbf{i} + a \sin \phi \sin \theta \, \mathbf{j} + a \cos \phi \, \mathbf{k},$$

where (ϕ, θ) is in

$$D = \left\{ (\phi, \theta) \mid 0 \leq \phi \leq \frac{\pi}{2}, \quad 0 \leq \theta \leq 2\pi \right\}.$$

From (11),

$$m(S) = \int_S z(x^2 + y^2) \, dS$$

$$= \int_D (a \cos \phi)(a^2 \sin^2 \phi)(a^2 \sin \phi) \, d\theta \, d\phi$$

$$= a^5 \int_0^{2\pi} d\theta \int_0^{\pi/2} \cos \phi \sin^3 \phi \, d\phi$$

$$= a^5 (2\pi) \left[\frac{\sin^4 \phi}{4} \bigg|_0^{\pi/2} \right] = \frac{\pi a^5}{2} \, .$$

By symmetry $M_{yz} = M_{xz} = 0$ (verify this by setting up the appropriate surface integrals), and

$$M_{xy} = \int_S z^2(x^2 + y^2) \, dS$$

$$= \int_D (a \cos \phi)^2 (a^2 \sin^2 \phi)(a^2 \sin \phi) \, d\theta \, d\phi$$

$$= a^6 \int_0^{2\pi} d\theta \int_0^{\pi/2} \cos^2 \phi \sin^3 \phi \, d\phi$$

$$= 2\pi a^6 \int_0^{\pi/2} \cos^2 \phi (1 - \cos^2 \phi) \sin \phi \, d\varphi$$

$$= 2\pi a^6 \left[\left(-\frac{\cos^3 \phi}{3} + \frac{\cos^5 \phi}{5} \right) \bigg|_0^{\pi/2} \right] = \frac{4\pi a^6}{15} \, .$$

The center of mass of S has coordinates $\bar{x} = \bar{y} = 0$ and

$$\bar{z} = \frac{\dfrac{4\pi a^6}{15}}{\dfrac{\pi a^5}{2}} = \frac{8a}{15} \, .$$

The moment of inertia of S about the z-axis is

$$I_z = \int_S z(x^2 + y^2)^2 \, dS$$

$$= \int_D (a \cos \phi)(a^4 \sin^4 \phi) a^2 \sin \phi \, d\theta \, d\phi$$

$$= a^7 \int_0^{2\pi} d\theta \int_0^{\pi/2} \cos \phi \sin^5 \phi \, d\phi$$

$$= 2\pi a^7 \left[\frac{\sin^6 \phi}{6} \, \Big|_0^{\pi/2} \right] = \frac{\pi a^7}{3} \, .$$

Many commonly occurring surfaces do not meet the requirements of Definition 6.1. For example, the surface of the parallelepiped in Fig. 6.9 is not smooth since no continuous normal vector can be defined so as to be continuous on its edges. The same can be said for the surface comprised of the hemisphere $x^2 + y^2 + z^2 = 1$ ($z \leq 0$) and the section of the paraboloid $z = 1 - x^2 - y^2$ for which $z \leq 0$, joined together by the unit circle on the xy-plane (Fig. 6.10).

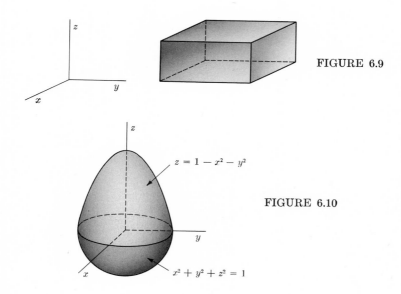

FIGURE 6.9

$z = 1 - x^2 - y^2$

FIGURE 6.10

$x^2 + y^2 + z^2 = 1$

Let us consider the surface of the parallelepiped further. Clearly it can be viewed as consisting of six smooth surfaces $S_1, \ldots, S_6$ (its faces) joined together. It is surely not surprising that we define

$$\int_S f \, dS = \sum_{i=1}^{6} \int_{S_i} f \, dS.$$

Definition 6.4   If f is a real valued function defined on a surface S made up of smooth sections $S_1, \ldots, S_k$, then

$$\int_S f \, dS = \sum_{i=1}^{k} \int_{S_i} f \, dS,$$

provided the integrals on the right exist.

Example 6.10   Let S be the surface shown in Fig. 6.10 and $f(x, y, z) = z$. We represent S_1, the upper half of S, by

$$\mathbf{X} = \mathbf{\Phi}_1(r, \theta) = r \cos \theta \, \mathbf{i} + r \sin \theta \, \mathbf{j} + (1 - r^2) \mathbf{k}$$

for (r, θ) in

$$D_1 = \{ (r, \theta) \mid 0 \le r \le 1, \quad 0 \le \theta \le 2\pi \}.$$

Then

$$\mathbf{N}_1(r, \theta) = 2r^2 \cos \theta \, \mathbf{i} + 2r^2 \sin \theta \, \mathbf{j} + r \mathbf{k}$$

and

$$dS = \mid \mathbf{N}_1(r, \theta) \mid dr \, d\theta = r \sqrt{1 + 4r^2} \, dr \, d\theta.$$

On S_1, $f(x, y, z) = z = 1 - r^2$; hence

$$\int_{S_1} f \, dS = \int_{D_1} (1 - r^2) \, r \sqrt{1 + 4r^2} \, dr \, d\theta$$

$$= \int_0^{2\pi} d\theta \int_0^1 (1 - r^2) \, r \sqrt{1 + 4r^2} \, dr.$$

We leave the evaluation of the integral to the student. The result is

(12)
$$\int_{S_1} f \, dS = \frac{\pi}{60} (25\sqrt{5} - 11).$$

The lower half S_2 of S can be parametrized by

$$\mathbf{X} = \mathbf{\Phi}_2(\phi, \theta) = \sin \phi \cos \theta \, \mathbf{i} + \sin \phi \sin \theta \, \mathbf{j} + \cos \phi \mathbf{k}$$

for (ϕ, θ) in

$$D_2 = \left\{ (\phi, \theta) \mid -\frac{\pi}{2} \le \phi \le 0, \quad 0 \le \theta \le 2\pi \right\}.$$

Then

$$\mathbf{N}_2(\phi, \theta) = \sin \phi \, (\sin \phi \cos \theta \, \mathbf{i} + \sin \phi \sin \theta \, \mathbf{j} + \cos \phi \, \mathbf{k})$$

and

$$dS = \sin \phi \, d\phi \, d\theta.$$

On S_2, $f(x, y, z) = \cos \phi$; hence

$$\int_{S_2} f \, dS = \int_{D_2} \cos \phi \sin \phi \, d\phi \, d\theta$$

$$= \int_0^{2\pi} d\theta \int_{-\pi/2}^0 \cos \phi \sin \phi \, d\phi$$

$$= 2\pi \left(\frac{\sin^2 \phi}{2} \bigg|_{-\pi/2}^0 \right) = -\pi.$$

From this and (12) it follows that

$$\int_S f \, dS = \frac{\pi}{60} (25\sqrt{5} - 71).$$

EXERCISES 6.6

In Exercises 1 through 11 compute the area of the given surface.

1. The part of the surface $z^2 = x^2 + y^2$ between the planes $z = 1$ and $z = 3$.

2. The spiral surface given by

$$\mathbf{X} = \mathbf{X}(u, v) = u \cos v \, \mathbf{i} + u \sin v \, \mathbf{j} + v \mathbf{k}$$

$$(0 \le u \le 1, \quad 0 \le v \le 2\pi).$$

3. The surface defined by

$$\mathbf{X} = \mathbf{X}(u, v) = (u + v)\mathbf{i} + (u - v)\mathbf{j} + u\mathbf{k}$$

$$(0 \le u \le 2, \quad 1 \le v \le 3).$$

4. The part of the sphere $x^2 + y^2 + z^2 = 1$ between the planes $z = 1/\sqrt{2}$ and $z = -1/\sqrt{2}$.

5. The part of the plane $x + y + z = 2$ inside the cylinder $x^2 + y^2 = 4$.

6. The part of the cylinder $x^2 + y^2 = 2x$ bounded above by the cone $x^2 + y^2 = z^2$ and below by the xy-plane.

7. The part of the sphere $x^2 + y^2 + z^2 = 1$ inside the upper nappe of the cone $x^2 + y^2 = z^2$.

8. The part of the cylinder $x^2 + y^2 = 4$ between the planes $z = x$ and $z = 2x + 3$.

9. The triangle determined by the plane

$$\frac{x}{a} + \frac{y}{b} + \frac{z}{c} = 1$$

and the coordinate planes.

10. The part of the cylinder $x^2 + z^2 = a^2$ inside the cylinder $y^2 + z^2 = a^2$.

11. The part of the cylinder $x^2 + y^2 = ax$ inside the sphere $x^2 + y^2 + z^2 = a^2$.

In Exercises 12 through 17 compute $\int_S f \, dS$ over the given surface.

12. $f(x, y, z) = x^2$; the cylinder defined by $x^2 + y^2 = a^2, |z| \leq 1$, including the top and bottom.

13. $f(x, y, z) = y$; the part of the plane $2x + y + z = 2$ inside the cylinder $x^2 + y^2 = 4$.

14. $f(x, y, z) = x^2$; the part of the paraboloid $z = 1 - x^2 - y^2$ for which $z \geq 0$.

15. $f(x, y, z) = x^2$; the part of the cone $x^2 + y^2 = z^2$ between the planes $z = 0$ and $z = 2$.

16. $f(x, y, z) = x$; the part of the sphere $x^2 + y^2 + z^2 = a^2$ inside the upper nappe of the cone $x^2 + y^2 = z^2$, together with the upper part of the cone contained within the sphere.

17. $f(x, y, z) = x$; the part of the cylinder $x^2 + y^2 = 1$ between the planes $z = 0$ and $z = x + 2$, together with the top and bottom.

In Exercises 18 through 22 compute the center of mass and I_z for the given surface S with density ρ.

18. $\rho(x, y, z) = \rho_0$; the surface of Exercise 17.

19. $\rho(x, y, z) = x^2$; the surface of Exercise 16.

20. $\rho(x, y, z) = x^2 + y^2$; the surface of Exercise 4.

21. $\rho(x, y, z) = x^2$; the surface of Exercise 5.

22. $\rho(x, y, z) = \sqrt{x^2 + y^2}$; the surface of Exercise 6.

THEORETICAL EXERCISES

T-1. Let $\mathbf{X} = \boldsymbol{\Phi}(u, v)$ $((u, v)$ in $D)$ and $\mathbf{X} = \boldsymbol{\Phi}_1(\xi, y)$ $((\xi, y)$ in $D_1)$ be equivalent parametrizations of an orientable surface S. Show that

$$\int_D f(\boldsymbol{\Phi}(u, v)) \left| \frac{\partial \boldsymbol{\Phi}}{\partial u}(u, v) \times \frac{\partial \boldsymbol{\Phi}}{\partial v}(u, v) \right| du\, dv$$

$$= \int_{D_1} f(\boldsymbol{\Phi}_1(\xi, y)) \left| \frac{\partial \boldsymbol{\Phi}_1}{\partial \xi}(\xi, y) \times \frac{\partial \boldsymbol{\Phi}_1}{\partial y}(\xi, y) \right| d\xi\, dy$$

if f is a scalar valued function for which either integral exists. *Hint:* See the proof of Theorem 6.1.

T-2. Let the surface S be given explicitly by $z = f(x, y)$ $((x, y)$ in $D)$.
  (a) Show that the area of S is given by

$$A(S) = \int\int_D \sqrt{1 + \left(\frac{\partial f}{\partial x}\right)^2 + \left(\frac{\partial f}{\partial y}\right)^2}\, dx\, dy.$$

  (b) If γ is the acute angle between the z-axis and the normal to S, show that

$$A(S) = \int\int_D \sec \gamma\, dx\, dy.$$

T-3. Let the surface S be given by $\mathbf{X} = \mathbf{X}(u, v)$ $((u, v)$ in $D)$ and define

$$E = \left|\frac{\partial \mathbf{X}}{\partial u}\right|^2, \qquad F = \frac{\partial \mathbf{X}}{\partial u} \cdot \frac{\partial \mathbf{X}}{\partial v}, \qquad G = \left|\frac{\partial \mathbf{X}}{\partial v}\right|^2.$$

Show that

$$A(S) = \int\int_D \sqrt{EG - F^2}\, du\, dv.$$

T-4. Prove the following theorem of Pappus:
  If a path of length L is revolved about the y-axis to obtain a surface S, then the area of S is $A(S) = 2\pi \bar{x} L$, where $(\bar{x}, \bar{y})$ is the centroid of the path. (Assume $x > 0$ on the path.)

T-5. Show that $A(T_{ij}) = |\mathbf{N}(u_{ij}, v_{ij})|\, (a_i - a_{i-1})(b_j - b_{j-1})$.

6.7   Surface Integrals of Vector Fields

We now consider surface integrals of vector functions over *orientable* surfaces in R^3.

Definition 7.1   An *orientable* surface S in R^3 is a surface which has a smooth parametric representation

$$\mathbf{X} = \mathbf{\Phi}(u, v) \qquad ((u, v) \text{ in } D)$$

that satisfies

(1)
$$\lim_{(u,v)\to(u_0,v_0)} \frac{\mathbf{N}(u, v)}{|\mathbf{N}(u, v)|} = \lim_{(u,v)\to(u_1,v_1)} \frac{\mathbf{N}(u, v)}{|\mathbf{N}(u, v)|}$$

whenever (u_0, v_0) and (u_1, v_1) are points on the boundary $B(D)$ of D which correspond to the same point on S; that is,

(2)
$$\mathbf{X}_0 = \mathbf{\Phi}(u_0, v_0) = \mathbf{\Phi}(u_1, v_1).$$

A point $\mathbf{X}_0$ on S which satisfies (2) with $(u_0, v_0) \neq (u_1, v_1)$ is a *multiple point* of S. From the definition of a smooth surface (specifically, part (b) of Definition 6.1, Section 6.6), (2) can be satisfied with $(u_0, v_0) \neq (u_1, v_1)$ only if both (u_0, v_0) and (u_1, v_1) are in $B(D)$.

Example 7.1   Consider the cylindrical surface S defined by $x^2 + y^2 = a^2$, $|z| \leq 1$, and represented parametrically by

(3)
$$\mathbf{X} = \mathbf{\Phi}(\theta, z)$$
$$= a \cos \theta \, \mathbf{i} + a \sin \theta \, \mathbf{j} + z\mathbf{k}$$

for (θ, z) in
$$D = \{(\theta, z) \mid 0 \leq \theta \leq 2\pi, \quad |z| \leq 1\}.$$

We have seen (Example 6.1, Section 6.6) that

$$\mathbf{N}(\theta, z) = a \cos \theta \, \mathbf{i} + a \sin \theta \, \mathbf{j};$$

hence (assuming that $a > 0$),

$$\frac{\mathbf{N}(\theta, z)}{|\mathbf{N}(\theta, z)|} = \cos \theta \, \mathbf{i} + \sin \theta \, \mathbf{j}.$$

To find the multiple points of S, we need only examine the image of $B(D)$ under (3). They are of the form $(a, 0, z_0)$ $(|z_0| \leq 1)$ since

$$\mathbf{\Phi}(0, z_0) = \mathbf{\Phi}(2\pi, z_0) = a\mathbf{i} + z_0\mathbf{k}.$$

However,

$$\lim_{(\theta,z)\to(0,z_0)} \frac{\mathbf{N}(\theta, z)}{|\mathbf{N}(\theta, z)|} = \lim_{(\theta,z)\to(2\pi, z_0)} \frac{\mathbf{N}(\theta, z)}{|\mathbf{N}(\theta, z)|} = \mathbf{i};$$

thus (1) is satisfied, and S is orientable.

Example 7.2 Consider the hemisphere S defined by $x^2 + y^2 + z^2 = a^2$, $z > 0$. We have seen (Example 6.2, Section 6.6) that S can be represented by

(4) $\mathbf{X} = \mathbf{\Phi}(\phi, \theta) = a \sin \phi \cos \theta\, \mathbf{i} + a \sin \phi \sin \theta\, \mathbf{j} + a \cos \phi\, \mathbf{k}$

for (ϕ, θ) in

$$D = \left\{ (\phi, \theta) \mid 0 \leq \phi \leq \frac{\pi}{2}, \quad 0 \leq \theta \leq 2\pi \right\},$$

and that

$$\mathbf{N}(\phi, \theta) = a^2 \sin \phi\, (\sin \phi \cos \theta\, \mathbf{i} + \sin \phi \sin \theta\, \mathbf{j} + \cos \phi\, \mathbf{k}).$$

Therefore (assuming $a > 0$),

$$\frac{\mathbf{N}(\phi, \theta)}{|\mathbf{N}(\phi, \theta)|} = \sin \phi \cos \theta\, \mathbf{i} + \sin \phi \sin \theta\, \mathbf{j} + \cos \phi\, \mathbf{k}.$$

Again we examine the image of $B(D)$ under (4) to find the multiple points of S. All points of the form $(0, \theta_0)$ map onto the "north pole" since

$$\mathbf{\Phi}(0, \theta_0) = a\mathbf{k};$$

however, for any such point,

$$\lim_{(\phi,\theta)\to(0,\theta_0)} \frac{\mathbf{N}(\phi, \theta)}{|\mathbf{N}(\phi, \theta)|} = \mathbf{k}.$$

Also, $\mathbf{\Phi}(\phi_0, 0) = \mathbf{\Phi}(\phi_0, 2\pi)$ and

$$\lim_{(\sigma,\theta)\to(\sigma_0,0)} \frac{\mathbf{N}(\phi, \theta)}{|\mathbf{N}(\phi, \theta)|} = \lim_{(\sigma,\theta)\to(\sigma_0,2\pi)} \frac{\mathbf{N}(\phi, \theta)}{|\mathbf{N}(\phi, \theta)|} = \sin \phi_0\, \mathbf{i} + \cos \phi_0\, \mathbf{k}.$$

Therefore S is orientable.

Example 7.3 Any smooth surface without multiple points is orientable since (2) then implies that $(u_0, v_0) = (u_1, v_1)$ and (1) follows automatically.

Example 7.4 The surface S defined by

$$\mathbf{X} = \mathbf{\Phi}(\theta, v)$$

$$= \left(1 + v \sin \frac{\theta}{2}\right) \cos \theta\, \mathbf{i} + \left(1 + v \sin \frac{\theta}{2}\right) \sin \theta\, \mathbf{j} + v \cos \frac{\theta}{2}\, \mathbf{k}$$

for (θ, v) in
$$D = \{(\theta, v) \mid 0 \leq \theta \leq 2\pi, \ |v| \leq 1\}$$
is called a *Möbius strip*. A model of S can be constructed by subjecting one end of a narrow strip of paper to a half twist and pasting it to the other end (Fig. 7.1). By straightforward calculation (Exercise T-1) it can be shown that

$$(5) \qquad \lim_{(\theta, v) \to (0,0)} \frac{\mathbf{N}(\theta, v)}{|\mathbf{N}(\theta, v)|} = -\lim_{(\theta, v) \to (2\pi, 0)} \frac{\mathbf{N}(\theta, v)}{|\mathbf{N}(\theta, v)|} = \mathbf{i}.$$

Since $\mathbf{\Phi}(0, 0) \neq \mathbf{\Phi}(2\pi, 0)$, S is not orientable. (Actually, we have not proved that the Möbius strip *cannot* be represented by a parametric function which has the properties of Definition 7.1. However, this can be shown.)

FIGURE 7.1

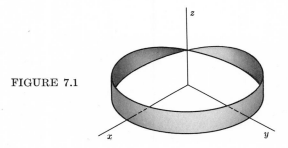

To each point $\mathbf{X}$ on an orientable surface S we can assign a unit vector $\mathbf{n}(\mathbf{X})$ as follows: if $\mathbf{X} = \mathbf{\Phi}(u, v)$ for (u, v) in D°, then

$$\mathbf{n}(\mathbf{X}) = \frac{\mathbf{N}(u, v)}{|\mathbf{N}(u, v)|},$$

while if $\mathbf{X} = \mathbf{\Phi}(u_0, v_0)$ for some (u_0, v_0) in $B(D)$, then

$$(6) \qquad \mathbf{n}(\mathbf{X}) = \lim_{(u, v) \to (u_0, v_0)} \frac{\mathbf{N}(u, v)}{|\mathbf{N}(u, v)|}.$$

Since $\mathbf{N}(u, v)$ is normal to S at $\mathbf{X} = \mathbf{\Phi}(u, v)$, it follows that $\mathbf{n}(\mathbf{X})$ is a unit normal to S at $\mathbf{X} = \mathbf{\Phi}(u, v)$. It follows from (1) that (6) is well defined even if $\mathbf{X}$ is a multiple point of S.

Example 7.5 For the cylindrical section of Example 7.1,

$$\mathbf{N}(\theta, z) = a \cos \theta \, \mathbf{i} + a \sin \theta \, \mathbf{j}$$
$$= x\mathbf{i} + y\mathbf{j};$$

hence

$$\mathbf{n}(\mathbf{X}) \;=\; \frac{x\mathbf{i} + y\mathbf{j}}{\sqrt{x^2 + y^2}}\;.$$

For the hemisphere of Example 7.2,

$$\mathbf{N}(\phi, \theta) \;=\; a \sin \phi\, \mathbf{X}(\phi, \theta),$$

and therefore

$$\mathbf{n}(\mathbf{X}) \;=\; \frac{\mathbf{X}}{|\mathbf{X}|} \;=\; \frac{x\mathbf{i} + y\mathbf{j} + z\mathbf{k}}{\sqrt{x^2 + y^2 + z^2}}\;.$$

We shall regard $\mathbf{n}$ as a vector field defined on S. It can be represented visually by drawing an arrow in the direction of $\mathbf{n}(\mathbf{X})$ at each point $\mathbf{X}$ of S (Fig. 7.2).

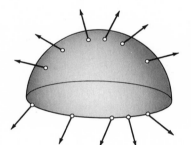

FIGURE 7.2

We shall call the vector field $\mathbf{n}$ the *orientation of S induced by the parametric representation* $\mathbf{X} = \mathbf{\Phi}(u, v)$. Along with $\mathbf{n}$, the vector $-\mathbf{n}$ is also an orientation of S. It is induced by $\mathbf{X} = \mathbf{\Psi}(v, u)$, where $\mathbf{\Psi}(v, u) = \mathbf{\Phi}(u, v)$ (Exercise T-2).

The student knows that some surfaces—such as hemispheres—have boundaries, while others—such as spheres—do not. We shall not attempt to give a precise definition of the boundary of a surface. Rather, we shall be guided by geometric intuition.

An orientable surface S which has no boundary is said to be *closed*. A closed surface determines a bounded set, *the inside of S*, and an unbounded set, the *outside* of S, such that any continuous curve with initial point inside and terminal point outside of S must pass through S. It is customary to designate the positive orientation of a closed surface to be the one for which the arrow representing $\mathbf{n}(\mathbf{X})$ is directed toward the outside of S at every point of S.

Theorem 7.1 Let $\mathbf{X} = \boldsymbol{\Phi}_1(u, v)$ and $\mathbf{X} = \boldsymbol{\Phi}_2(u, v)$ be equivalent smooth parametric representations of a surface S (Definition 6.3, Section 6.6), related by

$$\boldsymbol{\Phi}_2(g_1(u, v), g_2(u, v)) = \boldsymbol{\Phi}_1(u, v),$$

and let $\mathbf{n}_1$ and $\mathbf{n}_2$ be the orientations induced on S by $\boldsymbol{\Phi}_1$ and $\boldsymbol{\Phi}_2$, respectively. Then $\mathbf{n}_1 = \mathbf{n}_2$ if the Jacobian determinant $\partial(g_1, g_2)/\partial(u, v)$ is positive, and $\mathbf{n}_1 = -\mathbf{n}_2$ if the Jacobian determinant is negative.

We leave the proof of this theorem to the student (Exercise T-3).

Surface Integrals of Vector Fields

Let the surface S be oriented by the unit vector function $\mathbf{n}$ and suppose

$$\mathbf{F} = F_1\mathbf{i} + F_2\mathbf{j} + F_3\mathbf{k}$$

is a vector field defined on S. Then $\mathbf{F} \cdot \mathbf{n}$ is a scalar function on S, and the integral

(7) $$\int_S \mathbf{F} \cdot \mathbf{n}\, dS,$$

if it exists, is called the *surface integral of* $\mathbf{F}$ *over* S. If $\mathbf{X} = \mathbf{X}(u, v)$ ((u, v) in D) is a parametric representation of S which induces $\mathbf{n}$, then

$$\int_S \mathbf{F} \cdot \mathbf{n}\, dS = \int_D \left(\mathbf{F}(\mathbf{X}(u, v)) \cdot \frac{\mathbf{N}(u, v)}{|\mathbf{N}(u, v)|}\right) |\mathbf{N}(u, v)|\, du\, dv$$

$$= \int_D \mathbf{F}(\mathbf{X}(u, v)) \cdot \mathbf{N}(u, v)\, du\, dv$$

$$= \int_D \mathbf{F}(\mathbf{X}(u, v)) \cdot \left(\frac{\partial \mathbf{X}}{\partial u}(u, v) \times \frac{\partial \mathbf{X}}{\partial v}(u, v)\right) du\, dv.$$

Notice that (7) depends upon which of the two possible orientations of S occurs in the integrand. If S is closed it is to be understood, in the absence of a statement to the contrary, that $\mathbf{n}(\mathbf{X})$ is directed toward the outside of S. For a surface which is not closed the orientation must be stated explicitly.

Example 7.6 Let S be the sphere $x^2 + y^2 + z^2 = a^2$ ($a > 0$) and $\mathbf{F}(x, y, z) = x\mathbf{i} + y\mathbf{j}$, and suppose we wish to calculate $\int_S \mathbf{F} \cdot \mathbf{n}\, dS$, where $\mathbf{n}$

is the outward normal. We parametrize S by

(8) $\qquad X = \mathbf{X}(\phi, \theta)$

$\qquad\qquad = a \sin \phi \cos \theta \, \mathbf{i} + a \sin \phi \sin \theta \, \mathbf{j} + a \cos \phi \mathbf{k}$

for (ϕ, θ) in

$$D = \{ (\phi, \theta) \mid 0 \le \phi \le \pi, \;\; 0 \le \theta \le 2\pi \}.$$

From Example 7.5,

(9) $\qquad\qquad \dfrac{\mathbf{N}(\phi, \theta)}{\mid \mathbf{N}(\phi, \theta) \mid} = \dfrac{\mathbf{X}(\phi, \theta)}{\mid \mathbf{X}(\phi, \theta) \mid} ,$

which is a positive multiple of the position vector of the point $\mathbf{X} = \mathbf{X}(\phi, \theta)$ with respect to the origin. Hence the unit vector (9) points outward from the sphere S at every point (Fig. 7.3), and therefore (8) induces the positive orientation on S. Hence,

$$\int_S \mathbf{F} \cdot \mathbf{n} \, dS = \int_D (a \sin \phi \cos \theta \, \mathbf{i} + a \sin \phi \sin \theta \, \mathbf{j}) \cdot (a \sin \phi \, \mathbf{X}(\phi, \theta) \, d\phi \, d\theta$$

$$= \int_D [(a \sin \phi \cos \theta)^2 + (a \sin \phi \sin \theta)^2] a \sin \phi \, d\phi \, d\theta$$

$$= \int_D a^3 \sin^3 \phi \, d\phi \, d\theta = a^3 \int_0^{2\pi} d\theta \int_0^{\pi} \sin^3 \phi \, d\phi$$

$$= 2\pi a^3 \int_0^{\pi} (1 - \cos^2 \phi) \sin \phi \, d\phi$$

$$= 2\pi a^3 \left(-\cos \phi + \frac{\cos^3 \phi}{3} \right) \bigg|_0^{\pi} = 2\pi a^3 \left(\frac{4}{3} \right) = \frac{8\pi a^3}{3} .$$

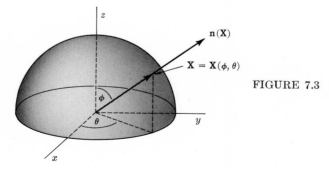

FIGURE 7.3

Example 7.7   Let S be the part of the paraboloid

$$z = f(x, y) = x^2 + y^2$$

which is bounded above by $z = 1$ (Fig. 6.8). It can be represented parametrically by

(10) $$\mathbf{X} = \mathbf{X}(x, y) = x\mathbf{i} + y\mathbf{j} + (x^2 + y^2)\mathbf{k},$$

where (x, y) is in

$$D = \{ (x, y) \mid x^2 + y^2 \le 1 \}.$$

Suppose we wish to integrate

$$\mathbf{F}(x, y, z) = y\mathbf{i} - x\mathbf{j} + z^2\mathbf{k}$$

over S, where the orientation is chosen so that $\mathbf{n}(\mathbf{X})$ has a negative z component. It is easy to verify that the representation (10) induces $-\mathbf{n}$; hence

$$\int_S \mathbf{F} \cdot \mathbf{n} \, dS = -\int_D \mathbf{F}(\mathbf{X}(x, y)) \cdot \mathbf{N}(x, y) \, dx \, dy$$

$$= -\int_D [y\mathbf{i} - x\mathbf{j} + (x^2 + y^2)^2\mathbf{k}] \cdot [-2x\mathbf{i} - 2y\mathbf{j} + \mathbf{k}] \, dx \, dy$$

$$= -\int_D (x^2 + y^2)^2 \, dx \, dy$$

Changing to polar coordinates yields

$$\int_S \mathbf{F} \cdot \mathbf{n} \, dS = -\int_0^{2\pi} d\theta \int_0^1 r^5 \, dr = (2\pi) \left(\frac{1}{6} \right) = -\frac{\pi}{3}.$$

Piecewise Orientable Surfaces

Many commonly occurring surfaces do not meet the requirements of Definition 7.1. For example, the surface of the parallelepiped in Fig. 7.4 is not orientable since it is not smooth, as noted in Section 6.6. The same can be said for the surface comprised of the hemisphere $x^2 + y^2 + z^2 = 1$ ($z \le 0$) and the section of the paraboloid $z = 1 - x^2 - y^2$ for which $z \le 0$, joined together by the unit circle on the xy-plane (Fig. 7.5).

Let us consider the surface of the parallelepiped further. Clearly it can be viewed as consisting of six orientable surfaces $S_1, \ldots, S_6$ (its faces)

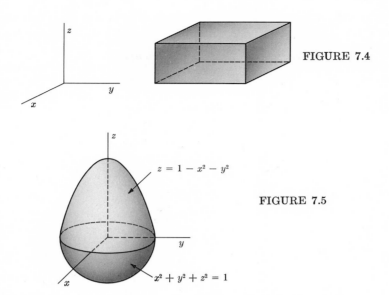

FIGURE 7.4

FIGURE 7.5

joined together. For surface integrals of scalar functions we defined

$$\int_S f \, dS = \sum_{i=1}^{6} \int_{S_i} f \, dS$$

for any real valued function f such that the integrals on the right exist. However, the situation is not so simple if we wish to define the integral of a vector field over S. To write simply

(11)
$$\int_S \mathbf{F} \cdot \mathbf{n} \, dS = \sum_{i=1}^{6} \int_{S_i} \mathbf{F} \cdot \mathbf{n} \, dS$$

would be nonsense, since there is as yet no orientation $\mathbf{n}$ defined over all six faces. We could remedy this by choosing orientations $\mathbf{n}_1, \ldots, \mathbf{n}_6$ for $S_1, \ldots, S_6$ and replacing the right side of (11) by

$$\sum_{i=1}^{6} \int_{S_i} \mathbf{F} \cdot \mathbf{n}_i \, dS.$$

However, this expression has in general $2^6 = 64$ distinct values, since each $\mathbf{n}_i$ can be chosen in two ways.

Clearly we must adopt a convention regarding the way in which the orientable pieces of surfaces such as those in Figs. 7.4 and 7.5 are to be joined together. To do this we first introduce the notion of a *directed boundary* of a surface.

Suppose an orientable surface S has a boundary which we denote by

$B(S)$. It can be shown that if S is represented by $\mathbf{X} = \boldsymbol{\Phi}(u, v)$ $((u, v)$ in $D)$ then every point on $B(S)$ is the image of a point on $B(D)$. However, there may be points on $B(D)$ whose images are not on $B(S)$. These statements are easily verified for the surfaces of Examples 7.1 and 7.2 (Exercise T-4).

If S has a boundary then $B(S)$ can be shown to consist of a finite number of piecewise smooth curves. For example, the boundary of the cylindrical surface of Example 7.1 consists of the two circles

$$C_1 = \{(x, y, z) \mid x^2 + y^2 = a^2, \quad z = 1\}$$

and

$$C_2 = \{x, y, z) \mid x^2 + y^2 = a^2, \quad z = -1\}.$$

Let C be a smooth curve in $B(S)$, parametrized by $\mathbf{X} = \boldsymbol{\Psi}(t)$. If, whenever $\dot{\boldsymbol{\Psi}}(t)$ exists, the vector $\mathbf{n}(\boldsymbol{\Psi}(t)) \times \dot{\boldsymbol{\Psi}}(t)$ points toward S, we say that C is *positively directed with respect to the orientation* $\mathbf{n}$, or simply *positively directed*. If every piecewise smooth curve in $B(S)$ is positively directed we say that $B(S)$ *is positively directed* (with respect to $\mathbf{n}$). In this case a person walking in the positive direction around $B(S)$ with his head in the direction of $\mathbf{n}(\mathbf{X})$ would always see S to his left (Fig. 7.6). (To see this, recall that $\dot{\boldsymbol{\Psi}}(t)$ is tangent to $B(S)$ at the point $\mathbf{X} = \boldsymbol{\Psi}(t)$, and points in the direction of motion.)

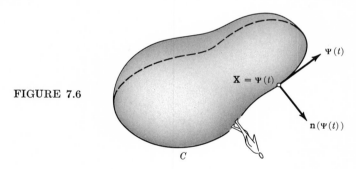

FIGURE 7.6

Example 7.8 Consider the hemisphere S of Example 7.2 with the orientation

$$\mathbf{n}(\mathbf{X}) = \frac{\mathbf{X}}{|\mathbf{X}|}$$

induced by (4). The boundary of S is the circle

$$C = \{(x, y, z) \mid x^2 + y^2 = a^2, \quad z = 0\}.$$

We define a direction on C by means of the parametrization

$$\boldsymbol{\Psi}(t) = a \cos t\,\mathbf{i} + a \sin t\,\mathbf{j} \qquad (0 \leq t \leq 2\pi).$$

Then the tangent vector

$$\dot{\Psi}(t) = -a \sin t \, \mathbf{i} + a \cos t \, \mathbf{j}$$

has the direction shown in Fig. 7.7a. Since

$$\mathbf{n}(\Psi(t)) = \frac{\Psi(t)}{|\Psi(t)|} = \cos t \, \mathbf{i} + \sin t \, \mathbf{j},$$

it follows that

$$\mathbf{n}(\Psi(t)) \times \dot{\Psi}(t) = \begin{vmatrix} \mathbf{i} & \mathbf{j} & \mathbf{k} \\ \cos t & \sin t & 0 \\ -a \sin t & a \cos t & 0 \end{vmatrix} = a\mathbf{k},$$

which points in the direction of S (Fig. 7.7a); therefore C is positively directed with respect to $\mathbf{n}$.

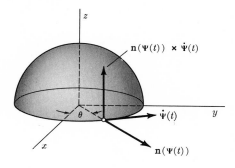

(a)

FIGURE 7.7

(b)

If we parametrize C by

(12) $\mathbf{X} = \mathbf{\Psi}_1(t) = a \sin t \, \mathbf{i} + a \cos t \, \mathbf{j}$ $(0 \le t \le 2\pi)$

then

$$\dot{\mathbf{\Psi}}_1(t) = a \cos t \, \mathbf{i} - a \sin t \, \mathbf{j},$$

$$\mathbf{n}(\mathbf{\Psi}_1(t)) \times \dot{\mathbf{\Psi}}_1(t) = \begin{vmatrix} \mathbf{i} & \mathbf{j} & \mathbf{k} \\ \sin t & \cos t & 0 \\ a \cos t & -a \sin t & 0 \end{vmatrix} = -a\mathbf{k},$$

which points away from S (Fig. 7.7b). Thus (12) does not direct C positively with respect to $\mathbf{n}$.

Example 7.9   The boundaries of the cylinders in Figs. 7.8a and 7.8b, traversed in the directions of the arrows, are positively directed with respect to the indicated orientations. In Fig. 7.8c, C_1 is positively directed, but C_2 is not.

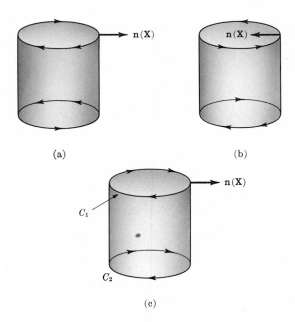

FIGURE 7.8

Definition 7.2   A surface S is *piecewise orientable* if it is orientable or can be constructed by joining together along sections of their boundaries a finite number of orientable surfaces $S_1, \ldots,$ S_k which have the following property: It is possible to choose orientations $\mathbf{n}_1, \ldots,$ $\mathbf{n}_k$ on $S_1, \ldots,$ S_k so that if γ_{ij} is any curve common to $B(S_i)$ and $B(S_j)$ $(i \neq j)$ then the positive direction of γ_{ij} with respect to S_i is the opposite of the positive direction of γ_{ij} with respect to S_j. If this is done we say that S is *oriented* and that $S_1, \ldots,$ S_k are *sections* of S.

Example 7.10   The surface S in Fig. 7.5 is orientable. In Figs. 7.9a and 7.9b, it has the outward (positive) and inward (negative) orientations, respectively; in Fig. 7.9c it is not oriented.

The orientation chosen for a section of an orientable surface determines the orientations for all of its sections. Thus, an orientable surface has exactly two orientations.

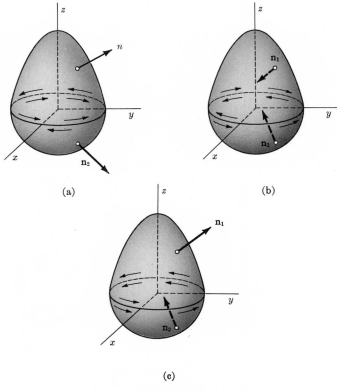

(a)   (b)

(c)

FIGURE 7.9

We can now define the integral of a vector field over a piecewise orientable surface.

Definition 7.3 Let S be a piecewise orientable surface with sections $S_1, \ldots, S_k$, and suppose that the sections are oriented in the manner described in Definition 7.2. Let $\mathbf{F}$ be a vector field defined on S. Then

$$\int_S \mathbf{F} \cdot \mathbf{n} \, dS = \sum_{i=1}^k \int_{S_i} (\mathbf{F} \cdot \mathbf{n}_i) \, dS$$

if the integrals on the right exist.

Example 7.11 Let S be the surface of Fig. 7.5. As in Example 6.10, Section 4.6, we represent S_1, the upper half of S, by

(13) $\qquad \mathbf{X} = \mathbf{\Phi}_1(r, \theta) = r \cos \theta \, \mathbf{i} + r \sin \theta \, \mathbf{j} + (1 - r^2) \mathbf{k}$

for (r, θ) in

$$D_1 = \{ (r, \theta) \mid 0 \leq r \leq 1, \quad 0 \leq \theta \leq 2\pi \},$$

and S_2, the lower half of S, by

$$\mathbf{X} = \mathbf{\Phi}_2(\phi, \theta) = \sin \phi \cos \theta \, \mathbf{i} + \sin \phi \sin \theta \, \mathbf{j} + \cos \phi \, \mathbf{k}$$

for (ϕ, θ) in

(14) $\qquad D_2 = \left\{ (\phi, \theta) \mid -\frac{\pi}{2} \leq \phi \leq 0, \quad 0 \leq \theta \leq 2\pi \right\}.$

Suppose we wish to integrate

$$F(\mathbf{X}) = \frac{y\mathbf{i} - y\mathbf{j} + \mathbf{k}}{\sqrt{x^2 + y^2}}$$

over S. The parametrizations (13) and (14) induce the outward (positive) orientations on S_1 and S_2; hence

(15) $\qquad \displaystyle\int_S \mathbf{F} \cdot \mathbf{n} \, dS = \int_{D_1} \mathbf{F}(\mathbf{\Phi}_1(r, \theta)) \cdot \mathbf{N}_1(r, \theta) \, dr \, d\theta$

$$+ \int_{D_2} \mathbf{F}(\mathbf{\Phi}_2(\phi, \theta)) \cdot \mathbf{N}_2(\phi, \theta) \, d\phi \, d\theta,$$

where $\mathbf{N}_1$ and $\mathbf{N}_2$ are defined by

$$\mathbf{N}_1(r, \theta) = 2r^2 \cos \theta \, \mathbf{i} + 2r^2 \sin \theta \, \mathbf{j} + r\mathbf{k}$$

and

$$\mathbf{N}_2(\phi, \theta) = \sin \phi \, (\sin \phi \cos \theta \, \mathbf{i} + \sin \phi \sin \theta \, \mathbf{j} + \cos \phi \, \mathbf{k}).$$

Easy calculations show that

$$\mathbf{F}(\boldsymbol{\Phi}_1(r, \theta)) \cdot \mathbf{N}_1(r, \theta) = 1$$

and

$$\mathbf{F}(\boldsymbol{\Phi}_2(\phi, \theta)) \cdot \mathbf{N}_2(\phi, \theta) = \cos \phi;$$

hence, from (15)

$$\int_S \mathbf{F} \cdot \mathbf{n} \, dS = \int_0^{2\pi} d\theta \int_0^1 dr + \int_0^{2\pi} d\theta \int_{-\pi/2}^0 \cos \varphi \, d\varphi$$

$$= 2\pi + 2\pi = 4\pi.$$

Example 7.12 Let V be the parallelepiped bounded by the planes $x = x_1$, $x = x_2$, $y = y_1$, $y = y_2$, $z = z_1$, and $z = z_2$ (Fig. 7.10). Let S be the surface of V with the positive (outward) orientation, and suppose $\mathbf{F} = F_1\mathbf{i} + F_2\mathbf{j} + F_3\mathbf{k}$ is a continuous vector function on S. Then

$$\int_S \mathbf{F} \cdot \mathbf{n} \, dS = \sum_{i=1}^6 \int_{S_i} \mathbf{F} \cdot \mathbf{n} \, dS,$$

where $S_1, \ldots, S_6$ are the faces of V.

Let S_1 be the face on which $x = x_2$; it can be represented by

$$\mathbf{X} = x_2\mathbf{i} + y\mathbf{j} + z\mathbf{k}$$

for (y, z) in

$$D = \{ (y, z) \mid y_1 \leq y \leq y_2, \quad z_1 \leq z \leq z_2 \}.$$

Clearly $\mathbf{n}_1 = \mathbf{i}$ and therefore

$$(16) \qquad \int_{S_1} \mathbf{F} \cdot \mathbf{n} \, dS = \int_D F_1(x_2, y, z) \, dy \, dz.$$

Similarly, the face S_2 on which $x = x_1$ can be represented by

$$\mathbf{X} = x_1\mathbf{i} + y\mathbf{j} + z\mathbf{k}$$

for (y, z) in D. Here $\mathbf{n}_2 = -\mathbf{i}$ and

$$(17) \qquad \int_{S_2} \mathbf{F} \cdot \mathbf{n} \, dS = -\int_D F_1(x_1, y, z) \, dy \, dz.$$

Now suppose $\mathbf{F}$ is defined and has continuous first partial derivatives on some open set in R^3 which contains V; then (16) and (17) imply that

$$(18) \quad \int_{S_1} \mathbf{F} \cdot \mathbf{n} \, dS + \int_{S_2} \mathbf{F} \cdot \mathbf{n} \, dS = \int_D (F_1(x_2, y, z) - F_1(x_1, y, z)) \, dy \, dz$$

$$= \int_D \left(\int_{x_1}^{x_2} \frac{\partial F_1}{\partial x}(x, y, z) \, dx \right) dy \, dz$$

$$= \int_V \frac{\partial F_1}{\partial x} \, dx \, dy \, dz.$$

If S_3 and S_4 are the faces for which $y = y_2$ and $y = y_1$, respectively, a similar argument yields

$$(19) \qquad \int_{S_3} \mathbf{F} \cdot \mathbf{n} \, dS + \int_{S_4} \mathbf{F} \cdot \mathbf{n} \, dS = \int_V \frac{\partial F_2}{\partial y} \, dx \, dy \, dz;$$

also,

$$(20) \qquad \int_{S_5} \mathbf{F} \cdot \mathbf{n} \, dS + \int_{S_6} \mathbf{F} \cdot \mathbf{n} \, dS = \int_V \frac{\partial F_3}{\partial z} \, dx \, dy \, dz,$$

where S_5 and S_6 are the faces for which $z = z_2$ and $z = z_1$, respectively. Adding (18), (19), and (20) yields

$$\int_S \mathbf{F} \cdot \mathbf{n} \, dS = \int_V \left(\frac{\partial F_1}{\partial x} + \frac{\partial F_2}{\partial y} + \frac{\partial F_3}{\partial z} \right) dV = \int_V \operatorname{div} \mathbf{F} \, dV.$$

This is a remarkable relationship. We shall explore its consequences in the next section.

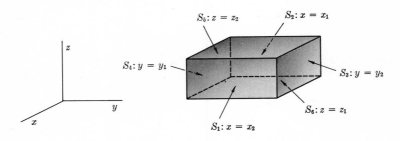

FIGURE 7.10

EXERCISES 6.7

In Exercises 1 through 14, compute $\displaystyle\int_S \mathbf{F} \cdot \mathbf{n} \, dS$ for the given $\mathbf{F}$ and S.

1. $\mathbf{F}(\mathbf{X}) = y\mathbf{i} - x\mathbf{j} + xy^2\mathbf{k}$; the surface of Exercise 1, Section 6.6; $\mathbf{n}(\mathbf{X})$ has a negative $\mathbf{k}$ component.

2. $\mathbf{F}(\mathbf{X}) = y\mathbf{i} - x\mathbf{j} + \mathbf{k}$; the surface of Exercise 2, Section 6.6, with the orientation induced by the given parametrization.

3. $\mathbf{F}(\mathbf{X}) = x^2\mathbf{i} + y^2\mathbf{j} + z^2\mathbf{k}$; the surface of Exercise 3, Section 6.6, with the orientation induced by the given parametrization.

4. $\mathbf{F}(\mathbf{X}) = \mathbf{X}$; the surface of Exercise 4, Section 6.6; $\mathbf{n}(\mathbf{X})$ pointing away from the origin.

5. $\mathbf{F}(\mathbf{X}) = (x^2 - y^2)\mathbf{i} + (y^2 - z^2)\mathbf{j} + (z^2 - x^2)\mathbf{k}$; the surface of Exercise 5, Section 6.6; $\mathbf{n}(\mathbf{X})$ has a positive $\mathbf{i}$ component.

6. $\mathbf{F}(\mathbf{X}) = (x - 1)\mathbf{i} + y\mathbf{j} + z\mathbf{k}$; the surface of Exercise 6, Section 6.6; $\mathbf{n}(\mathbf{X})$ points away from the axis of the cylinder.

7. $\mathbf{F}(\mathbf{X}) = x\mathbf{i}$; the surface of Exercise 7, Section 6.6; $\mathbf{n}(\mathbf{X})$ points away from the origin.

8. $\mathbf{F}(\mathbf{X}) = yz\mathbf{k}$; the surface of Exercise 8, Section 6.6; $\mathbf{n}(\mathbf{X})$ points toward the axis of the cylinder.

9. $\mathbf{F}(\mathbf{X}) = x\mathbf{i} + y^2\mathbf{j} + z\mathbf{k}$; the surface of Exercise 9, Section 6.6; $\mathbf{n}(\mathbf{X})$ points away from the origin. (Assume a, b, and $c > 0$.)

10. $\mathbf{F}(\mathbf{X}) = x\mathbf{i} + y\mathbf{j} + z^2\mathbf{k}$; the surface of Exercise 12, Section 6.6, positively oriented.

11. $\mathbf{F}(\mathbf{X}) = y\mathbf{i} - x\mathbf{j} + xyz\mathbf{k}$; the surface of Exercise 17, Section 6.6, positively oriented.

12. $\mathbf{F}(\mathbf{X}) = |\mathbf{X}|^2\mathbf{X}$; the surface consisting of

$$S_1 = \{\mathbf{X} \mid x^2 + y^2 + z^2 = 1, \quad x \geq 0, \quad y \geq 0, \quad z \geq 0\}$$

and

$$S_2 = \{\mathbf{X} \mid x^2 + y^2 \leq 1, \quad x \geq 0, \quad y \geq 0, \quad z = 0\},$$

oriented so that $n(\mathbf{X}) = \mathbf{k}$ on S_2.

13. $F(\mathbf{X}) = 2\mathbf{i} + 3\mathbf{j} + 4\mathbf{k}$; the surface consisting of

$$S_1 = \{\mathbf{X} \mid 0 \le z = 1 - x^2 - y^2\},$$

$$S_2 = \{\mathbf{X} \mid x^2 + y^2 = 1, \quad -1 \le z \le 0\}$$

and

$$S_3 = \{\mathbf{X} \mid x^2 + y^2 \le 1, \quad z = -1\},$$

oriented so that $\mathbf{n}(\mathbf{X}) = -\mathbf{k}$ on S_3.

14. $\mathbf{F}(\mathbf{X}) = x^2\mathbf{i} + y^2\mathbf{j} + z^2\mathbf{k}$; the surface of the parallelepiped

$$V = \{\mathbf{X} \mid 0 \le x \le 1, \quad 0 \le y \le 1, \quad 0 \le z \le 1\},$$

positively oriented.

THEORETICAL EXERCISES

T-1.   Verify Eq. (5) for the Möbius strip.

T-2.   Prove that if $\mathbf{n}$ is induced on S by $\mathbf{X} = \boldsymbol{\Phi}(u, v)$ then $-\mathbf{n}$ is induced by $\mathbf{X} = \boldsymbol{\Psi}(v, u)$, where $\boldsymbol{\Psi}(v, u) = \boldsymbol{\Phi}(u, v)$.

T-3.   Prove Theorem 7.1. *Hint*: See the proof of Theorem 6.1, Section 6.6.

T-4.   For the parametric representations of the cylindrical and hemispherical surfaces of Examples 7.1 and 7.2, verify that all points of $B(S)$ are images of $B(D)$. Show also that the images of some points of $B(D)$ are not in $B(S)$.

T-5.   For the surfaces of Exercises 7.1 through 7.9, specify the positive directions of the boundaries with respect to the given orientations.

6.8   The Divergence Theorem; Green's and Stokes's Theorems

In Example 7.12 of Section 6.7 we saw that

(1)
$$\int_S \mathbf{F} \cdot \mathbf{n} \, dS = \int_V \operatorname{div} \mathbf{F} \, dV$$

if S is the surface of a parallelepiped V with faces parallel to the coordinate planes, and $\mathbf{F} = F_1\mathbf{i} + F_2\mathbf{j} + F_3\mathbf{k}$ is a continuously differentiable vector field on an open set containing V. In this section we shall extend this result and consider some of its consequences.

The next theorem, which is known as *Gauss's theorem*, or the *divergence theorem*, extends (1) to more general regions.

Theorem 8.1   Let V be a region in R^3 enclosed by an orientable surface S, and let $\mathbf{n}$ be the outward normal on S. Suppose also that $\mathbf{F}$ is continuously differentiable in an open set containing V and S. Then

$$(2) \qquad \int_S \mathbf{F} \cdot \mathbf{n} \, dS = \int_V \operatorname{div} \mathbf{F} \, dV.$$

Proof.   We shall prove this theorem under the additional assumption that V is convex (Exercise T-12, Section 4.1), and simply indicate how it can be extended to more complicated regions. The convexity of V implies that any line which intersects S meets S in exactly two points, or is tangent to S. Let R_{xy} be the projection of V into the xy-plane (Fig. 8.1), and for each (x, y) in R_{xy} let $z_1(x, y)$ and $z_2(x, y)$ be the least and greatest values of z such that $(x, y, z_1(x, y))$ and $(x, y, z_2(x, y))$ are on S. Now V can be described as

$$V = \{ (x, y, z) \mid (x, y) \text{ in } R_{xy}, \quad z_1(x, y) \leq z \leq z_2(x, y) \},$$

and S can be divided into three parts:
  (a)   an upper surface S_2, given by

$$(3) \qquad \mathbf{X} = \mathbf{X}_2(x, y) = x\mathbf{i} + y\mathbf{j} + z_2(x, y)\mathbf{k} \qquad ((x, y) \text{ in } R_{xy});$$

  (b)   a lower surface S_1, given by

$$(4) \qquad \mathbf{X} = \mathbf{X}_1(x, y) = x\mathbf{i} + y\mathbf{j} + z_1(x, y)\mathbf{k} \qquad ((x, y) \text{ in } R_{xy});$$

  (c)   vertical sections on which the normal is parallel to the xy-plane.

It may happen that part (c) contains only boundary curves common to S_1 and S_2, as in Fig. 8.1; however, it may also contain sections of S with nonzero area, such as the shaded section in Fig. 8.2.

We shall first show that

$$\int_S (F_3\mathbf{k}) \cdot \mathbf{n} \, dS = \int_V \frac{\partial F_3}{\partial z} \, dV.$$

First we observe that

$$(5) \qquad \int_S (F_3\mathbf{k}) \cdot \mathbf{n} \, dS = \int_{S_2} (F_3\mathbf{k}) \cdot \mathbf{n} \, dS + \int_{S_1} (F_3\mathbf{k}) \cdot \mathbf{n} \, dS$$

even if S has sections of type (c) with positive area, since on such a section $\mathbf{n}$ would be perpendicular to $\mathbf{k}$. Let us assume that z_1 and z_2 are continuously differentiable; from (3) and (4),

$$(6) \qquad \mathbf{N}_2 = -\frac{\partial z_2}{\partial x}\mathbf{i} - \frac{\partial z_2}{\partial y}\mathbf{j} + \mathbf{k}$$

and

$$(7) \qquad \mathbf{N}_1 = -\frac{\partial z_1}{\partial x}\mathbf{i} - \frac{\partial z_1}{\partial y}\mathbf{j} + \mathbf{k}.$$

Since (6) induces the positive orientation on S_2 and (7) induces the negative orientation on S_1, it follows from (5) that

$$\int_S (F_3\mathbf{k})\cdot\mathbf{n}\,dS = \int_{R_{xy}} F_3(x, y, z_2(x, y))\mathbf{k}\cdot\mathbf{N}_2(x, y)\,dx\,dy$$

$$-\int_{R_{xy}} F_3(x, y, z_1(x, y))\mathbf{k}\cdot N_1(x, y)\,dx\,dy$$

$$= \int_{R_{xy}} F_3(x, y, z_2(x, y))\,dx\,dy - \int_{R_{xy}} F_3(x, y, z_1(x, y))\,dx\,dy.$$

$$= \int_{R_{xy}} \left(\int_{z_1(x,y)}^{z_2(x,y)} \frac{\partial F_3}{\partial z}(x, y, z)\,dz\right)dx\,dy.$$

Therefore

$$(8) \qquad \int_S (F_3\mathbf{k})\cdot\mathbf{n}\,dS = \int_V \frac{\partial F}{\partial z}\,dV.$$

A similar argument with x, y, z, and F_3 replaced by y, z, x, and F_1 yields

$$(9) \qquad \int_S (F_1\mathbf{i})\cdot\mathbf{n}\,dS = \int_V \frac{\partial F_1}{\partial x}\,dV.$$

Another permutation of the variables yields

$$(10) \qquad \int_S (F_2\mathbf{j})\cdot\mathbf{n}\,dS = \int_V \frac{\partial F_2}{\partial y}\,dV$$

(Exercise T-1), and (2) can be obtained by adding (8), (9), and (10).

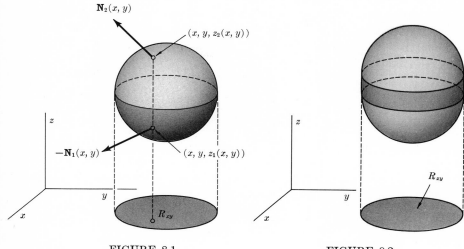

FIGURE 8.1   FIGURE 8.2

Now (2) can be extended to any region which can be broken down into convex subregions, and also to regions bounded by more than one surface, such as the punctured ball

(11) $$V = \{\mathbf{X} \mid 0 < \rho_1 < |\mathbf{X}| < \rho_2\}.$$

To see how this is accomplished for the punctured ball, we write

$$\int_V \operatorname{div} \mathbf{F} \, dV = \sum_{r=1}^{8} \int_{V_r} \operatorname{div} \mathbf{F} \, dV$$

where V_r is the part of V in the rth octant. Since V_r is convex,

(12) $$\int_{V_r} \operatorname{div} \mathbf{F} \, dV = \int_{\sigma_r} \mathbf{F} \cdot \mathbf{n} \, dS,$$

where σ_r comprises one spherical and three planar sections (Fig. 8.3). Every planar section appears twice among $\sigma_1, \ldots, \sigma_8$, with opposite signs; hence adding (12) for $r = 1, \ldots, 8$ yields

(13) $$\int_V \operatorname{div} \mathbf{F} \, dV = \int_{S_2} \mathbf{F} \cdot \mathbf{n} \, dS - \int_{S_1} \mathbf{F} \cdot \mathbf{n} \, dS,$$

where S_1 and S_2 are the positively oriented spheres

$$S_i = \{\mathbf{X} \mid |\mathbf{X}| = \rho_i\} \qquad (i = 1, 2).$$

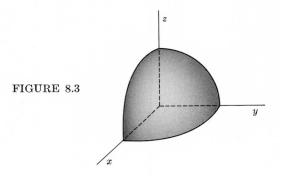

FIGURE 8.3

Example 8.1 Let V be the tetrahedron bounded by the coordinate planes and the plane $x + y + z = 1$ (Fig. 8.4), and let

$$\mathbf{F}(\mathbf{X}) = 3x^2\mathbf{i} + xy\mathbf{j} + z\mathbf{k}.$$

Then

$$(\text{div } \mathbf{F})(\mathbf{X}) = 6x + x + 1 = 7x + 1$$

and

$$\int_V \text{div } \mathbf{F}\, dV = \int_0^1 (7x + 1)\, dx \int_0^{1-x} dy \int_0^{1-x-y} dz$$

$$= \int_0^1 (7x + 1)\, dx \int_0^{1-x} (1 - x - y)\, dy$$

$$= \frac{1}{2} \int_0^1 (7x + 1)(x - 1)^2\, dx = \frac{1}{8}.$$

Let S_1, S_2, and S_3 be the faces of V in the xy-, yz- and zx-planes, respectively. Then

$$\int_{S_1} \mathbf{F} \cdot \mathbf{n}\, dS = \int_{S_1} (3x^2\mathbf{i} + xy\mathbf{j} + 0\mathbf{k}) \cdot (-\mathbf{k})\, dS = 0,$$

$$\int_{S_2} \mathbf{F} \cdot \mathbf{n}\, dS = \int_{S_2} (0\mathbf{i} + 0\mathbf{j} + z\mathbf{k}) \cdot (-\mathbf{i})\, dS = 0,$$

and

$$\int_{S_3} \mathbf{F} \cdot \mathbf{n}\, dS = \int_{S_3} (3x^2\mathbf{i} + 0\mathbf{j} + z\mathbf{k}) \cdot (-\mathbf{j})\, dS = 0.$$

The remaining face S_4 can be represented by

$$\mathbf{X} = \mathbf{X}(x, y) = x\mathbf{i} + y\mathbf{j} + (1 - x - y)\mathbf{k} \qquad (0 \le y \le 1 - x, \ \ 0 \le x \le 1);$$

hence
$$\mathbf{F}(\mathbf{X}(x, y)) = 3x^2\mathbf{i} + xy\mathbf{j} + (1 - x - y)\mathbf{k}$$
and
$$\mathbf{N}(x, y) = \mathbf{i} + \mathbf{j} + \mathbf{k}.$$
Therefore

$$\int_{S_4} \mathbf{F} \cdot \mathbf{n} \, dS = \int_0^1 dx \int_0^{1-x} (3x^2 + xy + 1 - x - y) \, dy$$

$$= \int_0^1 \left[\frac{(x - 1)^3}{2} - 3x^2(x - 1) \right] dx = \frac{1}{8}$$

and

$$\int_V \operatorname{div} \mathbf{F} \, dV = \int_S \mathbf{F} \cdot \mathbf{n} \, dS = \frac{1}{8},$$

as predicted by the divergence theorem.

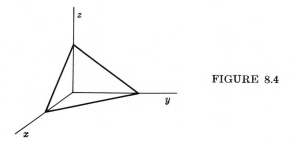

FIGURE 8.4

Example 8.2   Let $\mathbf{F}$ be the inverse square law force of gravity:

$$\mathbf{F}(\mathbf{X}) = -Gm_0 \frac{\mathbf{X}}{|\mathbf{X}|^3} \, ;$$

then div $\mathbf{F} = 0$ (Exercise T-9, Section 5.4) and therefore

$$\int_S \mathbf{F} \cdot \mathbf{n} \, dS = 0$$

if S is any closed surface. In particular, if V is the punctured ball (11), it follows from (13) that the integrals of $\mathbf{F}$ over any two concentric spheres about the origin are equal. To verify this, let S_ρ be the sphere defined by $|\mathbf{X}| = \rho$. The outward normal to S is

$$\mathbf{n}(\mathbf{X}) = \frac{\mathbf{X}}{|\mathbf{X}|},$$

and therefore

$$\int_{S_\rho} \mathbf{F} \cdot \mathbf{n} \, dS = -Gm_0 \int_{S_\rho} \left(\frac{\mathbf{X}}{|\mathbf{X}|^3} \right) \cdot \left(\frac{\mathbf{X}}{|\mathbf{X}|} \right) dS$$

$$= -Gm_0 \int_{S_\rho} \frac{|\mathbf{X}|^2}{|\mathbf{X}|^4} \, dS.$$

$$= -\frac{Gm_0}{\rho^2} \int_{S_\rho} dS = -\frac{Gm_0}{\rho^2} (4\pi\rho^2) = -4\pi Gm_0.$$

Green's Theorem in R^3

Let u and v be continuously differentiable scalar valued function in a region T in R^3 and let v_{xx}, v_{yy}, and v_{zz} exist and be continuous there. From Theorem 8.1 with

$$\mathbf{F} = u \, \boldsymbol{\nabla} v = u \frac{\partial v}{\partial x} \mathbf{i} + u \frac{\partial v}{\partial y} \mathbf{j} + u \frac{\partial v}{\partial z} \mathbf{k},$$

it follows that

$$(14) \qquad \int_S (u \, \boldsymbol{\nabla} v) \cdot \mathbf{n} \, dS = \int_V \text{div} (u \, \boldsymbol{\nabla} v) \, dV,$$

for any volume V in T with orientable surface S. Now

$$(15) \qquad \boldsymbol{\nabla} v \cdot \mathbf{n} = \frac{\partial v}{\partial \mathbf{n}},$$

where $\partial v / \partial \mathbf{n}$ stands for the directional derivative in the direction of the outward normal to S (Section 4.2). Direct calculation yields

$$\text{div} (u \, \boldsymbol{\nabla} v) = \boldsymbol{\nabla} u \cdot \boldsymbol{\nabla} v + u \, \boldsymbol{\nabla}^2 v$$

(Exercise T-12, Section 5.4), where $\boldsymbol{\nabla}^2 v$ is the Laplacian of v, defined in Section 5.4 as

$$(16) \qquad \boldsymbol{\nabla}^2 v = v_{xx} + v_{yy} + v_{zz}.$$

Using (15) and (16) we can rewrite (14) as

$$(17) \qquad \int_S u \frac{\partial v}{\partial \mathbf{n}} \, dS = \int_V (\boldsymbol{\nabla} u \cdot \boldsymbol{\nabla} v + u \, \boldsymbol{\nabla}^2 v) \, dV.$$

This relation is known as *Green's first theorem*.

If we also assume that u_{xx}, u_{yy}, and u_{zz} are continuous in T then we can interchange u and v in (17) to obtain

$$\int_S v \frac{\partial u}{\partial \mathbf{n}} \, dS = \int_V (\boldsymbol{\nabla} v \cdot \boldsymbol{\nabla} u + v \nabla^2 u) \, dV;$$

subtracting this from (17) yields

$$\int_S \left(u \frac{\partial v}{\partial \mathbf{n}} - v \frac{\partial u}{\partial \mathbf{n}} \right) dS = \int_V (u \nabla^2 v - v \nabla^2 u) \, dV,$$

which is known as *Green's second theorem*.

Green's theorems have important applications in the theory of *harmonic functions*; that is, functions which satisfy Laplace's equation,

$$\nabla^2 u = 0.$$

(See Section 5.4).

Stokes's Theorem

Suppose $\mathbf{F}$ has continuous second partial derivatives in a region T. Since $\text{div}(\mathbf{curl}\ \mathbf{F}) = 0$ (Example 11, Section 5.4), it follows from the divergence theorem that

(18)
$$\int_S (\mathbf{curl}\ \mathbf{F}) \cdot \mathbf{n} \, dS = 0,$$

if S is a closed orientable surface which encloses a volume contained in T.

Now suppose S is made up of two nonintersecting surfaces S_1 and S_2 with common boundary C. (For example, let S be the sphere $x^2 + y^2 + z^2 = 1$, C the circle $x^2 + y^2 = 1$, and S_1 and S_2 the upper and lower halves of S, as in Fig. 8.5.) Suppose S_1 and S_2 are oriented in opposite senses, so that C is positively directed with respect to both (Fig. 8.5). Then, from (18),

$$\int_{S_2} (\mathbf{curl}\ \mathbf{F}) \cdot \mathbf{n}_1 \, dS - \int_{S_1} (\mathbf{curl}\ \mathbf{F}) \cdot \mathbf{n}_2 \, dS = 0;$$

that is, the integrals of $\mathbf{curl}\ \mathbf{F}$ over all orientable surfaces in T with a given boundary C are equal. It is therefore reasonable that the common value of these integrals can be computed from a knowledge of $\mathbf{F}$ on C alone. The following theorem, which we shall not prove, tells how this can be done.

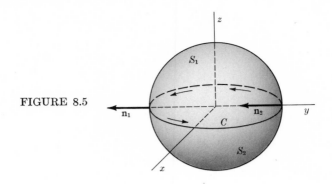

FIGURE 8.5

Theorem 8.2 (Stokes's Theorem)   Suppose S is an orientable surface and $\mathbf{F}$ is a continuously differentiable vector field on an open set containing S, and let $B(S)$ be the boundary of S. Then

$$(19) \qquad \int_S (\mathbf{curl\ F}) \cdot \mathbf{n}\, dS = \int_{B(S)} \mathbf{F} \cdot d\mathbf{X},$$

provided $B(S)$ is positively directed with respect to $\mathbf{n}$. If $B(S)$ consists of piecewise smooth curves $C_1, \ldots, C_n$, the right side of (19) is to be interpreted as

$$\int_{B(S)} \mathbf{F} \cdot d\mathbf{X} = \int_{C_1} \mathbf{F} \cdot d\mathbf{X} + \cdots + \int_{C_n} \mathbf{F} \cdot d\mathbf{X},$$

where $C_1, \ldots, C_n$ are positively directed with respect to S.

Example 8.3   Let S be the part of the sphere $x^2 + y^2 + z^2 = 1$ bounded by $z = 0$ and $z = \frac{1}{2}$, with the orientation shown in Figure 8.6, and let

$$F = z\mathbf{i} + x\mathbf{j} + y\mathbf{k}.$$

According to Stokes's theorem

$$(20) \qquad \int_S (\mathbf{curl\ F}) \cdot \mathbf{n}\, dS = \int_{C_1} \mathbf{F} \cdot d\mathbf{X} - \int_{C_2} \mathbf{F} \cdot d\mathbf{X},$$

where C_1 and C_2 are the directed circles shown in Fig. 8.6. Let us verify this.

We can represent S by

$$\mathbf{X} = \sin \phi \cos \theta\, \mathbf{i} + \sin \phi \sin \theta\, \mathbf{j} + \cos \phi \mathbf{k}$$

for (ϕ, θ) in

$$D = \left\{ (\phi, \theta) \ \middle|\ \frac{\pi}{3} \le \phi \le \frac{\pi}{2}, \quad 0 \le \theta \le 2\pi \right\}.$$

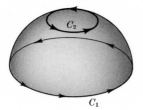

FIGURE 8.6

Then

$$(21) \quad \int_S (\mathbf{curl}\ \mathbf{F}) \cdot \mathbf{n}\ dS$$

$$= \int_D (\mathbf{i} + \mathbf{j} + \mathbf{k}) \cdot (\sin \phi\, \mathbf{X}(\phi, \theta))\ d\phi\ d\theta$$

$$= \int_{\pi/3}^{\pi/2} d\phi \int_0^{2\pi} (\sin^2 \phi \cos \theta + \sin^2 \phi \sin \theta + \sin \phi \cos \phi)\ d\theta$$

$$= 2\pi \int_{\pi/3}^{\pi/2} \sin \phi \cos \phi\ d\phi = \frac{\pi}{4}\ .$$

The circle C_1 can be represented by

$$\mathbf{X} = \cos t\, \mathbf{i} + \sin t\, \mathbf{j} \qquad (0 \le t \le 2\pi),$$

so that on C_1

$$\mathbf{F}(\mathbf{X}(t)) \cdot \mathbf{X}'(t) = (0\mathbf{i} + \cos t\, \mathbf{j} + \sin t\, \mathbf{k}) \cdot (-\sin t\, \mathbf{i} + \cos t\, \mathbf{j}) = \cos^2 t;$$

thus

$$(22) \qquad \int_{C_1} \mathbf{F} \cdot d\mathbf{X} = \int_0^{2\pi} \cos^2 t\ dt = \pi.$$

On C_2,

$$\mathbf{X} = \frac{\sqrt{3}}{2} \cos t\, \mathbf{i} + \frac{\sqrt{3}}{2} \sin t\, \mathbf{j} + \frac{1}{2} \mathbf{k}$$

and

$$\mathbf{F}(\mathbf{X}(t)) \cdot \mathbf{X}'(t) = \left(\frac{1}{2} \mathbf{i} + \frac{\sqrt{3}}{2} \cos t\, \mathbf{j} + \frac{\sqrt{3}}{2} \sin t\, \mathbf{k} \right) \cdot \left(-\frac{\sqrt{3}}{2} \sin t\, \mathbf{i} + \frac{\sqrt{3}}{2} \cos t\, \mathbf{j} \right)$$

$$= -\frac{\sqrt{3}}{4} \sin t + \frac{\sqrt{3}}{4} \cos^2 t;$$

hence

(23) $\qquad \int_{C_2} \mathbf{F} \cdot \mathbf{dX} = \int_0^{2\pi} \left(-\frac{\sqrt{3}}{4} \sin t + \frac{3}{4} \cos^2 t \right) dt = \tfrac{3}{4}\pi.$

Now (20) follows from (21), (22), and (23).

Green's Theorem in the Plane

We now obtain the following theorem, known as *Green's theorem in the plane*, as a special case of Stokes's theorem. (If one wished to prove Stokes's theorem, then it would be necessary to establish Green's theorem in the plane by other means, since the latter is an essential part of the proof of the former.)

Theorem 8.3 Let S be a region in the xy-plane bounded by piecewise smooth simple closed curves $C_1, \ldots, C_n$, no two of which intersect. Suppose further that P and Q are continuously differentiable functions in a region containing S and $C_1, \ldots, C_n$. Then

$$\int_S \left(\frac{\partial Q}{\partial x} - \frac{\partial P}{\partial y} \right) dx \, dy = \sum_{i=1}^{n} \int_{C_i} P dx + Q \, dy,$$

where each C_i is positively directed with respect to the orientation $\mathbf{n} = \mathbf{k}$ on S (Fig. 8.7); that is, the interior of S is to the left of C_i.

To obtain this result we take

$$\mathbf{F}(x, y, z) = P(x, y)\mathbf{i} + Q(x, y)\mathbf{j}$$

and parametrize S by

$$\mathbf{X} = x\mathbf{i} + y\mathbf{j} \qquad ((x, y) \text{ in } S);$$

FIGURE 8.7

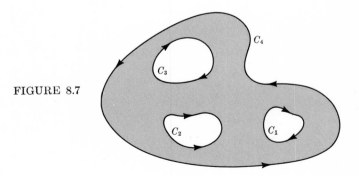

then $\mathbf{n}(\mathbf{X}) = \mathbf{k}$,

$$\text{curl } \mathbf{F} = \left(\frac{\partial Q}{\partial x} - \frac{\partial P}{\partial y}\right)\mathbf{k},$$

and the result follows from Stokes's theorem.

Example 8.4   Let us verify Green's theorem in the plane for $P(x, y) = x^2 + y$ and $Q(x, y) = 2x + 2y$, where S is the region outside the circle $x^2 + y^2 = \frac{1}{4}$ and inside the square with vertices at $(1, 1), (-1, 1), (-1, -1)$, and $(1, -1)$ (Fig. 8.8). In this case the double integral is simply the area of S:

$$(24) \qquad \int_S \left(\frac{\partial Q}{\partial x} - \frac{\partial P}{\partial y}\right) dx\, dy = \int_S (2 - 1)\, dx\, dy = 4 - \frac{\pi}{4}.$$

The boundary of S consists of five smooth segments, identified as $C_1, \ldots, C_5$ in Fig. 8.8. The integrals over these segments are as follows:

$$\int_{C_1} P\, dx + Q\, dy = -\int_0^{2\pi} \left[\left(\frac{1}{4}\cos^2 t + \frac{1}{2}\sin t\right)\left(-\frac{1}{2}\sin t\right)\right.$$
$$\left. + \left(\cos t + \sin t\right)\left(\frac{1}{2}\cos t\right)\right] dt = -\frac{\pi}{4};$$

$$\int_{C_2} P\, dx + Q\, dy = -\int_{-1}^1 (x^2 + 1)\, dx = -\tfrac{8}{3};$$

$$\int_{C_3} P\, dx + Q\, dy = -\int_{-1}^1 (-2 + 2y)\, dy = 4;$$

$$\int_{C_4} P\, dx + Q\, dy = -\int_{-1}^1 (x^2 - 1)\, dx = -\tfrac{4}{3},$$

and

$$\int_{C_5} P\, dx + Q\, dy = \int_{-1}^1 (2 + 2y)\, dy = 4.$$

Thus

$$\int_C P\, dx + Q\, dy = \sum_{i=1}^5 \int_{C_i} P\, dx + Q\, dy = 4 - \frac{\pi}{4},$$

which agrees with (24).

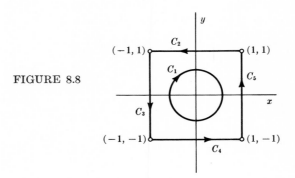

FIGURE 8.8

EXERCISES 6.8

In Exercises 1 through 11 verify the divergence theorem for the given vector function $\mathbf{F}$ and volume V.

1. $\mathbf{F}(\mathbf{X}) = x\mathbf{i} + y\mathbf{j} + z\mathbf{k}$; $V = \{\mathbf{X} \mid 0 \leq z \leq 4 - x^2 - y^2\}$.

2. $\mathbf{F}(\mathbf{X}) = x^2\mathbf{i} + y^2\mathbf{j} + z^2\mathbf{k}$; $V = \{\mathbf{X} \mid x^2 + y^2 \leq 4, \ 0 \leq z \leq x + 2\}$.

3. $\mathbf{F}(\mathbf{X}) = z\mathbf{i} + z\mathbf{j}$; $V = \{\mathbf{X} \mid \sqrt{x^2 + y^2} \leq z \leq 2\}$.

4. $\mathbf{F}(\mathbf{X}) = (x + y)\mathbf{i} + (y - z)\mathbf{j} + (x + y + z)\mathbf{k}$; the tetrahedron bounded by the plane $2x + y + z = 2$ and the coordinate planes.

5. $\mathbf{F}(\mathbf{X}) = x\mathbf{i} + y\mathbf{j} + z\mathbf{k}$; $V = \{\mathbf{X} \mid 0 \leq |\mathbf{X}| \leq a\}$.

6. $\mathbf{F}(\mathbf{X}) = x\mathbf{i} + y^2\mathbf{j} + z\mathbf{k}$; $V = \{\mathbf{X} \mid 2 + x^2 + y^2 \leq z \leq 6\}$.

7. $\mathbf{F}(\mathbf{X}) = x^2\mathbf{i} + y^2\mathbf{j} + z^2\mathbf{k}$; $V = \{\mathbf{X} \mid x_1 \leq x \leq x_2, \ y_1 \leq y \leq y_2,$
   $z_1 \leq z \leq z_2\}$.

8. $\mathbf{F}(\mathbf{X}) = \dfrac{x\mathbf{i} + y\mathbf{j} + z\mathbf{k}}{\sqrt{x^2 + y^2 + z^2}}$; $V = \{\mathbf{X} \mid 3 \leq |\mathbf{X}| \leq 4\}$.

9. $\mathbf{F}(\mathbf{X}) = x^2\mathbf{i} + y\mathbf{j} + z^2\mathbf{k}$; the solid bounded by the cylinder $x^2 + y^2 = 4$ and the planes $z = 1$ and $z = 3$.

10. $\mathbf{F}(\mathbf{X}) = 2x\mathbf{i} + 3y\mathbf{j} + z\mathbf{k}$; the solid bounded by the cylinder $y^2 + z^2 = 1$ and the planes $x = 0$ and $2x + y + z = 2$.

11. $F(\mathbf{X}) = x\mathbf{i} + y\mathbf{j} + z\mathbf{k}$; the solid bounded by the paraboloid $z = x^2 + y^2$, the cylinder $x^2 + y^2 = 9$, and the plane $z = 0$.

In Exercises 12 through 17 verify Stokes's theorem for the given function **F** and surface S.

12. $\mathbf{F}(\mathbf{X}) = z\mathbf{i} + x\mathbf{j} + y\mathbf{k}$; $\quad S = \{\mathbf{X} \mid 0 \leq z = 4 - x^2 - y^2\}$.

13. $\mathbf{F}(\mathbf{X}) = (x^2 + y)\mathbf{i} + yz\mathbf{j} + (x - z^2)\mathbf{k}$;

$\quad\quad S = \{\mathbf{X} \mid 2x + y + 2z = 2, \quad x \geq 0, \quad y \geq 0, \quad z \geq 0\}$.

14. $\mathbf{F}(\mathbf{X}) = x\mathbf{i} + z\mathbf{j} - y\mathbf{k}$; $\quad S = \{\mathbf{X} \mid |\mathbf{X}| = 2, \quad y \geq 0\}$.

15. $\mathbf{F}(\mathbf{X}) = x\mathbf{i} + y\mathbf{j}$; the part of the paraboloid $z = x^2 + y^2$ inside the cylinder $x^2 + y^2 = 4$.

16. $\mathbf{F}(\mathbf{X}) = y\mathbf{i} + x\mathbf{j} + z\mathbf{k}$; the part of the cylinder $x^2 + y^2 = 4$ between the planes $z = 1$ and $z = x + 4$.

17. $\mathbf{F}(\mathbf{X}) = (y + x)\mathbf{i} + (x + z)\mathbf{j} + z^2\mathbf{k}$; the part of the cone $z^2 = x^2 + y^2$ between the planes $z = 0$ and $z = 1$.

In Exercises 18 through 21 compute $\displaystyle\int_{\partial S} \mathbf{curl\ F \cdot n}\ dS$ by using Stokes's theorem. Orient S so that **n** has a nonnegative **k** component.

18. $\mathbf{F}(\mathbf{X}) = y\mathbf{i} + z\mathbf{j} + x\mathbf{k}$; S is the triangle with vertices $(1, 0, 0)$, $(0, 1, 0)$, $(0, 0, 1)$.

19. $\mathbf{F}(\mathbf{X}) = (x + y)\mathbf{i} + (y - z)\mathbf{j} + (x + y + z)\mathbf{k}$; S is the hemisphere $x^2 + y^2 + z^2 = a^2, z \geq 0$.

20. $\mathbf{F}(\mathbf{X}) = x\mathbf{i} + z\mathbf{j} + y\mathbf{k}$; S is the paraboloid $z = 4 - x^2 - y^2, \quad z \geq 0$.

21. $\mathbf{F}(\mathbf{X}) = x\mathbf{i} + y\mathbf{j}$; S is the surface $z = 1 - \sqrt{x^2 + y^2}, z \geq 0$.

In Exercises 22 through 29 verify Green's theorem in the plane for the given functions P and Q and the given region S.

22. $P(x, y) = x^2$, $\quad Q(x, y) = y^2$; $\quad S = \{(x, y) \mid 0 \leq x \leq 1, \quad 1 \leq y \leq 3\}$.

23. $P(x, y) = x$, $\quad Q(x, y) = y$; $\quad S = \{(x, y) \mid 1 \leq x^2 + y^2 \leq 4\}$.

24. $P(x, y) = 2y$, $\quad Q(x, y) = -3x$; the triangle bounded by the line $2x + y = 2$ and the coordinate axes.

25. $P(x, y) = y^2$, $\quad Q(x, y) = 2x$; the square with vertices $(1, 0)$, $(0, 1)$, $(-1, 0)$, and $(0, -1)$.

26. $P(x, y) = 2x + y, Q(x, y) = 3y + x$; the region bounded by $y = x^2$ and $y = x$.

27. $P(x, y) = y$, $Q(x, y) = 2x$;   the region between the circles $x^2 + y^2 = 4$ and $x^2 + y^2 = 9$.

28. $P(x, y) = -y$, $Q(x, y) = 2x$; $S = \{(x, y) \mid 0 \leq y \leq \sin x, \ 0 \leq x \leq \pi\}$.

29. $P(x, y) = 2y$, $Q(x, y) = 3x$; $S = \{(x, y) \mid \mid x \mid \leq y \leq 2\}$.

In Exercises 30 through 34 compute the line integral of $\mathbf{F}$ along C using Green's theorem. Orient C in the positive sense.

30. $\mathbf{F}(\mathbf{X}) = (x^2y + y^3 - \log x)\mathbf{i} + (xy^2 + x^2y + x^3 + \log y^2)\mathbf{j}$; C is the square with sides $x = 1$. $x = 2$, $y = 1$, $y = 2$.

31. $\mathbf{F}(\mathbf{X}) = (\sin y - x^2y)\mathbf{i} + (x \cos y + xy^2)\mathbf{j}$ over the circle $x^2 + y^2 = 1$.

32. $\mathbf{F}(\mathbf{X}) = (e^x \sin y - xy^2)\mathbf{i} + (e^x \cos y + x^2y)\mathbf{j}$ over the ellipse $x^2/4 + y^2/9 = 1$.

33. $\mathbf{F}(\mathbf{X}) = (x^2y + y^3x)\mathbf{i} + (y^4x^2 - x^3)\mathbf{j}$ over the boundary of the rectangle $1 \leq x \leq 3$, $0 \leq y \leq 1$.

34. $\mathbf{F}(\mathbf{X}) = (3x^2y^2 + \frac{1}{4}y^4 - 3xy)\mathbf{i} + (xy^3 + \frac{1}{3}x^3)\mathbf{j}$ over the boundary of the triangle with vertices $(0, 1)$, $(1, 0)$, and $(-1, 0)$.

THEORETICAL EXERCISES

T-1.   Verify Eq. (10).

T-2.   Let a region V be bounded by a closed orientable surface S, and suppose u is harmonic in an open set containing V and S. Show that

$$\int_S \frac{\partial u}{\partial \mathbf{n}} \, dS = 0$$

and

$$\int \frac{\partial (u^2)}{\partial \mathbf{n}} \, dS \geq 0,$$

where the inequality is strict if u is not constant.

T-3.   Let S and V be as in Exercise T-2, and suppose that u and v are both harmonic in an open set containing S and V. Show that

$$\int_S u \frac{\partial v}{\partial \mathbf{n}} \, dS = \int_S v \frac{\partial u}{\partial \mathbf{n}} \, dS.$$

Chapter 7

Series

7.1 Infinite Sequences

An *infinite sequence*, or more briefly a *sequence*, is a real valued function s defined on the set of integers greater than or equal to a given integer N. We denote the value of s for an integer n by s_n and write

$$(1) \qquad s = \{s_N, s_{N+1}, \ldots, s_{N+r}, \ldots\},$$

or more briefly

$$(2) \qquad s = \{s_n\}_N^\infty.$$

We often drop the symbol s altogether and say simply "the sequence $\{s_r\}_N^\infty \ldots;$" for example,

$$(3) \qquad \left\{\frac{1}{n}\right\}_1^\infty = \left\{1, \frac{1}{2}, \frac{1}{3}, \ldots\right\},$$

and

$$\{n\}_0^\infty = \{0, 1, 2, \ldots\}.$$

The real numbers $s_N, s_{N+1}, \ldots$ are called the *elements* of $\{s_n\}$. Usually, we are interested in the elements and the order in which they appear, but the particular initial value N chosen for the independent (subscript) variable is not important. It is always possible to renumber the elements in a sequence so that N can be replaced by zero, and we shall often do so as a matter of convenience. Thus, by defining $s_{N+r} = s'_r$ we can rewrite the right sides of (1) and (2) as

$$\{s'_0, s'_1, s'_2, \ldots\} = \{s'_r\}_0^\infty.$$

For example, (3) can be rewritten as

$$\left\{\frac{1}{n+1}\right\}_0^\infty = \left\{1, \frac{1}{2}, \frac{1}{3}, \ldots\right\}.$$

Henceforth, we shall write $\{s_r\}_0^\infty$ simply as $\{s_r\}$, reserving the notation $\{s_r\}_N^\infty$ for situations where it is necessary to specify N explicitly.

Earlier (Section 4.1), we defined the sum, difference, product, and quotient of two functions, as well as a constant multiple of a function. Nevertheless, we state the following definition for emphasis.

Definition 1.1   The *sum*, *difference*, and *product* of two sequences $\{s_n\}$ and $\{t_n\}$ are defined by

$$\{s_n\} + \{t_n\} = \{s_n + t_n\},$$
$$\{s_n\} - \{t_n\} = \{s_n - t_n\},$$

and

$$\{s_n\}\{t_n\} = \{s_n t_n\},$$

for those values of n for which s_n and t_n are both defined.

If $t_n \neq 0$ for all n sufficiently large, say $n \geq n_0 \geq 0$, then the *quotient of* $\{s_n\}$ *by* $\{t_n\}$ is

$$\frac{\{s_n\}}{\{t_n\}} = \left\{\frac{s_n}{t_n}\right\}_{n_0}^\infty.$$

If c is a real number, then

$$c\{s_n\} = \{c s_n\}.$$

Example 1.1   Let

$$\{s_n\} = \{n + 2\} = \{2, 3, 4, \ldots\},$$

and

$$\{t_n\} = \{n - 4\} = \{-4, -3, -2, \ldots\}.$$

Then

$$\{s_n\} + \{t_n\} = \{(n + 2) + (n - 4)\} = \{2n - 2\}$$
$$= \{-2, 0, 2, \ldots\},$$

and

$$\{s_n\} - \{t_n\} = \{(n+2) - (n-4)\} = \{6\}$$

$$= \{6, 6, 6, \ldots\}.$$

The product is given by

$$\{s_n\}\{t_n\} = \{(n+2)(n-4)\} = \{n^2 - 2n - 8\}$$

$$= \{-8, -9, -8, \ldots\},$$

while

$$\frac{\{s_n\}}{\{t_n\}} = \left\{\frac{n+2}{n-4}\right\}_5^\infty = \{7, 4, 3, \ldots\}$$

and

$$\frac{\{t_n\}}{\{s_n\}} = \left\{\frac{n-4}{n+2}\right\} = \left\{-2, -1, -\frac{1}{2}, \ldots\right\}.$$

Finally,

$$3\{s_n\} = \{3(n+2)\} = \{6, 9, 12, \ldots\}.$$

Bounded Sequences

Definition 1.2   A sequence $\{s_n\}$ is said to be

(a) *bounded* if there exists a number A such that $|s_n| \leq A$ $(n \geq 0)$;
(b) *bounded above* if there exists a number M such that $s_n \leq M$ $(n \geq 0$;
(c) *bounded below* if there exists a number m such that $s_n \geq m$ $(n \geq 0)$.

The numbers A, M, and m are called a *bound*, an *upper bound*, and a *lower bound*, respectively, for $\{s_n\}$.

Example 1.2   If $s_n = n + \dfrac{1}{n+1}$ then $\{s_n\}$ is bounded below $(s_n \geq 1)$, but unbounded above. If $s_n = -\left(n + \dfrac{1}{n+1}\right)$ then $\{s_n\}$ is bounded above $(s_n \leq -1)$, but unbounded below. The sequence defined by $s_n = \dfrac{(-1)^n}{n+1}$ is bounded $(|s_n| \leq 1)$, while the one defined by $s_n = (-1)^n n$ is not bounded above or below.

The following property of the real numbers, known as *completeness*, is needed for the study of convergence of sequences.

Completeness Axiom   Suppose S is a nonempty set of real numbers which is bounded above; that is, there is a number M such that every x in S satisfies $x \leq M$. Then there is a unique least real number M_0, called the *least upper bound of S*, such that $x \leq M_0$ for all x in S. We write $M_0 =$ l.u.b. S.

If $S = \{s_n\}$ and $\epsilon > 0$, then $s_m > M_0 - \epsilon$ for some m, for if $s_m \leq M_0 - \epsilon$ for all m, then $M_0 - \epsilon$ would also be an upper bound for $\{s_n\}$, which violates the definition of M_0.

Convergent Sequences

Usually we shall be interested in the behavior of a sequence $\{s_n\}$ as n becomes large; specifically, we shall be interested in whether a sequence *approaches a limit*, which we now define.

Definition 1.3   A sequence $\{s_n\}$ *has limit L* if for every $\epsilon > 0$ there is an integer N, which depends on ϵ, such that

$$(4) \qquad\qquad |s_n - L| < \epsilon \qquad \text{if} \quad n \geq N.$$

We say that $\{s_n\}$ *converges to L*, or *approaches L*, and write

$$\lim_{n \to \infty} s_n = L.$$

A sequence is said to be *convergent* if it has a limit, and *divergent* if it does not have a limit.

Theorem 1.1   If a sequence $\{s_n\}$ has a limit L, then L is its only limit; that is, the limit of a sequence is unique.

Proof.   Suppose L and L' are both limits of $\{s_n\}$. From Definition 1.3, corresponding to each $\epsilon > 0$ there are integers N and N' such that $|s_n - L| < \epsilon$ if $n \geq N$ and $|s_n - L'| < \epsilon$ if $n \geq N'$. Now we write

$$|L - L'| = |(L - s_n) + (s_n - L')|$$

$$\leq |L - s_n| + |s_n - L'|,$$

and take $n \geq \max(N, N')$ to obtain

$$| L - L' | < 2\epsilon.$$

Since ϵ can be taken as small as we wish, this implies that $L = L'$.

Definition 1.3 can be interpreted geometrically to mean that any arbitrarily small interval $(L - \epsilon, L + \epsilon)$ contains all but a finite number of the elements s_n.

Example 1.3 The sequence $\left\{ (-1)^n \dfrac{1}{n+1} \right\}$ approaches zero, and we write $\lim\limits_{n \to \infty} \dfrac{(-1)^n}{n+1} = 0$. To verify this, we observe that $\left| \dfrac{(-1)^n}{n+1} - 0 \right| = \dfrac{1}{n+1}$, which is less than ϵ if $n + 1 \geq 1/\epsilon$. Therefore (4) is satisfied if $N > (1/\epsilon) - 1$.

Example 1.4 Consider the sequence $\left\{ \dfrac{n}{n+1} \right\}$. We claim that

$$\lim_{n \to \infty} \frac{n}{n+1} = 1.$$

To verify this we calculate

$$\left| \frac{n}{n+1} - 1 \right| = \left| \frac{n - (n+1)}{n+1} \right| = \frac{1}{n+1},$$

and again (4) is satisfied if $N > (1/\epsilon) - 1$.

Example 1.5 The sequence $\{ (-1)^n \}$ is divergent. We show that this is so by assuming that $\lim\limits_{n \to \infty} (-1)^n$ exists and deducing a contradiction. Suppose $\lim\limits_{n \to \infty} (-1)^n = L$; then there is an integer N such that

$$(5) \qquad\qquad | (-1)^n - L | < \epsilon \qquad \text{for} \quad n \geq N.$$

Letting $n = 2m \geq N$ in (5) yields

$$(6) \qquad\qquad | 1 - L | < \epsilon,$$

while letting $n = 2m + 1 \geq N$ in (5) yields

$$(7) \qquad\qquad | -1 - L | = | 1 + L | \leq \epsilon.$$

However, (6) and (7) are mutually contradictory if $\epsilon < \frac{1}{2}$; hence $\{ (-1)^n \}$ is divergent.

From Definition 1.3, it follows that a convergent sequence is bounded (Exercise T-3), and that a constant sequence $\{c\}$ (that is, $s_n = c$ for all n) converges to c (Exercise T-4).

Theorem 1.2   Let $\lim\limits_{n \to \infty} s_n = S$ and $\lim\limits_{n \to \infty} t_n = T$. Then

(i) $\lim\limits_{n \to \infty} (s_n + t_n) = S + T$;

(ii) $\lim\limits_{n \to \infty} (s_n - t_n) = S - T$;

(iii) $\lim\limits_{n \to \infty} s_n t_n = ST$;

(iv) $\lim\limits_{n \to \infty} \dfrac{s_n}{t_n} = \dfrac{S}{T}$ if $T \neq 0$.

We prove (i) and (iii) and leave the proofs of (ii) and (iv) to the student (Exercises T-5 and T-6).

Proof of (i).   Given $\epsilon > 0$ there exist integers N_1 and N_2 such that

$$| s_n - S | < \epsilon \qquad \text{if} \quad n > N_1$$

and

$$| t_n - T | < \epsilon \qquad \text{if} \quad n > N_2.$$

These two inequalities are both satisfied if $n \geq N = \max(N_1, N_2)$. Therefore,

$$
\begin{aligned}
| (s_n + t_n) - (S + T) | &= | (s_n - S) + (t_n - T) | \\
&\leq | s_n - S | + | t_n - T | \\
&\leq \epsilon + \epsilon = 2\epsilon \qquad \text{if} \quad n \geq N.
\end{aligned}
$$

This proves (i), since replacing ϵ by any constant multiple of ϵ in (4) yields an equivalent definition of the limit of a sequence.

Proof of (iii).   Let ϵ and N be as in the proof of (i); then

$$
\begin{aligned}
| s_n t_n - ST | &= | (s_n - S)(t_n - T) + T(s_n - S) + S(t_n - T) | \\
&\leq | s_n - S | | t_n - T | + | T | | s_n - S | + | S | | (t_n - t) | \\
&\leq \epsilon^2 + | T | \epsilon + | S | \epsilon \leq \epsilon(| T | + | S | + 1)
\end{aligned}
$$

if $\epsilon \leq 1$ and $n \geq N$. This completes the proof of (iii).

Monotonic Sequences

A sequence $\{s_n\}$ in which each element is larger than its predecessor,

$$(8) \qquad s_{n+1} > s_n \qquad (n = 0, 1, \ldots),$$

is said to be *monotonically increasing*. If (8) is relaxed to allow equality,

$$(9) \qquad s_{n+1} \geq s_n \qquad (n = 0, 1, \ldots),$$

then $\{s_n\}$ is *monotonically nondecreasing*. If

$$(10) \qquad s_{n+1} < s_n \qquad (n = 0, 1, \ldots)$$

then $\{s_n\}$ is *monotonically decreasing*, while if

$$(11) \qquad s_{n+1} \leq s_n \qquad (n = 0, 1, \ldots),$$

then $\{s_n\}$ is *monotonically nonincreasing*.

A sequence satisfying any one of the inequalities (8) through (11) for all n is *monotonic*. Notice that if $\{s_n\}$ is monotonically increasing (nondecreasing), then $\{-s_n\}$ is monotonically decreasing (nonincreasing).

Example 1.6 The sequence

$$\left\{ \frac{n}{n+1} \right\} = \left\{ 0, \frac{1}{2}, \frac{2}{3}, \ldots \right\}$$

is monotonically increasing, while

$$\left\{ \frac{1}{n+1} \right\} = \left\{ 1, \frac{1}{2}, \frac{1}{3}, \ldots \right\}$$

is monotonically decreasing. The sequence

$$\{s_n\} = \{0, 0, 1, 1, 2, 2, \ldots, m, m, \ldots\},$$

defined by $s_{2m} = s_{2m+1} = m$, is monotonically nondecreasing, while

$$\{-s_n\} = \{0, 0, -1, -1, -2, -2, \ldots, -m, -m, \ldots\}$$

is monotonically nonincreasing.

The question of convergence of a monotonic sequence is easily settled, as shown by the following theorem.

Theorem 1.3 (a) A monotonically nondecreasing sequence is convergent if and only if it is bounded above, in which case the limit of the sequence equals its least upper bound. (b) A monotonically nonincreasing sequence is convergent if and only if it is bounded below, in which case the limit of the sequence equals its greatest lower bound.

Proof of (a). We remarked earlier that any convergent sequence is bounded (Exercise T-3); hence we prove only the converse. If $\{s_n\}$ is bounded above, it has a least upper bound M, from the completeness axiom. Since M is an upper bound for $\{s_n\}$, $s_n \leq M$ for all n; since M is the least upper bound, there exists for any $\epsilon > 0$ an N (which depends upon ϵ) such that $s_N > M - \epsilon$. (See the remark following the completeness axiom.) If $\{s_n\}$ is nondecreasing, then $s_n \geq s_N$ for $n \geq N$. Thus, we have the inequalities

$$M - \epsilon \leq s_N \leq s_n \leq M \qquad (n \geq N),$$

which imply that $|s_n - M| \leq \epsilon$ if $n \geq N$. Therefore, $\lim_{n \to \infty} s_n = M$, which completes the proof of (a).

We leave the proof of (b) to the student (Exercise T-7).

A Convergence Criterion for Sequences

Definition 1.3 suggests only one way to determine whether a sequence is convergent: Try to guess the limit L, and verify (4) for each $\epsilon > 0$. This method is unsatisfactory if $\{s_n\}$ does not have a limit, or has a limit which is difficult to guess. Moreover, it suffices in many applications to know that a limit does or does not exist, while its value is unimportant. It is therefore desirable to have a convergence test that can be applied directly to the elements of $\{s_n\}$, and does not require advance knowledge of the limit, which may not even exist. The following theorem provides such a test.

Theorem 1.4 A sequence $\{s_n\}$ is convergent if and only if, corresponding to each $\epsilon > 0$, there is an integer N, which depends upon ϵ, such that

$$(12) \qquad |s_n - s_m| < \epsilon \qquad \text{if} \quad n, m \geq N.$$

Proof. Suppose $\{s_n\}$ is convergent and $\lim_{n \to \infty} s_n = L$. Then for each $\epsilon > 0$ there exists an integer N such that $|s_r - L| < \epsilon/2$ if $r \geq N$.

If m and n are both greater than or equal to N, it follows that $|s_m - L| < \epsilon/2$ and $|s_n - L| < \epsilon/2$, and therefore

$$|s_n - s_m| \leq |(s_n - L) - (s_m - L)| \leq |(s_n - L)| + |s_m - L|$$

$$< \frac{\epsilon}{2} + \frac{\epsilon}{2} = \epsilon,$$

which proves that $\{s_n\}$ satisfies (12).

Now suppose $\{s_n\}$ satisfies (12). Then $\{s_n\}$ is bounded (Exercise T-3), say $|s_n| \leq K$, for $n = 0, 1, 2, \ldots$, and consequently so are the sequences

$$s^0 = \{s_0, s_1, s_2, \ldots\},$$

$$s^1 = \{s_1, s_2, s_3, \ldots\},$$

$$\vdots$$

$$s^r = \{s_r, s_{r+1}, s_{r+2}, \ldots\},$$

where s^r is obtained by omitting the first r elements from $\{s_n\}$. Let

$$\alpha_r = \text{l.u.b.}\{s_r, s_{r+1}, s_{r+2}, \ldots\} \qquad (r = 0, 1, 2, \ldots);$$

Then $\{\alpha_r\}$ is bounded below ($\alpha_r \geq -K$) and nonincreasing (why?), and therefore has a limit, by Theorem 1.3. Let $\lim_{r \to \infty} \alpha_r = \alpha$. We shall now show that $\lim_{n \to \infty} s_n = \alpha$. Let $\epsilon > 0$ be given. Then:

(i)   Since $\{s_n\}$ satisfies (12), there exists an integer N such that

$$|s_n - s_m| < \frac{\epsilon}{3} \qquad \text{if} \quad n, m \geq N.$$

(ii)   Since $\lim_{r \to \infty} \alpha_r = \alpha$, there exists a fixed $r_0 \geq N$ such that

$$|\alpha - \alpha_{r_0}| < \frac{\epsilon}{3}.$$

(iii)   Since $\alpha_{r_0} = \text{l.u.b.}\{s_{r_0}, s_{r_0+1}, s_{r_0+2}, \ldots\}$, there exists an integer $m \geq r_0$ such that $\alpha_{r_0} - (\epsilon/3) \leq s_m \leq \alpha_{r_0}$, which implies that

$$|s_m - \alpha_{r_0}| < \frac{\epsilon}{3}.$$

Now we can write

$$(13) \qquad |s_n - \alpha| = |(s_n - s_m) + (s_m - \alpha_{r_0}) + (\alpha_{r_0} - \alpha)|$$

$$\leq |s_n - s_m| + |s_m - \alpha_{r_0}| + |\alpha_{r_0} - \alpha|,$$

where each term is less than $\epsilon/3$ if $n \geq N$ and α_{r_0} and m are chosen as in (ii) and (iii). Since the left side of (13) depends only on n, it follows that $|s_n - \alpha| < \epsilon$ if $n \geq N$, which proves that $\lim_{n \to \infty} s_n = \alpha$.

We point out that condition (12) is quite strong, since it must hold for *all* m and n, no matter how far apart they may be, provided only that they both exceed N.

Example 1.7   The sequence $\{s_n\}$ with $s_n = \dfrac{1}{n+1}$ satisfies (12). To see this, suppose $n \geq m$; then

$$|s_n - s_m| = \left| \frac{1}{n+1} - \frac{1}{m+1} \right| = \frac{1}{m+1}\left| 1 - \frac{m+1}{n+1} \right|$$

$$< \frac{1}{m+1} < \epsilon$$

if $m + 1 > 1/\epsilon$. Therefore, $|s_n - s_m| < \epsilon$ if $n \geq m \geq N > 1/\epsilon$. (The requirement that $n \geq m$ is really no restriction at all; it merely means that we choose to designate the larger of n and m by n.)

Example 1.8   The sequence $\{s_n\} = \{n\}$ does not satisfy (12), since, for any integer N, we can make $(n - m)$ as large as we wish even while requiring that $n \geq m \geq N$.

Subsequences

A *subsequence* of an infinite sequence $\{s_n\}$ is a sequence of the form

$$\{s_{n_k}\} = \{s_{n_0}, s_{n_1}, s_{n_2}, \ldots\},$$

where $n_0 < n_1 < n_2 < \cdots$. Aside from the trivial case (corresponding to $n_0 = 0, n_1 = 1, \ldots, n_k = k, \ldots$) where the subsequence is identical with the original sequence, we can think of the subsequence $\{s_{n_k}\}$ as being obtained from $\{s_n\}$ by omitting some of the elements of the latter, and keeping the relative order of the remaining elements the same as it was originally.

Example 1.9   The sequence

$$\left\{ \frac{1}{2(k+1)} \right\} = \left\{ \frac{1}{2}, \frac{1}{4}, \frac{1}{6}, \ldots \right\}$$

is a subsequence of

$$(14) \qquad \left\{ \frac{1}{n+1} \right\} = \left\{ 1, \frac{1}{2}, \frac{1}{3}, \ldots \right\},$$

with $n_0 = 1, n_1 = 3, \ldots, n_k = 2k + 1, \ldots$. Another subsequence of (14) is

$$\left\{ \frac{1}{(k+1)^2} \right\} = \left\{ 1, \frac{1}{4}, \frac{1}{9}, \ldots \right\},$$

this time with $n_0 = 0, n_1 = 3, n_2 = 8, \ldots, n_k = (k+1)^2 - 1$.

Theorem 1.5 If a sequence $\{s_n\}$ converges to a limit L, then every subsequence of $\{s_n\}$ converges to L.

We leave the proof of this theorem to the student (Exercise T-8).

It is possible for a divergent series to have convergent subsequences, as in the next example.

Example 1.10 If

(15) $$\{s_n\} = \{(-1)^n\},$$

then the subsequences $\{s_{2n}\} = \{1\}$ and $\{s_{2n+1}\} = \{-1\}$ are both convergent. In fact, a subsequence $\{s_{n_k}\}$ of (15) is convergent if and only if the integers $\{n_k\}$ are, for sufficiently large k, either all odd or all even (Exercise T-9).

Divergence to Infinity

Definition 1.4 A sequence $\{s_n\}$ will be said to *diverge to* ∞ if, for every $\rho > 0$, there is an integer N such that $s_n > \rho$ for all $n \geq N$. If $\{s_n\}$ diverges to ∞ we shall write

$$\lim_{n \to \infty} s_n = \infty.$$

A sequence $\{s_n\}$ will be said to *diverge to* $-\infty$ if $\lim_{n \to \infty}(-s_n) = \infty$.

We emphasize that this definition merely points out a particular way in which a sequence may diverge; a sequence may be divergent without diverging to ∞ or $-\infty$.

Example 1.11 The sequences $\{n^2\}$ and $\{2^n\}$ diverge to ∞. Also,

$$\lim_{n \to \infty} \sqrt{n} = \infty.$$

Even though the sequence $\{s_n\}$ defined by

$$s_n = [1 + (-1)^n]n^3$$

has arbitrarily large elements, it does not diverge to ∞, since $s_{2r+1} = 0$ for all $r \geq 0$. However, the subsequence $\{s_{2r}\}$ does converge to ∞, since $s_{2r} = 2(2r)^3 = 16r^3$.

EXERCISES 7.1

In Exercises 1 through 7, write the first five terms of the sequence.

1. $\{2n + 1\}$. 2. $\{(-2k)^{k-1}\}$. 3. $\{(j + 1)^j\}$. 4. $\left\{\dfrac{1}{2n - 1}\right\}$.

5. $\left\{\dfrac{1}{2n}\right\}_1^\infty$. 6. $\left\{\dfrac{e^n}{n(n - 1)}\right\}_2^\infty$. 7. $\left\{\dfrac{1}{n}\sin\dfrac{n\pi}{4}\right\}_1^\infty$.

In Exercises 8 through 11, find $\{s_n\} + \{t_n\}$, $\{s_n\} - \{t_n\}$, and $c\{s_n\}$.

8. $s_n = 3n - 2$, $t_n = 2n + 1$, $c = -1$.

9. $s_n = 2n + 1$, $t_n = n - 2$, $c = 3$.

10. $s_n = e^n$, $t_n = e^{-n}$, $c = \tfrac{1}{2}$.

11. $s_n = \sin\dfrac{n\pi}{4}$, $t_n = \cos\dfrac{n\pi}{4}$, $c = 3$.

12. $s_n = n^2$, $t_n = n^2$, $c = -1$.

In Exercises 13, 14, and 15, state which of the sequences are bounded, bounded below, or bounded above.

13. (a) $\left\{\dfrac{3n + 2}{n - 1}\right\}_2^\infty$. (b) $\{[(-1)^n + 1]n^2 \sin(n + \tfrac{1}{2})\pi\}$.

(c) $\left\{\dfrac{n^2}{n - 1}\right\}_2^\infty$. (d) $\left\{\dfrac{2^n}{n + 1}\right\}$.

14. (a) $\{n \cos n\pi\}$. (b) $\left\{|n| \sin\dfrac{n\pi}{2}\right\}$.

(c) $\left\{\dfrac{n-1}{n+1}\right\}$. (d) $\{[(-1)^n+1]n\cos n\pi\}$.

15. (a) $\{n^2\sin n\pi\}$. (b) $\left\{1-\dfrac{1}{n+1}\right\}$.

(c) $n^2-n\sin\dfrac{n\pi}{2}$. (d) $\left\{\dfrac{1}{n}\sin\dfrac{n^2\pi}{2}\right\}$.

In Exercises 16 through 25, find the limit of the given sequence, if it exists, or state that there is no limit. In either case, prove your statement.

16. $\{1+(-1)^n\}$. 17. $\left\{\dfrac{2n-1}{n^2}\right\}_1^\infty$.

18. $\{\sqrt{n}\}$. 19. $\left\{\dfrac{n+1}{2n}\right\}_1^\infty$.

20. $\left\{\dfrac{1}{n}+\dfrac{1}{n-2}\right\}_3^\infty$. 21. $\{4\}$.

22. $\{(-1)^n\cos n\pi\}$. 23. $\left\{\dfrac{\sqrt{n^2+1}}{n+1}\right\}$.

24. $\left\{\dfrac{n(n+1)}{3n+2}\right\}$. 25. $\left\{\dfrac{n}{\sqrt{n^2+1}}\right\}$.

26. State which of the sequences of Exercise 13 are monotonic.

27. State which of the sequences of Exercise 14 are monotonic.

28. State which of the sequences of Exercise 15 are monotonic.

29. Let $s_n=\sin\dfrac{n\pi}{2}+\cos\dfrac{n\pi}{2}$. (a) Show that $\{s_n\}$ is divergent. (b) Show that the subsequences $\{s_{4r}\}$, $\{s_{4r+1}\}$, $\{s_{4r+2}\}$, and $\{s_{4r+3}\}$ are convergent, and find their limits.

30. Let $s_n=\dfrac{n-3}{n+3}$. (a) Show that the subsequence $\{s_{n_r}\}$, where $n_r=r^2$, has the same limit as $\{s_n\}$. (b) Repeat part (a) with $n_r=3r$.

THEORETICAL EXERCISES

T-1. Show that the least upper bound of a bounded set is unique.

T-2. *Prove:* If a set S is bounded below then there is a greatest real number m_0 (called *the greatest lower bound of* S) such that $x \geq m_0$ for all x in S. (*Hint:* Use the completeness axiom.)

T-3. Prove that an infinite sequence which satisfies Equation (12) is bounded. Then conclude that a convergent sequence is bounded.

T-4. If $s_n = c$ for all n, prove that $\lim_{n \to \infty} s_n = c$.

T-5. Prove (ii) of Theorem 1.2.

T-6. Prove (iv) of Theorem 1.2.

T-7. Prove (b) of Theorem 1.3.

T-8. Prove Theorem 1.5.

T-9. Show that a subsequence $\{s_{n_k}\}$ of $\{s_n\} = \{(-1)^n\}$ is convergent if and only if, for sufficiently large k, the integers $\{n_k\}$ are either all odd or all even.

T-10. Suppose $\lim_{n \to \infty} s_{2n} = L_1$ and $\lim_{n \to \infty} s_{2n+1} = L_2$. Show that $\lim_{n \to \infty} s_n$ exists if and only if $L_1 = L_2$, and that $\lim_{n \to \infty} s_n$ is equal to the common value of L_1 and L_2.

T-11. Suppose $s_n \leq t_n$ and $\{s_n\}$ and $\{t_n\}$ converge. Show that $\lim_{n \to \infty} s_n \leq \lim_{n \to \infty} t_n$.

T-12. Suppose $s_n \leq p_n \leq t_n$ and $\lim_{n \to \infty} s_n = \lim_{n \to \infty} t_n = L$. Show that $\lim_{n \to \infty} p_n = L$.

7.2 Infinite Series

A finite sum of real numbers

(1)
$$a_0 + a_1 + \cdots + a_n = \sum_{r=0}^{n} a_r$$

is easily understood, no matter how large n is; to obtain its value, we simply perform n additions. However, a sum of infinitely many terms

$$(2) \qquad a_0 + a_1 + \cdots + a_n + \cdots,$$

where the second "$\cdots$" indicates that we continue adding terms indefinitely, is another matter. Indeed, the phrase "adding terms indefinitely" is nonsensical in the absence of a precise definition.

An expression of the form (2) is called an *infinite series*, and is written more briefly as

$$(3) \qquad \sum_{r=0}^{\infty} a_r = a_0 + a_1 + \cdots + a_n + \cdots.$$

We call a_r the rth *term* of the series. To motivate the definition of the sum of an infinite series, we compute the sum (1) recursively:

$$s_0 = a_0,$$

$$s_1 = s_0 + a_1,$$

$$(4) \qquad s_2 = s_1 + a_2,$$

$$\vdots$$

$$s_n = s_{n-1} + a_n.$$

It can be shown that s_r is given explicitly by

$$(5) \qquad s_r = a_0 + a_1 + \cdots + a_r \qquad (0 \le r \le n),$$

and that s_n is the value of the sum (1).

We stopped the calculation (4) with s_n because there were no terms left to add to it. Now suppose $\{a_r\}$ is an infinite sequence of real numbers. We can associate with it a second sequence $\{s_r\}$, where s_r is defined by (5); we say that s_r is the rth *partial sum of the infinite series* $\sum_{r=0}^{\infty} a_r$.

Definition 2.1 If $\{s_r\}$ is the sequence of partial sums (4) of the infinite series $\sum_{r=0}^{\infty} a_r$ and

$$\lim_{r \to \infty} s_r = S$$

exists, we say that $\sum_{r=0}^{\infty} a_r$ is *convergent*, and that it *converges to the sum S.*

In this case we write

$$S = \sum_{r=0}^{\infty} a_r.$$

If $\{s_r\}$ is divergent, then we say that $\sum_{r=0}^{\infty} a_r$ is *divergent*.

Example 2.1 If $a_r = 0$ for $r > n$ then

(6) $$\sum_{r=0}^{\infty} a_r = a_0 + a_1 + \cdots + a_n + 0 + 0 + \cdots$$

and $s_{r+1} = s_r$ for $r \geq n$; therefore

$$\lim_{r \to \infty} s_r = s_n = a_0 + a_1 + \cdots + a_n.$$

It is natural to identify the infinite series (6) with the finite sum (1).

Example 2.2 Let α be a fixed real number, and consider the infinite series

(7) $$\sum_{r=0}^{\infty} \alpha^r = 1 + \alpha + \cdots + \alpha^r + \cdots,$$

which is called the *geometric series with ratio* α. Its nth partial sum is

(8) $$s_n = 1 + \alpha + \cdots + \alpha^n.$$

To find s_n in closed form, we multiply (8) by α to obtain

(9) $$\alpha s_n = \alpha + \alpha^2 + \cdots + \alpha^{n+1}$$

and subtract (9) from (8) to find that

$$(1 - \alpha)s_n = 1 - \alpha^{n+1};$$

hence, if $\alpha \neq 1$,

(10) $$s_n = \frac{1 - \alpha^{n+1}}{1 - \alpha}.$$

Since $\lim_{n \to \infty} \alpha^{n+1} = 0$ if $|\alpha| < 1$, we have

$$\sum_{r=0}^{\infty} \alpha^r = \lim_{n \to \infty} \frac{1 - \alpha^{n+1}}{1 - \alpha} = \frac{1}{1 - \alpha} \qquad \text{if} \quad |\alpha| < 1.$$

We shall see in Example 2.6 that the geometric series diverges if $|\alpha| \geq 1$. Thus the geometric series converges if and only if $|\alpha| < 1$.

It is often convenient to allow the index r in (3) to begin with an integer $r_0 \neq 0$. Therefore, we shall agree that the sum of the series

(11)
$$\sum_{r=r_0}^{\infty} b_r$$

is the limit of the sequence defined by

$$s_0 = b_{r_0},$$
$$s_1 = b_{r_0} + b_{r_0+1},$$
$$\vdots$$
$$s_n = b_{r_0} + b_{r_0+1} + \cdots + b_{r_0+n},$$

if the sequence has a limit. It is a trivial matter to rewrite (11) in the form (3) with

$$a_0 = b_{r_0}, a_1 = b_{r_0+1}, \ldots, a_n = b_{r_0+n}, \ldots.$$

For example,

$$\sum_{r=1}^{\infty} (-1)^{r-1} \frac{1}{r} = \sum_{r=0}^{\infty} (-1)^r \frac{1}{r+1}.$$

Theorem 2.1 If $\sum_{r=0}^{\infty} a_r$ and $\sum_{r=0}^{\infty} b_r$ exist, then so do $\sum_{r=0}^{\infty} (a_r + b_r)$, $\sum_{r=0}^{\infty} (a_r - b_r)$, and $\sum_{r=0}^{\infty} (ca_r)$, where c is a real number. Moreover,

$$\sum_{r=0}^{\infty} (a_r + b_r) = \sum_{r=0}^{\infty} a_r + \sum_{r=0}^{\infty} b_r,$$

$$\sum_{r=0}^{\infty} (a_r - b_r) = \sum_{r=0}^{\infty} a_r - \sum_{r=0}^{\infty} b_r,$$

and

$$\sum_{r=0}^{\infty} (ca_r) = c\left(\sum_{r=0}^{\infty} a_r\right).$$

We leave the proof of this theorem to the student (Exercise T-1).

It follows from Definition 2.1 that the convergence or divergence of a series is not altered if a finite number of terms at its beginning are changed. Thus, $\sum_{r=0}^{\infty} a_r$ and $\sum_{r=0}^{\infty} b_r$ are convergent or divergent together if $a_r = b_r$ for all

$r \geq m$ (Exercise T-2). In particular, taking $b_r = 0$ for $0 \leq r \leq m - 1$ and $b_r = a_r$ for $r > m$, we conclude that $\sum\limits_{r=0}^{\infty} a_r$ and $\sum\limits_{r=m}^{\infty} a_r$ are convergent or divergent together. Because of this, we shall write simply $\sum a_r$ when we are interested only in whether a series converges, and not in the actual value of its sum. Thus, the statement "$\sum a_r$ is convergent" means that $\sum\limits_{r=r_0}^{\infty} a_r$ is convergent for all r_0 such that a_r is defined for $r \geq r_0$.

Example 2.3 For each fixed p, all of the series

$$\sum_{r=r_0}^{\infty} \frac{1}{r^p} \qquad (r_0 = 1, 2, \ldots)$$

converge or diverge together. However, the expression $\sum\limits_{r=0}^{\infty} \frac{1}{r^p}$ is meaningless, since the general term is not defined for $r = 0$.

The next theorem provides a general convergence criterion.

Theorem 2.2 An infinite series $\sum a_r$ is convergent if and only if, for each $\epsilon > 0$, there exists an integer N such that

(12) $\qquad\qquad | a_n + a_{n+1} + \cdots + a_m | < \epsilon \qquad (m \geq n \geq N).$

Proof. Let s_n be the sequence of partial sums of $\sum\limits_{r=0}^{\infty} a_r$. If $m \geq n > 0$ then

$$a_n + a_{n+1} + \cdots + a_m = s_m - s_{n-1};$$

consequently, $\{s_n\}$ satisfies (12) of Section 7.1 if and only if the hypotheses of the theorem are satisfied. The proof now follows from Theorem 1.4, Section 7.1.

Intuitively, Theorem 2.2 means that $\sum a_r$ is convergent if and only if arbitrarily long sums of the form

$$a_n + a_{n+1} + \cdots + a_m = \sum_{r=n}^{m} a_r$$

are small, provided only that they start sufficiently far out.

Example 2.4 We know from Example 2.2 that the geometric series (7) is convergent if $| \alpha | < 1$. We can obtain a second proof of this result by observing that

(13) $$\sum_{n}^{m} \alpha^r = \alpha^n + \alpha^{n+1} + \cdots + \alpha^m = \alpha^n(1 + \alpha + \cdots + \alpha^{m-n})$$
$$= \alpha^n \frac{(1 - \alpha^{m-n+1})}{1 - \alpha}.$$

(See (10), with n replaced by $m - n$.) For $|\alpha| < 1$, (13) implies that

$$\left| \sum_{n}^{m} \alpha^r \right| < \frac{2|\alpha|^n}{1 - \alpha},$$

and therefore (12) is satisfied if N is chosen so large that

$$\frac{2|\alpha|^N}{1 - \alpha} < \epsilon.$$

Example 2.5 The *harmonic series*, $\sum \dfrac{1}{r}$, is divergent because no matter how large n is,

$$\sum_{n}^{2n} \frac{1}{r} = \frac{1}{n} + \frac{1}{n+1} + \cdots + \frac{1}{2n} > (n+1)\frac{1}{2n} > \frac{1}{2};$$

therefore (12) is violated for every n if we take $m = 2n$ and $\epsilon \le \frac{1}{2}$.

If $\sum a_r$ is convergent, then (12) with $n = m$ implies that $|a_n| < \epsilon$ if $n \ge N$. This yields the following necessary condition for convergence of an infinite series.

Theorem 2.3 If $\sum a_r$ is convergent, then

(14) $$\lim_{r \to \infty} a_r = 0;$$

thus, $\sum a_r$ must diverge if (14) is not satisfied.

We emphasize that (14) is a necessary, but not sufficient, condition for convergence of $\sum a_r$. That is, convergence of $\sum a_r$ implies (14), but (14) alone does not imply convergence of $\sum a_r$.

Example 2.6 The geometric series $\sum \alpha^r$ diverges if $|\alpha| \ge 1$, since then the sequence $\{\alpha^r\}$ does not converge to zero. The harmonic series $\sum \dfrac{1}{r}$ is divergent (Example 2.5), even though $\lim_{r \to \infty} \dfrac{1}{r} = 0$.

Series with Positive Terms

A series may diverge even though its sequence of partial sums is bounded.

For example $\sum_{r=0}^{\infty} (-1)^r$ is divergent (by Theorem 2.3) even though its partial sums satisfy $0 \le s_n \le 1$. However, a divergent series whose elements do not change sign cannot have bounded partial sums, as is shown by the following theorem.

Theorem 2.4 If $a_r > 0$ for $r = 0, 1, \ldots$ then $\sum a_r$ is convergent if and only if its sequence $\{s_n\}$ of partial sums is bounded above.

Proof. Since $s_{n+1} = s_n + a_{n+1} > s_n$, the sequence $\{s_n\}$ is increasing; hence $\{s_n\}$ converges if and only if it is bounded above (Theorem 1.3, Section 7.1).

Since $\sum_{r=0}^{\infty} a_r$ and $\sum_{r=r_0}^{\infty} a_r$ converge or diverge together, the hypothesis of Theorem 2.4 can be weakened to require only that $a_r > 0$ for $r = r_0$, $r_0 + 1, \ldots$, where r_0 is any nonnegative integer. Any convergence criterion for $\sum a_r$ will, of course, place conditions on the sequence $\{a_r\}$. However, it is not necessary for all elements of $\{a_r\}$ to satisfy the given condition; it suffices that they be satisfied for $r \ge r_0$ for some fixed $r_0 \ge 0$. (We say then that the conditions are met *for sufficiently large r.*) For convenience, we shall continue to state theorems with conditions on $\{a_r\}$ for all $r \ge 0$; however, the theorems will all continue to hold if their hypotheses are satisfied only for sufficiently large r.

Theorem 2.4 implies that a series of positive terms can diverge in only one way: Its partial sums must approach ∞. If $\sum_{r=0}^{\infty} a_r$ is an arbitrary series, possibly with infinitely many negative terms, such that $\lim_{n \to \infty} s_n = \infty$, then we shall write

(15) $$\sum a_r = \infty.$$

If $\lim_{n \to \infty} s_n = -\infty$, we shall write

(16) $$\sum a_r = -\infty.$$

Both (15) and (16) indicate that $\sum a_r$ diverges. We emphasize that a series such as $\sum (-1)^r$ may diverge without satisfying either (15) or (16).

We can also conveniently indicate convergence of a series whose terms are positive for all sufficiently large r by writing

$$\sum a_r < \infty .$$

Example 2.7 From our investigation of the geometric series, we know that

$$\sum 2^n = \infty ,$$

while

$$\sum \frac{1}{2^n} < \infty .$$

The series $\sum_{r=0}^{\infty} (-2)^r$ is divergent; however it does not satisfy either (15) or (16), since its partial sums

$$s_n = \sum_{r=0}^{n} (-2)^r$$

alternate in sign.

The next three theorems give useful tests for convergence of series with positive terms.

Theorem 2.5 (Comparison Test) Let $\sum a_r$ and $\sum b_r$ be series with positive terms and suppose that

(17) $$a_r \leq b_r.$$

Then:

(i) If $\sum b_r$ is convergent, so is $\sum a_r$.
(ii) If $\sum a_r$ is divergent, so is $\sum b_r$.

Proof. (i) Let $s_n = \sum_{r=0}^{n} a_r$ and $t_n = \sum_{r=0}^{n} b_r$. If $\sum b_r < \infty$ then $\{t_n\}$ is bounded above, by Theorem 2.4. Now (17) implies that $\{s_n\}$ is bounded above, since $s_n \leq t_n$, and Theorem 2.4 implies that $\sum a_r < \infty$. We leave the proof of (ii) to the student (Exercise T-3).

Example 2.8 Since

$$\frac{1}{2^r + r} < \frac{1}{2^r} \quad \text{and} \quad \sum \frac{1}{2^r} < \infty ,$$

it follows that

$$\Sigma \frac{1}{2^r + r} < \infty.$$

Since

$$\frac{1}{r} < \frac{1}{r - \frac{1}{2}} \quad (r \geq 1) \qquad \text{and} \qquad \Sigma \frac{1}{r} = \infty,$$

it follows that

$$\Sigma \frac{1}{r - \frac{1}{2}} = \infty.$$

Finally,

$$\frac{10r}{r^2 + 1} \frac{1}{3^r} < \frac{1}{3^r}$$

for r sufficiently large, and

$$\Sigma \frac{1}{3^r} < \infty;$$

therefore,

$$\Sigma \frac{10r}{r^2 + 1} \frac{1}{3^r} < \infty.$$

Theorem 2.6 (Integral Test) Let $\{a_r\}$ be a nonincreasing sequence of positive numbers, and suppose f is a continuous, positive, and nonincreasing function of x, for $1 \leq x \leq \infty$, such that $f(r) = a_r$, $r = 1, 2, \ldots$. Then

$$(18) \qquad \sum_{r=1}^{\infty} a_r < \infty$$

if and only if the improper integral $\int_{1}^{\infty} f(x)\, dx$ is convergent.

Proof. The relationship between $\{a_r\}$ and the graph of $y = f(x)$ is shown in Figure 2.1a. By comparing the area of the shaded rectangle in Figure 2.1b with the area under the curve $y = f(x)$ between $x = r - 1$ and $x = r$, we conclude that

$$(19) \qquad a_r \leq \int_{r-1}^{r} f(x)\, dx.$$

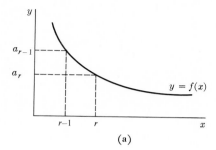

(a)

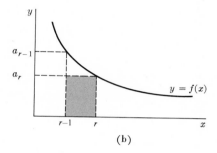

(b)

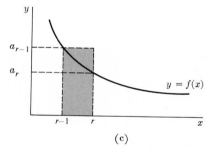

(c)

FIGURE 2.1

If $\displaystyle\int_1^\infty f(x)\ dx$ converges, then

$$\int_1^\infty f(x)\ dx = \lim_{n\to\infty} \int_1^n f(x)\ dx$$

$$= \lim_{n\to\infty} \sum_{r=2}^n \int_{r-1}^r f(x)\ dx;$$

hence, the series

$$\Sigma \int_{r-1}^{r} f(x) \, dx$$

is a convergent series of positive terms, and (18) follows from (19) and the comparison test (Theorem 2.5). On the other hand, comparing the area of the shaded rectangle in Figure 2.1c with the area under the curve yields

$$\int_{r-1}^{r} f(x) \, dx \le a_{r-1}.$$

Now (18) and the comparison test imply that

$$\Sigma \int_{r-1}^{r} f(x) \, dx < \infty,$$

which implies that $\int_{1}^{\infty} f(x) \, dx$ converges. This completes the proof.

Example 2.9   Consider the series $\sum_{1}^{\infty} \dfrac{1}{r^p}$. If $p \le 0$ the series diverges, by Theorem 2.3. If $p > 0$ then Theorem 2.6 can be applied, with $f(x) = x^p$. If $p = 1$ then

$$\int_{1}^{n} f(x) \, dx = \int_{1}^{n} \frac{dx}{x} = \log n.$$

If $p \ne 1$ then

$$\int_{1}^{n} f(x) \, dx = \int_{1}^{n} \frac{dx}{x^p} = \frac{1}{p-1}\left(1 - \frac{1}{n^{p-1}}\right).$$

Letting $n \to \infty$ we find that $\int_{1}^{\infty} f(x) \, dx$ exists if and only if $p > 1$. Hence, Theorem 2.6 yields

$$\Sigma \frac{1}{r^p} < \infty \qquad (p > 1),$$

and

$$\Sigma \frac{1}{r^p} = \infty \qquad (p \le 1).$$

Alternating Series

A series whose successive terms alternate in sign, such as $\sum \dfrac{(-1)^r}{r}$ or $\sum \dfrac{(-1)^r}{2^r}$, is called an *alternating* series. Such a series can be written as $\sum (-1)^r a_r$, where the terms $\{a_r\}$ are all of the same sign. The following theorem gives a useful sufficient condition for convergence of alternating series.

Theorem 2.7 (Alternating Series Test)   If $a_r > a_{r+1} > 0$ and $\lim\limits_{r \to \infty} a_r = 0$, then $\sum (-1)^r a_r$ is convergent. Moreover, if $S = \sum\limits_{r=0}^{\infty} (-1)^r a_r$ then

$$(20) \qquad |\, S - s_n \,| \;=\; \Big|\, S - \sum_{r=0}^{n} (-1)^r a_r \,\Big| \;<\; a_{n+1}.$$

Thus, an alternating series converges if the magnitudes of its terms form a nonincreasing series which approaches zero, and (20) gives a convenient estimate of the error committed in approximating the sum of the series by the nth partial sum.

Proof.   We first write the odd numbered partial sums in the form

$$s_{2n+1} = (a_0 - a_1) + (a_2 - a_3) + \cdots + (a_{2n} - a_{2n+1})$$

$$= s_{2n-1} + (a_{2n} - a_{2n+1}).$$

Since $a_{2n} > a_{2n+1}$, the sequence $\{s_{2n+1}\}$ is increasing. It is also bounded above, since

$$s_{2n+1} = a_0 - (a_1 - a_2) - \cdots - (a_{2n-1} - a_{2n}) - a_{2n+1} \le a_0$$

for all $n \ge 0$. Therefore, $\lim\limits_{n \to \infty} s_{2n+1} = S$ exists, by Theorem 1.3. Since $s_{2n} = s_{2n+1} + a_{2n+1}$ and $\lim\limits_{n \to \infty} a_{2n+1} = 0$, we have

$$\lim_{n \to \infty} s_{2n} = \lim_{n \to \infty} s_{2n+1} + \lim_{n \to \infty} a_{2n+1} = S + 0 = S.$$

It follows that $\lim\limits_{n \to \infty} s_n = S$ exists (Exercise T-10, Section 7.1).

To prove (20) we write

$$s_{n+2k+1} - s_n = (-1)^{n+1}[a_{n+1} - (a_{n+2} - a_{n+3}) - \cdots - (a_{n+2k} - a_{n+2k+1})].$$

Since $\{a_n\}$ is nonincreasing, we can infer from this that

$$|s_{n+2k+1} - s_n| \leq a_{n+1},$$

and therefore

$$(21) \qquad |S - s_n| \leq |S - s_{n+2k+1}| + |s_{n+2k+1} - s_n|$$

$$\leq |S - s_{n+2k+1}| + a_{n+1}.$$

However, $|S - s_{n+2k+1}|$ can be made as small as we wish by taking k sufficiently large, since $\lim_{k \to \infty} s_{n+2k+1} = S$ (Theorem 1.5, Section 7.1); hence

(20) follows from (21), and the proof of Theorem 2.7 is complete.

Example 2.10 The series

$$(22) \qquad \sum \frac{(-1)^r}{r}$$

meets the hypotheses of the alternating series test and is therefore convergent. Moreover, if $S = \sum_{r=1}^{\infty} \frac{(-1)^r}{r}$

then

$$\left| \sum_{r=1}^{n} \frac{(-1)^r}{r} - S \right| < \frac{1}{n+1}.$$

Also,

$$(23) \qquad \sum \frac{(-1)^r}{(r+1)(r+2)}$$

is convergent, and if

$$T = \sum_{r=0}^{\infty} \frac{(-1)^r}{(r+1)(r+2)}$$

then

$$\left| \sum_{r=0}^{n} \frac{(-1)^r}{(r+1)(r+2)} - T \right| < \frac{1}{(n+2)(n+3)}.$$

Absolute Convergence

It is possible for $\sum a_r$ to be convergent even though $\sum |a_r| = \infty$. In this case, $\sum a_r$ is said to be *conditionally* convergent. Thus (22) is a conditionally convergent series, from Examples 2.5 and 2.10. A stronger type of convergence is introduced in the following definition.

Definition 2.2 A series $\sum a_r$ is said to be *absolutely convergent* if the associated series of absolute values, $\sum |a_r|$, is convergent.

Example 2.11 Any convergent series of positive terms is absolutely convergent.

Example 2.12 The series (23) is absolutely convergent, since its associated series of absolute values,

$$\sum \frac{1}{(r+1)(r+2)}$$

is convergent by comparison with the known convergent series $\sum \frac{1}{r^2}$ (Example 2.9).

Theorem 2.8 An absolutely convergent series is convergent.

Proof. By the triangle inequality

(24) $|a_n + a_{n+1} + \cdots + a_m| \leq |a_n| + |a_{n+1}| + \cdots + |a_m|.$

If $\sum |a_r| < \infty$ and $\epsilon > 0$, then Theorem 2.2 implies that there is an integer N such that

$|a_n| + |a_{n+1}| + \cdots + |a_m| < \epsilon$ if $m \geq n \geq N.$

Therefore, from (24)

$|a_n + a_{n+1} + \cdots + a_m| < \epsilon$ if $m \geq n \geq N,$

and it follows, again from Theorem 2.2, that $\sum a_r$ is convergent.

A conditionally convergent series such as (22) owes its convergence, in part, to cancellation among its terms. Changing the signs of some of the terms of an absolutely convergent series may change the sum, but it will not alter the basic fact of convergence, which is due not to cancellation, but solely to the rapidity with which the terms approach zero.

Any convergence test applicable to series with positive terms can be used to test a series $\sum a_r$ for absolute convergence; one simply applies the test to $\sum |a_r|$. Another useful test for absolute convergence is the following.

Theorem 2.9 (The Ratio Test) Suppose $\sum a_r$ is a series of nonzero terms and

$$\lim_{r \to \infty} \left| \frac{a_{r+1}}{a_r} \right| = R.$$

Then

(i) $\sum a_r$ is absolutely convergent if $R < 1$;
(ii) $\sum a_r$ is divergent if $1 < R \leq \infty$.
(iii) If $R = 1$, no conclusion can be drawn; $\sum a_r$ may be divergent, conditionally convergent, or absolutely convergent.

Proof. If $R < 1$, choose R_0 so that $R < R_0 < 1$. From Definition 1.3, Section 7.1, there is an integer N such that

$$\left| \left| \frac{a_{r+1}}{a_r} \right| - R \right| < R_0 - R \qquad (r \geq N),$$

and therefore

(25) $$\left| \frac{a_{r+1}}{a_r} \right| < R_0 \qquad (r \geq N).$$

Thus, if $m \geq N + 1$,

$$|a_m| = \left| \frac{a_m}{a_{m-1}} \right| \left| \frac{a_{m-1}}{a_{m-2}} \right| \cdots \left| \frac{a_{N+1}}{a_N} \right| |a_N| < R_0^{m-N} |a_N|.$$

Since

$$\frac{|a_N|}{R_0^N} \sum R_0^m < \infty$$

(geometric series with ratio $R_0 < 1$), it follows from the comparison test that $\sum |a_r| < \infty$. This completes the proof of (i).

For (ii), let $1 < R \leq \infty$. By an argument similar to that which yielded (25) (Exercise T-4), there is an integer M such that

(26) $$\left| \frac{a_{r+1}}{a_r} \right| > 1 \qquad (r \geq M),$$

which implies that $|a_{r+1}| \geq |a_N|$, $r \geq M$. Hence, $\sum a_r$ diverges, since its terms do not approach zero.

To prove (iii), we need only give examples. This is done in Example 2.14.

Example 2.13 The series

$$\sum \frac{(-1)^r 10^r}{r!}$$

is absolutely convergent, since

$$\lim_{r \to \infty} \left| \frac{a_{r+1}}{a_r} \right| = \lim_{r \to \infty} \frac{\dfrac{10^{r+1}}{(r+1)!}}{\dfrac{10^r}{r!}} = \lim_{r \to \infty} \frac{10}{r+1} = 0.$$

So is

$$\sum \frac{r}{3^r}$$

convergent, since

$$\lim_{r \to \infty} \left| \frac{a_{r+1}}{a_r} \right| = \lim_{r \to \infty} \frac{\dfrac{r+1}{3^{r+1}}}{\dfrac{r}{3^r}} = \lim_{r \to \infty} \frac{1}{3} \left(1 + \frac{1}{r} \right) = \frac{1}{3}.$$

However,

$$\sum (-1)^r \frac{2^r}{r(r+1)}$$

is divergent, since

$$\lim_{r \to \infty} \left| \frac{a_{r+1}}{a_r} \right| = \lim_{r \to \infty} \frac{\dfrac{2^{r+1}}{(r+1)(r+2)}}{\dfrac{2^r}{r(r+1)}} = \lim_{r \to \infty} \frac{2r}{r+2} = 2.$$

Example 2.14 For the series,

(27)
$$\sum \frac{(-1)^r}{r^p},$$

$$\lim_{r \to \infty} \left| \frac{a_{r+1}}{a_r} \right| = \lim_{r \to \infty} \frac{r^p}{(r+1)^p} = \lim_{r \to \infty} \frac{1}{\left(1 + \dfrac{1}{r} \right)^p} = 1.$$

Thus, neither (i) nor (ii) of Theorem 2.9 applies. We have already seen that (27) is conditionally convergent if $p = 1$, absolutely convergent if $p > 1$, and divergent if $p < 1$.

EXERCISES 7.2

Find the sums of the geometric series in Exercises 1 through 4.

1. $1 + \dfrac{1}{2} + \dfrac{1}{4} + \cdots + \dfrac{1}{2^r} + \cdots$.

2. $\dfrac{1}{3} + \dfrac{1}{9} + \cdots + \dfrac{1}{3^r} + \cdots$.

3. $1 + \dfrac{1}{e} + \dfrac{1}{e^2} + \cdots + \dfrac{1}{e^r} + \cdots$.

4. $1 + \dfrac{1}{\sqrt{2}} + \dfrac{1}{2} + \cdots + \left(\dfrac{1}{\sqrt{2}}\right)^r + \cdots$.

Use Theorem 2.2 to determine whether the series in Exercises 5 through 8 are convergent.

5. $1 + \dfrac{1}{3} + \dfrac{1}{5} + \cdots + \dfrac{1}{2n+1} + \cdots$.

6. $1 + \dfrac{1}{4} + \dfrac{1}{9} + \cdots + \dfrac{1}{n^2} + \cdots$.

7. $1 + \dfrac{1}{6} + \dfrac{1}{6^2} + \cdots + \dfrac{1}{6^r} + \cdots$.

8. $\dfrac{1}{2} + \dfrac{2}{3} + \dfrac{3}{4} + \cdots + \dfrac{n}{n+1} + \cdots$.

Use the integral test to determine whether the series in Exercises 9 through 14 are convergent.

9. $1 + \dfrac{1}{2^3} + \dfrac{1}{3^3} + \cdots + \dfrac{1}{n^3} + \cdots .$

10. $\dfrac{1}{1.2} + \dfrac{1}{2.3} + \dfrac{1}{3.4} + \cdots + \dfrac{1}{n(n+1)} + \cdots .$

11. $\dfrac{2}{e} + \dfrac{4}{e^2} + \dfrac{6}{e^3} + \cdots + \dfrac{2n}{e^n} + \cdots .$

12. $\dfrac{1}{2} + \dfrac{1}{5} + \dfrac{1}{10} + \cdots + \dfrac{1}{n^2+1} + \cdots .$

13. $\dfrac{\log 1}{1} + \dfrac{\log 2}{2} + \cdots + \dfrac{\log n}{n} + \cdots .$

14. $1 + \dfrac{1}{\sqrt{2}} + \dfrac{1}{\sqrt{3}} + \cdots + \dfrac{1}{\sqrt{n}} + \cdots .$

Use the comparison test to determine whether the series in Exercises 15 through 18 are convergent.

15. $1 + \dfrac{1}{2^3+1} + \dfrac{1}{3^3+2} + \cdots + \dfrac{1}{n^3+n-1} + \cdots .$

16. $\dfrac{2}{e+1} + \dfrac{4}{e^2+1} + \dfrac{6}{e^3+1} + \cdots + \dfrac{2n}{e^n+1} + \cdots .$

17. $\dfrac{1}{3} + \dfrac{1}{5} + \cdots + \dfrac{1}{n^2-n} + \cdots .$

18. $-1 + \dfrac{1}{\sqrt{2}-2} + \dfrac{1}{\sqrt{3}-2} + \cdots + \dfrac{1}{\sqrt{n}-2} + \cdots .$

Use the ratio test to determine whether the series in Exercises 19 through 24 are convergent.

19. $\displaystyle\sum_{1}^{\infty} (-1)^r \dfrac{r^3}{3^r} .$ 20. $\displaystyle\sum_{r=1}^{\infty} \dfrac{2^r}{r!} .$ 21. $\displaystyle\sum (-1)^r \dfrac{r(r+3)}{5^r} .$

22. $\displaystyle\sum \dfrac{\log r}{2^r} .$ 23. $\displaystyle\sum \dfrac{1}{r!} .$ 24. $\displaystyle\sum \dfrac{r^r}{r!} .$

Determine whether the series in Exercises 25 through 34 are conditionally convergent, absolutely convergent, or divergent.

25. $\sum (-1)^r \dfrac{1}{2r+1}$.

26. $\sum (-1)^r \dfrac{1}{(3r+1)^2}$.

27. $\sum (-1)^{r+1} \dfrac{r}{2^{r-1}}$.

28. $\sum (-1)^{r+1} \dfrac{r+2}{2r+3}$.

29. $\sum (-1)^{r+1} \dfrac{r^2}{e^r}$.

30. $\sum (-1)^r \dfrac{\log r}{r}$.

31. $\sum (-1)^r \dfrac{r}{\log r}$.

32. $\sum (-1)^r \dfrac{1}{5^r}$.

33. $\sum (-1)^r \dfrac{1}{\sqrt{r}}$.

34. $\sum (-1)^r \dfrac{1}{r\sqrt{r}}$.

THEORETICAL EXERCISES

T-1.   Prove Theorem 2.1. (*Hint:* Work directly from Definition 2.1.)

T-2.   Give a formal proof, with careful definitions of the partial sums involved, that $\sum\limits_{r=0}^{\infty} a_r$ and $\sum\limits_{r=0}^{\infty} b_r$ are convergent or divergent together if $a_r = b_r$ for all $r \geq m$.

T-3.   Prove (ii) of Theorem 2.5.

T-4.   Verify Equation (26).

T-5.   Let $a_r \geq 0$ and $\sum a_r < \infty$. Suppose $\{b_r\}$ is a bounded sequence. Show that $\sum a_r b_r$ converges absolutely.

T-6.   Suppose $\sum a_r$ is absolutely convergent, and let $\{b_r\}$ be a sequence such that $\left\{\dfrac{b_r}{a_r}\right\}$ is bounded. Show that $\sum b_r$ is absolutely convergent.

T-7. Let $\sum\limits_0^\infty a_r = L$, and suppose $0 \le n_1 < n_2 < n_3 < \cdots$. Define

$$b_0 = (a_0 + \cdots + a_{n_1}),$$

$$b_1 = (a_{n_1+1} + \cdots + a_{n_2}),$$

$$\vdots$$

$$b_k = (a_{n_k+1} + \cdots + a_{n_{k+1}}).$$

Show that $\sum\limits_0^\infty b_r = L$.

T-8. Suppose $a_r > 0$, $b_r > 0$, and

$$\lim_{r \to \infty} \frac{a_r}{b_r} = L > 0.$$

Prove: $\sum a_r$ and $\sum b_r$ converge or diverge together.

7.3 Power Series

For $r = 0, 1, 2, \ldots$, let f_r be a real-valued function defined on a domain D of the real numbers. Then

$$f(x) = \sum_{r=0}^\infty f_r(x)$$

defines a real-valued function f, with domain consisting of those values of x for which the series on the right converges. A thorough study of functions defined by infinite series is beyond the scope of this book; however, we shall investigate an important class of functions that arise in this way.

Definition 3.1 An infinite series of the form

$$(1) \qquad\qquad f(x) = \sum_{r=0}^\infty a_r(x - x_0)^r,$$

where x_0 and $a_0, a_1, \ldots$ are real constants, is called a *power series in $x - x_0$*.

For a fixed value of x, the right side of (1) is simply a series of constants; therefore the results of Section 7.2 can be applied to power series. The nth *partial sum* of the series in (1) is the polynomial

$$(2) \qquad\qquad s_n(x) \;=\; \sum_{r=0}^{n} a_r(x - x_0)^r.$$

The domain of the function f defined by (1) is the set of values of x for which

$$f(x) \;=\; \lim_{n\to\infty} s_n(x)$$

exists. We shall also call the domain of f *the convergence set of* $\sum a_r(x - x_0)^r$.

Since $s_n(x_0) = a_0$ for $r = 0, 1, 2, \ldots$, it is clear that x_0 is in the convergence set of $\sum a_r(x - x_0)^r$, and $f(x_0) = a_0$. Our first objective is to determine precisely the nature of the convergence set of a power series.

Example 3.1 The series

$$\sum_{r=0}^{\infty} r!(x + 1)^r$$

diverges if $x \neq -1$, by the ratio test (Theorem 2.9, Section 7.2), since

$$\lim_{r\to\infty} \left| \frac{(r + 1)!(x + 1)^{r+1}}{r!(x + 1)^r} \right| = \lim_{r\to\infty} (r + 1) \,|\, (x + 1) \,| = \infty$$

if $x \neq -1$. The convergence set of the series $\sum_{r=0}^{\infty} r!(x + 1)^r$ consists of the single point $x = -1$.

Example 3.2 The series

$$\sum_{r=0}^{\infty} (x - 1)^r$$

is a geometric series with ratio $(x - 1)$ (Example 2.2, Section 7.2); hence it converges to

$$f(x) \;=\; \frac{1}{1 - (x - 1)} \;=\; \frac{1}{2 - x}$$

if $|\, x - 1 \,| < 1$, and diverges otherwise. Its convergence set is the open interval $(0, 2)$.

Example 3.3 The series

$$f(x) \;=\; \sum_{r=0}^{\infty} \frac{x^r}{r!}$$

converges for all x, by the ratio test, since

$$\lim_{r \to \infty} \frac{\dfrac{|x|^{r+1}}{(r+1)!}}{\dfrac{|x|^r}{r!}} = \lim_{r \to \infty} \frac{x}{r+1} = 0.$$

Its convergence set is the set of all real numbers.

Lemma 3.1   If the sequence $\{a_r \rho^r\}$ is bounded, then $\sum a_r(x - x_0)^r$ is absolutely convergent for $|x - x_0| < \rho$.

Proof.   Suppose $|a_r \rho^r| \le M$ for $r = 0, 1, \ldots$; then

$$(3) \qquad |a_r(x - x_0)^r| = |a_r \rho^r| \left| \frac{x - x_0}{\rho} \right|^r \le M \left| \frac{x - x_0}{\rho} \right|^r.$$

If $|x - x_0| < \rho$, then

$$\sum M \left| \frac{x - x_0}{\rho} \right|^r$$

is essentially a geometric series with ratio less than 1, and $\sum\limits_{r=0}^{\infty} a_r(x - x_0)^r$ is therefore convergent, from (3) and the comparison test (Theorem 2.5, Section 7.2).

Lemma 3.2   Let $x_1 \ne x_0$ and suppose $\sum a_r(x_1 - x_0)^r$ is convergent. Then $\sum a_r(x - x_0)^r$ is absolutely convergent if $|x - x_0| < |x_1 - x_0|$.

Proof.   If $\sum a_r(x_1 - x_0)^r$ converges, then $\{a_r |x_1 - x_0|^r\}$ is bounded (Exercise T-1), and the conclusion follows from Lemma 3.1, with $\rho = |x_1 - x_0|$.

Theorem 3.1   The power series

$$(4) \qquad\qquad f(x) = \sum_{r=0}^{\infty} a_r(x - x_0)^r$$

may diverge for all $x \ne x_0$, or it may converge absolutely for every x. If neither of these is the case, then there is an $R > 0$ such that (4) converges absolutely if $|x - x_0| < R$, and diverges if $|x - x_0| > R$. No general statement can be made regarding convergence at $x = x_0 + R$ and $x = x_0 - R$; (4) may converge absolutely or conditionally at one, both, or neither of these points.

Proof. Example 3.1 shows that (4) may diverge for all $x \neq x_0$. We leave it to the student to show that if (4) converges for all x, then the convergence is absolute (Exercise T-2). The remaining possibility is that (4) converges for some $x_1 \neq x_0$, and diverges for some x_2. In this case, let S be the set of positive numbers ρ such that (4) converges absolutely if $|x - x_0| < \rho$. Then S is bounded above by $|x_2 - x_0|$ and, by Lemma 3.2, S contains $|x_1 - x_0|$. From the completeness axiom S has a least upper bound R. We must show that R has the properties stated in the theorem. If $|x - x_0| < R$, there is a ρ_0 in S such that $|x - x_0| < \rho_0 < R$. Therefore, (4) is convergent, by definition of S. On the other hand, if $\sum a_r (\bar{x} - x_0)^r$ converges for some $\bar{x}$ such that $|\bar{x} - x_0| > R$, then Lemma 3.2 implies that every real number between R and $|\bar{x} - x_0|$ is also in S, which violates the definition of R. Consequently, R has the stated properties.

The number R is called the *radius of convergence* of the power series (4). To include the two extreme cases in this definition, we define $R = 0$ if (4) converges at $x = x_0$ only, and $R = \infty$ if (4) converges for all x. If $R > 0$ the convergence set is the open interval $(x_0 - R, x_0 + R)$, which is called the *interval of convergence* of (4).

Example 3.4 The radii of convergence of the power series in Example 3.1, 3.2, and 3.3 are 0, 1, and ∞, respectively. The series of Example 3.2 diverges at both endpoints, $x = 0$ and $x = 2$, of its interval of convergence.

Example 3.5 The series

$$(5) \qquad\qquad f(x) = \sum_{r=0}^{\infty} \frac{x^r}{r+1}$$

converges if $|x| < 1$ and diverges if $|x| > 1$, by the ratio test, since

$$\lim_{r \to \infty} \frac{\dfrac{|x|^{r+1}}{r+2}}{\dfrac{|x|^r}{r+1}} = \lim_{r \to \infty} \frac{r+1}{r+2} |x| = |x|;$$

thus $r = 1$. Setting $x = 1$ in (5) yields the divergent series

$$\sum \frac{1}{r+1},$$

(Example 2.5, Section 7.2); setting $x = -1$ yields the conditionally convergent series

$$\Sigma \frac{(-1)^r}{r+1}$$

(Example 2.10, Section 7.2). The interval of convergence of this series is $(-1, 1)$; the convergence set is $[-1, 1)$.

Example 3.6   The series

$$f(x) = \sum_{r=0}^{\infty} \frac{x^r}{(r+1)^2}$$

converges if $|x| < 1$ and diverges if $|x| > 1$, again by the ratio test; thus $r = 1$. (Verify.) Setting $x = 1$ yields

(6)
$$\sum_{r=0}^{\infty} \frac{1}{(r+1)^2},$$

which is absolutely convergent (Example 2.9, Section 7.2); setting $x = -1$ yields

$$\sum_{r=0}^{\infty} \frac{(-1)^r}{(r+1)^2}$$

which is absolutely convergent, by comparison with (6). The interval of convergence of this series is $(-1, 1)$; the convergence set is $[-1, 1]$.

Theorem 3.1 does not provide a practical way of determining R. The next theorem solves this problem for a large class of power series.

Theorem 3.2   The radius of convergence of

(7)
$$f(x) = \sum a_r(x - x_0)^r$$

satisfies

$$\frac{1}{R} = \lim_{r \to \infty} \left| \frac{a_{r+1}}{a_r} \right|$$

if the limit exists and is positive. If $\lim_{r \to \infty} \left| \frac{a_{r+1}}{a_r} \right| = 0$, then (7) converges for all x ($R = \infty$); if $\lim_{r \to \infty} \left| \frac{a_{r+1}}{a_r} \right| = \infty$, then (7) converges only if $x = x_0$ ($R = 0$).

Proof. Suppose

$$\lim_{r \to \infty} \left| \frac{a_{r+1}}{a_r} \right| = L,$$

where $0 < L < \infty$. Then, if $x \neq x_0$,

$$\lim_{r \to \infty} \left| \frac{a_{r+1}(x - x_0)^{r+1}}{a_r (x - x_0)^r} \right| = |x - x_0| \lim_{r \to \infty} \left| \frac{a_{r+1}}{a_r} \right|$$

$$= |x - x_0| \, L.$$

Now the ratio test (Theorem 2.9, Section 7.2) implies that (7) is absolutely convergent if $|x - x_0| L_1 < 1$, or divergent if $|x - x_0| L_1 > 1$. We leave the rest of the proof to the student (Exercise T-3).

Example 3.7 For the series

$$f(x) = \sum_{r=0}^{\infty} (r + 1)x^r,$$

$$\lim_{r \to \infty} \left| \frac{a_{r+1}}{a_r} \right| = \lim_{r \to \infty} \frac{r + 2}{r + 1} = 1;$$

hence $R = 1$.

Example 3.8 For the series

$$f(x) = \sum_{r=0}^{\infty} (-1)^r (r + 1)^2 \frac{(x + 6)^r}{2^r},$$

$$\lim_{r \to \infty} \left| \frac{a_{r+1}}{a_r} \right| = \lim_{r \to \infty} \frac{\dfrac{(r + 2)^2}{2^{r+1}}}{\dfrac{(r + 1)^2}{2^r}}$$

$$= \frac{1}{2} \lim_{r \to \infty} \left(\frac{r + 2}{r + 1} \right)^2$$

$$= \frac{1}{2} \lim_{r \to \infty} \left(1 + \frac{1}{r + 1} \right)^2 = \frac{1}{2};$$

hence $R = 2$.

Example 3.9 For the series

$$f(x) = \sum_{r=0}^{\infty} (-1)^r \frac{3^r (r+1)^2}{r!} (x-1)^r$$

$$\lim_{r \to \infty} \left| \frac{a_{r+1}}{a_r} \right| = \lim_{r \to \infty} \frac{\dfrac{3^{r+1}(r+2)^2}{(r+1)!}}{\dfrac{3^r(r+1)^2}{r!}}$$

$$= \lim_{r \to \infty} 3 \left(\frac{r+2}{r+1} \right)^2 \frac{1}{r+1}$$

$$= 0;$$

hence $R = \infty$.

Example 3.10 The series

(8) $$f(x) = \sum_{r=0}^{\infty} \frac{x^{2r}}{4^r(r+1)},$$

has no odd powers of x; hence $a_{2r+1} = 0$ for all r. Therefore, Theorem 3.2 does not apply, since $\dfrac{a_{2r}}{a_{2r+1}}$ is undefined. Nevertheless the ratio test still allows us to determine R, as follows:

$$\lim_{r \to \infty} \frac{\dfrac{x^{2r+2}}{4^{r+1}(r+2)}}{\dfrac{x^{2r}}{4^r(r+1)}} = \lim_{r \to \infty} \frac{x^2}{4} \left(\frac{r+1}{r+2} \right) = \frac{x^2}{4}.$$

Therefore, (8) converges if $\dfrac{x^2}{4} < 1$, and diverges if $\dfrac{x^2}{4} > 1$. Hence $R = 2$. (See Exercise T-4.)

Differentiation and Integration of Power Series

Lemma 3.3 If

(9) $$f(x) = \sum_{r=0}^{\infty} a_r (x - x_0)^r$$

has radius of convergence $R > 0$, then the power series

$$f_1(x) = \sum_{r=1}^{\infty} r a_r (x - x_0)^{r-1},$$

(10) $$f_2(x) = \sum_{r=2}^{\infty} r(r-1) a_r (x - x_0)^{r-2},$$

$$\vdots$$

$$f_n(x) = \sum_{r=n}^{\infty} r(r-1) \cdots (r-n+1) a_r (x - x_0)^{r-n},$$

which are obtained by repeatedly differentiating (9) term-by-term, all have radius of convergence R.

Proof.   For a fixed x in the interval of convergence, choose R_0 so that $|x - x_0| < R_0 < R$. Then $\sum_{r=1}^{\infty} |a_r| R_0^{r-1}$ is convergent (why?), and there is a number M such that

$$|a_r| R_0^r \leq M \qquad (r = 1, 2, \ldots).$$

Then

$$|r a_r (x - x_0)^{r-1}| \leq |a_r| R_0^{r-1} \left(r \left| \frac{x - x_0}{R} \right|^{r-1} \right)$$

$$\leq M r \rho^{r-1},$$

where $\rho = \left| \dfrac{x - x_0}{R} \right| < 1$. From Example 3.7, the series

$$M \sum_{r=1}^{\infty} r \rho^{r-1}$$

is convergent, and the comparison test implies that $f_1(x)$ is absolutely convergent. Since this holds for all x such that $|x - x_0| < R$, the radius of convergence of f_1 is at least R. We leave it to the student to show that it is not greater than R (Exercise T-5). Therefore, f_1 has radius of convergence R. Since f_2 is obtained from f_1 in the same way that f_1 was obtained from f, it follows that f_2 has radius of convergence R also. By repeating this argument, we find that $f, f_1, \ldots, f_n, \ldots$ all have radius of convergence R.

Theorem 3.3 A function f defined by a power series (9) with radius of convergence $R > 0$ has derivatives of all orders within its interval of convergence. The derivatives can be obtained by differentiating f term by term; that is

(11) $\qquad f^{(n)}(x) = f_n(x) \qquad (|x - x_0| < R, \quad n = 1, 2, \ldots),$

where f_n is defined by (10).

Proof. We shall first show that $f'(x_1) = f_1(x_1)$ if $|x_1 - x_0| < R$. Let $|x_1 - x_0| < R_0 < R$ and suppose $|x - x_0| < R_0$. Then $f_1(x_1)$ is convergent, by Lemma 3.3, and, from Theorem 2.1, Section 7.2,

(12) $\qquad \dfrac{f(x) - f(x_1)}{x - x_1} - f_1(x_1)$

$$= \sum_{r=0}^{\infty} a_r \left[\frac{(x - x_0)^r - (x - x_1)^r}{x - x_1} - r(x_1 - x_0)^{r-1} \right]$$

if $x \neq x_1$. Now define

(13) $\qquad\qquad\qquad p_r(x) = (x - x_0)^r.$

From Theorem 8.2, Section 4.8,

(14) $\qquad | p_r(x) - p_r(x_1) - p_r'(x_1)(x - x_1) | \leq \dfrac{M_r}{2}(x - x_1)^2,$

where M_r is an upper bound for p_r'' in the interval $x_0 - R_0 < x < x_0 + R_0$. Suppose $r \geq 2$; since

(15) $\qquad\qquad\qquad p_r'(x) = r(x - x_0)^{r-1}$

and

$$p_r''(x) = r(r - 1)(x - x_0)^{r-2},$$

we may take $M_r = r(r - 1)R_0^{r-2}$; then it follows from (13), (14), and (15) that

$$| (x - x_0)^r - (x - x_1)^r - r(x_1 - x_0)^{r-1}(x - x_1) | \leq \frac{r(r - 1)R_0^{r-2}}{2}(x - x_1)^2.$$

Dividing both sides by $x - x_1$ and substituting the resulting inequality into (12) yields

(16) $\qquad \left| \dfrac{f(x) - f(x_1)}{x - x_1} - f_1(x_1) \right| \leq \dfrac{|x - x_1|}{2} \sum_{r=2}^{\infty} r(r - 1) \, | a_r | \, R_0^{r-2}.$

The series on the right is convergent, from Lemma 3.3. (Specifically, because f_2 is convergent, and therefore *absolutely* convergent, for $| x - x_0 | < R$.) Hence, if

$$Q = \tfrac{1}{2} \sum_{r=2}^{\infty} r(r-1) \, | \, a_r \, | \, R_0^{r-2},$$

then (16) implies that

$$\left| \frac{f(x) - f(x_1)}{x - x_1} - f_1(x_1) \right| \le Q \, | \, x - x_1 \, |,$$

and letting x approach x_1 yields

$$f_1(x_1) = \lim_{x \to x_1} \frac{f(x) - f(x_1)}{x - x_1} = f'(x_1).$$

This proves (11) for $n = 1$. Repeated application of this result yields (11) for all n (Exercise T-6).

Since a differentiable function is continuous, Theorem 3.3 yields the following corollary.

Corollary 3.1   If $f(x) = \sum a_r (x - x_0)^r$ has radius of convergence $R > 0$, then f is continuous in its interval of convergence.

Example 3.11   We know that

(17)
$$\frac{1}{1 - x} = \sum_{r=0}^{\infty} x^r \qquad (| \, x \, | < 1).$$

Differentiating both sides yields

$$\frac{1}{(1 - x)^2} = \sum_{r=1}^{\infty} r x^{r-1} = \sum_{r=0}^{\infty} (r + 1) x^r \qquad (| \, x \, | < 1).$$

(Also, see Example 3.7.) Repeated differentiation of both sides of (17) yields

$$\frac{n!}{(1 - x)^{n+1}} = \sum_{r=n}^{\infty} r(r - 1) \cdots (r - n + 1) x^{r-n}$$

$$= \sum_{r=0}^{\infty} (r + n)(r + n - 1) \cdots (r + 1) x^r,$$

which can be rewritten as

(18)
$$\frac{1}{(1-x)^{n+1}} = \sum_{r=0}^{\infty} \binom{r+n}{n} x^r,$$

where

$$\binom{r+n}{n} = \frac{(r+n)!}{r!\,n!}$$

is the binomial coefficient. Each of the series (18) (for $n = 0, 1, 2, \ldots$) has radius of convergence $R = 1$.

Example 3.12 In Example 3.3 we saw that the series

(19)
$$f(x) = \sum_{r=0}^{\infty} \frac{x^r}{r!}$$

converges for all x. Differentiation yields

$$f'(x) = \sum_{r=1}^{\infty} \frac{x^{r-1}}{(r-1)!}$$

$$= \sum_{r=0}^{\infty} \frac{x^r}{r!}.$$

Therefore,
$$f'(x) = f(x),$$

which means that $f(x) = Ae^x$; since $f(0) = 1$ (from (19)), it follows that $A = 1$. We have now shown that

$$e^x = \sum_{r=0}^{\infty} \frac{x^r}{r!}.$$

Example 3.13 By the method of Example 3.10 (see also Exercise T-4) it can be shown that

(20)
$$C(x) = \sum_{r=0}^{\infty} (-1)^r \frac{x^{2r}}{(2r)!}$$

and

(21)
$$S(x) = \sum_{r=0}^{\infty} (-1)^r \frac{x^{2r+1}}{(2r+1)!}$$

converge for all x. Differentiating (20) yields

$$C'(x) = \sum_{r=1}^{\infty} (-1)^r \frac{x^{2r-1}}{(2r-1)!}$$

$$= \sum_{r=0}^{\infty} (-1)^{r+1} \frac{x^{2r+1}}{(2r+1)!} = -S(x),$$

and differentiating (21) yields

$$S'(x) = \sum_{r=0}^{\infty} (-1)^r \frac{x^{2r}}{(2r)!} = C(x).$$

It is also legitimate to integrate a power series term by term within its interval of convergence, as is shown by the following theorem.

Theorem 3.4   If

$$f(x) = \sum_{r=0}^{\infty} a_r (x - x_0)^r$$

has radius of convergence $R > 0$, then

$$F(x) = \sum_{r=0}^{\infty} \frac{a_r}{r+1} (x - x_0)^{r+1}$$

also has radius of convergence R. Moreover,

$$F(x) = \int_{x_0}^{x} f(t)\, dt.$$

The proof of this theorem is sketched in Exercises T-7 and T-8.

Example 3.14   Integrating the geometric series (17) yields

$$-\log(1 - x) = \int_0^x \frac{dt}{1 - t} = \sum_{r=0}^{\infty} \frac{x^{r+1}}{r+1} = \sum_{r=1}^{\infty} \frac{x^r}{r},$$

which provides a power series for $\log(1 - x)$, valid for $|x| < 1$:

$$\log(1 - x) = -\sum_{r=1}^{\infty} \frac{x^r}{r}.$$

Theorem 3.3 shows that a function f defined by a power series

$$(22) \qquad\qquad f(x) = \sum_{r=0}^{\infty} a_r (x - x_0)^r$$

with a positive radius of convergence has derivatives of all orders, which are given by

$$f^{(n)}(x) = \sum_{r=n}^{\infty} r(r-1)(r-n+1)a_r(x-x_0)^{r-n};$$

in particular

$$f^{(n)}(x_0) = n! \, a_n,$$

so that

$$a_n = \frac{f^{(n)}(x_0)}{n!}.$$

This means that the nth partial sum of (22) is actually the nth Taylor polynomial of f about x_0 (Definition 8.1, Section 4.8). This leads to the following definition.

Definition 3.2 Let f be a real-valued function of x which has derivatives of all orders at $x = x_0$. Then the power series

(23)
$$\sum_{r=0}^{\infty} \frac{f^{(r)}(x_0)}{r!}(x-x_0)^r$$

is called the *Taylor series of f about* $x = x_0$. In the special case where $x_0 = 0$, the series

$$\sum_{r=0}^{\infty} \frac{f^{(r)}(0)}{r!}x^r$$

is called *Maclaurin series of f*.

The Taylor series may converge only at $x = x_0$, or converge for some $x \neq 0$, but to a value different from $f(x)$. It is necessary to investigate the Taylor series (23) to determine whether it converges to $f(x)$. It may be possible to accomplish this by summing (23), or it may be necessary to investigate the error

$$R_n(x) = f(x) - T_n(x)$$

$$= f(x) - \sum_{r=0}^{n} \frac{f^{(r)}(x_0)}{r!}(x-x_0)^r.$$

Theorem 8.2 of Section 4.8 suggests a method for doing this, since it says that

(24)
$$|R_n(x)| \leq \frac{M_n(x, x_0)}{(n+1)!}|x-x_0|^{n+1},$$

where $M_n(x, x_0)$ is an upper bound for $|f^{(n+1)}(\xi)|$ as ξ varies over the interval with endpoints x and x_0.

Example 3.15   Let

$$(25) \qquad\qquad f(x) = \frac{1}{1-x};$$

then

$$(26) \qquad\qquad f^{(r)}(x) = \frac{r!}{(1-x)^{r+1}}.$$

Since $f^{(r)}(0) = r!$, the Taylor series for f about $x = 0$ is the geometric series

$$\sum_{r=0}^{\infty} x^r,$$

which we already know to be convergent to (25) for $|x| < 1$. This can also be seen by investigating the error term. Suppose $|x| < 1$ and consider

$$R_n(x) = f(x) - T_n(x) = \frac{1}{1-x} - 1 - x^2 - \cdots - x^n.$$

From (26), we can take

$$M_n(x, 0) = \frac{(n+1)!}{(1-|x|)^{n+2}} \qquad (-1 < x < 1),$$

so that (24) becomes

$$(27) \qquad\qquad |R_n(x)| < \frac{|x|^{n+1}}{(1-|x|)^{n+2}}.$$

Hence, $\lim_{n \to \infty} R_n(x) = 0$ if $|x| < 1$. If $|x| < \rho$, where ρ is fixed and $0 < \rho < 1$, then (27) can be replaced by

$$|R_n(x)| < \frac{\rho^{n+1}}{(1-\rho)^{n+2}} \qquad (0 \le |x| \le \rho),$$

which provides an upper bound for the maximum error over the interval $[-\rho, \rho]$.

A slightly different analysis of the same problem is given in Example 8.4, Section 4.8.

Example 3.16 In Example 8.3, Section 4.8, we saw that the Taylor polynomials about $x = 0$ for $f(x) = \cos x$ are

$$T_{2k}(x) = T_{2k+1}(x) = \sum_{r=0}^{k} (-1)^r \frac{x^{2r}}{(2r)!}.$$

Since

$$|f^{(2k+2)}(x)| = |\cos x| \leq 1,$$

we may take $M_{2k+1}(x, 0) = 1$, and (24) yields

$$|\cos x - T_{2k+1}(x)| = |R_{2k+2}(x)| \leq \frac{|x|^{2k+2}}{(2k+2)!};$$

hence

$$\lim_{n \to \infty} |\cos x - T_n(x)| = 0 \qquad (-\infty < x < \infty)$$

and

$$\cos x = \sum_{r=0}^{\infty} (-1)^r \frac{x^{2r}}{(2r)!} \qquad (-\infty < x < \infty).$$

Differentiation yields

$$-\sin x = \sum_{r=1}^{\infty} (-1)^r \frac{x^{2r-1}}{(2r-1)!} = \sum_{r=0}^{\infty} (-1)^{r+1} \frac{x^{2r+1}}{(2r+1)!},$$

so that

$$\sin x = \sum_{r=0}^{\infty} (-1)^r \frac{x^{2r+1}}{(2r+1)!} \qquad (-\infty < x < \infty).$$

We now see that the series $C(x)$ and $S(x)$ defined in Example 3.13 are the familiar functions $\cos x$ and $\sin x$.

EXERCISES 7.3

In Exercises 1 through 9, find the radius of convergence of the given power series. If $0 < R < \infty$, determine whether the series converges at the endpoints of the interval of convergence.

1. $\displaystyle\sum_{r=1}^{\infty} \frac{(-1)^r}{\sqrt{r}} (x+2)^r.$

2. $\displaystyle\sum_{r=1}^{\infty} \frac{(-1)^r}{r\sqrt{r}} (x-1)^r.$

3. $\displaystyle\sum_{r=0}^{\infty} (-1)^r \frac{\sqrt{r}}{2^r} (x+1)^r.$

4. $\displaystyle\sum_{r=0}^{\infty} \frac{\sqrt{r}}{r!} x^r.$

5. $\displaystyle\sum_{r=0}^{\infty} \frac{r!}{5^r} (x-6)^r.$

6. $\displaystyle\sum_{r=0}^{\infty} (-1)^r \frac{(x+3)^r}{6^r}.$

7. $\displaystyle\sum_{r=0}^{\infty} (-1)^r r^r x^r.$

8. $\displaystyle\sum_{r=0}^{\infty} \frac{(-1)^r}{2^r r^2 \sqrt{r}} (x-1)^{2r}.$

9. $\displaystyle\sum_{r=0}^{\infty} \frac{x^{3r+1}}{8^{3r+1}}.$

10. Let $f(x) = \displaystyle\sum_{r=0}^{\infty} \frac{x^{2r}}{2r!}$. Show that

$$f''(x) = f(x) \qquad (-\infty < x < \infty).$$

11. Use Theorem 3.3 and the geometric series to express

$$f(x) = \sum_{r=0}^{\infty} 3^r (r+1)(x-1)^{r+2}$$

in closed form.

12. Let $f(x) = \displaystyle\sum_{r=0}^{\infty} (-1)^r \frac{x^{2r}}{r!}$. Show that

$$f''(x) + 2f(x) + 2xf'(x) = 0 \qquad (-\infty < x < \infty).$$

13. Obtain a power series for $f(x) = \tan^{-1} x$, valid for $-1 < x < 1$, by integrating

$$\frac{1}{1+x^2} = \sum_{r=0}^{\infty} (-1)^r x^{2r}.$$

14. Use Theorem 3.4 to express

$$f(x) = \sum_{r=0}^{\infty} \frac{x^{2r}}{2r+1}$$

in closed form.

In Exercises 15 through 22, find the Taylor series for f about $x = x_0$. Show that the Taylor series converges to $f(x)$ within its interval of convergence.

15. $f(x) = \sin 2x$; $x_0 = 0$.

16. $f(x) = e^{x^2}$; $x_0 = 0$.

17. $f(x) = \cosh x$; $x_0 = 0$.

18. $f(x) = xe^x$; $x_0 = 0$.

19. $f(x) = \cos x$; $x_0 = \pi/4$.

20. $f(x) = e^{2x+1}$; $x_0 = -\frac{1}{2}$.

21. $f(x) = \dfrac{1}{x-5}$; $x_0 = 0$.

22. $f(x) = \dfrac{1}{x^2}$; $x_0 = 1$.

THEORETICAL EXERCISES

T-1. *Prove:* If $\sum a_r(x - x_0)^r$ converges, then $\{a_r(x - x_0)^r\}$ is bounded. (*Hint:* Use Theorem 2.3, Section 7.2.)

T-2. *Prove:* If $\sum a_r(x - x_0)^r$ converges for all x, then it converges absolutely for all x.

T-3. Complete the proof of Theorem 3.2.

T-4. *Prove:* The radius of convergence, R, of a power series $\sum b_r x^{2r}$ containing only even powers of x satisfies

$$\frac{1}{R^2} = \lim_{r \to \infty} \frac{b_{r+1}}{b_r}$$

if the limit on the right exists. (What is the radius of convergence if the limit is 0 or ∞?)

T-5. Show that the radius of convergence of $f_1(x) = \sum r a_r(x - x_0)^{r-1}$ is not greater than the radius of convergence of $f(x) = \sum a_r(x - x_0)^r$.

T-6.   Complete the proof of Theorem 3.3 by showing that $f^{(n)}(x) = f_n(x)$ for all n.

T-7.   Show that

$$F(x) = \sum_{r=0}^{\infty} \frac{a_r}{r+1} (x - x_0)^{r+1} \qquad \text{and} \qquad f(x) = \sum_{r=0}^{\infty} a_r (x - x_0)^r$$

have the same radius of convergence. (*Hint:* Use Theorem 3.3.)

T-8.   With F and f as defined in Exercise T-6, show that

$$F(x) = \int_{x_0}^{x} f(t)\, dt,$$

for all x in the interval of convergence of f. (*Hint:* Use the fundamental theorem of the calculus.)

Answers to Selected Problems

Chapter 1

Section 1.1, page 21

1. (a) $x = 2, \ y = -3$ (b) $x = 1, \ y = 2, \ z = -3$

3. (a) $x = 2, \ y = -2$ (b) $x = 1, \ y = 3$

5. (a) $\begin{bmatrix} 2 & 4 & 6 \\ -2 & 4 & 2 \\ 6 & 4 & 2 \end{bmatrix}$ (b) undefined   (c) $\begin{bmatrix} 1 & 0 & 2 \\ 0 & 0 & 0 \\ 3 & 1 & 0 \end{bmatrix}$

   (d) $\begin{bmatrix} 11 & 10 \\ 6 & -1 \end{bmatrix}$ (e) undefined

9. (a) 25   (b) 56

11. (a) $c_1 = 11, \ c_2 = 11, \ c_3 = 5, \ c_4 = 9$

13. (a) undefined   (b) $\begin{bmatrix} 0 & 10 & 6 \\ 4 & 3 & 5 \end{bmatrix}$ (c) $\begin{bmatrix} -1 & 6 \\ -2 & 10 \end{bmatrix}$

   (d) $\begin{bmatrix} 15 & 14 & -3 \\ 8 & 4 & 5 \end{bmatrix}$ (e) undefined   (f) $\begin{bmatrix} 23 & -10 \\ 20 & -8 \end{bmatrix}$

15. (a) $\begin{bmatrix} 2 & -1 \\ 4 & 4 \\ 1 & 2 \end{bmatrix}$ (b) $\begin{bmatrix} 3 & -3 \\ 6 & 9 \\ 6 & 3 \end{bmatrix}$ (c) $\begin{bmatrix} 3 & 4 \\ 10 & 8 \end{bmatrix}$

   (d) undefined   (e) $\begin{bmatrix} 1 & 5 \\ 20 & 4 \\ 8 & 0 \end{bmatrix}$

19. (a) $\begin{bmatrix} 1 & 1 \\ 2 & -3 \end{bmatrix} \begin{bmatrix} x \\ y \end{bmatrix} = \begin{bmatrix} 10 \\ 5 \end{bmatrix}$ (b) $\begin{bmatrix} 1 & -1 & 2 \\ 2 & 3 & -1 \\ 4 & 2 & 3 \end{bmatrix} \begin{bmatrix} x \\ y \\ z \end{bmatrix} = \begin{bmatrix} 5 \\ -3 \\ 8 \end{bmatrix}$

(c) $\begin{bmatrix} 1 & -1 & 2 & -1 \\ 2 & 2 & 3 & 0 \end{bmatrix} \begin{bmatrix} x \\ y \\ z \\ w \end{bmatrix} = \begin{bmatrix} 4 \\ 8 \end{bmatrix}$

Section 1.2, page 42

1. (a) $\begin{bmatrix} 4 & 1 & 2 & -3 \\ 2 & -1 & 3 & 1 \\ 1 & 0 & 1 & 2 \end{bmatrix}$ (b) $\begin{bmatrix} 1 & 0 & 1 & 2 \\ 2 & -1 & 3 & 1 \\ 0 & 3 & -4 & -5 \end{bmatrix}$

(c) $\begin{bmatrix} 3 & 0 & 3 & 6 \\ 2 & -1 & 3 & 1 \\ 4 & 1 & 2 & -3 \end{bmatrix}$

3. (a) $\begin{bmatrix} 1 & -1 & 2 & 0 \\ 0 & 1 & 6 & 3 \\ 0 & 0 & 1 & \frac{3}{13} \end{bmatrix}$ (b) $\begin{bmatrix} 1 & 0 & 0 & \frac{5}{13} \\ 0 & 1 & 0 & \frac{21}{13} \\ 0 & 0 & 1 & \frac{3}{13} \end{bmatrix}$

5. (a) $x = 1, \quad y = -1, \quad z = 2$ (b) no solution
(c) z, w arbitrary; $x = 3, \quad y = 1 + z + w$

7. (a) z arbitrary; $x = 5, \quad y = z - 1$ (b) no solution
(c) $x = 1, \quad y = -1, \quad z - 3.$

9. (i) $a^2 \neq 4$ (ii) $a = -2$ (iii) $a = 2$

11. (a) z arbitrary; $x = -3 - z, \quad y = 13, \quad w = 3$
(b) no solution

13. (a) $x = y = z = 0$
(b) x_4, x_5 arbitrary; $x_1 = \frac{1}{6}x_4 - x_5, \quad x_2 = -\frac{3}{4}x_4, \quad x_3 = \frac{1}{12}x_4$

Section 1.3, page 59

3. $\begin{bmatrix} -2 & 1 \\ 3 & -1 \end{bmatrix}$

5. (a) $\begin{bmatrix} 1 & 0 & 0 & 0 \\ 0 & 1 & 0 & 0 \\ 0 & 0 & 1 & -3 \\ 0 & 0 & 0 & 1 \end{bmatrix}$ (b) $\begin{bmatrix} 1 & 0 & 0 & 0 \\ 0 & 4 & 0 & 0 \\ 0 & 0 & 1 & 0 \\ 0 & 0 & 0 & 1 \end{bmatrix}$ (c) $\begin{bmatrix} 0 & 0 & 1 & 0 \\ 0 & 1 & 0 & 0 \\ 1 & 0 & 0 & 0 \\ 0 & 0 & 0 & 1 \end{bmatrix}$

7. (a) $\begin{bmatrix} 1 & 0 & 0 \\ 0 & 0 & 1 \\ 0 & 1 & 0 \end{bmatrix}$ (b) $\begin{bmatrix} 1 & 0 & 4 \\ 0 & 1 & 0 \\ 0 & 0 & 1 \end{bmatrix}$ (c) $\begin{bmatrix} 1 & 0 & 0 \\ 0 & \frac{1}{5} & 0 \\ 0 & 0 & 1 \end{bmatrix}$

9. $A^{-1} = \begin{bmatrix} \frac{1}{2} & -\frac{2}{3} & \frac{13}{12} \\ 0 & \frac{2}{3} & -\frac{5}{6} \\ 0 & \frac{1}{3} & -\frac{1}{6} \end{bmatrix} = E_7 E_6 \cdots E_1,$

where

$$E_1 = \begin{bmatrix} \frac{1}{2} & 0 & 0 \\ 0 & 1 & 0 \\ 0 & 0 & 1 \end{bmatrix}, \qquad E_2 = \begin{bmatrix} 1 & 0 & 0 \\ 0 & -1 & 0 \\ 0 & 0 & 1 \end{bmatrix}, \quad E_3 = \begin{bmatrix} 1 & 0 & 0 \\ 0 & 1 & 0 \\ 0 & 2 & 1 \end{bmatrix},$$

$$E_4 = \begin{bmatrix} 1 & 0 & 0 \\ 0 & 1 & 0 \\ 0 & 0 & -\frac{1}{6} \end{bmatrix}, \quad E_5 = \begin{bmatrix} 1 & -\frac{3}{2} & 0 \\ 0 & 1 & 0 \\ 0 & 0 & 1 \end{bmatrix}, \quad E_6 = \begin{bmatrix} 1 & 0 & -\frac{13}{2} \\ 0 & 1 & 0 \\ 0 & 0 & 1 \end{bmatrix},$$

$$E_7 = \begin{bmatrix} 0 & 0 & 0 \\ 0 & 1 & 5 \\ 0 & 0 & 1 \end{bmatrix}$$

11. (a) $\begin{bmatrix} 1 & 0 & 0 & 1 \\ 0 & 1 & 0 & -1 \\ 0 & 0 & 1 & -1 \\ 0 & 0 & 0 & 0 \end{bmatrix}$ (b) $\begin{bmatrix} 1 & 0 \\ 0 & 1 \end{bmatrix}, \begin{bmatrix} \frac{4}{7} & -\frac{3}{7} \\ -\frac{3}{7} & \frac{4}{7} \end{bmatrix}$

(c) $\begin{bmatrix} 1 & 0 & 0 & 0 \\ 0 & 1 & 0 & 0 \\ 0 & 0 & 1 & 0 \\ 0 & 0 & 0 & 1 \end{bmatrix}, \frac{1}{34} \begin{bmatrix} 13 & 1 & -4 & -6 \\ 4 & -18 & 4 & 6 \\ 6 & -10 & 6 & -8 \\ -13 & 33 & 4 & 6 \end{bmatrix}$

13. (a) $\begin{bmatrix} 1 & 0 & 0 & 0 \\ 0 & 1 & 0 & 0 \\ 0 & 0 & 1 & 0 \\ 0 & 0 & 0 & 1 \end{bmatrix}$, $\begin{bmatrix} -\frac{1}{10} & -\frac{1}{5} & 0 & \frac{1}{2} \\ -\frac{7}{5} & -\frac{9}{5} & 1 & 2 \\ \frac{21}{10} & \frac{11}{5} & -1 & -\frac{5}{2} \\ \frac{17}{10} & \frac{12}{5} & -1 & -\frac{5}{2} \end{bmatrix}$

(b) $\begin{bmatrix} 1 & 0 & 3 \\ 0 & 1 & -2 \\ 0 & 0 & 0 \end{bmatrix}$, singular.

(c) $\begin{bmatrix} 1 & 0 & 0 \\ 0 & 1 & 0 \\ 0 & 0 & 1 \end{bmatrix}$, $\frac{1}{13}\begin{bmatrix} -3 & 4 & 5 \\ 7 & -5 & -3 \\ 5 & 2 & -4 \end{bmatrix}$

15. (a) $\begin{bmatrix} 1 & 0 \\ 0 & 1 \end{bmatrix}$, $\frac{1}{5}\begin{bmatrix} -3 & 1 \\ 2 & 1 \end{bmatrix}$

(b) $\begin{bmatrix} 1 & 0 & 0 \\ 0 & 1 & 0 \\ 0 & 0 & 1 \end{bmatrix}$, $\begin{bmatrix} \frac{1}{2} & \frac{1}{2} & -\frac{1}{2} \\ \frac{1}{2} & -\frac{1}{10} & -\frac{1}{10} \\ -\frac{1}{2} & -\frac{3}{10} & \frac{7}{10} \end{bmatrix}$

(c) $\begin{bmatrix} 1 & 0 & 0 \\ 0 & 1 & 0 \\ 0 & 0 & 1 \end{bmatrix}$, $\begin{bmatrix} \frac{9}{7} & \frac{6}{7} & -1 \\ -\frac{5}{14} & -\frac{1}{14} & \frac{1}{2} \\ -\frac{1}{2} & -\frac{1}{2} & \frac{1}{2} \end{bmatrix}$

17. (a) $x = \frac{7}{6}$, $y = -\frac{3}{2}$, $z = \frac{5}{6}$.

(b) $x = \frac{1}{2}$, $y = -\frac{1}{2}$

19. (a), (b)

Section 1.4, page 86

1. (a) 4   (b) 5   (c) 4   (d) 2

3. (a), (b)

7. (a) -10 (b) -61 (c) 35

9. $D_1 = 5$, $D_2 = 10$, $D_3 = -5$

11. (a) -13 (b) 14   (c) -44

13. (a) 36   (b) 0   (c) 0

15. $A_{11} = -1,\quad A_{12} = 0,\quad A_{13} = 2,$
$A_{21} = 1,\quad A_{22} = 0,\quad A_{23} = -1,$
$A_{31} = -1,\quad A_{32} = 1,\quad A_{33} = 1$

17. (a) -3 (b) 0 (c) 47

19. (a) 20 (b) 36

21. $x = 2,\quad y = 0,\quad z = 2,\quad w = 1$

23. $\begin{bmatrix} \frac{1}{2} & \frac{1}{2} & -\frac{1}{2} \\ -\frac{1}{2} & \frac{1}{2} & \frac{1}{2} \\ \frac{1}{2} & -\frac{1}{2} & \frac{1}{2} \end{bmatrix}$

25. (a) $\begin{bmatrix} \frac{2}{5} & \frac{1}{5} \\ -\frac{3}{10} & \frac{1}{10} \end{bmatrix}$ (b) $\dfrac{1}{25}\begin{bmatrix} 4 & 3 & -5 \\ 6 & -8 & 5 \\ -3 & 4 & 10 \end{bmatrix}$

27. (a) $\begin{bmatrix} \frac{1}{2} & -\frac{1}{3} \\ 0 & \frac{1}{3} \end{bmatrix}$ (b) $\dfrac{1}{19}\begin{bmatrix} 7 & -6 & 3 \\ -9 & 5 & 7 \\ 4 & 2 & -1 \end{bmatrix}$

31. $-\frac{1}{3}$

Chapter 2

Section 2.1, page 103

5. (f) in Definition 1.1 is not satisfied.

7. (b), (c)

9. (a), (d)

11. (a), (d)

Section 2.2, page 119

1. (b), (d)

3. (a)

5. (b), (c), (d)

9. (a), (c), (d)

11. (a) $\begin{bmatrix} 0 \\ 0 \\ 0 \end{bmatrix} = 0 \begin{bmatrix} 1 \\ 1 \\ 2 \end{bmatrix} + 0 \begin{bmatrix} -2 \\ 3 \\ 4 \end{bmatrix}$ (c) $\begin{bmatrix} -1 \\ 1 \\ 9 \end{bmatrix} = -4 \begin{bmatrix} 1 \\ 2 \\ -3 \end{bmatrix} - 3 \begin{bmatrix} -1 \\ -3 \\ 1 \end{bmatrix}$

13. (a) $\cos 2t = \cos^2 t - \sin^2 t.$ (d) $\sec^2 t = 1 + \tan^2 t$

15. (c)

17. (a), (c), (d)

19. (a) $\mathbf{Y} = \frac{3}{2}\mathbf{X}_1 + \frac{1}{2}\mathbf{X}_2 - \frac{3}{2}\mathbf{X}_3$ (b) $\mathbf{Y} = \mathbf{X}_2$
    (c) $\mathbf{Y} = \frac{23}{9}\mathbf{X}_1 + \frac{3}{9}\mathbf{X}_2 - \frac{8}{9}\mathbf{X}_3$

21. (a) $\{p_1(t), p_2(t)\}$ (b) $\{p_1(t), p_2(t)\}$

23. $\left\{ \begin{bmatrix} 1 \\ -1 \\ 2 \end{bmatrix}, \begin{bmatrix} 0 \\ 1 \\ 0 \end{bmatrix}, \begin{bmatrix} 0 \\ 0 \\ 1 \end{bmatrix} \right\}$

25. $\left\{ \begin{bmatrix} 1 \\ 0 \\ 1 \end{bmatrix}, \begin{bmatrix} 1 \\ 2 \\ -3 \end{bmatrix}, \begin{bmatrix} 0 \\ 0 \\ 1 \end{bmatrix} \right\}$

Section 2.3, page 145

1. (a) yes   (b) yes   (c) yes   (d) yes   (e) no

3. (a) $\left\{ \begin{bmatrix} 1 \\ -1 \\ 1 \end{bmatrix} \right\}$ (b) $\left\{ \begin{bmatrix} 1 \\ 1 \\ 0 \end{bmatrix}, \begin{bmatrix} 1 \\ 0 \\ 1 \end{bmatrix} \right\}$ (c) no   (d) no

5. (a) kernel = zero subspace   (b) $\{2t^2, t, 1\}$
   (c) yes   (d) yes

7. (a) $\left\{ \begin{bmatrix} 2 \\ 1 \\ 4 \end{bmatrix} \right\}$ (b) $\left\{ \begin{bmatrix} 1 \\ -1 \end{bmatrix}, \begin{bmatrix} 2 \\ 2 \end{bmatrix} \right\}$ (c) no   (d) no

9. (a) kernel = zero subspace   (b) $\left\{ \begin{bmatrix} 1 \\ 1 \\ 0 \end{bmatrix}, \begin{bmatrix} 1 \\ -1 \\ 0 \end{bmatrix}, \begin{bmatrix} 0 \\ 0 \\ 1 \end{bmatrix} \right\}$

   (c) yes   (d) yes

11. (a) $\begin{bmatrix} 1 \\ 0 \\ 0 \end{bmatrix}$ (b) $\begin{bmatrix} 1 \\ -1 \\ -1 \end{bmatrix}$ (c) $\begin{bmatrix} 0 \\ \frac{1}{2} \\ \frac{3}{2} \end{bmatrix}$ (d) $\begin{bmatrix} 3 \\ 1 \\ -3 \end{bmatrix}$

13. $\begin{bmatrix} 1 & -1 & 1 \\ 0 & 0 & 1 \end{bmatrix}$ with respect to natural bases;

$\begin{bmatrix} 0 & 1 & -\frac{8}{3} \\ 1 & 1 & \frac{7}{3} \end{bmatrix}$ with respect to B and C; $\begin{bmatrix} 4 \\ 1 \end{bmatrix}$

15. $\begin{bmatrix} 1 & 0 \\ 0 & 1 \\ 0 & 0 \end{bmatrix}$

17. $\begin{bmatrix} 2 & -1 & 0 \\ 1 & 2 & 0 \\ 1 & 1 & 1 \end{bmatrix}$ with respect to natural bases;

$\begin{bmatrix} \frac{3}{2} & -\frac{1}{2} & 0 \\ \frac{1}{2} & \frac{5}{2} & 2 \\ \frac{1}{2} & -\frac{1}{2} & 1 \end{bmatrix}$ with respect to the basis B; $\begin{bmatrix} 5 \\ 0 \\ 4 \end{bmatrix}$

Section 2.4, page 163

1. $\left\{ \begin{bmatrix} 1 \\ 1 \\ -1 \\ 0 \end{bmatrix} \begin{bmatrix} -16 \\ -6 \\ 0 \\ 7 \end{bmatrix} \right\}$

3. $\left\{ \begin{bmatrix} 2 \\ -3 \\ 0 \\ 0 \\ 4 \end{bmatrix}, \begin{bmatrix} -2 \\ 1 \\ 0 \\ 2 \\ 0 \end{bmatrix}, \begin{bmatrix} 0 \\ 1 \\ 2 \\ 0 \\ 0 \end{bmatrix} \right\}$

5. 1

7. $\left\{ \begin{bmatrix} 2 \\ -1 \\ 1 \\ 0 \\ 0 \end{bmatrix}, \begin{bmatrix} -1 \\ -2 \\ 0 \\ 1 \\ 0 \end{bmatrix}, \begin{bmatrix} -3 \\ -2 \\ 0 \\ 0 \\ 1 \end{bmatrix} \right\}$ = basis for ker **L**;

$$\left\{ \begin{bmatrix} 1 \\ -1 \\ 0 \\ 2 \end{bmatrix}, \begin{bmatrix} 1 \\ 2 \\ 1 \\ -3 \end{bmatrix} \right\} = \text{basis for range } \mathbf{L}.$$

9. $\left\{ \begin{bmatrix} -2 \\ 1 \\ 1 \\ 0 \end{bmatrix}, \begin{bmatrix} -1 \\ 2 \\ 0 \\ 1 \end{bmatrix} \right\} = \text{basis for ker } \mathbf{L};$

$$\left\{ \begin{bmatrix} 1 \\ 2 \\ -1 \\ 2 \end{bmatrix}, \begin{bmatrix} 0 \\ 1 \\ 0 \\ 3 \end{bmatrix} \right\} = \text{basis for range } \mathbf{L}.$$

11. $\text{rank} = 3;$ $\begin{vmatrix} -2 & -1 & -5 \\ 3 & 2 & 8 \\ -1 & 1 & -1 \end{vmatrix} \neq 0$

13. $\text{rank} = 5,$ $\det \mathbf{A} \neq 0$

15. $\left\{ \begin{bmatrix} 0 \\ 5 \\ 0 \end{bmatrix}, \begin{bmatrix} 1 \\ 3 \\ -1 \end{bmatrix}, \begin{bmatrix} 1 \\ 1 \\ 2 \end{bmatrix} \right\}$

17. No

Section 2.5, page 181

1. (a) $\sqrt{30}$ (b) 0 (c) $\sqrt{2}$ (d) $\sqrt{23}$

3. (a) $2\sqrt{11}$ (b) $2\sqrt{17}$ (c) $2\sqrt{39}$

5. (a) -16 (b) 0 (c) -10 (d) 16

7. (a) $\sqrt{83}$ (b) $\sqrt{2}$ (c) $\sqrt{71}$ (d) $\sqrt{10}$

9. (a) $-\dfrac{16}{3\sqrt{66}}$ (b) 0 (c) $-\dfrac{10}{3\sqrt{66}}$ (d) $\dfrac{8}{\sqrt{90}}$

13. (a) $\begin{bmatrix} \dfrac{1}{\sqrt{5}} \\[4mm] \dfrac{2}{\sqrt{5}} \end{bmatrix}$ (b) $\begin{bmatrix} \dfrac{2}{\sqrt{29}} \\[4mm] -\dfrac{3}{\sqrt{29}} \\[4mm] \dfrac{4}{\sqrt{29}} \end{bmatrix}$ (c) $\begin{bmatrix} \dfrac{1}{\sqrt{7}} \\[4mm] \dfrac{2}{\sqrt{7}} \\[4mm] \dfrac{1}{\sqrt{7}} \\[4mm] -\dfrac{1}{\sqrt{7}} \end{bmatrix}$ (d) $\begin{bmatrix} 0 \\[2mm] 0 \\[2mm] \dfrac{1}{\sqrt{26}} \\[4mm] -\dfrac{5}{\sqrt{26}} \end{bmatrix}$

15. (c)

17. $\dfrac{1}{\sqrt{2}}\begin{bmatrix} 1 \\ 0 \\ 1 \end{bmatrix}$, $\dfrac{1}{\sqrt{6}}\begin{bmatrix} -1 \\ 2 \\ 1 \end{bmatrix}$

19. $\dfrac{1}{\sqrt{3}}\begin{bmatrix} 1 \\ 1 \\ 0 \\ 1 \end{bmatrix}$, $\dfrac{1}{\sqrt{5}}\begin{bmatrix} 2 \\ -4 \\ -1 \\ 2 \end{bmatrix}$, $\dfrac{1}{5\sqrt{6}}\begin{bmatrix} 8 \\ -1 \\ 6 \\ -7 \end{bmatrix}$

21. (a) $\dfrac{9}{\sqrt{6}}\mathbf{X_1} - \dfrac{1}{\sqrt{2}}\mathbf{X_2}$ (b) $\dfrac{5}{\sqrt{6}}\mathbf{X_1} + \dfrac{3}{\sqrt{2}}\mathbf{X_2} - \dfrac{1}{\sqrt{3}}\mathbf{X_3}$

   (c) $\dfrac{4}{\sqrt{6}}\mathbf{X_1} + \dfrac{2}{\sqrt{2}}\mathbf{X_2}$ (d) $\sqrt{6}\,\mathbf{X_1}$

Section 2.6, page 210

1. $(\lambda - 6)(\lambda + 2)$; $\lambda_1 = 6$, $\begin{bmatrix} 1 \\ 1 \end{bmatrix}$; $\lambda_2 = -2$, $\begin{bmatrix} 1 \\ -1 \end{bmatrix}$

3. $(\lambda - 1)(\lambda + 1)(\lambda + 2)$; $\lambda_1 = 1$, $\begin{bmatrix} 1 \\ 1 \\ 0 \end{bmatrix}$;

$$\lambda_2 = -1, \begin{bmatrix} 1 \\ 0 \\ 0 \end{bmatrix}; \quad \lambda_3 = -2, \begin{bmatrix} 1 \\ -2 \\ 1 \end{bmatrix}$$

5. No eigenvalues

7. $(\lambda - 3)^2(\lambda + 3)\lambda;$ $\lambda_1 = 3, \begin{bmatrix} 1 \\ 1 \\ -2 \\ 1 \end{bmatrix}, \begin{bmatrix} 1 \\ -2 \\ 1 \\ 1 \end{bmatrix};$

$$\lambda_2 = -3, \begin{bmatrix} -2 \\ 1 \\ 1 \\ 1 \end{bmatrix}; \quad \lambda_3 = 0, \begin{bmatrix} 1 \\ 1 \\ 1 \\ 2 \end{bmatrix}$$

9. $\lambda_1 = 0, \begin{bmatrix} 1 \\ 0 \\ 0 \end{bmatrix}; \quad \lambda_2 = 1, \begin{bmatrix} 0 \\ 1 \\ 0 \end{bmatrix}, \begin{bmatrix} 1 \\ 0 \\ 1 \end{bmatrix}$

11. $\lambda_1 = -3, \begin{bmatrix} 1 \\ 0 \\ 0 \end{bmatrix}; \quad \lambda_2 = -2, \begin{bmatrix} 6 \\ 1 \\ -1 \end{bmatrix}; \quad \lambda_3 = 6, \begin{bmatrix} 4 \\ 9 \\ 9 \end{bmatrix}$

13. $\lambda_1 = -2, \begin{bmatrix} 7 \\ -7 \\ -2 \end{bmatrix}; \quad \lambda_2 = 1, \begin{bmatrix} 0 \\ -2 \\ 1 \end{bmatrix}; \quad \lambda_3 = 5, \begin{bmatrix} 0 \\ 0 \\ 1 \end{bmatrix}$

15. $\lambda_1 = 9, \begin{bmatrix} -2 \\ 1 \\ 2 \end{bmatrix}; \quad \lambda_2 = -9, \begin{bmatrix} -2 \\ -2 \\ -1 \end{bmatrix}, \begin{bmatrix} -1 \\ 2 \\ -2 \end{bmatrix}$

17. $\lambda_1 = 0, \begin{bmatrix} -4 \\ 0 \\ 3 \end{bmatrix}; \quad \lambda_2 = 5, \begin{bmatrix} 3 \\ 5 \\ 4 \end{bmatrix}; \quad \lambda_5 = -5, \begin{bmatrix} -3 \\ 5 \\ -4 \end{bmatrix}$

19. $\lambda_1 = 3, \begin{bmatrix} 1 \\ 1 \\ 1 \end{bmatrix}; \quad \lambda_2 = -3, \begin{bmatrix} 1 \\ -1 \\ 0 \end{bmatrix}, \begin{bmatrix} 1 \\ 1 \\ -2 \end{bmatrix}$

21. $\lambda_1 = 2,$ $\dfrac{1}{\sqrt{2}}\begin{bmatrix} 1 \\ -1 \\ 0 \end{bmatrix}$; $\lambda_2 = -2,$ $\dfrac{1}{\sqrt{3}}\begin{bmatrix} 1 \\ 1 \\ -1 \end{bmatrix}$; $\lambda_3 = 4,$ $\dfrac{1}{\sqrt{6}}\begin{bmatrix} 1 \\ 1 \\ 2 \end{bmatrix}$

(a) $\dfrac{1}{8}\begin{bmatrix} -3 \\ -7 \\ 6 \end{bmatrix}$ (b) $\dfrac{1}{8}\begin{bmatrix} -5 \\ -9 \\ 10 \end{bmatrix}$ (c) $\dfrac{1}{2}\begin{bmatrix} 1 \\ -1 \\ 0 \end{bmatrix}$

25. $\dfrac{1}{\sqrt{2}}\begin{bmatrix} 1 & 1 \\ 1 & -1 \end{bmatrix}$

27. $\begin{bmatrix} 1 & 0 & 0 \\ 0 & \dfrac{1}{\sqrt{2}} & \dfrac{1}{\sqrt{2}} \\ 0 & \dfrac{1}{\sqrt{2}} & -\dfrac{1}{\sqrt{2}} \end{bmatrix}$

29. $\begin{bmatrix} \dfrac{1}{\sqrt{3}} & \dfrac{1}{\sqrt{2}} & \dfrac{1}{\sqrt{6}} \\ \dfrac{1}{\sqrt{3}} & -\dfrac{1}{\sqrt{2}} & \dfrac{1}{\sqrt{6}} \\ \dfrac{1}{\sqrt{3}} & 0 & -\dfrac{2}{\sqrt{6}} \end{bmatrix}$

31. $\begin{bmatrix} -\dfrac{4}{5} & -\dfrac{3}{5\sqrt{2}} & \dfrac{3}{5\sqrt{2}} \\ 0 & \dfrac{1}{\sqrt{2}} & \dfrac{1}{\sqrt{2}} \\ \dfrac{3}{5} & -\dfrac{4}{5\sqrt{2}} & \dfrac{4}{5\sqrt{2}} \end{bmatrix}$

33. $$\begin{bmatrix} \dfrac{1}{\sqrt{2}} & \dfrac{1}{\sqrt{3}} & \dfrac{1}{\sqrt{6}} \\[2em] -\dfrac{1}{\sqrt{2}} & \dfrac{1}{\sqrt{3}} & \dfrac{1}{\sqrt{6}} \\[2em] 0 & -\dfrac{1}{\sqrt{3}} & \dfrac{2}{\sqrt{6}} \end{bmatrix}$$

Chapter 3

Section 3.1, page 226

1. (a), (c), (d)

3. (a) $\dfrac{x_1 - 2}{0} = \dfrac{x_2 + 1}{-1} = \dfrac{x_3 - 3}{4} = \dfrac{x_4 - 3}{-3}$

(b) $\dfrac{x - 2}{1} = \dfrac{y - 1}{-1} = \dfrac{z}{2}$

5. $X = \begin{bmatrix} 1 \\ 2 \\ -1 \end{bmatrix} + t \begin{bmatrix} 4 \\ -2 \\ 5 \end{bmatrix}$

7. $X = \begin{bmatrix} 2 \\ 1 \\ 0 \\ 3 \end{bmatrix} + t \begin{bmatrix} 3 \\ 2 \\ -2 \\ 3 \end{bmatrix}$

9. $X = \begin{bmatrix} 4 \\ -2 \\ 5 \\ 2 \end{bmatrix} + t \begin{bmatrix} 0 \\ 1 \\ -2 \\ 4 \end{bmatrix}$

11. yes

13. $(-12, 16, -26)$

17. no

21. yes

23. $X = \begin{bmatrix} 3 \\ 5 \\ 3 \end{bmatrix} + t \begin{bmatrix} 19 \\ 73 \\ 8 \end{bmatrix}$

25. $\dfrac{-1}{\sqrt{14}\ \sqrt{39}}$

27. $\begin{bmatrix} 6 \\ -4 \\ 3 \end{bmatrix}$

29. $2x_1 - x_2 + 4x_3 - 5x_4 = 16.$

31. $3y - 6z = -15$

33. $X = \begin{bmatrix} 1 \\ 3 \\ -2 \\ 4 \end{bmatrix} + t \begin{bmatrix} 2 \\ 3 \\ 3 \\ -2 \end{bmatrix}$

Section 3.2, page 251

1. (a), (c)

3. (a) $\begin{bmatrix} 2 \\ -3 \\ 0 \end{bmatrix}$ (b) $\begin{bmatrix} 1 \\ \dfrac{-2}{3} \\ \dfrac{4}{7} \end{bmatrix}$ (c) $\begin{bmatrix} 0 \\ 2 \\ 0 \end{bmatrix}$ (d) $\begin{bmatrix} 3 \\ -1 \\ 1 \end{bmatrix}$

5. (a) $4\mathbf{i} + 8\mathbf{j} + 4\mathbf{k}$ (b) $20\mathbf{j} + 8\mathbf{k}$ (c) $-8\mathbf{i} + 11\mathbf{j} + \mathbf{k}$
   (d) $-5\mathbf{i} - 10\mathbf{j} - 6\mathbf{k}$

9. $3/\sqrt{2}$

11. $\frac{1}{2}\sqrt{362}$

13. $x - 5y - 2z = 15$

15. $X = \begin{bmatrix} 1 \\ -1 \\ 3 \end{bmatrix} + t \begin{bmatrix} 1 \\ -23 \\ -8 \end{bmatrix}$

19. $X = \begin{bmatrix} \dfrac{20}{3} \\ \dfrac{-5}{3} \\ 0 \end{bmatrix} + t \begin{bmatrix} -2 \\ 5 \\ 3 \end{bmatrix}$

21. distance $= 22/\sqrt{13}$; $\bar{X} = \left(\dfrac{25}{13}, -5, \dfrac{47}{13}\right)$

23. $3/\sqrt{5}$

25. $4/\sqrt{14}$

27. $4/\sqrt{6}$

29. $29x - 13y + 2z = 152$

31. $2y - z = 7$

33. $2x - 21y + 3z = -49$

35. $a[2(x - 2) + 3(y + 2)] + b[z - 5] = 0$; a and b arbitrary but not both zero

37. $x + 4y + 3z = 2$

39. 26

41. (a) $-25\mathbf{i} + 5\mathbf{j} - 20\mathbf{k}$ (b) $-2\mathbf{i} - 4\mathbf{j} - 2\mathbf{k}$

Section 3.3, page 271

1. $\mathbf{V}(t) = 3t^2\mathbf{i} + 2t\mathbf{j}$, $\mathbf{A}(t) = 6t\mathbf{i} + 2\mathbf{j}$,

$$\cos\theta = \frac{9t^2 + 2}{\sqrt{9t^2 + 4}\,\sqrt{9t^2 + 1}}, \qquad |\,\mathbf{V}(t)\,| = t\sqrt{9t^2 + 4},$$

$$\mathbf{T}(t) = \frac{3t\mathbf{i} + 2\mathbf{j}}{\sqrt{9t^2 + 4}}$$

3. $V(t) = e^t\mathbf{i} - \sin t\,\mathbf{j} + 3t^2\mathbf{k}$, $\mathbf{A}(t) = e^t\mathbf{i} - \cos t\,\mathbf{j} + 6t\mathbf{k}$

$$\cos\theta = \frac{e^{2t} + \sin t\cos t + 18t^3}{\sqrt{e^{2t} + \sin^2 t + 9t^4}\,\sqrt{e^{2t} + \cos^2 t + 36t^2}}$$

$$|\,\mathbf{V}(t)\,| = \sqrt{e^{2t} + \sin^2 t + 9t^4},$$

$$\mathbf{T}(t) = \frac{e^t\mathbf{i} - \sin t\,\mathbf{j} + 3t^2\mathbf{k}}{\sqrt{e^{2t} + \sin^2 t + 9t^4}}$$

7. $\dfrac{2t + 2t^3}{\sqrt{t^4 + 2t^2 + 2}}$; $\dfrac{A\mathbf{i} + B\mathbf{j} + C\mathbf{k}}{(t^4 + 2t^2 + 2)^{3/2}}$,

where
$$A = -t^4 - 2t^3 - 2t + 2, \qquad B = -t^4 - 2t^3 + 2t + 2,$$
$$C = 2t^3 + 4t^2$$

9. e^t; $\cos t \mathbf{i} - \sin t \mathbf{k}$

11. 1; $\dfrac{1}{(t+1)^2}\left[\dfrac{1-t}{\sqrt{2t}}\,\mathbf{i} - \mathbf{j} - \mathbf{k}\right]$

13. $T\left(\dfrac{\pi}{3}\right) = \dfrac{1}{2}\mathbf{i} - \dfrac{\sqrt{3}}{2}\mathbf{j}$; $N\left(\dfrac{\pi}{3}\right) = \dfrac{-\sqrt{3}}{2}\mathbf{i} - \dfrac{1}{2}\mathbf{j}$;

$B\left(\dfrac{\pi}{3}\right) = -\mathbf{k}$, $z = -2$

15. $T(0) = \dfrac{\mathbf{j}+\mathbf{k}}{\sqrt{2}}$, $N(0) = \mathbf{i}$, $B(0) = \dfrac{1}{\sqrt{2}}(\mathbf{j}-\mathbf{k})$; $y - z = 1$

17. $\frac{5}{2}$

19. $\kappa(t_0) = \frac{1}{2}$, $\rho(t_0) = 2$, $\tau(t_0) = \dfrac{1}{\sqrt{2}}$, $\mathbf{A}_T(t_0) = \mathbf{0}$,

$\mathbf{A}_N(t_0) = -\frac{1}{2}\mathbf{i} - \dfrac{\sqrt{3}}{2}\mathbf{j}$

21. $\kappa(t_0) = 2\sqrt{2}$, $\rho(t_0) = \dfrac{1}{2\sqrt{2}}$, $\tau(t_0) = 0$, $\mathbf{A}_T(t_0) = \mathbf{0}$,

$\mathbf{A}_N(t_0) = 2(\mathbf{i}+\mathbf{j})$

23. $\mathbf{A}_T(t) = \mathbf{0}$, $\mathbf{A}_N(t) = \dfrac{V_0^2}{r^2}\,\mathbf{X}(t)$

Section 3.4, page 282

3. both

5. both

7. neither

9. $\Phi'(t) = \begin{bmatrix} 2 \\ 4 \\ 4 \end{bmatrix}$

15. possible answers: $\Psi_1(s) = \begin{bmatrix} -\sin 2s \\ \cos 2s \\ \sin 2s \end{bmatrix}$ $\left(0 \leq s \leq \dfrac{\pi}{4}\right)$;

$\Psi_2(s) = \begin{bmatrix} -\sin s^2 \\ \cos s^2 \\ \sin s^2 \end{bmatrix}$ $\left(0 \leq s \leq \sqrt{\dfrac{\pi}{2}}\right)$

17. $s = t + 3$

19. $\mathbf{X} = \begin{bmatrix} 2\cos t \\ -3\sin t \end{bmatrix}$ $(\pi \leq t \leq 3\pi)$

21. $2(e^\pi - 1)$

Section 3.5, page 295

1. (a) $\left(10, \dfrac{11\pi}{6}, 2\right)$ (b) $\left(2\sqrt{2}, \dfrac{5\pi}{4}, 5\right)$

(c) $\left(1, \dfrac{\pi}{2}, 0\right)$ (d) $\left(6, \dfrac{\pi}{3}, -3\right)$

3. (a) $\left(2\sqrt{2}, \dfrac{\pi}{4}, \dfrac{\pi}{4}\right)$ (b) $\left(6, \dfrac{2\pi}{3}, \dfrac{\pi}{2}\right)$

(c) $\left(2, \dfrac{5\pi}{3}, \dfrac{5\pi}{6}\right)$ (d) $\left(2, \dfrac{5\pi}{6}, \dfrac{5\pi}{6}\right)$

5. (a) $\left(\sqrt{13}, \dfrac{\pi}{3}, \cos^{-1}\dfrac{2}{\sqrt{13}}\right)$ (b) $\left(5, \dfrac{2\pi}{3}, \cos^{-1} - \dfrac{3}{5}\right)$

(c) $\left(2\sqrt{5}, \dfrac{5\pi}{6}, \cos^{-1}\dfrac{2}{\sqrt{5}}\right)$ (d) $\left(2, \dfrac{3\pi}{4}, 3\right)$

7. (a) $\left(0, \dfrac{3}{\sqrt{2}}, -\dfrac{1}{\sqrt{2}}\right)$ (b) $(-3, -\sqrt{2}, 0)$

(c) $(-1, 2\sqrt{2}, 0)$ (d) $\left(1, \dfrac{3}{\sqrt{2}}, \dfrac{-7}{\sqrt{2}}\right)$

Section 3.6, page 306

1. hyperbolic paraboloid

3. hyperboloid of one sheet

5. hyperbolic cylinder

7. y axis

9. parabolic cylinder

11. sphere

13. hyperboloid of one sheet

15. ellipsoid

17. hyperbolic paraboloid

19. hyperbolic cylinder

21. $x' = x + 1,\quad y' = y - 1,\quad z' = z - 2;$ ellipsoid

23. $x' = x + 2,\quad y' = y - 3,\quad z' = z + 2;$ hyperboloid of two sheets

25. $x' = x - 2,\quad z' = z - 1;$ pair of lines

27. $x' = x + 2,\quad y' = y + 2,\quad z' = z + 1;$ elliptic paraboloid

29. $x' = x - 1,\quad y' = y - 2,\quad z' = z + 3;$ sphere

Chapter 4

Section 4.1, page 322

1. All (x, y) such that   (a) $x^2 + y^2 \geq 1$ (b) $xy \geq 0$
    (c) $xy \neq 0$ (d) $x^2 - y^2 \neq \pi n$ (n = integer)

   (e) $(x^2 - y^2) \neq (2n + 1)\,\dfrac{\pi}{2}$ (n = integer)   (f) all (x, y)

3. The common domain of $f - g$, $f + g$, and fg is the set of all (x, y)
    such that $x^2 + y^2 \leq 1$ and $x > y$; the domain of f/g is the set of all
    (x, y) such that $x^2 + y^2 \leq 1$, $x > y$ and $x - y \neq 1$.

5. $-\dfrac{1}{\cos(x+y)}$; the set of all (x, y) such that $x + y \neq (2n + 1)\dfrac{\pi}{2}$

 $(n = $ integer)   (b) -1

7. $\rho - |\mathbf{X}_1 - \mathbf{X}_0|$

9. (b)   has no limit points. The limit points of the other sets are the
points (x, y) such that   (a) $x^2 + y^2 \leq 1$ (c) $x^2 + y^2 \leq 1$
(d) $0 \leq x \leq y$ (e) x and y are arbitrary real numbers
(f) $x^2 + y^2 = 1$

11. (a)   9(b) and 9(f) have no interior points. The interior points of
the other sets are those (x, y) such that: 9(a) $x^2 + y^2 < 1$;
9(c) $x^2 + y^2 < 1, x \neq 0, y \neq 0$;   9(d) $0 < x < y$;
9(e) $(x, y) \neq (0, 0)$

 (b)   9(b) has no limit points. The limit points of the other sets which
are not interior points are (x, y) such that: 9(a) $x^2 + y^2 = 1$;
9(c) $x^2 + y^2 = 1, (x, y) = (x, 0) \, (|y| < 1)$, or $(x, y) = (0, y) \, (|y| < 1)$;
9(d) $x = 0, y \geq 0$ or $x = y \geq 0$;   9(e) $(x, y) = (0, 0)$;
9(f) $x^2 + y^2 = 1$

 (c)   The isolated points of 9(b) are $(0, 0)$ and $(1, 1)$; $(1, 2)$ is the
only isolated point of 9(f). The other sets have no isolated points.

13. (a)   8   (b)   no limit   (c)   0   (d)   2

15. (a)   0   (b)   2   (c)   no limit

17. 1

19. Each is continuous at every point of its domain.

21. All (x, y) such that $x^2 \geq y^2$

23. (a) (x, y, z) such that $3xyz \neq 1$ (b)   all (x, y)
(c)   all $(x, y) \neq (0, 0)$

25. (a)   0   (b)   1   (c)   0

Section 4.2, page 336

1. (a) $3\sqrt{2}$ (b) $\dfrac{7}{\sqrt{2}}$ (c) $\dfrac{1}{\sqrt{2}}$ (d) $\dfrac{1}{9\sqrt{2}}$

3. (a) $f_x(\mathbf{X}) = 4x, \quad f_y(\mathbf{X}) = 2y, \quad f_z(\mathbf{X}) = 1, \quad f_x(\mathbf{X}_0) = 4, \quad f_y(\mathbf{X}_0) = 2,$
$f_z(\mathbf{X}_0) = 1$;

 (b) $f_x(\mathbf{X}) = e^x \cos(x^2 + y^2 + z^2) - 2xe^x \sin(x^2 + y^2 + z^2), \quad f_y(\mathbf{X}) =$

$$-2y \sin(x^2 + y^2 + z^2), \quad f_z(\mathbf{X}) = -2z \sin(x^2 + y^2 + z^2),$$
$$f_x(\mathbf{X}_0) = -\sqrt{\pi}\, e^{\sqrt{\pi}/4} = f_y(\mathbf{X}_0), \quad f_z(\mathbf{X}_0) = 0$$

(c) $f_x(\mathbf{X}) = \dfrac{2x}{1 + x^2 + y + z}, \quad f_y(\mathbf{X}) = f_z(\mathbf{X}) = \dfrac{1}{1 + x^2 + y + z},$

$f_x(\mathbf{X}_0) = 0, \quad f_y(\mathbf{X}_0) = f_z(\mathbf{X}_0) = 1$

(d) $f_x(\mathbf{X}) = f_y(\mathbf{X}) = 1$

5. (a) $f_{xy}(-1,\ 2) = f_{yx}(-1,\ 2) = -14$ (b) $f_{xy}(-1,\ 2) = f_{yx}(-1, 2) = 2$

7. (a) $f_x(x, y, z) = ye^z - 3z \cos y,$
$f_y(x, y, z) = xe^z + 4yz + 3xz \sin y,$
$f_z(x, y, z) = xye^z + 2y^2 - 3x \cos y$

(b) $f_x(x, y) = \dfrac{4x^2 - 20x^3y + 4xy^2 - 15x^2y^3 - 6y^2}{(2x + y^2)^2},$

$f_y(x, y) = \dfrac{12xy + 5x^3y^2 - 4x^2y - 10x^4}{(2x + y^2)^2}$

(c) $f'(x) = 3$

11. (a) (x, y) such that $x = -y;$ (b) (x, y) such that $x^2 + y^2 = 1;$
(c) (x, y) such that $y \sin x = -1$

Section 4.3, page 350

1. (a) $a = 1, b = 0$ (b) $a = 3, b = 4$

3. (a) $(d_{X_0}f)\,(\mathbf{X} - \mathbf{X}_0) = 10(x + 1) + 2y, \qquad (d_{X_0}f)\,(\mathbf{X}_1 - \mathbf{X}_0) = 44$

(b) $(d_{X_0}f)\,(\mathbf{X} - \mathbf{X}_0) = \dfrac{2}{\sqrt{3}}\left(x - \dfrac{1}{\sqrt{3}}\right) - \dfrac{2}{\sqrt{3}}\left(y + \dfrac{1}{\sqrt{3}}\right)$

$$+ \dfrac{2}{\sqrt{3}}\left(z - \dfrac{1}{\sqrt{3}}\right),$$

$(d_{X_0}f)\,(\mathbf{X}_1 - \mathbf{X}_0) = -2.$

(c) $(d_{X_0}f)\,(\mathbf{X} - \mathbf{X}_0) = -\dfrac{\pi}{2}(x - 1) - \dfrac{\pi}{2}(y - 1) - \left(z - \dfrac{\pi}{2}\right),$

$$(d_{X_0} f)(\mathbf{X}_1 - \mathbf{X}_0) = -2\pi.$$

(d) $(d_{X_0} f)(\mathbf{X} - \mathbf{X}_0) = 10(z - 3);$ $(d_{X_0} f)(\mathbf{X}_1 - \mathbf{X}_0) = 140$

5. $\dfrac{1}{4} + \dfrac{5\pi\sqrt{3}}{72}$

7. (a) 0   (b) 1   (c) 2   (d) 2, 5, 6, 7, 9

9. (a) $d_{X_0} f = 10 \, dx + 2 \, dy$ (b) $d_{X_0} f = \dfrac{2}{\sqrt{3}} dx - \dfrac{2}{\sqrt{3}} dy + \dfrac{2}{\sqrt{3}} dz$

(c) $d_{X_0} f = -\dfrac{\pi}{2} dx - \dfrac{\pi}{2} dy - dz$ (e) $d_{X_0} f = 0.$

11. (a) $df = (9x^2 + 2xy + 1) \, dx + (x^2 + 1) \, dy$

(b) $df = \dfrac{2x \, dx + 2y \, dy + 2z \, dz}{x^2 + y^2 + z^2}$

(c) $df = -\sin xyz \, e^{\cos xyz}(yz \, dx + xz \, dy + xy \, dz)$

(d) $df = \dfrac{x^2 - y^2}{xy}\left(\dfrac{dx}{x} + \dfrac{dy}{y}\right)$

13. $d_{X_0}(2f) = \sqrt{2}(dx + dy + dz),$ $d_{X_0}(f + g) = 2\sqrt{2}(dx + dy + dz),$
$d_{X_0}(fg) = dx + dy + dz,$ $d_{X_0}(f/g) = 0$

15. (a) $\dfrac{2}{7\sqrt{3}}$ (b) $\dfrac{1}{\sqrt{3}}$ (c) $\dfrac{e^6}{\sqrt{3}}$ (d) $-\dfrac{1}{49\sqrt{3}}$

Section 4.4, page 357

1. (a) 0   (b) $-5\left(\dfrac{\pi}{2}\right)^{4/5}$ (c) 0   (d) 2   (e) $-\pi^2$

3. $\dfrac{1}{9\sqrt{2}}$

5. (a) 3   (b) 0

7. $e^y \log y \, (\tan y + \sec^2 y) + \dfrac{e^y \tan y}{y}$

Section 4.5, page 364

9. $18x + 16y + \sqrt{11}\,z = 61$

11. $2x + 3y - z = 6$

13. $2x - 8y - z = 9$

Section 4.6, page 373

1. (a) domain $= \{(1, 5, 1), (1, 1, 2), (1, 1, 1), (1, 1, 3)\}$;   range $=$
       $\{1, 2, 3\}$

    (c) domain $= \{(2, 7, 5), (1, 7, 5), (3, 6, 2), (7, 5, 3)\}$;   range $=$
       $\{1, 3\}$

    (d) domain $= \{(1, 2, 3), (2, 3, 4), (3, 4, 1), (4, 1, 2)\}$;   range $=$
       $\{1, 2, 3, 4\}$

3. (c) domain $= \{x | |x| \leq 1\}$;   range $= \{y | 0 \leq y \leq \pi\}$
    (d) domain $= \{x | |x| \leq 1\}$;   range $= \{y | \pi \leq y \leq 2\pi\}$

5. (a) domain $= \{(5, 1, 1), (1, 2, 2), (1, 1, 3), (1, 3, 2)\}$;   range $= \{1\}$
    (b) domain $= \{(1, 1, 1), (1, 2, 2), (1, 1, 2), (1, 3, 2)\}$;   range $= \{1\}$
    (d) domain $= \{(2, 3, 4), (3, 4, 1), (4, 1, 2), (1, 2, 3)\}$;   range $=$
       $\{1, 2, 3, 4\}$

7. (c) domain $= \{(x, z) | x^2 + z^2 \geq 1\}$;   range $= \{y | y \geq 0\}$

11. $g(0) = 1,\quad g'(0) = -\tfrac{2}{3};\quad h(0) = 2,\quad h'(0) = -\tfrac{10}{3}$

13. $g_x(1, 1, 1) = \tfrac{5}{9},\quad g_z(1, 1, 1) = \tfrac{4}{9},\quad g_w(1, 1, 1) = -\tfrac{8}{9}$

Section 4.7, page 383

1. (a) $\begin{bmatrix} a_1 \\ a_2 \\ \vdots \\ a_n \end{bmatrix}$ (b) $\begin{bmatrix} 0 \\ -2 \\ 4 \end{bmatrix}$ (c) $\begin{bmatrix} \frac{1}{6} \\ \frac{1}{6} \\ \frac{1}{6} \end{bmatrix}$ (d) $\begin{bmatrix} 2e \\ 0 \\ 0 \end{bmatrix}$

3. (a) $\dfrac{1}{\sqrt{a_1^2 + \cdots + a_n}} \begin{bmatrix} a_1 \\ a_2 \\ \vdots \\ a_n \end{bmatrix}$; $\sqrt{a_1^2 + \cdots + a_n^2}$

(b) $\dfrac{1}{\sqrt{5}}\begin{bmatrix} 0 \\ -1 \\ 2 \end{bmatrix}$; $2\sqrt{5}$ (c) $\begin{bmatrix} \dfrac{1}{\sqrt{3}} \\ \dfrac{1}{\sqrt{3}} \\ \dfrac{1}{\sqrt{3}} \end{bmatrix}$; $\dfrac{1}{2\sqrt{3}}$

(d) $\begin{bmatrix} 1 \\ 0 \\ 0 \end{bmatrix}$; $2e$

5. (a) 0 (b) $2e/\sqrt{3}$

9. $2x + 4y - 3z = -2$ (plane).

11. $\{(x, y, z) | x^2 + y^2 + z^2 = \pi\}$

13. $7x + 7y + z = 20$

15. $x = \sqrt{\pi}$

17. $x = 0$

19. $\begin{bmatrix} \dfrac{1}{\sqrt{2}} \\ -\dfrac{1}{\sqrt{2}} \end{bmatrix}$

21. $\begin{bmatrix} \dfrac{1}{\sqrt{2}} \\ \dfrac{1}{\sqrt{2}} \\ 0 \end{bmatrix}$

Section 4.8, page 396

1. (a) $e(1 + (t - 1) + \frac{1}{2}(t - 1)^2)$ (b) $1 + \frac{1}{2}(t - 1) - \frac{1}{8}(t - 1)^2$
   (c) $1 + \frac{1}{2}t^2$

3. $8 + 10(t - 1) + 5(t - 1)^2 + (t - 1)^3$

5. (a) 6   (b) 4   (c) 6   (d) 5

7. 1.6487

9. (a) $(d^0_{X_0} f)(\mathbf{X} - \mathbf{X}_0) = 0,$ $(d_{X_0} f)(\mathbf{X} - \mathbf{X}_0) = 2x + (y - 1),$
       $(d^2_{X_0} f)(\mathbf{X} - \mathbf{X}_0) = -4x^2 - 4x(y - 1) - (y - 1)^2$

    (b) $(d^0_{X_0} f)(\mathbf{X} - \mathbf{X}_0) = 0,$ $(d_{X_0} f)(\mathbf{X} - \mathbf{X}_0) = 0,$
       $(d^2_{X_0} f)(\mathbf{X} - \mathbf{X}_0) = 2(x^2 + y^2).$

11. (a) $(d^0_{X_0} f)(\mathbf{X} - \mathbf{X}_0) = 1,$ $(d_{X_0} f)(\mathbf{X} - \mathbf{X}_0) = 0,$
       $(d^2_{X_0} f)(\mathbf{X} - \mathbf{X}_0) =$

$$-\left[\left(x - \frac{\pi}{6}\right)^2 + \left(y - \frac{\pi}{6}\right)^2 + \left(z - \frac{\pi}{6}\right)^2 + 2\left(x - \frac{\pi}{6}\right)\left(y - \frac{\pi}{6}\right)\right.$$

$$\left. + 2\left(x - \frac{\pi}{6}\right)\left(z - \frac{\pi}{6}\right) + 2\left(y - \frac{\pi}{6}\right)\left(z - \frac{\pi}{6}\right)\right]$$

    (b) $(d^0_{X_0} f)(\mathbf{X} - \mathbf{X}_0) = 1,$ $(d_{X_0} f)(\mathbf{X} - \mathbf{X}_0) = x - 1,$
       $d^2_{X_0} f = y^2 + z^2$

13. (a) $2x + (y - 1) - 2x^2 - 2x(y - 1) - \frac{1}{2}(y - 1)^2$ (b) $x^2 + y^2$

15. $1 - \frac{1}{2}\left(x - \frac{\pi}{6}\right)^2 - \frac{1}{2}\left(y - \frac{\pi}{6}\right)^2 - \frac{1}{2}\left(z - \frac{\pi}{6}\right)^2 - \left(x - \frac{\pi}{6}\right)\left(y - \frac{\pi}{6}\right)$

$$-\left(x - \frac{\pi}{6}\right)\left(z - \frac{\pi}{6}\right) - \left(y - \frac{\pi}{6}\right)\left(z - \frac{\pi}{6}\right)$$

17. $6 + 2(x - 2) + 2(y - 3) + (x - 2)^2 - (x - 2)(y - 3)$

Section 4.9, page 409

1. (a) $f(-\frac{3}{2}) = -39,$ $f(1) = 11,$ $f(\frac{3}{2}) = 21/2$

    (b) $f(0) = 0,$ $f(1) = 1$

    (c) $f(1) = f(-1) = 0,$ $f(0) = 1$

    (d) $f(1) = f(-1) = 0,$ $f(0) = 1$

    (e) $f(x, -x) = 1$

    (f) $f(x, y) = 0$ if $xy = k\pi$ (k = integer); $f(x, y) = 1$ if $xy =$

$$(2k + 1)\frac{\pi}{2} (k = \text{integer})$$

3. (a) (0, 0)   (b) (0, 0), (0, 1), (0, −1), (1, 0), (−1, 0), (1, 1),

(−1, 1), (1, −1), (−1, −1)   (c) $\left(\dfrac{1}{e}, \dfrac{1}{e}\right)$ (d) (0, 0)

5. $\frac{7}{3}$

7. $l = w = (2v)^{1/3}, \quad h = (v/4)^{1/3}$

9. $\sqrt{10/3}$

11. $a/3, b/3, c/3$

13. (a) neither   (b) (0, 0) is a relative minimum; (1, 1), (−1, 1), (1,−1) and (−1, −1) are relative maxima; (0, 1), (0, −1), (1, 0) and (−1, 0) are not extreme points   (c) relative minimum (d) neither

15. $8abc/3\sqrt{3}$

17. (a) (0, 0) and (x, y) such that $(x^2 + y^2) = (2k + \frac{3}{2})\pi$ ($k = 0, 1, 2, \ldots$) are relative minima; (x, y) such that $x^2 + y^2 = (2k + \frac{1}{2})\pi$ ($k = 0, 1, 2, \ldots$) are relative maxima.   (b) (0, 0) is a critical point, but not an extreme point.   (c) (0, 0) is a relative minimum.   (d) (0, 0) is a critical point, but not an extreme point. (e) (0, 0) is a relative minimum.

Section 4.10, page 423

1. $l = \sqrt{A} = w$

3. $\left(\dfrac{2}{\sqrt{5}}, \dfrac{4}{\sqrt{5}}\right), \left(-\dfrac{2}{\sqrt{5}}, \dfrac{-4}{\sqrt{5}}\right)$

5. (0, 2, 2), (0, −2, 2)

7. $a/3, b/3, c/3$

9. (0, 2) and (0, −2) are closest; (0, 3) and (0, −3) are farthest

11. $8abc/3\sqrt{3}$

13. $f\left(\dfrac{10}{29}, \dfrac{15}{29}, \dfrac{-20}{29}\right) = \dfrac{25}{29}$

15. $\sqrt{2} - 1$

17. $Q(0, 0, 0, 1) = 1$

Chapter 5

Section 5.1, page 435

1. (a) $f_1(x, y) = \cos xy,\quad f_2(x, y) = \sqrt{x^2 + y^2 - 4},\quad f_3(x, y) = 1/xy;$
   $D = \{(x, y)\,|\,x^2 + y^2 \geq 4,\quad x \neq 0,\quad y \neq 0\}$

   (b) $f_1(x, y) = e^{x^2+y^2},\qquad f_2(x, y) = \dfrac{x}{y^2 - 1}\,;$

   $D = \{(x, y)\,|\,y \neq \pm 1\}$

   (c) $f_1(x, y) = x^2 + y^2 + z^2,\quad f_2(x, y) = x^2,\quad f_3(x, y, z) = -y^2;$
   $D = R^3.$

   (d) $f_1(x, y) = \sqrt{4 - x^2},\quad f_2(x, y) = \sqrt{4 - y^2},\quad f_3(x, y) = x + y;$
   $D = \{(x, y)\,|\,|x| \leq 2,\ |y| \leq 2\}.$

3. (a) $n = 2,\ m = 3$ (b) $n = 2,\ m = 2$
   (c) $n = 3,\ m = 3$ (d) $n = 2,\ m = 3.$

5. (a) $D = \{(x, y)\,|\,x^2 + y^2 \leq 2,\quad x \neq 0,\quad y \neq 0\}$

   (b) $\begin{bmatrix} \frac{9}{2} \\ 4 + \dfrac{\sqrt{3}}{2} \\ 1 \end{bmatrix},\quad \begin{bmatrix} -\frac{7}{2} \\ 4 - \dfrac{\sqrt{3}}{2} \\ 0 \end{bmatrix}$ (c) $\begin{bmatrix} 8 \\ 3 \\ 5 \end{bmatrix}$

7. (a) $\begin{bmatrix} -12 \\ 0 \\ 1 \end{bmatrix}$ (b) $\begin{bmatrix} -3x + 6y \\ 3x + 6y \\ 3x + 5y \end{bmatrix}$ (d) no

9. (a) $\begin{bmatrix} \frac{1}{2} \\ 2 \end{bmatrix}$ (b) $\begin{bmatrix} x - y \\ z \\ \sqrt{z} \end{bmatrix}$ (c) $D = \{(x, y, z)\,|\,z > 0\}$

11. (a) $\begin{bmatrix} \frac{1}{2} \\ 1 \end{bmatrix}$ (b) $\begin{vmatrix} \dfrac{\cos^2 y}{x^2} \\ x \end{vmatrix}$

13. (a) $\begin{bmatrix} \frac{1}{5} \\ -6 \\ \frac{12}{5} \end{bmatrix}$ (b) no limit (c) $\begin{bmatrix} 0 \\ 0 \\ 0 \end{bmatrix}$

15. (a) $\begin{bmatrix} 2 \\ -3 \end{bmatrix}$ (b) $\sqrt{2(x+z)^2 + 2y^2}$

17. (a) $\begin{bmatrix} x + y + z \\ 2x + z \\ x - y \end{bmatrix}$, $\begin{bmatrix} 7 \\ 9 \\ 2 \end{bmatrix}$

19. (a) $\{(x, y, z) | y \neq \pm 3, z \neq 0\}$
    (b) $\{(x, y) | x > y\}$ (c) $\{(x, y, z) | x \neq y, \quad x \neq z, \quad y \neq \pm z\}$

    (d) $\{(x, y) | x^2 - y^2 \neq (2k + 1) \dfrac{\pi}{2} \quad (k = \text{integer}), \quad |x| \geq |y|\}$

21. $\{(x, y) | x^2 + y^2 \leq 2, \quad x \neq 0, \quad y \neq 0\}$

23. R^2

25. $\{(x, y) | x \neq 0\}$,

$$\{(u, v) | u^2 + v^2 \neq 0, \quad v \neq (2k + 1) \dfrac{\pi}{2} \quad (k = \text{integer})\}$$

Section 5.2, page 449

1. (a) $\begin{bmatrix} \frac{1}{3} \, dx + \frac{2}{3} \, dy - \frac{2}{3} \, dz \\ 3 \, dx + 2 \, dy + 4 \, dz \end{bmatrix}$, $\begin{bmatrix} -8 \\ 3 \\ 10 \end{bmatrix}$

   (b) $\begin{bmatrix} dx - \quad dy \\ 12 \, dx + 2 \, dy \\ 8 \, dx + \quad dy \end{bmatrix}$, $\begin{bmatrix} 0 \\ 14 \\ 9 \end{bmatrix}$

   (c) $\begin{bmatrix} e^2 \, dx + e^2 \, dy + e^2 \, dz \\ \dfrac{1}{2\sqrt{2}} \, dx + \dfrac{1}{2\sqrt{2}} \, dy \\ \frac{3}{2} \, dx + \frac{3}{2} \, dy + \frac{3}{2} \, dz \end{bmatrix}$, $\begin{bmatrix} e^2 \\ 0 \\ \frac{3}{2} \end{bmatrix}$

3. (a) $\begin{bmatrix} dx + dy + dz \\ 2\,dx + 2\,dy + 4\,dz \end{bmatrix}$, $\begin{bmatrix} 2 \\ 6 \end{bmatrix}$

(b) $\begin{bmatrix} \dfrac{1}{\sqrt{2}}\,(dx + dy) \\[2ex] 2\,(dx + dy) \\[2ex] -\dfrac{1}{\sqrt{2}}\,(dx + dy) \end{bmatrix}$, $-\dfrac{\pi}{4\sqrt{2}}\begin{bmatrix} 1 \\ 2\sqrt{2} \\ -1 \end{bmatrix}$

(c) $\begin{bmatrix} dx \\ 0 \\ dx \end{bmatrix}$, $\begin{bmatrix} 1 \\ 0 \\ 1 \end{bmatrix}$

5. (a) $\begin{bmatrix} -\sin(x+y)\,(dx+dy) \\ \cos(x+y)\,(dx+dy) \end{bmatrix}$ (b) $\begin{bmatrix} dx \\ 2x\,dx \\ 3x^2\,dx \end{bmatrix}$

(c) $\begin{bmatrix} e^x \sin yz\,dx + ze^x \cos yz\,dy + ye^x \cos yz\,dz \\ ze^y \cos xz\,dx + e^y \sin xz\,dy + xe^y \cos xz\,dz \\ ye^z \cos xy\,dx + xe^z \cos xy\,dy + e^z \sin xy\,dz \end{bmatrix}$

7. (a) $\begin{bmatrix} e^x(y+z+w)\,dx + e^x(dy+dz+dw) \\ \cos(x+y+z+w)\,(dx+dy+dz+dw) \\ yz\,dx + xz\,dy + xy\,dz \end{bmatrix}$

(b) $\begin{bmatrix} -\cos(1-x-y)\,(dx+dy) \\ (2+y)\,dx + x\,dy \\ 3x^2\,dx + 2y\,dy \end{bmatrix}$

(c) $e^x\begin{bmatrix} (\sin x + \cos x)\,dx \\ (\cos x - \sin x)\,dx \end{bmatrix}$

9. (a) $\begin{bmatrix} 6\,dx + 24\,dy \\ 6\,dx + 3\,dy \end{bmatrix}$ (b) $\begin{bmatrix} 4\,dx + 7\,dy \\ 5\,dx + 5\,dy \end{bmatrix}$

(c) $\begin{bmatrix} -2\,dx + 19\,dy \\ -5\,dx - 10\,dy \end{bmatrix}$

11. (a) $\mathbf{g}(\mathbf{X}) = \begin{bmatrix} \frac{1}{3}x + \frac{2}{3}y - \frac{2}{3}z \\ -3x + 2y + 4z + 1 \end{bmatrix}$

(b) $\mathbf{g}(\mathbf{X}) = \begin{bmatrix} x - y - 1 \\ 12x + 2y - 17 \\ 8x + y - 8 \end{bmatrix}$

(c) $g(\mathbf{X}) = \begin{bmatrix} e^2(x + y + z - 2) \\ \dfrac{1}{2\sqrt{2}}(x + y + 2) \\ \frac{3}{2}x + \frac{3}{2}y + \frac{3}{2}z - 2 \end{bmatrix}$

Section 5.3, page 462

1. (a) $\begin{bmatrix} 4\,dx - 3\,dy + dz \\ dy + dz \end{bmatrix}$ (b) $\begin{bmatrix} 2\,dx \\ 2\,dx \end{bmatrix}$

3. (a) $\begin{bmatrix} 2 & -1 & 1 \\ 2 & -2 & 0 \end{bmatrix}, \begin{bmatrix} 1 & 1 \\ 1 & -1 \end{bmatrix}, \begin{bmatrix} 4 & -3 & 1 \\ 0 & 1 & 1 \end{bmatrix}, \begin{bmatrix} 4 & -3 & 1 \\ 0 & 1 & 1 \end{bmatrix}$

    (b) $\begin{bmatrix} 1 & 0 \\ 0 & 1 \end{bmatrix}, \begin{bmatrix} 2 & 0 \\ 2 & 0 \end{bmatrix}, \begin{bmatrix} 2 & 0 \\ 2 & 0 \end{bmatrix}, \begin{bmatrix} 2 & 0 \\ 2 & 0 \end{bmatrix}$

5. (a) $J\mathbf{U} = \begin{bmatrix} \dfrac{x}{\sqrt{x^2+y^2+z^2}} & \dfrac{y}{\sqrt{x^2+y^2+z^2}} & \dfrac{z}{\sqrt{x^2+y^2+z^2}} \\[3mm] \dfrac{-y}{x^2+y^2} & \dfrac{x}{x^2+y^2} & 0 \\[3mm] \dfrac{-xz}{\sqrt{x^2+y^2}(x^2+y^2+z^2)} & \dfrac{-yz}{\sqrt{x^2+y^2}(x^2+y^2+z^2)} & \dfrac{\sqrt{x^2+y^2}}{(x^2+y^2+z^2)} \end{bmatrix}$,

$J\mathbf{g} = \begin{bmatrix} v & u & 0 \\ 0 & w & v \\ w & 0 & u \end{bmatrix}$,

(b) $J_{\mathbf{X}_0}\mathbf{U} = \begin{bmatrix} \dfrac{1}{\sqrt{2}} & \dfrac{-1}{\sqrt{2}} & 0 \\[3mm] \dfrac{1}{\sqrt{2}} & \dfrac{1}{\sqrt{2}} & 0 \\[3mm] 0 & 0 & 1 \end{bmatrix}$, $J_{\mathbf{U}_0}\mathbf{g} = \begin{bmatrix} -\dfrac{\pi}{4} & 1 & 0 \\[3mm] 0 & 0 & -\dfrac{\pi}{4} \\[3mm] 0 & 0 & 1 \end{bmatrix}$,

$$
\mathbf{J_{x_0}h} = \begin{bmatrix} \dfrac{1}{\sqrt{2}}\left(1 - \dfrac{\pi}{4}\right) & \dfrac{1}{\sqrt{2}}\left(1 + \dfrac{\pi}{4}\right) & 0 \\[2ex] 0 & 0 & -\dfrac{\pi}{4} \\[2ex] 0 & 0 & 1 \end{bmatrix}
$$

(c) $\mathbf{d_{x_0}u} = \begin{bmatrix} \dfrac{1}{\sqrt{2}}\,(dx - dy) \\[2ex] \dfrac{1}{\sqrt{2}}\,(dx + dy) \\[2ex] dz \end{bmatrix}$, $\mathbf{J_{u_0}g} = \begin{bmatrix} -\dfrac{\pi}{4}\,dx + dy \\[2ex] -\dfrac{\pi}{4}\,dz \\[2ex] dz \end{bmatrix}$,

$$
\mathbf{J_{x_0}h} = \begin{bmatrix} \dfrac{1}{\sqrt{2}}\left(1 - \dfrac{\pi}{4}\right)dx + \dfrac{1}{\sqrt{2}}\left(1 + \dfrac{\pi}{4}\right)dy \\[2ex] -\dfrac{\pi}{4}\,dz \\[2ex] dz \end{bmatrix}
$$

7. (a) $\sqrt{3}, \sqrt{3}\,dt$ (b) $1, dt$ (c) $\dfrac{3 - 2\pi}{4}, \dfrac{3 - 2\pi}{4}\,dt$

9. (a) $\left(2 + \dfrac{1}{\sqrt{2}}\right)dr + (\sqrt{2} - 4)\,d\theta$

(b) $\begin{bmatrix} \left(\dfrac{1}{\sqrt{2}} - 2\right)dr - (\sqrt{2} + 4)\,d\theta \\[2ex] \dfrac{3}{\sqrt{2}}\,dr - \sqrt{2}\,d\theta \end{bmatrix}$ (c) $\begin{bmatrix} 8\,dr \\ 4\,dr \end{bmatrix}$

11. (a) $d\theta + dz$ (b) $\begin{bmatrix} dr + 2\,dz \\ 2\,dr - 2\,dz \end{bmatrix}$ (c) $\begin{bmatrix} d\theta \\ d\theta \end{bmatrix}$

13. (a) $\begin{bmatrix} -2\,d\theta \\ 2\,dr \end{bmatrix}$ (b) $\begin{bmatrix} \frac{1}{2}(dr - d\theta - d\phi) \\ \frac{1}{2}(dr + d\theta - d\phi) \\ dr + d\phi \end{bmatrix}$

   (c) $-\frac{1}{2}\,d\theta$

15. $dx + \frac{1}{5}\,dy$

17. $\frac{7}{5}, \frac{3}{5}, \frac{3}{25}$

19. $h_{uv} = g_{xx}x_u x_v + g_{xy}(x_u y_v + x_v y_u) + g_{yy}y_u y_v + g_x x_{uv} + g_y y_{uv}$;
     $h_{vv} = g_{xx}x_v^2 + 2g_{xy}x_v y_v + g_{yy}y_v^2 + g_x x_{vv} + g_y y_{vv}$

Section 5.4, page 477

1. (a) $ye^{xy} \sin z\,\mathbf{i} + xe^{xy} \sin z\,\mathbf{j} + e^{xy} \cos z\,\mathbf{k}$

   (b) $\dfrac{\mathbf{X}}{|\mathbf{X}|^2}$ (c) $z(x + 1)\mathbf{i} + x(x + \frac{1}{2})\mathbf{k}$

5. (a) $\dfrac{y}{x} + \dfrac{x}{y} + \dfrac{xy}{z}$ (b) x (c) $3x^2 + 2y + 3z^2$

   (d) $6(x + y + z)$ (e) $-[(xy)^2 + (yz)^2 + (xz)^2]\cos xyz$

7. (a) $z(x^2 + y^2)e^{xy}$ (b) $e^{yz}(y^2 + z^2 - 1)\cos x$

   (c) $\dfrac{2(3 + x^2 + y^2 + z^2)}{(1 + x^2 + y^2 + z^2)^2}$ (d) 4

   (e) $-[(xy)^2 + (yz)^2 + (xz)^2]\sin xyz$

11. (a) $-\frac{1}{2}x^2\mathbf{i} + (xy - z)\mathbf{j} + 2xz\mathbf{k}$
    (b) $2\mathbf{i} + \mathbf{j}$
    (c) $2z(ye^x - xe^y)\mathbf{i} + y(x^2 e^z - yze^x)\mathbf{j} + (z^2 e^y - x^2 e^z)\mathbf{k}$
    (d) $-(e^y \sin x + e^x \cos y)\mathbf{k}$

Section 5.5, page 488

1. $u = x + \frac{3}{4}y, \quad v = -\frac{5}{4}y;$ $\begin{bmatrix} 1 & \frac{3}{4} \\ 0 & -\frac{5}{4} \end{bmatrix}$

3. $\begin{bmatrix} 1 & 2 \\ -2 & 1 \end{bmatrix}, \quad \begin{bmatrix} -1 & 1 \\ 2 & 2 \end{bmatrix}, \quad \begin{bmatrix} 1 & \frac{3}{4} \\ 0 & -\frac{5}{4} \end{bmatrix}$

5. $\mathbf{f}(\mathbf{X}_0) = \begin{bmatrix} 1 \\ 1 \end{bmatrix}, \quad J_{\mathbf{X}_0}\mathbf{f} = \begin{bmatrix} 3 & -1 \\ 1 & 0 \end{bmatrix}, \quad d_{\mathbf{X}_0}\mathbf{f} = \begin{bmatrix} 3\,dx - dy \\ dx \end{bmatrix};$

   $\mathbf{g}(\mathbf{X}_0) = -\begin{bmatrix} 1 \\ 1 \end{bmatrix}, \quad J_{\mathbf{X}_0}\mathbf{g} = \begin{bmatrix} -3 & -1 \\ -1 & 0 \end{bmatrix}, \quad d_{\mathbf{X}_0}\mathbf{g} = \begin{bmatrix} -3\,dx - dy \\ -dx \end{bmatrix}$

7. 1

9. $\begin{bmatrix} 2 & -2 & 1 \\ -2 & 0 & -1 \\ -2 & 1 & -2 \end{bmatrix}$

11. (x, y), (x, u), (x, v), (y, u), (u, v)

Section 5.6, page 501

1. $\mathbf{f}^{-1}(u, v) = \begin{bmatrix} 3u - v \\ -2u + v \end{bmatrix}$, $\mathbf{Jf}^{-1}(u, v) = \begin{bmatrix} 3 & -1 \\ -2 & 1 \end{bmatrix}$

3. (b) $\begin{bmatrix} x \\ y \end{bmatrix} = \begin{bmatrix} \sqrt{\dfrac{2v - u}{3}} \\[4mm] \sqrt{\dfrac{2u - v}{3}} \end{bmatrix}$; $D = \{(u, v) \,|\, 0 \leq u \leq 2v, \;\; 0 \leq v \leq 2u\}$

(c) $\begin{bmatrix} x \\ y \end{bmatrix} = \begin{bmatrix} -\sqrt{\dfrac{2v - u}{3}} \\[4mm] \sqrt{\dfrac{2u - v}{3}} \end{bmatrix}$; $D = \{(u, v) \,|\, 0 \leq u \leq 2v, \;\; 0 \leq v \leq 2u\}$

(d) no

5. $\dfrac{1}{\sqrt{3}} \begin{bmatrix} -\frac{1}{2} & 1 \\[2mm] \dfrac{1}{\sqrt{13}} & \dfrac{-1}{2\sqrt{13}} \end{bmatrix}$

7. (b) $\begin{bmatrix} x \\ y \end{bmatrix} = \begin{bmatrix} 1 + \frac{2}{3}u - \frac{1}{3}v \\ 1 - \frac{1}{2}u + \frac{1}{2}v \end{bmatrix}$

9. $\dfrac{\partial(r, \theta, \phi)}{\partial(x, y, z)} = \begin{bmatrix} \cos\theta\sin\phi & \sin\theta\sin\phi & \cos\phi \\[2mm] \dfrac{-\sin\theta}{r\sin\phi} & \dfrac{\cos\theta}{r\sin\phi} & 0 \\[4mm] \dfrac{\cos\theta\cos\phi}{r} & \dfrac{\sin\theta\cos\phi}{r} & \dfrac{-\sin\phi}{r} \end{bmatrix}$

$\dfrac{\partial(x, y, z)}{\partial(r, \theta, \phi)} = \begin{bmatrix} \cos\theta\sin\phi & -r\sin\theta\sin\phi & r\cos\theta\cos\phi \\ \sin\theta\sin\phi & r\cos\theta\sin\phi & r\sin\theta\cos\phi \\ \cos\phi & 0 & -r\sin\phi \end{bmatrix}$

11. (a) $\begin{bmatrix} 1 & 0 & 0 \\ 0 & 1 & 0 \\ 0 & 0 & 1 \end{bmatrix}$ (b) $\begin{bmatrix} 1 & 0 & 0 \\ 0 & 0 & 1 \\ 0 & 1 & 0 \end{bmatrix}$

(c) $\begin{bmatrix} 0 & 0 & 1 \\ 0 & 1 & 0 \\ 1 & 0 & 0 \end{bmatrix}$

Chapter 6

Section 6.1, page 518

1. 12 3. 12 5. 10

7. 16 13. 48

15. 48 17. 1

Section 6.2, page 532

1. $\dfrac{135}{2}$ 3. $-\dfrac{32}{3}$ 5. $-\dfrac{5}{3}$ 7. -285 9. $\dfrac{114}{3}$

11. 324 13. $\dfrac{1}{27}(8e^9 + 7e^{-6})$ 15. 27 17. $\dfrac{3}{10}$

19. $\dfrac{5\sqrt{3}}{8} - \dfrac{3}{2} + \dfrac{\pi}{12}(1 + \sqrt{3})$ 21. $\dfrac{1}{8}$ 23. $\dfrac{32}{3}$ 25. 1

27. $\dfrac{1}{2} + \log 2$ 29. 72   31. $\dfrac{128}{21}$ 33. π 35. $\dfrac{1}{6}$

37. 16π

Section 6.3, page 552

1. (a) $\frac{1}{2} \tan^{-1} \frac{1}{2}$,   (b) $\sqrt{2} + \frac{9}{2} \sin^{-1} \frac{1}{3}$
   (c) $-\frac{1}{2}\sqrt{3} + \log(2+\sqrt{3})$ (d) $e^{\sqrt{3}} - e$

3. 24   5. $\dfrac{\pi}{4}(e^2 - 1)$ 7. 25π 9. 8π

11. (a) $\dfrac{\pi}{4}(e^4 - 1)$ (b) $\dfrac{\pi}{8} \log 5$

13. $\dfrac{93}{20}$ 15. $4 - \pi$ 17. $\dfrac{1}{3} + \dfrac{\pi}{2}$ 19. $\dfrac{8\pi}{3}$

21. $\dfrac{\pi}{2}(32\sqrt{3} - 45)$ 23. $\dfrac{4096\pi}{3}$ 25. $\dfrac{23{,}056\pi}{15}$

27. $2\pi ab$ 29. 0   31. 0   33. 8π

Section 6.4, page 576

1. (a) $18, -4, 5$ (b) $-5/\sqrt{14}$ (c) $(-\frac{4}{9}, \frac{5}{9}, 2)$
   (d) $41, 40, 9$ (e) $0, 10, -8$ (f) $\frac{409}{10}$

3. (a) $\frac{8}{3}c(b^3 - a^3)$ (b) $0, 0, 0$ (c) 0 (d) $(0, 0, 0), (0, 0, 0)$
   (e) $\frac{8}{45}c(3b^5 + 5c^2b^3 - 3a^5 - 5c^2a^3)$, $\frac{8}{45}c(6b^5 + 5c^2b^3 - 6a^5 - 5c^2a^3)$,
   $\frac{8}{5}c(b^5 - a^5)$
   (f) $0, 0, 0$ (g) $\frac{8}{45}c(2b^5 + c^2b^3 - 2a^5 - c^2a^3)$

5. (a) $\frac{2}{3}k\pi$ (b) $\frac{1}{3}k\pi, 0, 0$ (c) $-\dfrac{k\pi}{3\sqrt{3}}$

   (d) $(0, 0, \frac{1}{3}), (0, 0, \frac{1}{2})$ (e) $\frac{8}{15}k\pi, \frac{8}{15}k\pi, \frac{2}{3}k\pi$
   (f) $0, 0, 0$ (g) $\frac{8}{15}k\pi$

7. (a) $\dfrac{4}{3}k$ (b) $\dfrac{128}{105}k, \dfrac{128}{105}k, \dfrac{128}{105}k$ (c) $-\dfrac{152}{105}k$

(d) $\left(\dfrac{3}{4},\dfrac{3}{4},\dfrac{3}{4}\right)$, $\left(\dfrac{32}{35},\dfrac{32}{35},\dfrac{32}{35}\right)$ (e) $\dfrac{8}{3}k,\dfrac{8}{3}k,\dfrac{8}{3}k$

(f) $\frac{1}{3}k\pi,\frac{1}{3}k\pi,\frac{1}{3}k\pi$ (g) $\frac{1}{3}k(8-\pi)$

9. (a) $64k\pi$ (b) $\dfrac{768}{5}k\pi,0,0$ (c) $\dfrac{1344}{5\sqrt{11}}k\pi$

(d) $\left(0,0,\dfrac{9}{4}\right)$, $\left(0,0,\dfrac{12}{5}\right)$ (e) $\dfrac{4864}{9}k\pi,\dfrac{4864}{9}k\pi,\dfrac{2560}{9}k\pi$

(f) $0,0,0$ (g) $\dfrac{1280}{3}k\pi$

11. (a) $\frac{1}{2}abc^2k\pi$ (b) $0,0,0$ (c) 0 (d) $(0,0,0),(0,0,0)$
   (e) $\frac{1}{12}abc^2k\pi(b^2+2c^2)$, $\frac{1}{12}abc^2k\pi(a^2+2c^2)$, $\frac{1}{12}abc^2k\pi(a^2+b^2)$

(f) $0,0,0$ (g) $\dfrac{1}{156}abc^2k\pi(9a^2+4b^2+26c^2)$

13. $\frac{1}{12}ah\rho_0\pi(a^2+2h^2)$

15. (a) $\frac{1}{6}ka^2b^3$, $\frac{1}{6}ka^3b^2$ (b) 0 (c) $(a/2,b/2),(\frac{2}{3}a,\frac{2}{3}b)$

(d) $\frac{1}{8}ka^2b^4,\frac{1}{8}ka^4b^2,\frac{1}{9}ka^3b^3,\frac{1}{8}ka^2b^2(a^2+b^2)$ (e) $\dfrac{ka^4b^4}{36(a^2+b^2)}$

17. (a) $32k,\dfrac{32}{3}k$ (b) $-\dfrac{160}{3\sqrt{5}}k$ (c) $\left(1,\dfrac{10}{3}\right),\left(\dfrac{4}{3},4\right)$

(d) $\dfrac{448}{3}k,16k,\dfrac{704}{15}k,\dfrac{496}{3}k$ (e) $\dfrac{2128}{25}k$

19. (a) $\dfrac{5}{24}k,\dfrac{1}{2}k\left(\dfrac{1}{4}+\log 2\right)$ (b) $\dfrac{1}{24}k(12\log 2-17)$

(c) $\left(\dfrac{8}{3+6\log 2},\dfrac{5}{6+12\log 2}\right),\left(\dfrac{3}{10}+\dfrac{6}{5}\log 2,\dfrac{1}{2}\right)$

(d) $\dfrac{59}{480}k,\dfrac{3}{5}k,\dfrac{7}{30}k,\dfrac{347}{480}k$ (e) $\dfrac{41}{320}k$

21. (a) $\dfrac{1}{12}k,\dfrac{3}{35}k$ (b) $\dfrac{k}{420\sqrt{2}}$ (c) $\left(\dfrac{9}{20},\dfrac{9}{20}\right),\left(\dfrac{4}{7},\dfrac{5}{9}\right)$

(d) $\dfrac{3}{56}k, \dfrac{1}{18}k, \dfrac{3}{56}k, \dfrac{55}{504}k$ (e) $\dfrac{k}{1008}$

23.   (a) $0, \dfrac{5}{4}k\pi$ (b) $\dfrac{5}{4\sqrt{2}}k\pi$ (c) $\left(\dfrac{5}{6}, 0\right), \left(\dfrac{5}{6}, 0\right)$

(d) $\dfrac{21}{32}k\pi, \dfrac{49}{32}k\pi, 0, \dfrac{35}{16}k\pi$ (e) $\dfrac{35}{32}k\pi$

Section 6.5, page 608

1. $\dfrac{7}{15}$ 3.   0   5. $\dfrac{19}{6}$

7. $\dfrac{181}{4}$ 9. $-\pi ab$

11. $a^2\left(\dfrac{\pi}{3} - \dfrac{3}{8}\right) + \dfrac{\sqrt{3}}{6}\pi ab - \dfrac{\pi^2}{18}b^2$

13. $\dfrac{31}{2}$ 15. -2

17. $\dfrac{2}{3}e^3 + 2e + 2e^{-1} + e^{-3} - \dfrac{17}{3}$

19. $-\left(\dfrac{1}{3}x^3 + \dfrac{1}{2}xy^2 + \dfrac{1}{2}yz^2 + c\right)$

21.   not conservative

23. $-(x^2ye^z + xz^2e^y + y^2ze^x + c)$

25. $-(x\cos y + y\sin x + c)$

27.   6   29. $\dfrac{131}{2}$ 31. 16π

33. $\dfrac{\sqrt{2}}{26}(3e^{3\pi/4} + 2)$ 35. $\pi\rho R^3$

37. $\dfrac{5}{2}\rho_0 L$; center of gravity at a distance $\dfrac{2\sqrt{3}}{15}L$ from base, $\dfrac{5}{16}\rho_0 L^3$

Section 6.6, page 629

1. $8\pi\sqrt{2}$

3. $4\sqrt{6}$

5. $4\pi\sqrt{3}$

7. $\pi(2 - \sqrt{2})$

9. $\frac{1}{2}\sqrt{a^2b^2 + a^2c^2 + b^2c^2}$

11. $4a^2$

13. 0

15. $4\pi\sqrt{2}$

17. π

19. $\left(0, 0, \dfrac{27a}{160 - 85\sqrt{2}}\right), \dfrac{\pi a^6}{240}(128 - 81\sqrt{2})$

21. $(0, 0, 2), \dfrac{32}{3}\pi\sqrt{3}$

Section 6.7, page 647

1. 0

3. $\dfrac{104}{3}$

5. 0

7. $\dfrac{\pi}{12}(8 - 5\sqrt{2})$

9. $\dfrac{1}{12}abc(4 + b)$

11. 0

13. 0

Section 6.8, page 660

1. 24π

3. 0

5. $4\pi a^3$

7. $(x_2 - x_1)(y_2 - y_1)(z_2 - z_1)(x_1 + x_2 + y_1 + y_2 + z_1 + z_2)$

9. 40π

11. $\dfrac{243}{2}\pi$

13. $-\dfrac{13}{6}$

15. 0

17. 0

19. $-\pi a^2$

21. 0

23. 0

25. 4

27. 5π

29. 4

31. $\dfrac{\pi}{2}$

33. $-\dfrac{556}{15}$

Chapter 7

Section 7.1, page 674

1. $\{1, 3, 5, 7, 9, \ldots\}$ 3. $\{1, 2, 9, 64, 625, \ldots\}$

5. $\{\frac{1}{2}, \frac{1}{4}, \frac{1}{6}, \frac{1}{8}, \frac{1}{10}, \ldots\}$ 7. $\left\{\frac{1}{\sqrt{2}}, \frac{1}{2}, -\frac{1}{3\sqrt{2}}, 0, \frac{1}{5\sqrt{2}}, \ldots\right\}$

9. $\{3n - 1\}$, $\{n + 3\}$, $\{6n + 3\}$

11. $\left\{\sin\frac{n\pi}{4} + \cos\frac{n\pi}{4}\right\}$, $\left\{\sin\frac{n\pi}{4} - \cos\frac{n\pi}{4}\right\}$, $\left\{3\sin\frac{n\pi}{4}\right\}$

13. (a)   bounded   (b)   not bounded above or below
     (c)   bounded below   (d)   bounded below

15. (a)   bounded   (b)   bounded
     (c)   bounded below   (d)   bounded

17. 0   19. $\frac{1}{2}$ 21. 4   23. 1   25. 1

25. (a), (c)   27. (b)

Section 7.2, page 692

1. 2   3. $e/(e - 1)$ 5. divergent

7. convergent   9. convergent   11. convergent

13. divergent   15. convergent   17. convergent

19. convergent   21. convergent   23. convergent

25. conditionally convergent   27. absolutely convergent

29. absolutely convergent   31. divergent

33. conditionally convergent

Section 7.3, page 709

1. $R = 1$;   conditional convergence at $x = -1$, divergence at $x = -3$

3. $R = 2$;   divergence at $x = -3$ and $x = 1$

5. $R = 0$ 7. $R = 0$ 9. $R = 2;$ divergence at $x = \pm 2$

11. $f(x) = \dfrac{(x-1)^2}{(4-3x)^2}$

13. $\displaystyle\sum_{r=0}^{\infty} (-1)^r \frac{x^{2r+1}}{2r+1}$ 15. $\displaystyle\sum_{r=0}^{\infty} (-1)^r \frac{2^{2r+1}x^{2r+1}}{(2r+1)!}$

17. $\displaystyle\sum_{r=0}^{\infty} \frac{x^{2r}}{(2r)!}$ 19. $\dfrac{1}{\sqrt{2}}\displaystyle\sum_{r=0}^{\infty} (-1)^r \left(x - \frac{\pi}{4}\right)^r$

19. $\dfrac{1}{\sqrt{2}}\displaystyle\sum_{r=0}^{\infty} \frac{(-1)^r}{(2r)!}\left(x - \frac{\pi}{4}\right)^{2r} - \dfrac{1}{\sqrt{2}}\displaystyle\sum_{r=0}^{\infty} \frac{(-1)^r}{(2r+1)!}\left(x - \frac{\pi}{4}\right)^{2r+1}$

21. $-\displaystyle\sum_{r=0}^{\infty} \frac{x^r}{5^{r+1}}$

Subject Index

A

Absolute convergence, 689
Acceleration, 259
  normal component, 270
  tangential component, 270
Addition of matrices, 5
Additive inverse of a vector, 94
Adjoint of a matrix, 83
Affine function, 446
Alternating series test, 687
Angle between two vectors, 174
Arc length, 268, 280
Area
  of parallellogram, 239
  of triangle, 240
Augmented matrix, 20
Axis, 167

B

Ball, 311
Basis, 112
  coordinate vectors with respect to, 135
  natural, 114
  orthonormal, 234
  in R^n, 115
Bound, 665
  greatest lower, 676
  least upper, 666
  lower, 665
  upper, 665
Boundary, directed, 639
Boundary point, 326
Bounded function, 326

C

Cardioid, 544
Cauchy–Schwarz inequality, 172
Center of mass, 560, 562
Centroid, 562, 571
Chain rule, 355, 452

Characteristic
  equation, 189
  polynomial, 188
  value, 187
  vector, 187
Change of parameter, 278
  piecewise smooth, 279
  smooth, 279
Change of variable, 535, 537
Closed curve, 275
Coefficient matrix, 20
Cofactor, 77
Column rank, 152
Column space, 152
Component(s)
  of vector, 96
  of vector-valued function, 426
Comparison test, 683
Completeness axiom, 666
Composite function, 310, 429
Conditional convergence, 689
Connected, 590
  simply, 598
Conservative field, 469, 593
Continuity
  of composite functions, 320
  in direction of vector, 338
  of product, 320, 435
  of quotient, 320
  of real-valued function, 318
  of sum, 320, 435
Constraints, 412
Continuously differentiable
  curve, 276
  real-valued function, 354
Convergence criterion for sequences, 670
Convergence set, 696
Convergent
  absolutely, 689
  conditionally, 689
  sequence, 666
  series, 677